Proceedings of
The Rice Meeting

Proceedings of
The Rice Meeting

Volume 2

*1990 Meeting of the Division of Particles and Fields
of the American Physical Society*

Houston, Texas   3 – 6 January 1990

Editors
Billy Bonner and Hannu Miettinen
Department of Physics
Rice University, Houston

World Scientific
Singapore • New Jersey • London • Hong Kong

Published by

World Scientific Publishing Co. Pte. Ltd.

P O Box 128, Farrer Road, Singapore 9128

USA office: 687 Hartwell Street, Teaneck, NJ 07666

UK office: 73 Lynton Mead, Totteridge, London N20 8DH

THE RICE MEETING — 1990 MEETING OF THE DIVISION OF PARTICLES AND FIELDS OF THE AMERICAN PHYSICAL SOCIETY

ISBN 981-02-0258-X

Printed in Singapore by Loi Printing Pte. Ltd.

TABLE OF CONTENTS

Volume 1

Volume 2

QCD AND HADRON PHYSICS

NON-ACCELERATOR PHYSICS

HEAVY IONS / LATTICE

QCD AND HADRON PHYSICS

Conveners: K. Ellis, N. Isgur, & K. Meier

CDF Top Quark Searches

The CDF Collaboration[1]
presented by Robert Hollebeek
University of Pennsylvania

1990 DPF Conference, Houston

Introduction

The top quark has been the object of intensive searches during the past year using the CDF detector. This important ingredient in the standard model of particle physics however remains unseen. Searches for it or its discovery will eventually provide important constraints on the parameters of the standard model. Despite the fact that the top quark has not been found, evidence or prejudice for its existence is quite strong. The measurement of the b quark asymmetry [2,3] in e^+e^- annihilation indicates that the b quark has the expected asymmetry for a member of a left-handed doublet and right-handed singlet structure indicating that it should have a partner, namely the top quark. The measured value of the b quark asymmetry is

$$a^b = -1.08 \pm 0.29 \ .$$

If the b quark were alone in a singlet structure, one would expect zero for this parameter, while in the Standard Model, one expects -1. Other negative searches reported include results from AMY and Venus at Tristan [4]($M_t > 30.4 GeV$), MKII at Slac [5] ($M_t > 40.7 GeV$), UA1 and UA2 at CERN (M_t >61,69 GeV 95%CL) [6] and LEP (M_t >45.8 GeV). The searches described in this contribution extend these results to exclude the range

$$40 < M_t < 77 \qquad \text{excluded at 95 \% CL}$$

from an electron plus jets final state[8], and

$$28 < M_t < 72 \qquad \text{excluded at 95 \% CL}$$

from an electron muon final state[9].

t Quark Production

At Tevatron energies ($\sqrt{s} = 1.8 TeV$), the production of t quarks is dominated by the production of $t\bar{t}$ states. This situation differs from that found at the lower energies available at CERN where $W \rightarrow t\bar{b}$ also contributes significantly to t production. The difference is due to the faster rise of the t production cross section with energy relative to the W production cross section.[10] The expected behavior of the t cross section as a function of the top quark mass for these two energies including higher order corrections by Nason, Dawson, and Ellis [11] is well reproduced by the monte carlo Isajet, used later in these analyses for acceptance calculations. The standard model of t quark decays predicts that for a $t\bar{t}$ system, roughly 44% of the decays will be to fully hadronic final states with as many as 6 final state jets. While these rates are attractive, the calorimeter performance, the ability to reconstruct jets, and the copious QCD background in this channel limit our ability to utilize this mode. An additional 15% for each of the final states e, μ, τ, will consist of leptonic decays of one of the t quarks and the hadronic decay of the other, leading to a lepton, as many as four jets and missing E_t due to the neutrino from the decay of a virtual or real W. The remaining decay modes consist of 1% for ee,$\mu\mu$, and $\tau\tau$ pairs (each of which suffer from significant Drell-Yan backgrounds) and 2% for non-identical lepton pair final states. These leptons will be accompanied by as many as two additional jets and missing E_t . The searches reported here by the CDF collaboration were done using the electron plus jets final state and the electron plus muon final state. The electron plus jets mode has the advantage of higher rate than the $e\mu$ channel and can therefore probe heavier masses. The $e\mu$ channel has the largest di-lepton rate, does not suffer from Drell-Yan backgrounds, and can be used effectively at low masses where acceptances in the electron jet channel are uncertain.

A number of indirect constraints on the top quark mass are available. For example, the observation of $B_d\,\overline{B_d}$ mixing[?] implies that the b quark is inconsistent with an SU(2) singlet structure. Analysis of neutral current data and the W and Z masses [?] implies an upper bound on the top quark mass of approximately 190 GeV [12,13]. An additional indirect method involves the use of the ratio [15]

$$R = \frac{\sigma_W}{\sigma_Z} \frac{BR(W \to e\nu)}{BR(Z \to ee)}$$

where

$$\frac{BR(W \to e\nu)}{BR(Z \to ee)} = \left(\frac{\Gamma(W \to e\nu)}{\Gamma_{Wtot}}\right) / \left(\frac{\Gamma(Z \to ee)}{\Gamma_{Ztot}}\right) \ .$$

The ratio of cross sections for W and Z production can be predicted from theory as can the leptonic decay partial widths of the W and Z leaving a dependence in R on the t mass due to its influence on the total widths of the Z and W. While previous measurements of this ratio were thought to favor low top quark masses, more recent data give

$$\text{CDF} \quad 10.3 \pm 0.8 \pm 0.5$$

and are consistent with heavy top masses.

Electron Detection

The CDF analysis uses the following quantities to identify electrons: E/P (the ratio of the shower energy to the track candidate momentum), *had/em* (small leakage of energy from the electromagnetic to the hadronic sections of the calorimeter), $\Delta x, \Delta z$ (the shower to track matching parameters measured using strip chambers positioned at a depth of 6 radiation lengths within the electromagnetic calorimeter), χ_x^2, χ_z^2 (the shower profiles near shower max), and LSHR (a chi-square like parameter describing the shower sharing between adjacent calorimeter towers). Conversions are detected by reconstructing opposite sign pairs of tracks with low invariant mass and by using missing hits in the central VTPC (vertex time projection chambers) to identify conversions in the outer wall of the VTPC. Conversions in the outer wall of the VTPC can be used to test the track reconstruction for all conversions and to correct for inefficiencies in conversion detection.

The electron detection techniques can be tested by using the electrons from W decay, identified conversions, and electrons from the inclusive electron trigger. The distributions of the E/P and *had/em* parameters are shown in figure 1-2 for electron candidates with $E_t > 15\ GeV$.

A further cut called the border tower cut can be used to discriminate against electrons from parent b quarks. In this cut, the energy in towers immediately adjacent to the electron is summed and required to be less than 2 GeV. The number of towers summed in this manner depends on the shape of the electron cluster but ranges between 8 and 12. This cut requires the electron to be isolated and is efficient for heavy t quarks because the large mass of the t quark leads to increased separation on average between the decay electron and any associated jet activity.

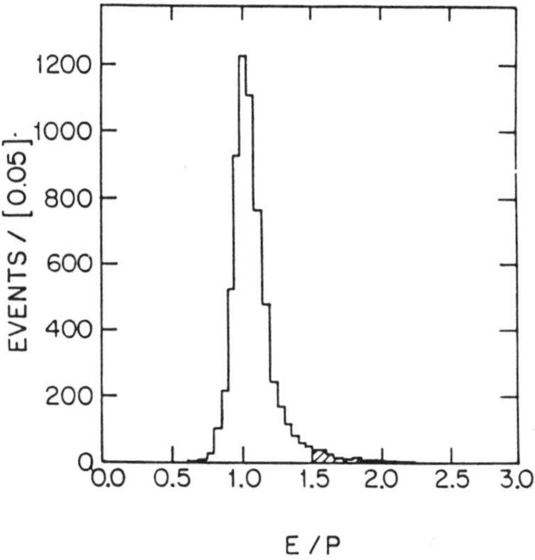

Figure 1: E/P distribution for electron candidates with $E_t > 15$ GeV

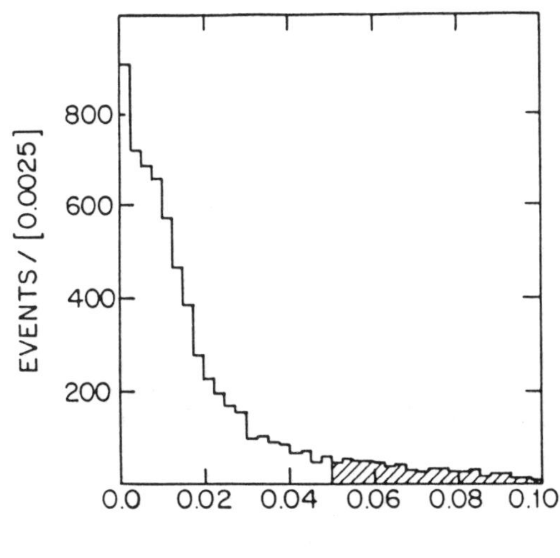

Figure 2: *had/em* distribution for electron candidates with $E_t > 15$ GeV

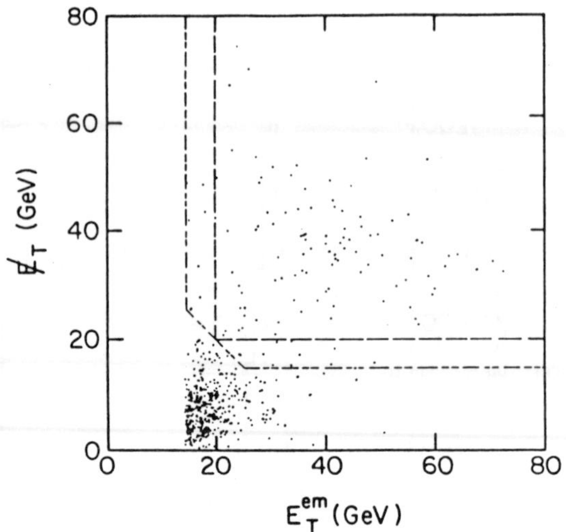

Figure 3: $\not{E}_T$ versus E_t for electron $+ \geq 2$ jet events.

Electron Plus Jets Search

The electron plus jets search proceeds by requiring a "golden" electron candidate, i.e. a cluster in the central electromagnetic calorimeter with $E_t > 12$ GeV, *had/em* less than 0.05, E/P less than 1.5, track to strip matches of less than 1.5cm in xy and 3cm in z, and good shower profiles. The distributions in figure 1 indicate that very little background remains in the sample from overlaps or charged pion interactions after these cuts. About 30% of the inclusive electrons above E_t of 12 GeV come from γ conversions and π^0 Dalitz decays. These are removed by requiring a track in the VTPC, and by removing any candidate with a second track which forms an invariant mass pair with mass less than 0.5 GeV. Electrons from Z^0 decays are removed by looking for a second EM cluster which forms an invariant mass greater than 70 GeV. At least two jets with $E_t > 10$ GeV and $|\eta| < 2$ are required.

The distribution of $\not{E}_T$ versus E_t shows (see figure 3) that the remaining sample of events comes primarily from events with large $\not{E}_T$ and large E_t due to W production, and lower $\not{E}_T$,E_t events which in general also show poor isolation and are consistent with the kinematic properties expected for events due to bottom production.

Two regions in the $\not{E}_T$, E_t plane are defined, one with $\not{E}_T > 20$ GeV and $E_t > 20$ and a second with $\not{E}_T$,$E_t > 15$ but $\not{E}_T + E_t > 40$ GeV. The looser cut is more efficient for low mass top, but lets in more background from light quarks. It is used to extend the search to lower top masses where the rates are large. The tighter cuts are used at higher masses. See figure 3.

The method used to search for top quark production in the electron plus 2 or more jet sample is to use the electron E_t and the missing transverse energy, $\not{E}_T$, to calculate a transverse mass. The transverse mass distribution is then compared to that expected from W and top production. The expected transverse mass distribution for W's is checked by studying the electron plus one jet sample which should be dominated by W production. The observed distribution for this sample is compared to a W+1jet prediction from the Papageno monte carlo [16] and found to be in excellent

M_{top}	a	$n_{t\bar{t}}$
40	$0.07 \pm 0.05 \pm 0.02$	130 ± 44
50	$0.06 \pm 0.05 \pm 0.03$	123 ± 31
60	$0.11 \pm 0.08 \pm 0.04$	101 ± 22
70	$0.00^{+0.12}_{-0.00} \pm 0.11$	43 ± 8
80	$0.00^{+0.27}_{-0.00} \pm 0.17$	32 ± 5

Table 1: Fitted value for parameter a and expected number of top events.

agreement. The detector resolution in transverse mass for the W+2jet sample, which is the major background to the top search, is similar to the resolution for the W+1jet sample.

The electron plus two or more jet sample has a transverse mass distribution which agrees with that expected for a pure W sample. The rate of events observed in this class is within 30% of the predicted rate from the Papageno monte carlo which is within the theoretical uncertainty of about 30-50%.[17] The transverse mass distribution for top decays where the top mass is below threshold for real W production, is quite different from that for W's. In general, it tends to peek at lower transverse masses. This fact is used to fit the observed distribution to the expected shape due to the sum of contributions from the W plus jets and from top. A binned maximum-likelihood fit is used to fit to the form

$$dN/dM_t = aT(M_t) + bW(M_t).$$

The results of the fits are shown in table 1 and figure 4 where the coefficients a and b would be 1 if the predicted source agreed in rate with the monte carlo prediction.

To interpret the fit as a limit for top production, the following systematic errors are included: 1% for electron energy calibration, 20% for jet energy scale uncertainties, 5% electron selection uncertainties, and 15% luminosity uncertainty. In addition, the fragmentation parameters of the top quark in the Isajet monte carlo have been varied [18], and uncertainties of 50% and 20% have been used for initial state radiation and the underlying event respectively. Each of these quantities is varied in the monte carlo, and the resultant effect on the fit (which varies with M_t) is taken as the systematic error due to this parameter. These errors are added in quadrature. The resulting upper limit for t production is shown in figure 5 where the top production cross section which has been assumed is the lower bound from Altarelli et al.[19] The results of the fit exclude the range

$$40 < M_t < 77 \quad \text{excluded} \quad 95\%c\ell$$

Additional information concerning the candidate events such as the jet-jet mass distribution for 2 jet events, and the distribution of the number of observed jets are also consistent with a pure W sample. Although these distributions have not been used in this analysis, they can be used to extend the sensitivity of the search in the future as can additional requirements such as additional jets or additional soft leptons..

Verification of Jet Detection Efficiencies

A number of techniques have been used to verify that the jet detection techniques are efficient and to verify the jet energy scale. Since the jet detection efficiency enters twice for a two jet event, the overall detection efficiency can be quite sensitive to errors in the jet energy scale. This sensitivity is less than might be expected for heavy M_t because the leading jets tend to be well over the 10 GeV threshold. The acceptance varies by 5% for example if the threshold is changed to 12 GeV. Monte carlo studies have shown that the $E_t > 10$ GeV threshold used in this analysis corresponds roughly

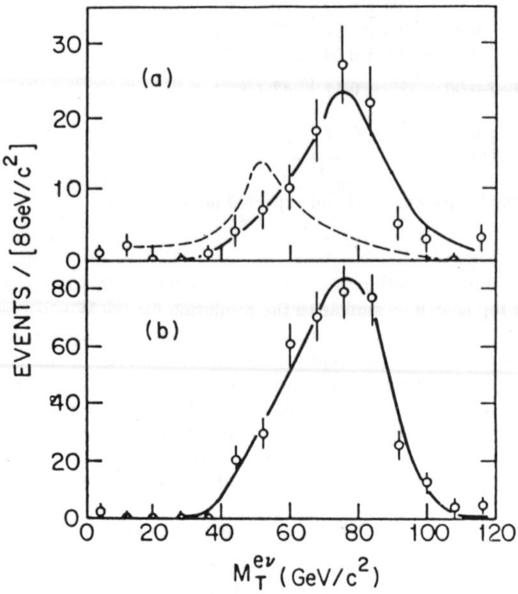

Figure 4: The transverse mass distributions for (a) 70 GeV top (dashed) and Papageno (solid) for ≥ 2 jets and (b) ≥ 1 jet.

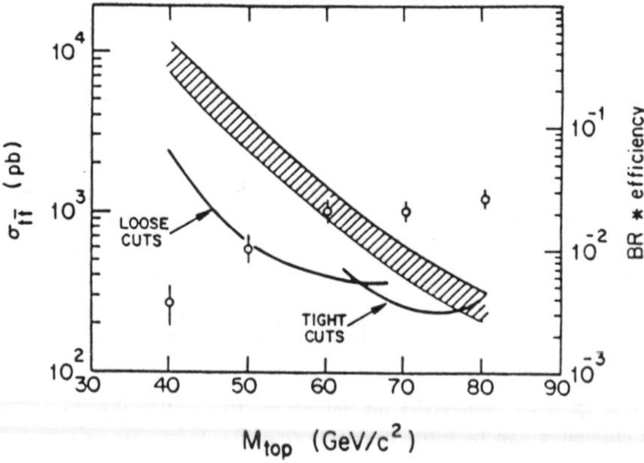

Figure 5: The 95% CL upper limit from the electron plus jets search and the acceptance (right hand scale)

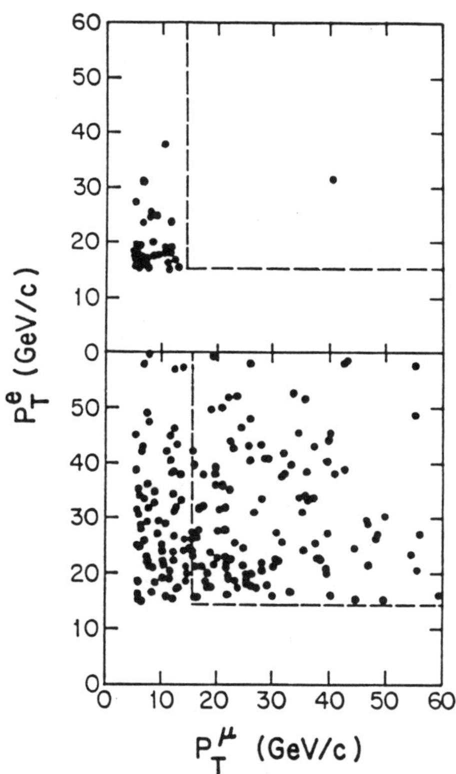

Figure 6: $e\mu$ (a) data and (b) 70 GeV top for 80 pb^{-1}

to a produced jet of 15 GeV E_t . The response of the detector depends on the charged to neutral ratio in the jet as well as the fragmentation spectra and the detector response. Fortunately, much is known about these parameters. We know for example from e^+e^- annihilation that the fragmentation of b quarks is not significantly different from other jets at these energies.

Three samples are used to check the jet response of the detector. The inclusive electron sample (dominated by electrons from b's), Z+jet events, and γ+jet events can all be used to balance the p_t of a known electron, Z or γ against a jet on the opposite side to determine the jet energy scale. The inclusive electron sample has 12,000 events with $E_t > 12$ GeV, and can be used for detailed studies not only of the jets, but also of the electron properties. We estimate the systematic uncertainty in the jet energy scale to vary from 13% to 5% over the range 18 to 180 GeV in observed energy for central jets. For lower energies, we use a conservative estimate of 20%.

$e\mu$ Analysis

The electron muon search for top uses the same electron sample discussed previously except that the electron E_t re-quirement is 12 GeV and the E/P cut is 1.4. Because backgrounds are smaller in this final state, the isolation cut is not applied. In the central region, $\eta < 0.65$, a track with impact parameter $< 0.5cm$, $p_t > 5$, which matches a muon chamber stub with $\Delta R\phi < 10cm$ is required. The difference in slope between the central track and the muon stub must be less than 0.1, and the calorimeter must have less than 2 GeV in the EM section and <6 GeV in the hadronic section. In the region $0.65 < \eta < 1.2$, the requirement of a muon stub is removed. In this case, the track p_t is required to be more than 10 GeV and the total calorimeter energy associated with the muon candidate less than 5 GeV in a cone of radius 0.4 about the track.

The events in the remaining sample cluster at low electron and muon E_t as would be expected for light quark production with the electron and muon being typically back-to-back. To reject these backgrounds, the electron and muon are both required to have $E_t > 15$ and to be of opposite sign. One event remains in the sample. (See figure 6.) Expected rates from background sources are 1 event from $Z \to \tau\tau$, 0.15 events from WW production, 0.05 events from WZ production, and of order 1 event or less from b quark jets though this rate is uncertain due to uncertainties in the b quark production cross section. The assumed systematic errors are 1% for electron calibration, 5% for electron selection efficiency, 15% luminosity uncertainty, and 20% acceptance uncertainty. The limit is determined by finding the mean of a Poisson distribution that when convoluted with the total systematic error yields a probability of 5% for observing zero or one event. This analysis excludes a top quark in the range $28GeV < M_t < 72GeV$ at 95%CL.

References

[1] Argonne, Brandeis, Chicago, FermiLab, Frascati,
    Harvard, Illinois, KEK, LBL, Pennsylvania, INFN,
    Purdue, Rockefeller, Rutgers, Texas A&M, Tsukuba,
    Tufts, and Wisconsin.

[2] The JADE Collaboration, DESY 88-154(1988).

[3] W.W.Ash et al.,Phys.Rev.Lett.58,1080(1987).

[4] Eva Law, AMY Collaboration, 1989 EPS Conference.

[5] MKII 1989 EPS Conference,Madrid.

[6] UA1 95% limits,1989 EPS Conference,
    T. Akesson et al., UA2 collab., CERN-EP/89-152(1989).

[7] D. Decamp et al., ALEPH collab., CERN-EP89-165(1989).
M. Akrawy et al., OPAL collab., CERN-EP/89-154(1989).

[8] F. Abe et al., CDF collab., Phys.Rev.Lett.**64**,142(1990).

[9] F. Abe et al., CDF collab., Phys.Rev.Lett.**64**,147(1990).

[10] see for example Barger and Phillips Collider Physics
figure 10.23.

[11] P.Nason,S.Dawson and R.K.Ellis,Nucl.Phys.B303,
607(1988).

[12] H.Albrecht et al.,Phys.Lett.192B,245(1987).

[13] M.Artuso et al.,Phys.Rev.Lett.62,2233(1989).

[14] U. Amaldi et al., Phys. Rev. **D36**,1385(1987). P. Langacker, Phys. Rev. Lett. **63**,1920(1989).

[15] P.Colas, D.Denegri, C.Stubenrauch, DPhPE 88-02.

[16] I.Hinchliffe,report in preparation,Papageno V2.71.

[17] M.Mangano,S.Parke Fermilab Pub 84/106-T(1989).

[18] The epsilon parameter was changed by a factor of 3.

[19] G.Altarelli,M.Diemos,G.Martinelli, P.Nason,
Nucl.Phys.B308,724(1988).

THE SEARCH FOR TOP IN UA2

UA2 Collaboration

Presented by
S. L. Singh

University of Cambridge,
Cambridge, CB3 0HE, UK.

ABSTRACT

During 1988 and 1989 the upgraded UA2 detector collected data corresponding to an integrated luminosity of 7.5 pb^{-1} from p$\bar{\text{p}}$ collisions at a centre of mass energy of 630 GeV. A search has been performed for the top quark (t) or a member of a hypothetical fourth family (b') by looking for their semi-electronic decays. No evidence has been found for these particles. Using production rates and the decay branching ratios from the Standard Model, this analysis implies that the top quark mass is greater than 69 (71) GeV, and that the b$'$ mass is greater than 54 (57) GeV, at 95(90)% confidence.

1   INTRODUCTION

The existence of the top quark (t) is essential if the Standard Model is to remain consistent with experimental data. Quark families consist of left-handed fields forming doublets under SU(2) and right-handed singlet fields. Quark classification into left-handed weak isospin doublets prevents various processes, notably flavour-changing neutral curents, from occurring at rates in contradiction to experiment; the existence of the bottom quark (b) implies the existence of its partner, the top quark.

Calculations indicate that the top quark mass must be less than 200 GeV [1], if the Standard Model is to remain consistent with the body of experimental data. Experimental searches at LEP and SLC have excluded top quark masses below 44.5 GeV at the 95% confidence level.

The new Antiproton Accumulator Complex allowed the CERN p$\bar{\text{p}}$ Collider to run at a peak luminosity of $3.10^{30}\text{cm}^{-2}\text{s}^{-1}$. During 1988 and 1989 the upgraded UA2 detector collected a data sample corresponding to 7.5 pb^{-1}, at a centre of mass energy, $\sqrt{s}$=630 GeV. Using 7.1 pb^{-1} of this data sample, a search has been performed for top quark production followed by semi-electronic decay.

In the next section the UA2 detector will be described in the context of searching for a top signature. The following section discusses top production and decay at the CERN p$\bar{\text{p}}$ Collider. After applying a set of selection criteria to the data in section 4, a comparison will be made between the final data sample, a background estimate and the estimated top signal. In section 7 upper limits on the cross-section for top production will be estimated, in order to extract lower mass limits for the top quark. Mass limits will also be set for a hypothetical b$'$quark, assuming that b$'$ production is similar to top quark pair production.

2 THE UA2 DETECTOR

Between 1985 and 1987 the UA2 detector underwent a major upgrade, which involved the addition of calorimeter endcaps, in order to improve its missing transverse momentum resolution, and the introduction of a new central detector, in order to improve tracking and electron identification. Much of the relevant detail regarding the UA2 detector has been given at this conference by G.Blaylock and J.Incandela in their talks about W/Z physics at UA2. Consequently, only a brief outline of the detector will be given here, with some extra detail concerning electron identification in the top analysis.

The central calorimeter remains largely unchanged from the original UA2 [2]. It has full azimuthal coverage and covers the region 40-140° in θ. It has 240 electromagnetic and hadronic cells, each subtending 10° in θ and 15 ° in ϕ. The endcap calorimeter sections (endcaps) [3] extend the coverage from $\pm$ 1-3 in pseudorapidity (η), or from 40°- 6° in θ as measured from the beam axis. Each endcap consists of 12 modules containing 16 cells.

Pattern recognition in the calorimeter involves a clustering algorithm which accociates cells with a common edge if their energy transverse to the beam is greater than 400 MeV. Clusters were identified as being electromagnetic if the distribution of the energy deposition was compatible with that expected from a single electron shower.

2.1 The Central Detector

The UA2 detector has neither a magnetic field nor muon detection. Therefore good electron identification is essential, and the new central detector was designed with special emphasis on this requirement. What follows is a guide to how each subdetector is used to identify electrons and reject backgrounds.

Figure 1 shows the upper half of a lateral cross-section through the central detector. The subdetectors have coverage down to approximately 20° with respect to the beam axis. The diagram is not drawn to scale, but each detector is represented and its response to an electron and various backgrounds is indicated.

An electron leaving the interaction vertex encounters the inner silicon layer (ISI) [4] first. Each inner silicon pad covers 2mm along the z-axis and 30° in azimuth, and a minimum ionising particle (MIP) such as the electron will deposit what is referred to as a one MIP signal. The electron then passes through the jet vertex detector (JVD) [5] until it reaches the outer silicon layer (OSI), which has a similar granularity to the inner layer. Once again a one MIP signal is deposited.

The transition radiation detector (TRD) [6] consists of two chambers made up of radiators and proportional chambers. Electrons passing through the radiators emit X-rays which then give rise to signals in the chamber.

The scintillating fibre detector (SFD) [7] has 8 stereotriplet layers of fibres read out by photomultiplier tubes and CCDs. The first 6 layers provide tracking, and then 1.5 radiation lengths of lead cause the electron to shower before reaching the 2 preshower layers. Shower development then continues into the electromagnetic compartment of the calorimeter. Energy leakage into the first hadronic compartment should be low, and the energy profile should be compatible with that expected from an isolated electron shower. The SFD is used to identify electrons using track/preshower matching. This involves matching the position of the track in the first 6 layers with the point of shower commencement in the preshower section. The SFD achieves a resolution of 0.4mm in R.ϕ and 1.1mm in z.

A hadron provides one MIP signals in both silicon layers and a track in the JVD and the SFD. The lack of any significant preshower in the SFD means that the hadron fails the track/preshower match. In addition, the shower in the calorimeter ought to give significant leakage into the hadronic cells.

Conversions provide two electrons which create the necessary track/preshower match and an

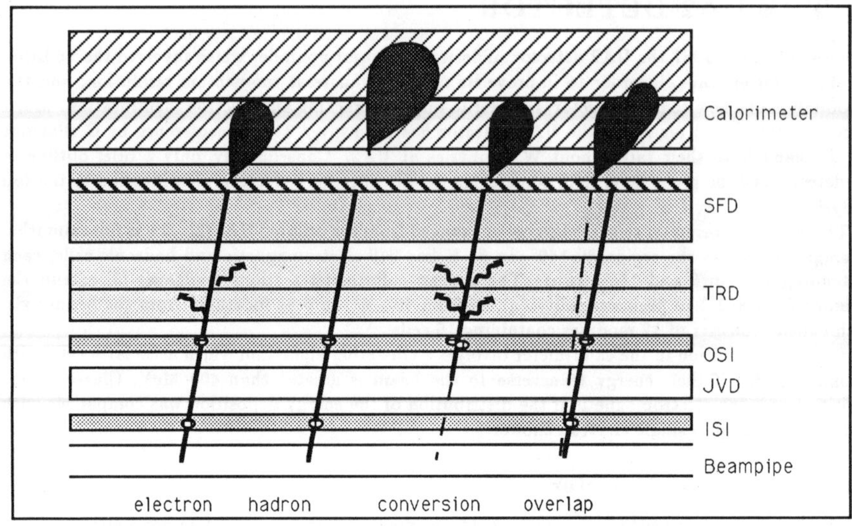

Fig.1 Particle Identification with the Central Detector

acceptable calorimeter shower profile. The silicon layers are then the strongest means of rejection. Conversions before the OSI give 2 MIP signals in one or both silicon layers. A silicon hit is required on every track.

The overlap is a low energy hadron giving a single MIP in both silicon layers, accompanied by a neutral pion or a photon which does not convert until after the OSI. The calorimeter cluster can still be classified as electromagnetic if the hadron does not cause excessive leakage into the hadronic compartment, owing to its low energy. Hence, the only means of rejection is the SFD track/preshower match. The preshower is caused by the photon whereas the track is that of the hadron. Unless the hadron and photon are sufficiently close together they will not be accepted as being an electron candidate.

3 TOP PRODUCTION AND DECAY

Top production occurs at the SPS via two processes;

Weak interaction, $p\bar{p} \to W+X$, $W \to t\bar{b}$ or $b\bar{t}$

Strong interaction, $p\bar{p} \to t\bar{t}+X$.

The production cross-section for the weak process can accurately be derived from the cross-section for W production and decay to an electron as measured by UA2 [8]. The main errors in the weak cross section estimate come from the statistical uncertainty in the number of Ws, and the uncertainty in the mass of the W, which affects the phase space available for top production. The uncertainty on the mass of the W leads to a 5% systematic uncertainty on the branching ratio of $W \to t\bar{b}$, for a top mass of 60 GeV. To be conservative corrections due to higher order QCD processes are set to 1.0, as these are derived from incomplete theoretical calculations.

The strong production cross-section has been evaluated [9] using the full next-to-leading order

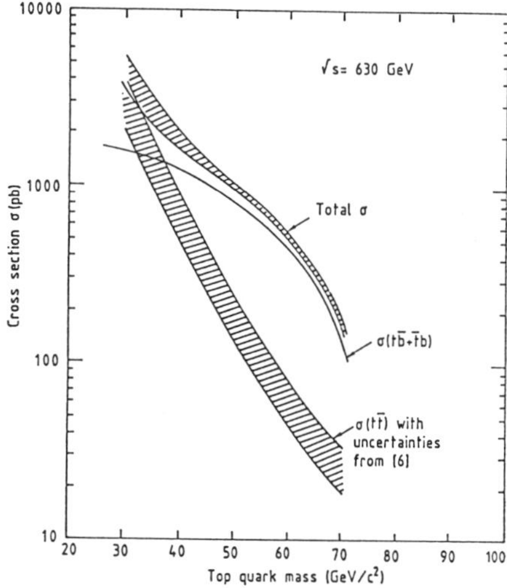

Fig.2 Cross-sections for top production in p$\bar{\text{p}}$ interactions at $\sqrt{s}$=630 GeV.

calculation of Ref.[10]. Figure 2 shows production cross-section as a function of top mass for both the strong and weak processes. The band indicates the uncertainty associated with the strong production cross-section, and in order to be conservative in this analysis, the lower limit has been used to calculate expected top signals. This lower limit is approximately 0.7 of the central value independent of the top mass.

The dominant decay mode for both strong and weak production results in a multi-jet state, but this is swamped by the QCD background. Consequently, this analysis looks for the semi-electronic decay channel (t$\rightarrow$ beν) which has a branching ratio of approximately 1/9. The signature is one electron with high transverse momentum, large missing transverse momentum resulting from the neutrino, and at least one jet from either the semi-electronic decay or the associated quark.

4 DATA SELECTION

The 1988 data sample corresponds to 2.7 pb^{-1}. All events have just one electromagnetic cluster with a transverse momentum, $p_T > 12$ GeV. The 1989 data sample is equivalent to 4.4 pb^{-1}. These events have a raw missing transverse momentum above 9.5 GeV and a calorimeter cluster with transverse momentum above 6 GeV, in addition to the electromagnetic cluster.

The analysis commences by selecting electron candidates employing the criteria discussed in section 2. The efficiency for each stage of electron identification is estimated using electrons from W candidates and testbeam studies, but owing to the additional complexity of top events the efficiencies have been reduced accordingly. For masses from 30 GeV to 70 GeV, the estimated electron selection efficiency increases from 35.3% to 41.4% for t$\bar{\text{b}}$, and decreases from 36.1% to 33.8% for t$\bar{\text{t}}$. Events were rejected if the high p_T electron was in the endcap regions of the calorimeter, where QCD background is still high but where electrons from top quarks are less likely to be found.

The corrected missing transverse momentum ($\not{p}_T$) distribution for the events in the 1988 data sample with an electron candidate has a peak at low $\not{p}_T$ which is caused by the QCD background where one arm of a 2-jet event passes the electron selection criteria. Most of the QCD background is rejected with a cut, $\not{p}_T > 15.0$ GeV.

Events are rejected unless there is at least one jet present with transverse momentum greater than 10 GeV. This cut is efficient for top, and it reduces the W$\rightarrow$ eν contribution significantly. To reduce the efficiency for QCD background to survive this cut, the jet is rejected if $|\eta| > 2.2$.

Finally, a cut is applied against the back-to-backness of the electron and the jet, such that they must be separated by less than 160° in azimuth. This rejects both QCD where one of the jets fakes an electron, and Z$\rightarrow$ e$^-$e$^+$, where one of the two electrons is mis-identified as a jet.

The transverse mass distribution of the identified electron and neutrino ($M_T{}^{e\nu}$) for the 137 events remaining after all the cuts have been applied can be seen later in figure 5, where it is eventually used to extract a mass limit.

5 BACKGROUND PROCESSES

5.1 $W \to e\nu$ and $W \to \tau\nu, \tau \to e\nu\nu$

W events contribute to the final sample if there is associated jet production in addition to the electron and neutrino emitted at high transverse momentum. The EKS Monte Carlo programme [11] was used to model the $M_T^{e\nu}$ distribution of W events, and the results are shown for W$\to$ eν and W$\to\tau\nu$ in fig.3. Although the EKS Monte Carlo programme accurately predicts the transverse mass distribution, the number of events expected has a high uncertainty. Hence, absolute normalisation is achieved by assuming that the 105 events in the final data sample with $M_T^{e\nu} > 50$ GeV are all W events, and then comparing this region with the EKS Monte Carlo. This leads to an expectation of 148.5 ± 14.5 W events in the full transverse mass distribution. The W background dominates in this analysis, and with this contribution alone any significant top contribution can be excluded.

5.2 $Z \to e^+ e^-$

In this case one electron is mis-identified as a jet. The difference in calorimeter response with respect to electrons and hadronic jets gives rise to missing momentum. The cut on the back-to-backness of the electron and jet reduces the Z contribution from 15 ± 3 to 1.7 ± 0.5 events.

5.3 $Z \to \tau\tau, \tau \to e\nu\nu, \tau \to \nu + X$

The EKS Monte Carlo programme was employed to estimate this background. Owing to the softer transverse momentum spectrum of the electron relative to the $Z \to e^+ e^-$ process, the contribution is only 0.8 ± 0.3 events.

5.4 $p\bar{p} \to b\bar{b}+$ x, b$\to$ eνc

This process is similar to the top signature for which we are searching, but the transverse momentum spectra of the electron, neutrino and jet are all softer. The $b\bar{b}$ background was estimated using the Eurojet Monte Carlo programme [12] and the production cross-section as measured by UA1 [13]. Few $b\bar{b}$ events passed the trigger requirements, and even fewer had high missing transverse momentum, or an electron which survived the electron selection criteria, owing to their non-isolated nature. All these factors result in an estimate of only 1.0 ± 0.6 events contributing to the final sample.

5.5 Jets mis-identified as electrons

QCD jets can fake electrons, and in addition give rise to $\not{p}_T$, resulting from incorrect weighting of energy in the calorimeter. Distributions of samples dominated by QCD events (with poor electron quality or low $\not{p}_T$) were used to quantify QCD contamination in the final sample. This resulted in a background estimate of 2.4 ± 1.5 events.

The final background estimate for all processes is 154.4 ± 14.6 events. The estimate for the transverse mass region 15-50 GeV is 26.2 ± 3.4 events.

6 ESTIMATION OF THE TOP SIGNAL

6.1 Monte Carlo Simulation of Top events

In order to determine the expected number of top events, the Eurojet Monte Carlo programme was used to obtain an acceptance and transverse mass distribution. The Eurojet simulation includes the matrix elements for higher order tree level processes in heavy quark production ($O(\alpha_s^2)$ for t$\bar{b}$ and

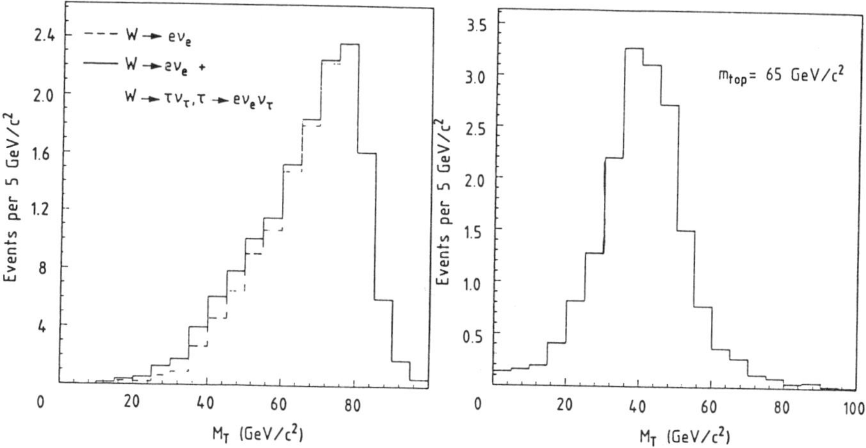

Fig.3 $M_T{}^{e\nu}$ distribution for W events   Fig.4 $M_T{}^{e\nu}$ distribution for 65 GeV top.

$O(\alpha_s^3)$ for $t\bar{t}$). The lack of a calculation complete to higher orders should underestimate the top acceptance as such diagrams will only serve to increase the number of jets in the final state.

After hadronisation into a top meson or baryon, top quark decay was simulated, such that the branching fractions used were those expected in the Standard Model. The fraction of the top quark momentum carried by the top hadron was drawn from the parameterisation of Ref.[14]. Light quarks were fragmented following the parameterisation of Field and Feynman [15].

In order to estimate acceptance, events were put through a full simulation of the UA2 calorimeter, and then treated as much as possible in the same way as the real data.

6.2   Systematic Errors in Acceptance

The uncertainty from simulating the underlying event is reduced by replacing the Eurojet underlying event by the energy pattern of minimum bias events as measured in UA2. Experience with W events shows that reasonable simulation of the underlying event can be achieved by superimposing two minimum bias events. By comparing the acceptance in this case with that resulting from superimposing one and three minimum bias events, the systematic error due to the underlying event could be estimated. This was found to be $\pm4\%$ for $t\bar{b}$ and $\pm2\%$ for $t\bar{t}$ for a top mass of 65 GeV.

The response of the calorimeter to low energy hadrons is uncertain and so, in order to give a pessimistic acceptance, the response curve was adjusted to provide the lowest response consistent with testbeam data. Allowing for uncertainty in the absolute energy scale of the calorimeter leads to a worst case where the acceptance is reduced by 5% for $t\bar{b}$ and 2% for $t\bar{t}$, for a top quark mass of 65 GeV.

The parameters used in the fragmentation functions were varied within limits consistent with the observed energy flow in jets with transverse energy about 10 GeV as measured in UA2. In the worst case the loss in acceptance was 2% for both $t\bar{b}$ and $t\bar{t}$.

A lower limit can now be set on the acceptances by setting each of the above parameters to the most pessimistic values.

For top quarks in the mass range from 30 to 70 GeV the acceptance increases from 1.8(1.5)% to 15.3(13.6)% for $t\bar{b}$ and 1.9(1.7)% to 35.5(32.3)% for $t\bar{t}$. The number in brackets indicates the lower limit. The expected number of events from both processes in the final data sample falls from 39.2(26.2) to 11.5(8.5)in the full transverse mass range.

530

The errors on the integrated luminosity $(7.1\pm0.5\ \mathrm{pb^{-1}})$, on the observed number of $W\to e\nu$ decays, on the electron cut efficiency and on the number of Monte Carlo events have not been included in the numbers just quoted. Instead, they will be treated as independent Gaussian errors in the determination of the top quark mass limit. Figure 4 shows the predicted $M_T^{\,e\nu}$ distribution for a top mass of 65 GeV. Not only do we expect a significant number of events, a lower limit of 13.6, but they also peak at a lower transverse mass compared to the dominant background source.

7 EXTRACTING MASS LIMITS

A mass limit was obtained by comparing the $M_T^{\,e\nu}$ distribution of the data with that expected from background sources alone, and with the addition of a top signal. Before describing this likelihood fitting method, the strength of the analysis can be demonstrated by using a simple argument [16] based on Poisson statistics. Taking the case of a 65 GeV top quark, we would expect to see a lower limit of 10.6 events in addition to the predicted background in the $M_T^{\,e\nu}$ region 15-50 GeV. Comparing this to our observed number of events and accounting for the error in the background estimate, the existence of a 65 GeV top quark can be excluded at the 98.6% confidence level.

For the likelihood method, a fit is performed to the observed events with two free parameters, corresponding to the fraction of the event sample due to top and to W events. Remaining events are attributed to the remaining background sources. Transverse mass distributions for top and W events were derived from Eurojet and EKS Monte Carlo programmes respectively. The exact distribution for other background sources was uncertain, but the final result is insensitive to this uncertainty. The total likelihood function normalisation was constrained, within Poisson statistics, to the total number of 137 observed events in the data sample. The 90 and 95% confidence limits on the number

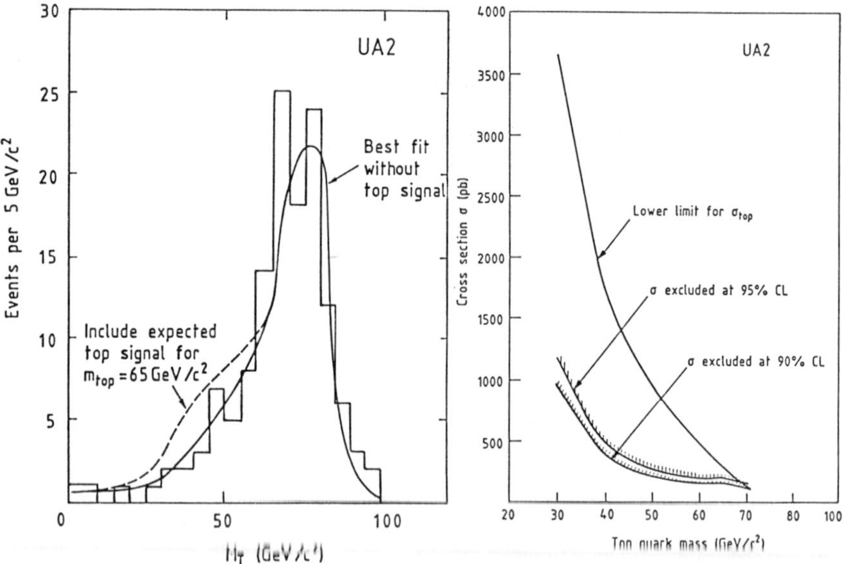

Fig.5 Best fit (full curve) to the $M_T^{\,e\nu}$ distribution with no top signal. The lowest expected contribution from 65 GeV top quark is added (dashed curve).

Fig.6 Lower limit for the top production cross-section and the 90 and 95% CL excluded cross-sections as a function of top quark mass.

of top events in the sample were obtained by integrating the likelihood distribution over all possible values of the signal. Fig.5 shows the case of a 65 GeV top quark. The solid line shows the best fit to the data with no signal, and the dotted line is the best fit with the inclusion of a lower limit contribution of 13.4 events from top. This hypothesis can be excluded at the 99.0% CL.

Allowing for acceptances, the fit method can be used to exclude production cross-sections at the 90 and 95% CL. Figure 6 shows the excluded region and the expected lower limit for top production as a function of mass. The existence of the top quark with a mass between 30 and 69 (71)GeV can be excluded at the 95(90)% confidence level.

The search for the top quark can be adapted in order to extract a mass limit for the b' quark, a member of a hypothetical fourth family. It can be assumed that the partner of the b' quark is too heavy to allow producton via the electroweak process, thus restricting the production to b' pair production via the strong process. Without correcting for the slightly different decay matrix element for b', this analysis implies $m(b')>53$ (56) GeV at the 95 (90)% confidence level. If the correct matrix element for b' decay is used, and it is assumed that the b' branching ratio to a charm quark and a virtual W boson is 100%, then this implies $m(b')>54$ (57)GeV at the 95 (90)% confidence level.

8 CONCLUSIONS

With 7.1 pb^{-1} of data accumilated at the CERN SPS Collider, a search has been performed for production and decays of the top quark and b' quark. The final data sample after applying selection criteria shows no evidence for such processes. This implies lower limits for the mass of these quarks, such that at the 95% confidence level $m(top)>69$ GeV and $m(b')>54$ GeV. At the 90% confidence level $m(top)>71$ GeV and $m(b')>56$ GeV.

REFERENCES

[1] For example:J. Ellis and G.L. Fogli, Phys.Lett. B231 (1989) 189, and references therein:
     P. Langacker, Phys. Rev. Lett. 63 (1989) 1920.

[2] A. Beer et al., Nucl. Inst. and Meth. 224 (1984) 360.

[3] F.Alberio et al., The Electron, Jet, and The Missing Transverse
     Energy Calorimetry of the Upgraded UA2 Experiment at the CERN
     Collider, to be published in Nucl. Inst. Meth..

[4] R. Ansari et al., Nucl. Inst. Meth. A279 (1989) 388.

[5] F. Bosi et al., CERN-EP/89-82 (1989).

[6] R. Ansari et al., Nucl. Inst. Meth. A263 (1988) 51.

[7] J. Alitti et al., Nucl. Inst. Meth. A279 (1989) 364.

[8] UA2 Collaboration, H. Plothow-Besch, Proc. of the 1989 Europhys.
     Conf. on High Energy Phys. (Madrid), to be published in Nucl. Phys.

[9] G. Alterelli et al., Nucl. Phys. B308 (1988) 724.

[10] P.Nason, S. Dawson and R.K. Ellis, Nucl. Phys. B303 (1988) 607.

[11] S.D. Ellis, R. Kleiss, and W.J. Stirling, Phys.Lett.154B(1985)435;
     F.A. Berends et al., Phys.Lett. 224B (1989) 237.

[12] A. Ali and B. van Eijk, Proc. 5th Top. Work. on Proton-Antiproton
     Coll. Phys., St Vincent, Aosta, Italy (1985);
     B. van Eijk, CERN-EP/85-121 (1985); A. Ali et al., CERN-TH.4523/86 (1986);
     A. Ali et al., Nucl. Phys. B292 (1987) 1.

[13] UA1 Collaboration, C.Albajar et al., Z.Phys. C37 (1988) 489.

[14] C. Peterson et al., Phys. Rev. D27 (1983) 105.

[15] R.D. Field and R.P. Feynman, Nucl. Phys. B136 (1978) 1.

[16] Particle Data Group, Review of Particle Properties, Phys.Lett. 204B (1988) 81.

Properties of Inclusive W,Z Events
in $\bar{p}p$ Collisions at 1.8 TeV

CDF Collaboration*
Presented by T.Watts, Rutgers University.

Abstract

A preliminary measurement of the properties of W and Z production along with accompanying jets has been made in $\bar{p}p$ collisions at 1.8 TeV using the CDF detector at Fermilab. Distributions of jet multiplicity, and boson E_T, with and without selection on jet multiplicity, were obtained. Agreement was found with perturbative QCD predictions.

Introduction

The production of W and Z bosons, along with hard jets, in $\bar{p}p$ collisions provides a quantitative test of perturbative QCD. In addition, such events can show contributions from new physics such as heavy quark or Higgs particle decays. The recent CDF run, at Tevatron energy and with integrated luminosity of 4.4 pb^{-1}, provided a sample of W,Z events with larger statistics than has been available before. Results are shown for the boson decays $W \to e\,\nu$ and $Z \to e^+e^-$.

The data was taken in the CDF detector [1]. Important parts of the detector for this analysis were: scintillation counter planes (BBC) upstream and downstream of the interaction region which were used to record interactions in the trigger and as a monitor of $\bar{p}p$ luminosity; vertex time projection chambers (VTPC) around the interaction region which were used to determine the z coordinate of the collision; central tracking drift chamber (CTC) and axial magnetic field; electromagnetic and hadron calorimeters which covered polar angles to within 2° of the beam pipe; proportional chambers with wire and strip readout embedded in the EM calorimetry to give accurate EM shower positions.

Readout of data was triggered by a particle in one side of the BBC counters, by EM clusters in the calorimetry with E_T above 12 GeV, an associated track momentum above 6 GeV/c, and a ratio of transverse energy in hadron and EM parts of the cluster less than 0.125. The events were also filtered by an online farm of microprocessors which tested shower shape *inter alia*. About 4500 W candidates were collected and 350 Z candidates.

Energy scales of the detector components were calibrated with test beam data and with the mass peaks from $K^0_s \to \pi^+\pi^-$, $J/\psi \to \mu^+\mu^-$, and $\Upsilon \to \mu^+\mu^-$ [2]. Calorimeter energies were compared to track data, and dijet E_T balancing was used to compare calorimeters at different polar angles.

*The CDF Collaboration consists of the following institutions: Argonne National Laboratory, Brandeis University, University of Chicago, Fermi National Accelerator Laboratory, Laboratori Nazionale di Frascati (INFN), Harvard University, University of Illinois, National Laboratory for High Energy Physics (KEK), Lawrence Berkeley Laboratory, University of Pennsylvania, INFN University and Scuola Normale Superiore of Pisa, Purdue University, Rockefeller University, Rutgers University, Texas A&M University, University of Tsukuba, Tufts University, and University of Wisconsin.

Analysis and Results

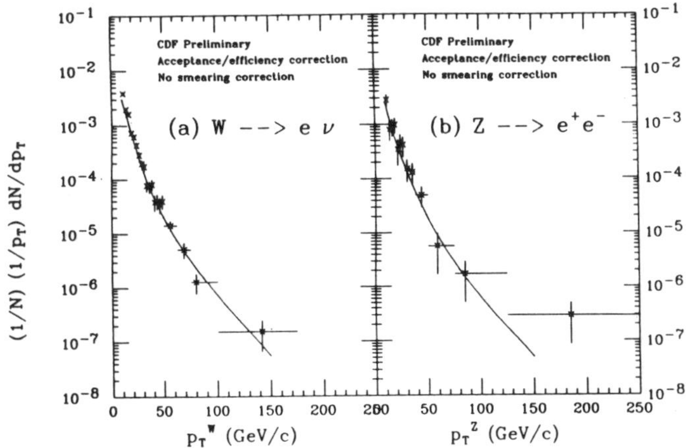

Figure 1: Inclusive p_T spectra for (a) W and (b) Z events. The solid line is the theoretical calculation of Bawa et al., and Martin et al.[8]

Stringent cuts were made for one central electron in all events[4]. These cuts included $E_T \geq$ 20 GeV, isolation of the EM cluster, shower shape tests, match of cluster and track momenta and of positions to a few cm, and fiducial cuts to avoid cracks and bad towers in calorimetry.

Hadron clusters were found in the calorimetry with a fixed cone size of 0.7 in η-ϕ space [5]; these jets were required to have $E_T \geq$ 10 GeV. From calorimeter responses to individual particles in test beam data, and from Monte Carlo simulation of jet fragmentation and detector responses, a correction was applied to jet energies [5]. At low jet energies this is strong correction, e.g. 10 GeV after correction becomes 15 GeV.

The missing E_T in the W events was required to be $\geq$ 20 GeV and large compared to its expected fluctuation. In order to apply calorimeter energy corrections, the E_T sum over all towers was separated into 3 components: electron cluster, jet clusters, and underlying event contribution from the rest of the calorimeter towers. Jets were corrected as above, and the underlying event multiplied by 1.2, a factor derived from analysis of the underlying event contribution in Z events.

The W was identified by calculating the transverse mass of electron and missing E_T, and requiring this mass $\geq$ 40 GeV/c² [3]. For the Z, the second electron passed less stringent cuts and was also allowed up to $| \eta | \leq$ 2.2; the cut on the mass of two electrons was 75 through 105 GeV/c². After these cuts and those discussed above, 2685 W events remained and 220 Z events.

The inclusive distributions of p_T of both W and Z are shown in Figure 1. These plots show shape only and have a background in the mass cuts estimated to be 4% not yet subtracted. The W distribution is expected to have a background which could reach 20% at p_T above 80 GeV/c, and the effects of "smearing" caused by calorimeter resolution have not yet been unfolded; this smearing could change the W distribution by 15% at p_T above 80 GeV/c. The data points are corrected for acceptance and efficiency. The W and Z distributions are very similar and are compared to an $\mathcal{O}(\alpha_s^2)$ QCD calculation [8] which used the MRSB parton distribution. At this stage of the data, agreement is quite acceptable.

The fraction of W or Z events with n jets, where n=0,1,2,3,4, is shown in Figure 2. Again W and Z distributions are very similar and the relation $(\sigma_n/\sigma_0) \simeq (\sigma_1/\sigma_0)^n$ [10] is approximately satisfied. Comparison is made to a QCD calculation with up to 3 jets [9] with parton E_T above 15 GeV, and to a Monte Carlo event generator [7] which gives up to 2 jets. This latter comparison used ISAJET for fragmentation and underlying event simulation [6], and used the CDF detector simulation; generated events were passed through the data analysis chain and the plots normalized by luminosity. Both predictions agree with the data.

Figure 3 investigates the agreement between data and perturbative QCD in more detail by showing the p_T distribution for W bosons in events with 1 and 2 accompanying jets. The QCD model compared is again the PAPAGENO/ISAJET/CDFSIM chain and comparison and cuts are made in uncorrected p_T.

In summary, perturbative QCD predictions for the production of W and Z bosons accompanied by jets shows acceptable agreement at this stage of the data.

References

[1] Abe, F. *et al.*, Nucl. Inst. and Meth. **A271**, 387 (1988).

[2] Abe, F. *et al.*, Phys. Rev. Lett. **63**, 720 (1989)

[3] Abe, F. *et al.*, Phys. Rev. Lett. **62**, 1005 (1989);

[4] Proudfoot, J., Proceedings of the Workshop on Calorimetry for the Superconducting Supercollider, Tuscaloosa, Alabama, 12-17 March 1989; ANL-HEP-CP-89-40.

[5] Abe, F., *et al.*, Phys. Rev. Lett. **62**, 613 (1989).

[6] Paige, F. and Protopopescu, S.D., ISAJET Monte Carlo V6.21, BNL Report No. BNL38034 (1986).

[7] Hinchliffe, I., PAPAGENO (Parton-level event generator) Monte Carlo program.

[8] Bawa, A.C. and Stirling, W.J., Phys. Lett. **B203**, 172 (1988); Martin, A.D. *et al.*, Z.Phys. **C42**, 277 (1989).

[9] Berends, F.A. *et al.*, University of Leiden Preprint, Preprint-89-0318 (1989)

[10] Ellis, S.D., Kleiss, R., and Stirling, W.J., Phys. Lett. **154B**, 435 (1985).

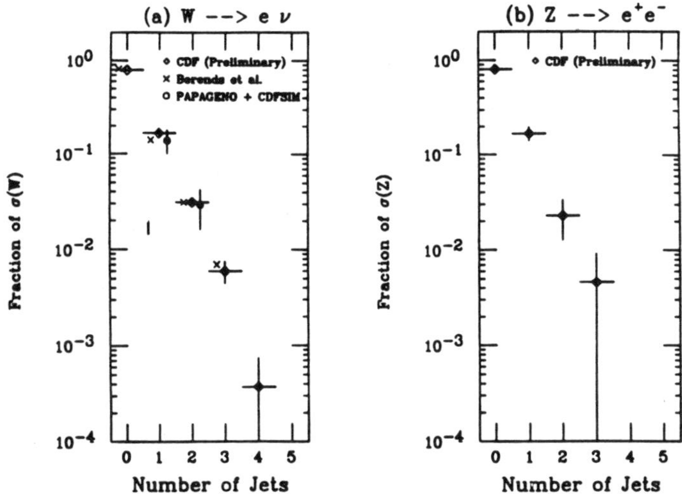

Figure 2: Jet multiplicity distributions for (a) W and (b) Z events. Comparison is made to theoretical predictions of Berends et al. [9], and PAPAGENO/ISAJET/CDFSIM [7].

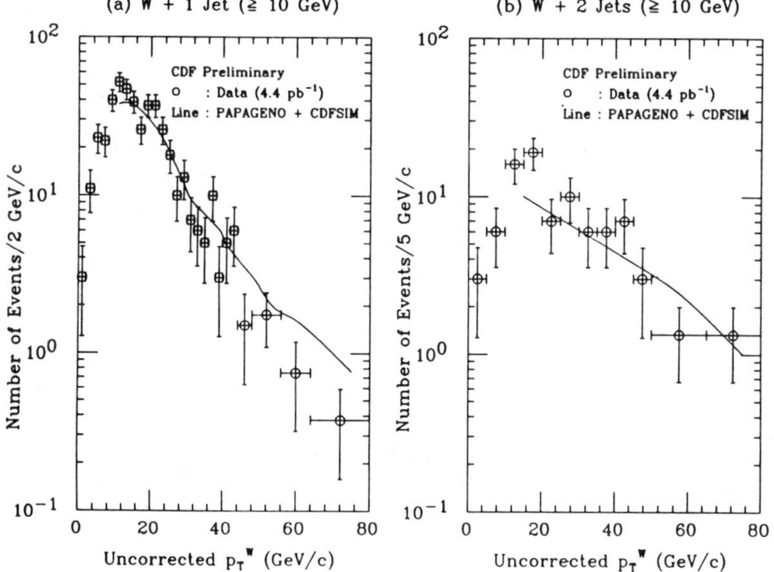

Figure 3: (a) W p_T spectrum for W + 1-jet events. (b) W p_T spectrum for W + 2-jet events. The solid curves are predictions from PAPAGENO/ISAJET/CDFSIM [7].

THEORY OF THE W/Z TRANSVERSE MOMENTUM DISTRIBUTION

Richard J. Gonsalves
Department of Physics and Astronomy
State University of New York at Buffalo
Buffalo, New York 14260, U.S.A.

ABSTRACT

QCD predictions for the inclusive production of W and Z bosons with fixed transverse momentum Q_T in $p\bar{p}$ collisions at the Tevatron are reviewed in the light of recently completed calculations of the full $\mathcal{O}(\alpha_s^2)$ QCD radiative corrections.

1. INTRODUCTION

The Standard Model makes rather precise predictions for the production of W and Z bosons with invariant mass-squared $Q^2 = M_W^2$ or M_Z^2 as a function of transverse momentum Q_T at Tevatron energies, $\sqrt{S} \approx 2$ TeV. The total cross sections are $\sigma_W \approx 15$ nb and $\sigma_Z \approx 5$ nb. The typical momentum fractions of the annihilating partons in the underlying Drell-Yan mechanism are of order $x \approx \sqrt{\tau} = \sqrt{(Q^2/S)} \approx 0.04$. At these x values, gluons contribute roughly 30% of the events at large Q_T via the Compton process $qG \to W + X$: this is a dramatic qualitative difference from CERN Collider energies at which the dominant subprocess is $q\bar{q}$ annihilation. Thus the large Q_T distribution is a potentially useful probe of the gluon density in the proton. Since large values of Q_T probe short distances, the Q_T distribution is a good place to look for new physics. Possible signatures include anomalously large numbers of events with lepton + jet(s) + missing transverse energy ($\not{E}_T$) in the W distribution, and monojets or jets + $\not{E}_T$ in the Z distribution.

Perturbative QCD may be used to obtain precise testable predictions [1,2] for $d\sigma/dQ_T$ in the range 20 GeV $\lesssim Q_T \lesssim$ 250-300 GeV: the upper limit values are determined by Tevatron luminosity, and the lower limit is roughly the momentum at which large infrared logarithms of Q^2/Q_T^2 begin to spoil the convergence of the perturbation series. In the range 2 GeV $\lesssim Q_T \lesssim$ 20 GeV these infrared logarithms must be summed to all orders in α_s to yield a "Sudakov" form factor which causes the distribution to peak at ≈ 10 GeV. Theory seems to indicate [3] that this Sudakov region is cleanly separated from the region below ≈ 2 GeV where perturbatively intractable parton intrinsic momentum effects dominate.

2. CHOOSING RENORMALIZATION AND FACTORIZATION SCALES

The Lagrangian of QCD depends in principle on only one parameter $\Lambda \approx 0.2$ GeV if quark masses and the CP-violating θ parameter can be ignored in making predictions at high energies. In practise, the wave function of the proton cannot be calculated with sufficient accuracy to allow predictions to be made at high energies, and one must rely on the QCD-improved parton model formula

$$\frac{d\sigma}{dQ_T} = \sum_{ab} \int dx_a dx_b \, f_a(x_a, M^2) f_b(x_b, M^2) \frac{d\sigma^{ab}}{dQ_T^2}(\alpha_s(\mu^2), \mu^2, M^2) \,, \tag{1}$$

where σ^{ab} is the parton level distribution for partons with momentum fractions $x_{a,b}$ from which ultraviolet and collinear infrared divergences have been subtracted. This formula depends on the physical parameter Λ in an explicitly calculable way through the effective coupling

$$\frac{\alpha_s(\mu^2)}{2\pi} = \frac{6}{(33 - 2n_f)\ln(\frac{\mu^2}{\Lambda^2})} - \frac{\beta_2 \ln\ln(\frac{\mu^2}{\Lambda^2})}{\beta_1^3 \ln^2(\frac{\mu^2}{\Lambda^2})} \,, \tag{2}$$

as well as implicitly in an incalculable way through the parton densities $f_a(x, M^2)$. The renormalization scale μ is a remnant of the process of discarding ultraviolet divergent contributions "at scale μ" according to some arbitrary prescription (taken to be $\overline{\text{MS}}$ renormalization in what follows). The factorization scale M is likewise a remnant of the process of discarding collinear singularities according to some arbitrary prescription (taken to be $\overline{\text{MS}}$ factorization in what follows). In principle, the parameters μ and M are unphysical, unrelated to one another, and not specified by the theory in terms of any measured energy scale E_s (which can for example be chosen to be Q_T, $\sqrt{Q^2}$, $\sqrt{Q^2 + Q_T^2}$ or $\sqrt{S}$). Indeed, proofs of renormalizability, and factorization theorems imply that

$$\mu\frac{\partial}{\partial\mu}\left(\frac{d\sigma}{dQ_T}\right)_{\text{exact}} = 0 \,, \qquad M\frac{\partial}{\partial M}\left(\frac{d\sigma}{dQ_T}\right)_{\text{exact}} = 0 \,. \tag{3}$$

In practise, this exact scale independence is spoiled if the perturbation series in α_s is truncated at any finite order. This unfortunate fact is signalled by the appearance in σ^{ab} of powers of $\ln(\mu^2/E_s^2)$ and $\ln(M^2/E_s^2)$ which are not precisely compensated by the μ dependence of α_s and the M dependence of $f(x, M^2)$. While the potentially destabilizing logarithms can be made small by choosing

$$\zeta_\mu \equiv \frac{\mu^2}{E_s^2} \approx 1 \,, \qquad \zeta_M \equiv \frac{M^2}{E_s^2} \approx 1 \,, \tag{4}$$

the theory makes absolutely no more precise determination of $\zeta_{\mu,M}$. This spurious scale dependence is of course precisely cancelled by higher (uncalculated) orders of perturbation theory. The logically optimal (but obviously unworkable) solution would be to choose scales which minimize the remainder of the perturbation series. Various strategies for guessing the unknown remainder have been suggested. Two such strategies will be discussed here. The first is (a simplified version of) Stevenson's "Principle of Minimal Sensitivity" [4] in which the unphysical scales are fixed at a local extremum or saddle point of $d\sigma/dQ_T$ in the variables ζ:

$$\zeta_{\mu,M}^{\text{SP}}(Q_T): \quad \zeta_\mu\frac{\partial}{\partial\zeta_\mu}\left(\frac{d\sigma}{dQ_T}\right)_{\text{NLO}} = 0 \,, \quad \zeta_M\frac{\partial}{\partial\zeta_M}\left(\frac{d\sigma}{dQ_T}\right)_{\text{NLO}} = 0 \,, \tag{5}$$

where SP stands for saddle point and the subscript NLO signifies that the parton level cross section is computed through next-to-leading order in perturbation theory and the parton densities have likewise been evolved using the next-to-leading order Altarelli-Parisi equations. Two remarks may be made concerning this choice of scales: First, there is the question of the existence and uniqueness of a saddle point. The answer to this question is not known, but in practise, there does usually seem to exist a fairly well defined saddle point at physically reasonable values of ζ. Second, even if a unique saddle point does exist, there is no guarantee that using the saddle point scales yields the best estimate of the uncalculated remainder of the series. The NLO cross section at the saddle point has the same derivatives with respect to ζ as the exact cross section, but not necessarily the same magnitude. A second strategy for choosing the unphysical scales ζ may be called the effective scales (ES) prescription, which is a simplified version of Grunberg's "Method of Effective Charges" [5]. One would like to be able to choose the scales such that the NLO approximation is equal in magnitude to the exact distribution. In the absence of any information about the remainder of the series, we choose scales such that the leading order (LO) cross section is equal in magnitude to the cross section computed through next-to-leading order:

$$\zeta^{\rm ES}_{\mu,M}(Q_T): \quad \left(\frac{d\sigma}{dQ_T}\right)_{\rm LO} = \left(\frac{d\sigma}{dQ_T}\right)_{\rm NLO}. \tag{6}$$

This equation determines a line of possible effective scale choices in the $\zeta_\mu - \zeta_M$ plane.

The choice of scales was studied by Bawa and Stirling [6] for the non-singlet (i.e., valence quark) contributions to the Q_T distribution at $\sqrt{S} = 630$ GeV. They chose $E_s = Q_T$ and set $\zeta_M = \zeta_\mu$, and found that the one dimensional analogs of Eqs. 5 and 6 yielded $\zeta^{SP} \simeq \zeta^{ES} \simeq 0.16$ for $Q_T > 15$ GeV. The situation at $\sqrt{S} = 1.8$ TeV, where gluon initiated contributions cannot be neglected is more complicated. It was found in Refs. [1,2] that if ζ_M and ζ_μ were set equal, then the one dimensional version of Eq. 5 had no physically acceptable solution for $Q_T \lesssim 80$ GeV. We have since studied this problem [7] by allowing the two scales to vary independently of one another, as has been done for prompt photon production by Aurenche et al., [8]. We find that there is a well defined saddle point in the $\zeta_M - \zeta_\mu$ plane. Results of this analysis for W production and using the parton densities of Martin, Roberts and Stirling are shown in Fig. 1. The saddle point scales are seen to move away from the diagonal $\zeta_M = \zeta_\mu$ line for lower values of Q_T. It is interesting that there is an apparent clustering of ES choices at $\zeta_M \approx \zeta_\mu \approx .1$, and that the SP scales move toward this region for large Q_T.

What is the correct choice of scales? Fig. 2 illustrates what we believe is the answer to this question. In this figure are plotted the leading order (LO) and the full (NLO) distributions evaluated at the naive choice of scales $\zeta_M = \zeta_\mu = 1$, and at the saddle point scales. It is apparent that the LO distribution is quite sensitive to variations in scale. Inclusion of the radiative corrections considerably decreases this sensitivity and makes the choice of scales irrelevant! If experimental errors were to become smaller than the difference between the NLO curves in Fig. 2, the only rigorous way to improve the theoretical prediction would be to compute the $\mathcal{O}(\alpha_s^3)$ radiative corrections, thus further narrowing the difference and rendering a choice of scales unnecessary at this increased resolution.

3. OTHER SOURCES OF THEORETICAL UNCERTAINTY

The main emphasis of this review has been on renormalization and factorization scale ambiguities. In this section we briefly discuss two important additional theoretical issues.

Parton densities

The most significant contribution to the theoretical uncertainty at large Q_T arises from different choices of input parton densities. For consistency, next-to-leading order parton density parametrizations should be used in conjunction with the next-to-leading parton level cross sections of Refs. 1,2. Ref. 1 presents results using the parametrizations of Martin, Roberts and Stirling [9]. Cross sections derived using the MRSB and MRSE sets typically differ from one another by less than 20% for all $Q_T > 20$ GeV. Refs. 2 present results using the parametrizations of Diemoz et al. The different DFLM sets yield distributions that agree to within $\pm 15\%$. Taking both analyses into account, QCD with 5

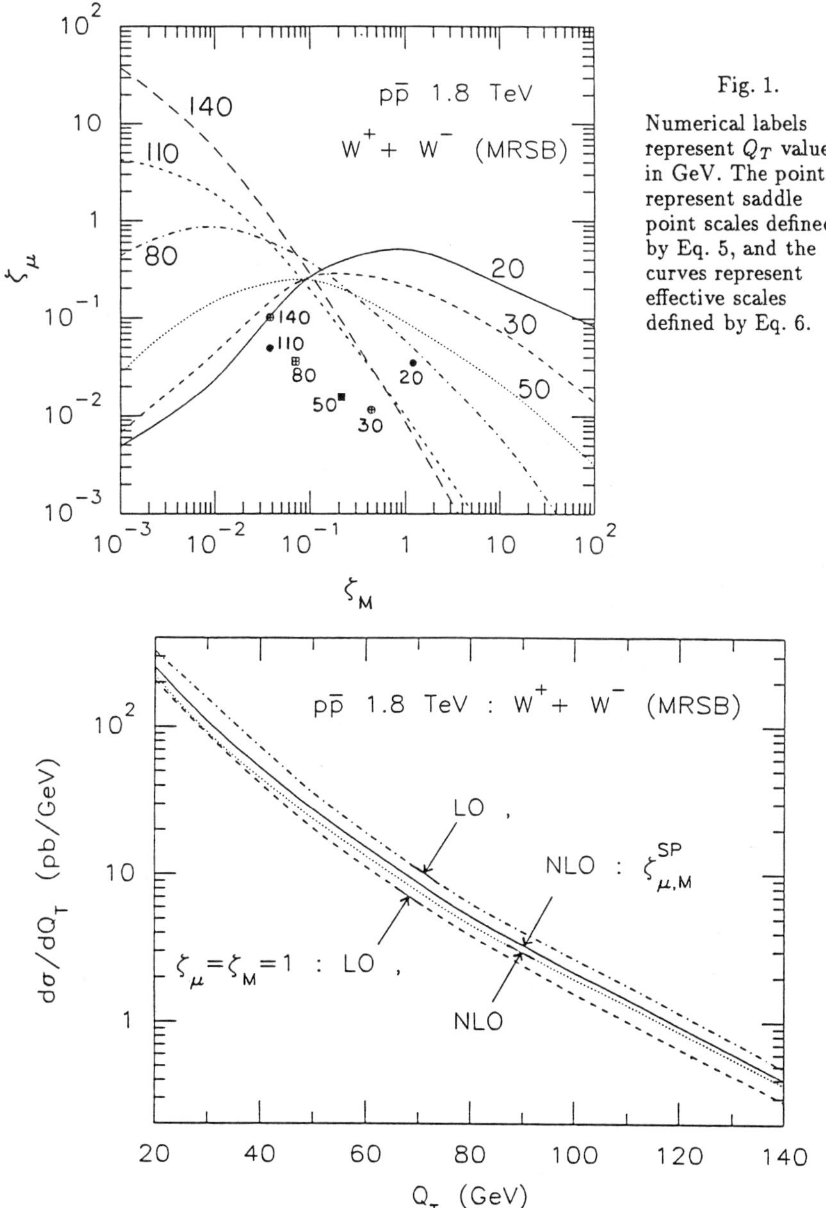

Fig. 2. Including $\mathcal{O}(\alpha_s^2)$ contributions (NLO) makes a choice of scales unnecessary.

540

light flavors yields absolute predictions for the Q_T distribution with $Q_T > 20$ GeV with a conservative uncertainty of no more than about 20-25%.

Sudakov logarithms

The predictions of QCD for $Q_T \lesssim 20$ GeV are less certain on account of the occurrence in the perturbation series of terms of the type $\alpha_s^n \ln^m(Q^2/Q_T^2)/Q_T^2$, $m \leq 2n - 1$. These terms arise from gluon radiation off the initial state partons. They tend to suppress the emission of a W with very small Q_T, and in fact drive the cross section negative. This unphysical result must be dealt with by summing these logarithms to all orders. It has been shown, principally by Collins and Soper, that this can be done consistently [10]. Various perturbative coefficients required for this analysis have been computed through $\mathcal{O}(\alpha_s^2)$ for the non-singlet distribution [11], and a prescription for matching the exponentiated logarithms at small Q_T with the perturbative large Q_T tail has been given [3]. The principal result of a rather involved analysis is that the distribution should peak at $Q_T \approx 10$ GeV. However, a full analysis including gluon initiated events has not been performed.

4. CONCLUSIONS

QCD makes firm and reliable predictions for the W and Z transverse momentum distributions at large Q_T. Any deviation from these predictions would constitute a firm signal for new physics beyond the Standard Model. The inclusion of the full $\mathcal{O}(\alpha_s^2)$ QCD radiative corrections makes the choice of unphysical renormalization and factorization scales almost irrelevant. Accurate experimental measurement of the Q_T distribution can provide a means of probing the quark and especially the gluon densities in the proton. Further theoretical work needs to be done on summing Sudakov logarithms in the region of small $Q_T \lesssim 20$ GeV.

ACKNOWLEDGMENTS

I would like to thank R.K. Ellis and the organizers of DPF90 for the opportunity to present these results. I thank J. Pawłowski, C.F. Wai and C.M. Hung for their collaborative contributions. This work was supported in part by an NSF grant, PHY 87-13231.

REFERENCES

1. R.J. Gonsalves, J. Pawłowski and C.-F. Wai, Phys. Rev. D **40**, 2245 (1989).
2. P.B. Arnold and M.H. Reno, Nucl. Phys. **B319**, 37 (1989); P. Arnold, R.K. Ellis and M.H. Reno, Phys. Rev. D **40**, 912 (1989).
3. G. Altarelli et al., Nucl. Phys. **B246**, 12 (1984); Z. Phys. C **27**, 617 (1985).
4. P.M. Stevenson, Phys. Rev. D **23**, 2916 (1981); P.M. Stevenson and H.D. Politzer, Nucl. Phys. **B277**, 758 (1986).
5. G. Grunberg, Phys. Rev. D **29**, 2315 (1984).
6. A.C. Bawa and W.J. Stirling, Phys. Lett. **203B**, 172 (1988).
7. R.J. Gonsalves, J. Pawłowski and C.-F. Wai, UB-HET 90/3, submitted to Phys. Lett.
8. P. Aurenche, R. Baier, M. Fontannaz and D. Schiff, Nucl. Phys. **B286**, 509 (1987).
9. A.D. Martin, R.G. Roberts and W.J. Stirling, Phys. Rev. D **37**, 1161 (1988).
10. J.C. Collins, D.E. Soper and G. Sterman, Nucl. Phys. **B250**, 199 (1985).
11. C.T.H. Davies, W.J. Stirling and B.R. Webber, Nucl. Phys. **B256**, 413 (1985); J. Kodaira and L. Trentadue, Phys. Lett. **112B**, 66 (1982).

Charm Photoproduction

M. V. Purohit

Physics Dept., Princeton University

Princeton, NJ 08544

Abstract

Results on the photoproduction of 10000 charmed particles from the 10^8 recorded triggers of Fermilab experiment E691 have been analyzed in the photon-gluon fusion model. We find that the total cross-section, its rise with energy, and the p_T^2 and x_F distributions can be explained by a high mass for the charm quark ($m_c = 1.74^{+0.13}_{-0.18}$ GeV/c^2) and a soft gluon distribution ($G(x) \sim (1\text{-}x)^n$ where n=7.1$\pm$2.2).

I. Introduction

Predictions of perturbative QCD can be used to extract information on the structure of hadrons at fixed target energies. In particular, the quark structure functions have been measured in deep inelastic scattering experiments with lepton probes. These experiments are not directly sensitive to the gluon structure function, $G(x)$ which enters only as a correction in deep inelastic scattering. However, other processes such as hadro- and photo-production of heavy quarks and prompt photon production are directly sensitive to $G(x)$. In a paper published last year we described the measurement of the total photoproduced charm cross-section, its dependence on energy and the p_T^2 and x_F distributions of charmed mesons.[1] In this talk we describe the extraction of the gluon structure function and the charm quark mass from these measurements using the photon gluon fusion model[2] (PGF) and data from Fermilab experiment E691.[3] Next-to-leading order corrections have recently been calculated[4] and are included in the analysis below.

II. Data Analysis

As is to be expected, the total charm production cross-section is the quantity most sensitive to the charm quark mass. Further, it does not suffer from the uncertainties associated with fragmentation. Using the next to leading order results we find that the charm quark mass is determined to be $1.6^{+0.3}_{-0.1}$ GeV/c^2. Using the rise of the total cross-section with energy (we had determined the ratio of cross-sections at 100 and 200 GeV to be 1.96 ± 0.24) we find that $n_g = 8.8 \pm 2.3$.

The p_T^2 distribution is sensitive to both the charm quark mass and the shape of the gluon structure function. Using this distribution (see fig. 1) we find that $m_c = 1.8^{+0.2}_{-0.4}$ GeV/c^2. The shape of the gluon structure function is parameterized as $(1-x)^{n_g}$ where we determine n_g to be 5^{+5}_{-1} and m_c to be $1.8^{+0.2}_{-0.4}$. The x_F distribution is not very sensitive to the PGF model parameters, but we use it to determine the systematic errors from fragmentation uncertainties. This can be done because the distribution is sensitive to different forms of fragmentation functions. These are illustrated in fig. 2 and the differences between them are used to estimate the systematic error.

In conclusion, the results of fits to $\sigma_{c\bar{c}}$, its rise with energy, the x_F distribution and the p_T^2 distribution are tabulated in table 1. Also listed are the results of a combined fit. The most reliable result is the combined fit including data only for the region $p_T^2 > 2$ GeV^2/c^2 which yields $m_c = 1.74^{+0.13}_{-0.18}$ GeV/c^2 and $n_g = 7.1 \pm 2.2$ at $Q^2 = 20$ GeV^2. The value of m_c should be useful in similar perturbative QCD dynamics, e.g., in neutrino-induced charm production provided those calculations are also performed to next-to-leading order.

References

[1] J. C. Anjos et al., Phys. Rev. Lett. **62**, 513 (1989).

[2] L. M. Jones, and H. W. Wyld, Phys. Rev. **D17**, 759 (1978).
M. A. Shifman, A. I. Vainstein and V. I. Zakharov, Phys. Lett. **65B**, 255 (1976).

[3] J. R. Raab et al., Phys. Rev. **D37**, 2391 (1988) and additional references therein.

[4] R. K. Ellis and P. Nason, Nucl. Phys. **B312**, 551 (1989).

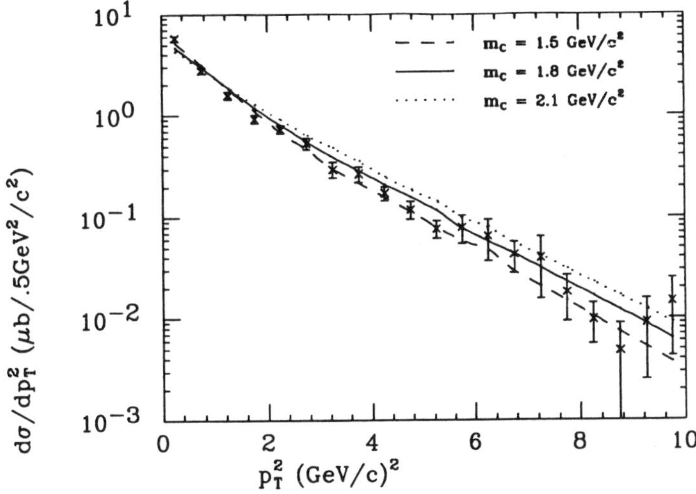

1. The p_T^2 dependence of the cross-section on Be for production of charm events ($d\sigma_{c\bar{c}}/dp_T^2$). The curves are for different values of m_c, the charm quark mass.

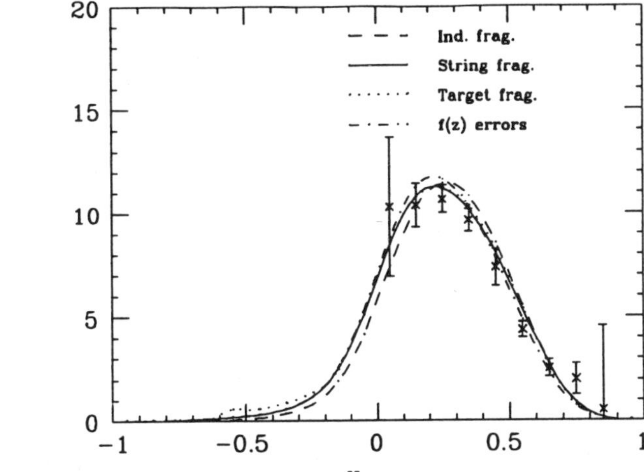

2. The x_F dependence of the cross-section on Be for production of charm events ($d\sigma_{c\bar{c}}/dx_F$). The curves are for string fragmentation (solid), independent fragmentation (dashed), for the case where the diquark carries a fixed fraction of the remnant nucleon momentum (dotted) and for the fragmentation function parameters changed by one standard deviation (dotdashed).

Preliminary Results of Elastic J/ψ Photoproduction Cross Section Measurement At Fermilab E687[*]

R. Yoshida
Department of Physics and Astronomy
Northwestern University, Evanston, IL 60208
for the E687 COLLABORATION[†]

1 Introduction

Elastic photoproduction cross section of J/ψ's has been measured, through the decay channel into two muon, at Fermilab E687 using the highest energy photon beam thus far.

2 E687 Spectrometer

The beamline and the spectrometer are described elsewhere in these proceedings[1]. Inner Muon Detectors, used in obtaining the events, consist of 3 arrays of scintillation counters and 4 arrays of 2 inch diameter proportional tubes and cover 0 to ± 30 mr, polar angle approximately.

The acceptance for $J/\psi \to \mu^+\mu^-$ was small below J/ψ lab momenta of 100 GeV/c and went to zero below ~ 50 GeV/c. We were, however, able to obtain a good sample of higher energy events.

3 Elastic J/ψ analysis

A minimum bias trigger was based on trigger scintillator signals consistent with a photon interaction with at least 2 charged particles in the spectrometer. A gap in the downstream counters discriminates against e^+e^- pairs. The second level muon trigger was the muon scintillation counters giving signal consistent with 2 muons. Events with muon trigger were stripped off the data tapes and reconstructed separately. Events were then selected with following criteria: There must be 2 oppositely charged tracks coming from a vertex. (A third track was allowed if it did not come from the vertex.) The muon counter hits must

---

[*]Supported by U.S. DOE and NSF and by INFN

[†]E687 COLLABORATION: University of Colorado; Fermilab; University of Illinois; INFN Bologna; INFN Frascati; INFN Milano; Northwestern Universiy; University of Notre Dame; INFN Pavia

[1]See paper by J. Wilson in these proceedings

be associated with the tracks(Muon ID). Finally, the invariant mass of the dimuon must be greater than 1 GeV.

We calculate, for events above invariant mass of 2.5 GeV the error on the mass measurement, $\sigma(M)$, event by event, and cut on three sigma of the Gaussian fit to the quantity $(M_0 - M)/\sigma(M)$ where M is the measured mass, and M_0 is the nominal mass of the J/ψ.

Further, a cut on extremely low momentum transfer is applied to remove the remaining Bethe-Heitler dimuons and the coherent part of the cross section. A more detailed t analysis is in progress. This cut removes $\sim 15\%$ of the events. The events surviving above cuts are called 'elastic incoherent' J/ψ's. (257 events)

Two independent normalization methods are used.

1. Determine photon flux from scaler count from a beam counter, measured electron spectrum and the measured electron energy loss spectrum.

2. Normalize to the calculable Bethe-Heitler dimuon background $(M_{\mu^+\mu^-} > 1GeV)$

   - A Bethe-Heitler dimuon Monte Carlo was written to calculate the acceptance and the high-mass cross sections

Photon flux was calculated from both methods and shows very good agreement.

4 Preliminary Results

Elastic J/ψ Photoproduction Cross Section
(Bethe-Heitler Normalization)

Energy($E_{J/\psi}$)(GeV)	σ(per nucleon[2])(nb)	Error(stat. only)(nb)
121	14.1	± 3.2
177	19.7	± 4.1
223	17.2	± 3.8
272	21.7	± 4.8
324	29.7	± 7.8
374	25.4	± 12.6

Fit to a straight line: slope $= 0.05 \pm 0.02$ nb/GeV ; $\chi^2 = 1.19$; $Q = 0.88$

5 In Progress

Still in progress are determination of systematic errors, analysis of the inelastic J/ψ sample, and also determination of the open charm cross section and its energy dependence.

---

[2] Assumes $A^{0.94}$ dependence

Observation of the Semileptonic Decay
$D^+ \to \overline{K}^{*0} \mu^+ \nu_\mu$ (+ charge conjugate)

Steve Culy

University of Colorado

The E687 Collaboration

Abstract

We report the observation of events that are consistent with the decay $D^+ \to \overline{K}^{*0} \mu^+ \nu_\mu$ (+ c. c.). Evidence that these events come from charmed particle decay is demonstrated by the asymmetry between events containing $K^- \pi^+ \mu^+$ and $K^+ \pi^- \mu^+$ final states. We also find no significant evidence of the non-resonant decay $D^+ \to (K^- \pi^+)_{NR} \mu^+ \nu_\mu$.

1. Introduction

E687 is a photoproduction experiment performed with the Wide Band Spectrometer at Fermilab. This spectrometer will not be discussed here, since it is described in detail elsewhere in these proceedings. Instead, we will report on the semileptonic analysis of data taken during the first run of the experiment. First, we will describe the analysis techniques used to isolate the signal $D^+ \to K^- \pi^+ \mu^+ \nu_\mu$, to demonstrate that these techniques do not bias our results. We will then present the results and our conclusions, followed by a brief statement on work in progress on semileptonic decays in E687.

2. Overview of Analysis

This analysis begins with the formation of D^+ meson "candidate" vertices from charged-particle tracks identified as K, π, and μ, with the K and π required to be of opposite charge. These tracks are refit using a constrained-vertex algorithm, and the vertex is rejected if $\chi^2_{DOF} > 2.0$, if the total $K\pi\mu$ momentum is less than 40 GeV, or if the invariant mass of the $K\pi\mu$ tracks ($M_{K\pi\mu}$) is greater than the nominal mass of the D^+ meson ($M_D = 1.87$ GeV). The remaining tracks in the event are then used to construct candidates for the primary photon interaction vertex. From the mass difference $(M_D^2 - M_{K\pi\mu}^2)$, the maximum possible transverse neutrino momentum can be calculated. This in turn can be used (in conjunction with the measured $K\pi\mu$ momentum) to generate an upstream "cone of (kinematic) acceptance" for primary vertex candidates. Events which have no acceptable primary vertex (within vertexing uncertainty) are rejected. In addition, the distance of separation between the primary and secondary is required to be 7 times the combined vertexing uncertainty.

3. Backgrounds from Other Charm Decays

Backgrounds from other charm decays are of interest because they contribute to the asymmetry between $K^- \pi^+ \mu^+$ and $K^+ \pi^- \mu^+$ events, so an accurate measurement of the asymmetry requires that corrections be made for these decays. The most significant of these is the decay $D^{*+} \to D^0 \pi^+ \to (K^- \mu^+ \nu_\mu) \pi^+$. If the momementum of the D^0 is calculated by balancing the transverse momentum (assuming the mass of the D^0), the mass of the D^{*+} can be calculated. For this analysis, we require that this mass be greater than 2.02 GeV.

Another significant background is $D^+ \to K^- \pi^+ \pi^+$, where one π^+ decays to $\mu^+ \nu_\mu$. We can correct for this by recalculating $M_{K\pi\mu}$ under the assumption that the μ track has the π mass. Events are rejected if $M_{K\pi\pi} > 1.80$ GeV.

4. Results

The results of this analysis are displayed in Figure 1. These are histograms of the $K\pi$ invariant mass for the $K^-\pi^+\mu^+$ (right sign) and $K^+\pi^-\mu^+$ (wrong sign) events, where both plots include the charge conjugate events as well. The maximum-likelihood fit imposed is the sum of a Breit-Wigner resonance and a background with the expected shape (determined from Monte Carlo) for the decay $D^+ \rightarrow (K^-\pi^+)_{NR}\mu^+\nu_\mu$. In the right-sign case, the mean, width, and yield of the resonance are all allowed to vary; in the wrong-sign case, the mean and width are fixed to the world-average values for the $K^{*0}(892)$ resonance.

For the right-sign case, we find 76 ± 9 K^{*0} events and 35 ± 14 non-resonant events. The wrong-sign case yields 12 ± 7 resonant events and 23 ± 9 non-resonant events. Combining these results, we find a charm-decay consistent asymmetry of 64 ± 11 for events containing a K^{*0} and an asymmetry of 12 ± 16 for non-resonant events. These results are very preliminary, and they represent only 30% of the estimated yield from the current E687 data sample. All errors quoted are only statistical. Systematic errors will be forthcoming as we study the results of subsequent analyses.

5. Work in Progress

It is clear that this analysis is far from complete. We are now working to perform this analysis on the remainder of the data sample, while using our Monte Carlo to measure the spectrometer acceptance for this state. We plan to determine the branching ratio for this state by measuring $\Gamma(D^+ \rightarrow \overline{K}^{*0}\mu^+\nu_\mu)/\Gamma(D^+ \rightarrow K^-\pi^+\pi^+)$, and we should also be able to measure the polarization of the $\overline{K}^{*0}$ meson as well.

We also plan to extend our semileptonic analysis to other charm decays. First, we will extend our lepton particle ID to identify electron semileptonic decays (such as $D^+ \rightarrow \overline{K}^{*0}e^+\nu_\mu$). Afterwards, we will concentrate on extracting signals for the following states: $D^0 \rightarrow K^-l^+\nu_l$, $D_s^+ \rightarrow \phi l^+\nu_l$, and $\Lambda_c^+ \rightarrow Xl^+\nu_l$ (exclusive states).

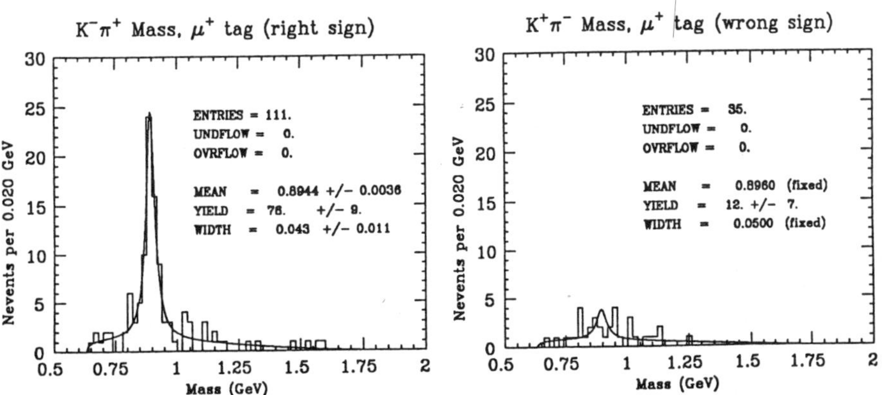

Figure 1. Invariant mass of $K\pi$ tracks in events containing $K^-\pi^+\mu^+$ and $K^+\pi^-\mu^+$ vertices. The mean, yield, and width values quoted represent the parameters for the Breit-Wigner resonance in the maximum-likelihood fit.

HADROPRODUCTION OF CHARM
AT FERMILAB E769

G.A. Alves,[1] J.C. Anjos,[1] J.A. Appel,[2] S.B. Bracker,[5] L.M. Cremaldi,[3]
R.L. Dixon,[2] D. Errede,[7] H.C. Fenker,[2],[a] C. Gay,[5] D.R. Green,[2] R. Jedicke,[5]
D. Kaplan,[4],[b] P.E. Karchin,[8] S. Kwan,[2] I. Leedom,[4] L.H. Lueking,[2]
G.J. Luste,[5] P.M. Mantsch,[2] J.R.T. de Mello Neto,[1] J. Metheny,[6]
R.H. Milburn,[6] J.M. de Miranda,[1] H. da Motta,[1] A. Napier,[6]
A.B. de Oliveira,[1],[c] A.C. dos Reis,[1] S. Reucroft,[4] W.R. Ross,[8]
A.F.S. Santoro,[1] M. Sheaff,[7] M.H.G. Souza,[1] W.J. Spalding,[2]
C. Stoughton,[2] M.E. Streetman,[2] D.J. Summers,[3] Z. Wu[8]

(The Tagged-Particle Spectrometer Collaboration)

[1] Centro Brasileiro de Pesquisas Físicas, Rio de Janeiro, Brazil
[2] Fermi National Accelerator Laboratory, Batavia, Illinois 60510, USA
[3] University of Mississippi, Oxford, Mississippi 38677, USA
[4] Northeastern University, Boston, Massachusetts 02115, USA
[5] University of Toronto, Toronto, Ontario M5S 1A7, Canada
[6] Tufts University, Medford, Massachusetts 02155, USA
[7] University of Wisconsin, Madison, Wisconsin 53706, USA
[8] Yale University, New Haven, Connecticut 06520, USA

Presented by

Lee Lueking

Fermilab, Batavia, Ill. 60510, U.S.A.

Abstract

Experiment E769 at Fermilab obtained charm data during the 1987-88 Fixed
Target running period with a 250 GeV tagged hadron beam incident on thin target
foils of W,Cu,Al, and Be. From analysis of 25% of the recorded 400M trigger sample
we have explored the Feynman x, p_t^2 and the atomic number dependence of charm
quark production for D^+ and D^0 mesons.

1. INTRODUCTION

Experiment E769 has completed its first analysis pass on 25% of the data which
was taken during the 1987-88 Fixed Target running period. This data was collected
using a 250 GeV tagged hadron beam incident on a segmented foil target with four
nuclear materials: Be, Al, Cu, and W. The total data set consists of 400 M triggers,

with about 130 M negative beam events; 85% π, and 15% K and 240 M positive beam triggers; 40% π, 30% K, and 30% p. The principal focus of the experiment is to measure the characteristics of hadroproduction of charm, including total charm cross section, Feynman x, Pt, A dependence and beam particle type dependence.

The most important processes contributing to the hadroproduction of charm are gluon-gluon fusion and quark-antiquark annihilation. The contributions to the charm cross section of these processes were calculated to lowest order perturbative QCD by Ellis and Quigg, resulting in predictions for production characteristics [1]. Recent calculations have been made including the next to leading order terms, showing that $\sigma_{c\bar{c}}(\alpha_s^2+\alpha_s^3)/\sigma_{c\bar{c}}(\alpha_s^2) \approx 3$ [2]. Measurement of the charm production can be used to test values of the charm quark mass and structure functions employed in this calculation. For hard scattering there is no reason for a nuclear A dependence on the cross section other than $\sigma_{c\bar{c}} = A^1$ which is assumed in the theory. Any special effects when the final charm particle contains one of the valence quarks of the incident beam particle, the so-called "leading particle effect", are also not included in this model.

The current status of fixed target research in this area is marked by difficulty and success. Hadroproduction of charm is more difficult than photoproduction in that the charm cross section relative to the total cross section is about 5 times lower. Also, track multiplicities are higher, further increasing backgrounds. Despite these problems, the results for the $\sigma_{c\bar{c}}$ are in good agreement with theory [3]. Early measurements of the $\sigma_{c\bar{c}} \propto A^\alpha$ dependence indicate that $\alpha < 1$, and there is reason to believe α may depend on the variables x_F and p_t. Leading particle effects have been reported, but are not established.

2. APPARATUS

The E769 detector is the Tagged Photon Spectrometer described elsewhere [4] with several improvements. The detector consists of 13 planes of Silicon Microstrip vertex Detectors (SMD), 35 planes of drift chambers, two threshold Cerenkov counters, and electromagnetic and hadronic calorimeters. The following enhancements were made to the E691 photoproduction apparatus for this data: 1) beam tracking for the incident hadron, 2) additional $25\mu m$ silicon immediately downstream of the target, 3) Y measurement 2mm PWC's before the first analysis magnet, 4) improved data acquisition capable of logging 400 events/second with a 40% dead time [5]. The silicon vertex detector provides $20\mu m$ x-y and $300\mu m$ z vertex resolution with an angular acceptance of ± 100 mr. A trigger based on the transverse energy in the forward calorimeters is used to enhance the charm sample.

Consisting of 10 Be, 5 Al, 3 Cu and 4 W foils, the target represents 2% of a nuclear interaction length and was designed with the high Z materials located most upstream, to minimize multiple scattering effects. the Be, Al, and Cu foils were $250\mu m$ thick, and the W foils were $100\mu m$ in thickness. The foils are placed at 1.2 mm intervals in the beam direction and simultaneously exposed to the incident 250 GeV hadrons.

550

The secondary hadron beam was tagged using two devices, a differential Cerenkov counter, and a transition radiation detector. The Differential Isochronous Self-focusing Cerenkov counter (DISC) was 50 % efficient tagging kaons at a pion contamination level of less than 5% and it was employed in this capacity for most of the run. However, we recorded nearly 60M triggers with the DISC used to trigger on protons only. A 24 module TRD with polypropylene radiators was employed to tag pions with a 95% efficiency and a contamination level of $< 3\%$ from protons or kaons [6].

3. RESULTS

The portion of the data reported on here consists of about 100M negative beam triggers which are principally pions. From this sample of the data, we have extracted charm signals for the modes $D^+ \to K^-\pi^+\pi^+$ and $D^0 \to K^-\pi^+$ [7]. The charm signals are extracted by forming vertex candidates using the SMD track information. From this list of vertices for each event the primary vertex is defined to be the one with the largest number of attached tracks, and a secondary decay vertex candidate is chosen based on several criteria. First, the significance of the primary to secondary separation, $SDZ = \Delta z/(\sigma_{z_p}^2 + \sigma_{z_s}^2)^{1/2}$, must be greater than 12. For each track in the secondary candidate the transverse distance to the primary, b_p, and to the secondary, b_s, are determined. A parameter "RATIO" is then defined as the product of the ratios $\Pi_{i=1}^n b_s/b_p$, with n being the number of tracks in the secondary decay mode. The value of RATIO must be less than 0.006 for the 3 body decays, and .06 for the two body decays. The impact parameter of the momentum vector of the reconstructed D secondary decay with respect to the primary vertex must be less than 80μm for the 3 body decays and 100μm for the two body decays. The decay vertex location must be isolated by more than 60μ from additional tracks in the event for the charged D's. The sum of p_t^2 of the D candidate decay tracks relative to the D direction must be greater than $.5 \ GeV^2/c^2$. For the $D^+ \to K\pi\pi$ and $D^0 \to K\pi$, modes respectively signal sizes of 684 ± 42 and 496 ± 39 events are observed (Fig. 1).

From the $K\pi\pi$ and $K\pi$ decay mode samples a preliminary analysis has been performed to extract the x_F, p_t^2 distributions, and the A dependence. Acceptance corrected x_F distributions for the D^+ and D^0, are fit to $(1 - x_F)^n$. The value for n from the two fits is $n = 3.6 \pm 0.4$ for the D^+ (Fig. 2a), and $n = 4.3 \pm 0.8$ for the D^0 (Fig. 2b). Shown in Fig. 3 are acceptance corrected p_t^2 distributions, with the fit representing $e^{-bp_t^2}$. For the D^+ and D^0 samples the value for b are $b^. = 0.95 \pm 0.1$ and $b = 0.87 \pm 0.1$ respectively. The A dependence is shown in Fig. 4a and b for the D^+ and D^0 signals and the fits represent A^α dependence. The values for α are $\alpha = 0.95 \pm 0.07$ for the D^+ and $\alpha = 1.00 \pm 0.1$ for the D^0. In all of the analyses the errors quoted are statistical only; the systematic errors are not yet determined. Also, there may be a systematic correction due to the effect of the Et trigger which has not been made pending further study.

Our plans for the future include completion of our reconstruction, analysis of more

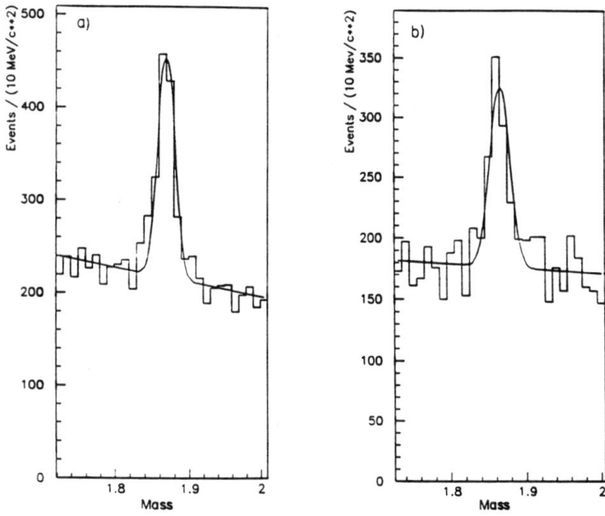

Figure 1: Charm signals for the decay modes (a) $D^+ \rightarrow K^-\pi^+\pi^+$; 683 ± 42 signal events and (b) $D^0 \rightarrow K^-\pi^+$; 496 ± 39 signal events.

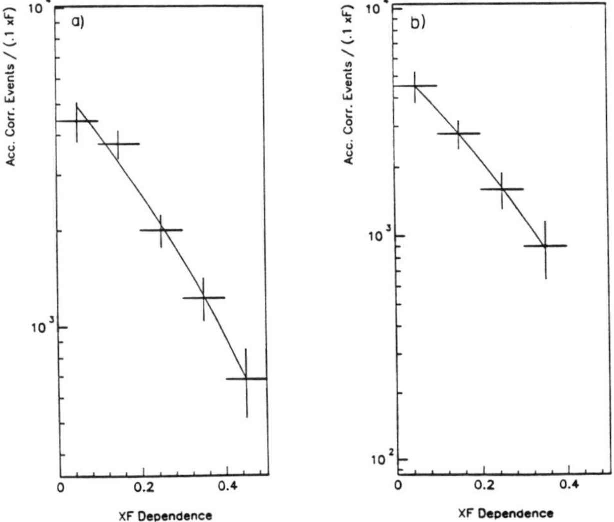

Figure 2: Acceptance corrected x_F distribution for the (a) $D^+ \rightarrow K^-\pi^+\pi^+$ and (b) $D^0 \rightarrow K^-\pi^+$ modes with the fit representing $(1-x_F)^n$, the value of n is $n = 3.6 \pm 0.4$ and $n = 4.3 \pm 0.8$ respectively.

552

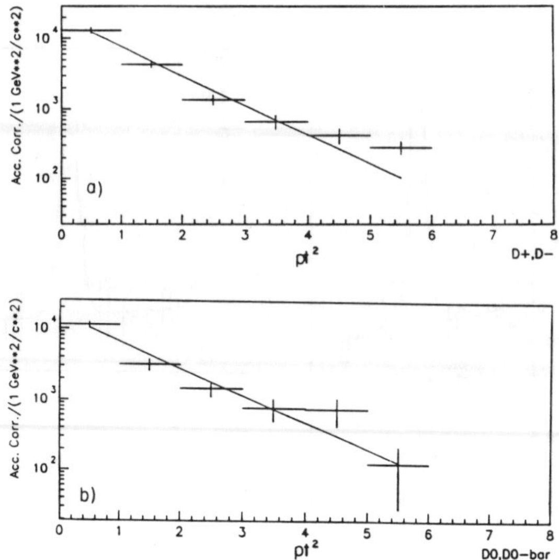

Figure 3: Acceptance corrected p_t^2 distributions, with the fit representing $e^{-bp_t^2}$ for (a) $D^+ \rightarrow K^-\pi^+\pi^+$ and (b) $D^0 \rightarrow K^-\pi^+$ modes; $b = 0.95 \pm 0.1$ and $b = 0.87 \pm 0.1$ respectively.

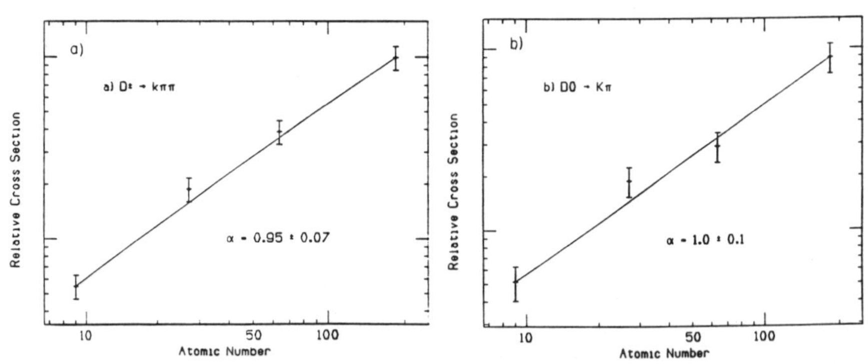

Figure 4: Nuclear A dependence for the (a) D^+ and (b) D^0 signals; the fits represent A^α. The values for α are $\alpha = 0.95 \pm 0.07$ for the D^+ and $\alpha = 1.0 \pm 0.1$ for the D^0.

charm states and modes, measurement of the total and differential charm cross sections, and use of the beam tagging information. We have currently reconstructed nearly 90% of our raw data and hope to complete this processing by summer. Our analysis should proceed quickly and will include the $D^0 \rightarrow K\pi\pi\pi$ decay mode, additional charmed meson states such as the D^*, D_s and charmed baryon states including Λ_c. We will search for incident particle type related physics with the tagged hadrons in the positive beam.

4. SUMMARY

A preliminary analysis has been performed on approximately 25% of our 400 M trigger data sample. This sample has yielded signals for the decay modes for $D^+ \rightarrow K\pi\pi$ and $D^0 \rightarrow K\pi$ of 683 $\pm$ 42 and 496 $\pm$ 39 events respectively. For these modes distributions for x_F, p_t^2, and the nuclear A dependence have been extracted. Work is continuing to understand systematic errors and corrections, and to include more charmed states including charm-strange mesons and charmed baryons.

We gratefully acknowledge the assistance we received from the Fermilab staff and the support staffs at the various universities. This research was supported by the U.S. Department of Energy, the National Science and Engineering Research Council of Canada, and the Brazilian Conselho Nacional de Desenvolvimento Científico e Tecnológico.

REFERENCES

[a]Present address: SSC Laboratory, Dallas, Texas 75239.

[b]Present address: University of Oklahoma, Norman, Oklahoma 73019.

[c]Present address: University of Cincinnati, Cincinnati, Ohio 45221

[1] R .K. Ellis, C. Quigg, FN-445 (1987).

[2] S .Dawson, R.K. Ellis, P. Nason, Nuclear Physics $\underline{B303}$ (1988) 607.

[3] G . Altarelli, et al., Nuclear Physics $\underline{B308}$ (1988) 724; E.L. Berger,ANL-HEP-CP-89-107.

[4] V .K. Bharadwaj et al., Nucl. Inst. Meth. $\underline{155}$ (1978) 411; V.K. Bharadwaj et al., Nucl. Inst. Meth. $\underline{228}$ (1985) 283; D.J. Summers, Nucl. Inst. Meth. $\underline{228}$ (1985) 290; P. Karchin et al., IEEE Trans. Nucl. Sci. $\underline{32}$ (1985) 612; J.A. Appel et al., Nucl. Inst. Meth. $\underline{A243}$ (1986) 361; D. Bartlett et al., Nucl. Inst. Meth. $\underline{A260}$ (1987) 55.

[5] D . Errede et al., FERMILAB-Conf-88/180-E; D. Errede et al., IEEE Trans. Nucl. Sci. $\underline{36}$ (1989) 106.

[6] S . Bracker, C. Gay, IEEE Trans. Nucl. Sci. $\underline{34}$ (1987) 870.

[7] T hroughout this paper, refrences to decays include both the particles and their charge conjigates.

Xe/D$_2$ Cross-section Ratio at Low x_{Bj}
from Muon Scattering at 490 GeV/c

Stephen R. Magill
University of Illinois at Chicago
for the
E665 Collaboration at Fermilab[1]

Abstract

First measurements from the E665 experiment at Fermilab on the relative cross-sections of 490 GeV/c muons scattered from deuterium and xenon targets are presented. The scattered muons were in the kinematic range $Q^2 > 0.1(GeV/c)^2$ and very low x_{Bj} ($0.001 < x_{Bj} < 0.1$). Triggering on events in this kinematic region was accomplished with a special Small Angle Trigger which projected individual beam muons to form a veto region 30 meters from the target. Using this trigger, muons with scattered angles as small as 0.5 milliradians were detected.

1   E665 at FNAL

E665 at FNAL is the highest energy muon deep inelastic scattering experiment in the world. This experiment was designed to provide physics information on a variety of topics, including:

- Structure Functions and ratio of Structure Functions for nucleons and nuclei at low x_{Bj}.

- Current jet fragmentation with identified hadrons.

- Target jet fragmentation.

- Event structure relative to the virtual photon axis.

- Exclusive vector meson production.

The apparatus is shown in Figure 1 and includes the following features :

- E_μ from 100 GeV to 500 GeV.

- Open geometry with $\sim 4\pi$ hadron detection.

- Two large field volume superconducting dipole magnets.

- Particle identification system including Time-of-Flight, two Threshold Cerenkov Counters, and a Ring-Imaging Cerenkov Counter.

- Electromagnetic Calorimeter for neutral energy detection.

- Streamer Chamber vertex detector for the target fragmentation region.

- Independent scattered muon triggers based on minimum scattering angles of 0.5 mrad (Small Angle Trigger - SAT) and 3 mrad (Large Angle Trigger - LAT).

- Targets : Xenon (8.5 g/cm^2), Deuterium (16 g/cm^2), Hydrogen (7 g/cm^2).

The first data-taking period for this new apparatus began in July 1987 and ended in February 1988. For more information on the E665 apparatus, see reference [2].

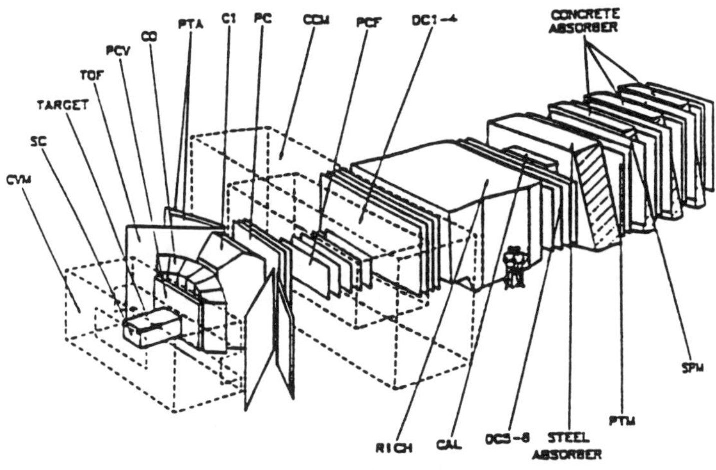

Figure 1: E665 Experimental Apparatus

2 Deep Inelastic Scattering Kinematics

Figure 2 shows the single virtual photon exchange Feynman diagram for muon-induced deep inelastic scattering. Included are the definitions and calculations of kinematical variables applicable in this analysis. The calculation of Q^2 includes the mass of the muon since the sample of low x_{Bj} is obtained at Q^2 down to 0.1 $(\text{GeV}/\text{c})^2$.

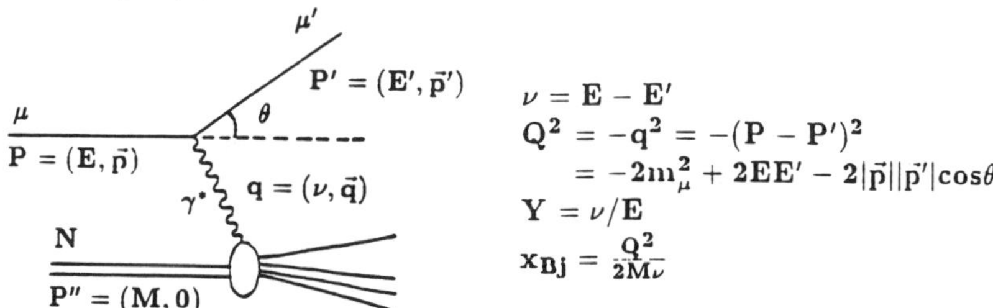

Figure 2: DIS Single γ^* Exchange Diagram

The Feynman diagrams which represent radiative contributions to the muon vertex of order α^2 are shown in Figure 3. The vacuum polarization and vertex correction diagrams are target independent and so do not contribute to the total correction in the ratio of cross-sections. However, the diagrams corresponding to the bremsstrahlung process contribute in 3 different ways. The coherent process, in which the nucleus remains intact, and the quasi-elastic process, in which the nucleus breaks up into its constituent nucleons only, are strongly A dependent and, therefore, must be considered as corrections in the ratio of cross-sections. The inelastic contribution, in which the nucleon breaks up, depends in part on the structure function of the target nucleon, and is, therefore, weakly A dependent. In a heavy target, such as xenon, the coherent and quasi-elastic contributions dominate the total radiative correction at $y = \nu/E$ values greater than 0.5.

The contribution of the dominant radiative processes has been estimated by two independent

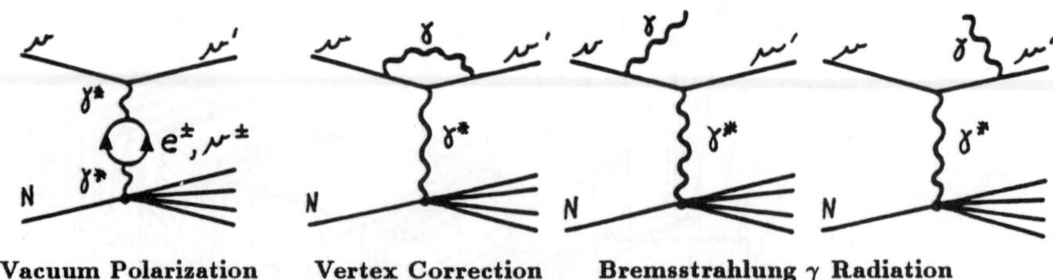

Figure 3: Radiative Corrections to Muon Vertex

methods : 1) a radiative correction Monte Carlo program based on an exact calculation [3]; and 2) use of the E665 electromagnetic calorimeter to eliminate radiative events from the deep inelastic scattered sample.

3 Event Sample - Kinematic Cuts

The distribution of events in the SAT trigger sample is shown in Figure 4. Note the position in the

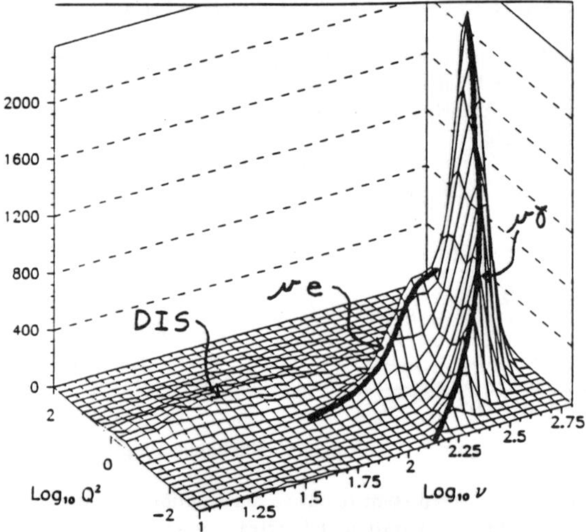

Figure 4: Log Q^2 vs Log ν Distribution of SAT Xenon Target Events

Q^2 vs ν event plane of the deep inelastic scatters, the elastic scattered $\mu - e$ events at constant x_{Bj}, and the $\mu - \gamma$ bremsstrahlung events primarily at high ν and low Q^2. The cuts used to define the current event sample are as follows : $Q^2 > 0.1\ (GeV/c)^2$; $\nu > 40\ GeV$; $x_{Bj} > 0.001$; $y < 0.75$; and are chosen to reduce the amount of electromagnetic background. In addition, a loose target position cut on the muon vertex was made.

4 Corrections and Systematic Errors

The data have been corrected for target acceptance, reconstruction efficiency of the scattered muon, empty target subtraction, and radiative events (unless the electromagnetic calorimeter was used to cut these events). Radiative corrections represent the largest correction to the raw ratio ($\sim 25\%$ at $x_{Bj} = 0.001$). Using the electromagnetic calorimeter, E665 has been able to compare the results of radiative corrections from the two methods mentioned above, showing that they are consistent within statistical errors. The following table shows current estimates of the systematic errors due to the above corrections as well as from beam normalization and target density.

Summary of Systematic Errors					
Source	Estimate	Source	Estimate	Source	Estimate
Beam Norm.	0.7 %	Trigger Acc.	2.8 %	MT Target Sub.	1.0 %
Target Density	0.4 %	Radiative Corr.	1.4 %	μ Recon.	4.0 %

5 Results

Figure 5 shows the ratio σ_{Xe}/σ_{D2} as a function of x_{Bj}. The data show that the cross-section per

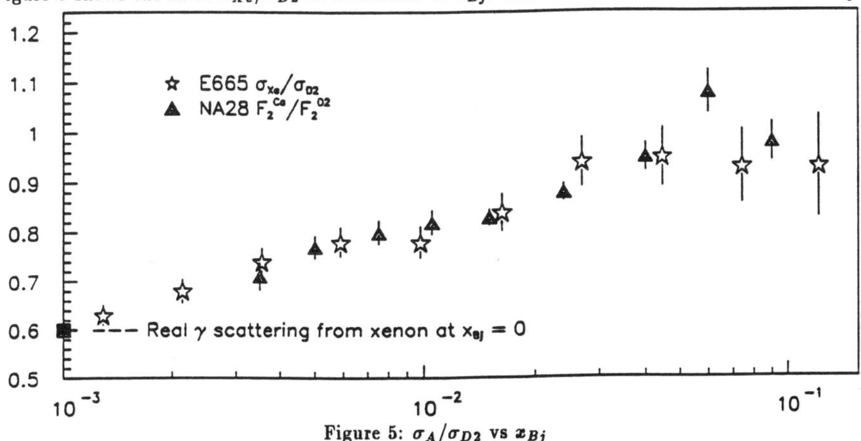

Figure 5: σ_A/σ_{D2} vs x_{Bj}

nucleon for muon scattering on Xe is suppressed relative to that for D_2 in the region $0.001 < x_{Bj} < 0.1$. This phenomena, called "shadowing", has been seen in γ - nucleus scattering at high energies [4] as well as in virtual photon scattering at low and high energies [5,6]. Recently, CERN experiment NA28 [6] has made measurements on carbon and calcium nuclei in much the same kinematic region as in this analysis. In Figure 5, data from E665 xenon and NA28 calcium are shown with statistical error bars only, indicating agreement between the two results in the shadowing region, and showing that E665 extends the kinematical range studied by NA28. Note that the data are consistent with an estimated real photon point (0.6 for xenon for $A^{eff} = 131^{0.9}$) [4].

In Figure 6, the Q^2 variation of σ_{Xe}/σ_{D2} for two x_{Bj} regions is presented with a comparison of NA28 calcium data. Once again, only statistical error bars are shown. The E665 data significantly extend the Q^2 range, and show that there is no Q^2 dependence of the σ_{Xe}/σ_{D2} ratio. This behavior indicates that the shadowing phenomenon observed is not consistent with naive vector dominance, but rather with models which predict shadowing as a result of parton recombination in heavy nuclei at low x_{Bj} [7] or with generalized vector dominance models [8].

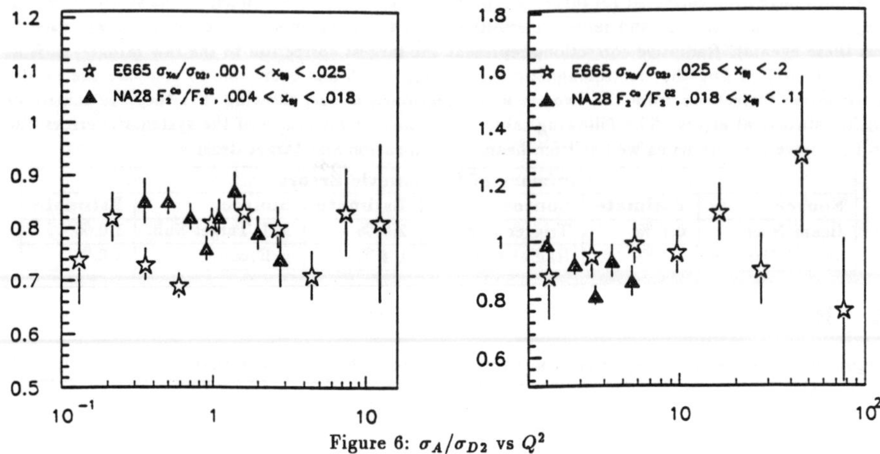

Figure 6: σ_A/σ_{D2} vs Q^2

6 Conclusion

In conclusion, the following points can be made from this analysis :

- Shadowing is seen in the ratio σ_{Xe}/σ_{D2} as a function of x_{Bj}.

- The dependence of σ_{Xe}/σ_{D2} on Q^2 is flat, indicating that the observed shadowing is of partonic origin. This is observed over the Q^2 range of $0.1 < Q^2 <\sim 30(GeV/c)^2$.

References

[1] Argonne, U.C. San Diego, Cracow, Fermilab, Freiburg, Harvard, U.I.C., Maryland, MIT, MPI Munich, Washington, Wuppertal, Yale.

[2] M. R. Adams, et. al.; FERMILAB-Pub-89/200-E;1989.

[3] L. W. Mo and Y. S. Tsai; Rev. Mod. Phys.; 41 (1969); Y. S. Tsai; SLAC-PUB-848; (1971).

[4] for example : D. O. Caldwell, et. al.; Phys. Rev. Lett.; 42 (1979).

[5] for example : J. Eickmeyer, et. al.; Phys. Rev. Lett.; 36 (1976).

[6] for example : M. Arneodo, et. al.; Nucl. Phys.; B333 (1990).

[7] F. E. Close, et. al.; Phys. Rev.; D40 (1989).

[8] for example : O. L. Bilchak, et. al.; Phys. Lett., B214 (1988).

A Next-to-leading-logarithm Monte Carlo Calculation for Direct Photon Production[*]

J. F. Owens

Department of Physics B-159, Florida State University

Tallahassee, Florida 32306

ABSTRACT

A method for calculating direct photon production in the next-to-leading-logarithm approximation is described. The method uses a mix of analytic and Monte Carlo integration techniques for performing the required phase space integrations. The flexibility afforded by the use of these techniques allows a variety of observables to be easily calculated. Some examples are discussed, including the photon plus jet cross section.

Observations of direct photon production in hadronic collisions have the potential to yield valuable information concerning the gluon distribution in the proton.[*] When combined with measurements of other processes, such as deep-inelastic lepton-nucleon scattering or the production of high-mass lepton pairs, it is also possible to place constraints on the various quark distributions.[4] However, most of the existing direct photon measurements have been for the invariant cross section, the calculation of which involves a convolution of parton distributions. It has long been realized that more precise information concerning the gluon distribution could be obtained from data for the double inclusive photon plus jet cross section[5] since, in the leading-logarithm approximation, the arguments for both parton distributions would be fixed by the kinematics of the observed jet and photon. On the theoretical side, existing next-to-leading-logarithm calculations for large momentum transfer processes have also focussed on the single particle, jet, or photon inclusive cross section. The information contained in the correlation between the various partons in the process is lost when the unobserved partons are integrated over the available phase space. On the other hand, it is difficult to know what observable should be calculated ahead of time, since there are many variations in jet definitions, acceptance corrections, etc. Ideally, one would like to have a method of performing next-to-leading-logarithm calculations which can be easily adapted to

---

[*] Work supported in part by the U. S. Department of Energy.

[*] See Refs. (1, 2, and 3), and references therein, for reviews.

the particular experimental situation being studied. In this report a new calculation of direct photon production is described which explicitly includes the point-like photon subprocesses through $O(\alpha\alpha_s^2)$. The various two-body and three-body phase space integrations are performed using a mix of analytic and Monte Carlo methods, thereby allowing one to calculate observables which involve correlations between the photon and another parton. A variety of observables can be calculated simultaneously by forming suitable histograms within the same program.

In a typical leading-logarithm calculation for high-p_T photon production, there are two different types of production mechanisms to be considered. The first consists of the $O(\alpha\alpha_s)$ subprocesses $gq \to \gamma q$ and $q\bar{q} \to \gamma g$ and is referred to as the direct or pointlike component. The second mechanism arises because quarks participating in a hard scattering process can radiate photons. This is taken into account by introducing photon fragmentation functions for both quarks and gluons. These can be obtained by solving the appropriately modified Altarelli-Parisi equations.[6] For this photon bremsstrahlung component the hard scattering subprocesses are $O(\alpha_s^2)$ while the photon fragmentation function is $O(\alpha/\alpha_s)$ so that the results are of the same order as for the direct photon case. Both of these constitute all orders calculations, but only the leading-logarithm from each of the terms beyond $O(\alpha\alpha_s)$ is retained. In order to improve the precision of the predictions, the next step is to include completely the effects of the various $O(\alpha\alpha_s^2)$ subprocesses involving a single photon. Specifically, the one-loop contributions to $gq \to \gamma q$ and $q\bar{q} \to \gamma g$ and the three-body subprocesses $qq' \to \gamma qq'$, $q\bar{q}' \to \gamma q\bar{q}'$, $q\bar{q} \to \gamma q'\bar{q}'$, $qq \to \gamma qq$, $q\bar{q} \to \gamma q\bar{q}$, $gq \to \gamma qg$, $gg \to \gamma q\bar{q}$, and $q\bar{q} \to \gamma gg$ have been included in this calculation using expressions obtained from Refs. (7, 8). In principle, there are additional $O(\alpha\alpha_s^2)$ terms which come from $O(\alpha_s^3)$ $2 \to 3$ subprocesses convoluted with photon fragmentation functions. These terms provide a correction to the bremsstrahlung component. However, the bremsstrahlung component does not provide a significant contribution except in the small x_T region. Therefore, the neglect of these additional terms is justified for those experiments which do not go to very small values of x_T. Furthermore, some experiments use isolation cuts in the process of defining the photon sample, thereby reducing the role of the bremsstrahlung terms. It is worth noting, however, that this question will become important as new data at small x_T become available from the various high energy collider experiments.

The calculation proceeds along much the same lines as the usual inclusive single photon calculation, in which one integrates over the unobserved final state partons. Such integrations can be performed analytically at the parton level. For a suitably defined inclusive observable, the infrared singularities associated with the one-loop contributions will cancel against the soft singularities associated with the three-body tree graphs. Furthermore, the hard collinear singularities associated with the initial partons or the produced photon can be factorized and included in the relevant parton distribution or fragmentation functions, respectively. The remaining hard collinear singularities will cancel against corresponding singularities from the one-loop graphs. If, however, one wants to constrain one or more of the final state partons to be in a singularity-free region of phase space, then the integration regions are restricted and Monte Carlo methods offer an alternative to the purely analytic approach. The basic challenge, then, is to find a way of ensuring that all of the required cancellations can take place within the context of a Monte Carlo calculation. In order to discuss the technique for isolating the various singularities, let the four-vectors of the two-body and three-body subprocesses be labelled by $p_1 + p_2 \rightarrow p_3 + p_4$ and $p_1 + p_2 \rightarrow p_3 + p_4 + p_5$, respectively, and define the following Lorentz scalars $s_{ij} = (p_i + p_j)^2$ and $t_{ij} = (p_i - p_j)^2$. Briefly, the calculation proceeds via the following steps. The ultraviolet singularities associated with the one-loop contributions are regulated using the method of dimensional regularization[9] and subtracted using the $\overline{\text{MS}}$ scheme.[10] Similarly, dimensional regularization is used in treating the infrared, soft, and collinear divergences. Next, two cut-off parameters, δ_s and δ_c, are introduced whose purpose is to allow the separation of the regions of phase space which contain the singularities. For the three-body subprocesses, the soft singularities are associated with the phase space region where one final state gluon becomes soft. The soft region is defined to be that where the relevant parton energy in the subprocess rest frame becomes less than $\delta_s \sqrt{s_{12}}/2$. If δ_s is chosen to be sufficiently small, then the relevant three-body subprocesses can be evaluated using the soft-gluon approximation wherein the the gluon energy is set to zero in the numerator of the expression. The resulting expression is then easily integrated over the soft region of phase space. At this stage, this integrated soft piece contributes to the two-body part which contains the one-loop terms. The soft and infrared singularities can then be cancelled explicitly. Next, the collinear regions of phase space are defined to be those where any invariant (s_{ij} or t_{ij}) becomes smaller in magnitude than $\delta_c s_{12}$. If δ_c is chosen to be sufficiently small, then in each collinear region the relevant subprocess can be evaluated using the leading-pole approximation. The result is easily integrated in n-dimensions, thereby explicitly displaying the collinear singularities. These are then factorized and included in the relevant structure functions or cancelled with corresponding singularities in the two-body expressions. At this point, the remainder of the three-body phase space contains no singularities and the subprocesses can be evaluated in four dimensions.

The calculation now consists of two pieces – a set of two-body contributions and a set of three-body contributions. Each set consists of finite parts, and all singularities having been cancelled, subtracted, or factorized. However, each part depends separately on the two theoretical cut-offs δ_s and δ_c. Each by itself has no intrinsic meaning. However, when the two- and three-body contributions are combined to form a suitably inclusive observable, $e.g.$, an inclusive single photon invariant cross section, all dependence on the cut-offs cancels. It turns out that the answers are stable against variations of these cut-offs over quite a wide range, which is as it should be. The cut-offs merely serve to distinguish the regions where the phase space integrations are done by hand from those where they are done by Monte Carlo. When the results are added together, the precise location of the boundary between the two regions is not relevant. The results reported below are stable to reasonable variations in the cut-offs, thus providing a check on the calculation.

The method described above has been used to perform a variety of next-to-leading-logarithm calculations including the photoproduction[11] of jets, symmetric di-hadron production,[12] and direct photon production.[13,14] Space does not permit a detailed description of the results for these various processes, so the interested reader is instead referred to the relevant publications just listed. Instead, I would like to summarize some of the features learned from such calculations.

One question which repeatedly comes up when discussing next-to-leading-logarithm calculations is the choice of scales. In each of the cases calculated to date we have found that the next-to-leading-logarithm calculations are, in general, less sensitive to the choice of scale. Detailed studies have been made taking the factorization and renormalization scales to be the same. The reduced scale dependence is most noticeable in the intermediate x_T region; for very small values of x_T the sensitivity is comparable to that of the leading-logarithm calculation. From a practical standpoint this means that thhe theoretical uncertainty due to the remaining scale dependence is often reduced by including the next-to-leading-logarithm corrections which, after all, was one of the reasons for performing the calculation in the first place. Moreover, the remaining scale dependence means that there is no unique value for the ratio of the leading-log and next-to-leading-log results, *i.e.*, there is no unique "K-factor". Furthermore, this ratio is different for different observables. As a case in point, consider the photon plus jet cross section. The amount which the three-body matrix elements contribute to the final result depends in part on the fiducial cuts used to define the recoiling jet, the jet clustering or coalescence algorithm, and the photon isolation criteria, for example. Since the three-body contribution is positive definite, changing these factors alters the ratio of the $O(\alpha\alpha_s)$ and $O(\alpha\alpha_s^2)$ terms. This variation would be missed in any approximate approach which simply rescaled the two-body leading-log contributions.

In some experiments an isolation cut is placed on electromagnetic triggers as part of the definition of single photon events. Events are rejected which have more than a certain amount of hadronic energy in a cone about the electromagnetic trigger direction. It is difficult to precisely duplicate this type of cut if the various associated parton four-vectors have already been integrated over. However, in the present case it is possible to duplicate the effects of such cuts, at least at the parton level.

One limitation of the present calculation is that fragmentation is not included, in the sense of an event generator. The Monte Carlo procedure is used simply to perform integrations over limited regions of phase space. The weights generated do not correspond to individual weight one events and, indeed, the individual contributions to the histograms are not positive definite – only their sum should be. Finally, all different flavor contributions and subprocesses are summed over for each kinematic configuration, making it difficult to put in a specific model for fragmentation. Note, however, that it is relatively easy to put in the effects of collinear fragmentation functions. Therefore, this technique in its present form is not suited for studying global event structures. Rather, it should simply be viewed as a means for calculating inclusive observables in a flexible manner.

The results already published reproduce the known good agreement between perturbative QCD and data for direct photon production. As new data for additional types of observables are obtained, this technique will help to improve the precision of the theoretical calculations which are used to extract information on parton distributions and also to test the underlying theory.

REFERENCES

1) J. F. Owens, Rev. Mod. Phys. **59**, 465 (1987).

2) E. Berger, E. Braaten, and R. D. Field, Nucl. Phys. **B239**, 52 (1984).

3) T. Ferbel and W. R. Molzen, Rev. Mod. Phys. **56**, 181 (1984).

4) P. Aurenche, R. Baier, M. Fontannaz, J. F. Owens, and M. Werlen, Phys. Rev. **D39**, 3275 (1989).

5) L. Cormell and J. F. Owens, Phys. Rev. **D22**, 1609 (1980).

6) R. J. DeWitt *et al.*, Phys. Rev. **D19**, 2046 (1979); Phys. Rev. **20**, 1751(E) (1979).

7) P. Aurenche, R. Baier, A. Douiri, M. Fontannaz, and D. Schiff, Nucl. Phys. **B286**, 553 (1987).

8) The expressions in Ref. 6 for the $qq \to qq\gamma$ and $qq' \to qq'\gamma$ subprocesses contain a number of typographical errors where the variable a_3 appears in place of a_5. The original preprint, LPTHE Orsay 86/24, contains the correct expressions.

9) G. 't Hooft and M. Veltman, Nucl. Phys. **B44**, 1972 (189).

10) W. A. Bardeen, A. J. Buras, D. W. Duke, and T. Muta, Phys. Rev. **D18**, 1978 (3998).

11) H. Baer, J. Ohnemus, and J. F. Owens, Phys. Rev. **D40**, 2844 (1989).

12) L. Bergmann, FSU-HEP-890215, Ph. D. Dissertation.

13) H. Baer, J. Ohnemus, and J. F. Owens, Phys. Lett. **B234**, 127 (1990).

14) H. Baer, J. Ohnemus, and J. F. Owens, Phys. Rev. D, in press.

Event Structure and Cross Sections for Single Photon Production at 530 GeV/c

S. Mani for the E706 collaboration

DELHI–FERMILAB–MICHIGAN STATE–MINNESOTA–NORTHEASTERN
OKLAHOMA–PENNSYLVANIA STATE–PITTSBURGH–ROCHESTER

Fermilab experiment E-706 is a study of direct photons and associated hadrons produced at high transverse momenta in hadronic interactions. During the 1987-88 fixed target run of the Tevatron, the spectrometer recorded events produced by 530 GeV/c secondary beams and triggered by electromagnetic showers with $P_T > 3.5$ GeV/c. Preliminary measurements of inclusive cross sections for direct photons and studies of the accompanying jets are presented.

Introduction

Direct photons are produced in hadronic collisions as a result of hard scatterings of constituent partons[1]. Since the Compton process ($q + g - > \gamma + q$) provides the dominant contribution, the cross sections are uniquely sensitive to the gluon structure function. The availability of next to leading order QCD calculations[2] renders even more significance to these measurements.

The E-706 spectrometer (figure 1) is designed to study direct photon production in the P_T range of 4-10 GeV/c. The MW beamline was used to deliver 530 GeV/c secondary beams with π^-/K^- and p/π^+ tagging provided by an upstream Cerenkov counter. A lead and liquid argon calorimeter[3] (LAC) with finely segmented read out in r and ϕ views served as the photon detector. It also generated P_T triggers with appropriate sums of the deposited energies in the r-view. A large acceptance charged particle spectrometer[4] provided measurements on the accompanying hadrons. The tracking was performed by 8 silicon strip detectors in front of a dipole magnet and 16 proportional wire counters between the magnet and the LAC. The integrated luminosity for the 1987-88 run was 1 pb^{-1}.

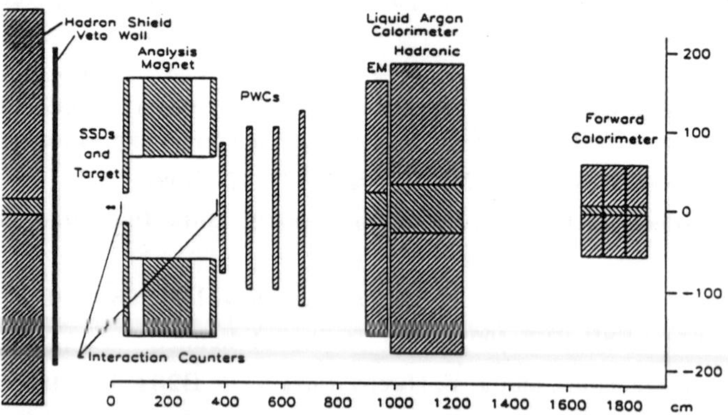

Figure 1. The E-706 Spectrometer.

Cross Sections

The upper curve in figure 2 shows the invariant mass spectrum for all photon pairs with a combined P_T greater than 3.5 GeV/c. The peaks corresponding to the π^0 and η mesons are clearly visible above a relatively small background that has an apparent exaggeration due to the logarithmic scale used on the vertical axis. The lower curve is obtained by demanding that the energy asymmetry $(A = |E_{\gamma_1} - E_{\gamma_2}|/(E_{\gamma_1} + E_{\gamma_2}))$ for the pair be less than 0.75. This reduces the background even further. For spinless mesons like the π^0 or the η the energy asymmetry distribution should be uniform. Figure 3 is a plot of the asymmetry of pairs in the π^0 mass band(105-165 MeV) from which the asymmetry of the pairs in the side-bands (75-100 MeV and 170-195 MeV) has been subtracted. The departure from a uniform population at values above $A = 0.75$ is due to inefficient detection of low energy photons. The prediction(dotted curve) of our Monte Carlo(MC) program reproduces the general shape of the asymmetry distribution. The cross sections for the production of π^0 and η mesons have been calculated by employing the asymmetry cut and correcting for the losses[5].

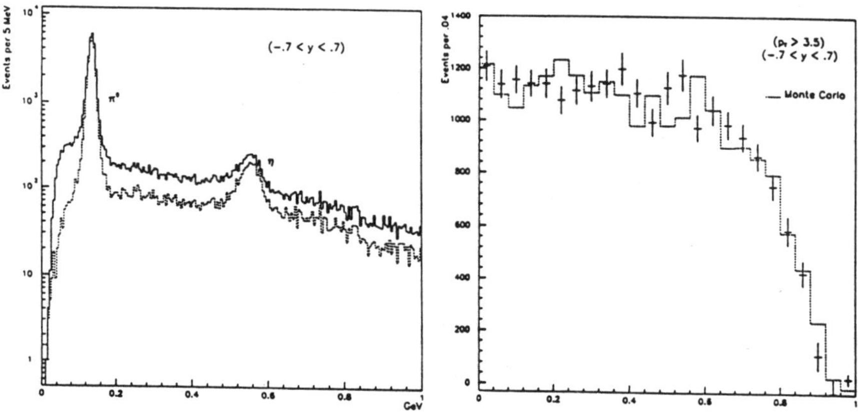

Figure 2. $\gamma\gamma$ invariant mass spectrum.

Figure 3. $A = |E_{\gamma_1} - E_{\gamma_2}|/(E_{\gamma_1} + E_{\gamma_2})$ distribution with MC overlay.

The procedure for calculating the direct photon cross sections is as follows. First, the $\gamma\gamma$ pairs that lie within the π^0 or η mass band and satisfy the $A < 0.75$ cut are removed from the event sample. Second, a cross section is calculated for the remaining *single* γ sample. Finally, a subtraction is performed, based on the prediction of our MC for the contamination of this sample by asymmetric π^0 and η mesons, thereby yielding the cross section for *direct* photon production. These cross sections for incident π^- and proton beams on a beryllium target are shown in figures 4a and 4b respectively. The measurements, however, remain preliminary while we attempt to further refine our MC subtractions.

Event Structure

The hadrons recoiling from the trigger particle exhibit a definite jet structure. Figure 5a is a distribution of the angular differences in the transverse plane between the trigger particle and all the charged particles in the event. The peak at π radians demonstrates that the tracks cluster in a direction opposite to that of the trigger. This in itself can be

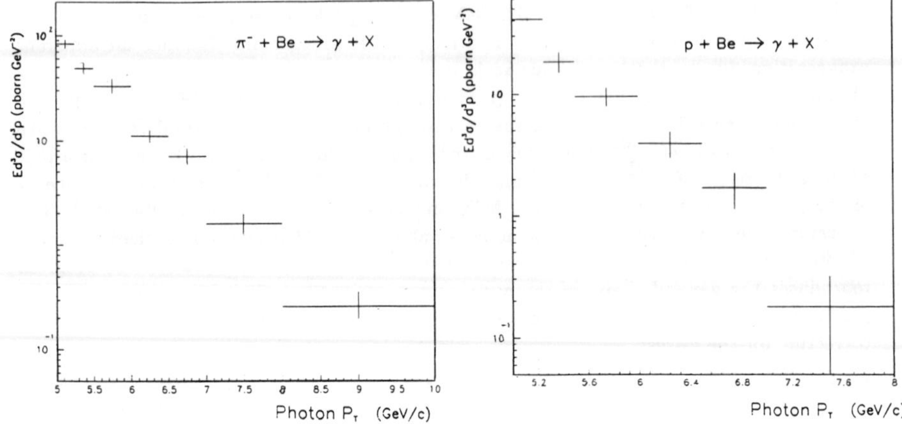

construed to be a statement of momentum conservation. However, clustering of tracks in rapidity space as well would indicate jet production. This is shown in figure 5b which is a histogram of the difference in the pseudo-rapidities of the two highest P_T particles in the hemisphere opposite to that of the trigger direction. A subtraction of accidentals has been performed on this plot using track combinations from uncorrelated events.

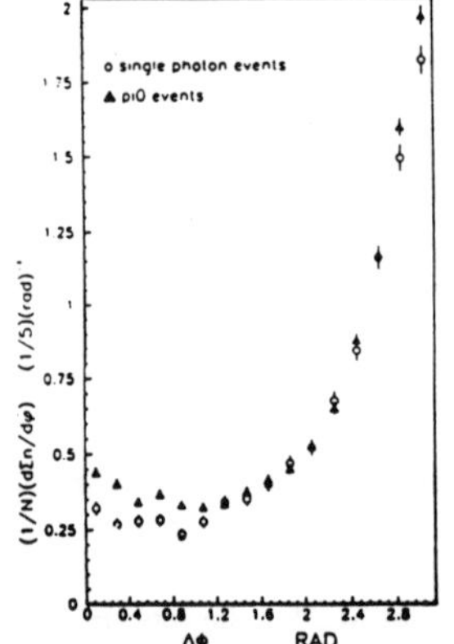

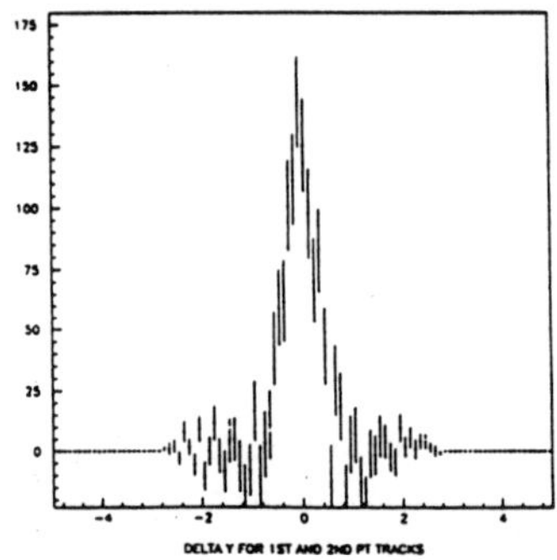

Figure 5a. ϕ difference distributions.

Figure 5b. Distribution of pseudo-rapidity differences of leading particles.

A reconstruction of the jet direction in these events was done using an algorithm quite similar to the one used by WA-70 [9]. The initial seed for the jet direction is determined by taking the particle with the highest P_T within a sector defined by $\delta\phi > 2.0$ with respect to the trigger direction. A cut of $P_T > 500$ MeV/c is imposed on this leading particle. Both charged tracks and photon showers are included in this definition. Now all the particles are assigned a set of three weights along the trigger, leading particle and beam directions according to the formula, $w_i = \cos\theta_i e^{-5q_{t_i}}$, where θ_i and q_{t_i} are the angle and the transverse momentum with respect to the ith axis. Each particle is now assigned to the jet direction for which it has the highest weight and the momenta of the resulting groups are summed to define jet vectors. Two more iterations are performed wherein the weights for each particle are recalculated with respect to the jet axes and they are regrouped to form new jets. An event is considered to have jet structure if the final recoil jet has at least two particles and the ratio $|P_{T_{jet}}|/|P_{T_\gamma}|$ is greater than 0.2. This algorithm was tested on events that were generated by ISAJET[7] and propagated through simple detector simulation. The variables of interest for theoretical calculations[8] of triple differential cross sections are the P_T and the rapidity of the direct photon and η_{jet}, the pseudo-rapidity of the recoil jet. The generated photons and recoil jets were restricted to a rapidity range of $|0.7|$. Figure 6a shows the jet finding efficiency versus P_{T_γ} and η_{jet}. In figure 6b the difference between the reconstructed and generated η_{jet} is shown. Again, emphasis should be laid on the preliminary nature of this algorithm.

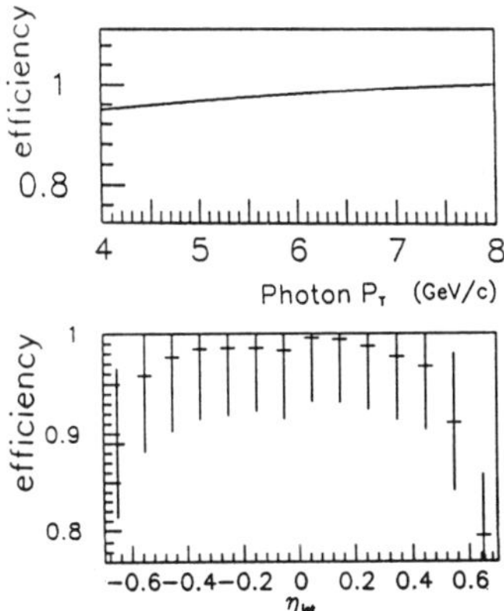

Figure 6a. Jet finding efficiency.

Figure 6b. Jet algorithm resolution of η_{jet}

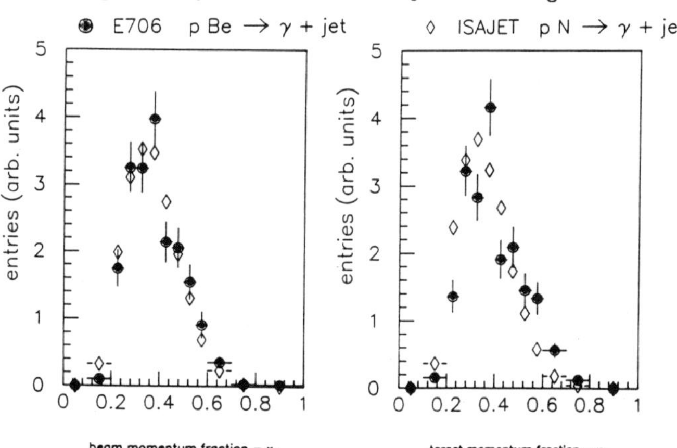

Figures 7a. and 7b Reconstructed parton momentum fractions.

If we make the assumption that the leading order processes dominate the production of direct photons and that the effects of intrinsic k_t are small, we can scale the P_T of the recoil jet to equal that of the trigger jet. This also compensates for any loss of particles that were part of the jet. Using the rescaled jet vectors we can solve the two-body kinematics to obtain the momentum fractions of the beam and target partons. The area normalized distributions for the proton induced events are shown in figure 7. The contamination due to π^0 and η mesons has been subtracted from these distributions. The data, however, are uncorrected for trigger inefficiencies. Also shown is a distribution obtained by following a similar procedure on ISAJET generated events. The Duke- Owens set I structure functions were used for event generation. Simulation of k_t effects is done by randomly introducing small boosts in the transverse direction in the photon-jet system. A beryllium nucleus was simulated by appropriately mixing protons and neutrons in the target. The shape of the data and ISAJET distributions appear to agree with each other.

Conclusions

The inclusive cross sections for direct photon production by proton and π^- beams incident on beryllium have been measured. Preliminary studies of the recoil jet have been made and compared to ISAJET.

Acknowledgements

This work is supported in parts by the DOE, NSF and UGC (India).

References

1. J. Owens, Rev. of Mod. Physics 59 (1987) p.465

2. P. Aurenche, *et al.*, Nucl. Phys. B297 (1988) p.661

3. F. Lobkowicz, *et al.*, NIM A235 (1985) p.332

4. E. Engels Jr., *et al.*, NIM A279 (1989) p.272 ; E. Engels Jr., *et al.*, NIM A253 (1987) p.523 ; E. Engels Jr., *et al.*, NIM A226 (1985) p.59

5. Due to space limitations the π^0 cross sections are not shown here. See, for example, G. Ginther, *et al.*, "Preliminary Results from Fermilab Experiment E706 - A Study of Direct Photon Production in Hadronic Collisions"*Rencontres de Moriond, New Results in Hadronic Interactions* Les Arcs, France, 1989 (published by Editions Frontières), p.181.

6. M. Bonesini, *et al.*, Z. Phys C - Particles and Fields 44 (1989) p.71

7. ISAJET version 6.21, F. Paige and S. Protopopescu, Brookhaven National Lab, Upton, NY 11973, private communication.

8. See Jeff Owens' contribution to this conference.

THE PROTON "SPIN CONTENTS": Current Status & Perspectives

Ta-Pei Cheng[*]
Department of Physics, University of Missouri, St. Louis, MO 63121

Ling-Fong Li
Department of Physics, Carnegie-Mellon University, Pittsburgh, PA 15213

Abstract

The present status of the phenomenological and theoretical interpretation of the EMC result on the polarized deep inelastic scattering is reviewed. We focus our discussion on the possibility of a significant gluonic contribution to the proton spin *via* the axial anomaly. We contrast the variant perspectives on this question: the viewpoint that stresses the interpretation in terms of the parton distributions *vs.* the one that concentrates on the matrix elements of local operators. Some remarks concerning the validity of OZI rule for the strange quark are also included.

I. Introduction

In this report we shall discuss various issues raised by the EMC result concerning the intrinsic structure of the proton spin. In Sec. II we review the interpretation of the EMC data as indicating a significant proton matrix element of the strange quark axial vector current, which almost cancel the u, d contributions. In Sec. III we discuss the suggestion that these matrix elements $\Delta q'$ may contain significant gluonic fraction Δg because of axial anomaly:

$$\Delta q' = \Delta q - \frac{\alpha_S}{2\pi}\Delta g \tag{1}$$

where Δg & Δq are the spin-dependent distributions of the gluon and the quark (of a particular flavor q), respectively. Arguments for the anomaly as local probe of the gluonic distribution, and some features of infrared collinear regularization of the perturbative QCD calculations will be reviewed. In Sec. IV we emphasize the advantage, for the problem at hand, of working directly with the current divergence instead of the axial vector current itself. By taking the forward proton matrix element of the axial anomaly equation, we obtain

$$\Delta q' = \Delta Q - \frac{\alpha_S}{2\pi}\Delta G \tag{2}$$

where ΔQ & ΔG are defined directly in terms of the matrix elements of local operators $2m\bar{q}i\gamma_5 q$ and $TrGG$, respectively. The second viewpoint allows us to see more clearly how the decoupling theorem is enforced for the heavy quark contribution $\Delta q' \to 0$. This approach also naturally suggests a useful comparison with the anomaly equation for the trace of the energy momentum tensor. The proton matrix element of this equation yields a sum rule for the proton mass. In this connection the problem of pion-nucleon sigma term with its possible interpretation as indicating a significant proton matrix element of the strange quark scalar-density will be briefly recalled. In Sec. V we show how ΔG can be estimated using the current divergence equation, and its peculiar feature of large isospin violation will be analyzed. In Sec. VI, we summarize and contrast the various perspectives of this "spin content" problem. Our approach suggests that the quark model OZI rule is not generally applicable for the strange quark, and future investigation as prompted by such a suggestion is briefly outlined.

---

[*] Presenter of the talk at the *DPF90* meeting.

II. The EMC result & its interpretations

Let us recall that the EMC collaboration measured the longitudinal spin-spin asymmetry in the deep inelastic muon-proton scattering [1]. The asymmetry measurements directly yield information on the spin-dependent structure function $g_1(x,Q^2)$. Through the operator product expansion one can show that the first moment of g_1 is then related to the proton matrix element of the axial vector current (weighted by the quark charge-squared) [2]. Including the leading QCD radiative correction to the Wilson coefficient function [3], we have

$$\int_0^1 dx\, g_1^{\,p}(x) = \frac{1}{2}\left[\frac{4}{9}\Delta u' + \frac{1}{9}\Delta d' + \frac{1}{9}\Delta s'\right]\left[1 - \frac{\alpha_S}{\pi}\right] \tag{3}$$

with

$$<p,s|\bar{q}\gamma_\mu\gamma_5 q|p,s> = \Delta q' s_\mu \tag{4}$$

where s_μ is the covariant spin vector of the proton. In the simple parton model one would naturally identify $\Delta q'$ with the polarized quark density.

The baryon matrix elements of the axial vector current also appear in the neutron and hyperon semileptonic decays. One can then express, through flavor SU(3), two of the quark densities in terms of the axial vector F and D coefficients:

$$\int_0^1 dx\, g_1^{\,p}(x) = \frac{1}{18}(9F - D + 6\Delta s')\left[1 - \frac{\alpha_S}{\pi}\right] \tag{5}$$

Since the proton does not have strange valence quark, it seems reasonable to conjecture that $\Delta s' \simeq 0$, expressing a kind of OZI rule [4]. This yields the Ellis-Jaffe sum rule [5] with r.h.s. of Eq. (5) being *0.171*. This has become the benchmark for experimenters and phenomenologists alike. What is considered so surprising about the EMC result [1] is that it deviates significantly from this baseline expectation.

$$\int_0^1 dx\, g_1(x) = 0.126 \pm 0.010 \pm 0.015 \tag{6}$$

Using SU(3) to express the EMC result in terms of the Δq's, we have

$$\Delta u' = 0.77, \qquad \Delta d' = -0.49, \qquad \Delta s' = -0.16 \tag{7}$$

which sum up to
$$\Delta\Sigma' = 0.12$$

Here we have used $F = 0.47$ & $D = 0.81$, which are the best-fit values of all the neutron and hyperon β decay measurements [6, 7]. This represents a significant downward revision, from 0.63 down to 0.58, of the F-to-D ratio widely used up until recently [8]. We also note that, because not all measured F & D parameters are consistent with each other, by emphasizing any particular set of data the resultant Δq's can vary somewhat [9]. However we expect that the uncertainties not to exceed twenty percent, and the conclusions that $\Delta s'$ is non-negligible will stand.

It has also been pointed out [6] that the EMC result is consistent with the existing data on low energy elastic $\nu p \to \nu p$ scattering [10], which is sensitive to a different combination of $\Delta u'$, $\Delta d'$, and $\Delta s'$.

The results of a significant $\Delta s'$ and a suppressed $\Delta\Sigma'$ seem to deviate from the naive quark model expectation, which would predict a negligible contribution from the strange quark term $\Delta s' \simeq 0$ and it would have the proton spin given completely by the valence quarks $\Delta\Sigma' \simeq \Delta u' + \Delta d' = 1$. Of course such an expectation is perhaps too simple, since we know from the deep inelastic scattering (DIS) momentum sum rule that the gluons carry about half of the proton momentum. Therefore it is likely that the situation will change considerably when the gluon degrees of freedom are introduced: while gluons do not contribute to the additive quantum numbers like electric charge, and strangeness, etc. there is no reason to expect the gluon spin and orbital angular momentum not to be an important factor in the discussion of the proton spin. Still, because of the simplicilty of the nonrelativistic quark model (and many of its successes), the EMC result of $\Delta s'$ & $\Delta \Sigma'$ had people calling it *a proton spin crisis*. Several different interpretations of the EMC data have been advanced. Even a breakdown of perturbative QCD has been suggested [11]. (For arguments against such an interpretation, see Refs. [9, 12].) We list below some of the less radical suggestions:

(1) Data not yet in the scaling region ?

It has been suggested that perhaps the Q^2 values reached in the EMC experiment are still not large enough to be in the scaling region [13]: one observes that the $Q^2 = 0$ limit of the first moment is fixed by the Drell-Hearn-Gerasimov sum rule, and this implies a very different value for the r.h.s. of the integral (in fact a sign-change). However, this is not a viable option, since data over a sizable range of Q^2 exist and no appreciable Q^2-dependence is visible. Actually it is remarkable that the scaling curve represents, in the sense of Bloom and Gilman [14], a very good average over local fluctuations down to $Q^2 \simeq 0.5 \ GeV^2$. (And, one can even understand the change of sign in the DHG sum rule as reflecting the N^* dominance in the low Q^2 region [15].)

(2) Unreliable extrapolation to $x = 0$?

The small $x \equiv Q^2/2M\nu$ limit corresponds to the high energy, $\nu \to \infty$, limit of the virtual Compton scattering. The EMC extrapolation is quite consistent with the standard Regge expectation [12], and thus $g_1(x)$ should be regular in the $x = 0$ limit. Clearly if this expectation turns out to be incorrect and there is some sort of singular behavior, then the resulting moment integral will be significantly changed [16]. However there is not a hint of such a singular behavior in the data, and theoretically the Pomeron-Pomeron-cut $(x l n x)^{-1}$ behavior is not expected here either.

(3) Large SU(3) breaking ?

Unlike the vector charges which are protected from large symmetry breaking corrections by the Ademollo-Gatto theorem [17], the axial vector charges may not obey SU(3) relations all that well. Again there is no such way out. Several hyperon beta decays have been measured, and one finds that SU(3) symmetry actually works remarkably well. In fact the uncertainties in the F & D matrix elements result mostly from experimental systematic errors [7]. Also one can show that the baryon axial charges are not sensitive to large SU(3) breakings in the baryon wave functions [18].

(4) Gluonic contribution via axial anomaly ?

Perhaps the most interesting suggestion for revising the straightforward interpretation of the EMC result is that, because of the presence of axial anomaly [19, 20], $\Delta q'$ not only represents the quark contribution to the proton spin but has a significant fraction coming from the gluon polarization as well, see Eq. (1), where Δq & Δg are idendified with the spin-dependent quark and gluon distributions, respectively:

$$\Delta q = \int_0^1 dx \ [q^+(x) - q^-(x) + \bar{q}^+(x) - \bar{q}^-(x)] \quad and \quad \Delta g = \int_0^1 dx[g^+(x) - g^-(x)]. \quad (8)$$

The remarkable claim is that one can in a perturbative QCD calculation single out a subset of gluon reducible diagrams with a small coefficient proportional to $\alpha_S(Q^2)$. This possibility was first raised by Efremov and Teryaev [21], and then Altarelli and Ross [22] correctly derived the above spin density relation. Thorough discussions have also been given by Carlitz, Collins, and Mueller [23, 24]. Further discussion as to the validity of this relation has been given in [7, 25]. We shall suggest [26] that current divergence equations and decoupling theorem can provide another perspective for clarifying some of the perplexities in this problem.

III. Gluonic contribution via axial anomaly

The axial anomaly under discussion is the one associated with the global axial U(1) symmetry of QCD. Namely, because the triangle diagram of the (quark) axial vector current with two gluon legs is linearly divergent and any cut-off scheme will either break the vector or the axial-vector symmetry, the regularization scheme that keeps the vector current conservation necessarily leads to the anomaly term in the divergence of the axial vector current [19]. For our purpose it is more convenient to think all this in terms of each flavor separately. Thus, we have

$$\partial^\mu \bar{q}\gamma_\mu \gamma_5 q = 2m_q \bar{q}i\gamma_5 q + \frac{\alpha_S}{2\pi}Tr G\tilde{G} \tag{9}$$

The key point to keep in mind is that anomaly comes from regulating the ultraviolet divergence.

The first moment of the spin dependent structure function, as measured in the inelastic eletroproduction in the scaling limit, is related, through the operator product expansion, to the proton matrix element of the local axial vector current. In the parton model, this can be evaluated by taking the axial vector current between the quark states and multiplying it by the probability of finding the quark in the target proton. Or, equivalently, this is the matrix element $<q|j_{5\mu}|q>$ for the quark state with helicity aligned to the proton helicity times the spin-dependent quark distribution Δq as given in Eq. (8). What AR have shown [22], and is corroborated later by more detailed discussion by CCM [23], is that there is another potentially important contribution to this proton matrix element of the axial vector current: it is the spin-dependent gluon distribution Δg times the gluonic matrix element of the axial current $<g|j_{5\mu}|g>$. The last factor is just the triangle diagram for the anomaly which, with $<q|j_{5\mu}|q>$ normalized to one, yields a coefficient of $-\alpha_S/2\pi$. Thus, we have schematically:

$$<p|j_{5\mu}|p> = <q|j_{5\mu}|q>\Delta q + <g|j_{5\mu}|g>\Delta g \tag{10}$$

or

$$\Delta q' = \Delta q - \frac{\alpha_S}{2\pi}\Delta g$$

Let us emphasize that the operator product expansion approach to DIS itself is not under challenge. What is under discussion is the contribution to the proton matrix element of the local operator of quark axial vector current. With regard to the anomaly contribution to this matrix element, we make the following observations:

(1) $\alpha_S \Delta g$ can be large

The QCD correction term $\alpha_S \Delta g$ in $\Delta q'$ can be large because Δg grows with Q^2. It is easy to see that helicity is not conserved in a tree diagram representing gluon bremsstrahlung by a quark. While quark helicity does not change, there is a gain of the gluon helicity. Of course what is conserved is the total angular momentum, for which one has to include the orbital contribution [27]. As it turns out for large Q^2, Δg goes as lnQ^2 [28], and, since the strong coupling $\alpha_S \propto (lnQ^2)^{-1}$, the product $\alpha_S \Delta g$ is Q^2-independent at the leading order. In effect, one can regard $\alpha_S \Delta g$ as being of the zeroth order, and thus potentially it can be substantial.

We can state this more precisely in terms of the Altarelli & Parisi equation [29]. The evolution of the moment of the structure function is controlled by the moment of the 'splitting function' which can be calculated from QCD vertices. For $t \equiv lnQ^2$, we have [22]

$$\frac{d}{dt}\begin{pmatrix} \Delta\Sigma \\ \Delta g \end{pmatrix} = \alpha_S(t)\begin{pmatrix} 0 & 0 \\ A_{qg} & b \end{pmatrix}\begin{pmatrix} \Delta\Sigma \\ \Delta g \end{pmatrix} \tag{11}$$

The zero entries reflect the Q^2 independent nature of the quark helicity distribution $\Delta\Sigma$. The key point is that the (g,g) entry in the evolution matrix is proportional to the leading coefficient in the QCD β-function: $d\alpha_S/dt = -b\alpha_S{}^2$. Because to this order $\Delta\Sigma$ is t-independent, thus we can ignore the $\alpha_S(t)\Delta\Sigma$ term for large Q^2, and obtain

$$\frac{d}{dt}\Delta g(t) \simeq b\alpha_S(t)\Delta g(t)$$

This implies that

$$\alpha_S\frac{d\Delta g}{dt} + \frac{d\alpha_S}{dt}\Delta g = \frac{d}{dt}(\alpha_S\Delta g) \simeq 0 \tag{12}$$

(2) Anomaly as a local probe of the gluon distribution

In the parton model language, one can represent the probing of the *quark* distribution by the tree diagram. The phenomenological manifestation is the 1-jet process corresponding to a hard quark. As for the gluon, we have box diagrams. Since this involves a integration over all values of momentum it generally does not represent a pointlike probe. However the finite result $-\alpha_S/2\pi$ extracted above in the perturbation calculation comes from the high $k_T{}^2 \simeq Q^2$ region of the loop integration [23]. Because only the high k_T region contributes, this loop graph is effectively a pointlike coupling. Furthermore, because the quark lines are hard, the phenomenological signal of gluon parton contribution will be 2-jet processes.

Also, had the low momentum scales contributed, the running coupling would not be small and the perturbative approach could not be trusted. Fortunately since the present situation only involves hard quark lines, we do have a small coupling $\alpha_S(Q^2)$, as guaranteed by asymptotic freedom, and the relevant perturbation calculation is reliable.

In the operator product expansion language, the above box graph generates in the scaling limit the triangle diagram of the axial vector current matrix element. And the above discussion is compatible with our understanding that anomaly comes from the ultraviolet behavior of the loop integration. Again this means that one can view the hard quark triangle loop as effetively pointlike. Thus we have a rather subtle situation: One would think that there was no pointlike gluon contribution to the first moment because there was no (color) gauge invariant twist-two gluon operator, but the above calculations suggest that, the gluon contribution via the anomaly being effectively pointlike, we can think of the local quark axial vector current operator $j_{5\mu}$, as also containing a gluon part:

$$j_{5\mu} = \tilde{j}_{5\mu} - \frac{\alpha_S}{2\pi}k_\mu \tag{13}$$

where k_μ is a twist-two local gluonic operator.

This possibility has long been discussed in the context of the anomaly equation. The anomalous divergence itself can be written as the four-divergence of a gluonic current:

$$\partial^\mu k_\mu = -Tr\tilde{G}G, \qquad with \qquad k_\mu = -\epsilon_{\mu\nu\lambda\sigma}Tr\left[A^\nu G^{\lambda\sigma} - \frac{2}{3}A^\nu A^\lambda A^\sigma\right] \tag{14}$$

Thus the decomposition of Eq. (13) is possible, $\tilde{j}_{5\mu}$ being conserved in the chiral limit. Indeed it is tempting to suggests that Eq. (1) can be trivially obtained from the current decompostion (13) if the forward matrix elements of the currents k_μ & $j_{5\mu}$ are to be idendified with Δg & Δq just as $j_{5\mu}$ defines $\Delta q'$ in Eq. (3). If this could be done, we would have then simple operator matrix element expressions of the spin-dependent parton density functions Δq & Δg. But the topological current k_μ is not (color) gauge invariant. Related to this, the matrix elements of k_μ & $j_{5\mu}$ have massless ghost-pole terms and the corresponding zero momentum transfer limits are singular. Thus it seems that we cannot go beyond the partonic definitions of Δq & Δg as given in Eqs. (10, 8).

To be more specific we write out the nonforward matrix elements, with the momentum difference of the final and initial proton states being $t_\mu \equiv p'_\mu - p_\mu$, as follows:

$$<p',s|j_{5\mu}|p,s> = \bar{N}(p')[\gamma_\mu\gamma_5 G_1(t^2) - t_\mu\gamma_5 G_2(t^2)]N(p) \tag{15a}$$

$$<p',s|\tilde{j}_{5\mu}|p,s> = \bar{N}(p')\left[\gamma_\mu\gamma_5\tilde{G}_1(t^2) - t_\mu\gamma_5\tilde{G}_2(t^2)\right]N(p) \qquad (15b)$$

$$<p',s|k_\mu|p,s> = \bar{N}(p')[\gamma_\mu\gamma_5 H_1(t^2) - t_\mu\gamma_5 H_2(t^2)]N(p) \qquad (15c)$$

In the forward $t_\mu \to 0$ limit, the massless ghost-pole terms in the gauge variant current matrix elements must cancel because the gauge invariant current j_5 is nonsigular even in the chiral limit [7, 30-34]:

$$R_q - \frac{\alpha_S}{2\pi}R_g = 0, \qquad with \quad \tilde{G}_2(t^2) \simeq \frac{2MR_q}{t^2} \quad and \quad H_2(t^2) \simeq \frac{2MR_g}{t^2} \qquad (16)$$

(M is the proton mass.) The forward matrix elements of Eq. (13) are then related

$$\Delta q' = \tilde{G}_1(0) - \frac{\alpha_S}{2\pi}H_1(0) \qquad (17)$$

where we have used $G_1(0) = \Delta q'$. Indeed it is tempting to recover Eq. (1) by making the idendifications of

$$\Delta q = \tilde{G}_1(0) \quad and \quad \Delta g = H_1(0). \qquad (18)$$

However this is not warranted because we do not expect $H_1(0)$ & $\tilde{G}_1(0)$ to be invariant under large gauge transformations. Thus, for Δq & Δg, it is not clear how to go beyond the partonic definitions [as given in Eqs. (10, 8)], to the local operator matrix element definitions. We shall return to this problem of expressing quark and gluon contributions in terms of operator matrix elements in Secs. IV and VI.

(3) Regulating the infrared collinear divergence

In the chiral limit ($m = 0$) with the gluon on-shell ($p^2 = 0$), the box- and triangle-graphs have infrared collinear divergences. The problem of a proper regularization has been studied in detail by CCM [23]. They note that the final result (symbolically represented by Δ_5 below) depends sensitively on the form of the IR regulation:

$$\Delta_5 = -\frac{\alpha_S}{2\pi} \quad for \quad \frac{m^2}{p^2} \to 0 \qquad (20a)$$

$$\Delta_5 = 0 \quad for \quad \frac{p^2}{m^2} \to 0 \qquad (20b)$$

CCM point out that the first limit (20a) is the appropriate one: gluons inside a confined hadron is expected to be off-shell by an amount characterized by the QCD confinement scale $p^2 \simeq \Lambda^2$ of a few hundred MeV^2, which is large compared to the u, d, s quark masses. Furthermore the result (20a) comes from high k_T region of the loop integration, in agreement with our expectation that the anomaly results from UV regularization of the triangle graph.

The zero of the second limit (20b) results from a cancellation of this high k_T term by an equal and opposite low k_T contribution. As we have already pointed out, the high k_T part corresponds to hard quark lines and thus a 2-jet process. It is compatible with our expectation of a scattering off the gluon parton. The low k_T contribution to the scatterings does not produce 2-jet events. Thus this factor should really be counted as (higher order) part of Δq. In short, the second form of IR regularization is not the physically relevant one.

For different viewpoints on this question of infrared sensitivity, see Refs. [7, 25].

IV. The perspective from current divergence equations & decoupling theorem

(1) Axial anomaly equation and the proton spin

The problems of gauge invariance and infrared regularization prompt us [26] to suggest that a useful approach to the problem of gluonic contribution via the axial anomaly is to work directly with current divergences. (For previous discussions of the axial U(1) Ward idenditiy in connection with the proton spin problem, see Refs. [7, 30-35].) The current divergences are expected to be less IR sensitive, and gauge invariance is also manifest.

For the gauge invariant axial vector current $j_{5\mu}$, becuase there is no massless Goldstone boson even in the chiral limit, it is trivial to go from the forward matrix element of the current to that of the divergence: Denoting the proton (axial) spin-vector and pseudoscalar density by $s_\mu \equiv N\gamma_\mu\gamma_5 N$ and $v \equiv Ni\gamma_5 N$, respectively, we have

$$<p,s|j_{5\mu}|p,s> = s_\mu \Delta q' \quad \rightarrow \quad <\partial j_5> = 2Mv\Delta q' \tag{21}$$

For the gauge variant currents $\tilde{j}_{5\mu}$ and k_μ, even though the current matrix elements are singular, those of the divergences are regular and manifestly gauge invariant:

$$<\partial \tilde{j}_5> = <p,s|2m\bar{q}i\gamma_5 q|p,s> \equiv 2Mv\Delta Q \tag{22a}$$

$$<\partial k> = <p,s|-G\tilde{G}|p,s> \equiv 2Mv\Delta G \tag{22b}$$

From the expressions in Eq. (15) and the definitions in Eq. (16), we have

$$\Delta Q = \tilde{G}_1(0) - R_q \quad and \quad \Delta G = H_1(0) - R_g. \tag{23}$$

Namely, the (large) gauge non-invariant parts on the r.h.s. cancel to yield gauge invariant ΔQ & ΔG. Taking the forward proton matrix elements of the familiar anomaly equation (9), we obtain Eq. (2) very similar to that of Eq. (1).

Given the expressions (22) for ΔQ & ΔG in terms of the quark and gluon fields, we can naturally interpret Eq. (2) as another way of separating the quark and gluon contributions to the matrix element of the axial vector current. The interpretation of ΔQ & Δq as the "intrinsic quark contributions" is justified as in the free-field limit we have: $(\Delta q')_0 = (\Delta Q)_0 = (\Delta q)_0$.

Because Eqs. (1) and (2) have similar quark-gluon decompositions, we can view their relation as follows: Eq. (2) is just the Eq. (1) with certain constant being shifted from the first to the second term on the r.h.s.[*]

$$\sigma_q = \Delta q - \Delta Q, \quad \sigma_g = \Delta g - \Delta G \quad with \quad \sigma_q - \frac{\alpha_S}{2\pi}\sigma_g = 0. \tag{24}$$

Since the $\sigma's$ in the quark and gluon factors should always cancel, it really does not matter which set, $(\Delta q, \Delta g)$ or $(\Delta Q, \Delta G)$, one uses in the discussion of the $\Delta q'$ properties. (Of course, this is not the case when one wishes to discuss the separate quark or gluon contributions.) In this sense, one can look upon Eq. (1) as being the forward matrix element of the axial anomaly equation. The main advantage of Δq & Δg is their direct parton interpretations, while that of ΔQ & ΔG is their simple expression in terms of the (manifestly gauge invariant) field operators.

One of the advantages of having the explicit operator representations of (22) is that it allows us to see more clearly the source of 'IR regulation ambiguity' discussed in Sec. III.(3). Because we now view the relation Eq. (1) as the forward proton matrix elements of the anomaly equation Eq. (9), it is straightforward to see how a heavy quark ($q = c, b, t$) contribution to the proton spin 'decouples'. According to the Appelquist-Carrazzone theorem [36] $\Delta q'$ for heavy quarks should be suppressed by powers of the quark masses. However from our Eq. (2) for $\Delta q'$ we see that ΔQ does *not* vanish in the $m \rightarrow \infty$ limit, because when we integrate out the heavy quark field the pseudoscalar density for the heavy quark gives rise to a dimension-four operator:

$$m\bar{q}i\gamma_5 q \rightarrow -\frac{\alpha_S}{4\pi} Tr G\tilde{G} + O(m^{-2}) \tag{25}$$

---

[*] G. Altarelli, private communication.

Consequently, to the leading order of this 'heavy quark expansion' [37], the quark and gluon pseudoscalar density terms cancel in the divergence equation. Thus, the gluonic contribution via anomaly is needed to enforce the physical requirement of decoupling of heavy quarks.

It is easy to see how this cancellation comes about. Recall that anomaly term results from regulating the UV divergence. We can think that this is accomplished by introducing a 'regulating fermion' with a large mass. This of course breaks the chiral symmetry, and it gives rise to a mass term in the divergence of the axial vector current. Integrating out this heavy fermion we can then transform this naive divergence term into the anomaly factor. In every step it contributes the same as our original heavy quark. The difference is that for the regulating fermion loop there is an extra minus sign. This accounts for the cancellation.

Given our previous comment on the equivalence of $(\Delta Q, \Delta G)$ and $(\Delta q, \Delta g)$ in any calculation of $\Delta q'$, this cancellation due to decoupling is the same one encountered above in the discussion of the IR collinear regularization, Eq. (20). This shows once again that the $p^2/m^2 \to 0$ limit is not the appropriate one for the u, d, s quarks. (They are not heavy quarks on the QCD confinement scale.)

Viewed in this perspective, it should also be clear that there is no reason to expect that the strange quark fraction $\Delta s'$ be suppressed. Although $m_s >> m_u, m_d$ it is still not a heavy quark, and there is no reason to expect that the approximation of Eq. (25) is valid for the strange quark. (After all, treating the s quark as a light quark in chiral SU(3) and kaon PCAC does lead to reasonable results.) In this sense the EMC result finding $\Delta s'$ to be comparable to $\Delta u', \Delta d'$ should not really be looked upon as being so puzzling. This is certainly not a crisis.

(2) Trace anomaly equation and the proton mass

Given the new perspectives of current divergence equations and decoupling theorem, it is natural to make a comparison of another static property of the proton: its mass. Here the relevant current is the dilation current with a divergence given by the trace of the energy momentum tensor.

$$\partial^\mu D_\mu = \theta^\mu{}_\mu = \Sigma m_i \bar{q}_i q_i - \left[11 - \frac{2}{3}n_f\right]\frac{\alpha_s}{8\pi} Tr G^{\mu\nu} G_{\mu\nu} \tag{26a}$$

Again, the naive divergence is due to the quark masses, and there is also an anomaly term, the *trace anomaly* [38].[*] When taken between the proton states, this divergence equation yields

$$M = <m_u \bar{u}u + m_d \bar{d}d + m_s \bar{s}s> + \sum_{heavies} <m_i \bar{q}_i q_i> - \left[9 - \frac{2}{3}n_h\right]\frac{\alpha_s}{8\pi}<TrGG> \tag{26b}$$

where n_h is the number of heavy quarks. Again the heavy quarks decouple through the cancellation of the quark mass terms and their contributions to the anomaly. As for the surviving term, it has been tempting to say that the $<m\bar{q}q>$ terms are negligible because $m_u, m_d, \& m_s$ are small. Thus one ends up with the proton mass being almost all given by the gluon contribution. This view has been advocated by Shifman, Vainshtein, and Zakharov [39].

Actually there is experimental information on the quark contributions. The u, d term is measured by the πN sigma term (ignoring a small isospin violation correction)

$$<N|m_u \bar{u}u + m_d \bar{d}d|N> \simeq 60 \; MeV. \tag{27}$$

And just as we can deduce the strange quark contribution from the EMC result, here one can use SU(3), the octet baryon mass differences, and the canonical quark mass ratio to get the fraction

---

[*] Trace anomaly equation has also been discussed in Ref. [7] albeit for the purpose of illustrating the 'ambiguity' of idendifying a gluonic contribution (see our comments in Sec. VI.(2)). Also see [40], for another discussion.

$$\frac{2<N|\bar{s}s|N>}{<N|\bar{u}u + \bar{d}d|N>} \simeq 0.47. \qquad (28)$$

This is the so called '$\sigma_{\pi N}$ problem' [41]: nucleon seems to have surprisingly significant amount of the strange quark sea. But from the perspective of the decoupling theorem it really is not so odd. Incidently with such interpretation of the $\sigma_{\pi N}$ value, one finds that the proton mass is about equally shared between the quarks and the gluons [42].

V. Estimating ΔG & ΔQ

In the context of divergence equation, we can use the standard current algebra approach of Goldstone pole saturation to estimate ΔG and ΔQ [35] as defined in Eq. (22).

$$\Delta u' = \frac{m_u}{M} v_u - \frac{\alpha_s}{2\pi} \Delta G$$

$$\Delta d' = \frac{m_d}{M} v_d - \frac{\alpha_s}{2\pi} \Delta G \qquad (29)$$

where we have written out ΔQ in terms of the quark masses and the pseudoscalar density matrix elements, $v_q N i \gamma_5 N = <p|\bar{q} i \gamma_5 q|p>$. One can solve for ΔG and $v_u + v_d$ (the "unknowns") if we regard as known quantities: $\Delta u'$, $\Delta d'$ (as given *via* the EMC result) and $v_u - v_d$ (*via* saturation of the nonsinglet channel by the Goldstone poles). Let us first recall that the Goldberger-Treiman relation [43] can be derived in the charged channel by the $\pi^\pm$ pole-dominance of the pseudoscalar density. Taking the nucleon matrix element of the divergence equation,

$$\partial^\mu \left[\bar{u} \gamma_\mu \gamma_5 d \right] = (m_u + m_d) \bar{u} i \gamma_5 d$$

one obtains

$$2 M g_A = 2 f_\pi g_{\pi NN} + \mu_\pm \qquad (30)$$

where $\mu_\pm$ denotes the correction to the $\pi^\pm$ pole-dominance, and is the correction to g_A as given by the Goldberger-Treiman relation. Repeating this for the neutral isovector channel,

$$\partial^\mu \left[\bar{u} \gamma_\mu \gamma_5 u - \bar{d} \gamma_\mu \gamma_5 d \right] = 2 m_u \bar{u} i \gamma_5 u - 2 m_d \bar{d} i \gamma_5 d$$

we have

$$2 M g_A = 2 f_\pi g_{\pi NN} + \mu_0 + (m_u - m_d)(v_u + v_d) \qquad (31)$$

Comparing these two expressions for g_A one concludes that the singlet density is small, on the order of *correction* to the GT expression for g_A. If we assume $\mu_0 = \mu_\pm$, the isosinglet pseudoscalar density vanishes [41], and this leads to

$$\frac{\alpha_s}{2\pi}(\Delta G)_p \simeq -0.32 \quad and \quad \sum_Q (\Delta Q)_p \simeq -0.84 \qquad (32)$$

If one repeats the same calculation for the neutron, one finds a very different result:

$$\frac{\alpha_s}{2\pi}(\Delta G)_n \simeq +0.04 \quad and \quad \sum_Q (\Delta Q)_n \simeq +0.24 \qquad (33)$$

Thus although the operator $Tr G\tilde{G}$ is manifestly iso-symmetric, its nucleon matrix elements violate isospin symmetry strongly.

If one goes back through the calculation, one finds that the isospin violation comes in the form of

$$\left[\frac{m_u - m_d}{m_u + m_d}\right](\Delta u' - \Delta d'). \tag{34}$$

It is not difficult to see what this represents: the $(\Delta u' - \Delta d')$ factor is the isovector π^0 axial vector coupling to the nucleon, and it gives opposite sign contributions to $(\Delta G)_p$ and $(\Delta G)_n$, respectively. This term comes about as follows: Because we are *not* setting $m_u = m_d$, the isospin violating term in the QCD Hamiltonian causes a $\pi^0\eta$ mixing. This brings about the π^0 pole in the matrix element of the isosinglet operators, with a mixing angle $\sim(m_u - m_d)$. Such a pion pole contribution is enhanced by the smallness of the pion mass, as indicated by the $(m_u + m_d)$ factor in the denominator.

With such a large isospin violation one should worry that this is not compatible with our original assumption of flavor SU(3) symmetry in extracting $\Delta q'$ from the EMC data. But, it turns out that the pion pole contributions (*i.e.* the large isospin violations) in the gluon ΔG and the quark mass terms ΔQ cancel out. The axial vector matrix elements $\Delta u'$ and $\Delta d'$ respect isospin symmetry to a high degree. All this was worked out long ago by Gross, Treiman & Wilczek [45] who pointed out this cancellation as a consequence of the Sutherland theorem [46].

Hatsuda [32] using an effective Lagrangian approach showed that to the leading order in the large N (number of colors) expansion, ΔQ & ΔG are given in terms of the η', η, & π^0 pole contributions. Again they mostly cancel in the combination of Eq. (2) and thus the sum of the axial vector current matrix elements $\Delta\Sigma'$ is given by the η' nucleon coupling. Thus the EMC result of Eq. (7) implies that η' essentially decouples from the nucleon. Veneziano [33] has argued that this relation between $\Delta\Sigma'$ and $g_{\eta'NN}$ is independent of the large N expansion, and is a generalized Goldgerger-Treiman relation in the singlet channel. That $\Delta\Sigma' \simeq 0$ and the decoupling of η' from the nucleon is first derived in the Skyrmion model by Brodsky, Ellis, and Karliner [47].

VI. Summary & discussion

(1) Proton matrix element of the axial vector current

This is perhaps the least controversial part of the discussion on the problem of the proton spin contents:

(i) The extraction of $\Delta q'$ (for the flavor q) from the EMC data should be dependable. In particular the result of a significant $\Delta s'$ should be taken seriously. As for the sum $\Delta u' + \Delta d' + \Delta s'$, it is most likely suppressed, perhaps not as strongly as indicated in the original analysis, see Eq. (7).

(ii) Since the connection of axial vector current matrix element to the measurement of spin asymmetry in the deep inelastic scattering is not in question, we have every reason to expect that the Bjorken sum rule [2] to hold when the corresponding measurements on the neutron target are carried out in future experiments. (Namely, one should keep in mind that the discussion in Sec. V of large isopin violation concerns the separate quark and gluon contributions rather than $\Delta q'$ itself, which enters the sum rule directly.)

(iii) By the same reasoning, at sufficient high Q^2 (above the heavy quark thresholds) the first moment of $g_1(x)$ will in principle receive contributions from $\Delta c'$, $\Delta b'$, & $\Delta t'$, respectively. However by our decoupling argument, independent of the specific quark-gluon decompositions, matrix elements of such heavy quark local operators should be suppressed by inverse powers of the heavy quark masses. The key question that determines whether a quark is heavy or not is the following: what's the relevant scale for the proton matrix element of a local operator? Is it the QCD confinement scale $\simeq 1$ GeV? Or, is it the Q^2 of the virtual photon? One would think that the issue of the relative sizes of the quark mass vs. Q^2 has to do with the Wilson coefficients (with its flavor thresholds) multiplying the local operator. Once the heavy flavor threshold is passed, one then gets a clear-cut connection between the first moment and the matrix element of the heavy quark bilinear. If the relevant scale is the confinement radius, such matrix elements should all be suppressed,

$$\Delta c' \simeq \Delta b' \simeq \Delta t' \simeq 0, \tag{35}$$

independent of the value Q^2 (besides it's sufficiently above the corresponding quark threshold). Several authors offer a different prediction on this point [15, 48].

(2) The quark and gluon contributions: variant perspectives

In this report we have focused our discussions on the quark and gluon contributions to $\Delta q'$. It has been pointed out that there are two closely related, yet different, decompositions of $\Delta q'$:

$$\Delta q' = \Delta q - \frac{\alpha_s}{2\pi}\Delta g \tag{36}$$

$$\Delta q' = \Delta Q - \frac{\alpha_s}{2\pi}\Delta G \tag{37}$$

where Δq and Δg are spin-dependent parton distributions, as given in Eq. (8), and ΔQ and ΔG are matrix elements of local operators bilinear in the quark and gluon fields, as given in Eq. (22). Thus we can regard the two equations (36, 37) as being related by shifting a piece from the quark to the gluon term, as in the manner of Eq. (24). In fact a plausible expectation is that $(\Delta q, \Delta g)$ corresponds to those parts of $(\Delta Q, \Delta G)$ that can be reached in perturbative QCD, while the shifted constants (σ_q, σ_g) are purely non-perturbative (instanton interactions, ghost poles, *etc.*). Efremov, Soffer, and Törnqvist [34] have pointed out that the result obtained by Veneziano [33] should really be interpreted as the ghost-pole residue in the chiral limit; we can then interpret this to mean:

$$\sum_q \sigma_q = n_f \frac{\alpha_s}{2\pi}\sigma_g = \frac{F_{\eta'}}{2M}g_{\eta'NN} \tag{38}$$

where the decay constant $F_{\eta'} \simeq 287\ MeV$. Because ΔQ is explicitly proportional to the quark mass, it vanishes in the chiral limit, leading through Eq. (24) to:

$$\sum_q \Delta Q = 0, \quad or \quad \Delta u + \Delta d + \Delta s = \frac{F_{\eta'}}{2M}g_{\eta'NN}. \tag{39}$$

However one must be very careful in making such an estimate. As the calculation in Sec. V showed that $\Sigma\Delta Q$ is not particularly small and it is probably dominated by the non-perturbative σ_q, therefore using such an approximation to determine $\Sigma\Delta q$ is not likely to be reliable (besides the difficulty of an uncertain $g_{\eta'NN}$).

We also comment that that if one does not set the quark masses to zero, chiral and isospin symmetry breaking will bring about mixings between η' with η and π^0 [32, 33]. The pion pole contribution is the source of the large isospin violation discussed in the last section. Thus this large isospin violation is associated with σ_q & σ_g, which cancel out in axial vector matrix elements $\Delta q'$, Eq. (24). This is again in agreement with the result of ref. [45], and with our expectation that there should not be such large isospin violation in the parton distributions Δq & Δg. However it does not mean that this isospin violation is unphysical because there are situations where it is ΔQ or ΔG that is the physical observable. For example ΔQ's enter directly in the pseudoscalar Higgs boson nucleon couplings [44, 49], ΔG can in principle contribute to the neutron electric dipole moment, etc.

We should emphasize that the existence of these two variant approaches Eqs. (36, 37) does not imply there are unavoidable ambiguities so that it is meaningless to talk about any separate quark and gluon contributions. (For such a viewpoint, see ref. [7].) Both decompositions are meaningful because in each case the quark and gluon terms can be physically defined. Each decomposition emphasizes a particular aspects of the spin content problem; each definition has its own advantage: Clearly Δq & Δg emphasizes the parton aspect of the problem, while ΔQ & ΔG are directly given by matrix elements of gauge invariant local operators, which are related to a variety of physically measurable quantities.[†]

(3) OZI rule is not generally applicable for the strange quark

The decoupling theorem suggests that there is no *a priori* reason to expect that $\Delta s'$ should be particularly small. In fact, because strange quark is not a heavy quark on the QCD confinement scale, it suggests that OZI rule should not generally be applicable to the strange quark. (For an alternative view, suggesting that there is no strange quark OZI rule for baryons, as *vs.* mesons, see [50].) The nucleon matrix elements of strange operators generally should not be suppressed. (For recent discussions of such matrix elements, see refs. [6, 7].) Of course this does not mean that there is no physical instance in which the nonstrange hadron matrix element of some strange quark bilinear is suppressed. However, only for the truly heavy quarks c, b, t, OZI rule is generally applicable, (*i.e.* it is really a *rule*) because of asymptotic freedom and the decoupling theorem. To put it in another way, what we are suggesting is that for the case of strange quark bilinears there should be a "new baseline of expectations": generally we should expects their matrix elements to be comparable to those of the u, d operators, *except* for cases where there are specific symmetry or dynamic causes for their suppression.

For example, the OZI rule for the vector mesons (ϕ, ω) system may wll be explainable by vector meson dominance of the strange quark vector current, which has zero forward matrix element because it measures the net strangeness quantum number of the hadron state.

$$<nonstrange\ hadron\,|\,\bar{s}\gamma_\mu s\,|\,nonstrange\ hadron> = 0 \tag{40}$$

Such a picture is also compatible with the observed feature of 'channel-dependence' of the OZI rule for the strange quark. Recall that while there is OZI suppression for the vectors (ϕ, ω) the pseudoscalar system of (η, η') deviates significantly from ideal mixing. A challenge for future investigations will be to discover the specific mechanism for the suppression of the strange quark contribution to the DIS momentum sum rule, especially in view of its large contribution to the nucleon mass in the energy momemtum trace sum rule as discussed in this report.

Acknowledgement

It is a pleasure to thank G. Altarelli, D. Kaplan and A. Mueller for discussions, and P. Herczeg for alerting us to a possible numerical error in the preliminary version of this report. Our work is supported in part by the National Science Foundation (PHY-8907949) and in part by the Department of Energy (DE-AC02-76ER03066).

References

[1] EMC collaboration, J. Ashman, *et al.*, Phys. Lett. B **206**, 3641 (1988), CERN-EP 89-73.

[2] J. D. Bjorken, Phys. Rev. **148**, 1467 (1966), Phys. Rev. D **1**, 1976 (1970).

[3] J. Kodaira, *et al.*, Nucl. Phys. B **165**, 129 (1979).

[4] S. Okubo, Phys. Lett. **5**, 165 (1963); G. Zweig, CERN Report No. 8419-TH412 (1964); I. Iizuka, K. Okada, and O. Shito, Prog. Theor. Phys. **35**, 1061 (1966).

[5] J. Ellis and R. L. Jaffe, Phys. Rev. D **9**, 1444 (1974).

[6] D. B. Kaplan and A. Manohar, Nucl. Phys. B **310**, 527 (1988).

[7] R. L. Jaffe and A. Manohar, MIT preprint CTP-1706 (1989).

[8] M. Bourquin *et al.*, Z. Phys. C **21**, 27 (1983).

[9] F. E. Close, Phys. Rev. Lett. **64**, 361 (1990).

[†] The failure to distinguish ΔG from Δg in our paper [35] has caused some confusion.

[10] L. A. Ahren *et al.*, Phys. Rev. D **35**, 785 (1987).

[11] G. Preparata and J. Soffer, Phys. Rev. Lett. **61**, 1167 (1988); G. Preparata, P. Radcliffe, and J. Soffer, Milan Report No. IRN 1939408 (1989).

[12] J. Ellis and M. Karliner, Phys. Lett. B **213**, 73 (1988).

[13] M. Anselmino, B. L. Ioffe, and E. Leader, NSF-ITP-88-94 (1988).

[14] E. Bloom and F. Gilman, Phys. Rev. Lett. **25**, 1140 (1970).

[15] G. Altarelli and W. J. Stirling, CERN-TH-5249 (1988).

[16] F. E. Close and R. G. Roberts, Phys. Rev. Lett. **60**, 1471 (1988).

[17] M. Ademollo and R. Gatto, Phys. Rev. Lett. **13**, 264 (1964).

[18] H. J. Lipkin, Phys. Lett. B **230**, 135 (1989).

[19] S. L. Adler, Phys. Rev. **177**, 2426 (1969); J. S. Bell and R. Jackiw, Nuovo Cimento A **51**, 47 (1969).

[20] R. L. Jaffe, Phys. Lett. B **193**, 101 (1987).

[21] A. V. Efremov and O. V. Teryaev, Dubna Report No. E3-88-287 (1988).

[22] G. Altarelli and G. Ross, Phys. Lett. B **212**, 391 (1988); Altarelli & Stirling, Ref. [16].

[23] R. D. Carlitz, J. C. Collins, and A. F. Mueller, Phys. Lett. B **214**, 229 (1988); Argonne preprint ANL-HEP-CP-89-69, to be published in Proc. "Rencontre de Moriond", Les Arcs (1989).

[24] A. F. Mueller, Columbia Univ. preprint CU-TP-442 (1989).

[25] G. T. Bodwin and J. Qiu, Argonne preprint ANL-HEP-PR-89-83 (1989).

[26] T. P. Cheng and L. F. Li, to be published.

[27] P. Ratcliffe, Phys. Lett. B **192**, 180 (1987).

[28] J. Babcock, E. Monsay, and D. Silvers, Phys. Rev. D **19**, 1483 (1978); J. Qiu, G. Ramsey, D. Richards, and D. Silvers, *ibid*. **39**, 361 (1989).

[29] G. Altarelli and G. Parisi, Nucl. Phys. B **126**, 298 (1977).

[30] H. Fritzsch, Phys. Lett. B **229**, 122 (1989).

[31] S. Forte, Phys. Lett. B **224**, 189 (1989); Saclay preprint SPhT-89-106.

[32] T. Hatsuda, Stony Brook preprint (1989), to be published in Nucl. Phys.

[33] G. Veneziano, CERN-Th 5050 (1989).

[34] A. V. Efremov, J. Soffer, and N. A. Törnqvist, Phys. Rev. Lett. **64**, 1495 (1990).

[35] T. P. Cheng and L. F. Li, Phys. Rev. Lett. **62**, 1441 (1989).

[36] T. Appelquist and J. Carrazzone, Phys. Rev. D **11**, 2856 (1975).

[37] A. F. Vainshtein, V. I. Zakharov, and M. A. Shifman, JETP Lett. **22**, 55 (1975); E. Witten, Nucl. Phys. B **104**, 445 (1976).

[38] R. Crewther, Phys. Rev. Lett. **28**, 1421 (1972); M. Chanowitz and J. Ellis, Phys. Lett. B **40**, 397 (1972); Phys. Rev. D 7, 2490 (19730; S. L. Adler, J. C. Collins, and A. Duncan, *ibid*. **15**, 1712 (1977); J. C. Collins, A. Duncan, and S. D. Joglekar, *ibid*. **16**, 438 (1977); N. K. Nielsen, Nucl. Phys. B **120**, 212 (1977).

[39] M. A. Shifman, A. I. Vainshtein, and V. I. Zakharov, Phys. Lett. B **78**, 443 (1978).

[40] A. Patel, CERN-TH 5381 (1989).

[41] T. P. Cheng, Phys. Rev. D **13**, 2161 (1976).

[42] T. P. Cheng, Phys. Rev. D **38**, 2869 (1988).

[43] M. L. Goldberger and S. B. Treiman, Phys. Rev. **110**, 1178 (1958).

[44] A. A. Anselm *el al.*, Phys. Lett. B **152**, 116 (1985); H. Y. Cheng, *ibid*., **219**, 347 (1989).

[45] D. Gross, S. B. Treiman, and F. Wilczek, Phys. Rev. D **19**, 2188 (1979).

[46] D. Sutherland, Phys. Lett. **23**, 384 (1966).

[47] S. Brodsky. J. Ellis, and M. Karliner, Phys. Lett. B **206**, 309 (1988).

[48] J. Ellis, M. Karliner, and C. T. Sachrajda, CERN-TH 5471 (1989).

[49] T. P. Cheng and L. F. Li, Phys. Lett. B **234**, 163 (1990).

[50] J. Ellis, E. Gabathuler, and M. Karliner, Phys. Lett. B **217**, 173 (1989).

Understanding Quark Flow in High Momentum Transfer Exclusive Reactions

C. White, H. Courant, G. Fang[a], K. Heller, K. Johns[b], M. Marshak, M. Shupe[b]
University of Minnesota, Minneapolis, MN 55455

R. Appel, D. Barton, G. Bunce
A. Carroll, S. Gushue, M. Kmit, D. Lowenstein, Y. Makdisi
Brookhaven National Laboratory, Upton, NY 11973

X. Ma, J. Russell
Southeastern Massachusetts University, N. Dartmouth, MA 02747

S. Heppelmann
Pennsylvania State University, University Park, PA 16802

A 5.9 GeV/c secondary beam of pions, kaons, and protons directed into a liquid hydrogen target has been used to study high momentum transfer exclusive reactions of the form $A+B \rightarrow C+D$. The high sensitivity of this experiment has allowed the differential cross section for 19 two body exclusive reactions to be measured around 90 degrees in the center of mass frame. These high statistic measurements confirm the conclusion of an earlier 10 GeV/c experiment[1] which found that the relative magnitudes of the cross sections are consistent with the dominance of the quark interchange diagram.

1. Introduction

Exclusive scattering has been predicted by Brodsky and Farrar[2], Matveev et al[3], and others to follow a scaling rule, provided that the scattering occurs with a momentum transfer sufficiently high to probe the individual quarks. The differential cross section should scale as: $d\sigma/dt \propto s^{-n}$, where n=8 for meson-baryon scattering and n=10 for baryon-baryon scattering. Dimensional scaling has been experimentally observed.[4] BNL experiment 838 measured differential cross sections within the observed dimensional scaling region. We therefore believe that we have achieved a hard scatter of the valence quarks, thus allowing perturbative quantum chromodynamics to be applicable. In the PQCD picture, the amplitude calculations for two body exclusive reactions involve thousands of Feynman diagrams; however, the diagrams can be broken up into four basic groups: gluon exchange (GEX), quark interchange (QIN), quark annihilation (ANN), and a combination of quark interchange and annihilation (COMB). By simultaneously measuring a large number of reactions, the contributions to the cross section from the various groups can be tested. Furthermore, a comparison of common reactions between the 6 GeV/c data and 10 GeV/c data, obtained earlier from the same apparatus, provides a good test for dimensional scaling.

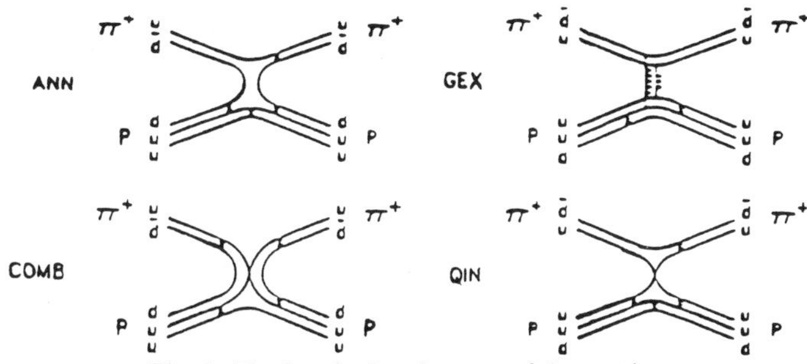

Fig. 1. The four basic valence quark interactions.

2. Apparatus

The apparatus had been used for several previous experiments and is discussed in detail elsewhere[5]. The principal component was a spectrometer arm that pivoted for ease in positioning. The arm tracked charged particles, resolving their trajectories to less than one milliradian, and measured the particles' momentum to better than 1%. The spectrometer magnet's horizontal aperture accepted tracks at $27^\circ \pm 3^\circ$ with respect to the beam line. This corresponded to $\mid cos\theta_{cm} \mid < .11$ for πp elastics. The azimuthal acceptance was $\pm 7.5^\circ$ for $cos\theta_{cm} = 0$. An array of large proportional chambers tracked the second final state product, or its charged decay products. Two differential Cherenkov counters provided beam particle identification while threshold Cherenkov counters identified spectrometer pions.

3. Analysis

Data analysis began with the extraction of the six elastic reactions. Simple kinematic cuts were used to isolate the elastics from the background. The cuts included: coplanarity, opening angle, and a good vertex within the target volume. pp elastics were separated exclusively with these cuts, while Cherenkov identification was needed to separate πp and Kp elastics. A measure of the πp contamination within the Kp elastic signal was obtained from the TDC distribution of the beam Cherenkov kaon rings. A small level of contamination was observed and subtracted. Proton anti-proton annihilation into two mesons is similiarly constrained. The opening angles for these reactions are distinct and thus they can be easily extracted from the data. Six $\overline{p} + p \to \pi^+ + \pi^-$ and two $\overline{p} + p \to K^+ + K^-$ event candidates were observed without background.

Extracting resonant reactions was a bit more difficult. The precise kinematics that apply to elastic events cannot be used due to the decay of the final state particle. The extraction relied heavily upon Cherenkov identification and vertex resolution. The missing mass squared was then plotted for each reaction. The signal shape was determined from Monte Carlo calculations while the background distibution was parameterized with a cubic polynomial of the form:

$$\sum_{n=1}^{3} P_n(X - X_0)^n + B_0$$

The fitting process added a large systematic error into the cross section due to the variability of the signal size when various fitting parameters were altered. The overall systematic error was dominated by the fitting. A few of the final state products possess decay modes which allowed for improved separation of the signal from the background. The original trajectories for Σ^+ and Δ^+ were projected into the side chambers. The decays $\Sigma^+ \rightarrow p + \pi^0$ and $\Delta^+ \rightarrow p + \pi^0$ produced shallow angle charged tracks such that a cone cut about the projected track accepted the proton track while excluding a large portion of the background. This cut improved the quality of the fit and thus reduced the systematic error involved with the fitting process.

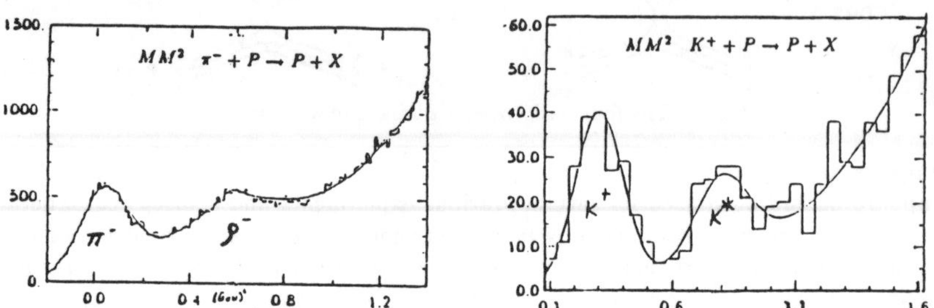

Fig. 2. Missing mass squared plots reveal two elastic peaks as well as ρ^- and $K^*(892)$ resonances.

4. Preliminary Results

A comparison of the pp elastic cross section was then made with previous experiments as a check on our absolute normalization. Figure 3 clearly indicates that our results agree with the earlier measurements. It should also be noted that all of our cross sections were measured simultaneously, such that normalization and systematic errors were similiar, allowing us to make comparisons between the various reactions. A compilation of all 19 cross sections is shown in Figure 4. The figure also indicates the groups of diagrams that contribute in the PQCD cross section calculation for each reaction.

First, the significant variation in elastic cross sections indicates that gluon exchange (GEX) does not dominate the cross section amplitude for two-body exclusive reactions about 90 degrees in the CM frame. Second, at 6 GeV/c the total cross sections for pp and $\bar{p}p$ scattering are about the same, yet the differential cross sections at 90 degrees differ by a factor of 50. Similiarly, the K^+p and K^-p elastic cross sections vary by an order of magnitude. Examination of the contributing diagrams suggests that quark interchange (QIN) dominates the interaction when allowed. Further evidence for QIN dominance is seen in hyperon production and in the kaon resonance $K^*(892)$. In both cases, the amplitude of the reaction which permits QIN was observed to be a factor of 10 greater than the corresponding reaction which does not permit QIN.

Vector meson production provides further evidence that this PQCD picture applies. The negative and positive rho reaction cross sections agree nearly exactly with the corresponding elastic cross section. This is presumably true since the contributing diagrams are the same, except for GEX which already has been shown to have a small contribution. Similarly, the cross sections for $K^\pm + p \rightarrow p + K^*(892)$ match the Kp elastic cross sections, which again possess the same contributing diagrams, except for GEX. The reactions $\bar{p} + p \rightarrow 2$ mesons reinforce the suggestion that the annihilation channels are suppressed. If one considers crossing, the two reactions can be considered to be the ANN and COMB contributions to the elastic cross sections. The small values for these cross sections indicate that ANN and COMB do not contribute significantly to the elastic cross sections. An anomaly in the data occurs when one examines delta production. Regardless of the contributing diagrams, the cross section for the four delta reactions were observed to be nearly equal. Further study will hopefully shed some light on this point. Finally, table 1 tabulates the exponential scaling factors observed between the 6 GeV/c and the 10 GeV/c data. Our results are in reasonable agreement with the predicted dimensional scaling rule: $d\sigma/dt \propto s^{-n}$.

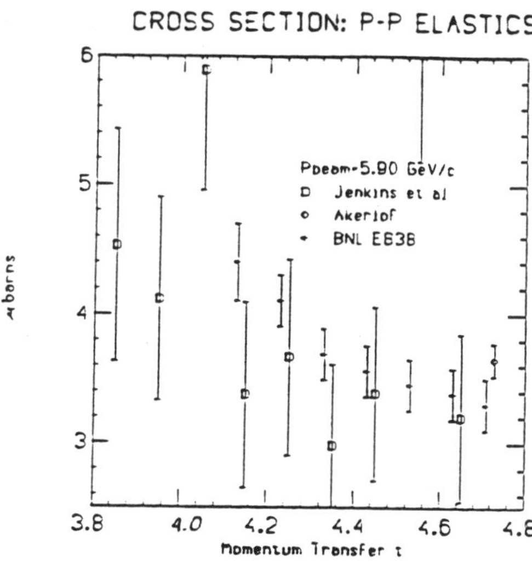

Fig. 3. Comparison of pp elastic cross sections

reaction	scaling factor n
$\pi^+ + p \to p + \pi^+$	$6.7 \pm .5$
$\pi^- + p \to p + \pi^-$	$8.2 \pm .6$
$K^+ + p \to p + K^+$	$8.5 \pm .6$
$K^- + p \to p + K^-$	7.0 ± 1.5
$\pi^+ + p \to p + \rho^+$	$7.7 \pm .7$
$\pi^- + p \to p + \rho^-$	$8.8 \pm .7$
$\pi^+ + p \to \pi^+ + \Delta^+$	$7.0 \pm .6$
$\pi^- + p \to \pi^+ + \Delta^-$	≥ 11.5
$p + p \to p + p$	$9.3 \pm .4$
$\overline{p} + p \to p + \overline{p}$	≥ 8.5

Table 1

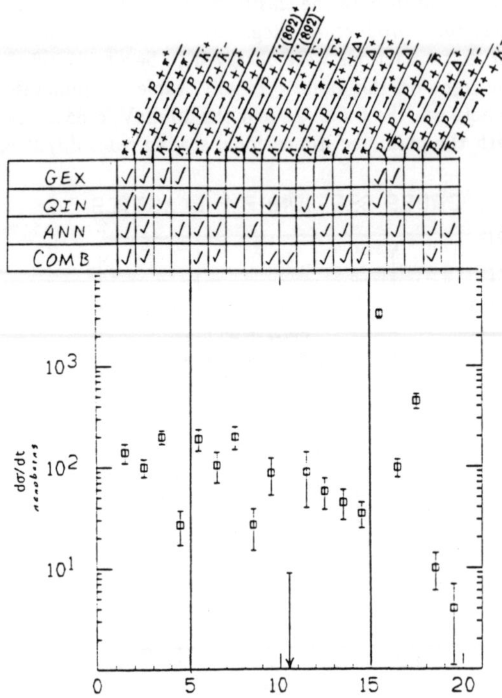

Fig. 4. Summary of the 19 measured cross sections about $\cos\theta_{cm}=0$.

*Supported in part by the DOE and NSF

[a]Current Address: Harvard University, Cambridge, Massachusetts 02139

[b]Current Address: University of Arizona, Tucson, Arizona, 85721.

[1]B. Baller et al., Phys. Rev. Lett. **60**, 1118 (1988)

[2]S. Brodsky and G. Farrar, Phys. Rev. D **11**, 1309 (1975)

[3]V. Matveev, R. Muradyan, A. Tavhelidze, Lett. Nuovo Cimento **7**, 719 (1973)

[4]K. Jenkins et al., Phys. Rev. D. **21**, 2445 (1980)

[5]G. Blazey et al., Phys. Rev. Lett. **55**, 1820 (1985)

[6]C. Akerlof et al., Phys. Rev. **159**, 1138 (1967)

RECENT EXPERIMENTAL RESULTS FROM TWO–PHOTON COLLISIONS

A. M. Eisner

Intercampus Institute for Research at Particle Accelerators
Physics Department, University of California, La Jolla, CA 92093

1. INTRODUCTION

Two-photon collisions have provided extensive information (a) on the formation of $C = +1$
resonances, including spin-one mesons (accessible if one photon, γ^*, is virtual) and glueball
candidates (not expected to couple strongly to
$\gamma\gamma$); and (b) on strong interaction dynamics
(e.g., the photon structure function, F_2^γ), taking
advantage of the relatively simple nature of the
initial state. Figure 1 illustrates the two-photon
process at an e^+e^- collider. When photon i is
tagged by detecting the corresponding $e^\pm$, let
$Q^2 \equiv -q_i^2$. W is the $\gamma\gamma$ center-of-momentum-
frame energy.

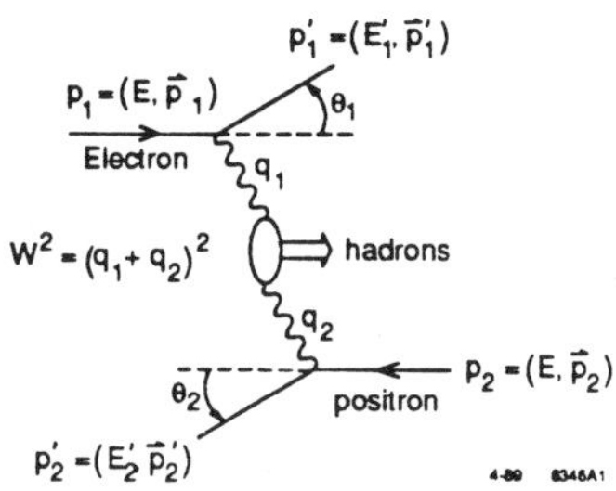

Fig. 1: The two-photon reaction in e^+e^- collisions. Shown
are lab frame four-momenta
and angles.

My focus will be on several processes bearing upon the Quark Parton Model
(QPM) or perturbative QCD, using measurements from PEP and PETRA:

- Topology of inclusive final states [TPC/Two–Gamma]. Emphasis is on the transition between hadronlike and pointlike behavior of the initial photons, with p_T
  and Q^2 the handles for accessing the pointlike domain.

- Charmed-meson production [TASSO]. Comparisons to the QPM are relevant because of the high c-quark mass.

- $\gamma\gamma \to \pi^0\pi^0$ above $2\,\text{GeV}$ [JADE]. QCD predictions come into play for high p_T.

- $\eta\gamma\gamma^*$ and $\eta'\gamma\gamma^*$ form factors [TPC/Two–Gamma]. Q^2 is the parameter of interest.

The latter two items are examples of processes predicted[1] to be sensitive to both
hard scattering (perturbative QCD) and meson wave functions.

2. FINAL STATE TOPOLOGY ("JETS ") IN $\gamma\gamma \to$ hadrons

The Q^2 dependence of TPC/Two–Gamma data[2] on the total $\gamma\gamma$ cross section
is well described for $W > 3\,\text{GeV}$ by a sum of QPM plus Vector Meson Dominance
(VMD) contributions. For $Q^2 < 5\,\text{GeV}^2$, Generalized VMD (GVMD) also works;
while PLUTO data[3] were better described by QPM plus GVMD. If a minimum p_T
is imposed on the pointlike part, to avoid double-counting,[4] the fit to TPC/Two–
Gamma data is further improved (see Fig. 5 of Ref. 2).

Topologically, in the $\gamma\gamma$ c.m. frame a hadronlike (VMD) contribution is characterized by limited p_T with respect to the $\gamma\gamma$ axis and can be represented by two low-p_T "jets." The QPM leads to two back-to-back jets at any angle; while perturbative QCD corrections can yield ≥ 3 jets. For the analysis to be described, VMD is modeled by $\gamma\gamma \rightarrow q\bar{q}$, with quark p_T^2 distributed $\propto exp\{-bp_T^2\}$. For both VMD and the QPM, the $q\bar{q}$ undergoes Lund-model hadronization.

TPC/Two–Gamma[5] has analyzed events in both untagged and single-tagged data samples, requiring at least four charged tracks and a minimum visible W of 4.0 and 3.5 GeV for the two samples. Using the detected hadrons to define the $\gamma\gamma$ c.m., each event was forced into a two-jet interpretation by a thrust algorithm. The $\gamma\gamma$ axis was taken to be the e^+e^- axis for the untagged data, and the tagged γ direction for the tagged data. The result was then a reconstructed jet p_T for each event. Monte Carlo events were put through the same procedure. Hadron distributions in the data with respect to the jet axis were compatible with the Monte Carlo models. The QPM contribution has fixed normalization, while the VMD contribution was normalized at the lowest p_T; both $b=2$ and $b=5$ give adequate fits there. Figure 2 shows the ratio $R'_{\gamma\gamma}$ of data (after background subtractions) to the resulting Monte Carlo prediction for $b=2$. There is a substantial excess of events at $1 < p_T < 3$ GeV, as observed earlier by PLUTO.[6] (The preliminary systematic uncertainty is largest, but still $\leq 30\%$, near 2 to 3 GeV.) The excess decreases for $Q^2 > 0$.

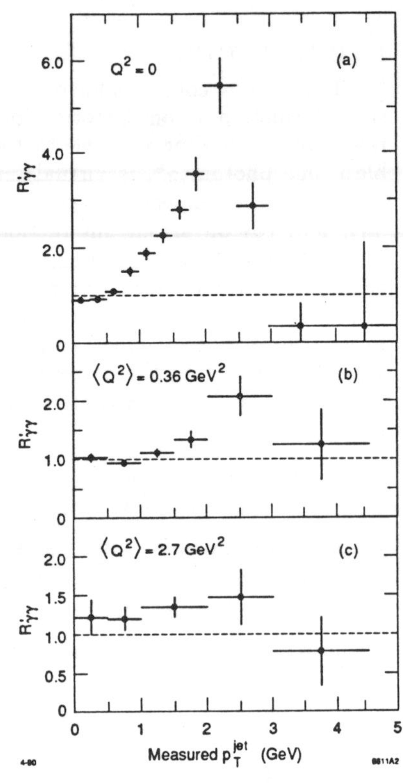

Fig. 2. TPC/Two–Gamma ratios of background-subtracted data to VMD + QPM vs the reconstructed jet p_T: (a) untagged and (b)–(c) tagged. Error bars are statistical only.

In the range $2 < p_T < 3$ GeV, the VMD contribution for any plausible b is negligible. Figure 3 shows that the thrust distribution there is more spherical than implied by the QPM. A possible three-jet contribution[7] has also been calculated, but it accounts for only a small fraction of the excess. Even if it is rescaled to the entire excess, the total thrust distribution is still less spherical than the untagged data.

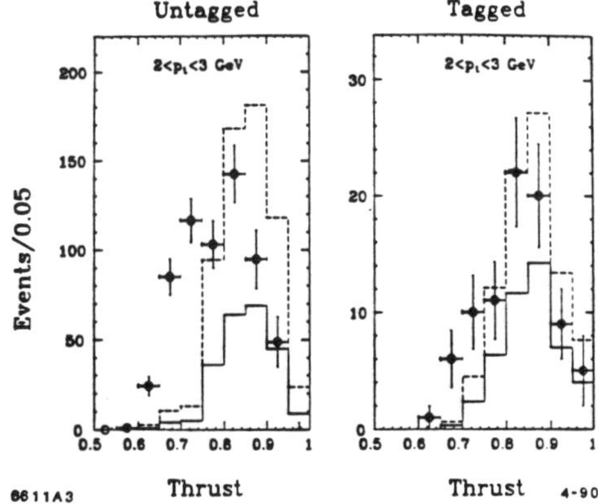

Fig. 3. TPC/Two–Gamma thrust distributions (statistical errors only) for $2 < p_T^{\text{jet}} < 3$ GeV. The solid histogram is the QPM prediction; the dashed histogram is the same normalized to the data.

3. $\gamma\gamma$ PRODUCTION OF CHARM

TASSO has recently analyzed data on inclusive $\gamma\gamma$ production of $D^{*\pm}$ (detected via $D^{*+} \to D^0\pi^+$, with $D^0 \to K^-\pi^+$ or $K^-\pi^+\pi^+\pi^-$) and has reported a first observation of $\gamma\gamma \to D^0\bar{D}^0$. The result[8] for $\sigma(e^+e^- \to e^+e^-D^{*\pm}X)$ is 97 ± 29 pb. A QPM estimate, based upon a pointlike process $\gamma\gamma \to c\bar{c}$ with $m_c = 1.6$ GeV, yields a cross section of ≈ 46 pb. (See Ref. 8 for assumptions. Acceptance calculations used this process, followed by Lund-model fragmentation.) Earlier JADE results were likewise high compared to such a model; preliminary TPC/Two–Gamma results, while also high, were consistent with the model.

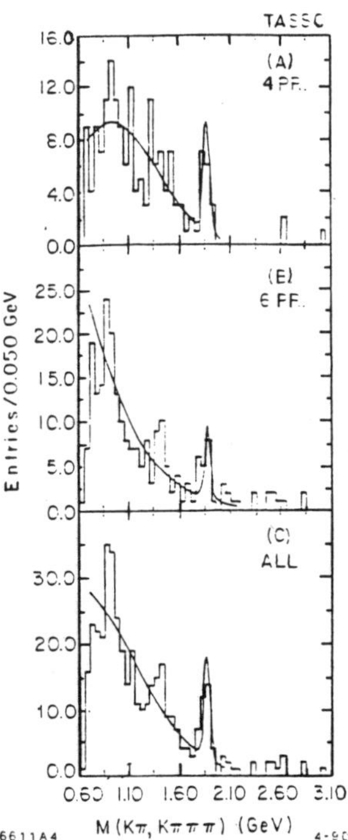

Fig. 4: TASSO's distributions of (a) $M(K\pi)$ recoiling against a $D^0 \to K\pi$ or $K\pi\pi^0$ candidate (with the π^0 undetected); (b) $M(K\pi)$ or $M(K3\pi)$, respectively, recoiling against a $D^0 \to K3\pi$ or $K\pi$ candidate; (c) the sum of (a) and (b). See Ref. 8 for cuts and for the fits used to provide the curves.

Evidence for $D^0\bar{D}^0$ is clearly seen in Fig. 4, which shows the mass of one $K\pi$ or $K\pi\pi$ combination when the other is in a D^0 window. Because events with undetected low-energy $\gamma's$ or $\pi^{0's}$ could pass Σp_T cuts, the measurement is actually sensitive to $\sigma(\gamma\gamma \to D^0\bar{D}^0) + \sigma(\gamma\gamma \to D^{*0}\bar{D}^0) + c.c. + \sigma(\gamma\gamma \to D^{*0}\bar{D}^{*0})$. Assuming a 1:6:9 mix of these final states, and using Monte Carlo acceptance calculations, TASSO obtains $\sigma(\gamma\gamma \to D^0\bar{D}^0 + X) = 56 \pm 13 \pm 12\,(syst)$ nb for visible W within 1 GeV above threshold. They estimate that at least 53 ± 13 nb of this is due to prompt production of neutral meson (D, D^*) pairs. If the corresponding prompt *charged* meson pairs are produced with the same 1:6:9 ratio, a Monte Carlo calculation allows translating

590

the inclusive D^{*+} measurement into a $\gamma\gamma$ cross section of $61 \pm 17\,\mathrm{nb}$ (statistical error only) for prompt production of charged meson pairs. If so, then the charged and neutral cross sections for visible W near the charm threshold are similar. This would favor a $\gamma\gamma \to c\bar{c}$ mechanism over other possible diagrams; but the cross sections are substantially larger than predicted.

4. $\gamma\gamma \to \pi^0\pi^0$ AT $W > 2$ GeV

Brodsky and Lepage[1] have predicted the $\gamma\gamma$ production of meson pairs at high p_T in terms of QCD hard scattering and meson wave functions. Measured values[9] of $\sigma(\gamma\gamma \to \pi^+\pi^-)$ vs W lie above the predictions for $W < 2\,\mathrm{GeV}$ [perhaps due to a tail from the $f_2(1270)$], with agreement better at higher W. More recent predictions,[10] using improved nonasymptotic wave functions, are somewhat closer. Both QCD models predict that $\pi^0\pi^0$ production is at least ten times smaller than $\pi^+\pi^-$.

JADE[11] is reporting new results on the 4γ final state, including a first measurement of $\gamma\gamma \to \pi^0\pi^0$ at $W > 2\,\mathrm{GeV}$. The data show a clean $\pi^0\pi^0$ signal (dominated by the f_2), with $a_0(980)$ and $a_2(1320) \to \eta\pi^0$ also evident. No evidence is found for $f_4(2050) \to \pi^0\pi^0$; indeed, a helicity analysis of events with $1.82 < W < 2.22\,\mathrm{GeV}$ is compatible with spin-2, helicity-2 (the f_2 tail?) or spin-0, but not with spin-4. $\Gamma_{\gamma\gamma}(f_4) < 1.1\,\mathrm{keV}$ at the 95% C.L.

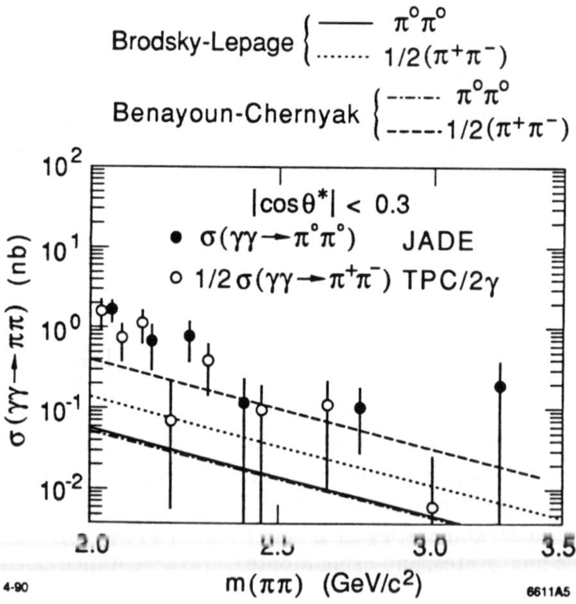

Fig. 5: $\sigma(\gamma\gamma \to \pi\pi)$, with QCD predictions of Ref. 1 [using model (c) of their Fig. 3] and Ref. 10. Here θ^* is the c.m. angle of one π.

The results are shown in Fig. 5, with the $\pi^+\pi^-$ measurements[9] (scaled by 1/2). From the rough equality seen there, JADE concludes that the data can be explained as isospin-0 formation. Also shown are predictions from two QCD models[1,10] The $\pi^0\pi^0$ data lie well above QCD predictions, and hence are likely due to some other mechanism. The $\pi^+\pi^-$ data are compatible with either a QCD model, or an isospin-0 hypothesis, or (given the limited precision) with some combination. Improved statistics for both processes will be needed to reach a firmer conclusion.

5. $\gamma\gamma$ FORM FACTORS OF PSEUDOSCALAR MESONS

The coupling of a pseudoscalar meson P to two photons may be written[12] $i\sqrt{X}\,F_P(q_1^2, q_2^2)$. Then the two-photon width of P is

$$\Gamma_{\gamma\gamma}(P) = \frac{m_P^3}{64\pi}\,F_P^2(0,0) \quad , \tag{1}$$

and the single-tagged cross section for a narrow P is

$$\sigma(\gamma^*\gamma \to P) = \frac{\pi\sqrt{X}}{4}\,F_P^2(Q^2)\,\delta(W^2 - m_P^2) \quad . \tag{2}$$

Two models for the form factor F_P are (i) naive VMD, using a ρ or ϕ pole, and (ii) a QCD-inspired form[1] which interpolates between a current-algebra value at $Q^2 = 0$ and a QCD prediction at $Q^2 \to \infty$:

$$F_P(Q^2) = \frac{\alpha}{\pi f_P}\left(1 + \frac{Q^2}{8\pi^2 f_P^2}\right)^{-1} \quad . \tag{3}$$

TPC/Two–Gamma[13] has measured F_η and $F_{\eta'}$ using a detected final state of $\pi^+\pi^-\gamma$. Monte Carlo calculations imply that 77% of η events were due to a $\pi^+\pi^-\pi^0$ decay with only one γ detected, while 23% were from the direct $\pi^+\pi^-\gamma$ decay; 7% of η' events were from $\pi^+\pi^-\eta$, $\eta \to \gamma\gamma$ with one γ detected, while 93% were from the $\rho^0\gamma$ channel. Figure 6 shows the results; systematic uncertainties are $\approx 12\%$. Fits of these $F_P(Q^2)$ to $A/[1 + (Q^2/\Lambda_P^2)^n]$, using the world-average $\Gamma_{\gamma\gamma}$ values at $Q^2 = 0$, gave $n = 1.09 \pm 0.29$ for η and $n = 1.08 \pm 0.15$ for η'. Fits with fixed $n = 1$, as per both models for F_P, yield $\Lambda_\eta = 0.70 \pm 0.08\,\text{GeV}$ and $\Lambda_{\eta'} = 0.85 \pm 0.07\,\text{GeV}$. Thus F_η is consistent with ρ-dominance, $F_{\eta'}$ is less so. Both Λ_P's are quite close to the values from $(dF/dQ^2)/F$ near $Q^2 = 0$ in quark loop models,[14] used for the QCD curves in Fig. 6.

How good is the interpolation formula? One can use Eq. (3) with the measured Q^2 dependence (Λ_P values) to extract $f_\eta = 79 \pm 9\,\text{MeV}$ and $f_{\eta'} = 96 \pm 8\,\text{MeV}$;

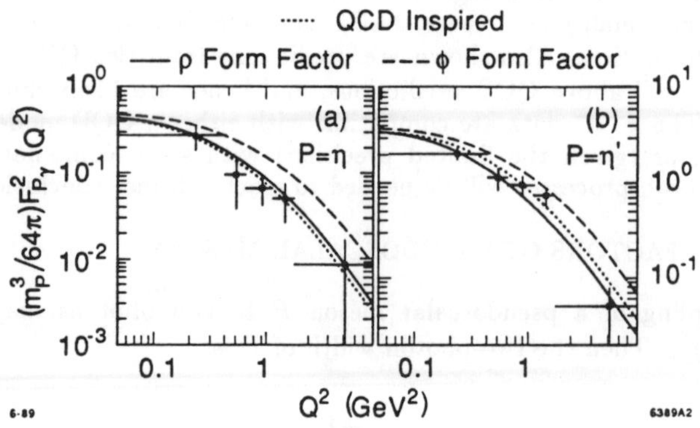

Fig. 6: TPC/Two–Gamma data[13] for $m_P^3 F_P^2(Q^2)/64\pi$, with statistical errors.

while the well-known values of $\Gamma_{\gamma\gamma}$ imply [via Eqs. (1) and (3)] $f_\eta = 91 \pm 4\,\mathrm{MeV}$ and $f_{\eta'} = 76 \pm 3\,\mathrm{MeV}$. These are close, but not in complete agreement. With better statistics, particularly at higher Q^2, it should be possible to use $Q^2 F_P(Q^2)$ with Eq. (16) of Ref. 1 to help constrain the meson wave functions.

In conclusion, there are a number of $\gamma\gamma$ processes (including the four considered here) for which meaningful comparisons with QPM or QCD predictions are possible in principle. We have seen some examples of approximate agreement, as well as some intriguing differences. An order-of-magnitude more events will allow considerable headway to be made in these studies.

REFERENCES

1. S. J. Brodsky and G. P. Lepage, *Phys. Rev.* **D24**, 1808 (1981).
2. H. Aihara et al., to be published in *Phys. Rev. D* (May 1990).
3. Ch. Berger et al., *Z. Phys.* **C26**, 353 (1984).
4. J. H. Field, F. Kapusta, and L. Poggioli, *Phys. Lett.* **181B**, 362 (1986); *Z. Phys.* **C36**, 121 (1987).
5. H. Aihara et al., paper on $\gamma\gamma$ jets, in preparation.
6. Ch. Berger et al., *Z. Phys.* **C33**, 351 (1987).
7. N. Arteaga-Romero et al., *Z. Phys.* **C32**, 105 (1986).
8. W. Braunschweig et al., DESY Report 90–012 (1990).
9. H. Aihara et al., *Phys. Rev. Lett.* **57**, 404 (1986).
10. M. Benayoun and V. L. Chernyak, *Nucl. Phys.* **B329**, 285 (1990).
11. T. Oest et al., DESY Report 90–025 (1990).
12. G. Köpp, T. F. Walsh, and P. Zerwas, *Nucl. Phys.* **B70**, 461 (1974).
13. H. Aihara et al., *Phys. Rev. Lett.* **64**, 172 (1990).
14. J. Bijnens, A. Bramon, and F. Cornet, *Phys. Rev. Lett.* **61**, 1453 (1988).

MULTIQUARK EFFECTS AND MULTIQUARK FRAGMENTATION

K. E. LASSILA AND A. N. PETRIDIS

Physics Department and Ames Laboratory, Iowa State University, Ames, IA 50011

C. E. CARLSON

Physics Department, College of William and Mary, Williamsburg, VA 22901

U. P. SUKHATME

Physics Department, University of Illinois, Chicago IL 60680

ABSTRACT

We examine effects of multiquark clusters in nuclei ,emphasizing their influence on Drell-Yan experiments with pion beams. We briefly mention their relevance to proton beam results presented at this conference and note the predicted "shadowing". We discuss a theoretical framework for calculating multiquark fragmentation based on our previous analysis of diquark data and examine relevant backward proton production in experiments with neutrinos incident on nuclei.

1. INTRODUCTION

Experiments by the European Muon Collaboration (EMC) have shown that quarks in nuclei behave differently from those in free nucleons.[1] Many effects contribute to this difference between bound and free nucleons; it is important to understand them all because nuclear target data give us much of the input from which the nucleon structure functions are determined. Here, we study the contributions from multiquark[2] clusters to this difference, but first we discuss other contributions to develop a feeling for how they are related to one another: (i) Fermi motion is probably the most studied and well known. The Fourier transform of the wave function for a nucleon in the nucleus determines the momentum distribution of the nucleon in a bound state.[3] This produces a rapid rise in the ratio R(A/D) of the deep inelastic lepton scattering structure function $F_2(x)$ measured on nuclear target A divided by that measured on a deuterium target D, where x is the fraction of the total nucleon momentum carried by the parton. (ii) Multiquark cluster probabilities can be calculated as the overlap of one nucleon wave function with others,[4] and the behavior of a particular quark or gluon in such an enlarged system (cluster) will necessarily be different than in a single nucleon. (iii) Since pions are exchanged between nucleons, the deep inelastic process will occasionally probe pions and consequently produce a distortion which must likewise be taken into account.[5] (iv) Bound state properties of the probed nucleon in the nucleus result in a rescaling of the definition of the parton momentum fraction distribution.[5] (v) Overlapping of the parton wave functions in adjacent nuclei can occur when the partons have very small x values. This effect has been referred to as shadowing.[5] (vi) Q^2 rescaling effects for a bound nucleon have also been shown to contribute in certain kinematic regions.[5] (vii) It has been argued that nucleon correlations contribute to the EMC effect.[5] It would seem most unlikely that any one of the above effects could by itself account for the remarkable structure of R(A/D). All of them have some contribution, often in only one kinematic regime.

It is important that all be understood along with their ranges of overlap. We will explore the contributions of (ii), multiquark clusters, and place emphasis on the difference of these effects from (vii), correlations. We find that effects in backward production of protons and pions in neutrino reactions which had been attributed to correlations can be understood logically as the by-products of multiquark fragmentation. Our treatment of multiquark clusters will follow the approach in Ref. 2, where deviations of R(A/D) from unity were shown to result from differences in the regions of z in which valence $V(z)$ or ocean $O(z)$ quarks in N quark clusters make their contributions. For a generic N quark cluster we write

$$O_N(z) = A_N(1-z)^{a_N}, \quad V_N(z) = B_N \sqrt{z}(1-z)^{b_N}, \quad G_N(z) = C_N(1-z)^{c_N}, \tag{1}$$

for the ocean, valence, and gluon distributions. Here, z is the scaled momentum fraction, e.g., $z = x$ for the 3q nucleon and $z = x/2$ for a 6q two nucleon color singlet cluster since a given quark can have a maximum value $x = 2$. The six parameters in Eq. (1) were determined from normalization to N valence quarks, momentum conservation constraints, and by examining data for $N = 2$ (pion) and $N = 3$ (nucleon) clusters, from which the ratio of momentum in the ocean to that in valence quarks could be fixed as independent of N. With the physically reasonable requirement that $a_N > b_N$, it is logical to take a_6 to be approximately in the range 11 - 13. This[2] led to a satisfactory representation of the ratio R(A/D) which could be described by the three parameter sets

$$A = A(3,9,9,11), B = B(3,9,10,11), C = C(3,9,10,13), where X = X(b_3, a_3, b_6, a_6). \tag{2}$$

Each of these three sets of values A, B, C led to values of R(A/D) at $x \simeq 0$ which were less than unity. In numerical evaluations these three sets of parameters will be specified as A, B, C.

2. DRELL-YAN EFFECT WITH INCIDENT PIONS

If nucleons in nuclei form quark clusters, there should be evidence of them in $q\bar{q}$ annihilation to lepton pairs in hadron-nucleus collisions. We will examine the NA10 data from CERN[6] from this point of view, and will find consistency with the measured x distribution and with the nucleon number (A) dependence of Drell-Yan pairs produced in pion-nucleus collisions. For now, we limit our present analysis to inclusion of 6q (two nucleon) clusters (though in principle 9q, 12q, etc., clusters could be included). The value of x_1, the momentum fraction of the incident pion carried by the $\bar{q}$ is kinematically selected large enough that contributions from the pion sea can be neglected when comparing with experiment. We have performed this calculation both for isoscalar and nonisoscalar targets. When clusters are present the usual isoscalar corrections given by the difference in proton and neutron number are not strictly correct. The fraction f (in the notation of Ref. 2) of clusters in the nucleus can be shown to influence this correction and we have been able to take it into account in the following calculations. The ratio of differential cross sections for dilepton pairs produced from tungsten W to that produced from deuteron targets is

$$R_\pi(x) \equiv \frac{d\sigma(q\bar{q} \to \mu\mu, W)/dx}{d\sigma(q\bar{q} \to \mu\mu, D)/dx}$$
$$= \frac{A[4(2f_p V_p{}^u + f_n V_p{}^u) + 4(f_{pp} V_{pp}{}^u + \frac{1}{2} f_{nn} V_{pp}{}^u + f_{pn} V_{pn}{}^u) + 5(f_3 S_p{}^u + f S_{pp}{}^u)]}{(1+f)[4(2V_p{}^u + V_p{}^d) + 10 S_p{}^u]}. \quad (3)$$

In Eq.(3), A for tungsten is 184, f is the probability of a $6q$ cluster as in Ref. 2, f_n (f_p) is the probability for a $3q$ neutron (proton) cluster, f_3 is $f_n + f_p = 1 - f$, and the V_n (V_p) are the appropriately normalized valence distribution functions for the neutron (proton) going as $[1-x]^4$ ($[1-x]^3$) as $x \to 1$. The remaining quantities, except for the sea quark distributions S (given by $\frac{1}{5}O_N$), are related to the cluster distributions and probabilities which are constructed from V_6 and O_6 in Eq.(1) in the same fashion that V_3 is related to V_n and V_p. The probabilities for (pp), (pn), and (nn) clusters likewise are found by simple counting, given f. The calculation for the isoscalar case proceeds as in Ref. 2, and leading to a limiting form of Eq.(3) which is plotted for cases A,B,C in Fig. 1a while the curves following from Eq.(3) are plotted in Fig. 1b with data[6] for $R_\pi(x)$ vs x.

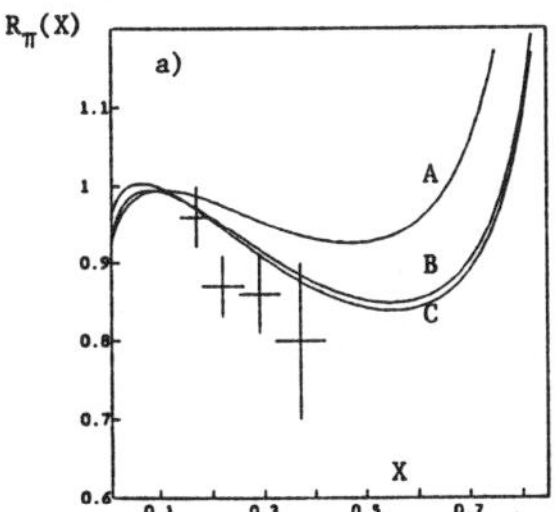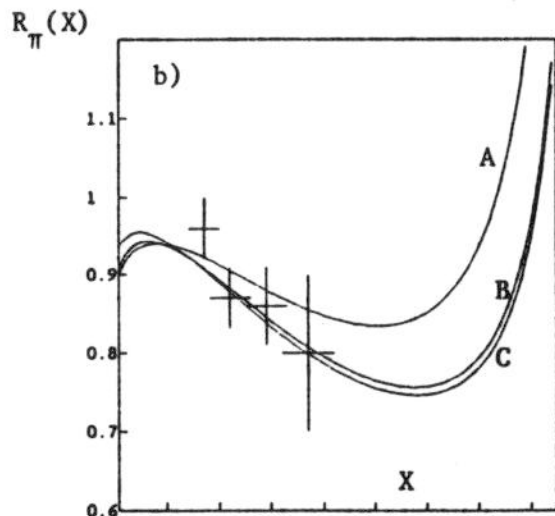

Fig. 1. a) Predictions for $R_\pi(x)$ *vs* x calculated assuming W is an isoscalar target. b) Predictions for $R_\pi(x)$ *vs* x calculated with complete nonisoscalar corrections included in Eq.(3).

We can integrate the x distributions discussed above to deduce the A^α dependence of the dilepton cross section. The prediction for α for the three cases (A,B,C) with the nonisoscalar corrections given in Eq.(3) is (0.92, 0.91, 0.90) in the range of the data[6], $0.14 \leq x \leq 0.42$. For the isoscalar calculation, the three cases all give values of $\alpha \simeq 0.98$, whereas the result given in Ref. 6 is 0.976.

A final calculation we shall mention here is that for the ratio of dilepton pairs produced on tungsten by incident positive pions compared with negative incident pions. This ratio peaks at small x and falls rapidly at intermediate x values. None of the paramenters apparent above in the nucleon or cluster structure functions make any difference to this behavior and it is very much like recently published data.[7] This result is illustrated in Fig. 2a.

3. DILEPTON PAIRS FROM PROTONS ON NUCLEI

Calculations have been done which are relevant for the discussion of the results of Fermilab experiment E772 which detected Drell-Yan lepton pairs produced in proton beams on various nuclear targets. We noted earlier[8] that such calculations had been done in the cluster model and presented some predictions. It seems of much interest to superimpose the smallest x that could be read from Brown's transparencies[9]; this comparison with the published curve and small x data is shown in Fig. 2b. All targets showed "shadowing" as predicted.

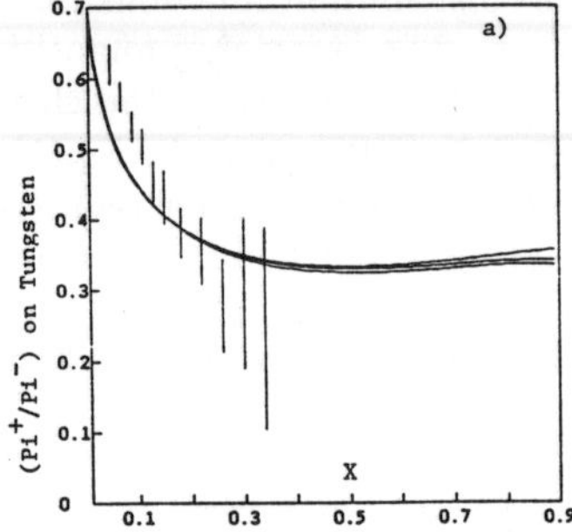
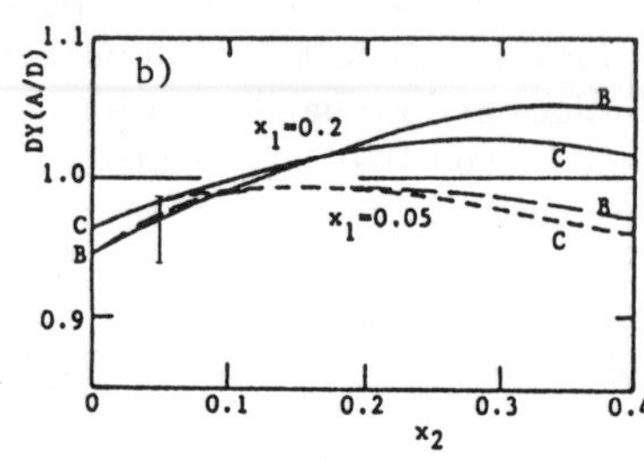

Fig. 2. a) Drell-Yan ratio on W for incident π^+ to π^-. b) Drell-Yan ratio for protons on Fe compared with small x data from E772 shown at this conference by Brown (these proceedings).

4. MULTIQUARK FRAGMENTATION

An interaction on a nucleon target producing an event via a hard QCD process will nessarily have slow fragmentation products from the color antitriplet left behind whenever a valence quark initiating the (hard) jet is ejected. Our[10] initial analysis of relevant experiments on nucleon targets led to information on the fragmentation of the residual two quark, or diquark, system. It has proved possible to streamline these earlier computer calculations of the coupled integral equations for the two quark fragmentation to analytical Mellin transforms. This, in itself, is of considerable interest as new experiments, e.g., E665 at Fermilab, will yield more data than was available earlier. But, additionally this Mellin transform solution can be extended to more complicated color triplets or antitriplets, e.g., the $5q$ system left behind when a quark is ejected from a $6q$ cluster. A summary of an application of these considerations follows.

5. BACKWARD PRODUCTION OF PROTONS AND PIONS By NEUTRINOS

Multiquark clusters lead naturally to the appearance of backward particle production in various reactions including neutrino-nucleus inelastic scattering.[11] (Backward means with respect to the incoming neutrino direction; the latter is substantially parallel to the incoming W direction.) The mechanism is simple: The virtual W enters and strikes a quark that is moving very fast in the same direction as the W and drives it out of the cluster. If this struck quark is carrying more than one nucleon's share of the momentum in a light cone or infinite momentum frame, the $5q$ residuum must be moving backwards in the nucleus rest frame, and it is just a matter of recombination to produce a backward moving nucleon or meson. To be more explicit, we use light-cone variables, choosing $q^+ = q^0 + q^3 = 0$, where q is the W momentum. This means that the W enters from more or less the $-\hat{z}$ direction so that the large x quarks move forward in the nucleus rest frame. The cross section for producing a backward moving baryon (say) is a convolution of the cross section for scattering from a $6q$ cluster times a fragmentation functions $D_{B/5q}(y)$ which describes how the $5q$ residuum turns into a baryon with momentum fraction y of the residuum. Thus, the cross section

$$\frac{d\sigma}{dE'd\Omega d\alpha} = \sigma_m[W_2^{(6)}(x, Q^2) + 2W_1^{(6)}(x, Q^2)tan^2\frac{\theta}{2}](2 - x)^{-1}D_{B/5q}(\frac{\alpha}{2 - x}),$$

where, $x = Q^2/2m_N\nu$ is the momentum fraction of the struck quark, and α is the momentum fraction[12] of the baryon with respect to the whole nucleus normalized to one at-rest nucleon, $0 \le \alpha, x \le 2$. In this view, $\alpha > 1$ gives backward baryons. In general, smaller x gives more chance for finding backward baryons, and $x > 1$ gives no chance for finding backward baryons. Applying the counting rules to Sect. 4 suggests $D_{B/5q}(y) \propto (1 - y)^3$ for large y; and, this leads to a qualitatively plausible description of the α dependence of the data.

A more detailed account of this work will be submitted for publication.

This research was supported in part by the U.S. Department of Energy, Contract No. W-7405-82, Office of Energy Research (KA-01-01), Division of High Energy and Nuclear Physics.

REFERENCES

1. See, e.g., J. Ashman et al., *Phys. Lett.* **202B**, 603 (1988) or R. G. Arnold et al., *Phys. Rev. Lett.* **52**, 727 (1984).

2. K. E. Lassila and U. P. Sukhatme, *Phys. Lett.* **209B**, 343 (1988) and C. E. Carlson and T. J. Havens, *Phys. Rev. Lett.* **51**, 261 (1983).

3. A. Bodek and J. L. Ritchie *Phys. Rev.D* **23**, 1070 (1981); **24**, 1400 (1981).

4. M. Sato et al., *Phys. Rev.C* **33**, 1062 (1986).

5. For this effect and others see E. L. Berger and J. Qiu, *Topical Conf. on Nuclear Chromodynamics*, Ed. J. Qiu and D. Sivers (World Scientific, Teaneck NJ, 1988) 119 or L. L. Frankfurt and M. I. Strikman, *Phys. Rep.* **160** 235 (1988).

6. P. Bordalo et al., *Phys. Lett.* **193B**, 368 (1987).

7. J. G. Heinrich et al., *Phys. Rev. Lett.***63**, 356 (1989).

8. K. E. Lassila and U. P. Sukhatme, *Nuclear and Particle Physics on the Light Cone*, Ed. M Johnson and L. Kisslinger (World Scientific, NJ 1989) 115.

9. C. Brown, talk presented at DPF90, Rice University, these proceedings.

10. U. P. Sukhatme, K. E. Lassila, and R. Orava, *Phys. Rev.D* **25**, 2975 (1982).

11. See e.g., E. Matsinos et al., *Z. Phys. C* **44**, 79 (1989) and references therein.

12. L. L. Frankfurt and M. I. Strikman, *Phys. Lett.* **69B**, 93 (1977).

Hadronic Decays of Ψ' and J/Ψ

STEPHEN S. PINSKY

Department of Physics, The Ohio State University

Columbus OH 43210

Experimentally the ratio of all exclusive Ψ' to J/Ψ hadronic decays are constant for all channels except for the final states K^*K, $\rho\pi$, $\eta\gamma$ and $\eta'\gamma$. I calculate the ratio for these final states and show that the vector pseudoscalar final states are unique. The hadronic decays of the Ψ' to these channels is closely related the hindered M1 transition $\Psi' \to \eta_c\gamma$. They are in effect a generalized hindered M1 transition. The OZI forbidden processes for both the J/Ψ and Ψ' decays are treated in identical ways and no rapid energy variation or fine tuning is needed. A corollary to the OZI rule is proposed stating that OZI forbiddens transition in the vector channel do not change the radial quantum number.

Introduction

The ratio of hadronic Ψ' to J/Ψ decays has attracted attention since 1982 [1]. A universal ratio

$$Q_{hadron} = \frac{B(\Psi' \to \text{hadrons})}{B(J/\Psi \to \text{hadrons})} \approx .14 \tag{1}$$

appears to exist for all hadronic final states except K^*K, $\rho\pi$, $\eta\gamma$ and $\eta'\gamma$. For these vector-pseudoscalar final states Q is much smaller ($<.0045$ for $\rho\pi$). Over the years there have been a number of proposals to explain the hadronic charmonium-decay puzzle [2,3,4,5]. Some previous explanations [2,4,5] have relied on a glueball mass finely tuned to the J/Ψ mass, which has the effect of decreasing Q by increasing the J/Ψ decay rate. These models also employ different OZI mechanisms for the J/Ψ and Ψ' decays. Another explanation based on sequential fragmentation [3] reduces Q by reducing the Ψ' rates for some channels. The model however, does not select out just the vector-pseudoscalar channel. Yet other models suggest that it is the two body meson final states that are unique[5].

The prediction of a universal ratio can easily be obtained by drawing a direct parallel between the decay of orthopositronium to three photons and the decay of J/Ψ or Ψ' to three massless gluons[7]. One easily finds

$$\frac{B(\Psi' \to ggg)}{B(J/\Psi \to ggg)} = \left[\frac{\alpha_s(\psi')}{\alpha_s(J/\Psi)} \right]^3 \frac{B(\Psi' \to e^+ e^-)}{B(J/\Psi \to e^+ e^-)} = .135 \pm .023 \; . \tag{2}$$

Of course ggg is not a measurable final state, however these gluons will materialize as light hadrons. Therefore it is reasonable to equate the exclusive decay of charm to massless gluons to the inclusive decay to all light hadrons. This is particularly plausible for multiparticle final states since the gluons like to produce quark- antiquark pairs. Remarkably for almost all final states the experimental date[8] respect this universal ratio.

$$Q_{p\bar{p}} = .086 \pm .024 \qquad Q_{\bar{p}p\pi^0} = .140 \pm .063 \qquad Q_{\bar{p}p\pi^+\pi^-} = .151 \pm .041$$
$$Q_{5\pi} = .095 \pm .027 \qquad Q_{K^+K^-\pi^+\pi^-} = .22 \pm .09 \qquad Q_{7\pi} = .13 \pm .07 \qquad (3)$$
$$Q_{\gamma f_2} = .09 \pm .0 \qquad Q_{B_1\pi} = .09$$

The Model

Consider the decays of J/Ψ and Ψ′ → η′γ or ηγ. The OZI forbidden transitions in these processes occur in the pseudoscalar channel as shown in Fig. 1. The details of the transition are not important to the calculations so I will simple call this OZI forbidden transition amplitude F_p. The only thing that is important about F_p is that it does not generate configuration mixing and that it connects single particle states. Thus we have, for example,

$$\Gamma(J/\Psi \to \eta\gamma) = \left(p_\eta^\Psi\right)^3 \frac{F_p^2}{\left(M_{\eta_c}^2 - M_\eta^2\right)^2} \frac{\Gamma(J/\Psi \to \eta_c\gamma)}{\left(p_{\eta_c}^\Psi\right)^3}, \qquad (4)$$

with a similar formula and graph for the Ψ′ decay and for the η′ final state. Calculating Q, everything but the phase space factors cancels out giving

$$\frac{B(\Psi' \to \eta\gamma)}{B(J/\Psi \to \eta\gamma)} = \frac{\left(p_\eta^{\Psi'}\right)^3}{\left(p_\eta^{J/\Psi}\right)^3} \frac{\left(p_{\eta_c}^{J/\Psi}\right)^3}{\left(p_{\eta_c}^{\Psi'}\right)^3} \frac{B(\Psi' \to \eta_c\gamma)}{B(J/\Psi \to \eta_c\gamma)} = .2\% . \qquad (5)$$

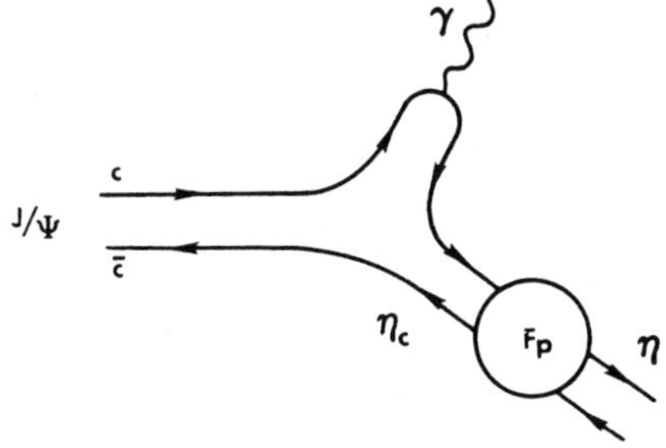

Figure 1. Charmonium decay to γ and a light pseudoscalar

This is a factor of ten below present experimental limits. The result is the same for η'. The reason for this very small ratio is that the decay $\Psi' \to \eta_c \gamma$ is a hindered M1 transition[6]. If there is a pseudoscalar glueball with a significant coupling to η_c' it will, by factorization, induce configuration mixing (mixing of radial quantum numbers). The ι meson is the leading candidate for such a glueball. However, it is not seen in radiative Ψ' decays [9], so it will not induce configuration mixing or modify (5).

To calculate the decay of J/Ψ or Ψ' to a vector-pseudoscalar final state consider the process shown in Fig. 2. The amplitude for the OZI transition from J/Ψ to a light vector meson is indicated by F_V and from Ψ' to a light radial vector meson is indicated by $F_{V'}$. Therefore it is the light quark state, either a vector meson or a radial vector mesons that finally decays to the vector-pseudoscalar final state with a coupling strength G_{vvp} or $G_{v'vp}$. The strength of the VVP coupling is well know from $\dfrac{G_{\omega\rho\pi}}{\sqrt{12\pi}} = 1.4\,(\text{GeV})^{-1}$. This phenomenological model, with some additional details, has been used to calculate the decay of J/Ψ to all vector and pseudoscalar final states [10].

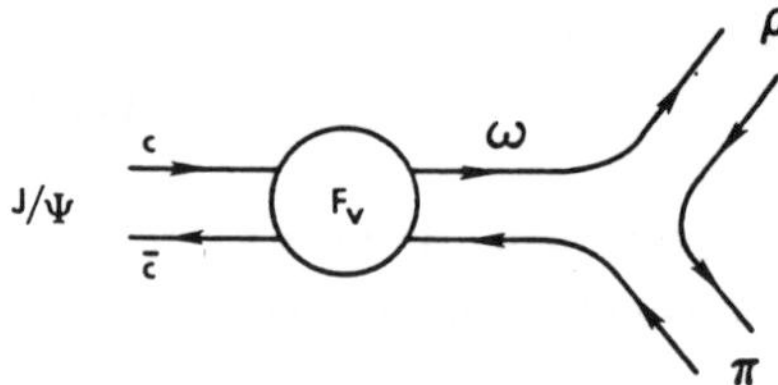

Figure 2. Charmonium decay to a light vector and pseudoscalar

The coupling $G_{\omega'\rho\pi}$ can be obtained from the decay of $\Psi' \to \eta_c \gamma$ [9]. Using the vector-dominance model to relate $\Psi' \to \eta_c \gamma$ to $\Psi' \to \Psi \eta_c$ we find

$$\Gamma\left(\Psi' \to \eta_c \gamma\right) = 6.8 \times 10^{-4}\,\text{Mev} = p^3 \frac{G^2_{\Psi'\Psi\eta_c}}{12\pi}\,\zeta^2_\psi \tag{6}$$

where the vector-dominance coupling[10] is $\zeta_\psi = 2.48 \times 10^{-2}$. We find therefore

$$\frac{G_{\omega'\rho\pi}}{\sqrt{12\pi}} = \frac{\sqrt{2}\,G_{\Psi'\Psi\eta_c}}{\sqrt{12\pi}} = .092\,(\text{GeV})^{-1}, \tag{7}$$

showing the effect of the generalized hindered M1 transition.

It should be noted that the vector-dominance model gives excellent results for $\Gamma(\Psi \to \gamma\eta_c)$. This indicates that it is not necessary to include all the radially excited charm states in the vector-dominance relation.

The model suggested by Fig. 2 for the J/Ψ decays gives

$$\Gamma\left(J/\Psi \to \rho\pi\right) = p^3 \frac{6\, F_v^2}{\left(M_\Psi^2 - M_\omega^2\right)^2}\frac{G_{\omega\rho\pi}^2}{12\pi}. \tag{8}$$

The analogous formula and graph for the Ψ' decay may be obtained using (8) and Fig. 2 with the J/Ψ replaced by Ψ' and ω replaced by ω'. Using eqns. (8) and the analogous formula for Ψ' we find

$$B\left(\Psi' \to \rho\pi\right) = 1.47 \frac{\Gamma_T(\Psi)}{\Gamma_T(\Psi')} B\left(J/\Psi \to \rho\pi\right)\left[\frac{G_{\omega'\rho\pi}}{G_{\omega\rho\pi}}\right]^2 \frac{F_{v'}^2}{F_v^2} . \tag{9}$$

(This same result holds for all the vector-pseudoscalar final states.) We finally get

$$B\left(\Psi' \to \rho\pi\right) = 2.2 \times 10^{-5} \frac{F_{v'}^2}{F_v^2}. \tag{10}$$

The OZI forbidden transition amplitudes F_V and $F_{V'}$ are hard to calculate. However since the OZI process are similar we expect the amplitudes to be similar. From other OZI process my best estimate[11] is $F_V^2/F_{V'}^2 = .3$. Then we get $B\left(\Psi' \to \rho\pi\right) = .68 \times 10^{-5}$. The direct photon process competes with the OZI forbidden transition [10]. It is a negligible effect for the J/Ψ decay but for the Ψ' decay it generates configuration mixing and therefore gives a rate slightly larger than the OZI process. The most conservative upper limit for the decay of $\Psi' \to \rho\gamma$ is obtained by adding the direct photon process coherently to the OZI process. Thus

$$B\left(\Psi' \to \rho\pi\right) = 4 \times 10^{-5} \tag{11}$$

The calculation of the limit for K^*K and the other vector pseudoscalar final states is very similar to the above calculation and therefore gives a very similar limit for Q.

This phenomenological estimate is to be compared with the recent results from Mark III[8] $B\left(\Psi' \to \rho\pi\right) < 5.7 \times 10^{-5}$ at a 90% CL. or $(2.1 \pm 1.4 \pm .4) \times 10^{-5}$ if the two events in the $\rho\pi$ bands are real.

Summary

In summary therefore, the phenomenological estimates of the decays of J/Ψ and Ψ' to K*K, $\rho\pi$, $\eta\gamma$ and $\eta'\gamma$ described above are consistent with the existing data. The calculation involved no fine tuning or rapid energy variation and the J/Ψ and Ψ' decays are treated the same way. The fundamental reason that Q_{pv} is so small is that the Ψ' decays are related to the hindered M1 decays. This generalized hindered M1 transition mechanism will apply to other heavy quark systems The other crucial ingredient is the corollary to the OZI rule forbidding configuration mixing in the vector channel.

Acknowledgements

I would like to thank S. Meshkov, H.Lipkin, L. M. Simmons, and G. Karl for useful conversations. I grateful to the Aspen Center for Physics for their hospitality

References

1. M.E.B Franklin, Ph. D, Thesis, Stanford University, 1982 (unpublished); M.E.B. Franklin, et al., . Rev. Lett. 51, 963 (1983) . G. Trilling, J. Phys. (Paris) Colloq. 43 (1982) C3-81; E. Bloom, J. Phys (Paris) Colloq 43, (1982) C3- 407.

2. W. S. Hou and A. Soni Phys. Rev. Lett. 50, 569, (1983).

3. G. Karl, W. Roberts, Phys. Lett., 144B, 263, (1984).

4. S. J. Brodsky, G. P. Lepage, and S. F. Tuan,Phys Rev. Lett, 59, 621(1987).

5. M. Chaichian, and N. A. Tornqvist Helsinki Preprint HU-TFT-88-15.

6. E. Eichten, K. Gottfried, T. Kinoshita, K. D. Lane, and T. M. Yan, Phys Rev. D 21 ,203 (1980).

7. L. Kopke, and N. Wermes Phys Reports 174, 67(1989).

8. W. Toki, 1989 International Symposium on Heavy Quark Physics, Cornell University, Ithaca NY June 13-17 1989.

9. Particle Data Group, Phys. Lett. B204 , (1988)

10. S. Pinsky, Phys. Rev D31, 1753 (1985)

11. J. F. Bolzan, K. A. Geer, W. F. Palmer and S. Pinsky, Phys. Rev. Lett. 35, 419 (1975); J. F. Bolzan, K. A. Geer, W. F. Palmer and S. Pinsky, Phys. Lett 59B, 351 (1975); W. F. Palmer and S. Pinsky, Phys. Rev. D14, 1916 (1976); J. F. Bolzan, W. F. Palmer and S. Pinsky, Phys. Rev. D14, 1920 (1976); D14, 3202 (1976); W. F. Palmer and S. Pinsky, Phys. Rev. D 22, 223 (1980).

The One-Jet Inclusive Cross Section at Order α_s^3

Stephen D. Ellis
Department of Physics, FM-15
University of Washington, Seattle, WA 98195, USA

Zoltan Kunszt
Institute of Theoretical Physics
Eidgenossosche Technische Hochschule, CH-8093 Zürich, Switzerland

Davison E. Soper
Institute of Theoretical Science
University of Oregon, Eugene, OR 97403, USA

Abstract

The next-to-leading order inclusive jet cross section is calculated. At this order, one is first able to see the dependence of the cross section on the size of the cone used to define the jet. The calculated cross section has a much reduced dependence on the renormalization scale μ, indicating a reduced theoretical uncertainty compared to the Born calculation. A first comparison with experimental data is exhibited.

Introduction

In this talk, I discuss the next-to-leading order cross section for making one jet plus anything in high energy collisions at hadron colliders. In preliminary work reported earlier[1], we included only gluons in the calculation. We have now completed the calculation including quarks as well as gluons, so we are able to compare the calculation to experimental data.

Let me begin with the motivation for a higher order jet calculation. One of the dramatic features of the recent data [2] from hadron colliders, both at CERN and Fermilab, is the obvious appearance of hadron jets [3] as a characteristic feature of a sizeable fraction of the final states. These jets are an essential tool for organizing and analyzing the data and are potentially important as a test of our quantitative understanding of the underlying strong interaction theory, QCD. This is particularly true if one wants to look [4] for a breakdown of the standard model. One possible source of deviation from the standard model that might be seen in jet physics would be a composite structure of the particles now thought to be elementary. Another would be new, heavy, strongly interacting particles such as axial gluons.

One therefore would like to analyze the scattering of quarks and gluons at the largest p_T scale possible. In order to get to very large p_T in the face of falling cross sections as $p_T \to \infty$ one needs to use the strong interactions rather than electroweak interactions. The signal for hard parton-parton scattering via the strong interactions is jet production. The most straightforward jet cross section is that for the inclusive production of a jet. Thus we look at the cross section $d\sigma/dE_T d\eta$ for the production of a jet plus anything else, where the jet carries transverse energy E_T $[= E\sin(\theta)]$ at pseudorapidity η $[= -\ln\tan(\theta/2)]$.

In earlier theoretical studies [5], only incomplete QCD matrix elements at order α_s^3 were available. Recently the full order α_s^3 matrix elements in $4 - 2\epsilon$ dimensions have been calculated by Ellis and Sexton [6]. The present talk gives a brief summary of the results of an analysis of jet cross sections using these full matrix elements.

Related calculations [7] have been performed by Aversa, Chiappetta, Greco and Guillet. These calculations include quarks as well as gluons, but so far are limited to very small sizes of the cone that is used to define a jet.

Importance of the Jet Definition

At the Born level, the one jet inclusive cross section is computed using graphs like that shown at the right. It is important to note that, strictly speaking, the cross section at this level is not a jet cross section but is rather just the parton cross section. Not only is this parton cross section strongly μ^2 dependent but it has no jet definition dependence to match that in the experimental cross section and, in fact, is not finite at higher orders in perturbation theory unless a finite jet size is introduced.

Graphs for Born cross section.

Some higher order graphs.

At order α_s^3 graphs like those above are allowed and there is an explicit dependence on the jet definition. In the calculation, one must decide when two partons count as two jets and when they count as one. The calculation at this order allows us to account for the power of the "experimental microscope" to resolve one parton into two. It is exactly this careful treatment of the finite size of the jet which renders the jet cross section finite at all orders in perturbation theory in analogy to what happens for similar quantities in e^+e^- physics [8]. Also, when two partons do count as one jet, one must define the resulting jet axis and jet E_T. In the experimental measurement, the differences between jet definition algorithms are now expected to matter, in that they can change the measured cross section at the same level as the α_s^3 corrections in the theory.

We thus need a good jet defintion that will help us to look for new physics at short distances, while washing out effects of long distance physics. The following properties should be considered in deciding on a definition: (1) It should be simple to implement in an experimental analysis. (2) It should be simple to implement in a calculation of first corrections. (3) It should be defined at any order of perturbation theory. (4) It should yield finite cross sections at any order of perturbation theory. (5) The jet cross section should be insensitive to hadronization. (6) The cross section should agree with the parton cross section in the infinite energy limit (*i.e.* at the Born level of calculation). (7) It should respect the symmetries of high energy hadron collisions. Thus we require that the definition should be invariant under rotations about the beam direction. It should also be covariant under Lorentz boosts along the beam direction: under a boost through hyperbolic angle ω, the rapidity η of each jet should transform to $\eta + \omega$, but the jet E_T should be unaffected. This boost covariance requirement is suggested by thinking of hadron beams as really being parton beams, with a "broad band" spectrum of partons. Then the velocity of the detector as viewed from the parton-parton center-of-mass frame should not affect the measurement. We note that this requirement is satisfied when one uses transverse momenta and rapidities instead of energies and angles in the jet definition.

Jet definition we use

The definition we use is as follows [9]. Let the calorimeter consist of cells i, in which transverse energies $E_{T,i}$ are measured. Define a jet cone of radius R in η-ϕ space, $R = \sqrt{\Delta\eta^2 + \Delta\phi^2}$. The E_T of the (trial) jet is then

$$E_{T,J} = \sum_{i \text{ in cone}} E_{T,i} \; . \tag{1}$$

The jet axis is defined by the weighted averages:

$$\eta_J = \frac{1}{E_{T,J}} \sum_{i \text{ in cone}} E_{T,i} \, \eta_i \; , \tag{2}$$

$$\phi_J = \frac{1}{E_{T,J}} \sum_{i \text{ in cone}} E_{T,i} \, \phi_i \; . \tag{3}$$

Finally, the cone axis must agree with the jet axis determined by eqs. (2) and (3). If it does not, one simply iterates the process until there is agreement.

In our calculation, this algorithm is applied to the final state consisting of two or three final state partons. At most two of these partons can be in the measured jet, so it is easy to write the condition that partons fall inside a cone centered on the jet axis in terms of the angles and transverse momenta of the partons.

R-Dependence

As a first illustration of the cross section $d\sigma/dE_T d\eta$ at order α_s^3 we exhibit the dependence of the calculated cross section on the cone size R. This is illustrated in Figure 1, where the inclusive jet cross section is plotted versus R for $\eta = 0, E_T = 100$

GeV and $\sqrt{s} = 1800$ GeV. For comparison the *R-independent* Born cross section (with the one loop form of the running coupling) is also plotted. This dependence of the cross section on R is an intrinsic feature of QCD. It will be important to test it experimentally.

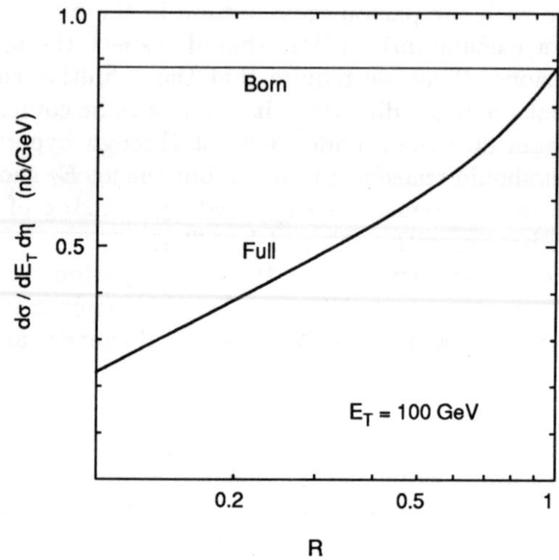

Figure 1. Inclusive jet cross section, $d\sigma/dE_T d\eta$, versus the cone size R for $\sqrt{s} = 1800$ GeV, $E_T = 100$ GeV, $\eta = 0$ and $\mu = 0.5E_T$.

Scale dependence

We now turn to the dependence of the jet cross section on the scale μ. This scale appears as the argument of the running coupling constant, $\alpha_s(\mu)$, and as the evolution parameter in the parton structure functions, $G(x, \mu)$. The scale μ arises as an artifact of the truncation of the perturbation series. Any residual dependence on it is a reminder of the missing higher order contributions and is not physical. In Figure 2, the inclusive jet cross sections at both lowest order and order α_s^3 are plotted versus μ, again for $\eta = 0, E_T = 100$ GeV and $\sqrt{s} = 1800$ GeV, but now with $R = 0.6$. Clearly the Born cross section is a monotonic function of μ while the higher order result is quite stable in the range around $\mu/E_T \sim 0.5$. In the range $0.5 \lesssim \mu/E_T \lesssim 1$ the Born cross section varies by approximately 40% while the higher order result varies only by about 5%. In this sense the inclusion of the higher contributions has greatly reduced the theoretical uncertainty due the arbitrary choice of μ.

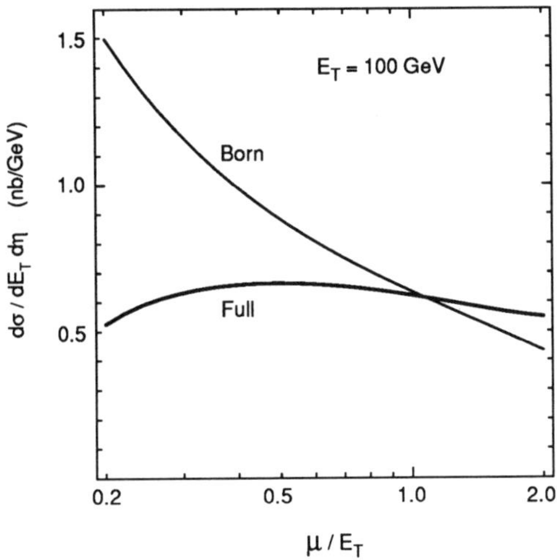

Figure 2. Inclusive jet cross section, $d\sigma/dE_T\,d\eta$, versus the ratio μ/E_T for $\sqrt{s} = 1800$ GeV, $E_T = 100$ GeV, $\eta = 0$ and $R = 0.6$.

Dependence on E_T

Finally, we compare the calculated inclusive jet cross section to data. Figure 3 exhibits the data from CDF[2] for $R = 0.6$ and $\sqrt{s} = 1800$ versus E_T. The data are averaged over the angular range $0.1 \leq |\eta| \leq 0.7$. In the original CDF publication[2], the experimental values for E_T were adjusted in an attempt to account for the part of the original parton's energy that fell outside the cone and for "background" energy (corresponding to the general energy level in the calorimeter) that fell inside the cone. It is important to recognize that the status of such corrections is different when comparing to the higher order cross section. The order α_s^3 perturbative contribution to both corrections is, in fact, correctly *included* in the calculation. Thus we have attempted to remove these corrections from the data plotted in Figure 3 by subtracting 1 GeV from each quoted jet E_T. There are, of course, residual corrections. We include these in our estimate of the theoretical uncertainty described below. There are also some uncertainties related to differences between the CDF jet definition and ours, particularly with respect to the definition of the jet E_T. The magnitude of these uncertainties has not been investigated, although we do not expect them to be large. They are not included in our error estimate.

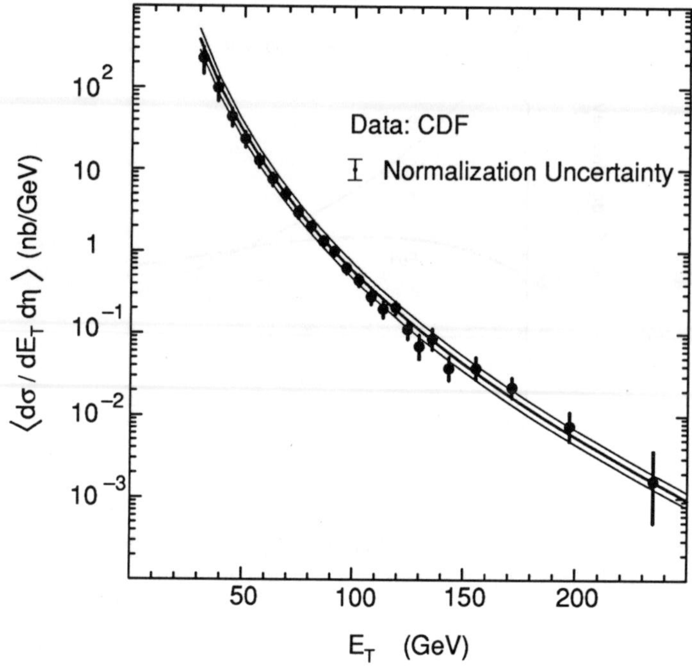

Figure 3. Inclusive jet cross section, $d\sigma/dE_T d\eta$, averaged over $0.1 < |\eta| < 0.7$ versus E_T at $\sqrt{s} = 1800$ GeV for $R = 0.6$ and $\mu = 0.5E_T$. The data are from the CDF Collaboration[2]. The experimental error bars include both the energy dependent systematic errors and the statistical errors as given in Ref. 2. The magnitude of the energy independent systematic uncertainty is indicated by the single, isolated error bar.

The theoretical curve in Figure 3 corresponds to $R = 0.6$ and $\sqrt{s} = 1800$ and is averaged over η in the same way as the data. We have used the set B structure functions of Martin, Roberts and Stirling[10] with $\Lambda_{\overline{\rm MS}}^{5\ \rm flavors} = 130$ MeV to generate the results in Figure 3. (This corresponds to $\Lambda_{\overline{\rm MS}}^{4\ \rm flavors} = 200$ MeV. For Figures 1 and 2, we have used an older set of results with a slightly different strong coupling: $\Lambda_{\overline{\rm MS}}^{5\ \rm flavors} = 200$ MeV.)

To estimate the overall theoretical fractional uncertainty $\Delta\sigma/\sigma$ we include three contributions: (1) an uncertainty of the form $6\,{\rm GeV}/E_T$ corresponding to effects such as contamination from the background event. (2) a perturbative contribution to the theoretical uncertainty of the form $5\,\alpha_s^2$. (3) a 20% uncertainty arising from the uncertainty in the parton distribution function used. We have added these uncertainties in quadrature to determine the width of the theoretical curve shown in Figure 3.

We note that the agreement between the theoretical curve and experiment is quite good. No evidence for a deviation from QCD is seen.

This work was supported in part by the US Department of Energy, Contracts No. DE-AS06-88ER40423 and DE-FG06-85ER40224. We thank our many experimental and theoretical colleagues for useful discussions, especially S. Behrends, J.C. Collins, R.K. Ellis, J. Huth, S.E. Kuhlmann, F.I. Olness and A. Para.

REFERENCES

[1] S. D. Ellis, Z. Kunszt and D. E. Soper, *Proc. XXIV International Conference on High Energy Physics*, Munich, August, (1988); Phys. Rev. Lett. 62, 726 (1989); Phys. Rev. D40, 2188 (1989).

[2] For jet data from UA1 see, G. Arnison, *et al.*, *Phys. Lett.* **172B**, 461 (1986); from UA2 see J. Appel, *et al.*, *Phys. Lett.* **160B**, 349 (1985); from CDF see F. Abe, *et al.*, *Phys. Rev. Lett.* **62**, 613 (1988). For a recent review of the CERN data see P. Bagnaia and S. D. Ellis, *Ann. Rev. Nucl. Part. Sci.* **38**, 659 (1988).

[3] For a recent discussion of the theoretical properties of jets and their role in Collider Physics see S. D. Ellis, *Proceedings of the 1987 Theoretical Advanced Study Institute in Elementary Particle Physics*, Santa Fe, July 1987, ed. R. Slansky and G. West, p. 274, (World Scientific, Singapore, 1988) and also R. K. Ellis and W. G. Scott, *Proton Antiproton Collider Physics*, ed. G. Altarelli and L. Di Lella, p. 131, (World Scientific, Singapore, 1989).

[4] See, for example, the discussion in V. Barnes, *et al.*, Proceedings of the Workshop on *Experiments, Detectors and Experimental Areas for the Supercollider*, Berkeley, July, 1987, p. 235, ed. R. Donaldson and M. G. D. Gilchriese (World Scientific, Singapore, 1988).

[5] R. K. Ellis, M. A. Furman, H. E. Haber and I. Hinchliffe, Nucl. Phys. B173, 397, (1980); M. A. Furman, Nucl. Phys. B197, 413, 1982, W. Furmański and W. Słomiński, Jagellonian preprint TPJU-11/81 (1981).

[6] R. K. Ellis and J. C. Sexton, Nucl. Phys. B269, 445 (1986).

[7] P. Chiappetta, *Proc. XXIV International Conference on High Energy Physics*, Munich, August, (1988); Phys. Lett. 210B, 225 (1988); Phys. Lett. 211B, 465 (1988); Nucl. Phys. B327, 105 (1989); preprint LAPP-TH-255/89.

[8] See G. Sterman and S. Weinberg, Phys. Rev. Lett. 39, 1436 (1977).

[9] See the contribution by D .E. Soper, *Proceedings of the XIXth International Symposium on Multiparticle Dynamics*, Arles, France, June, 1988.

[10] A. D. Martin, R. G. Roberts and W. J. Stirling, *Mod. Phys. Lett.* A4, 1135 (1989).

EXPERIMENTAL REVIEW OF STRUCTURE FUNCTION MEASUREMENTS

Ewa Rondio, EMC/NMC Collaboration,
Warsaw University, POLAND

Abstract

The status of structure functions $F_2(x_{Bj}, Q^2)$ and $R(x_{Bj}, Q^2)$ from different experiments and their consistency is discussed. New results on the ratio of neutron to proton structure functions and its implications are presented. Short discussion of Λ_{QCD} determinations as well as quark and gluon parameterizations are also included.

1 INTRODUCTION

Deep inelastic lepton scattering has proven to be a very useful tool to look inside the nucleon. The basic graph describing this interaction in lowest order is presented schematically below.

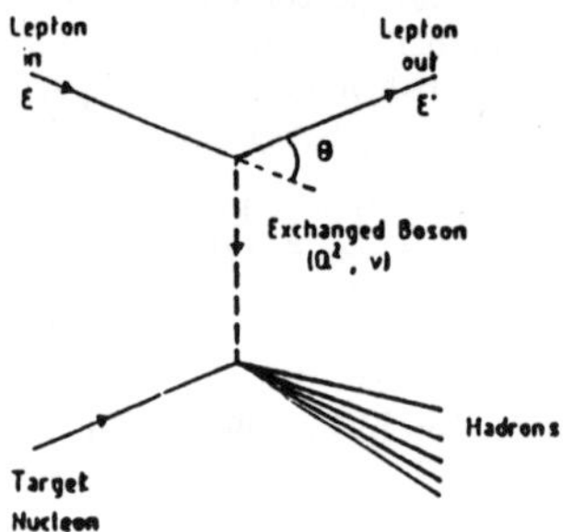

The most important variables describing this interaction are: the four momentum transfer squared $-Q^2$, the energy transfer $\nu = E - E'$ and the Bjorken scaling variables $x_{Bj} = \frac{Q^2}{2M\nu}$, $y = \frac{\nu}{E}$. At fixed beam energy two of them are independent and sufficient to define interaction kinematics. In charge lepton scattering the measured variables are momenta of incoming and outgoing lepton and the scattering angle θ, in neutrino experiments - outgoing lepton momentum and total energy of produced hadrons. From the number of interactions in each kinematical interval one gets a double differential cross-section which can be expressed in terms of structure functions:

$$\frac{d^2\sigma^{\mu,e}}{dx\,dy} = \frac{4\pi\alpha^2 s}{Q^4}[xy^2 F_1 + (1-y)F_2] \quad (1)$$

or

$$\frac{d^2\sigma^{\nu,\bar{\nu}}}{dx\,dy} = \frac{G^2 s}{2\pi}[xy^2 F_1 + (1-y)F_2 \pm (y - \frac{y^2}{2})xF_3] \quad (2)$$

where s is the total energy squared in the lepton-nucleon center of mass system. Introducing value R defined as the cross-section ratio of longitudinally and transversely polarized virtual intermediate bosons .

$$R = \frac{\sigma_L}{\sigma_T} = \frac{(1 + \frac{4M^2 x^2}{Q^2})F_2 - 2xF_1}{2xF_1}, \quad (3)$$

the total cross-section can be expressed by R and F_2 for charged leptons and these two plus F_3 functions for neutrino scattering. In the parton model F_2 and F_3 have a simple interpretation in terms of quark distributions.

2 EXPERIMENTAL DATA

Data which will be discussed below comes from the following experiments:

CHARGED LEPTONS

electrons

- SLAC E140 + combined analysis of all previous data from SLAC [2],[3] (scattering on elementary as well as nuclear targets).

  The main differences with respect to previous analysis are:

  - improvements in radiative correction procedure (up to 5% change)

- new determination of 8 GeV spectrometer acceptance (2% change)

- new parameterization of $R(x_{Bj}, Q^2)$

- extended range in x_{Bj} and Q^2.

- overall normalization uncertainty of the result reduced to $\pm 2\%$.

muons

- BCDMS (CERN-NA4) muon-hydrogen [17], muon-deuterium [18] at beam energies 100, 120, 200 and 280 GeV. Comparison of neutron and proton structure functions is also made [19].

- EMC (CERN-NA2) muon-hydrogen [12], muon deuterium [13] at the same energies as for BCDMS. Muon-deuterium results at low x_{Bj} and low Q^2 (NA28) [15].

- NMC (CERN-NA37) muon-hydrogen and muon-deuterium scattering at beam energy of 280 GeV giving the ratio of neutron to proton structure function with small systematic errors [24].

NEUTRINOS

- CDHSW (CERN) wide band beam final results from scattering on iron [8], determination of Λ_{QCD}, $R(x_{Bj}, Q^2)$ and parton distribution, global fit to the double differential cross section.

- CCFR preliminary results from scattering on iron [11].

- BEBC (CERN-WA21,WA25) wide band beam results on hydrogen and deuterium from two experiments giving the ratio of valance quark distributions: WA21 [6] from "classical" fits with QCD evolution, WA21+WA25 combined analysis globally fitting the quark distributions to the ratio of data distributions from hydrogen and deuterium in x_{Bj}, y and Q^2 [7].

3 NEW RESULTS ON $R(x_{Bj}, Q^2)$.

The measurement of R is difficult and requires data taken at different beam energies. The relative normalisation of datasets play a very important role. There are not many results and

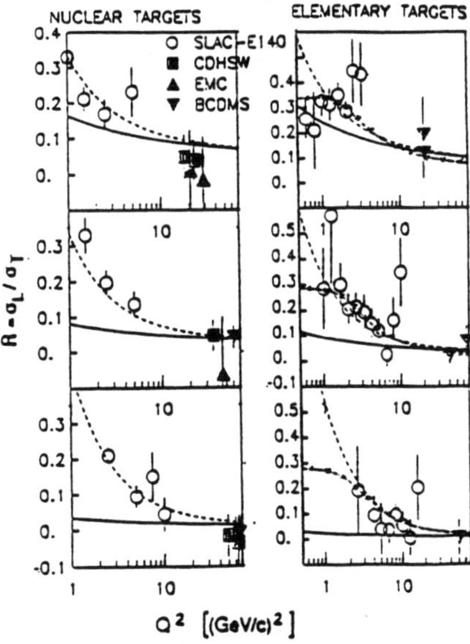

Figure 1: Experimental data on R from measurements on nuclear and elementary targets.

they cover a narrower range in x_{Bj} and Q^2 than absolute F_2. New results published last year came from scattering on elementary and nuclear targets and they are presented in fig. 1. The dashed line represents perturbative QCD predictions, the solid line QCD with target mass correction and asterisks fit to the data (for elementary targets only) based on a phenomenological model. At high Q^2, description by pure perturbative QCD is good, for $Q^2 \leq 5$ GeV^2 the target mass effect becomes important. A phenomenological model was introduced to fit better elementary target data [3] in full Q^2 range. For further analysis of SLAC hydrogen and deutrium data this model was used.

Does R depend on target material? Recent measurements from SLAC give the following results:

$$R^A - R^D = 0.001 \pm 0.018 \pm 0.016 [2] \quad (4)$$

for A being Fe and Au and

$$R^d - R^p = 0.002 \pm 0.009 \pm 0.010 [3] \quad (5)$$

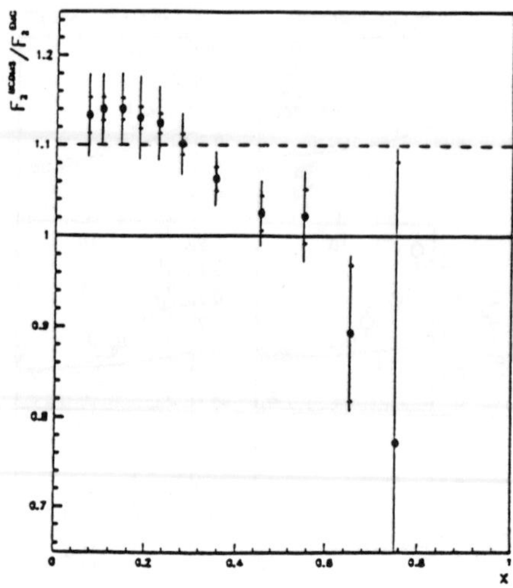

Figure 2: Comparison of F_2 structure functions for hydrogen as measured in two muon experiments: BCDMS and EMC.

From these one can assume that in the region of this measurement ($x_{Bj} \geq 0.1$ and $Q^2 \leq 16$ GeV^2) such dependence is very small or does not exist at all.

4 STRUCTURE FUNCTION $F_2(x_{Bj}, Q^2)$

The results from charged lepton scattering on elementary targets will be discussed in this section, concentrating on comparison and consistency of different datasets.

Figure 2 shows the ratio of values of F_2 structure functions averaged over Q^2 in x_{Bj} intervals from two muon experiments at CERN [12],[17]. Clear discrepancies are visible. At low x_{Bj} it is possible to bring both datasets to agreement by a normalization shift, at high x_{Bj} errors are big and this makes the situation unclear. For further discussion normalization was selected to be fixed by the SLAC data [3] and the following shifts were applied : BCDMS -2% and EMC $+8\%$. Figure 3 presents three datasets after the nor-

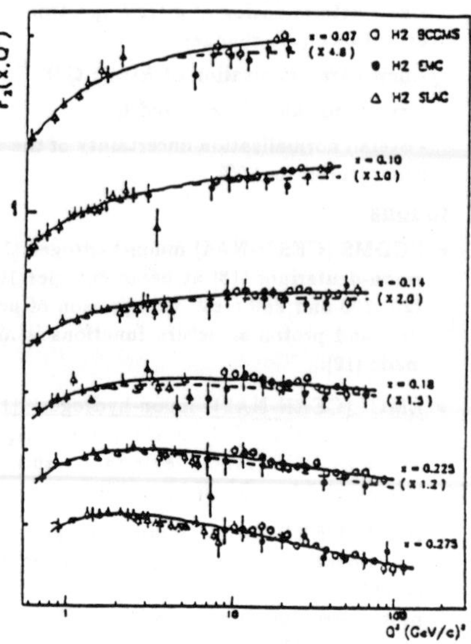

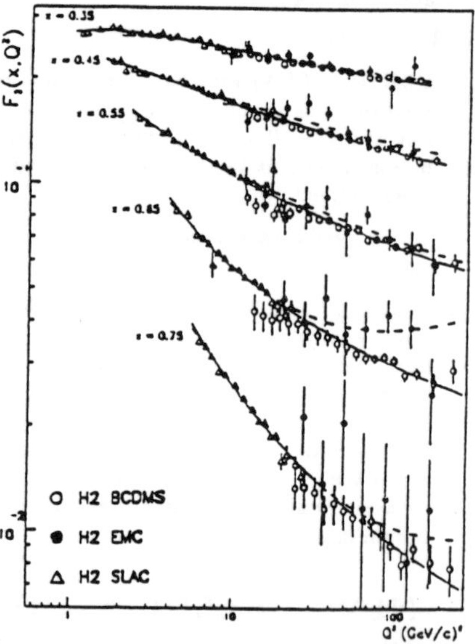

Figure 3: Experimental data on F_2 structure function from highest statistics experiments after normalisation corrections (see text). Fitted lines are for SLAC and BCDMS(solid) or SLAC and EMC (dashed)

malization correction. Required shifts are comparable with quoted systematic errors on the normalization. For this plot all data were processed with consistent parameterization for R. This picture shows quite a reasonable agreement in x_{Bj} intervals below 0.4. The same size of relative normalization shift seems to be necessary also for deuterium data. The EMC-NA2 [13] deuterium data, after such shift, come to good agreement with the NA28 deuterium data [15] at low x_{Bj} and the same is true for EMC results from iron [14] in comparison with CDHS [9].

To estimate possible areas of discrepancies, still left in F_2 structure function measurements from charge lepton scattering, one can try to fit different pairs of datasets and compare results. In fig. 3 results of such procedure is also presented [26]. At higher x_{Bj}, differences between SLAC and BCDMS can be seen in overlapping Q^2 regions. These discrepancies can be removed by the change of BCDMS data corresponding to 1.3 quoted systematic error. In this region the BCDMS data are dominated by systematic errors coming from energy and magnetic field calibration. For $x_{Bj} \leq 0.6$ residual differences are below 6% whereas at high x_{Bj} the comparison is not conclusive due to statistical errors for the EMC data and dominant systematic errors for BCDMS in this region. Similar conclusions can be reached when F_2 data on deuterium are considered.

Using all existing data for structure functions one can get the best fit for $F_2(x_{Bj},Q^2)$. Such an attempt was made by Y. Mizuno [27] from the New Muon Collaboration also using low energy experimental data covering the resonance region. For hydrogen the following data sets were used: data from resonance region [5], SLAC [3], EMC [12], BCDMS [17], CHIO [16] and also NMC ratio F_2^n/F_2^p [24] together with fitted function for F_2^d. Results are shown in fig 4. The procedure of the fit allows a free variation of structure function normalization between different datasets. The best fit was obtained with the following changes in the normalization: EMC +10%, BCDMS 0%, SLAC +2% and CHIO −5%. Relative shifts between EMC, BCDMS and SLAC are the same as obtained in the previous comparison (see fig.3). F_2 determination in a wide kinematic range reduces uncertainty on the radiative corrections coming from

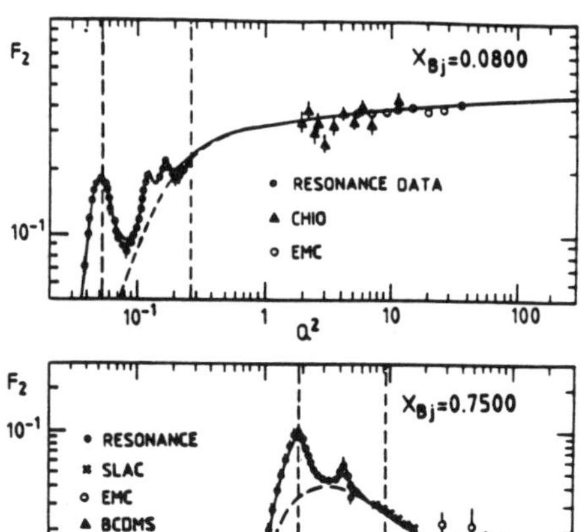

Figure 4: Results of the common fit to all existing data on F_2 hydrogen in two selected x_{Bj} bins.

regions not measured in the experiment.

4.1 THE NEW RESULTS ON F_2^d IN LOW x_{Bj}, LOW Q^2 RANGE

Recently new experimental data in the region of x_{Bj} from 0.002 to 0.17 and Q^2 from 0.2 to 8 GeV^2 have been published by the CERN muon experiment NA28 (from the EMC collaboration) [15]. Figure 5 presents very interesting result from this analysis showing that the structure function F_2, of deuterium, is not dependent on x_{Bj} below $x_{Bj}=0.1$ at low values of Q^2. This behaviour was predicted in the Regge exchange (Pomeron) approximation so the experimental result supports this kind of picture. Looking at the Q^2 dependence of F_2^d averaged over x_{Bj}, linear behaviour in $\log Q^2$ is observed and the line extrapolates to zero at $Q^2=0$.

Figure 6 shows the comparison of the data at $x_{Bj}=0.005$ with different models. The prediction from Generalised Vector Dominance Model [23] is

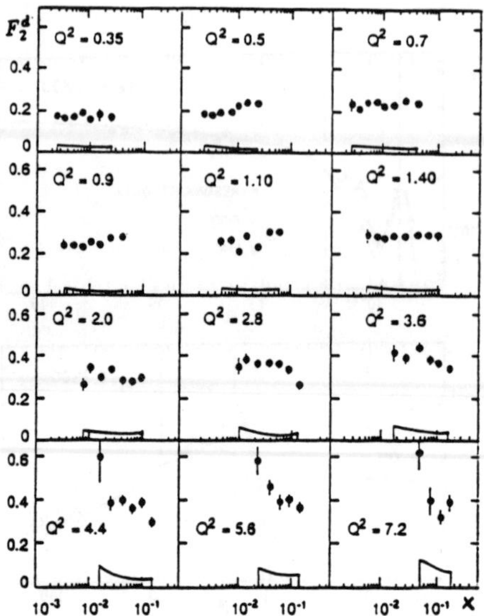

Figure 5: F_2^d as function of x_{Bj} in different Q^2 intervals

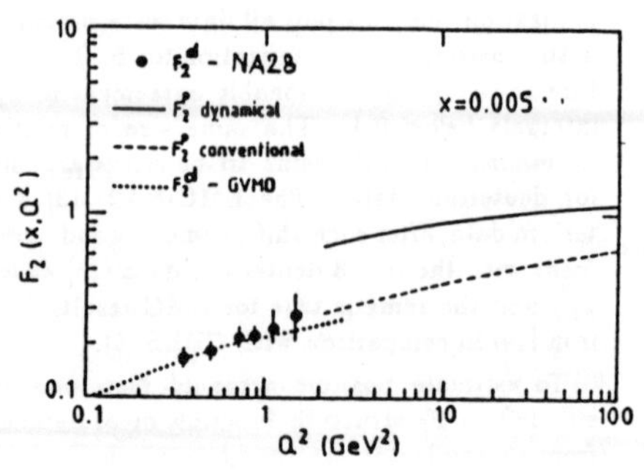

Figure 6: F_2^d at $x_{Bj}=0.005$

in good agreement with the data, extrapolation of F_2 obtained from conventional quark distributions [22] to low Q^2 also agrees quite well while F_2 from the dynamical evolution model [20] clearly predicts too high values. (It should be noticed that the next version of the dynamical model [21] published after this conference, is able to describe this data well.) This kind of discrepancy may lead to very big differences in the kinematic region of HERA experiments.

5 SUMMARY OF Λ_{QCD} DETERMINATIONS

New results on Λ_{QCD} determination have been presented by two neutrino experiments CDHSW and CCFR and by the muon experiment – BCDMS. The results from the CDHSW experiment came from the next-to-leading order fit to all data with Q^2 above 2GeV^2. In the small x_{Bj} region the data for xF_3 do not agree with perturbative QCD predictions, however systematic

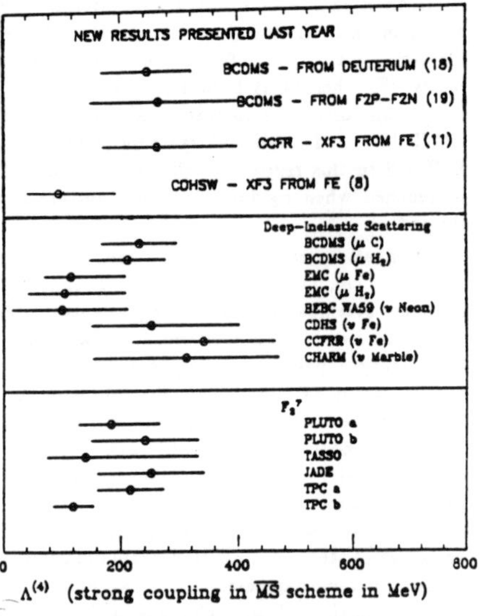

Figure 7: Measurements of Λ_{QCD} new measurements and summary from Particle Data Group.

errors in this region are too large to consider it as significant [8]. For reasonably large Q^2 (above 10 GeV^2) the agreement is good allowing determination of Λ_{QCD}.

The preliminary results presented by the CCFR [11] collaboration also came from fits to the structure function xF_3. Here the agreement of data with QCD predictions is satisfying but statistical errors are bigger.

The BCDMS presented two methods of obtaining Λ_{QCD} from muon scattering on hydrogen and deuterium. The results came from non-singlet fits of F_2 structure function results from scattering on deuterium and from analysis of the difference of the structure function F_2 for proton and neutron: $F_2^p - F_2^n$. It is important that all BCDMS determination of Λ_{QCD}, coming from different targets and different methods of calculation, are consistent.

All these results together with older ones [1] are summarised in figure 7.

6 PARTON DISTRIBUTIONS

6.1 F_2^n/F_2^p RATIO AND IT'S IMPLICATIONS

New results from F_2^n/F_2^p ratio from NMC and BCDMS muon scattering were presented. The NMC (NA37) collaboration performed measurements on hydrogen and deuterium targets simultaneously [24], covering x_{Bj} range (0.004 − 0.8) and Q^2 range (1 − 190 GeV^2). This method allows good control of systematic effects. Results from BCDMS come from separate measurements of hydrogen and deuterium in the range $x_{Bj}=(0.07 - 0.8)$ and $Q^2=(8 - 230$ GeV^2) [19]. Figure 8 shows results from both experiments together with earlier data from SLAC [4] and EMC [13]. Bands on the picture show systematic errors. Except for the EMC results, data have not been corrected for Fermi smearing. NMC and BCDMS taken together give results with small statistical and systematic errors in broad x_{Bj} range. These very accurate data provide a strong constraint for parton distribution parameterizations showing best agreement with [25]. The observed difference, relative to the SLAC result, may be due to lower Q^2 in this experiment

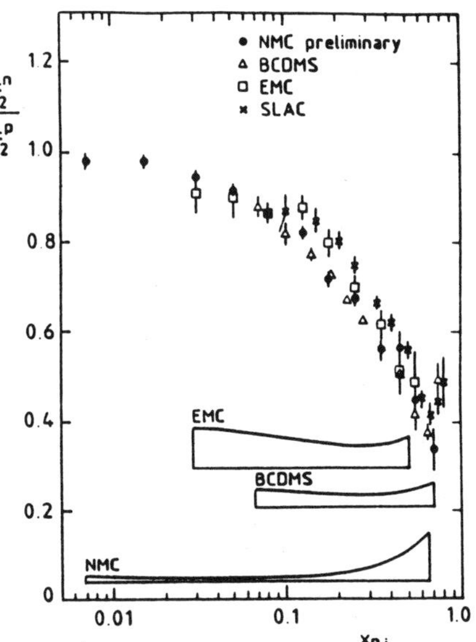

Figure 8: Structure function ratio F_2^n/F_2^p. Not corrected for nuclear effects.

(1 − 16 GeV^2). New results from SLAC reanalysis in a broader kinematic range are expected very soon.

Figure 9 shows results of F_2^n/F_2^p ratio from BCDMS after correction for Fermi smearing together with corresponding results derived from BEBC neutrino and antineutrino data on the valance up to down quark ratio. Neutrino results were obtained in two ways. One is WA21 analysis of νp and $\bar{\nu} p$ data [6] where no smearing correction is necessary. The other is a combined study of WA21 and WA25 data [7] where D/H cross section ratios are fitted while the number of valance quarks and the gluons momentum are constrained. The latter data are systematically higher. At x_{Bj} above 0.3 where valance partons dominate, this data can be directly compared to the BCDMS results on F_2^n/F_2^p. The muon data suggest somewhat stronger x_{Bj} dependence.

6.2 QUARK DISTRIBUTIONS

New results on quark distributions are coming from neutrino experiments on nuclear targets and

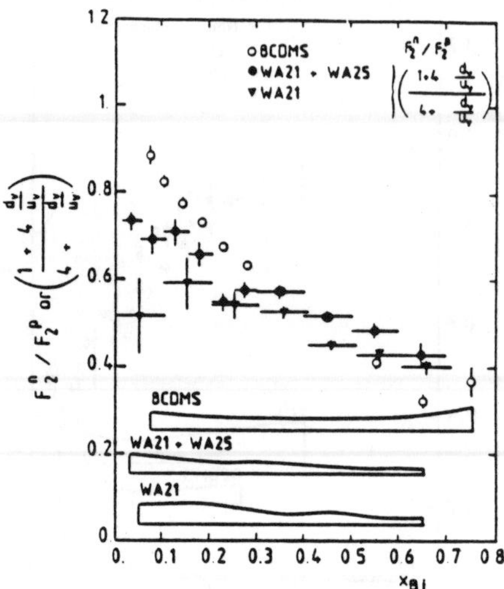

Figure 9: The valance up to down quark ratio from WA21 and WA25 compared with F_2^n/F_2^p ratio from BCDMS as a function of x_{Bj}.

were presented by CDHSW [8] from measurements of F_2 and F_3 which correspond to the sum and difference of quark and antiquark distributions (see fig.10).

From the fits to BEBC [7] data one also gets parameterizations of individual quark distributions. This should allow better constraints in the global parameterizations reducing considerably the big spread of predictions for high Q^2 soon accessible by experiments at HERA.

6.3 GLUON DISTRIBUTIONS

The situation is even less clear for gluons. The description of gluon distribution is obtained from QCD analysis introducing strong correlation between parameters of gluon distribution and Λ_{QCD} in the fit. The new results from CDHSW [8] gives much softer gluon distribution than obtained previously in this experiment [10].

In fig.11 different results of QCD fits are compared. At low x_{Bj} not much is known. New measurements here may come from J/Ψ muoproduction, measured by the NMC collaboration in the near future.

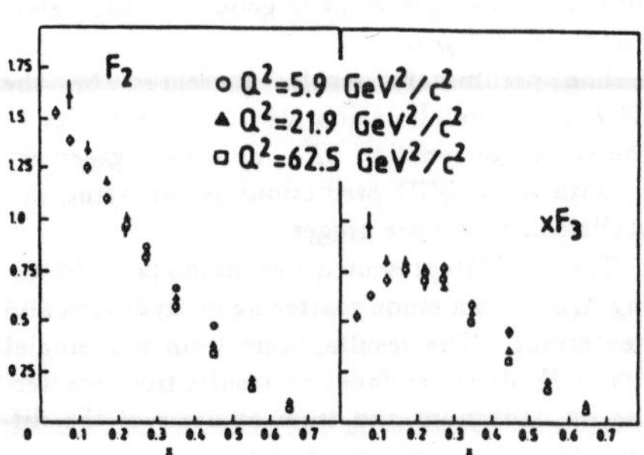

Figure 10: $F_2(x_{Bj},Q^2)$ and $xF_3(x_{Bj},Q^2)$ structure functions measured in CDHSW at three different values of Q^2

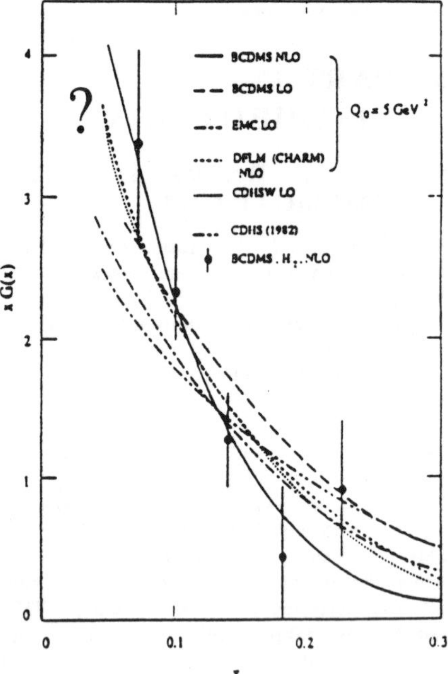

Figure 11: Gluon distributions as a function of x_{Bj} estimated from different experimental data.

7 SUMMARY

The new parameterization of $R = \frac{\sigma_L}{\sigma_T}$ in low Q^2 and $x_{Bj} \geq 0.1$ was obtained from SLAC data. The R values seem to be higher than QCD predictions even with target mass correction.

F_2 structure function results from charged leptons scattering can produce consistent picture at low and medium x_{Bj} after applying normalization shifts. The size of this shift, estimated in different ways, gives the same value and is compatible with quoted normalization uncertainty. At high x_{Bj}, differences between SLAC and BCDMS data can be understood in terms of systematic errors dominating BCDMS data there.

Results on the deuterium structure function F_2 at low x_{Bj} and low Q^2 are in agreement with predictions from the Pomeron exchange.

New measurements of the F_2^n/F_2^p ratio made by NMC and BCDMS are very accurate in complementary x_{Bj} ranges, covering together $x_{Bj} = (0.07\text{-}0.8)$. They provide constraint for parton distribution parameterization allowing to exclude some of them.

The new determinations of Λ_{QCD} give consistent value of about 200 MeV.

References

[1] Review of Particle Properties Phys.Lett.*B204* (1988).

[2] SLAC, S.Dasu et al, Phys.Rev.Lett.*25* (1988), 2591.

[3] SLAC, L.W.Whitlow et al., preprint SLAC-PUB-5100 (1989).

[4] SLAC, A.Bodek et al.,Phys.Rev.*D20* (1979), 1471.

[5] F.W.Brasse et al.,Nucl.Phys.*B110* (1976), 413.

[6] G.T.Jones et al.,Oxford Univ. rep. OUNP-89-18, (1989).

[7] G.T.Jones et al.,"Preliminary results on the ratio $d/u(x)$.." paper submitted to the 1989 Int.Symp. on Lepton and Photon Interactions at High Energies, Stanford Univ. 1989.

[8] CDHSW, J.P.Berge et al., preprint CERN-EP/89-103 (1989).

[9] CDHS, H.Abramowicz et al., Z.Phys.*C17* (1983), 283.

[10] CDHS, H.Abramowicz et al., Z.Phys.*C12* (1982), 289.

[11] CCFR, E.Oltman et al.,Nevis preprint no.1417, July 1989.

[12] EMC, J.J.Aubert et al.,Nucl.Phys.*B259* (1985), 189.

[13] EMC, J.J.Aubert et al.,Nucl.Phys.*B293* (1987), 740.

[14] EMC, J.J.Aubert et al.,Nucl.Phys.*B272* (1986), 158.

[15] EMC, M.Arneodo et.al. preprint CERN-EP/89-121. submitted to Nucl.Phys.B

[16] CHIO, A.B.Gordon et al., Phys.Rev.*D20* (1979), 2645.

[17] BCDMS, A.C.Benvenuti et al., Phys.Lett. *223B* (1989), 485.

[18] BCDMS, A.C.Benvenuti et al., Phys.Lett. *237B* (1990), 592 and preprint CERN-EP/89-171.

[19] BCDMS, A.C.Benvenuti et al., Phys.Lett. *237B* (1990), 599.

[20] M.Gluck,R.M.Godbole and E.Reya, Z.Phys. *C41* (1989), 667.

[21] M.Gluck,E.Reya and A.Vogt, Univ.of Dortmund,preprint DO-TH 89/20.

[22] M.Diemoz et al.,Z.Phys. *C39* (1988), 21.

[23] J.Kwiecinski and B.Badelek, Z.Phys. *C43* (1989), 251.

[24] NMC: presented by J.Nassalski at Europhysics Conference on High Energy Physics, Madrid,1989.

[25] A.D.Martin,R.G.Roberts and W.J.Stirling, Phys.Lett. *206B* (1988), 327.

[26] M.Virchaux (BCDMS) private communication.

[27] Y.Mizuno (NMC) private communication.

RECENT RESULTS ON JET PRODUCTION AT HADRON COLLIDERS

The CDF Collaboration[1]
The UA2 Collaboration[2]

Presented by James F. Patrick

Fermi National Accelerator Laboratory

Batavia IL 60510

USA

ABSTRACT

An overview of preliminary results on jet production from the recent high luminosity runs of the CDF and upgraded UA2 detectors is presented. Results include the inclusive single jet E_t distribution, the angular distribution of two jet events, and kinematic distributions in 3 and 4 jet events. The data are compared to QCD predictions and are found to be in reasonable agreement with those predictions. Results of a search by the UA2 collaboration for for W and Z decays into quark pairs are presented, this shows a signficant signal (5 standard deviations) over the QCD background.

1 Introduction

The standard picture of jet production in $p\bar{p}$ collisions is of hard scattering according to QCD between constituents (quarks and gluons) within the incident nucleons, followed by fragmentation of the final partons into the observed jets. Measurements made at the CERN $S\bar{p}pS$ collider of jet production at $\sqrt{s} = 630$ GeV are well described by this standard picture and have provided important tests of QCD [1]. These results included the discovery of jet structure in hadronic collisions, and measurements of the inclusive single jet E_t distribution, the two jet angular distribution, and the topology of multijet events. Results from the first CDF run in 1987 extended these results to a new energy regime, $\sqrt{s} = 1.8$ TeV [2]. Comparison of the single jet E_t spectrum to QCD predictions has placed the most stringent limit to date on a possible compositeness scale Λ_c for quarks, 750 GeV. Since the initial CERN results, the UA2 detector has been signficantly upgraded[3]. Following this upgrade, an integrated luminosity of 7.8 pb^{-1}, nearly a factor of 10 more than in the original runs, has been recorded. Also, the CDF detector has recorded 4.7 pb^{-1} at $\sqrt{s} = 1.8$ TeV. This report presents preliminary results on jet production from this data.

For n jet production, the differential cross section may be written:

$$\frac{d\sigma^n}{dx_1 dx_2} = \sum_{processes} \int d\hat{\sigma}^n F_1(x_1, Q^2) F_2(x_2, Q^2) \frac{dx_1}{x_1} \frac{dx_2}{x_2} \tag{1}$$

[1]Argonne - Brandeis - Chicago - Fermilab - Frascati - Harvard - Illinois - KEK - LBL - Pennsylvania - Pisa - Purdue - Rockefeller - Rutgers - Texas A&M - Tsukuba - Tufts - Wisconsin
[2]Bern - Cambridge - CERN - Heidelberg - Milano - Orsay - Pavia - Perugia - Pisa - Saclay - CERN

where $F_i(x_i)$ are nucleon structure functions which reflect the probability of obtaining a parton with momentum fraction x_i, Q^2 is the momentum transfer in the scattering, $d\hat{\sigma}^n$ is the differential cross section for production of an n parton final state from the incident partons, and the sum is over possible processes involving quarks and gluons. Thus from a knowledge of the nucleon structure functions determined from deep inelastic scattering experiments, and the subprocess cross sections computed using QCD, predictions can be made regarding production of n jets final states. These predictions would be modified, for example, if quarks had substructure. This would be exhibited as an excess of jets at high E_t in the inclusive jet E_t distribution [5].

2 Detectors

Most relevant for jet detection, both the UA2 and CDF detectors contain finely segmented calorimetry covering a large solid angle. The UA2 apparatus was substantially upgraded during the years 1985 to 1987. Among other modifications, end cap detectors were added so that the range $-3 < \eta < +3$ is now covered by calorimeters using scintillator as the sampling medium. The central calorimeter covers the range $-1 < \eta < +1$, and is segmented into 240 cells subtending $10°$ in $\Delta\theta$ and $15°$ in $\Delta\phi$. The calorimeter is longitudinally segmented into an electromagnetic compartment, where lead is the absorber, and two hadronic compartments, where iron is the absorber. In the end cap region, the cells in the pseudorapidity interval $1.0 < |\eta| < 2.5$ have one electromagnetic and one hadronic compartment with a segmentation of 0.2 units in $\Delta\eta$ and $15°$ in $\Delta\phi$. The region $|\eta| > 2.5$ consists of only a hadronic compartment, the $\Delta\phi$ segmentation here is $30°$. The event z vertex is determined by a time of flight method using a matrix of scintillation counters in the rapidity range $2.3 < |\eta| < 4.1$.

The CDF detector [4] contains calorimeters covering the pseudorapidity range $-4.2 \leq \eta \leq 4.2$ and the entire azimuthal range constructed in a projective tower geometry. Lead and steel plates are used as the absorbing medium in the electromagnetic and hadronic sections respectively. The calorimetry is divided into central ($|\eta| \leq 1.1$), endplug ($1.1 \leq |\eta| \leq 2.2$) and forward ($2.2 \leq |\eta| \leq 4.2$) sections. The central section uses plastic scintillator as the sampling medium. Here the tower segmentation is 0.1 units in $\Delta\eta$ and $15°$ in $\Delta\phi$. In the endplug and forward sections, proportional chambers are used as the sampling medium. The segmentation is 0.1 units in $\Delta\eta$ and $5°$ in $\Delta\phi$. The event vertex is is determined by a set of 8 time projection chambers surrounding the beam pipe.

3 Cluster Algorithms

The UA2 group uses several different algorithms to define clusters of calorimeter energy depending on the specific analysis goal. In the first algorithm, clusters are structures of adjacent calorimeter cells each having an energy deposition > 400 MeV. While this method has good resolution in space and good two jet resolving power, energy collection is limited to the central core of the jet. This method is used for the two and four jet studies described below. The second algorithm starts with the basic clusters described above, then merges clusters within a cone $\cos\omega > 0.2$ about a primary seed cluster. This algorithm has improved energy collection efficiency at the expense of two jet resolving power, and is used in the measurement of the inclusive jet E_t distribution. For reasons of execution speed, the online trigger uses a "window" algorithm. This combines energy in rectangular windows of $70° \times 75°$ in the θ, ϕ plane. The search for W, Z $\rightarrow$ jets uses a cone algorithm, which refines the window algorithm by including only energy in a cone $\eta^2 + \phi^2 < 0.64$ about a window jet centroid.

The CDF clustering algorithm starts by forming "preclusters", contiguous towers each with $E_t \geq 1$ GeV. All towers with $E_t \geq 0.1$ GeV within some radius $\Delta R = \sqrt{\Delta\eta^2 + \Delta\phi^2}$ (typically set to

0.7) in $\eta - \phi$ space with respect to the E_t weighted centroid of the precluster are collected and the centroid of the cluster computed. Towers within the cone defined by the new centroid are collected, this process is iterated until the list of towers in the cluster is stable. Only clusters with $E_t \geq 10$ GeV and $|\eta| \leq 3.5$ are considered for subsequent analysis.

4   Energy Corrections

The calorimeters in both detectors are non-compensating, having significantly non-linear response for low energy particles. Whereas jets are composed of a number of particles of potentially widely varying momentum, the response as a function of particle momentum must be well understood to infer the energy of the initial partons from that of the observed jets. For the UA2 experiment, the calorimeters were calibrated by exposing them to test beam particles of known momentum. The central electromagnetic and hadronic compartments were calibrated using electrons and pions respectively of momentum 40 GeV. The endcap calorimeters were exposed to beams of pions, protons, and electrons covering the momentum range from 300 MeV to 150 GeV. Whereas the measurements of response and resolution as a function of particle momentum in the endcap are well reproduced by calculations done with the GEANT [7] simulation package, this simulation was used to infer the response of the central calorimeter where only measurements at a single beam energy were available.

The CDF experiment measured the calorimeter response in a test beam of electrons and pions over the energy range 10 GeV to 150 GeV. For lower energy particles, isolated tracks in minimum bias events were used. The particle momentum was measured using the central tracking chamber, immersed in a 1.4 T magnetic field. This was compared to the energy in the calorimeter region pointed to by the track.

Both experiments track the time dependence of the calorimeter response using radioactive sources and light flashers. In addition, the UA2 calorimeter has been reexposed to test beams periodically since its initial construction.

To infer the produced jet energy from that observed, a Monte Carlo program that includes a QCD based model of jet production and a detector simulation is tuned to reproduce the observed calorimeter response, and particle multiplicity and (in the case of CDF) momentum distributions. The correction is determined as a function of observed jet energy, with uncertainties reflecting uncertainties in the model and detector response. There are corrections for cracks, leakage, contribution from the underlying event, and energy outside the clustering cone in addition to those due to non-linear calorimeter response. In the CDF case, the total jet energy correction ranges from $50 \pm 12\%$ for a jet with $E_t = 30$ GeV to $15 \pm 5\%$ for a jet with $E_t = 300$ GeV.

5   Inclusive Jet E_t Distribution

Measurement of the inclusive single jet E_t distribution is restricted to the central region ($|\eta| < 0.85$ for UA2, $0.1 < |\eta| < 0.7$ for CDF) in both experiments. This region has best known energy scale and resolution, and therefore the least systematic uncertainty. In addition to correcting observed jet energies as described above, the steeply falling distribution was corrected for resolution effects. Figure 1a) shows the results of the UA2 measurement, at present the uncertainty is estimated to be approximately 50%. The distribution is consistent with a lowest order QCD calculation. The present results are also in agreement with results from pre-upgrade runs, demonstrating the successful tracking of the calorimeter response. Figure 1b) shows the preliminary CDF measurement, based on an integrated luminosity of 865 nb^{-1}. Uncertainties due to the energy corrections, smearing correction, and luminosity range from $\pm 70\%$ at $E_t = 25$ GeV to $\pm 35\%$ at $E_t = 250$ GeV. Again, the

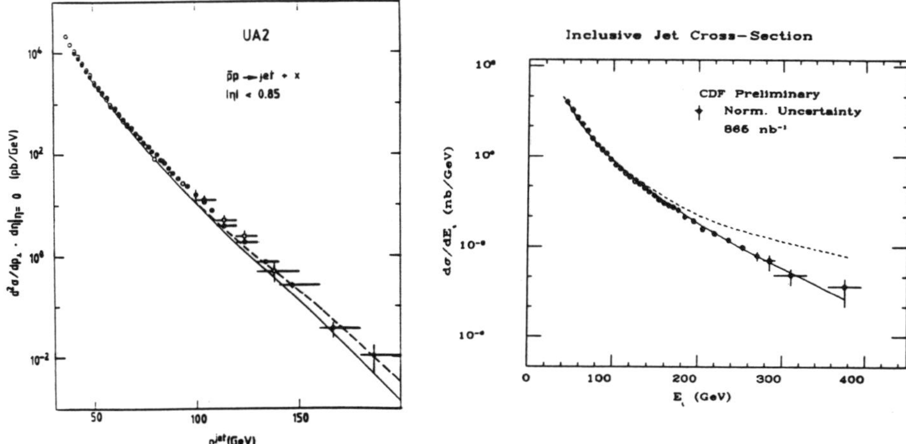

Figure 1: The inclusive single jet E_t distribution measured by UA2 (left) and CDF (right). The solid line is a calculation based on leading order QCD. The dashed line is a QCD prediction modifed by a quark compositeness scale of 750 GeV.

results are consistent with the QCD calculation. There is a significant theoretical uncertainty due to uncertainties in structure functions and the definition of the Q^2 scale. The new data strongly support the previously published limit on a possible quark compositeness scale of 750 GeV. Further work is needed on understanding the energy and smearing corrections and the associated uncertainties at high E_t before quoting a new limit on a possible compositeness scale. Also, non leading order calculations of this distribution have been presented at this conference. These will allow a more stringent test of QCD and reduce the theoretical uncertainty in deriving limits on quark compositeness.

6 Two Jet Mass Distribution

The CDF group has measured the invariant mass distribution for two jet events. This provides a further test of QCD, and also could exhibit modifications due to production of high mass states decaying into two jets. Events used in this analysis were required to be back to back in the transverse plane within $20°$, and satisfy $0.5 \times |\eta_1 + \eta_2| < 0.4$ and $0.5 \times |\eta_1 - \eta_2| < 0.4$. The latter cuts insure that both jets are contained in the central calorimeter. Results are shown in Figure 2 together with a range of QCD predictions based on various choices of structure functions and Q^2 scales. The uncertainty is approximately $\pm 35\%$. Also shown is the curve expected including a composite interaction among quarks with scale $\Lambda_c = 750$ GeV. The curve is consistent with QCD, within the uncertainties. Determination of a precise limit on Λ_c, and a search for structure is in progress.

7 Two Jet Studies

The UA2 experiment has performed a study of an exclusive selection of two jet events. The results are based on an integrated luminosity of 3.1 pb^{-1}. In each event, jets are sorted in order of decreasing

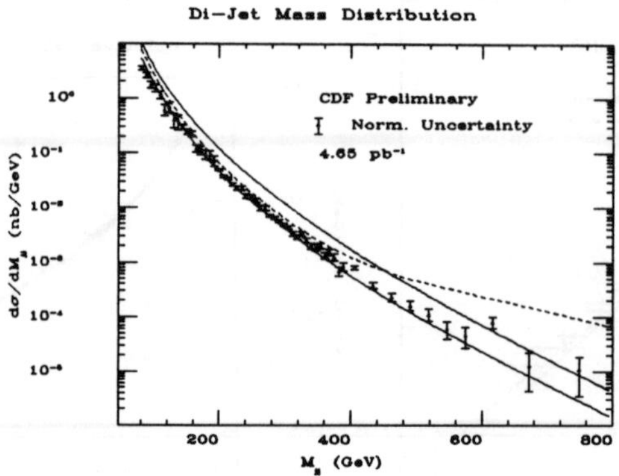

Figure 2: The two jet invariant mass distribution as measured by CDF. The band enclosed by solid lines are calculations based on leading order QCD with different assumptions regarding Q^2 scale and structure functions. The dashed line is a QCD prediction modifed by a quark compositeness scale of 750 GeV.

transverse energy and labeled 1, 2, 3, etc. Then a sample of 32000 events is selected with the following criteria:

- $E_{t,1,2} > 35$ GeV; $E_{t,3} < 10$ GeV,

- $p_t^\xi < 10$ GeV; $p_t^\eta < 50$ GeV

where p_t^ξ and p_t^η are the components of the transverse momenta of the jet pair projected on the bisectors of the transverse momenta. These cuts selected well balanced two jet events without large contributions from soft radiation not detected as a third jet. Figure 3a) characterises the transverse energy flow around the jet axes of the two leading jets in the lab system. In this figure, the measured transverse energy has been integrated over the psuedorapidity range $-2 < \eta < 2$ and binned in $\Delta\phi$ bins corresponding to the cell structure of $15°$ width. The two jet structure is clearly visible. The data are compared to the PYTHIA (version 4.8) [8] simulation, and are found to be in good agreement. The full line in the figure compares the data to PYTHIA with the initial state radiation switched off. While this approach describes well the width of the jets, it fails to account for the transverse energy detected at large angles to the jet axis. Finally, the angular distribution of the two jets in their center of mass system has been measured, this is shown in Figure 3b). The PYTHIA simulation again exhibits good agreement with the data.

8 Three Jet Studies

CDF has studied the kinematics of 3 jet final states. QCD allows either incoming or outgoing partons to radiate additional partons. If the radiated parton is hard and at large angle, it will be visible as a separate jet in the final state. There are 4 independent variables describing the kinematics. These

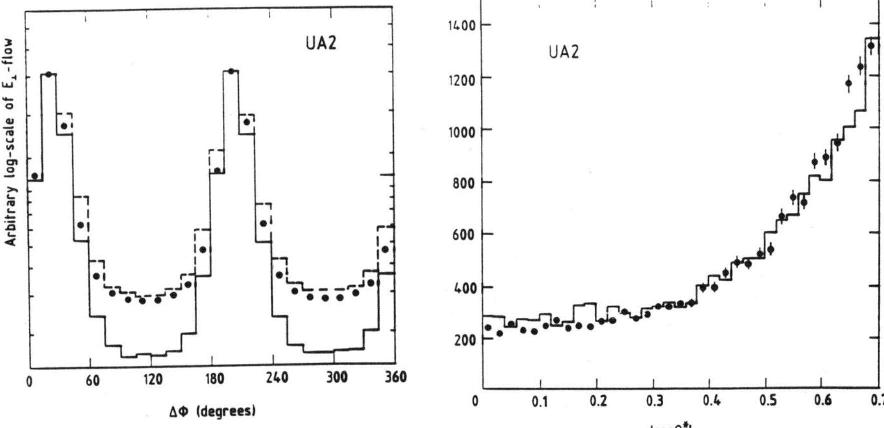

Figure 3: a) Distribution of the transverse energy flow in two jet events. b) Distribution of the center of mass angular distribution for two jet events as measured by UA2. In both plots, the histogram is from a simulation based on the PYTHIA model.

are taken to be θ^*, the angle of the leading jet with respect to the beam axis; ψ, the angle between the planes containing the beam and leading jet (designated parton 3), and the plane containing the two subleading jets (partons 4 and 5), and the jet energies scaled to the center of mass energy, $x_i = 2p_i/\sqrt{\hat{s}}$ $(x_3 + x_4 + x_5 = 2)$.

Events are selected with at least three jets each with at least 10 GeV in transverse energy. The event is boosted to the center of mass frame, and the jets ordered by energy in this frame. The following additional selection criteria are imposed to produce an event sample for which the detector acceptance is approximately uniform:

- $|\cos\theta^*| \leq 0.726$,

- $\sqrt{\hat{s}} \geq 200$ GeV,

- $30° \leq \psi \leq 150°$,

- $x_3 \leq 0.9$.

Following application of the cuts, 4973 events remain from 2 pb^{-1}. Figure 4 shows the distribution of the energy fraction variables x_3 and x_4, together with Monte Carlo predictions based on QCD (Papageno) [9] and also Monte Carlo predictions assuming a phase space distribution for the three jets. The agreement of the data with the QCD calculation is good. The data do not appear consistent with the phase space distribution.

9   A Study of Four Jet Events

The yield and phase space density of four jet events provide additional tests of perturbative QCD. Also, a new production mechanism could play a role, the production of two jets each from two independent pairs of incoming partons [10]. While this process is a major motivation for four jet

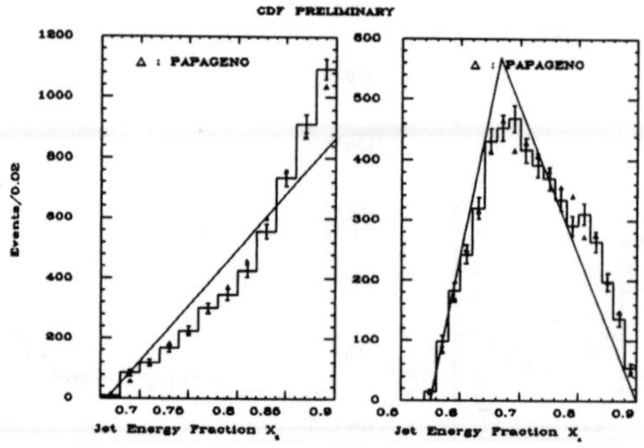

Figure 4: Distributions of the energy fraction variables x_3 and x_4 in 3 jet events as measured by CDF. The histogram shows the data, the triangles are a Monte Carlo calculation based on QCD. The solid lines are predictions of a phase space Monte Carlo.

physics, the analysis to date concentrates on a comparison with leading order QCD (order α_s^4). Events for this analysis were required to satisfy:

- $E_{t1,2} > 30$ GeV; $E_{t3,4} > 15$ GeV,
- $|\eta| < 2$ for all four jets,
- opening angle (jet1, jet2) $> 143°$.

The selected data sample corresponding to an integrated luminosity of 3.1 pb^{-1} consists of 857 such events. Figure 5 presents the observed distributions of invariant mass and sphericity of the 4 jet system together with the normalized prediction from a leading order QCD calculation. These distributions are well understood in terms of the QCD calculation and in particular do not exhibit any structure or strong deviation QCD. A search for a possible multiparton component to the 4 jet events is in progress.

10 Search for W, Z → jets

The intermediate vector bosons are expected to decay into quark-antiquark pairs with the branching ratios: $\Gamma(Z^\circ \to q\bar{q})/\Gamma(Z^\circ \to ee) \approx 20$; $\Gamma(W \to q\bar{q})/\Gamma(W \to e\nu) \approx 6$. Observation of these signals in the two jet mass spectum provides a test of the standard model predictions for these decay modes. In addition, it can serve as a test case for W and Z detection via the two jet decay mode as has been proposed for Higgs and other particle searches at future hadron colliders. The two jet final state produced by W and Z decay is experimentally very similar to two jet events originating from QCD parton interactions. Thus the search must be performed by measuring a localized peak structure on the smooth two jet mass spectrum originating from QCD processes. The QCD background is expected to dominate over the W, Z processes by about two orders of magnitude.

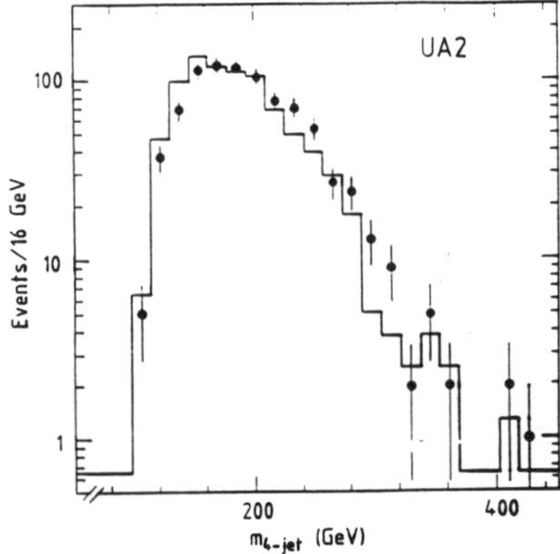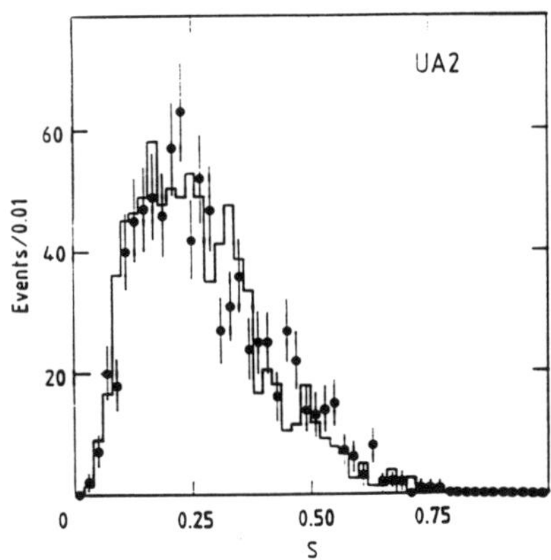

Figure 5: Distributions of the invariant mass and sphericity of the jets in 4 jet events as measured by UA2. The histogram is based on a leading order QCD calculation.

For this analysis, a special trigger was used to record approximately 10^7 events each with two jets with $E_t > 13$ GeV (or $E_t > 10$ GeV prescaled by a factor of 8). The offline event selection required two jets satisfying $|\cos\theta| < 0.6$, well contained within the central calorimeter where the signal/background is expected to be most favorable. Also, the z coordinate of the event vertex was required to be < 20 cm from the center of the detector. Additional cuts rejected $Z \to e^+e^-$ decays, jets with insufficient longitudinal containment, and events with a third jet with > 20 GeV of transverse energy. There were 4.5×10^6 event remaining after these cuts.

Figure 6 shows the resulting 2 jet mass distribution with a vertical scale weighted by $(m/100)^6$. This procedure largely removes the steep slope from the background distribution making the structure in the W,Z mass region much more apparent. A fit to the distribution was made assuming a double Gaussian form for the signal, $m_Z/m_W = 1.14$, and $\sigma \times B(Z^0 \to q\bar{q})/\sigma \times B(W \to q\bar{q}) = 0.43$. The function used was

$$f \times (m^\alpha e^{-\beta m} e^{-\gamma m^2} + \text{Signal}(\text{size}, m_W, \sigma_m/m)). \tag{2}$$

The results of the fit give 5620 ± 1130 as the number of events, 78.9 ± 1.5 GeV as the reconstructed W boson mass, and $9.3 \pm 2.0\%$ as the mass resolution.

These reconstructed signal parameters have been compared to the expectations obtained using the PYTHIA event generator combined with a calorimeter simulation. The resulting mass and resolution are consistent with those observed. The number of events expected from the simulation is 4250 ± 150, about 1.2 standard deviations below the observed number. However this simulation does not account for deviations from the simple double Gaussian shape, or possible interference between the weak interaction signal with the QCD background.

11 Summary

Preliminary results on jet physics from the recent high luminosity runs of the CDF and upgraded UA2

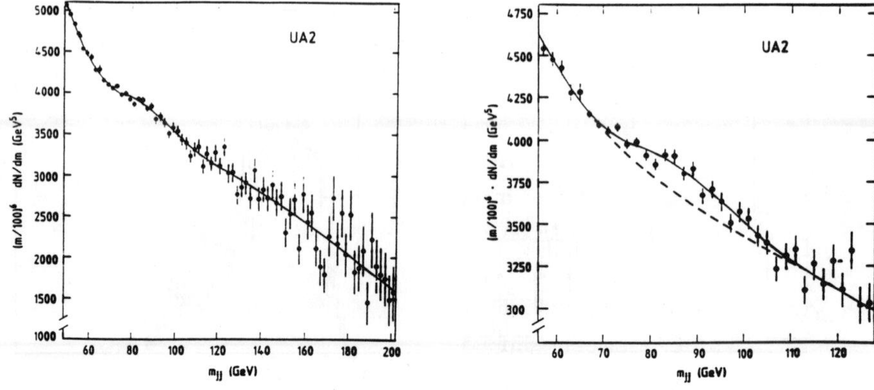

Figure 6: Two jet mass distribution measured by UA2.

detector have been presented. The inclusive single jet E_t distribution, and various distributions in 2, 3, and 4 jet events appear to be well described by QCD predictions. The UA2 group has observed a 5 standard deviation signal in the 2 jet invariant mass distribution consistent with W and Z decays to quark jets.

References

[1] M. Banner, *et. al.* (UA2 collaboration), Phys. Lett. <u>118B</u>, 203 (1982).
G. Arnison, *et. al.* (UA1 collaboration), Phys. Lett. <u>123B</u>, 115 (1983).
L. DiLella, Ann. Rev. Nucl. Part. Sci. <u>35</u>, 107 (1985).
P. Bagnaia, and S. D. Ellis, Ann. Rev. Nucl. Part. Sci. <u>38</u>, 659 (1988).

[2] F. Abe, *et. al.*, Phys. Rev. Lett. <u>62</u>:613 (1989).
F. Abe, *et. al.*, Phys. Rev. Lett. <u>62</u>:3020 (1989).
F. Abe, *et. al.*, Phys. Rev. Lett. <u>64</u>, 157 (1990).

[3] C. N. Booth, Proc. 6th Topical Workshop on Proton-Antiproton Collider Physics, Aachen, eds. Eggert, K. *et. al.* World Scientific, Singapore, 381 (1987).

[4] F. Abe, *et. al.*, Nucl. Inst. Meth. <u>A271</u>:387 (1988).

[5] E. Eichten, *et. al.*, Phys. Rev. Lett. <u>50</u>:811 (1983).

[6] D. Soper, these proceedings.

[7] R. Brun, *et. al.*, CERN-DD/EE/84-1.

[8] H. Bengtsson, and T. Sjostrand, LUTP 87 2, UCLA 87 001.

[9] I. Hinchliffe, report in preparation.

[10] N. Paver, and D. Treleani, Nuovo Cimento <u>70A</u>, 215 (1982), <u>73A</u>, 392 (1983).

Multihadronic Events in e^+e^- Collisions Below Z^0

Kaori Maeshima

AMY Collaboration, University of California, Davis

ABSTRACT

Some of the selected topics on multihadronic final states in e^+e^- Collisions below Z^0 are discussed here with emphasis on the TRISTAN results. At the TRISTAN accelerator in KEK, Japan, we have studied e^+e^- collisions at $\sqrt{s}$ between 50 and 64 GeV. The topics discussed here include: comparison of several different Monte Carlo models and data, observation of the difference in quark and gluon jets, observation of the Non-Abelian nature of QCD -- (running coupling constant via 3-jet ratio (R_3), evidence of 3 gluon couplings). The TRISTAN experiments' combined R_{had} measurement continued to be slightly higher than the standard model prediction using the newly established precise Z^0 mass value, 91.1 GeV/c^2.

1. Comparison Between Different Monte Carlo Models and Data, and Event Shape Variables.

Although QCD has become almost universally accepted as the theory of the strong interaction, the transition from partons to the observable hadrons is not well understood. However in the previous ten years, many studies of hadron production in e^+e^- annihilation have been performed by PETRA and PEP experiments $(12 < \sqrt{s} < 43$ GeV$)$, and the models have been refined to obtain better agreement with experimental results. We have studied the general properties of multi-hadron final states in the TRISTAN energy range.[1] Event shape, particle flow, and inclusive particle distributions have been measured. The data have been corrected for the effects of detector acceptance and initial state radiation and compared with the LUND Parton-Shower (PS),[2,3] LUND matrix element,[4] and Webber[5,6] (BIGWIG 4.3) QCD + fragmentation models. At TRISTAN energies, with the parameter values tuned for PEP and PETRA energies, LUND ME shows some large deviations and demonstrates significant difficulties in reproducing the experimental data for most of the event shape distributions. On the other hand, both the Webber and the LUND PS models provide a good description of the data. In general, the Webber model gives a fairly good description of the P_T^{in} - related quantities, but reproduces less well the P_T^{out} -related distributions. The LUND PS, not only, is in reasonable agreement with P_T^{in} - related distributions, but also yields an accurate reproduction of P_T^{out} quantities. Figure 1 (a) and

(b) show the AMY data of P_T^{in} and P_T^{out} distribution of charged particles (with respect to the sphericity axis in the event plane) respectively, as compared with predictions of the models with the parameters tuned by Mark II experiment at $\sqrt{s} = 29$ GeV.

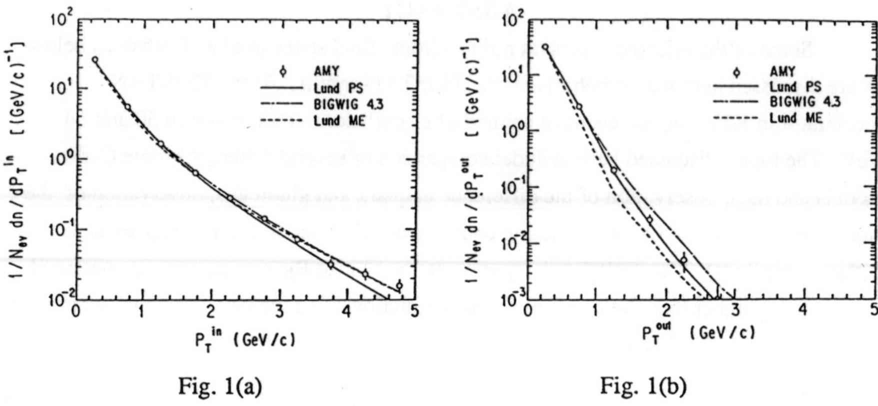

Fig. 1(a)

Fig. 1(b)

The Figure 2 show the observed charge multiplicity distribution (AMY)[7]. The solid line in the Figure 2 (a) shows LUND parton-shower model prediction and Figure 2 (b) shows the prediction from LUND ME model. LUND PS model agrees with AMY data better than the LUND ME model.

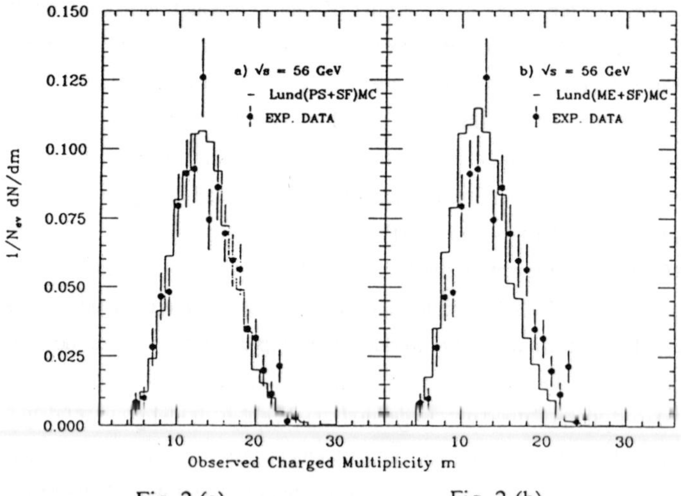

Fig. 2 (a)

Fig. 2 (b)

The total χ^2 of the fits following the LUND Parton-shower approach is the lowest among the three models examined.

Examining various event shape variables by comparing to Monte Carlo data is also a very effective way of searching for heavy flavour quark production. In Figure 3, some of the examples are shown. Figure 3 (a) is the AMY thrust distribution at $\sqrt{s}$ = 61.4 GeV, Figure 3 (b) shows the VENUS sphericity distribution at $\sqrt{s}$ = 61.4 GeV, and Figure 3 (c) shows the TOPAZ aplanarity distribution at $\sqrt{s}$ =52 and 60 GeV respectively. Solid lines are predictions from 5 flavour LUND 6.3 (PS) Monte Carlo. Dashed and dotted lines are for the case of when the heavy quarks were produced. Data shown are consistent with 5 flavour productions.

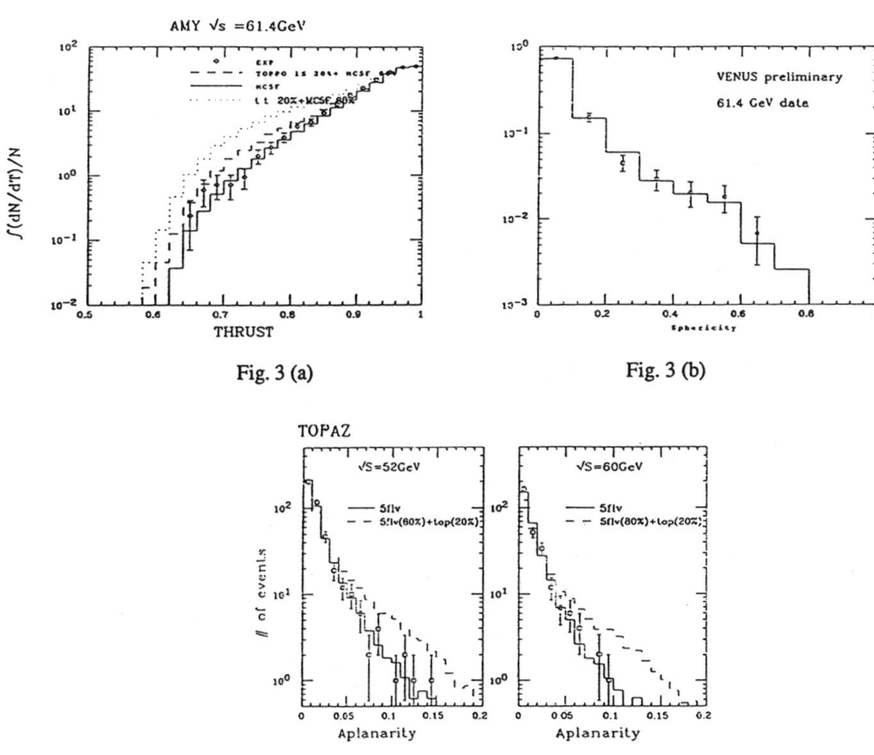

Fig. 3 (a) Fig. 3 (b)

Fig. 3 (c)

2. Comparison Between Quark and Gluon Jets

In the theory of QCD, the strong force is mediated by the exchange of massless vector gluons between quarks. The first clear evidence of the existence of gluons was observed at PETRA in 1979. Since QCD predicts gluons to have a larger colour factor than quarks, they fragment into softer particles than quarks, hence we expect softer structure in gluon jets that that in quark jets.

In PETRA and PEP experiments, there has been a considerable amount of experimental effort to see the differences between quark and gluon induced jets in the past years, however unambiguous difference between them has not yet been established. For example; HRS (PEP) looked at multiplicities of two types of jets and found no significant difference.[8] Some other experiments looked at another quantity, transverse momentum, and JADE (PETRA) group observed some positive difference,[9] TPC (PEP) observed less clear difference,[10] and CELLO (PETRA) saw no significant difference.[11] Mark II and TASSO groups used symmetric three jet events and two jet events at $\sqrt{s} = 2/3$ of that of three jet data to compare the jets at the similar energy range. The Mark II group observed some indication of the difference,[12] however TASSO group did not see any significant difference[13] in the same observable.

At TRISTAN, being at higher energy than PEP and PETRA, we expect the original parton nature of the jets to be more significant. The AMY group has studied the difference of quark and gluon jets by analysing three jet events[14]. Jets are formed by means of the jet-clustering algorithm introduced by the JADE group[15]. Since in three jet events, gluons are produced by Bremsstrahlung from one of the pair produced quarks, gluon jets tends to be the lowest energy jet amongst the three jets. Monte Carlo study shows that approximately two-third of the jets in the jet-3 sample (jet-1, jet-2, and jet-3 are defined in the order of the energy of jets. Jet-3 is the lowest energy jet) are gluon jets at TRISTAN energy range. The results are shown in Figure 4. The core energy fraction, ξ, is defined as the fraction of the visible jet energy , E_{vis}^{jet}, that is contained in a cone of half angle $60°/(E_{cal}^{jet})^{1/2}$, (where E_{cal}^{jet} in GeV) that is coaxial with the jet direction. The solid lines are the predictions from independent fragmentation models where quarks and gluons fragment the same way. They indicate that the data analysis is not biasing to give artificial differences between jet-3 and jet-1 + jet-2 samples. Since QCD predicts that gluons fragments softer, we expect ξ (the core energy fraction) of quark jets to be higher than that of the gluon jets.

Also the rapidity distribution of the leading particle of the quark jets is expected to be higher than the gluon jets. The dashed curves are predictions from the Parton-Shower Monte Carlo. The data clearly demonstrates the difference between the quark and gluon enriched jet samples.

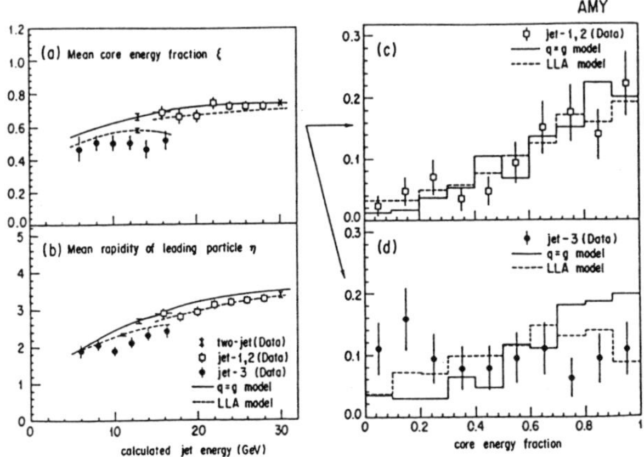

Fig. 4

3. Experimental Evidence for the Non-Abelian Nature of QCD

Non-Abelian nature of QCD was studied at TRISTAN. A comparison of the three-jet event fraction at TRISTAN to the fraction at PETRA shows that the QCD coupling strength α_s decreases with increasing Q^2. In addition, measurements of the angular distributions of four jet events show evidence for the triple-gluon vertex. Details of these analysis results were discussed in the plenary TRISTAN talk by Nagashima in this DPF90 conference, hence I refer the details to his proceeding and also the reference 16.

The evidence of the running nature of α_s was shown by S. Bethke (OPAL, LEP) by looking at R_3 in this conference, to extend the E_{cm} to the mass of Z^0. The data is consistent with the QCD prediction (running α_s), and not consistent with constant α_s. Mark II (PEP and SLC data), recently analysed data to determine α_s.[17] They used the differential jet multiplicity distribution. This method has an advantage of being free from possible bias caused by having the same event enter in different data points in R_3 analysis. The differential jet multiplicity distribution is compared with the second order the QCD prediction and α_s is determined to be $0.123 \pm 0.009 \pm 0.005$ at $\sqrt{s}$ around 91 GeV and $0.149 \pm 0.002 \pm 0.007$ at $\sqrt{s} = 29$ GeV. The running of α_s between these two center of mass energies is consistent with the QCD prediction in this method as well.

632

4. R VALUE in e+ e- COLLISIONS BELOW Z^0

R value in $e^+ e^-$ collisions is defined as;

$$R \equiv \frac{\sigma_{(ee \to \text{multi-hadrons})}}{\sigma_{(ee \to \mu\mu)}\text{lowest order point like QED calculation}}$$

Experimentally, R_{had} is measured by following;

$$R_{experimental} = \frac{N_{had} - N_{bkg}}{\varepsilon \, (1+\delta) \int Ldt \, \sigma_{(ee \to \mu\mu)}}$$

where

N_{had}	= number of hadronic events observed,
N_{bkg}	= number of estimated background events in N_{had},
ε	= efficiency of detecting N_{had},
$(1+\delta)$	= radiative correction,
$\int Ldt$	= luminosity (measured experimentally by small angle bhabha events)
$\sigma_{(ee \to \mu\mu)}$	= the lowest order QED cross section of ee -> $\mu\mu$.

The standard Model prediction is:
in the lowest order quark parton model,

$$R_{\text{ lowest order}} = 3\Sigma \, e_q^2 \quad \Rightarrow \quad \frac{11}{3} \qquad \text{for 5 flavours.}$$

In addition to the lowest order quark parton calculation, we need to correct R by higher order QCD effects due to gluon radiation ($\approx$ 5% at TRISTAN energy range). At this energy range, contribution from annihilation via Z^0 becomes comparable to annihilation via virtual gamma. The contribution from annihilation through Z^0 raises R by about 30% at TRISTAN energy range. The standard model prediction on R is very sensitive to the mass of Z^0 around TRISTAN energy region, however, the mass of Z^0 is so well measured now by 4 LEP experiments[18], Mark II[19] at SLC, and CDF[20] at FNAL in the past half year, that the measured error bar on R is much larger than the uncertainty in the prediction caused by the error on Z^0 mass. Therefore, we do not use the mass Z^0 as a free parameter to fit the data any more. The 3 TRISTAN experiments separately, and also combined plus other e+ e- R results are shown in Figure 5 (a) and (b)[21]. Though the statistical significance is less than 2 sigma and small, the data points are almost consistently higher than standard model predictions. That effect is true not only for TRISTAN data but also in PEP and PETRA R measurements as well. The systematic errors (mainly come from luminosity measurements) were studied carefully and are estimated to be 3.5%(AMY), 5.4%(TOPAZ), and 4.2%(VENUS). An interesting thing to point out is that while measured R_{had} tends to be high, $R_{\mu\mu}$ tends to be measured lower compared to the standard model using the same normalization constants. Figure 6 show the plot of $R_{\mu\mu}$. Because of the lack of statistical significance, we can not make any conclusive statement about the validity of the existing standard model which was used to compare R values. However it is curious to think what could cause this effect if it were real. It was pointed out by a few theorists that introducing a second Z boson, $(Z^{0\prime})$ with certain properties, the effect (high R_{had} and low $R_{\mu\mu}$) can be explained.[22]

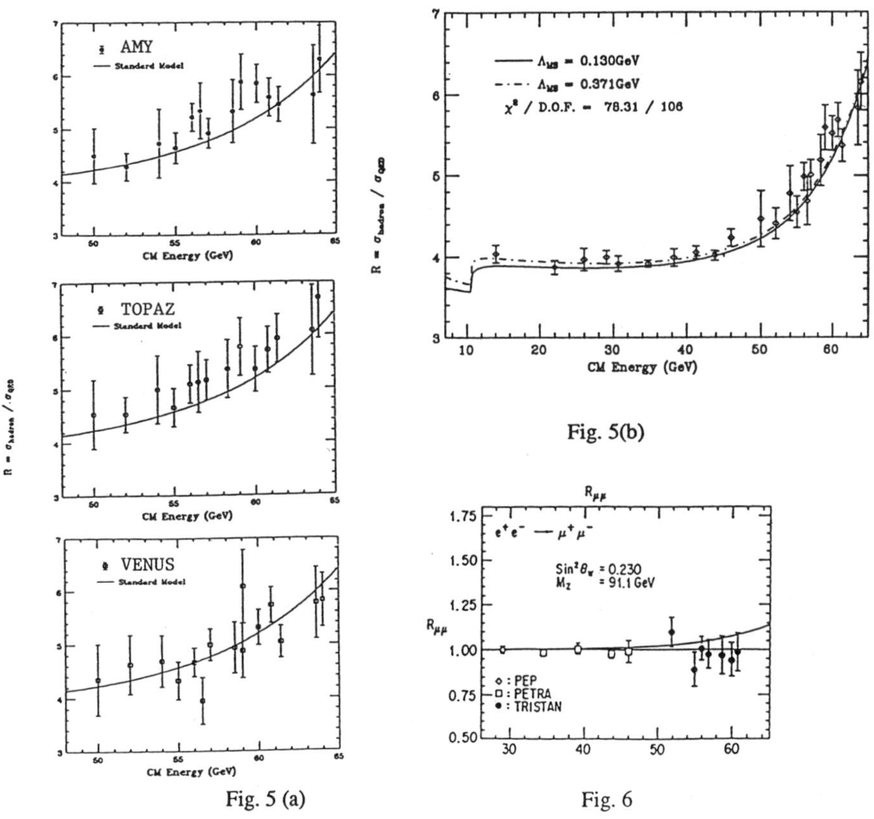

Fig. 5(b)

Fig. 5 (a)

Fig. 6

CONCLUSION

At TRISTAN accelerator in KEK, Japan, we have studied $e^+ e^-$ collisions at $\sqrt{s}$ between 50 and 64 GeV since 1987. We compared several different Monte Carlo models and data on multi hadronic events, and found that LUND 6.3 (with parton shower and string fragmentation model) describes our data very well, LUND 6.2 matrix element model had obvious difficulties to describe some of the quantities, and the Webber model also describes the data fairy well but not as good as LUND 6.3. We observed differences in quark and gluon jets quite clearly in the core energy fraction variable. We also observed the Non-Abelian nature of QCD -- (running coupling constant via 3-jet ratio (R_3), evidence of 3 gluon couplings). Our combined R_{had} measurement continued to be slightly high when we compare the experiment with the standard model with the newly established precise Z^0 mass, 91.1GeV.

REFERENCES

1. T. Sjostrand and M.Bengtsson, Comput. Phys. Commun. **43** (1987) 367.
2. M.Bengtsson and T. Sjostrand, Phys. Lett. **185B** (1987) 435.
3. T. Sjostrand, Comput. Phys. Commun. **39** (1986) 347.
4. G. Marchesini and B. Webber, Nuc. Phys **B238** (1984) 1; B. Webber, Nuc. Phys. B238 (1984) 492.
5. G. Marchesini and B. Webber, Nuc. Phys **B310** (1988) 461
6. Y.K.Li et al.(AMY), KEK preprint 89-149
7.H.W. Zheng et al.(AMY), KEK preprint 90 - 5. Submitted to Phys. Rev. D
8. M.Derrick et al. (HRS), Phys. Lett. **B165** (1985) 449
9. W. Bertel et al. (JADE) Phys. Lett. **123B** (1983) 460
10. R.J. Madaras et al.(TPC) Proceeding, Rencontre de Moriond, 1986
11. CELLO collaboration. Contribution paper. International Lepton Photon Conf. 1987
12. A. Peterson et al. (MARKII), phys. Rev. Lett. **55** (1985) 1954
13. W. Braunschweig et al. (TASSO), WIS-89/7/89-PH (unpublished)
14. Y.K.Kim et al. (AMY) Phys. Rev. Lett. **63** (1989) 1772
15. W. Bartel et al. (JADE) Z. Phys. C **33** (1986) 23
16. I.H. Park et al. (AMY) Phys. Rev. Lett. **62** (1989) 1713
17. S.Komamiya et al. (MarkII) SLAC-PUB_5137, LBL-27999, November 1989
18. ALEPH collaboration Phys. Lett. **B231**, number 4 (1989) 519
    DELPHI collaboration Phys. Lett. **B231**, number 4 (1989) 539
    LEP3 collaboration Phys. Lett. **B231**, number 4 (1989) 509
    OPAL collaboration Phys. Lett. **B231**, number 4 (1989) 530
19. MARKII collaboration Phys. Rev. Lett. **63**, number 7 (1989) 724
20. CDF collaboration Phys. Rev. Lett. **63**, number 7 (1989) 720
21. T. Kumita et al.(AMY) KEK preprint 89 - 2100. Submitted to Phys. Rev. D
22. K. Hagiwara et al. KEK preprint 89-57

PARTIAL WAVE ANALYSES OF $J/\psi \to \gamma K\overline{K}\pi$ AND $J/\psi \to \gamma\eta\pi^+\pi^-$

J.J. DRINKARD

Institute for Particle Physics, University of California

Santa Cruz, California 95064

ABSTRACT

The preliminary results of isobar-model partial wave analyses of the reactions $J/\psi \to \gamma K\overline{K}\pi$ and $J/\psi \to \gamma\eta\pi^+\pi^-$ are presented. The $K\overline{K}\pi$ and $\eta\pi^+\pi^-$ systems are studied in the respective 3-body mass intervals 1.35-1.6 GeV/c^2 and 1.2-1.5 GeV/c^2, using data corresponding to 5.8 million produced J/ψ events collected with the Mark III detector at SPEAR. This study addresses the presently unresolved status of the meson spectrum near 1.4 GeV/c^2, a region of considerable interest in the search for non-$q\overline{q}$ states. Results of the $\eta\pi^+\pi^-$ analysis show evidence for $f_1(1285)$ and $\eta(1400)$ production, where both states decay through $a_0(980)\pi$. The $K\overline{K}\pi$ system is best described by a mixture of $J^{PC} = 0^{-+}$ and $J^{PC} = 1^{++}$ amplitudes, corresponding to $a_0(980)\pi$ (S-wave), $K^*\overline{K} + c.c.$ (P-wave) and $K^*\overline{K} + c.c.$ (S-wave) decay modes. The mass dependence of these partial waves is consistent with a multi-resonance interpretation for the $\eta(1430)$.

1. INTRODUCTION

The historic reputation of $\eta(1430)$ [formerly $\iota(1440)$] as a gluonium candidate is built largely on early observations[1] of $J/\psi \to \gamma\eta(1430)$, $\eta(1430) \to K\overline{K}\pi$ at a rate considered to be the highest of all (gluon-mediated) J/ψ radiative decays. A clear understanding of this state's production properties, J^{PC}, and decay modes has yet to be achieved, however.[2]

An early analysis[3] of the process $J/\psi \to \gamma K^+K^-\pi^0$ determined that $\eta(1430)$ is a state which decays primarily through the $J^{PC} = 0^{-+}$ $a_0\pi$ channel, but no such state was observed in radiative J/ψ decay into $\eta\pi\pi$ (a contradiction, since the a_0 has a large $\eta\pi$ decay mode). More recent analyses[4] of $J/\psi \to \gamma K\overline{K}\pi$ show an $\eta(1430)$ signal distribution which is markedly asymmetric and poorly described as an S-wave Breit-Wigner resonance. While the asymmetry is plausibly the result of a large observed $J^{PC} = 0^{-+}$ K^*K mode,[4] the possibility of multiple states cannot be ruled out, and is here addressed for the first time in a full amplitude analysis of the Mark III data. Specifically, we present below the preliminary results of an isobar-model partial wave analysis (PWA) of the reaction $J/\psi \to \gamma K\overline{K}\pi$ in both the $K_S^0 K^{\pm}\pi^{\mp}$ and $K^+K^-\pi^0$ modes.

We also present preliminary results of a PWA of the complementary process, $J/\psi \to \gamma\eta\pi^+\pi^-$.

2. PWA OF $J/\psi \to \gamma K \overline{K} \pi$

Details regarding the Mark III detector and the selection of the $K_S^0 K^\pm \pi^\mp$ and $K^+ K^- \pi^0$ channels are not provided here; they may be found in reference 6. The PWA, which is performed separately for each channel, is described as follows. A model is formulated using a set of interfering amplitudes, the relative magnitudes and phases of which are estimated in a maximum (log) likelihood fit. An invariant tensor formalism[6] is used to construct amplitudes for the process $J/\psi \to \gamma X$, where the system X is in a well defined J^{PC} state and decays through a specified two-body channel. All $0^{-+}, 1^{++}$ and 1^{-+} amplitudes accessible to the $a_0\pi$ and $K^* K$[7] channels are considered in the initial analysis. A relativistic P-wave propagator[8] (with m_{K^*} and Γ_{K^*} taken from reference 2) is used to parametrize the K^* mass dependence, and a coupled-channel parametrization[9] is used for the $a_0(980)$. Finally, an isotropic "background" term, corresponding to three-body phase space, is added incoherently to the likelihood.

Results: $\gamma K_S^0 K^\pm \pi^\mp$

Independent fits are performed in ten 25 MeV/c^2 $K_S^0 K^\pm \pi^\mp$ mass bins spanning the interval 1.35-1.6 GeV/c^2. Various combinations of amplitudes are fitted separately, and log likelihoods are compared to determine the dominant amplitudes. The $1^{-+} K^* K$ and $1^{++} K^* K$ D-wave amplitudes are found not to contribute significantly to the likelihood, and are thus excluded at this point. Due to limited charged track acceptance and statistics, our technique cannot distinguish unambiguously between a 0^{-+} $a_0\pi$ amplitude and a 1^{++} $a_0\pi$ amplitude with polarization such that a uniform $cos\theta_{a_0}$[10] distribution is produced (we hereafter refer to this circumstance as "flat-polarization"). Consequently, we cannot rule out the possibility of a flat-polarized 1^{++} $a_0\pi$ component in the data which is attributed to the 0^{-+} $a_0\pi$ intensity, and we fit only to the $1^{++} K^* K$ (S-wave), $0^{-+} K^* K$ (P-wave) and 0^{-+} $a_0\pi$ (S-wave) amplitudes. The resulting intensities, normalized to the number of produced $J/\psi \to \gamma K_S^0 K^\pm \pi^\mp$ events, appear in Fig. 2.

Fit stability is tested by repeating the PWA many times, each time using a series of variations in selection cuts, a_0 parametrization,[11] parameter initialization values, fit normalization, and mass binning. Relative $0^{-+}/1^{++}$ intensities vary at most by 20% under this testing program. The sensitivity of the results to the level of incoherent phase space allowed in the fits is tested as follows. The PWA is repeated many times under various conditions, which include (i) fixing the phase space term to various fractions, or (ii) adding extra amplitudes parametrized by varying mixtures of $0^{-+}, 1^{++}, 2^{++}, 2^{-+}$ terms involving (as applicable) the decays $K^* K, a_0\pi$ and $a_2\pi$. In each test, the dominant features of the fit remain essentially unchanged; the only significant difference is an arbitrary reduction of the incoherent phase space fraction.

Fig. 3 shows a comparison in two-body invariant mass projections for the three mass regions commensurate with intensity peaks in the $a_0\pi$ S-wave intensity (1.4-1.425 GeV/c^2), the K^*K S-wave intensity (1.425-1.45 GeV/c^2) and the K^*K P-wave intensity (1.475-1.5 GeV/c^2), and indicates that reasonable fits have been achieved. Similarly good agreement is observed in the angular production variables.

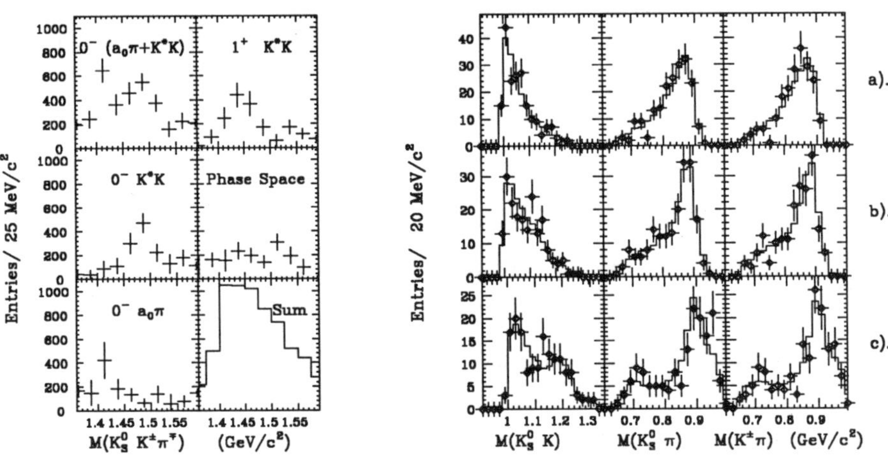

Fig. 2 (left): Acceptance-corrected PWA results for $1.35 < M_{K_S^0 K^\pm \pi^\mp} < 1.60$ GeV/c^2. **Fig. 3** (right): Distributions in two-body invariant masses for the data (points) and Monte Carlo (histograms) for the $K_S^0 K^\pm \pi^\mp$ mass bins (a) 1.4-1.425 GeV/c^2, (b) 1.425-1.45 GeV/c^2, and (c) 1.475-1.5 GeV/c^2.

Results: $\gamma K^+ K^- \pi^0$

An analysis of the $K^+ K^- \pi^0$ channel, using the same procedure as described above, yields results which are consistent with those from the $K_S^0 K^\pm \pi^\mp$ channel; the K^*K P-wave, K^*K S-wave and $a_0\pi$ S-wave are the dominant amplitudes, and the measured moduli and phases generally fall within one standard deviation of agreement with those measured in the $K_S^0 K^\pm \pi^\mp$ analysis.

3. PWA RESULTS: $\gamma\eta\pi^+\pi^-$

The PWA of this channel was performed in an independent analysis,[13] in which all 0^{-+} and 1^{++} amplitudes accessible to the $a_0\pi$ and $f_0(1400)\eta$ channels were used in the initial fits, and two modes (corresponding to the $\pi^+\pi^-\pi^0$ and $\gamma\gamma$ decay modes of the η) were fitted separately, in 20 MeV/c^2 $\eta\pi^+\pi^-$ mass bins. Preliminary results for the $\eta \to \gamma\gamma$ mode, appearing as intesitities normalized to the number of observed events, are shown in Fig. 4 (acceptance does not vary appreciably with mass). The fit to the $\eta \to \pi^+\pi^-\pi^0$ channel was performed with only the $a_0\pi$ amplitudes (it was not possible to obtain stable fits with more terms), and yielded results which were in very good agreement with the results from the $\eta \to \gamma\gamma$ mode.

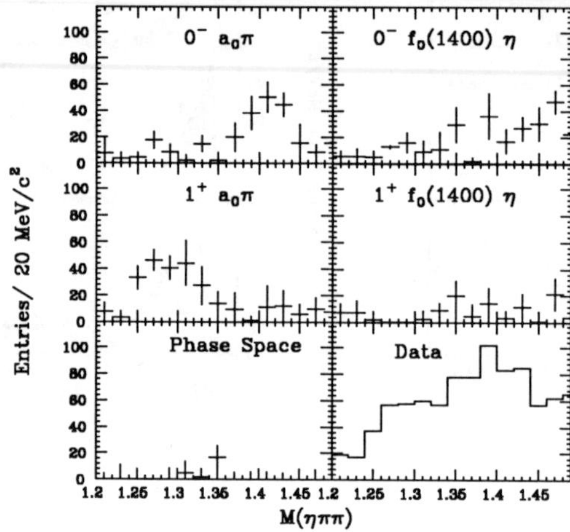

Fig. 4 : PWA results for $1.2 < M_{\eta\pi^+\pi^-} < 1.5$ GeV/c^2.

4. DISCUSSION

As there are no slowly varying amplitudes available for referencing in the above analyses, we can neither prove nor disprove the possibility that any of the peaks observed in Figures 2 and 4 is resonant. Still, the very presence of peaks in the spectra is highly suggestive. The intensity peaks in Fig. 2, for example, show that three states may contribute to the $\eta(1430)$ signal; a 0^{-+} $a_0\pi$ state near 1.410 GeV/c^2, a 1^{++} K^*K state near 1.44 GeV/c^2, and a 0^{-+} K^*K state near 1.48 GeV/c^2. In the $\eta\pi^+\pi^-$ analysis, the 1^{++} intensity peak (Fig. 4) is evidence for $f_1(1285)$ production, while the 0^{-+} $a_0\pi$ intensity peak near 1.4 GeV/c^2 appears to confirm recent reports[12] of a 0^{-+} $a_0\pi$ state, $\eta(1400)$, in this same mass region. The excess of 1^{++} events near 1.32 GeV/c^2 is presently not understood,[14] and is still under study. Further analysis of these results, including resonance parameter and branching ratio estimation, will appear in forthcoming publications.

Summarizing, we have shown preliminary results which demonstrate that the $\eta(1430)$ structure observed in $J/\psi \rightarrow \gamma K\overline{K}\pi$ contains a small, but significant, $a_0\pi$ component at low mass [which may be compatible with $\eta(1400)$ in the $\eta\pi^+\pi^-$ mode], and a dominant contribution from K^*K. The latter is split evenly between the 0^{-+} and 1^{++} channels. In a similar PWA of the process $J/\psi \rightarrow \gamma\eta\pi^+\pi^-$, we find evidence for two states: $f_1(1285)$ and $\eta(1400)$.

REFERENCES

1. D. Scharre *et al.*, Phys. Lett. **97B**, 329 (1980).

2. See pg. 294 in M. Aguilar-Benitez *et al.* (Particle Data Group) Phys. Lett. **B204**, 1 (1988).

3. C. Edwards *et al.*, Phys. Rev. Lett. **49**, 259 (1982).

4. W.J. Wisniewski, Proceedings of the 2nd Int. Conf. on Hadron Spectroscopy, Tsukuba, Japan, Apr 16-18, 1987; T.H. Burnett, Proceedings of the BNL Workshop on Glueballs, Hybrids and Exotic Hadrons, Upton, N.Y., Aug 29 - Sep 1, 1988.

5. All references to K^*K imply $K^*\overline{K} + \overline{K}^*K$.

6. J.Drinkard Ph.D. Thesis, UC Santa Cruz, SCIPP-90/04 (unpublished) Mar. 1990.

7. The K^*K amplitudes are constructed as G-parity eigenstates, with G=C=+1.

8. J.D. Jackson, Il Nuovo Cimento, Vol XXXIV No.6, Dec 16, 1964.

9. S.M. Flatté, Phys. Lett. **63B**, 224, 1976.

10. θ_{a_0} is defined in the $K\overline{K}\pi$ frame as the polar angle of the $K\overline{K}$ system with respect to the $K\overline{K}\pi$ production axis.

11. An alternative parametrization due to Tornqvist et. al. was used for this purpose. See: N. A. Tornqvist, Ann. Phys. **123**, 1 (1979); M. Roos and N. A. Tornqvist, Z. Phys. **C5**, 205 (1980); N. A.Tornqvist, Acta Phys. Pol. **B16**, 503 (1985).

12. A. Ando *et al.*, Phys. Rev. Lett. **57**, 1296 (1986). T.Tsuru, Proceedings of the 2nd Int. Conf. on Hadron Spectroscopy, Tsukuba, Japan, Apr 16-18, 1987.

13. C.A. Heusch and M.Burchell, Proceedings of the 3rd Int. Conf. on Hadron Spectroscopy (to be published) Ajaccio, France. Sept 23-27, 1989.

14. The events near 1.32 GeV/c^2 have no clear explanation; they are too high in mass to be associated with the relatively narrow $f_1(1285)$ ($\Gamma \sim 25$ MeV/c^2).

THE REACTION $\pi^- p \to \pi^0 \pi^0 n$ NEAR THRESHOLD AND CHIRAL SYMMETRY BREAKING

K.D. LARSON, B. BASSALLECK, J.R. HALL, D.M. WOLFE
University of New Mexico, Albuquerque, NM 87131

J. LOWE
University of New Mexico, Albuquerque, NM 87131 and
University of Birmingham, Birmingham B15 2TT, UK

M.D. HASINOFF, G. KOCH, A.J. NOBLE, M. SEVIOR, C.E. WALTHAM
University of British Columbia, Vancouver, BC V6T 2A3, Canada

J.P. MILLER, B.L. ROBERTS, T.M. WARNER
Boston University, Boston, MA 02215

M. SAKITT
Brookhaven National Laboratory, Upton, NY 11973

W.J. FICKINGER, D.K. ROBINSON
Case Western Reserve University, Cleveland, OH 44106

D. HORVATH
KFKI Budapest and Triumf, Vancouver, BC V6T 2A3, Canada

N.W. TANNER
Oxford University, Oxford OX1 3RH, UK

1. INTRODUCTION

Quantum Chromodynamics has proved to be very successful at explaining high-energy strong-interaction phenomena. However, QCD is very difficult to apply directly at lower energies. Consequently, approximations to QCD have been sought that would allow the calculation of low-energy strongly interacting processes. One such model is Chiral Perturbation Theory. Here, one builds a chirally symmetric theory of QCD in which the quark masses are zero and then uses perturbation theory to extrapolate to the case of broken chiral symmetry. Gasser and Leutwyler[1] have used Chiral Perturbation Theory to predict the values of the s-wave scattering lengths for the $\pi - \pi$ interaction.

The total cross sections for the reactions $\pi N \to \pi \pi N$ near threshold have long been known to be very sensitive to the $\pi - \pi$ scattering lengths. However, to extract the scattering lengths, measurements must be made close to threshold for all the different charge channels. In addition, other diagrams not involving the scattering lengths also contribute to the $\pi N \to \pi \pi N$ cross sections[2]. Measurements of all the charge channels will sort out these effects and enable the scattering lengths to be determined.

In all low-energy exeriments published or in progress[3], the only data available for the $\pi N \to \pi \pi N$ reactions were for channels involving at least one charged pion in the final state[3]. These data are insufficient to obtain unambiguous values for the scattering lengths or to study chiral symmetry breaking. We report here the first measurement close to threshold of the total cross section for $\pi^- p \to \pi^0 \pi^0 n$.

2. EXPERIMENT

The experiment was carried out at the C8 beam line of the Brookhaven AGS. A momentum-analysed π^- beam was incident on a liquid-hydrogen target, located at the center of the crystal box, a 396-element array of NaI detectors. This array provides approximately 2π coverage for detection of γ—rays from the target. Two stages of charge-veto detectors between the target and the NaI rejected events with charged particles in the final state. A scintillator hodoscope in the beam just upstream of the target measured the beam-particle position in the dispersion plane of the last bending magnet, providing additional momentum analysis of the beam.

Since it is difficult to calculate the absolute acceptance of the crystal box with adequate accuracy, two physics triggers were recorded during the measurements, to enable cross sections to be extracted:

(a) A trigger to detect γ-multiplicity-4 events, from $\pi^-p \to \pi^0\pi^0n$, and

(b) A γ-multiplicity-2 trigger to detect $\pi^-p \to \pi^0n$. The cross section for this reaction is well known, and is used as a basis for normalisation of the cross section for reaction (a).

3. RESULTS

Measurements were made at ten beam momenta between 250 and 400 MeV/c. Analysis of the tapes shows very clean signals for each of the above reactions, after

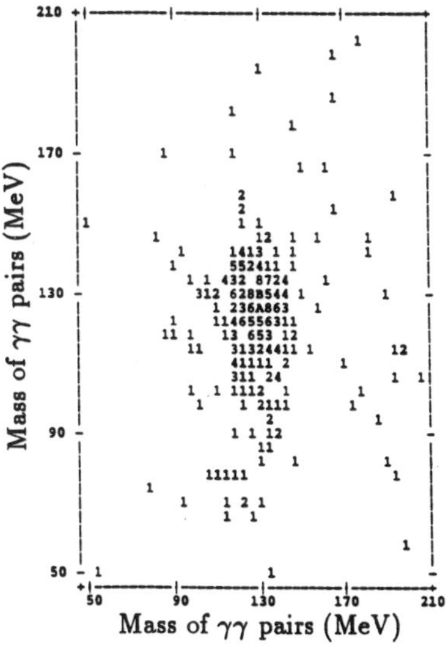

Fig. 1. Mass of gamma pairs from four-γ events at an incident π^- momentum of 290 MeV/c.

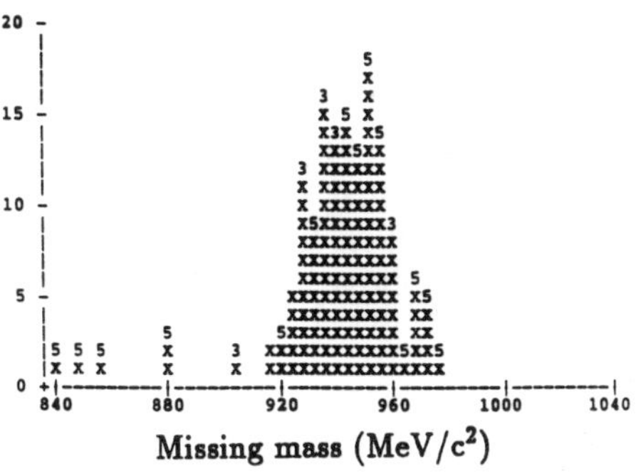

Fig. 2 Missing-mass spectrum for 4-γ events at a beam momentum of 290 MeV/c.

subtraction of the target-empty background. Fig. 1 shows a scatter plot of the invariant masses of γ pairs for multiplicity-4 events, at a beam momentum of 290 MeV/c. The $\pi^0\pi^0$ events can be seen clearly. The missing-mass spectrum for γ-multiplicity-4 events at this momentum is shown in fig. 2. Both multiplicity-2 and multiplicity-4 spectra showed clean peaks at the neutron mass. The cross sections for $\pi^- p \to \pi^0\pi^0 n$ were extracted from the ratios of the numbers of events in these two peaks. All acceptances were calculated using the simulation program GEANT.

A significant background arises from the channel $\pi^- p \to \pi^- \pi^0 p$ followed by the final-state π^- stopping in liquid hydrogen, giving $\pi^- p \to \pi^0 n$ at rest. The contribution from this was also simulated, and gives rise to a correction to the results of about 10%.

The cross sections are being fitted to an expression of the form

$$\sigma(\pi^- p \to \pi^0\pi^0 n) \;=\; 1/2 \times |a|^2 \times Q^2 \times (phasespace)$$

where Q is the center-of-mass incident pion momentum. The amplitude a is taken to be linear in the center-of-mass kinetic energy, a form which has been found to give an adequately good fit to data on other charge states[3]. Results from the comparison with other charge states will be presented.

In some other channels[3] an enhancement has been observed in the mass plot of the final-state $\pi - \pi$ system at a mass of about 400 MeV. In the present experiment, the incident momentum was not high enough to probe this mass region, though there is evidence that at the highest beam momenta measured, the $\pi-\pi$ mass distributions are distorted from a phase-space distribution by the presence of a high-mass resonance.

REFERENCES

1. J. Gasser and H. Leutwyler, Phys. Lett. **B125**, 325 (1982).

2. R. Arndt et al., Phys. Rev. **D20**, 651 (1979); E. Oset and M.J. Vincente-Vacus, Nucl. Phys. **A446**, 584 (1985).

3. Bjork et al., Phys. Rev. Letts. **44**, 62 (1980); G. Kernel et al., Phys. Letts. **B216**, 244 (1989) and Phys. Letts. **B225**, 198 (1989), R. Baran et al., PSI Progress Report (1989), M. Sevior et al, private communication.

Flavor and Higher Symmetries
in Hadrons Containing Both Heavy and Light Quarks

D.B. LICHTENBERG

Department of Physics, Indiana University

Bloomington, Indiana 47405

ABSTRACT

The symmetry of flavor-SU(3) and higher symmetries incorporating flavor-SU(3), including flavor-spin-SU(6) and the supersymmetry SU(6/21), are discussed for hadrons containing both heavy and light quarks. The symmetry properties are those of the light quarks alone, with the heavy quarks being treated as singlets whose principal role is to contribute to the mass and spin of the hadron. Mechanisms and qualitative features of the symmetry breaking are also pointed out.

1. SYMMETRY AND SUPERSYMMETRY

Many properties of different hadrons have been related to one another in terms of broken symmetries. We shall be concerned here with three of those symmetries: flavor-SU(3) symmetry,[1] the symmetry SU(6) of flavor and spin,[2] and the supersymmetry SU(6/21).[3] With the advent of the quark model and the theory of quantum chromodynamics (QCD), it has been possible to understand the reasons for the existence of these symmetries and also the reasons for their breaking.

The symmetry SU(3) arises from the fact that QCD is a theory of color interactions, and different flavors of quarks have the same color properties. The breaking of SU(3) arises from the fact that the different quark flavors u, d, and s, have different masses, giving rise both to kinematical and dynamical symmetry breaking. The breaking of SU(6) arises from the fact that, according to QCD, the quark interactions are spin dependent.

The supersymmetry SU(6/21) is a symmetry between mesons and baryons. It arises from the fact that any two quarks in a baryon are in a color $\bar{3}$ state, the same color state as that of an antiquark in a meson. Because the forces between two colored states depend mainly on their color configuration, the interaction of a quark with an antiquark in a meson is similar to the interaction of a quark with a pair of quarks, or diquark,[4] in a baryon. The symmetry is broken because the mass and spin of a diquark is different from the mass and spin of an antiquark, and also because a diquark is not a point particle. Calculations in lattice QCD confirm[5] the approximate (but not exact) equality of the quark-antiquark and quark-diquark potentials.

The idea of generalizing SU(3)-flavor to SU(4), SU(5), and SU(6) to include the heavy quarks c, b, and t is not useful, except to count states, because these higher flavor symmetries are badly broken by the large masses of these quarks. It follows that the flavor-spin and supersymmetries containing flavor-SU(n) ($n > 3$) are also not useful.

There is a better way to consider hadrons containing heavy quarks. We regard the heavy quarks as singlets of SU(3), SU(6), and SU(6/21). Then, in hadrons containing both heavy and light quarks, we look at the symmetry properties of the light quarks alone. The principal role of the heavy quarks is to contribute to the mass and spin of the hadron. This view is analogous to considering the symmetry properties of the electrons in an atom, while regarding the nucleus as adding mass and spin. This approximation improves with increasing mass of the heavy quarks. Hadrons containing heavy and light quarks have been considered by many authors from other points of view. Here we merely cite three related papers,[6] from which additional references may be traced.

2. HADRON MULTIPLETS

Because heavy quarks are SU(3)-flavor singlets, mesons containing a light quark and antiquark come in mixed SU(3) octets and singlets, whereas mesons containing a light quark and a heavy antiquark come in SU(3) triplets. Likewise, whereas baryons containing only light quarks come in SU(3) octets and decuplets, baryons containing one heavy quark come in sextets and antitriplets, and baryons containing two heavy quarks come in triplets. Hadrons containing only heavy quarks are SU(3) singlets.

Let us for simplicity consider only ground-state mesons and baryons and make the approximation that these states contain no orbital angular momentum. In flavor-spin SU(6), a meson containing a light quark and a heavy antiquark has the symmetry of the light quark, and so belongs to a **6** multiplet. A baryon containing one heavy quark has the symmetry of a symmetric (in space and spin) diquark[4] belonging to a **21** multiplet. A baryon containing two heavy quarks has the symmetry of a single light quark just like a meson, and again belongs to a **6** of SU(6), while a baryon containing no light quarks is a singlet.

We denote a light quark (u, d, s) by q, a light diquark (uu, ud, dd, us, ds, ss) by D, and a heavy quark (c, b, t) by Q. First we consider hadrons containing only one heavy quark. The meson multiplets $\bar{q}Q$ belong to SU(3)-flavor $\bar{\mathbf{3}}$ multiplets and flavor-spin-SU(6) $\bar{\mathbf{6}}$ multiplets. The baryon multiplets DQ belong to SU(3)-flavor **6** multiplets and $\bar{\mathbf{3}}$ multiplets, the former having spin 1 and the latter spin 0. Together, these states belong to **21** multiplets of SU(6). The states of $\bar{q}Q$ and DQ all belong to **27** supermultiplets of SU(6/21). There are three different supermultiplets (qQ, DQ), depending on whether Q is c, b, or t.

Turning to hadrons containing two heavy quarks, there are the baryons qQQ, which belong to an SU(3) triplet and to a **6** of SU(6). These baryons do not belong in a supermultiplet of SU(6/21) with any ordinary mesons, but rather in a $\mathbf{\overline{27}}$ multiplet with the manifestly exotic mesons $\bar{D}QQ$.

3. SYMMETRY BREAKING

It is not enough to point out the existence of the symmetries SU(3), SU(6), and SU(6/21) in hadrons containing both light and heavy quarks. It is also important to discuss how these symmetries are broken.

It is well-known that SU(3) is broken by the mass differences between the u, d, and s quarks. The largest of these mass differences is about 200 MeV in many models and leads to typical mass differences of about the same amount in hadrons containing only light quarks. On the other hand, in mesons containing one heavy quark, the symmetry breaking arising from the difference in light quark masses is only about 100 MeV, as can be seen in the tables of the Particle Data Group[8] and as has already been pointed out by Martin.[6] The experimental evidence is not yet in for baryons containing one heavy quark, but we can expect smaller SU(3) breaking in the baryon case also.

The supersymmetry SU(6/21) is not very good in hadrons containing only light quarks, because a baryon containing three quarks of a given flavor has a mass more than 3/2 as great as a meson containing only two.[7] However, in mesons and baryons containing one heavy quark, the supersymmetry breaking ought to be smaller, at least fractionally, as the main contribution to the mass comes from the heavy quark.

Because of the spin of the heavy quark, each meson $\bar{q}Q$ consists of two states, one with spin 0 and the other with spin 1. Because of the finite masses of the c and b quarks, the color-hyperfine interaction[9] breaks the degeneracy of these states. In the case of the c quark, the meson of spin 1 is about 140 to 145 MeV heavier than its spin-0 counterpart, while in the case of the b quark, the corresponding mass difference is about 50 to 55 MeV.[8]

In the case of the baryons DQ containing a spin 1 diquark, there exist spin doublets (3/2,1/2) arising from the spin of the heavy quark. There is not yet any experimental data concerning the magnitude of the breaking of the degeneracy between corresponding states. However, based on a comparison of mesons and baryons containing only light quarks and on qualitative considerations concerning the mass-dependence of the color-hyperfine splitting,[10] we can confidently predict that the breaking is substantially less than in the meson case. It follows that the splitting is less than the pion mass, and, as a consequence, that the spin-3/2 baryons belonging to the SU(3) sextet cannot decay strongly into their spin-1/2 counterparts.

The baryons containing a spin-0 diquark have spin 1/2. They are split from the baryons containing a spin-1 diquark by the color-hyperfine interaction of light quarks. As a consequence, these splittings are rather large. The splittings are made even larger by the fact that the heavy spectator quark shrinks the baryon wave function.[10] The result is that both the spin-3/2 and spin-1/2 baryons of the

SU(3) sextet should be able to decay strongly into their counterparts in the SU(3) antitriplet. There exists evidence for this, in that the Σ_c (uuc) has a mass which is 168 MeV greater than that of the Λ_c (udc), and decays strongly into the Λ_c by pion emission.[8] This is in contrast to the case of the ordinary Σ containing only light quarks, which is only 77 MeV heavier than the Λ, and cannot decay strongly. The $\Sigma_b - \Lambda_b$ splitting, which is not known experimentally, should be even greater than the $\Sigma_c - \Lambda_c$ splitting. The members of the antitriplets Dc and Db should all decay weakly.

One should not think that because the Σ_c-Λ_c splitting is much larger than the Σ-Λ splitting that SU(3) is more badly broken in baryons DQ than in baryons containing only light quarks. The Σ_c and Λ_c belong to *different* SU(3) multiplets (**6** and **$\bar{3}$** respectively), while the Σ and Λ belong to the *same* SU(3) multiplet (**8**).

The *ssc* and *ssb*, which belong to SU(3) sextets, have no flavor counterparts in SU(3) antitriplets. The spin-3/2 members should decay electromagnetically to the spin-1/2 members, which should decay weakly.

We now turn to hadrons containing two heavy quarks, and first consider the baryons qQQ. These are SU(3) triplets, and, as with the mesons $\bar{q}Q$, the SU(3) symmetry should be less badly broken than in SU(3) multiplets containing only light quarks. If QQ are cc or bb, then the ground state of the QQ pair has spin 1 in order to satisfy the Pauli principle. These heavy spin-1 diquarks can combine with the light quark to form a multiplet with spin 3/2 and another with spin 1/2. We estimate the splitting of these multiplets by the color-hyperfine interaction to be less than a pion mass so that none of these spin-3/2 multiplets can decay strongly.

In the case that QQ are cb, the heavy diquark may have either spin 1 or spin 0. The spin-1 case is similar to that in which the heavy diquark is cc or bb; the spin-0 case leads to a another baryon multiplet with spin 1/2. This second spin-1/2 multiplet is lighter than the first because of the color-hyperfine interaction. Again we estimate that the color-hyperfine energy is too small to allow these states to decay strongly.

It is tempting to predict the existence of a $\overline{2}1$ multiplet of manifestly exotic mesons $\bar{D}QQ$ as SU(6/21) partners of the heavy baryons qQQ. What makes this prediction unreliable is that, whereas the color configuration of a baryon containing three quarks is unique, the color configuration of an exotic meson containing two quarks and two antiquarks is not. This leaves open the possibility for an exotic meson $\bar{D}QQ$ to have in its color wave function the two-meson state $\bar{q}Q$-$\bar{q}Q$. To the extent that this is so, the $\bar{D}QQ$ state can simply "fall apart" into two mesons. Therefore, it cannot be stated with any confidence that there exist quasi-stable $\bar{D}QQ$ exotic mesons.

In conclusion, we have discussed the multiplet structure of hadrons containing both light and heavy quarks with respect to the groups flavor-SU(3), flavor-spin-SU(6), and the supergroup SU(6/21). We have also considered mechanisms for breaking these symmetries and have given qualitative estimates of the magnitude of the breaking in several cases. These estimates lead to predictions as to whether strong decays of certain states are energetically allowed or not.

ACKNOWLEDGEMENTS

We should like to thank Alan Kostelecký, Steve Gottlieb, and Enrico Predazzi for helpful discussions. This work was supported in part by the U.S. Department of Energy.

REFERENCES

1. M. Gell-Mann and Y. Ne'eman, *The Eightfold Way* (W. A. Benjamin, New York, 1964).
2. F. Gürsey and L. Radicati, Phys. Rev. Lett. 13, 173 (1964).
3. H. Miyazawa, Prog. Theor. Phys. 36, 1266 (1966); Phys. Rev. 170, 1586 (1968); S. Catto and F. Gürsey, Nuovo Cimento 86, 201 (1985); 99, 685 (1988).
4. D. B. Lichtenberg, Proc. Workshop on Diquarks, eds. M. Anselmino and E. Predazzi (World Scientific, Singapore, 1989), p. 1, and references contained therein.
5. H. B. Thacker, E. Eichten,and J. C. Sexton, Nucl. Phys. B (Proc. Suppl.) 4, 234 (1988).
6. J. D. Bjorken, Hadrons Spectroscopy—1985, Ed. by S. Oneda (American Institute of Physics, New York, 1985), p. 390; A. Martin and J. M. Richard, Phys. Lett. B 185, 426 (1987); A. Martin, Proc. Workshop on Heavy Flavors: Status and Perspectives, Erice, June 1988 (to be published).
7. S. Nussinov, Phys. Rev. Lett. 51, 2081 (1983); J. P. Ader, J. M. Richard and P. Taxil, Phys. Rev. D 25, 2370 (1982); J. M. Richard, Phys. Lett. B 139, 408 (1984).
8. Particle Data Group: G.P. Yost et al., Phys. Lett. B 204, 1 (1988).
9. A. de Rújula, H. Georgi, and S. L. Glashow, Phys. Rev. D 12, 147 (1975).
10. D. B. Lichtenberg, Phys. Rev. D 35, 2183 (1987).

A Preliminary Measurement of the Polarization of Ω^- and Ξ^- Hyperons Produced by 800 GeV Protons

J. Duryea, G. Guglielmo, K. Heller, K. Johns[a], M. Shupe, K. Thorne
Dept. of Physics, University of Minnesota, Minneapolis, MN 55455

T. Diehl, S. Teige, G. Thomson, Y. Zou
Dept. of Physics, Rutgers–The State University, Piscataway, NJ 08854

P. M. Ho, M. J. Longo, A. Nguyen
Dept. of Physics, University of Michigan, Ann Arbor, MI 48109

C. James, K. B. Luk[b], R. Rameika
Fermilab, Batavia, IL 60510

A preliminary measurement of the polarization of the Ω^- and Ξ^- hyperons inclusively produced by 800 GeV protons has been made. A non-zero Ξ^- polarization has been measured while the Ω^-'s do not appear to be polarized. Also, there are indications that Ξ^- polarization is smaller in magnitude than Ξ^0 polarization.

1. Introduction

Since the discovery that Λ hyperons produced by protons were polarized at high energies, polarization measurements have also been made on most of the other stable baryons. A non-zero polarization has been measured for the $\Lambda, \Xi^0, \Xi^-, \Sigma^+, \Sigma^0$, and Σ^- hyperons while a polarization consistent with zero has been found for the $\overline{\Lambda}$ and the proton. [1] Experimental evidence shows that Λ polarization is approximately independent of energy from 12 GeV to 2000 GeV equivalent lab energy but depends strongly on x_F and p_T.[1]

Many models attempt to predict and explain this hyperon polarization. In most of these models it is necessary that the hyperon has a common quark with the incident proton; Ω^-'s then would not be expected to be polarized. Also since the quark content of Ξ^-'s and Ξ^0's is similar it might be expected that P_{Ξ^-} behaves like P_{Ξ^0}. A model by DeGrand and Miettinen predicts $P_{\Xi^-} = P_{\Xi^0}$.[2]

2. Apparatus

In experiment E756 at Fermilab Ω^-'s and Ξ^-'s were produced with a beam of 800 GeV protons which struck a 1/4 interaction length beryllium target at a production angle of 2.4 mrad. The particles passed through a curved channel imbedded in a large dipole magnet, which served to select a negatively charged beam. The hyperons were were then detected through the decays: $\Omega^- \to \Lambda + K^-$ and $\Xi^- \to \Lambda + \pi^-$ in a charged particle spectrometer. Because the topologies of these decays are almost identical, the Ξ^-'s were detected and analyzed the same way as the Ω^-'s.

Figure 1 shows the spectrometer which consisted of nine MWPC's and a set of four silicon strip detector planes in each view. An analyzing magnet served to measure the momentum of the daughter particles. The trigger required no hit in the veto counters (V1 and V2), hits in both S1 and S2, and a signal in the multiplicity (M) counter corresponding to at least two but less than five charged tracks. The downstream part of the trigger looked for the characteristic "V" of the Λ by triggering on the right half of C8 and the left half of C9. This spectrometer was also used to study a sample of Ξ^-'s and Ω^-'s polarized using a spin transfer technique with which a preliminary measurement of the magnetic moment of the Ω^- has been made. [3]

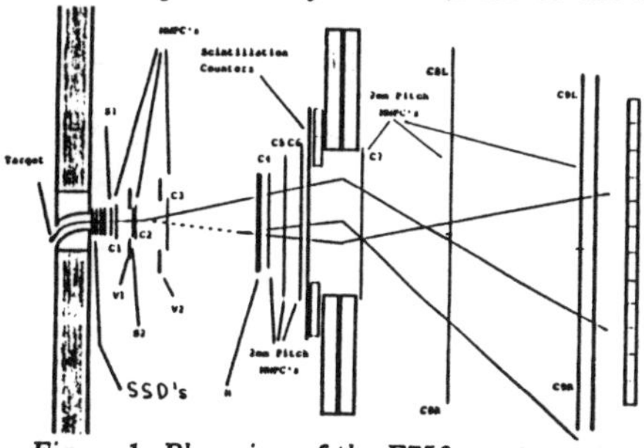

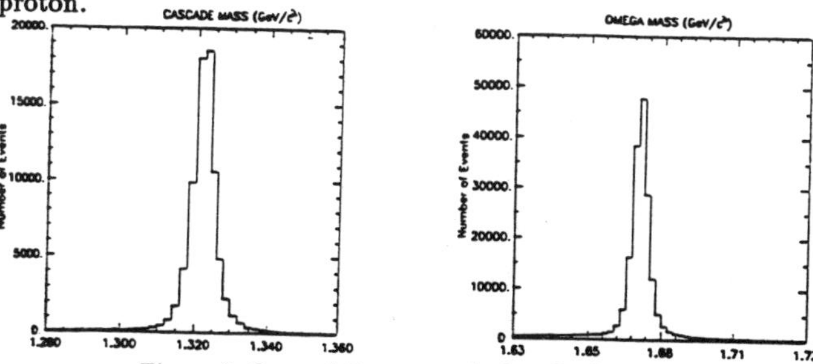

Figure 1. Plan view of the E756 spectrometer.

3. Analysis

The polarization of the Ξ^-'s or Ω^-'s is measured by first finding the polarization of the daughter Λ since P_Ω and P_Ξ can easily be related to P_Λ.[4,5] The polarization of the Λ is measured by examining the distribution of the daughter proton in the rest frame of the Λ.[5]

4. Results

A preliminary polarization analysis was done on a sample of 100,000 Ω^-'s and 400,000 Ξ^-'s; this represents 100% of the total Ω^- sample and about 5% of the total Ξ^- sample. Figure 2 shows two mass plots for the Ω^-'s and a fraction of the Ξ^-'s before final cuts. Figure 3 shows the Ω^- and Ξ^- polarization as a function of momentum. While there is clearly a non-zero polarization for the Ξ^-'s, P_Ω appears to be consistent with zero, as predicted by a model requiring a polarized hyperon to have a common valence quark with the incident proton.

Figure 2. Ω^- and Ξ^- masses before final cuts.

If the Ξ^- and Ξ^0 polarization is approximately energy independent like Λ polarization then P_{Ξ^-} can be compared with P_{Ξ^0} as long as the data points have the same p_T and x_F. All of the existing Ξ^0 polarization results are from data taken with a 400 GeV proton beam.[6] In order to compare to the E756 data it is necessary to find a sample of Ξ^0's taken at a 4.8 mrad production angle so that p_T and x_F are matched. Unfortunately no such data exist; however Ξ^0 data does exist at production angles of 7.2 mrad and 3.5 mrad.[6] and it is reasonable to expect that P_{Ξ^0} measured at 5.0 mrad would fall in between P_{Ξ^0} measured at these two angles. Figure 4 is a plot of these data and the 800 GeV Ξ^- polarization as a function of x_F.

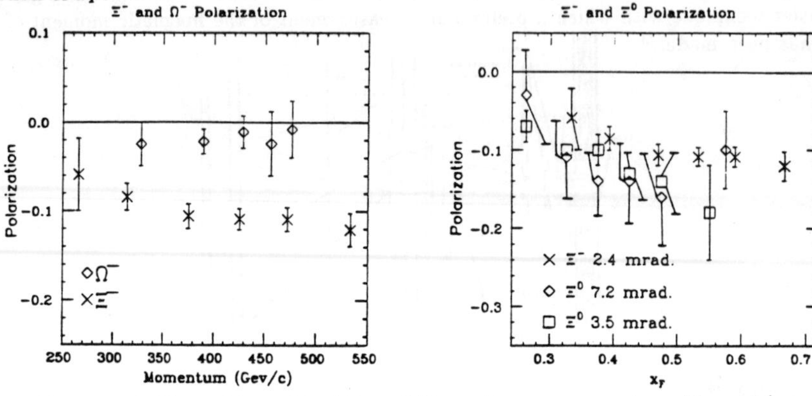

Figures 3-4. Figure 3 is a plot of P_Ω and P_Ξ vs. hyperon momentum. Figure 4 is a plot of P_Ξ^- and P_Ξ^0 versus x_F for Ξ^-'s at 800 GeV and 2.4 mrad and Ξ^0's at 400 GeV and production angles of 7.2 mrad and 3.5 mrad.

4. Discussion

It appears from Figure 4 that P_{Ξ^0} is systematically larger in magnitude than P_{Ξ^-} although the error bars on the Ξ^0 data do not allow for a definite determination. Any real difference between the polarization of Ξ^-'s and Ξ^0's would contradict the model by DeGrand and Miettinen.[2] It is clear that a more precise measurement of Ξ^0 polarization is necessary in order to check this.

[a]Current Address: Department of Physics, University of Arizona, Tucson AZ, 85721.

[b]Current Address: Department of Physics, University of California, Berkeley CA, 94720.

[1]K. Heller, Journal de Physique **46**, C2-121 (1985)

[2]T.A. DeGrand, and H. Miettinen Phys. Rev. D **24**, 2419 (1981)

[3]T. Diehl, Ph.D. thesis, Rutgers University, 1990.

[4]K. B. Luk et al., Phys. Rev. D **38**, 19 (1988)

[5]R. E. Marshak, Riazuddin, C. Ryan, *Theory of Weak Interactions in Particle Physics* (Wiley) p.513-517

[6]K. Heller et al., Phys. Rev. Lett. **51**, 2025 (1983)

A SUMMARY OF THE RESULTS FROM LASS AND THE FUTURE OF STRANGE QUARK SPECTROSCOPY[*]

D. ASTON,[1] N. AWAJI,[2] T. BIENZ,[1] F. BIRD,[1] J. D'AMORE,[3]
W. DUNWOODIE,[1] R. ENDORF,[3] K. FUJII,[2] H. HAYASHII,[2] S. IWATA,[2]
W. JOHNSON,[1] R. KAJIKAWA,[2] P. KUNZ,[1] Y. KWON,[1] D. LEITH,[1] L. LEVINSON,[1]
J. MARTINEZ,[3] T. MATSUI,[2] B. MEADOWS,[3] A. MIYAMOTO,[2] M. NUSSBAUM,[3]
H. OZAKI,[2] C. PAK,[2] B. RATCLIFF,[1] P. RENSING,[1] D. SCHULTZ,[1] S. SHAPIRO,[1]
T. SHIMOMURA,[2] P. SINERVO,[1] A. SUGIYAMA,[2] S. SUZUKI,[2] G. TARNOPOLSKY,[1]
T. TAUCHI,[2] N. TOGE,[1] K. UKAI,[4] A. WAITE,[1] S. WILLIAMS[1]

[1]*Stanford Linear Accelerator Center, Stanford University,*
[2]*Department of Physics, Nagoya University,*
[3]*University of Cincinnati,*
[4]*Institute for Nuclear Study, University of Tokyo*

ABSTRACT

A brief summary is presented of results pertinent to strange quark spectroscopy derived from high statistics data on K^-p interactions obtained with the LASS spectrometer at SLAC. The present status of strange meson spectroscopy is briefly reviewed, and the impact of the proposed KAON Factory on the future of the subject considered.

1. INTRODUCTION

This paper summarizes the results on strange quark spectroscopy obtained from an exposure of the Large Aperture Superconducting Solenoid (LASS) spectrometer at SLAC to a K^- beam of 11 GeV/c. The spectrometer and relevant experimental details are described elsewhere.[1,2] The raw data sample contains $\sim$ 113 million triggers, and the resulting useful beam flux corresponds to a sensitivity of 4.1 events/nb. The acceptance is approximately uniform over almost the full 4π solid angle.

[*] Work supported in part by the Department of Energy under contract No. DE-AC03-76SF00515; the National Science Foundation under grant Nos. PHY82-09144, PHY85-13808, and the Japan U.S. Cooperative Research Project on High Energy Physics.

Invited talk presented by W.Dunwoodie at DPF90, Houston, Texas, January 3-6, 1990

652

2. Ω^{*-} SPECTROSCOPY

Although the Ω^- was discovered in 1962, claims for the observation of Ω^{*-} resonances have been made only recently.[3,4,5] Preliminary evidence for an Ω^{*-} of mass ~ 2.26 GeV/c^2 was reported in the present experiment.[3] Subsequently, evidence for states with masses of 2251 and 2384 MeV/c^2 produced in Ξ^- interactions in beryllium was published.[4] More recently,[5] the full data sample from the present experiment has yielded clear evidence for the production of an Ω^{*-} resonance of mass 2253 ± 13 and width 85 ± 40 MeV/c^2, thus confirming the preliminary result.[3] The state is observed in the $\Xi^*(1530)\overline{K}$ decay mode, as shown in fig.1a, and the corresponding inclusive cross section is estimated to be 630 ± 180 nb. It is reasonable to consider this result as confirmation of the 2251 MeV/c^2 state of ref. 4, however there is no evidence for the production of the state of mass 2384 MeV/c^2 in the K^-p data.

The study of inclusive Ω^- production in the present experiment has resulted in the $\Omega^-\pi^+\pi^-$ mass distribution of fig.1b.[6] A clear peak is observed at ~ 2.47 GeV/c^2. Interpreting this as being due to the production of a new Ω^{*-}, the fit denoted by the solid curve yields mass and width estimates of 2474 ± 12 and 72 ± 33 MeV/c^2, respectively, and the signal has $\sim 5.5\sigma$ significance.

In summary, the existence of an Ω^{*-} of mass ~ 2250 MeV/c^2 decaying primarily to $\Xi^*(1530)\overline{K}$ seems confirmed; moreover, the mass and principal decay mode are quite consistent with theoretical expectations.[7,8] The status of the state at 2384 MeV/c^2 is uncertain, since it is not observed in the present experiment. Finally, the new Ω^{*-} at ~ 2470 MeV/c^2 needs confirmation from other experiments; theoretical predictions concerning possible states in this mass range would also be of interest.

3. K^* SPECTROSCOPY

A principal objective in studying meson spectroscopy is the precise definition of the level structure of the anticipated $q\bar{q}$ meson states. In this regard, the strange sector is particularly favored since it is free, not only from the need for isoscalar-isovector separation, but also from confusion resulting from the possible production of e.g. $K\overline{K}$ molecule and/or glueball states. The quark model level diagram is expected to take the form[9] illustrated qualitatively in fig.2 for the charmonium states.[10] Here S and L denote total quark spin and orbital angular momentum, respectively, and C indicates the charge conjugation parity of the resulting meson state. The para-charmonium levels are singlets, whereas the ortho-charmonium levels, other than ^{3}S, are triplets separated in mass due to spin-orbit interaction. Within each column, the lowest-lying state is the ground state, and the higher states correspond to radial excitations of this state. By convention, the leading orbital excitations are the ground states with largest total angular momentum, J, and, for the triplet levels, the remaining states

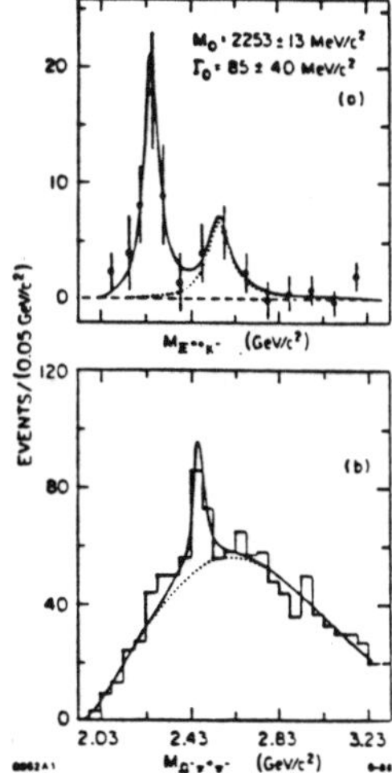

Fig.1.(a) The inclusive $\Xi^{*0}(1530)K^-$ mass distribution; the solid curve is the result of a fit using two Breit-Wigner line-shapes. (b) The inclusive $\Omega^-\pi^+\pi^-$ mass distribution; the solid curve is the result of a fit using a Breit-Wigner line-shape and a polynomial background.

Fig.2. The quark model level diagram for the charmonium states; the mass scale is only qualitative.

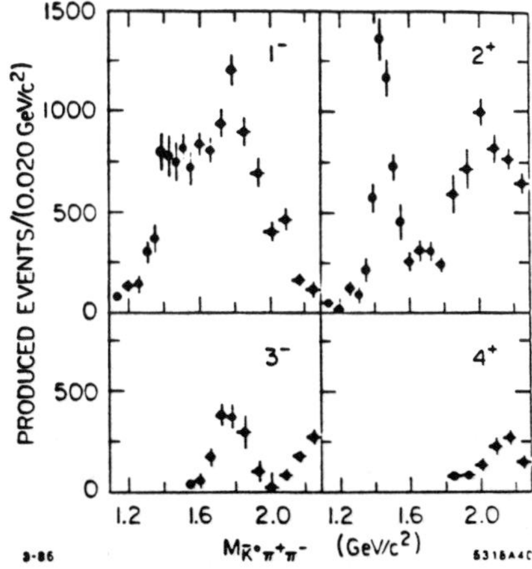

Fig.3. The natural spin-parity intensity distributions obtained from the amplitude analysis of reaction (1).

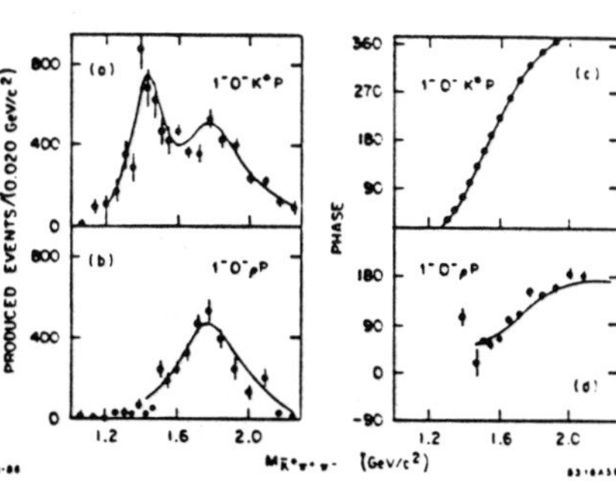

Fig.4. The individual isobar contributions to the 1^- distribution of fig.3; the curves are described in ref. 12.

are termed the underlying states. In $e^+ e^-$ collisions, the ^3S tower of states is readily accessible, while information on the leading orbital excitations with $J \geq 3$ is lacking, and that on the corresponding underlying states is sparse. Results of a complementary nature are obtained from hadronic interaction experiments, where production of the leading orbital states predominates, and precise information on the underlying and radially- excited states requires careful amplitude analyses of high statistics data.

To date the present experiment has provided such information primarily on the natural parity strange meson (i.e. K^*) states. The resulting light quark level structure is related to the nature of the long-range (i.e. confining) part of the $q\bar{q}$ interaction, and thus, in principle, provides information on non-perturbative QCD. In practice, it enables us to learn about QCD potential models incorporating relativistic corrections.[11]

The results relevant to K^* spectroscopy in the present experiment are obtained from amplitude analyses of the $\overline{K}^0 \pi^+ \pi^-$ system in the reaction

$$K^- p \to \overline{K}^0 \pi^+ \pi^- n \tag{1}$$

and of the $K^- \pi^+$, $\overline{K}^0 \pi^-$ and $K^- \eta$ systems produced in the reactions

$$K^- p \to K^- \pi^+ n \tag{2}$$

$$K^- p \to \overline{K}^0 \pi^- p \tag{3}$$

and

$$K^- p \to K^- \eta\, p \ ; \tag{4}$$

these analyses are discussed in detail in refs. 12-15 respectively, and so only the main features will be summarized here.

The analysis of reaction $(1)^{[12]}$ yields the intensity distributions of fig.3 for the natural spin-parity (J^P) states of the $\overline{K}^0 \pi^+ \pi^-$ system. Signals corresponding to the $\overline{K}^*_2(1430)$, the $\overline{K}^*_3(1780)$ and the $\overline{K}^*_4(2060)$ are observed in the 2^+, 3^- and 4^+ distributions, respectively. The 2^+ distribution exhibits a second peak at ~ 2 GeV/c^2, while the 1^- distribution has a shoulder at ~ 1.4 GeV/c^2 followed by a peak at ~ 1.8 GeV/c^2. The $\overline{K}^*(892)\pi$ and $\overline{K}\rho(770)$ contributions to the 1^- spectrum are shown in figs.4a,b; the curves result from a description in terms of two Breit-Wigner(BW) resonances. The lower mass state (M=1420±17, Γ=240±30 MeV/c^2) approximately decouples from the $K\rho$ channel, and its production characteristics indicate weak coupling to $K\pi$ also; it is most readily interpreted as the first radial excitation of the $\overline{K}^*(892)$. The second state (M=1735±30, Γ=423±48 MeV/c^2) may be the underlying member of the ^3D ground state; however its mass is greater and its

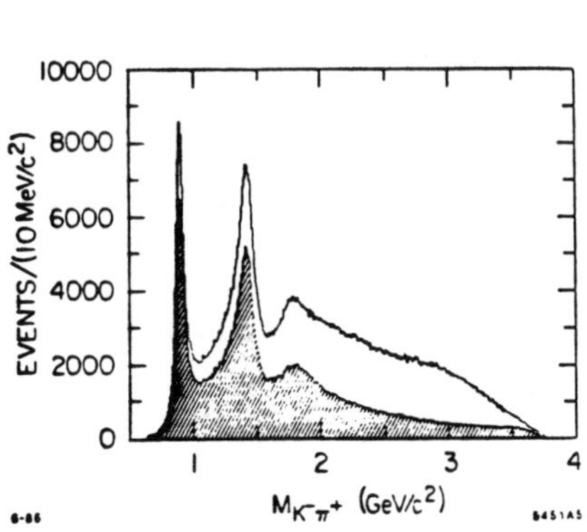

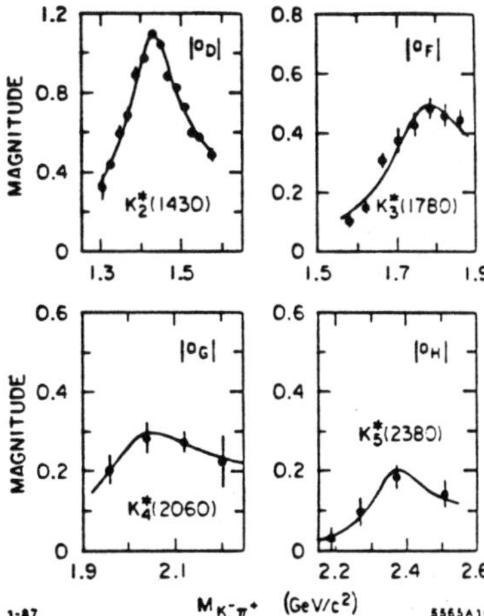

Fig.5. The raw $K^-\pi^+$ mass distribution for reaction (2); the shaded region corresponds to events with $M(\text{n }\pi^+) \geq 1.7$ GeV/c^2.

Fig.6. The mass dependence of the amplitudes of the leading orbital states obtained from reaction (2).

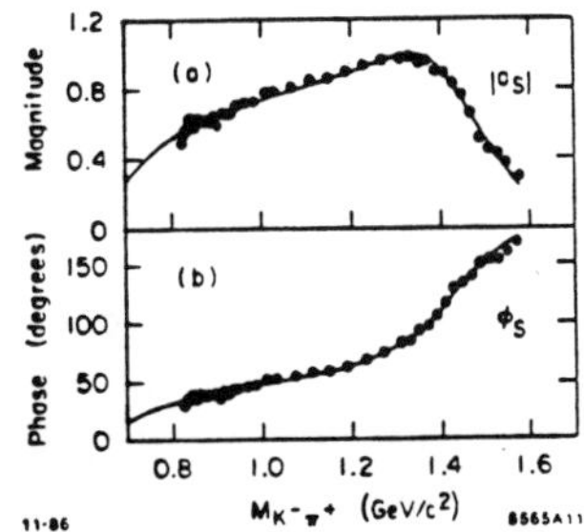

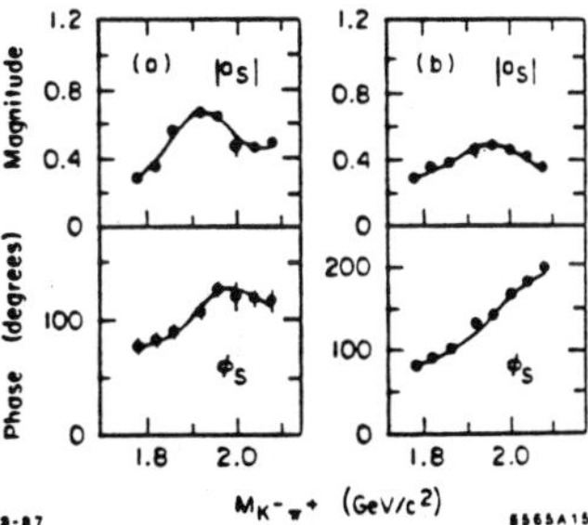

Fig.7. The behavior of the S wave $K\pi$ scattering amplitude up to 1.6 GeV/c^2 from reaction (2); the curves are described in ref.13.

Fig.8. The behavior of the S wave $K\pi$ scattering amplitude solutions above 1.8 GeV/c^2 from reaction (2); the curves are described in ref.13.

width significantly larger than that obtained from reaction $(2)^{[13]}$ so that it could be a mixture of this state and the second radial excitation of the $\overline{K^*}(892)$.

A similar fit to the 2^+ data yields mass and width estimates of 1973 ± 33 and 373 ± 93 MeV/c^2 respectively for the higher mass state, which corresponds most probably to the first radial excitation of the $\overline{K^*}(1430)$,$^{[12]}$ although it may also be a partner to the $\overline{K^*}_4(2060)$ in the F wave ground state triplet.

The raw $K^-\pi^+$ mass spectrum for reaction (2) is shown in fig.5; the shaded region corresponds to events with $M(n\ \pi^+) \geq 1.7$ GeV/c^2. In addition to an elastic BW amplitude describing the $\overline{K^*}(892)$ (cf.fig.9), the amplitude analysis$^{[13]}$ yields the resonant D-H wave signals of fig.6, thereby extending the observed leading orbital states to $J^P = 5^-$. The corresponding behavior of the underlying S wave amplitude up to 1.6 GeV/c^2 is shown in fig.7. The curves correspond to an effective range parametrization plus a BW amplitude ($M=1412 \pm 4 \pm 5$, $\Gamma=294 \pm 10 \pm 21$ MeV/c^2), and the resultant S wave is approximately elastic up to ~ 1.5 GeV/c^2. The state described by the BW is the $^3P_0$ partner of the $\overline{K^*}(1430)$. Above 1.8 GeV/c^2, there are two S wave solutions, as shown in fig.8. Both resonate in the 1.9-1.95 GeV/c^2 region, have width ~ 0.2 GeV/c^2 and elasticity ~ 0.5. In the quark model such a state can only be the first radial excitation of the $^3P_0$ ground state.

Finally, the behavior of the P wave $\overline{K}\pi$ scattering amplitude from reaction (2) is shown in fig.9 for the region up to 1.8 GeV/c^2. The solid curve provides a satisfactory description in terms of three BW resonances:$^{[13]}$ (i) the $\overline{K^*}(892)$; (ii) a resonance at ~ 1.4 GeV/c^2 of elasticity ~ 0.07 which is consistent with the radial excitation of the $\overline{K^*}(892)$ observed in reaction (1); and (iii) a resonance ($M=1677 \pm 10 \pm 32$, $\Gamma=205 \pm 16 \pm 34$ MeV/c^2) of elasticity ~ 0.39 which is readily interpreted as the $^3D_1$ ground state.

The production mechanism for reactions (1) and (2) is predominantly unnatural parity, helicity zero exchange. In contrast, the $\overline{K}^0\pi^-$ system in reaction $(3)^{[14]}$ is produced mainly by natural parity, helicity one exchange (cf. fig.10). Nevertheless, a description of the mass dependence of the resulting amplitudes and relative phases requires the same P, D and F wave resonance structure found for reactions (1) and (2). This is illustrated in fig.11, where the solid curves correspond to three P wave resonances, two D wave resonances, and one F wave resonance. Clearly, these structures are required in order to reproduce the relative phase motion, and the parameter values obtained from reaction (3) agree well with those obtained from reactions (1) and (2).

In general, theoretical predictions$^{[11]}$ are in quite good agreement with these measurements. However, the $^3P_0$ ground state is predicted ~ 170 MeV/c^2 low, and the $^3S_1$ first radial excitation is predicted ~ 160 MeV/c^2 high with respect to the measured mass value. Also, for the $^3S_1$ and $^3P_0$ levels the expected splitting between the

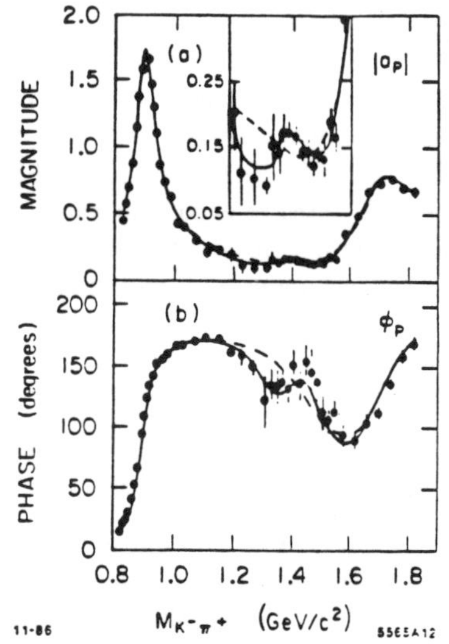

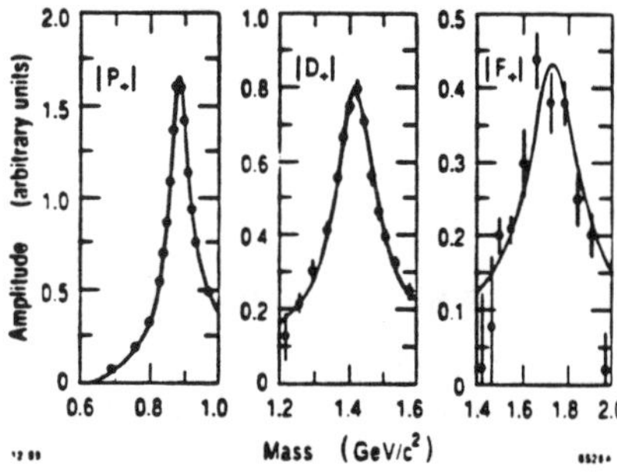

Fig.9. The behavior of the P wave $\overline{K}\pi$ scattering amplitude up to 1.8 GeV/c² from reaction (2); the curves are described in ref.13.

Fig.10. The mass dependence of the leading $\overline{K}^*$ amplitudes from reaction (3); the subscript indicates that production is via helicity one natural parity exchange in the t channel.

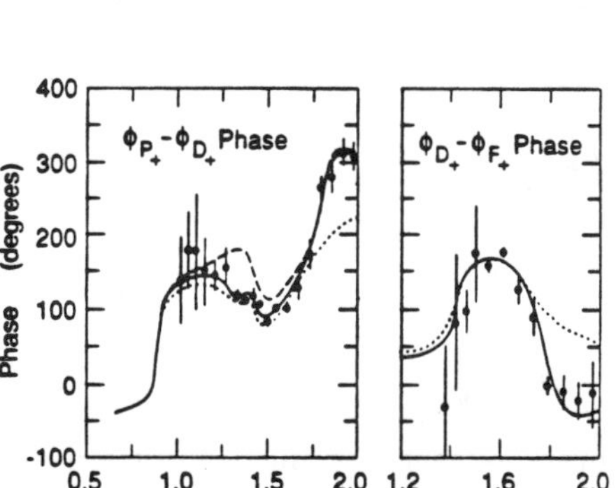

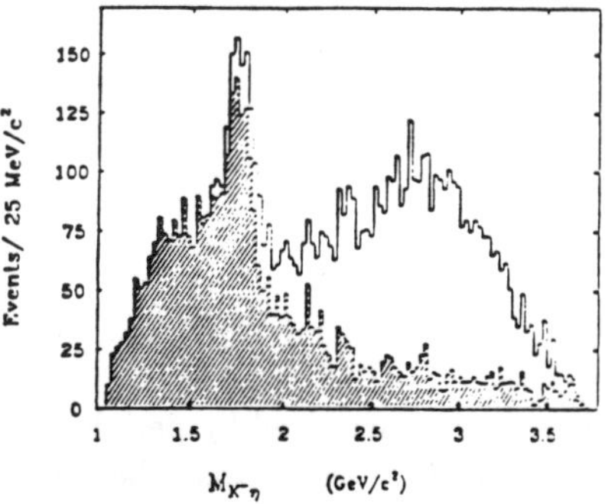

Fig.11. The mass dependence of the relative phase between the P-D and D-F amplitudes of fig.10; the solid curves are described in the text; the dotted curve is obtained if the second D wave resonance is excluded; the dashed curve is obtained if the small P wave resonance at ~ 1.4 GeV/c² is excluded.

Fig.12. The $K^-\eta$ mass distribution from reaction (4); the shaded region corresponds to $M(\eta p)\geq 2.0$ GeV/c² and $M(K^-p) \geq 1.85$ GeV/c².

ground state and first radial excitation is $\sim$650-680 MeV/c^2, whereas it is observed to be $\sim$500-550 MeV/c^2. It should be noted also that for the $^3S_1$ states this splitting is predicted to increase from the charmonium to strange meson sector; it appears actually to decrease by $\sim$70 MeV/c^2.

Reaction (4) has been measured with large statistics for the first time in the present experiment, and the resulting $K^-\eta$ mass spectrum is shown in fig.12. An amplitude analysis of the $K^-\eta$ system[15] has established that the peak at $\sim$1.8 GeV/c^2 results from production of the $\overline{K}^*_3(1780)$ (cf.fig.13). This, together with the results of a similar analysis of reaction (3),[14] yields a branching fraction to $\overline{K}\eta$ of 7.7$\pm$1.0%, in accord with SU(3). SU(3) predicts also that K^* states of even spin couple only weakly to $K\eta$. This is confirmed by the D wave intensity distribution of fig.13. No $\overline{K}^*_2(1430)$ signal is observed, and a 95% confidence level upper limit on the branching fraction is established at 0.45%. This value is an order of magnitude smaller than that obtained from previous low statistics measurements.

4. $s\bar{s}$ SPECTROSCOPY

The processes most relevant to the study of $s\bar{s}$ (i.e.strangeonium) spectroscopy are those in which a $K\overline{K}$ or $K\overline{K}\pi$ system is produced against a recoil hyperon. The objectives in studying such processes are basically the same as for the strange sector. However, in recent years candidate glueball states which also couple to these meson systems have emerged from the study of J/ψ decay. Consequently, it has become increasingly important to define the $s\bar{s}$ level structure in order to distinguish those states which may be of a different dynamical origin. In the present experiment, the relevant reactions which have been studied are

$$K^-p \to K^0_S K^0_S \Lambda_{seen} \tag{5}$$

$$K^-p \to K^- K^+ \Lambda_{seen} \tag{6}$$

and

$$K^-p \to K^0_S K^\pm \pi^\mp \Lambda_{seen} \ . \tag{7}$$

They are characterized by peripheral (i.e.quark exchange) production of the meson system, and it follows that the creation of glueball states should be disfavored in such processes. Although the data samples are considerably smaller than for reactions (1)-(4), amplitude analyses have been performed,[2,16,17,18] the results of which are summarized here.

The raw $K\overline{K}$ mass distributions for reactions (5) and (6) (fig.14) differ, since for the former only even spin states are produced. In addition, for reaction (6), the mass

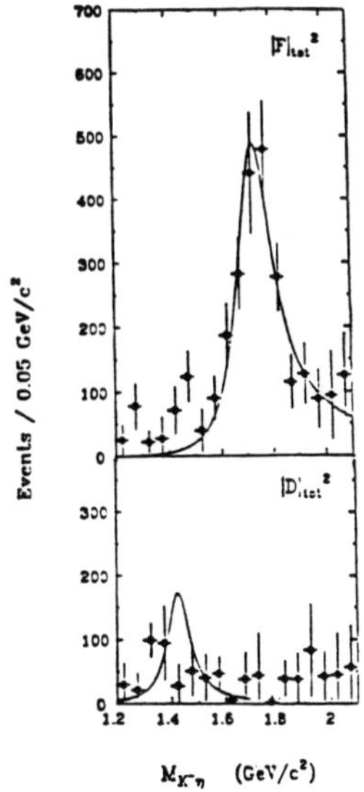

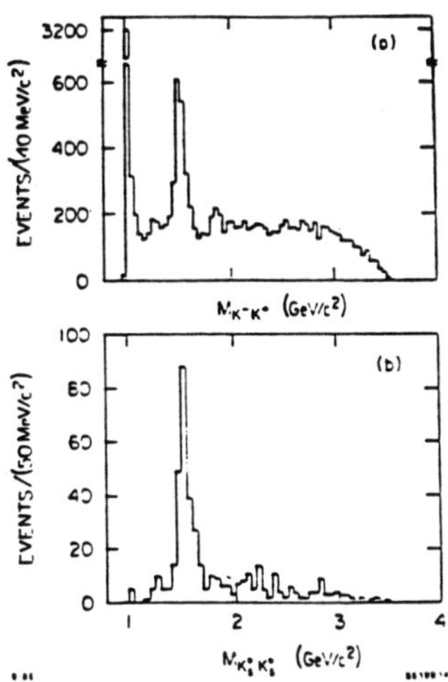

Fig.13. The total F and D wave intensity distributions from reaction (4): the F wave curve corresponds to a $\overline{K}^*_3$ (1780) BW; the D wave curve indicates the 95% confidence level limit on $\overline{K}^*_2$ (1430) production.

Fig.14. The $K\overline{K}$ mass projections for reactions (5) and (6).

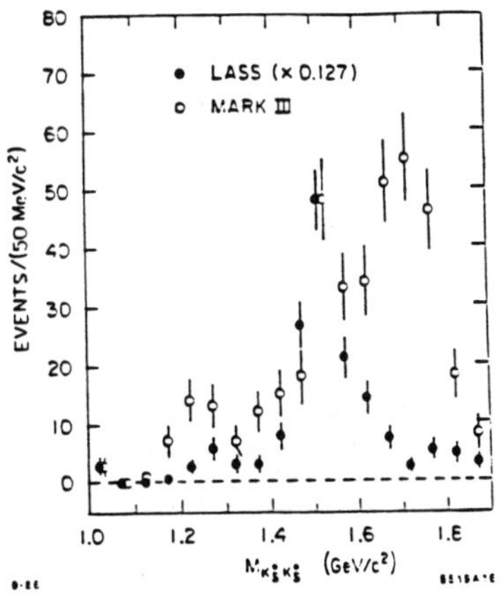

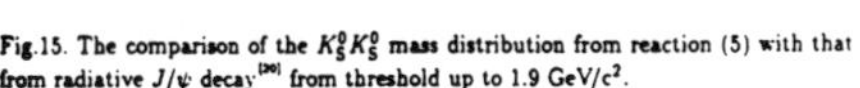

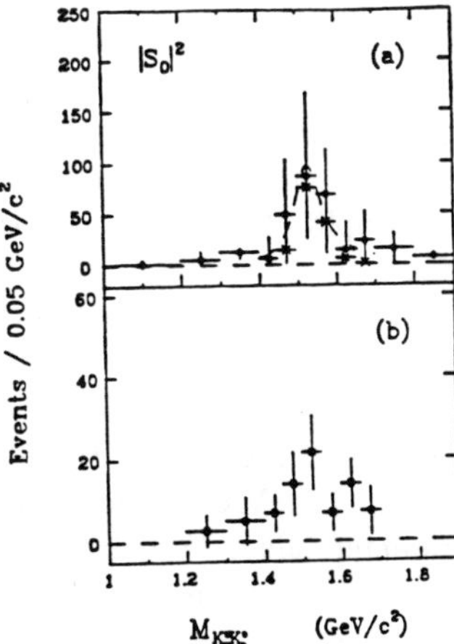

Fig.15. The comparison of the $K^0_S K^0_S$ mass distribution from reaction (5) with that from radiative J/ψ decay[30] from threshold up to 1.9 GeV/c^2.

Fig.16. (a) The S wave intensity distribution from the amplitude analysis of reaction (5). (b) The S wave intensity distribution from the amplitude analysis of reaction (6) at 8.25 GeV/c.[33]

region above ~ 1.7 GeV/c^2 is dominated by the reflection of diffractive production of low mass ΛK^+ systems.[19] In fig.15, the $K_S^0 K_S^0$ mass spectrum from reaction (5) is compared to the MARK III data on radiative J/ψ decay[20] in the mass region below 1.9 GeV/c^2 (the LASS data have been scaled to the MARK III data at the peak of the $f_2'(1525)$). Both spectra show a small, but intriguing, threshold rise,[2,21] followed by activity in the $f_2(1270)/a_2(1320)$ region and then the large $f_2'(1525)$ peak. At higher mass, the MARK III spectrum is dominated by the $f_2(1720)$. There is no evidence for any such signal in the LASS distribution, and the upper limit on the production cross section is 94 nb. at the 95% confidence level; it should be noted also that there is no evidence for $f_2(1720)$ production in reaction (7) in the present experiment[16] (cf.fig.21f). These results indicate that the $f_2(1720)$ is indeed a strong candidate glueball state.

The amplitude analysis of reaction (5)[2] yields the S wave intensity distribution of fig.16a. Although the uncertainties are large, the data seem to peak in the range ~ 1.5-1.6 GeV/c^2. A similar distribution (fig.16b) has been obtained in an analysis of reaction (6) at 8.25 GeV/c,[22] and there are indications of S wave structure at this mass from the data on reaction (6) in the present experiment. These results suggest the existence of a 0^+ state in this mass region which is naturally interpreted as a partner of the $f_2'(1525)$ in the ^{3}P ground state. It is not clear whether this state might be identified with the $f_0(1590)$,[23] although the latter appears not to couple strongly to $K\overline{K}$. An immediate consequence of this observation is that the $f_0(975)$, which is usually assigned to this multiplet, may well be a weakly bound $K\overline{K}$ system,[24] or be of some other non-$q\,\overline{q}$ origin. This is discussed at greater length in ref.2.

For the $K\overline{K}$ system in reaction (6), the acceptance corrected spherical harmonic moments ($t_L^M = \sqrt{4\pi}N\langle Y_{LM}\rangle$; L $\leq$ 8, M=0) in the t-channel helicity frame are shown in fig.17 for the mass region 1.68-2.44 GeV/c^2, and $t' \leq 0.2$ (GeV/c)^2. It should be noted that amplitudes with spin J can contribute to moments with L$\leq$2J. There is a peak in the mass spectrum, t_0^0, at ~ 1.86 GeV/c^2, and similar structure is present for all moments with L $\leq$ 6, but is absent for L $\geq$ 7. This, together with the absence of such a signal for reaction (5),[2] indicates the presence of a J$^{PC} = 3^{--}$, mostly $s\,\overline{s}$ state. A detailed analysis of this region[17] yields the total F wave intensity distribution of fig.18a for $t' \leq 1.0$(GeV/c)^2; the fitted curve gives BW mass and width 1855 $\pm$ 22 and 74 $\pm$ 67 MeV/c^2, respectively. A similar fit to the mass spectrum (fig.18b) gives parameter values 1851 $\pm$ 9 and 66 $\pm$ 29 MeV/c^2, in agreement with previous measurements from the mass distribution only.[25,26] It should be noted that the spin of this state, which is listed as the $\phi_J(1850)$,[10] is established by the LASS data, and that the mass and width have been estimated for the first time on the basis of an amplitude analysis.

A significant J^P=3$^-$ signal (fig.21g) has been obtained in the $\phi_3(1850)$ region

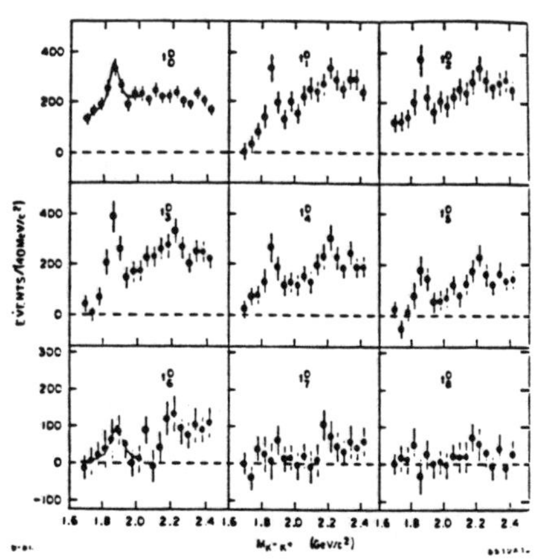

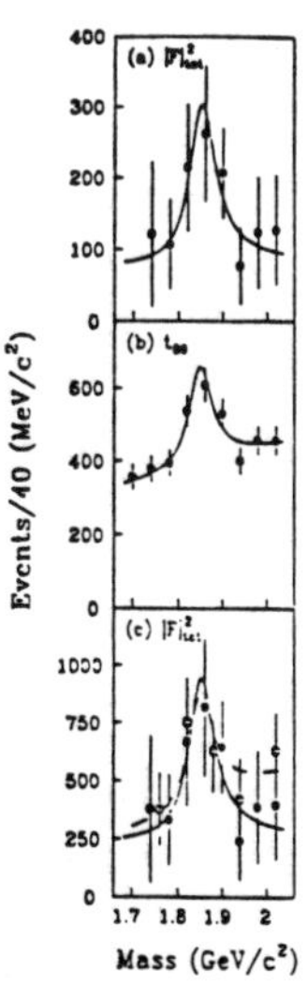

Fig.17. The mass dependence of the unnormalized spherical harmonic moments of the K^-K^+ system from reaction (6) in the region 1.68-2.44 GeV/c^2 and $t' \leq 0.2(\text{GeV/c})^2$

Fig.18. (a) The total F wave intensity distribution, and (b) the corresponding acceptance-corrected mass distribution from reaction (6) for $t' \leq 1.0(\text{GeV/c})^2$ in the ϕ_3 mass region. (c) The comparison of the F wave intensity distributions, after all corrections, from reaction (6) (solid dots) and reactions (7) (open dots). The curves in (a)-(c) correspond to fits using a BW line shape plus linear background term.

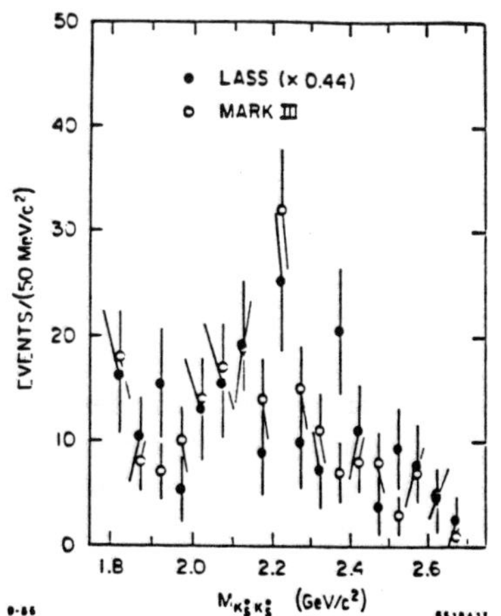

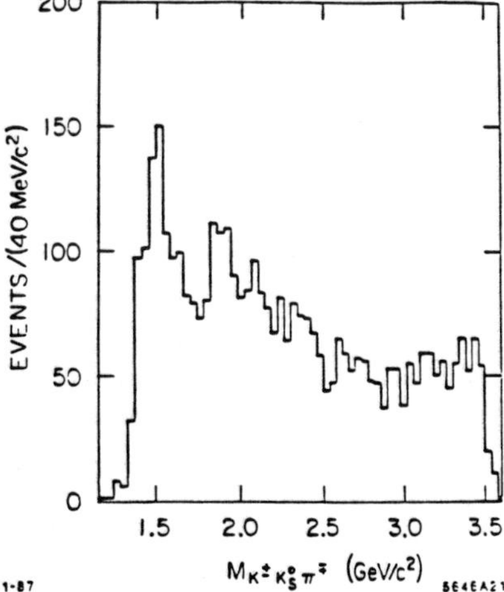

Fig.19. The comparison of the $K_S^0 K_S^0$ mass distribution from reaction (5) with that from radiative J/ψ decay[20] in the mass range 1.8-2.7 GeV/c^2.

Fig.20. The combined raw $K\overline{K}\pi$ mass distributions from reactions (7).

from the partial wave analysis of reaction (7).[16] This signal, after all corrections, is shown in fig.18c (open dots) in comparison with that from reaction (6) (solid dots); the branching ratio obtained is $BR[(\phi_3(1850) \to (K^*\overline{K} + c.c.))/(\phi_3(1850) \to K\overline{K})] = 0.55^{+0.85}_{-0.45}$, in agreement with theory.[11]

The $\phi_3(1850)$ is interpreted as belonging to the $^3D_3$ quark model nonet which also includes the $\rho_3(1690)$, the $\omega_3(1670)$, and the $K_3^*(1780)$. Using current mass values,[10] the mass formula[27] yields an octet-singlet mixing angle of $\sim 30°$. This indicates that the multiplet is almost ideally mixed, and that the $\phi_3(1850)$ is an almost pure $s\bar{s}$ state, in accord with its production characteristics.

The analysis of the ϕ_3 region[17] has shown that the peaks observed in fig.17 are due to interference between the resonant F wave and the approximately imaginary amplitude describing the diffractive production of the low mass ΛK^+ system. In fig.17, there is also a small signal in every moment with $L \geq 1$ at ~ 2.2 GeV/c^2; no such signal is observed for $L \geq 9$. This indicates the existence at this mass of a small, resonant 4^+ amplitude which interferes with the large, imaginary, diffractive background, just as for the ϕ_3. An analysis based on this interpretation[18] yields mass and width values of $2209 ^{+17}_{-15}$ and $60 ^{+107}_{-57}$ MeV/c^2 for this $J^{PC} = 4^{++}$ state. These values are consistent with those obtained by MARK III for the X(2220),[20] and with a signal in the $\eta\eta'$ mass spectrum observed by the GAMS collaboration.[28] In addition, the $K\overline{K}$ mass distribution from reaction (5) in this region has been shown[2] to be very similar to that observed for radiative J/ψ decay (fig.19). This suggests that the X(2220), which has been conjectured to be a glueball state, may instead be a member of the quark model $^3F_4$ ground state nonet.

The raw $K\overline{K}\pi$ mass spectrum for reactions (7) (fig.20) exhibits a small signal at the $f_1(1285)$, followed by a rapid rise at $K\overline{K}^*$ threshold to a peak at ~ 1.5 GeV/c^2, and a second peak at ~ 1.9 GeV/c^2. The low mass structure is similar to that observed at 4.2 GeV/c.[29] An amplitude analysis[16] yields the partial wave intensity distributions of fig.21. The low mass peak is associated primarily with 1^+ waves, while the 1.9 GeV/c^2 bump is due mainly to 2^- and the 3^- contribution discussed previously. Only K^* or $\overline{K}^*$ isobar production is important; amplitudes involving the $a_0(980)$ are negligibly small. The 1^+ intensity at low mass shows a pronounced asymmetry in favor of the $\overline{K}^*$ isobar, and also exhibits K^*-$\overline{K}^*$ interference, which is destructive near threshold, and constructive at ~ 1.5 GeV/c^2. This suggests the existence of two 1^+ states of opposite G-parity in this region. The corresponding J^{PG} amplitude combinations yield 1^{++} and 1^{+-} production intensity distributions (fig.22) which are well described as BW resonances of mass 1.53 ± 0.01 and 1.38 ± 0.02 GeV/c^2 respectively, with corresponding widths 0.10 ± 0.04 and 0.08 ± 0.03 GeV/c^2. The 1^{++} state confirms an earlier observation,[29] while the 1^{+-} state is new. If these are isoscalars,[29] the lower mass state, the $h_1(1380)$, completes the $J^{PC}=1^{+-}$ ground state

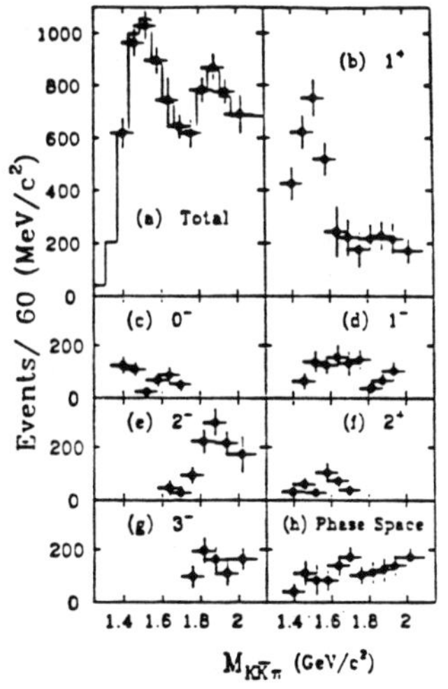

Fig.21. The intensity distributions corresponding to the partial wave decomposition for the $K\bar{K}\pi$ system in reactions (7).

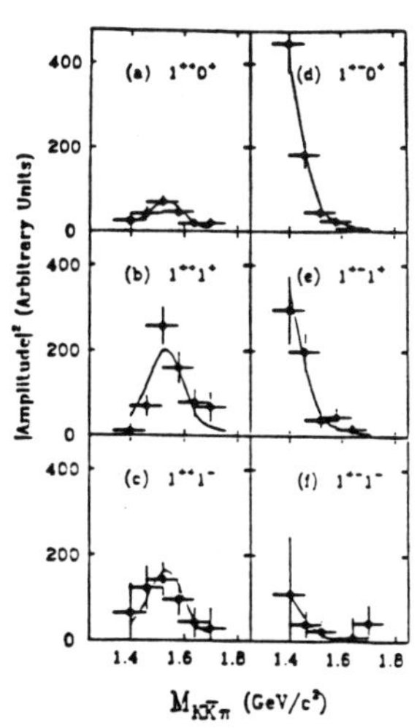

Fig.22. The production amplitude intensity distributions from reactions (7) for the $K\bar{K}^*$ and $\bar{K}K^*$ G-parity eigenstate combinations; (a)-(f) are labelled by $J^{PG}M^\eta$, where M is the helicity and η the naturality of the t-channel exchange.

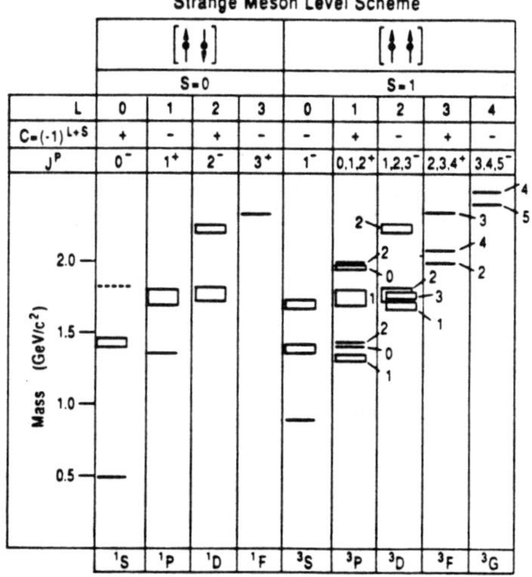

Fig.23. The quark model level diagram summarizing the status of strange meson spectroscopy; the C parity is that of the neutral, non-strange members of the relevant SU(3) multiplet.

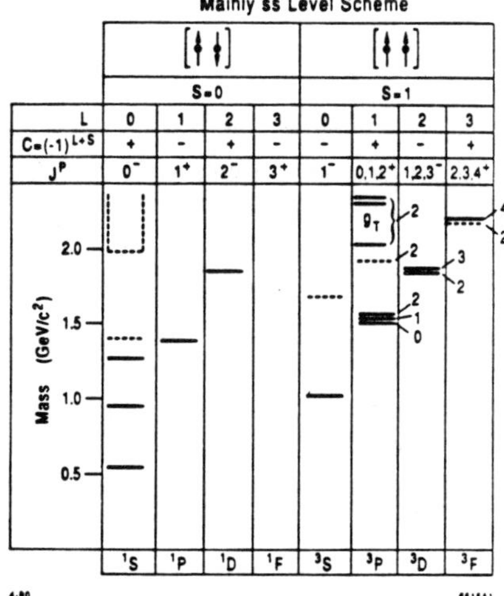

Fig.24. The quark model level diagram summarizing the status of the strangeonium states.

nonet, while the higher mass state, the $f_1(1530)$, is a strong candidate to replace the $E/f_1(1420)$ in the corresponding 1^{++} nonet. Use of the $f_1(1530)$ in this nonet gives a mixing angle $\sim 55°$; this implies that the $f_1(1530)$ is mainly $s\bar{s}$ and that the $f_1(1285)$ has little $s\bar{s}$ content, in agreement with their production characteristics. This is not the case when the E is used, since its production properties are not consistent with the predicted large $s\bar{s}$ content. Finally, a previous analysis[30] of the 1^{++} nonet made use of the E mass and width to predict an a_1 mass of ~ 1.47 GeV/c^2; when the $f_1(1530)$ mass and width are used instead, the predicted mass is ~ 1.28 GeV/c^2, in much better agreement with the accepted value.[10]

The 1^+ states observed in the present experiment satisfactorily complete the ground state $^1P_1$ and $^3P_1$ quark model nonets. Furthermore, the small spin-orbit interaction implied when the new information on the 0^+ is also taken into account contrasts markedly with the corresponding behavior in the charmonium sector. The status of the E meson as a $q\bar{q}$ state is dubious. However, it may be that its proximity to $K\overline{K}^*$ threshold is significant in that it could be a $K\overline{K}^*$ molecule or four-quark state of some kind.

5. STATUS OF THE SPECTROSCOPY

A somewhat optimistic summary of the observed $q\bar{q}$ levels in the strange meson sector is presented in fig.23. This includes results from the present experiment, together with information derived from analyses of the charged $K\pi\pi$, $K\phi$ and $\Lambda\bar{p}/\overline{\Lambda}p$ systems.[10] Of the unnatural parity levels with J^P other than 0^-, only the $J^P = 1^+$ ground state has been resolved into the states corresponding to different quark total spin; consequently, the remaining such entries in the left and right halves of fig.23 are identical. The second radial excitation in the 0^- tower is shown as a dashed line since, of the two experiments analyzing $K\phi$,[31,32] only one finds a significant 0^- amplitude.[31]

Some comments regarding the level splittings compared to those predicted have been made already in section 3. In a more empirical vein, the level structure appears to depend in first approximation only on principal quantum number value; i.e. there is approximate mass degeneracy of states having different spin-parity but the same value of the (orbital+radial) excitation quantum numbers. A notable exception involves the S wave ground states; this suggests that the spin-spin interaction is large only in this configuration.

For the triplet states, the spin orbit interaction, as measured by the mass difference between the leading state and the other natural parity member of the triplet, appears to increase approximately linearly as (J-1), where J is the spin of the leading state. This might be expected from simple angular momentum considerations. However, the behavior of the unnatural parity triplet member does not conform to such a pattern at all. For the ground state triplets, the unnatural parity state relative to the leading state is ~ 100 MeV/c^2 low for the P wave, about equal for the D wave, and ~ 250 MeV/c^2 high for the F wave, if the reported 3^+ state is the ground state. This in fact, may not be the case, since for the G wave the difference is only ~ 100 MeV/c^2; however the behavior is still anomalous. In addition, for the first radially excited P wave triplet, the unnatural parity state appears to be at least twice as far below the leading state as it is in the ground state triplet. It would appear difficult to reproduce such behavior in a simple potential model, and so it would be of interest to attempt to fit the structure observed in fig.23 by means of models such as that of reference 11.

A similar summary is presented in fig.24 for the mainly $s\bar{s}$ sector. Information on radial excitations is sparse, and in this regard the 0^- states observed by Mark III in $\phi\phi$ and $K^*\overline{K^*}$[33,34] and the 2^+ states observed in hadronic $\phi\phi$ production[35] have been somewhat arbitrarily included; it is difficult to assess the extent to which they might really belong in such a picture. The spin-spin interaction in the S wave ground state again appears to be large. Furthermore, the limited information available on the ground state triplets indicates that the spin-orbit interaction is rather weak and does not exhibit the anomalous behavior observed for the strange sector. Clearly, much experimental work needs to be done in order to bring the information on the $s\bar{s}$ sector even to the level of that on the strange sector.

6. THE FUTURE: KAON

The LASS experiment has demonstrated:

(i) the effectiveness of a programmatic approach to the study of meson spectroscopy;

(ii) the feasibility of obtaining very large statistics in a fixed target experiment by running a 4π acceptance spectrometer in interaction mode;

(iii) that the corresponding computing requirements can be met by means of a combination of large mainframe and microprocessor farm availability;

(iv) that the resultant large data samples for individual reactions can be analyzed in a relatively straightforward manner;

(v) the value of being able to analyze many different channels in the same experiment under conditions of uniform and well-understood acceptance;

(vi) that such a program need not involve a large number of physicists and graduate students.

In the next ten years, it would be desirable to push such a program to a significantly higher statistical level, while maintaining the uniform acceptance which characterizes the LASS experiment. This would imply an experiment of something like five billion triggers. Such experiments are already contemplated at Fermilab, (e.g. E791) and it seems that the data processing, storage and retrieval needs can be met. The level structure of fig. 23 would be significantly refined by such a program, and that of fig.24 would be carried beyond the present-day understanding of the strange sector. In addition, a programmatic study using pion beams should be carried out in order to similarly advance the understanding of the isovector and non-$s\bar{s}$ sectors.

The first requirement for the successful implementation of such a program is that it should form a significant component of the overall program of the lab. in which it is situated; significant long-term commitment of lab. resources is involved. There should be strong in-house involvement in the experiments for purposes of continuity and support, but there should also be an active and enthusiastic external user group. Constructive in-house theoretical interest would also be desirable. The facilities should include a high-intensity r.f. separated beam in order that data can be acquired in six months to a year of running. The spectrometer should be an upgraded LASS-type detector; it should have better resolution, photon detection, silicon vertex detector, etc., and should be able to cope with data transfer rates of at least five megabytes/sec. There should be powerful mainframe computing on-site together with a sizable microprocessor farm and extensive data storage and retrieval capability; a network of user-friendly low-end work-stations should also exist.

The most likely place for most of these ingredients to come into being on the indicated time-scale is at TRIUMF by means of the KAON Factory project. The present plans call for an r.f. separated beam line in the 6-20 GeV/c range providing $\sim 10^7$ K's per spill. They are favoring a spectrometer facility which is something of a cross between LASS and the Crystal Barrel detector at LEAR. Furthermore, this spectrometer facility figures prominently in the planned program of the lab., thereby indicating strong in-house commitment. There appears to be enthusiasm for such a facility among the user community, and development of the computing capabilities of the lab. along the lines indicated above is being considered.

It seems clear that the future of strange quark spectroscopy in particular, and indeed of light quark spectroscopy in general, is tied to the existence of the KAON Factory. The programmatic approach to the study of meson spectroscopy over the last twenty years or so at SLAC has proven its worth. However, there is much that remains to be understood, and it would seem that this greater understanding can only be achieved by means of the higher level, long-term program embodied in KAON.

7. CONCLUSION

The present experiment has made significant contributions to the various areas of strange quark spectroscopy discussed above. Data analysis is continuing, and new results will be forthcoming on inclusive Ξ^- and Ξ^* production, the $K^-\omega p$ final state, and the production of meson systems such as $K^*\overline{K^*}$, $\phi\phi$, $\phi\pi$, and $\pi^+\pi^-$ against a recoil Λ. Indeed, preliminary results from some of the hypercharge exchange reactions have already been presented.[36]

Future advances in the area of light quark spectroscopy require the kind of programmatic approach which has been adopted at SLAC over the last twenty years or so, but at a significantly higher statistical level and including experiments of a similar nature with incident pions. It is probable that such broad advances can be made only by means of an upgraded LASS-type detector located at the KAON Factory, and consequently that the future of the subject is intimately bound to the approval and construction of this facility.

REFERENCES

1. D. Aston et al., The LASS Spectrometer, SLAC-REP-298 (1986).

2. D. Aston et al., Nucl. Phys. B301, 525 (1988).

3. D. Aston et al., contributed talk given by B. Ratcliff in session P09 at the International Europhysics Conference on High Energy Physics, Bari, Italy, July 1985; see S. Cooper, Rapporteur Talk in the Proceedings of this Conference, p. 947 and SLAC-PUB-3819.

4. S.F. Biagi et al., Z. Phys. C31, 33 (1986).

5. D. Aston et al., Phys. Lett. 194B, 579 (1987).

6. D. Aston et al., Phys. Lett. 215B, 799 (1988).

7. N. Isgur and G. Karl, Phys. Rev. D19, 2653 (1979).

8. K.T. Chao, N. Isgur and G. Karl, Phys. Rev. D23, 155 (1981).

9. W. Grotrian, Graphische Darstellung der Spektren von Atomen und Ionen mit ein, zwei und drei Valenzelektronen (Springer, Berlin, 1928).

10. Particle Data Group, Phys. Lett. 204B (1988).

11. S. Godfrey and N. Isgur, Phys. Rev. D32, 189 (1985).

12. D. Aston et al., Nucl. Phys. B292, 693 (1987).

13. D. Aston et al., Phys. Lett. 180B, 308 (1986); Nucl. Phys. B296, 493 (1988).

14. P.F. Bird, Ph.D. Thesis, SLAC Report 332 (1988).

15. D. Aston et al., Phys. Lett. 201B, 169 (1988).

16. D. Aston et al., Phys. Lett. 201B, 573 (1988).

17. D. Aston et al., Phys. Lett. 208B, 324 (1988).

18. D. Aston et al., Phys. Lett. 215B, 199 (1988).

19. D. Aston et al., SLAC-PUB-4202 / DPNU-87-08 (1987).

20. R.M. Baltrusaitis et al., Phys. Rev. Lett. 56, 107 (1986).

21. K.L. Au et al., Phys. Rev. D35, 1633 (1987).

22. M. Baubillier et al., Z. Phys. C17, 309 (1983).

23. D. Alde et al., Phys. Lett. 201B, 160 (1988), and references therein.

24. J. Weinstein and N. Isgur, Phys. Rev. Lett. 48, 659 (1982); Phys. Rev. D27, 588 (1983), and references therein.

25. S. Al-Harran et al., Phys. Lett. 101B, 357 (1981).

26. T. Armstrong et al., Phys. Lett. 110B, 77 (1982).

27. M. Gell-Mann, Caltech report TSL-20 (1961) unpublished; S. Okubo, Prog. Theor. Phys. (Kyoto) 27, 949 (1967).

28. D. Alde et al., Phys. Lett. 177B, 120 (1986).

29. Ph. Gavillet et al., Z. Phys. C16, 119 (1982).

30. R.K. Carnegie et al., Phys. Lett. 68B, 287 (1977).

31. T. Armstrong et al., Nucl. Phys. B221, 1 (1983).

32. D. Frame et al., Nucl. Phys. B276, 667 (1986).

33. G. Eigen, CALT-68-1595 (1989).

34. Z. Bai et al., SLAC-PUB-5159 (1990).

35. A. Etkin et al., Phys. Lett. 201B, 568 (1988).

36. D. Aston et al., SLAC-PUB-5150 (1989).

NON-ACCELERATOR PHYSICS

Convener: B. Sadoulet

IMPROVED LIMIT ON THE MASS OF $\bar{\nu}_e$ FROM THE BETA DECAY OF MOLECULAR TRITIUM

T.J. Bowles, R.G.H. Robertson, J.L. Friar,
G.J. Stephenson, Jr., D.L. Wark, and J.F. Wilkerson,
Los Alamos National Laboratory, Los Alamos, NM 87545.

D. A. Knapp,
Lawrence Livermore National Laboratory,
Livermore, CA 94550.

ABSTRACT

We report a new upper limit of 13.4 eV (95% confidence level) on the mass of the electron antineutrino from a study of the shape of the beta spectrum of free molecular tritium. This result appears to be inconsistent with a reported value for the mass of 26(5) eV. The electron neutrino is evidently not massive enough to close the universe by itself.

1. INTRODUCTION

In 1981, a group at the Institute for Theoretical and Experimental Physics (ITEP) in Moscow reported[1] from their study of the tritium spectrum that n_e had a mass of 35 eV, with revolutionary implications for particle physics and cosmology. More recent ITEP work[2] has reduced this value slightly to 26(5) eV, with a "model-independent" range of 17 to 40 eV. Fritschi et al.[3] found in a similar type of experiment at the University of Zürich an upper limit of 18 eV, and other measurements have neither confirmed nor contradicted these works.[4,5] Both the Zürich and ITEP experiments have very high statistical accuracy, and the difference between the two results must be a consequence of systematic effects. A probable origin for such effects is the solid source materials used, for which molecular structure calculations are difficult to carry out to the necessary precision.

Unlike other experiments presently in operation, our experiment at the Los Alamos National Laboratory makes use of a gaseous source of T_2 to capitalize on the simplicity of the two-electron system. When a triton decays to ^{3}He, the daughter atom can be left in an excited final state. The resulting energy spread impressed on the outgoing beta must be very precisely calculated (at the 1% level) if serious errors in interpreting the data are to be avoided. Such calculations can be carried out with some confidence for atomic and molecular tritium, but with less certainty for solid sources like those used by ITEP and Zürich. Use of a gaseous source also confers the advantages of minimal and well understood energy-loss corrections, and no backscatter corrections.

2.   SYSTEM IMPROVEMENTS

In an earlier paper[5] we described our apparatus briefly and reported the initial result obtained with it: $n_e < 27$ eV at 95% confidence level (CL). Since this first result, we have made a number of improvements[21], the principal one being the replacement of the simple single-element proportional counter in the spectrometer with a 96-pad Si microstrip detector array.

3.   DATA ACQUISITION, REDUCTION, AND ANALYSIS

The beta spectrum is formed by setting the spectrometer to analyze a fixed momentum (equivalent to an energy of 23 keV) and scanning the accelerating voltage on the source. Data voltages are repeated in random order with a frequency that reflects the parts of the spectrum most significant in determining the neutrino mass. Before and after a tritium data set, the 17820-eV K-conversion line of $^{83}Kr^m$ is recorded two or three times to determine the instrumental resolution and energy scale.

Each pad of the silicon microstrip detector receives counts corresponding to a slightly different momentum, the total range being about 100 eV in energy from one end of the detector to the other. The data are thus organized by summing counts from the same pad numbers on each wafer to form 12 spectra, each independently calibrated by a $^{83}Kr^m$ spectrum similarly formed. The "raw" tritium spectra can be compared to the theoretical spectrum modified by corrections for energy loss, instrumental resolution, apparatus efficiency, and the final-state spectrum. The neutrino mass and its variance is then estimated[5] from a plot of the sum of Ξ^2 for all pads against m_ν^2.

Electrons lose energy by inelastic scattering as they spiral through the source gas. Monte Carlo calculations, together with the differential energy cross sections[4-7] and measurements of the source density determine the probability of an electron interacting to be 8.5%, and the number of interactions per decay is 9.1%, an indication of the generally small scattering probability and the very minor role played by plural interactions.

Measurement of the instrumental resolution is accomplished by circulating $^{83}Kr^m$ (from the decay of ^{83}Rb) through the source and recording the nominally monoenergetic K-conversion line at 17820(3) eV. Conversion lines are accompanied by shakeup and shakeoff satellites, and, rather than rely on calculations for their positions and intensities, we have carried out a K-shell photoionization measurement on Kr at the Stanford Synchrotron Radiation Laboratory.[8,9] Excellent agreement between the shapes of the spectra is obtained when the slightly better-resolution photoionization spectrum is convoluted with a Gaussian to match the internal-conversion data. The Gaussian variances from these data for the 12 spectra from the silicon microstrip detector averaged about 120 eV^2. Other contributions to the tritium linewidth not contained in the Kr calibration, such as Doppler broadening and zero point motion and molecular vibrations of the T_2 molecule, are negligible.[11]

The small variation of apparatus efficiency with acceleration voltage introduces a spectral distortion that can influence the neutrino mass derived. It is customary to parameterize this with empirically determined linear and quadratic correction terms a_1 and a_2 in the spectrum. From Monte Carlo calculations, we find that the spectrometric efficiency is strictly linear, with $a_1 = -2.0(3) \times 10^{-5}$ eV^{-1}, in reasonable agreement with the value $-2.6(2) \times 10^{-5}$ derived by fitting the spectrum to a linear term only. Rather than rely on Monte Carlo calculations, and because linear and quadratic corrections produce different neutrino masses, we consider both linear and quadratic terms in the fit separately, adopt a mean value, and treat the total spread in neutrino mass squared (105 eV^2) as a systematic uncertainty.

Experimental tests of a number of possible sources of systematic error were conducted. In two different tests, trapped ions were searched for, and we set a limit of 5 x 10^{-4} on the ratio of ions or metastables to neutrals, corresponding to an excess variance of order 0.2 eV^2. Tests to search for electrons scattered into the beam from the walls (which are highly contaminated with tritium) indicated no contribution, with a limit set at a level of 10^{-4} of the source strength.

The final-state spectrum (of the THe^+ ion) has the most important influence on the tritium spectrum. Calculations have been reported for the decay of T_2 in the sudden approximation.[10,13-16] The uncertainty on neutrino mass of the different final state calculations is rather small.

The largest uncertainty due to final state effects is uncertainty about validity of the sudden approximation. Williams and Koonin[17] (WK) have claimed that the rescattering contributions (i.e., the interaction of the beta directly with orbital electrons) were less than 10^{-3} in the case of the atom. However, their treatment is limited and their argument that higher partial waves were unimportant was incorrect.[22] Calculations by Friar[18] indicate that m_ν^2 shifts upward by 3 eV^2, and we have doubled this contribution to make an estimate for the T_2 molecule. Friar also shows that higher ℓ contributions fall off very rapidly for bound states, and we find from a semiclassical model that this is true for the continuum as well. We estimate the total effect to be not more than about 3 times the p-wave, but the need for a complete continuum calculation for the T_2 molecule is manifest.

4. RESULTS

The tritium data set obtained (13400 measurements) spanned an interval of about 9 days and led to spectra totalling 8000 counts in the last 100 eV, of which 1400 were background. The spectra covered the range 16500-19200 eV. The maximum-likelihood procedure described earlier[5] was used to obtain values for m_ν^2, E_0, amplitude, background, a_1 and a_2. These results, and their 1-s statistical uncertainties, are listed in Table I.

Table I. Results from analysis of August, 1988 data; uncertainties are one standard deviation statistical.

	a_1 only	a_2 only	
m_ν^2	-250(90)	-145(90)	eV^2
E_0	18567.2	18569.2	eV
a_1	-2.6(2) x 10^{-5}	--	eV^{-1}
a_2	--	-8.8(5) x 10^{-9}	eV^{-2}
χ^2 (av.)	945.8	945.1	(892 D.O.F)

In Table II we list the estimated uncertainties (1-s) in m_ν^2 from all sources. We have not at this time considered all relevant contributions to the uncertainty in the endpoint energy, E_0.

Table II. Contributions (eV^2) to the uncertainty in m_ν^2 at one standard deviation.

Analysis:	
Statistics	90
Beta monitor statistics, dead time	5
Energy Loss:	
18% in theoretical spectrum shape:	15
5% Uncertainty in source density	4
Resolution	
Width	12
Skewness	6
Tail	15
Final States	
Differences between theories	8
Region above truncation point	10
Rescattering	20
Apparatus Efficiency	
Linear vs Quadratic	105
Total	143

We take the conservative viewpoint that we do not know which is the correct description of the curvature, and that the two choices are but a selection from a large variety of possible efficiency functions. Our best estimate is then the average of the a_1 and a_2 fits, and the uncertainty associated with efficiency correction is the difference between them.

In Figure 1, we plot the residuals for the fit near the endpoint for m_ν = 0 and 30 eV, from which it may be seen qualitatively that a 30-eV mass is rejected. That conclusion is borne out quantitatively when all uncertainty components are considered.

As discussed elsewhere,[19] setting confidence levels on quantities physically forbidden from having negative values is a complex issue, especially when the measured value (through normal statistical fluctuations) falls in the non-physical regime. Our result (from the average of the fits with a_1 and a_2 separately) is m_ν^2 = -198(143) eV^2, which would have arisen in less than 8% of trials from a neutrino mass greater than 0 eV. To derive a confidence level on the mass is less straightforward. The Particle Data Group sets forth[20] a Bayesian prescription that is at least well-defined, if not rigorously justified. On that basis, we find a 95% confidence level upper limit on the neutrino mass of 13.4 eV.

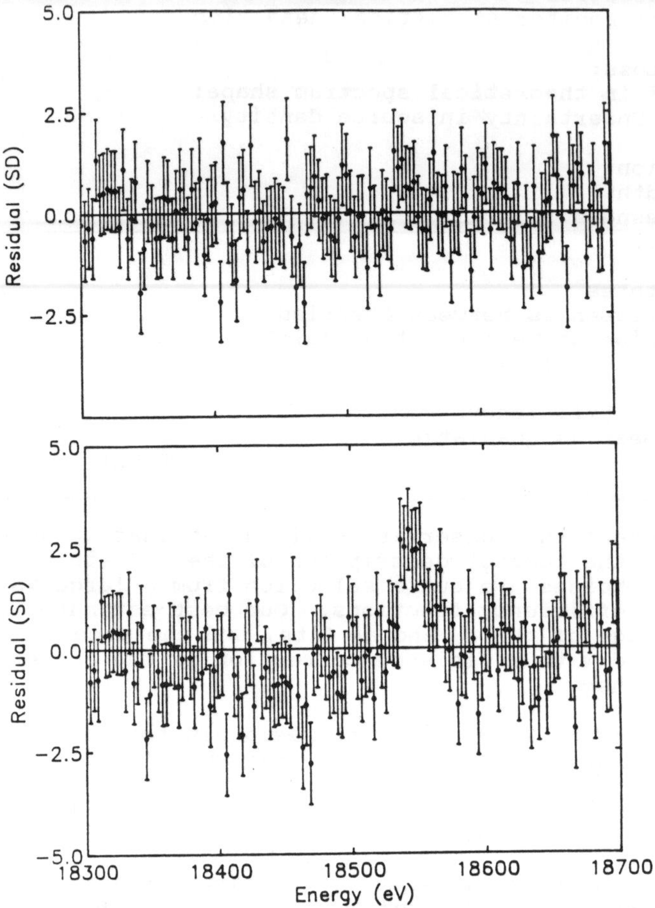

Fig.1. Residuals in fits to neutrino masses of 0 (top) and 30 eV (bottom). All other parameters including a_1 and a_2 have been allowed to vary.

5. CONCLUSIONS

The results of our experiment are entirely inconsistent with the ITEP result, 17 to 40 eV. We are not able to speculate usefully on the exact source of this disagreement, but we have noted how demanding this type of experiment is on the precision of all correction factors. Uncertainties in energy loss, backscattering, and final-state effects are presumably substantially larger for solid sources than for free molecular T_2. We note, finally, that this low limit on the mass of n_e should permit a new analysis of the neutrino data from the supernova SN1987a largely free of the time-dispersive effects introduced by neutrino mass, and that n_e is incapable of closing the universe by itself.

We gratefully acknowledge the assistance of M. Anaya in this work.

REFERENCES

1.   Lubimov, V. A., Novikov, E. G., Nozik, V. Z., Tret'yakov, E. F., Kozik, V. S., Phys. Lett. 94B, 266 (1980).
2.   Boris, S., et al., Phys. Rev. Lett. 58, 2019 (1987); Boris, S. D., et al., Pis'ma Zh. Eksp. Teor. Fiz. 45, 267 (1987) [Sov. Phys. JETP Lett. 45, 333 (1987)].
3.   Fritschi, M., et al., Phys. Lett 173B, 485 (1986).
4.   Kawakami, H., Nisimura, K., Ohshima, T., Shibata, S., Shoji, Y., et al. Phys. Lett. 187B, 198 (1987).
5.   Wilkerson, J. F., et al., Phys. Rev. Lett. 58, 2023 (1987).
6.   Liu, J. W., Phys. Rev. A 7, 103 (1973).
7.   Inokuti, M., Rev. Mod. Phys. 43, 297 (1971).
8.   Bartlett, R. J., Brown, G. S., Bowles, T. J., Crasemann, B., Henderson, J., Knapp, D. A., Robertson, R. G. H., Schaphorst, S. Sorenson, S. L., Trela, W. J, Wark, D. L., and Wilkerson, J. F. (unpublished).
9.   Robertson, R. G. H., et al., Proceedings of the VIIth Moriond Workshop, Les Arcs, France, January 1989 (to be published).
10.   Kolos, W., et al., Phys. Rev. A 31, 551 (1985).
11.   Cohen, J. S., private communication.
12.   Staggs, S. T., et al., Phys. Rev. C 39, 1503 (1989).
13.   Martin, R. L. and Cohen, J. S., Phys. Lett. 110A, 95 (1985).
14.   Fackler, O. et al., Phys. Rev. Lett. 55, 1388 (1985).
15.   Agren, H. and Carravetta, V., Phys. Rev. A 38, 2707 (1988).

16.   Kaplan, I. G. and Smelov, G. V., in Proceedings of the
International Symposium on Nuclear Beta Decays and Neutrino,
edited by T.Kotani, H. Ejiri and E. Takasugi (World
Scientific, Singapore 1986), p. 347.
17.   Williams, R. D. and Koonin, S. E., Phys. Rev. C $\underline{27}$,
1815 (1983)
18.   Friar, J. L. private communication.
19.   Robertson, R. G. H. and Knapp, D. A., Ann. Rev. Nucl.
Part. Sci. $\underline{38}$, 185 (1988).
20.   Particle Data Group, Phys. Lett. $\underline{170B}$, 1 (1986).
21.   Wilkerson, J. F., et al., Proceedings of the VIIIth
Moriond Workshop, Les Arcs, France, January 1990 (to be
published).

NEUTRINO OSCILLATIONS

D. Hywel White
Los Alamos National Laboratory
Los Alamos, NM 87545

ABSTRACT

The present status of searches for neutrino oscillations is described at accelerators, reactors and in neutrino sources in the atmosphere. There have been searches for $\nu_e \to \nu_\mu$, $\nu_\mu \to \nu_\tau$, ν_μ and ν_e disappearance. There are no compelling signals for neutrino oscillation at this time.

INTRODUCTION

In this paper we shall review the present evidence from searches for neutrino oscillations. There is no clear evidence at present for neutrino oscillations and so the discussion will center on the limits that are set by experiments to date. Experiments come in two categories, those that search for the appearance of a second neutrino flavor in a nearly pure beam, and those who verify the variation of flux from the neutrino source to see if anomalous variation indicates that neutrinos of the dominant flavor are disappearing from the beam.

The standard view of neutrino types is that there exists three flavors, ν_e, ν_μ , and ν_τ: flavor is approximately conserved and the experiments that are reviewed here search for deviations in beam constituents as a function of distance from the source. The probability that neutrinos of one flavor will be observed in a beam of another flavor is

$$P = \sin^2(2\alpha) \sin^2(1.27 \ \Delta m^2 \ l/E)$$

where $\sin^2(2\alpha)$ is the mixing angle between dominant production and minority appearance, Δm^2 is the mass difference squared between the two eigen states in a situation where only two flavors are involved in eV^2, l is the distance from source to point of detection in km, and E is the neutrino energy in GeV. Three state mixing has been discussed at length for example by Marciano[1]. Experiments are designed to measure the probability of appearance P or at least to set a limit on P. If we write $P = x \sin^2 by$, where $b = 1.27 \ l/E$ then for a fixed value of P the curve traced out by coordinates x and y bound the region of exclusion in the $\sin^2(2\alpha) \ \Delta m^2$ plane. In practice experiments have finite resolution in b so that the region of exclusion is given by

$$x = 2P / (1 - \cos(2by) \sin (2by\delta/(2by\delta))$$

approximating the resolution function by a rectangle of width δ.

This curve is illustrated in Fig. 1 and will be familiar to anyone who has heard a talk on neutrino oscillations. The maximum sensitivity (x is Minimum) is when $\cos 2by = -1$, $x = P$ when $\Delta m^2 = \pi E/2l$; for 1 GeV and 1 km $\Delta m^2 = 1.57 \ eV^2$. At high Δm^2, $x = 2P$. These rules of thumb will allow the casual observer to interpret the results of experiments in different energy and distance regimes.

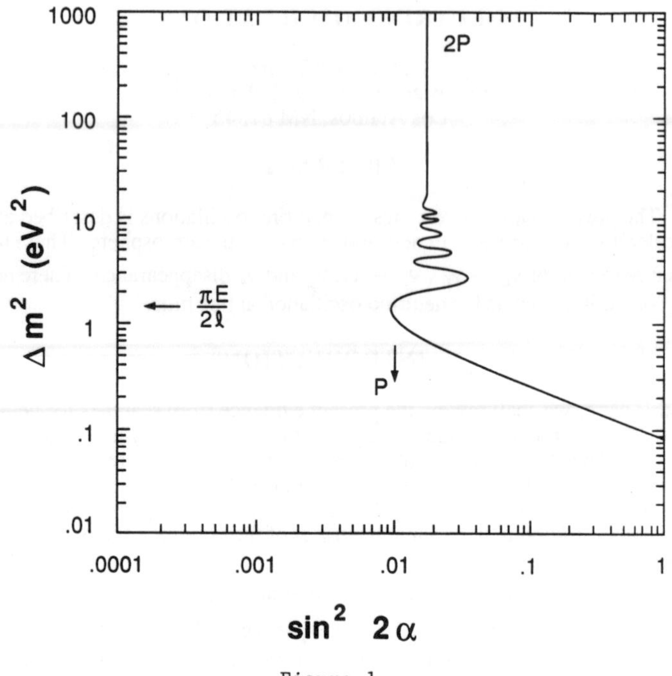

Figure 1

DECAY IN FLIGHT BEAMS, ν_e APPEARANCE

For decay in flight beams at accelerators the most recent limits are given in Table I.

		$< rdE_\nu >$	ℓ	ℓ/E	2P	Ref
BNL	E734	1300	120	0.1	3.4×10^{-3}	2
	CIJH	1300	800	0.8	2.0×10^{-2}	3
CERN	BEBC	1100	825	0.9	2.3×10^{-2}	4
Los Alamos	E764	150	30	0.45	1.5×10^{-2}	5

TABLE I

The region of exclusion of these experiments is shown in Fig. 2.

We have omitted two experiments which observed excess electron events at a statistically marginal level (2.5 σ), CERN PS 191 and BNL E 816[6] although each time with very different numbers of events. It is now believed that these anomalies may be real and ascribed to insufficient knowledge of the low energy region of the neutrino spectrum where the situation is complicated and difficult to measure precisely.

PION DECAY AT REST, $\bar{\nu}_e$ APPEARANCE

In a heavy element beam stop the only decays are

$$\pi^+ \to \mu^+ + \nu_\mu \qquad\qquad \mu^+ \to e^+ + \nu_e + \bar{\nu}_\mu$$

because π^- are 99% absorbed in the target no $\bar{\nu}_e$ are produced, (only 1% decay in flight). A search for the reaction

$$\bar{\nu}_e + p \to e^+ + n$$

giving electrons between 30 and 50 MeV provides a sensitive way to search for $\bar{\nu}_e$ appearance, presumably from $\bar{\nu}_\mu \to \bar{\nu}_e$ oscillations. Experiment E645 at Los Alamos[7] has given a limit shown in Table II and in Fig. 2.

TABLE II

		$< E_\nu >$	1	1/E	2P
Los Alamos	E645	40	26	0.65	1.3×10^{-2}

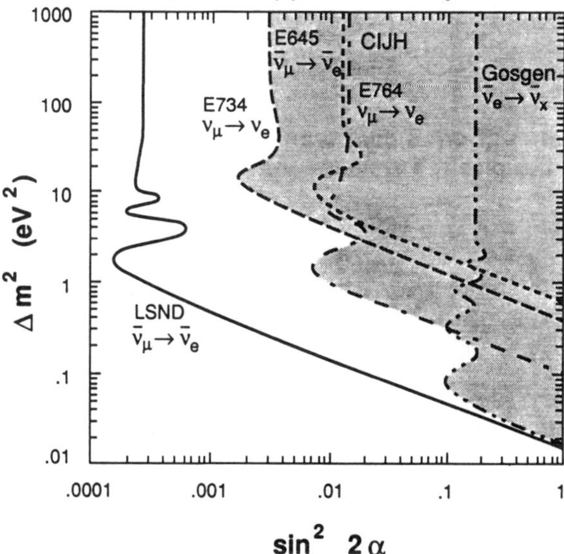

Figure 2

There is an experiment under construction at Los Alamos (LSND) to improve this limit by increasing the mass of the detector by a factor of ten and acceptance by a factor of five. A low decay in flight efficiency also gives a small ν_e background in the beam, so the decay in flight limit for $\nu_\mu \to \nu_e$ is also expected to be significantly improved over those in Table I[8]. The expected limits are also shown in Fig. 2.

DECAY IN FLIGHT, ν_τ APPEARANCE

Experiments that search for anomalous electron events to limit oscillations can also be used to set limits on $\nu_\mu \to \nu_\tau$ oscillations because of the decay mode $\tau^+ \to e^+ + \nu_e + \bar{\nu}_\tau$ gives excess electrons. The sensitivity of these searches is less than $\nu_\mu \to \nu_e$ directly because of the branching ratio of τ decays to electrons (~4%). A direct search has been conducted at Fermilab (E531)[9] detecting ν_τ events by observation of a "kink" near the production vertex in emulsion. There is very low background since ν_τ are expected only from charm decay J/ψ, D_s. The sensitivity of the measurement is limited by the counting rate for neutrino events in the emulsion stack and the threshold energy needed to create τ charged current events, the region of exclusion is shown in Fig. 3. A new experimental proposal[10] has emerged using a proposed Fermilab new injector at 130 GeV which gives significantly more neutrino flux that otherwise at Fermilab. The proposal uses improved event detection to ease the scanning problem with scintillating fibers between emulsion stacks and the apparatus all immersed in the magnetic field of the old 15' bubble chamber magnet. The oscillation $\nu_\mu \to \nu_\tau$ is particularly attractive as a way to solve the dark matter problem with a ν_τ mass in the range of tens of electron volts together with relatively light ν_μ and ν_e.

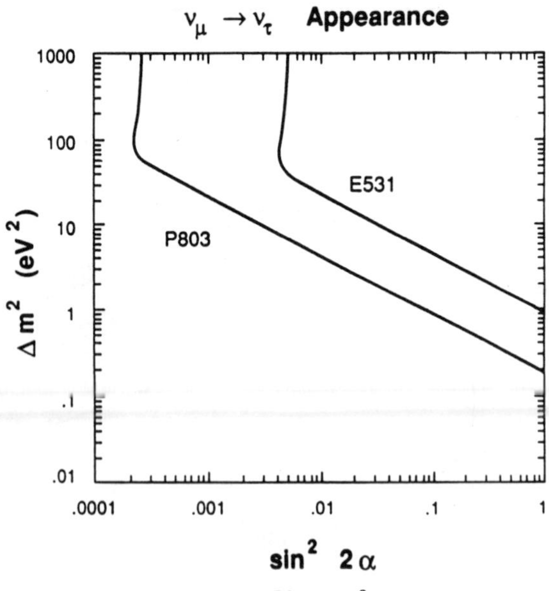

Figure 3

ν_μ DISAPPEARANCE

The principal interest in ν_μ disappearance experiments at this time is in the possibility of oscillation to ν_τ. Direct searches are limited in Δm^2 mass range by the neutrino energy needed to produce charged current events. No such limit is imposed on disappearance experiments. Unfortunately the systematic problems of understanding neutrino beams seem to limit the sensitivity of the searches to a few percent mixing. There are three experiments CCFR (Fermilab)[11], and CDHSW[12], CHARM (CERN)[13]. Beam source at Fermilab was a normal focused beam $<E_\nu> \sim 20$ GeV and the experiment utilized detectors at 110m and 700m. At CERN an unfocused beam from the PS was used with detectors at 120m and 900m. The exclusion curve looks somewhat like the curve described in the introduction, with the addition that at high Δm^2 the sensitivity falls off because finite experimental resolution means that the many oscillations that would occur between the detectors cannot be resolved. The limits achieved by these experiments are shown in Fig. 4.

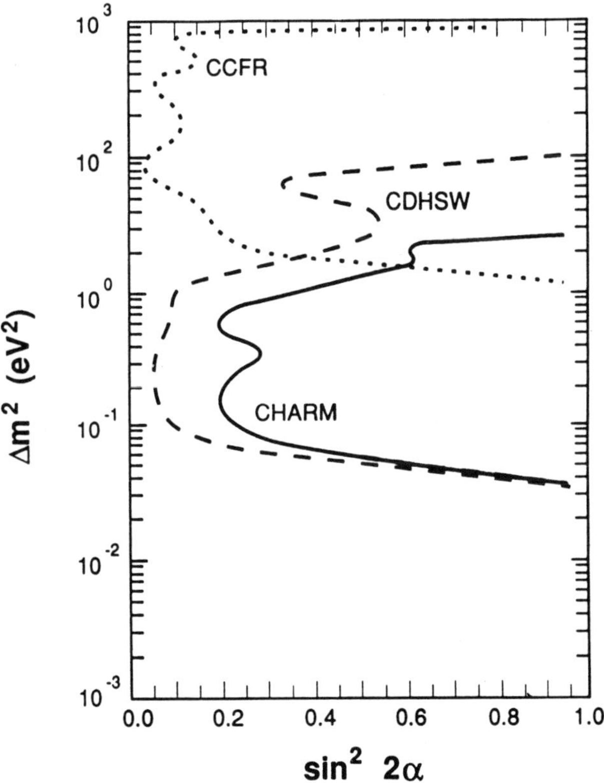

Figure 4

$\overline{\nu}_e$ DISAPPEARANCE

Fission products from a reactor are neutron rich and so when the fission products decay $\overline{\nu}_e$ are produced. The energy of the neutrinos is generally low (a few MeV) but the flux is copious. Searches for neutrino oscillations in the disappearance mode have been made at reactors, the two most recent being at the Bugey reactor in France and the Gosgen reactor in Switzerland[14]. The systematic problems that are common these experiments is that the source of neutrinos is large physically and the intensity distribution changes with time. Reactors are monitored in detail and the conviction is that the luminosity distribution is well known as a function of time. Measurements of the reaction

$$\overline{\nu}_e + p \rightarrow e^+ + n$$

are made observing the positron directly, together with the subsequent capture of the neutron after thermalization in ^{3}He counters. The sum of the positron and the neutron proton mass difference gives the neutrino energy with good precision. The neutrino spectrum is measured at two distances giving a limit on $\overline{\nu}_e$ disappearance. There were conflicting results at first from the positron spectrum measured in a single detector at Bugey[15], but now it it accepted that the dependence of flux on distance is compatible with no oscillations and the limit from the Gosgen measurement is also shown in Fig. 2. Initial studies for a 1000 ton detector using Gadolinium loaded liquid scintillator to observe thermal neutrons through the 8 MeV capture gamma rays are underway to extend the Δm^2 range down to 10^{-4} eV^2 with a power reactor.

ATMOSPHERIC NEUTRINOS

The large proton decay detectors at Frejus, IMB and Kamioka[16] have observed events from neutrinos produced in atmospheric interactions of cosmic rays. The initial cosmic ray interactions are high in the atmosphere to that the dilute medium allows pion and muon decays to be complete. Because pion (+) decays produce a ν_μ and muon (+) decays produce ν_e and $\overline{\nu}_\mu$ the ratio of muon induced charged current events should be approximately twice those from ν_e. A detailed Monte Carlo gives 1.7 ± 0.15 for the ratio of muon induced vs electron induced events. This is agreed to by all three experiments. The experimental data are shown in Table III.

TABLE III

Detector	Exposure	Total Events	Data/MC
IMB	3.77	401	1.00 ± 0.05
Frejus	1.53	197	0.97 ± 0.07
Kamioka	2.80	265	0.83 ± 0.05

The exposure is in kiloton years. Frejus and Kamioka have published momentum distributions for muon like and electron like events in accord with Monte Carlo. An up down ratio for neutrinos is measured with significantly different l/E but at present the ratios are not statistically compelling. Here we have a three and one half standard deviation effect in one experiment and agreement in the others. One of the Kamioka collaborators who was present at the meeting was quoted "At the moment, the totality of atmospheric neutrino data from the Frejus, IMB, and Kamioka detectors provide no compelling evidence for neutrino oscillations". That is probably where this issue should rest but if you want a space to watch...

SUMMARY

After more that a decade of work there remains no clear evidence for neutrino oscillations of any type from accelerators or reactors. The area of exclusion on the Δm^2 - $\sin^2(2\alpha)$ plot is large, but the conviction remains that any further experiments must be capable of extending this region by a major factor and be capable of keeping systematic effects well under control. Even so there are a number of well conceived experiments that propose to continue the struggle.

REFERENCES

1. W. J. Marciano, Proceedings of BNL Neutrino Workshop, **BNL 52079** (1987).
2. L. A. Ahrens et al., *Phys. Rev. D* **31** 2732 (1985).
3. B. Blumenfeld et al., "Search for $\nu_\mu \to \nu_e$ Oscillations," *Phys. Rev. Lett.* **62,** 2237 (1989).
4. C. Angelini et al., *Phys. Lett.* **179B,** 307 (1986).
5. T. Dombeck et al., *Phys. Lett. B194,* 591, (1987).
6. P. Astier et al., "A Search for Neutrino Oscillations," Los Alamos National Laboratory document LA-UR-89-3319 and CERN-EP/89-128 (to be published in Nuclear Physics, September 1989).
7. L. S. Durkin et al., "Limits on $\overline{\nu}_\mu \to \overline{\nu}_e$ Oscillations," Phys. Rev. Lett. 61, 1811 (1988).
8. X.-Q. Lu et al., "Proposal to Search for Neutrino Oscillations with High Sensitivity in the Appearance Channels $\nu_\mu \to \nu_e$ and $\overline{\nu}_\mu \to \overline{\nu}_e$," Los Alamos National Laboratory document LA-UR-89-3764 (December 1989).
9. N. Ushida et al., *Phys. Rev. Lett.* **57,** 2173 (1989).
10. Fermilab Proposal, P803.
11. I. E. Stockdale et al., *Phys. Rev. Lett.* **52** 1384 (1984).
12. F. Dydak et al., Phys. Lett. **134B** 281 (1984).
13. F. Bergsma et al., Phys. Lett. **142B** 103 (1984).
14. G. Zacek et al., *Phys. Rev. D* **34,** 2624 (1986).
15. J. F. Cavaignac et al., *Phys. Lett.* **B148,** 387 (1984).
16. A. K. Mann, Proceedings of 9th International Conference on Physics in Collision.(1989).

EXPERIMENTAL SEARCHES FOR DARK MATTER PARTICLES

David O. Caldwell

Physics Department, University of California, Santa Barbara, CA 93106

ABSTRACT

The particle which constitutes probably more than 90% of the mass of the universe is something unknown in the Standard Model of particle physics. There are few constraints on what it may be, and hence broad-ranging searches are needed. A combination of non-accelerator experiments, particularly those using Ge and Si ionization detectors, and recent accelerator experiments, especially at SLC and LEP, have eliminated as dark matter wide classes of formerly popular candidate particles, such as weak isodoublet neutrinos more massive than ~ 30 eV/c^2.

1. INTRODUCTION

While one of the most important problems confronting astrophysics and cosmology is understanding what constitutes most of the mass of the universe, this quest could be of equal importance for particle physics, since this dominant particle is something outside the presently successful Standard Model of particle physics. A very new field at the boundary of astro and particle physics is the experimental search for this particle. Even though this work has been going on for a short time, already many types of candidate particles have been ruled out or severely restricted and wide ranges of particle masses and interaction cross sections have been eliminated.

There is considerable evidence for gravitational effects which are not accounted for by visible mass. In galaxies the tangential velocity, v, remains constant out to a radius, r, about an order of magnitude beyond the luminous mass, M, whereas from $GmM/r^2 = mv^2/r$, $v \alpha r^{-1/2}$ would be expected. Thus, an unobserved mass increasing linearly as r must exist. On this basis, the ratio of mass density to the density required to close the universe is $\Omega = 0.05$–0.10. On larger scales, $\Omega = 0.2$–0.3 is required by, for example, the random peculiar velocities of galaxies which remain in a cluster, yet include velocities exceeding the apparent escape velocity corresponding to the observed total mass.[1]

2. THE NEED FOR NON-BARYONIC DARK MATTER

The success of nucleosynthesis theory in predicting the abundance in the universe of ^{4}He, ^{3}H, ^{2}H, and ^{7}Li, as well as three flavors of neutrinos, gives confidence that Ω from baryons lies between 0.02 and 0.11.[2] Observed baryons give only $\Omega \approx 0.007$, indicating some baryonic dark matter. This is probably in the form of Jupiters (stars too small to have initiated nuclear burning) or black holes outside the range of detectability ($< 10^{-6}$ or $> 10^6$ solar masses). While it is difficult to reconcile the nucleosynthesis limit with dark matter (DM) needed on very large scales, the main argument for non-baryonic DM is the apparent necessity to have $\Omega = 1$, assuming a zero cosmological constant. Unless there is severe fine tuning in the early universe, Ω would have been driven toward 0 or ∞, unless its value was the only time-stable value of unity, which gives a flat universe. Inflation theory provides a reason for $\Omega = 1$, and some observations support such a large value.[3]

Non-baryonic DM appears to be necessary also to explain galaxy formation. If at the time the DM particles dropped out of equilibrium with other particles they were relativistic, this "hot" DM seems not to have been sufficient for this task, unless cosmic strings exist. If instead the DM was moving slowly, this "cold" DM would enhance initial fluctuations to seed galaxies.

3. CANDIDATE DARK MATTER PARTICLES

Of early importance as a hot DM candidate was the electron neutrino, since ν's are present at a level of $\sim 10^2$/cm^3 everywhere, and the right mass (~ 30 eV/c^2) to give $\Omega = 1$ had been reported[4] in a ^{3}H β decay experiment. Now the limit[5] on that mass from the same type of experiment is 13 eV/c^2, but such a light neutrino is likely to be a Majorana particle, and the limit[6] on its mass from neutrinoless $\beta\beta$ decay is 1–2 eV/c^2. While massless in the Standard Model, the μ and τ neutrinos are viable hot DM candidates. There may be a problem with dwarf galaxies not being able to hold enough DM in the form of such light neutrinos, but this is controversial.

Although even lighter than such neutrinos, axions are a cold DM candidate. Resulting from a possible solution to the strong CP problem, the axion's existence is at least motivated by a reason having nothing to do with DM. If it exists and has a mass $\sim 10^{-5}$ eV/c^2, it could give $\Omega = 1$. The lack of evidence for energy transport by axions in stellar cooling and from supernova SN1987a limits axion mass to $\lesssim 10^{-3}$ eV/c^2. Two experiments[7] tried to detect axions by looking for the conversion of axions to photons in a magnetic field. So far, these difficult experiments lack a factor $\sim 10^2$ in sensitivity for seeing the "invisible" axions.

More accessible experimentally are weakly interacting massive particles, or WIMPs. These can be either Majorana particles with spin-dependent (axial) couplings or Dirac particles with predominantly spin-independent (vector) couplings. To be the main component of DM, a Majorana particle must be in a narrow range of masses (e.g., $\sim$ 5–8 GeV/c^2 for a weak isodoublet neutrino), as determined by the annihilation rate of these particles when they were in thermal equilibrium. Dirac particles can have a very wide mass range, however, since an initial particle-antiparticle asymmetry may have existed, enabling the annihilation rate to be adjusted suitably.

4. TECHNIQUES IN THE SEARCH FOR WIMPS

While most of the rest of this discussion will concern non-accelerator searches for WIMPs, accelerator experiments have contributed recently. In particular, the recent determinations of the width of the Z^0 at SLC and LEP provide important restrictions on weak isodoublet neutrinos. The limitation to three generations of light neutrinos means that such Majorana neutrinos, which could have been DM only in the small mass range given above, are eliminated completely. The SLC/LEP result also rules out the possible 4–10 GeV/c^2 mass range for weak isodoublet Dirac neutrinos which had not yet been reached by the Ge non-accelerator experiments to be described below.

The accelerator results also superseded the theoretically rather uncertain limits which were placed on heavy neutrinos by proton decay detectors.[8] The events for which they searched were ν_e and ν_μ (> 2 GeV) coming from the direction of the sun. DM particles captured in the sun could annihilate with each other, producing ν's in the cascade of decays starting with heavy quarks. Some limits[9] on supersymmetric neutralinos were obtained if the Higgs were sufficiently light, with similar limits[10] coming from scattering experiments discussed next. While proton-decay detector searches produced some useful results, including ruling out sneutrinos, attempts at observing annihilations of DM in space into γ rays, positrons, or antiprotons have not set any useful limits so far.

The broadest search has been with semiconductor ionization detectors. As the earth moves through the all-pervasive, thermally agitated galactic halo of DM particles, these could strike a nucleus of a detector, giving a small recoil energy (~ 10 keV). Only a fraction (typically $\frac{1}{5}$–$\frac{1}{3}$) of that energy goes into ionization, and the signal rate is very small, $\sim$ a count/keV per kg of detector per day. New levels of low energy thresholds with unprecedentedly low backgrounds had to be achieved. Record low backgrounds in the MeV region had been obtained with underground Ge detectors searching for $\beta\beta$ decay in ^{76}Ge, but altering these detectors to go to very low energies showed new background problems. Now the main remaining backgrounds appear to be those induced by cosmic ray bombardment when the detector and its surrounding materials were above ground. Examples are ^{3}H, a spallation product, and Ga and Zn x-rays which probably come from ^{70}Ge (n, 3n) $\rightarrow$ ^{68}Ge $\rightarrow$ ^{68}Ga (plus other steps in the decay chain).

These experiments are particularly sensitive to Dirac particles with their vector couplings, since these would exhibit nuclear coherent scattering, giving usually cross sections proportional to the square of the number of neutrons. All of the limits given below for Ge detectors assume such nuclear coherence, although it is sometimes provided by scalar particle exchange.

Three $\beta\beta$ experiments are now searching for DM. Two journal articles have appeared, but better results have been reported at many conferences. The group[11] most recently joining this work is that of Caltech, Neuchatel, and Paul Scherrer Institute. They have a very low threshold and hence exclude Dirac neutrinos with the standard weak interaction in the range of 10 GeV/c^2 to 1 TeV/c^2. The first group[12] to do this work utilized the Pacific Northwest Laboratories and South Carolina data. They now have better backgrounds and similar limits are about 12 GeV/c^2 to 4 TeV/c^2.

5. RESULTS FROM THE UCSB/LBL/UCB Ge EXPERIMENT

The group[13] with the largest data sample is that of the University of California (Santa Barbara and Berkeley) and Lawrence Berkeley Laboratory, and hence their results will be used here to illustrate both the points made above and the extent of the exclusion of DM by these data. Signals from one of two 0.9 kg detectors show in Fig. 1 electronic noise at the lowest energies followed by X-ray peaks due to identified radioactivities, such as those discussed above. The form of a signal due to a 10 GeV/c^2 Dirac neutrino, also shown, had to be transformed from a nuclear recoil energy to observable ionization. The efficiency of the slow, heavy nucleus in producing ionization had been determined[14] accurately by scattering neutrons from Ge down to an equivalent electron energy of 2 keV ($\approx$ 10 keV recoil energy), and these results agree very well with theory.[15]

The Ge data have been used to produce the exclusion plot of Fig. 2. To obtain the limits, first a fit is made to the gaussian electronic noise, all known radioactivity backgrounds, and the unresolved superposition of β and Compton spectra at larger energy (parameterized by a roughly flat constant term plus an exponential). The maximum of the likelihood function is found with respect to all these variables, giving a χ^2 compatible with the number of degrees of freedom. Next, at each mass point, the DM cross section is increased from zero to whatever value gives a 95% confidence level rejection of the DM hypothesis. Since the two detectors have different resolution and background shapes, they are fitted separately and their χ^2 added to produce Fig. 2.

The DM signals for Figs. 1 and 2 are generated by convoluting an s-wave scattering cross section with a Maxwellian velocity distribution (v_{rms} = 270 km/s) for the halo, truncated at the escape velocity (1300 km/s). The earth velocity (230 km/s) is added vectorially with a random angle in the convolution integral. At each particle mass it is assumed that only that particle is responsible for the galactic halo density of 0.3 GeV/c^2 $\cdot$ cm^3 = 5 $\times$ 10^{-25} g/cm^3. For large masses there is a loss of coherence because the momentum transfer becomes too big to leave the nucleus in its ground state, and a suitable correction has been made for that effect. While there are uncertainties in the astrophysical parameters, they have very little effect on the high-mass results of interest here.

The top boundary of the exclusion plot, calculated by Monte Carlo, is from the slowing of particles in the 600 m.w.e. of rock above the detector. However, all DM particles with very large cross sections on Ge are excluded[16] by a combination of observations, including data from this experiment[17] taken without the active NaI and passive Pb shielding and also near the earth's surface.

Particles having the normal weak, spin-independent interaction would lie along the line in Fig. 2 labelled "Dirac ν." Such particles are excluded between 10 GeV/c^2 and $\sim$ 3 TeV/c^2. The significance of the results will be discussed in Sec. 8.

6. COSMIONS

One possible type of particle which is searched for with more sensitivity using a detector made of Si rather than Ge is the so-called Cosmion. The Cosmion was hypothesized[18] to be not only DM but also to explain the apparent difference of a factor of $\sim$ 3 between the observations[19] of the ^{8}B neutrinos coming from the sun and the rate calculated from solar models.[20] The Cosmion DM could scatter in the sun, be captured and fall in toward the core, from which its circulation to larger radii could cool the core by the $\sim$ 10% needed to reduce the ^{8}B neutrino flux. It is impressive that the cooling requires an admixture of Cosmions which is 10^{-11} of the 10^{57} baryons in the sun, and the flux of DM and the age of the sun are such that 10^{46} Cosmions could have been accumulated by now.[21] The Cosmion mass should be $\gtrsim$ 4 GeV/c^2 so that they do not rapidly evaporate from the sun, but $\lesssim$ 10 GeV/c^2 so that, after thermalizing, they are not confined so near the solar center that they are inefficient in energy transport. The cooling and capture requirements also necessitate that the cross section for interactions with protons or helium be roughly two orders of magnitude larger than the weak cross section. While their scattering cross section should be high, their annihilation rate must be low and the lifetime very long to maintain an adequate density of Cosmions.

The 8 proposed Cosmions models[21,22] differ mainly in the mechanisms to provide the extra interaction and suppress the annihilation, but recent accelerator experiments eliminate many of these, leaving those in which Cosmions couple only to light quarks. The published results[13] of the Ge experiment ruled out Cosmions of mass > 9 GeV/c^2, and Fig. 2 extends this limit down to 7 GeV/c^2. To get to lower masses, Si detectors are desirable[23] because the lighter nucleus gives a larger recoil energy, more of that energy goes into ionization, and lower energy thresholds can be obtained. These advantages were sufficient to reduce limits to $\sim$ 4 GeV/c^2 using a small quantity of unselected Si with large backgrounds in what was supposed to be a test run. While four Si detectors of 18 g each replaced two Ge detectors in the UCSB/LBL/UCB apparatus, data from only the one of these with the lowest threshold and background were used, and its spectrum is shown in Fig. 3.

Two maximum likelihood analyses of the data of Fig. 3 were done, either of which produces the exclusion plot of Fig. 4. One was similar to that described above for Ge using a fit to known backgrounds up to 225 keV, and the other used a linear fit to the background in the small region 1.1–1.5 keV. The hypothesized Cosmion signal was generated as in the Ge case, except that the ionization efficiency had to be obtained in a separate experiment[24] by neutron scattering. The dashed lines in Fig. 4 indicate the region[25] of mass and cross section for coherently scattering Cosmions, if they are to solve the solar ν problem. All presently viable Cosmion models would be within this boundary. For the astrophysical parameters given above, the regions for possible Cosmions not excluded by this experiment are shown cross hatched. There is considerable dependence of the lower mass limits on v_{rms}, as indicated by changes in that quantity of $\pm$10% shown in Fig. 4.

7. SIGNIFICANCE OF THE RESULTS

As Fig. 4 shows, there is very little probability that Cosmions are the solution to the DM and solar ν problems. New detectors are being constructed to remove all doubts on this question.

Also essentially eliminated as DM is any particle more massive than $\sim$ 30 eV/c^2 having coherent interactions (i.e., vector or scalar) with nuclei which are at least as strong as the standard weak interaction. This includes isodoublet Dirac neutrinos, although there is some dispute about the upper mass limit. If the view[26] is accepted that with the breakdown of perturbation theory at $\sim$ 1 TeV/c^2 the strongly interacting ν would not exist as a free, stable state, these are eliminated. Also ruled out are technibaryons, the lightest one[27] of which (2.6 TeV/c^2) should have sufficient stability and be an excellent dark matter candidate.[28]

Eliminated as DM also are electron neutrinos of any mass, sneutrinos, micro-charged shadow matter, and μ and τ Majorana neutrinos more massive than $\sim$ 30 eV/c^2. Some restrictions are placed on weak isosinglet neutrinos associated with a Z$'$, which is constrained[29] to be > 1.4 TeV/c^2 by nucleosynthesis requirements, making the neutrino harder to detect. Accelerator and non-accelerator experiments impose complementary restrictions on the lightest supersymmetric particle, and the most likely candidate now appears to be a bino heavier than about 40 GeV/c^2.

In short, dark matter must be some particle which is extremely difficult to detect, such as a very light neutrino, a very weakly interacting heavy isosinglet neutrino or supersymmetric particle, or an axion. This presents a severe challenge for future experiments.

ACKNOWLEDGEMENTS

Thanks are due the author's collaborators in these experiments, particularly B. Magnusson, M.S. Witherell (UCSB); F.S. Goulding, A.R. Smith (LBL); B. Sadoulet, A. Da Silva (UCB); J. Rich, M. Spiro (Saclay). The work was partially supported by the U.S. Department of Energy.

REFERENCES

1. F. Zwicky, *Helv. Phys. Acta.*, **6**, 110 (1933).

2. K.A. Olive et al., *Phys. Lett.* **B236**, 454 (1990).

3. E.D. Loh and E.J. Spillar, *Ap. J.* **307**, L1 (1986); A. Meiksin and M. Davis, *Ap. J.* **91**, 191 (1986); A. Yahil, D. Walker, and M. Rowan-Robinson, *Ap. J.* **301**, L1 (1986); J.V. Villumsen and M.A. Strauss, *Ap. J.* **322**, 37 (1987).

4. V.A. Lubimov in "Massive Neutrinos in Astrophysics and Particle Physics," ed. O. Fackler and J. Tran Thanh Van (Editions Frontières, France 1986), p. 441; *Phys. Lett.* **B94**, 266 (1980).

5. T.J. Banks et al., this volume.

6. D.O. Caldwell et al., *Phys. Rev. Lett.* **59**, 419 (1987); Proc. of 12th Int. Workshop on Weak Interactions and Neutrinos, Ginosar, Israel (1989).

7. W. Wuensch et al., *Phys. Rev.* **D40**, 3153 (1989); P. Sikivie, N. Sullivan, and D. Tanner, experiment in progress.

8. IMB Collaboration, *Phys. Lett.* **B188**, 388 (1987) and *Ap. J.* **315**, 420 (1987); Fréjus Collaboration, "New and Exotic Phenomena", eds. O. Fackler and J. Tran Thanh Van, (Editions Frontières France, 1987), p. 215; Kamioka Collaboration, Univ. of Tokyo preprint ICR-192-89-9 (1989, unpublished).

9. G. Gelmini, P. Gondolo, and E. Roulet, preprint SISSA 88 EP89 (1989, unpublished).

10. R. Barbieri, M. Frigeni, and G.F. Giudice, *Nucl. Phys.* **B313**, 725 (1989).

11. F. Boehm et al., in "Theoretical and Phenomenological Aspects of Underground Physics" (l'Aquila, Italy, 1989, to be published).

12. S.P. Ahlen et al., *Phys. Lett.* **195**, 603 (1987).

13. D.O. Caldwell et al., *Phys. Rev. Lett.* **61**, 510 (1988).

14. C. Chasman, "Penetration of Charged Particles in Matter — A Symposium", Nat. Acad. of Sciences (1970), p. 16, and references therein.

15. J. Lindhard et al., Kgl. Dan. Vidensk. Selsk. Mat.-Fys. Medd. **33**, 14 (1963); *loc. cit.* **33**, 10 (1963).

16. G.D. Starkman et al., (to be published in *Phys. Rev. D*, 1990).

17. This UCSB/LBL/UCB result is not yet in written form.

18. R.L. Gilliland et al., *Ap. J.* **306**, 703 (1986) and references therein.

19. J.W. Rowley, B.T. Cleveland, and R. Davis, Jr., "Solar Neutrinos and Neutrino Astronomy," eds. M.L. Cherry et al., AIP Conf. Proc. 126 (Am. Inst. of Physics, NY, 1985), p. 1; K.S. Hirata et al., *Phys. Rev. Lett.* **63**, 16 (1989).

20. J.N. Bahcall and R.K. Ulrich, *Rev. Mod. Phys.* **60**, 297 (1988); S. Turck-Chièze et al., *Ap. J.* **335**, 415 (1988).

21. G.B. Gelmini, L.J. Hall, and M.J. Lin, *Nucl. Phys.* **B281**, 726 (1987).

22. S. Raby and G.B. West, *Nucl. Phys.* **B292**, 793 (1987); *Phys. Lett.* **B194**, 557 (1987); *Phys. Lett.* **B200**, 547 (1988); *Phys. Lett.* **B202**, 47 (1988); G.B. Ross and G.C. Segrè, *Phys. Lett.* **B197**, 45 (1987); G.F. Giudice and E. Roulet, *Phys. Lett.* **B219**, 309 (1989); J.G. Bartlett, M. Gleiser, and J. Silk, Inst. for Theor. Phys. preprint NSF-ITP-89-132 (1989, unpublished).

23. B. Sadoulet et al., *Ap. J.* **324**, L75, (1988).

24. G. Gerbier et al., Saclay preprint (1990, unpublished).

25. A. Gould, Inst. for Adv. Study preprint IASSNS-AST 89/49 (to be published in Ap. J., 1989); Y. Giraud-Héraud et al., to appear in Proc. of the Int. Conf. "Inside the Sun" (Versailles, France, 1989).

26. K. Griest and M. Kamionkowski, *Phys. Rev. Lett.* **64**, 615 (1990).

27. S.M. Barr, B.S. Chivukula, and E. Farhi, Inst. for Theor. Phys. preprint NSF-ITP-90-27 (1990, unpublished).

28. S. Nussinov, *Phys. Lett.* **B165**, 55 (1985) and **B179**, 103 (1986); R.S. Chivukula and T.P. Walker, *Nucl. Phys.* **B329**, 445 (1990).

29. M.C. Gonzalez-Garcia and J.W.F. Valle, Univ. of Valencia preprint FTUV/89-48 (1989, unpublished).

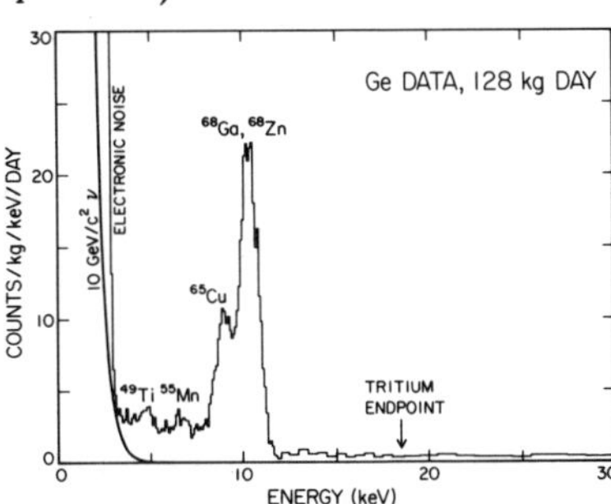

Figure 1. Energy (ionization) in a Ge detector and the expected shape of a 10 GeV/c² Dirac ν.

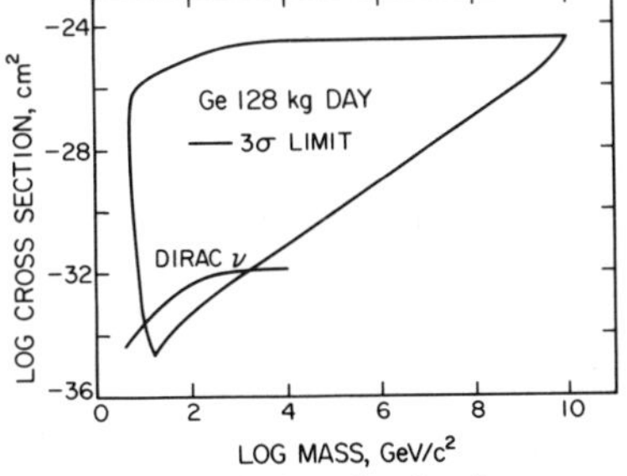

Figure 2. Exclusion plot for the mass and elastic cross section on Ge for dark matter particles. The weak interaction cross section is indicated by "Dirac ν."

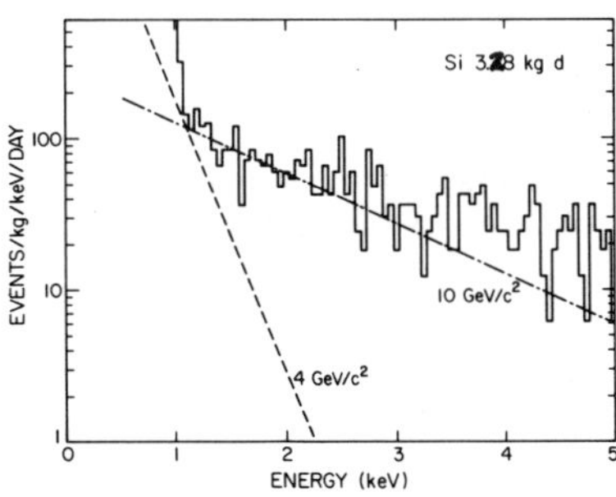

Figure 3. Energy (ionization) in a Si detector and signals from 4 and 10 GeV/c² Cosmions.

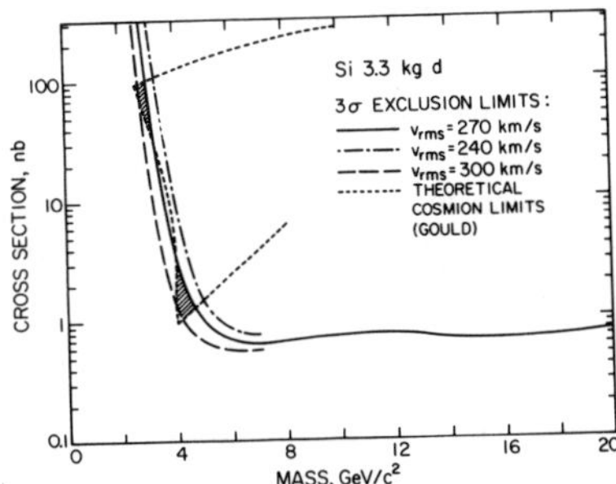

Figure 4. Exclusion plot for the mass and elastic cross section on Si for dark matter particles. Cosmions would lie within the dashed lines, leaving only the cross-hatched areas.

The UMC Extensive Air Shower Array: Results and Prospects

J. Matthews

Department of Physics, University of Michigan
Ann Arbor, Michigan, USA 48109

ABSTRACT

The Utah-Michigan-Chicago array addresses the need for greater measurement sensitivity in searches for ultra-high energy gamma rays. Results from the Utah-Michigan experiment demonstrate the power of muon rejection. Even if photons interact in unconventional ways, the combination of CASA and the large muon array will measure these effects.

1. INTRODUCTION

There have been many attempts to observe UHE ($> 10^{14}$eV) gamma rays from celestial sources since the reported observation of Cyg X-3 by the Kiel group in 1983[1]. However, results so far have been inconclusive with most positive reports having weak statistical significance[2].

The Utah-Michigan-Chicago (UMC) Air Shower Array is designed to provide a definitive answer to the question of whether UHE sources exist. The technique increases the size and improves the resolution of the ground array method and rejects hadron showers using large-scale muon detection. I survey here some recent results from the first stage of the experiment, the Utah-Michigan array, which demonstrate the power of muon rejection.

2. APPARATUS AND SHOWER ANALYSIS

The UMC experiment (Figure 1) consists of the Chicago Air Shower Array (CASA) and a large buried array for muon detection located around the Fly's Eye II detector at Dugway, Utah (40°N, 113°W, 870 g/cm^2). CASA is 1089 scintillator stations spread uniformly over a 480m×480m grid. Fast timing provides angular resolution of $\lesssim 0.5°$; a sophisticated triggering scheme records EAS above 10^{14} eV at a rate of 20 sec^{-1}. The muon array components[3] are 2.5 m^2 plastic scintillator sheets arranged in banks of 64 adjacent counters, buried at a depth of 3m. The total coverage of 16 banks (1024 counters with 2560 m^2 total area) far exceeds that of any other operating array. Measurements of muon lateral distributions and also with a test arrangement of buried counters at two depths confirm that electromagnetic punch-through to the muon counters is negligible when they are buried to this depth.

The size, resolution, and hadron rejection abilities will result in 15 γ-showers day^{-1} from Cyg X-3 (with a signal to noise factor of 33) if the source intensity is like that reported by the Kiel experiment. Put another way, a 12σ excess over background would emerge in one month if the source intensity was only 10^{-1} the Kiel report.

The Utah-Michigan array (Figure 1) is the first stage of the UMC experiment and operated from late 1987 through February 1990. There are 33 surface stations, each with four plastic scintillators, arranged over an area of radius 100m.

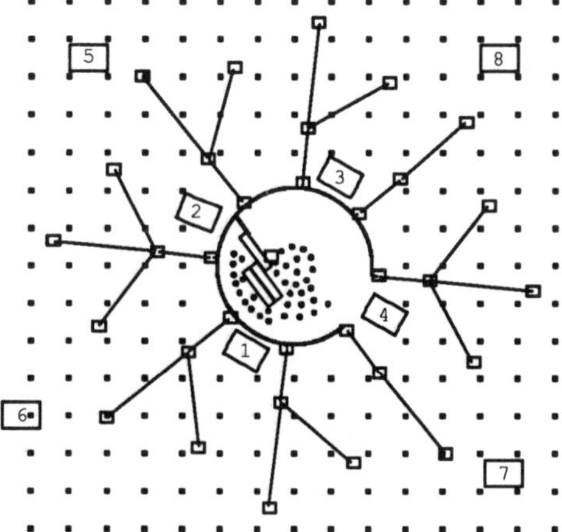

Figure 1. Layout of the central region of the UMC experiment showing the Utah-Michigan array. Grid marks are 15 m spacings and represent the location of CASA detectors under installation. The numbered rectangles indicate eight 64-counter muon banks. The smaller connected rectangles show positions of the 33 unit Utah surface array. The Fly's Eye II and electronics trailers are at the center. The full UMC array extends beyond that shown in the diagram and will include a total of sixteen muon banks and 1089 CASA stations.

The eight muon banks (512 counters totalling 1280 m^2) is the largest muon detector of any existing or previous air shower experiment. Triggering, shower reconstruction, and resolution are fully described elsewhere[4], but I will highlight features relevant to the present discussion.

The location and direction of the shower axis are found by fitting the pulse heights and arrival times of the surface counter hits. The total size $N \approx N_e$ and muon size N_μ are computed from maximum likelihood fits of surface data to an NKG function[5] and muon counter hits to a Greisen muon density function.[6]

The directional resolution $\delta\theta$ is defined such that 72% of events from a point source will reconstruct within $\delta\theta$ of the source direction. This definition maximizes $S/\sqrt{B}$ for a signal S in the presence of a uniform background B. For cores within 100m of the center of the array and $N > 10^4$, $\delta\theta = 3°$. Systematic pointing error is negligible, determined by comparison to data obtained by a tracking air-Cerenkov telescope operated in coincidence with the arrays.

Calculations predict 98% of γ-ray induced showers will have less than one-tenth the mean number of muons in hadron showers with similar N_e and zenith angle. We define the relative muon content of an air shower as $R_\mu = \log_{10}(N_\mu) - \langle\log_{10}(N_\mu)\rangle$. The expected number of muons is determined from fits to the data:

$$\langle\log_{10}(N_\mu)\rangle = -0.98 + 0.62\sec(\theta) + 0.82\log_{10}N$$

We accept showers meeting the μ-poor criterion $R_\mu < -1$ as γ-ray candidates.

3. MUON REJECTION AND DIFFUSE GAMMA RAYS

The relative muon content R_μ of showers taken in 1988 and 1989 is shown in Figure 2. Showers from within 10° of the galactic plane are distinguished from non-galactic showers. Rejecting events with $R_\mu > -1$ retains only about 1/200 showers with $\log_{10}N > 4.5$; the rejection factor improves to 1/2500 for larger showers ($\log_{10}N > 5.5$).

A possible "calibration" source of diffuse γ-rays might be expected from interactions of ordinary cosmic rays with interstellar gas and dust. Inspection of Figure 2 shows no excess μ-poor showers from the galactic plane when compared to non-galactic data. The expected number of background showers with $R_\mu < -1$ is computed by scaling the nongalactic sample so that the two sets have the same number of events with $R_\mu > 0$. The ratio of the integral flux of γ-rays above about 200 TeV to the ordinary hadronic cosmic ray flux is $< 8 \times 10^{-5}$ (90% CL). This limit is more sensitive by two orders of magnitude than previous experiments[7] with less extensive or no muon coverage.

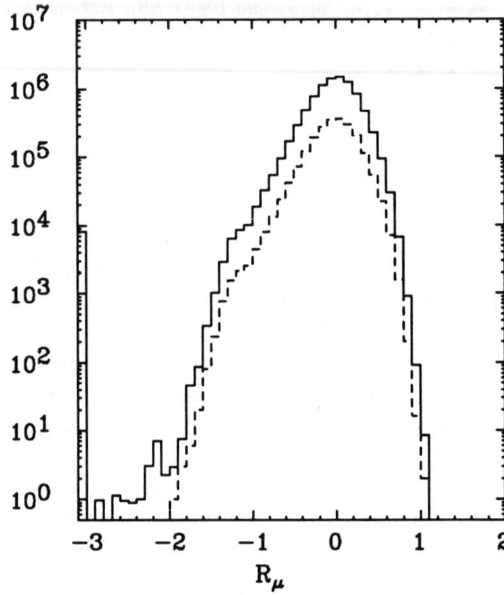

Figure 2. Relative muon content R_μ of showers. Dashed histogram contains showers from within 10° of the galactic plane, non-galactic data is the solid line. Nongalactic events have been slightly weighted according to zenith angle in order to match the overall zenith angle distribution of galactic events. There is no excess of muon poor showers from the galactic plane. Showers with no recorded muons are plotted as underflows at the left. The shapes of the distributions are in accord with that expected for proton induced showers.

4. CYG X-3 AND OTHER POINT SOURCES

Table 1 displays results of a search for excess showers from the direction of Cyg X-3. The γ-ray energy threshold E_0 is defined as the energy at which our acceptance of γ showers has attained 25% of its maximum value; this is very nearly the median energy of detected γ rays from an E_γ^{-1} integral spectrum. The background is determined using the measured rate of all off-source events in local coordinates to predict the rate of ordinary events from the source direction as it moves across the sky. The data are shown with and without cuts on the muon content.

We find no excess activity from the source, with or without a μ-poor cut. The power of muon rejection is evident in Table 1. The sample is reduced by factors of $10^{-2} - 10^{-3}$ and flux limits are improved by a factor of ten. The flux limits obtained for μ-poor showers are approximately ten times lower than previously reported observations[2] at these energies (see Figure 3). We have also sorted these data according to muon content and find no excesses in any interval of R_μ.

Table 1. Cyg X-3 Observations from 4 April 1988 to 3 August 1989

$\log_{10}N$	$E_o(eV)$	Observed Events	Expected Background	Flux(90%CL) $(cm^{-2}sec^{-1})$
4.5–6.0	2×10^{14}	35188	35480	$< 1.3 \times 10^{-13}$
	(μ-poor)	238	244	$< 1.5 \times 10^{-14}$

We have searched in a similar manner data taken during the large radio outbursts from Cyg X-3 in June and July 1989[8]. We find no evidence for any enhanced emission from the source on timescales from 15 minutes to 1 day during this period. Flux limits are a factor of five below those reported during similar radio flares in October 1985 by the Baksan experiment[9].

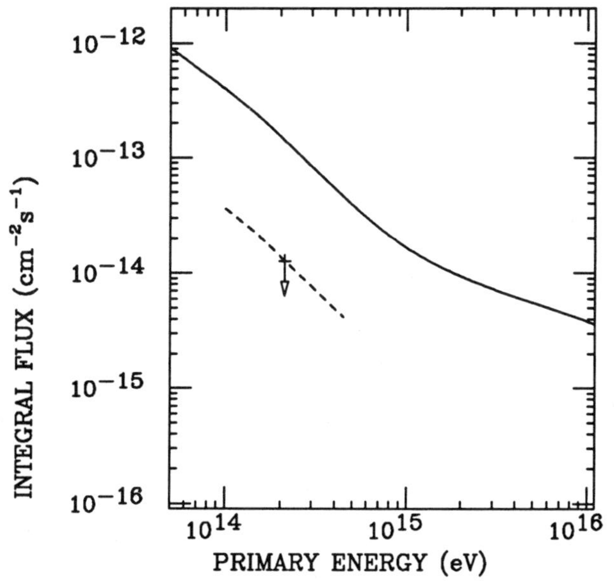

Figure 3. Upper limits on the γ-ray flux of Cygnus X-3 from the Utah-Michigan array using muon-poor showers with $\log_{10}N > 4.5$. (from Table 1). The dashed line indicates the flux upper limit (90% CL) and the arrow is drawn at approximately the median energy of γ-rays which would have triggered the detector (see text). The solid line approximates some of the earlier results and limits obtained above 10^{14} eV[2]. The dip in the E_γ^{-1} spectrum corresponds to the expected loss of UHE γ rays which interact with the 3° cosmic background radiation.

Similarly negative results have been obtained in searches for emission from Her X-1[9] and a whole sky survey for sources[10]. We have also analyzed data from the Crab Nebula[11]. This object is emerging as the leading candidate for a standard candle of VHE gamma rays. The persistent emmission reported by at least two groups[12] has an apparently softer spectrum than that presumed from Cyg X-3. Consequently, our flux limits are about twenty times higher than these measured flux levels extrapolated to 200 TeV.

5. OUTLOOK AND SUMMARY

The Chicago Air Shower Array will permit flux sensitivity substantially below previous reports of Cyg X-3, Her X-1, and perhaps approaching extrapolated levels for the Crab Nebula. CASA offers eight times the target area and improves the angular resolution by a factor of four over the existing apparatus. Using $S/\sqrt{B}$ as a figure of merit, we anticipate a factor of ten improvement from these factors alone. Furthermore, the energy threshold will be somewhat lower due to the higher density of counters.

The first 529 units of the large array are now in place with data taking begun in February 1990. The remaining 560 units are scheduled for installation in 1990 along with the final eight muon patches.

In summary, large scale muon detection and surface measurements will offer a definitive resolution of the current puzzle regarding UHE gamma ray emission from astrophysical objects. If photons interact conventionally, muon rejection provides demonstrated sensitivity to fluxes well below those previously reported. Should there be new physics and/or anomalous interactions, the UMC array will be more sensitive than prior attempts and the muon structure will be examined in detail.

REFERENCES

[1] M. Samorski and W. Stamm, Astrophys. J.**268**, L17-21,(1983); M. Samorski and W. Stamm, Proc. 18th Int. Cosmic Ray Conf. **11**, 244 (Bangalore, 1983).

[2] For reviews see R.J. Protheroe, Proc. 20th Int. Cosmic Ray Conf.**8**, 21 (Moscow, 1987); W. Hermsen et al.,Astron. Astrophys. **175**, 141 (1987).

[3] D. Sinclair, Nucl. Instrum. Meth. **A278**,583 (1989).

[4] G.L. Cassiday et al., Proc. 21st Int. Cosmic Ray Conf., paper HE 3.4-1 (Adelaide, 1990).

[5] K. Greisen, Prog. Cosmic Ray Phys. **3**, 1 (1956); E.J. Fenyves et al., Phys. Rev. **D37**,649 (1988).

[6] K. Greisen, Ann. Rev. Nucl. Sci. **10**, 63 (1960).

[7] R.W. Clay, R.J. Protheroe, P.R. Gerhardy, Nature (London) **309**, 687 (1984).

[8] G.L. Cassiday et al., Phys. Rev. Lett. **63**,2329 (1989); D. Ciampa et al., Proc. 21st Int. Cosmic Ray Conf., paper OG 4.1-12 (Adelaide, 1990)

[9] V.V. Alekseenko et al., Proc. 20th Int. Cosmic Ray Conf., **1**, 229 (Moscow, 1987); V.S. Berezinsky, Nature (London) **334**, 506 (1988).

[10] D. Ciampa et al., Proc. 21st Int. Cosmic Ray Conf., paper OG 4.2-9 (Adelaide, 1990).

[11] D. Ciampa et al., Proc. 21st Int. Cosmic Ray Conf., papers OG 4.0-19, OG 1.7 3 (Adelaide, 1990).

[12] S.C. Corbato et al., Proc. 21st Int. Cosmic Ray Conf., paper OG 4.3-11 (Adelaide, 1990).

[13] T.C. Weekes et al., Astrophys. J. **342**, 379 (1989); C. Akerlof et al., Proc. 21st Int. Cosmic Ray Conf., paper OG 4.3-1 (Adelaide, 1990).

Astronomy at Ultra-High Energies: Results from the CYGNUS Experiment

Todd J. Haines
University of Maryland

The CYGNUS Collaboration

D.E. Alexandreas,[1] R.C. Allen,[1] D. Berley,[2]§ S.D. Biller,[1] R.L. Burman,[3] R. Cady,[4]
C.Y. Chang,[2] B.L. Dingus,[2] G.M. Dion,[1] R.W. Ellsworth,[5] J.A. Goodman,[2] T.J. Haines,[2]
C.M. Hoffman,[3] P. Kwok,[2] J. Lloyd-Evans,[3]* X-Q. Lu,[1] D.E. Nagle,[3] M.E. Potter,[3]
V.D. Sandberg,[3] M.J. Stark,[2] R.L. Talaga,[2]‡ P.R. Vishwanath,[1]†
G.B. Yodh[1], and W. Zhang[3]

[1] *The University of California, Irvine, California 92717*
[2] *The University of Maryland, College Park, Maryland 20742*
[3] *Los Alamos National Laboratory, Los Alamos, New Mexico 87545*
[4] *The University of Notre Dame, Notre Dame, Indiana 46556*
[5] *George Mason University, Fairfax, Virginia 22030*

ABSTRACT

The CYGNUS experiment is composed of an air-shower array and muon detectors, located in Los Alamos, NM, and operating at energies above about 50 TeV. Recent results include a search for emission from Cygnus X-3 during the radio outbursts of June and July, 1989, preliminary results from a search for diffuse emission from the galactic plane, and preliminary results from a search for emission from possible northern hemisphere point sources, both known and unknown.

The interest in ultra-high energy astronomy has blossomed in the last several years: there is now at least one experiment on nearly every continent in the world. Further, the motivation for these efforts is twofold: an intense interest in the astrophysical mechanisms capable of generating particles of such high energies, and in the particle physics that can be uniquely explored at these energies. It is with these motivations that the CYGNUS collaboration is exploring this unique field.

The salient features of the CYGNUS experiment, which has operated nearly continuously since April, 1986, will be highlighted here; further discussions of the properties of the experiment, as well as previous results, can be found in ref. [1]. The experiment is located at an altitude of 2,130 m, corresponding to an atmospheric overburden of 800 g/cm^2, in Los Alamos, NM (106.3°W, 35.9°N). The air-shower array is composed of 106 scintillation counters, each counter having an area of about 1 m^2, deployed over an area of about 20,000 m^2; there is typically 15 m between counters. The direction of each event, determined from the timing information from each counter, is reconstructed with a typical uncertainty of 0.8°; the maximum systematic error in directional reconstruction is determined to be less than 0.3°, small enough to be of no practical concern. No requirement is made on the core position of the events. The experiment has a roughly cubical muon detector, 6 m on a side, that presents an effective area of about 44 m^2 to a typical air shower event; a muon must have a minimum energy of about 2 GeV to penetrate the shielding around the muon detector.

During the week of 30 June, 1989, a layer of lead, about one radiation length thick, was placed above each detector. The lead produces a lower trigger threshold and a larger effective area (area over which triggers are recorded) of the array by converting some of the photons in the shower. This results in both a change in data collection rate (about twice the event rate) and a change in the average number of detected muons per event. The number of detected muons per event decreased from typically 2.0 to 1.65 after the lead was installed.

More recently, an additional 96 counters have been deployed over an area of about 60,000 m^2. The spacing between the new counters varies from 20 m for the region nearest the original array to 30 m for the region furthest from the original array. The reason for the graded approach is to expand the high-energy capabilities of the CYGNUS experiment in a smooth fashion. This expanded array has a sensitivity to point sources about a factor of 2 to 4 times larger than the original array. An additional muon detector, composed of 30 scintillation counters of total area 70 m^2, is buried near the center of the new array under an overburden of about 710 g/cm^2, corresponding to a muon energy threshold of about 1.5 GeV. Finally, the anticoincidence shield from another neutrino detector, LAMPF experiment 645, is soon to be added to the data collection system, providing an additional 80 m^2 of muon detection near the center of the old array.

A powerful check of the absolute pointing accuracy and angular resolution of the array can be made by observing the deficit of cosmic rays from the direction of the sun and moon. Since cosmic rays of these energies are not effected by the solar or terrestrial magnetic fields, the shadowing caused by absorption on the sun and moon should be observable as a deficit around these objects, which each are about 0.25° across. A search for the sun and moon shadow produces a combined 4σ deficit from a total of over 200,000 events within 5° of the sun and moon, consistent with the number of cosmic ray events and the size of the objects. A maximum likelihood analysis of the deficit yields an average angular resolution of 0.8°. This is the first time the shadows of the sun and moon have been observed and used to establish the angular resolution of an air shower array.

In 1989, Cygnus X-3 had two large radio outbursts, the first began on about 2 June and the second on 21 July. In fact, Cygnus X-3 was first discovered as a γ-ray source at energies above ~ 1 TeV in observations made about one week after the first observation of a large radio outburst in September, 1972. Several groups reported signals at both ~ 1 TeV and ≥ 100 TeV energies[2] during the last such outburst of October, 1985. Most of these reports were of emission, lasting from minutes to days, during the first several days after the outburst; other experiments reported possible excesses lasting approximately one month around the time of the radio flare.

The data are analyzed on a day-by-day basis, each day defined to be the time from which Cygnus X-3 rises until it sets in the sky. The signal region is chosen to maximize the significance of a point-source signal; the total bin width is 2.3°in declination by 3.0°in right ascension and should contain 71% of the total signal. The daily background is estimated from the average of 5 background bins on each side of the source bin in right ascension.

This simple technique accounts for daily variations in the operating characteristics of the array from, for example, changes in the local air pressure. An appropriate cut in hour angle is made to ensure equal exposure for the source and background bins.

It is expected that showers initiated by γ-ray primaries should be deficient in muons when compared with proton primaries; a cut requiring muon deficient events should thus enhance the sensitivity to γ-ray point sources. Nevertheless, evidence exists to the contrary.[1] Thus, all observed events were analyzed; events with no detected muons were considered separately. This second set of data is a subset of the first so that the results from the two analyses are not independent.

The calculation of the effective area, needed to calculate all of the fluxes given here, is based partially on the observed cosmic ray rate and partially on a Monte Carlo calculation of the acceptance of the array. The motivation for this is to make the fluxes less prone to errors in the Monte Carlo simulation due to, for example, uncertainties in the exact trigger condition for the array. For each data set, the effective area for cosmic ray showers is simply calculated using the known cosmic ray flux and the observed number of background events. A Monte Carlo calculation is used to correct for the change in array response due to the different energy spectra for background and source events, and due to the different longitudinal development of photon and hadron induced showers. This increases in the effective area by a factor of 3.0, mostly due to the assumed hard spectrum, for the data considered here. The Monte Carlo calculation predicts the absolute cosmic ray trigger rate of the array to within $\pm$ 50% for a variety of different trigger conditions.

The daily excess number of events is shown in Figure 1; occasional gaps, the most notable of which was the day of the first burst, are due to experimental hardware failures. Taken together, there were 4347 events in the source region with 4338 ± 21 expected background (2255 observed and 2254 ± 15 expected for events with no detected muons). A 90% confidence level (CL) upper limit on the flux from Cygnus X-3 during this time is $3.0\times10^{-13}\mathrm{cm}^{-2}\mathrm{s}^{-1}$ for all data ($2.2\times10^{-13}\mathrm{cm}^{-2}\mathrm{s}^{-1}$ for data with no detected muons); all flux limits here are for the integral flux above 50 TeV assuming an integral power-law spectral index of -1.0.

The phaseograms for the different data sets were constructed in an effort to search for emission correlated with the 4.8 hr period of Cygnus X-3[3]. There is no evidence of such a correlation.

There is no evidence for either pulsed or unpulsed emission during the week following either of the radio bursts. The 90% CL upper limit for the flux for the week after the first burst is $2.5\times10^{-12}\mathrm{cm}^{-2}\mathrm{s}^{-1}$ ($1.0\times10^{-12}\mathrm{cm}^{-2}\mathrm{s}^{-1}$ for events with no detected muons) and for the week after the second burst it is $8.4\times10^{-13}\mathrm{cm}^{-2}\mathrm{s}^{-1}$ ($6.2\times10^{-13}\mathrm{cm}^{-2}\mathrm{s}^{-1}$).

The day with the largest excess, UT 23 July 1989 (JD 2447730.5), is about two days after the second burst, with 104 from the source bin when only 77.3 were expected. The probability for this excess to have occurred from a statistical fluctuation in the entire data set is about 18%. However, if the fact that this day is only two days after the second burst is considered important to establish a correlation with the burst, the probability is about 1%. Most of our excess is concentrated during two hours (about UT 7:30 to UT 9:30). If the excess during this day is considered as evidence for a burst, it would correspond to a flux of about $5.3\times10^{-12}\mathrm{cm}^{-2}\mathrm{s}^{-1}$ averaged over the entire day and about $1\times10^{-11}\mathrm{cm}^{-2}\mathrm{s}^{-1}$ during the two hours. The 4.8 hr phase of Cygnus X-3 during these two hours was between 0.8 to 0.2.

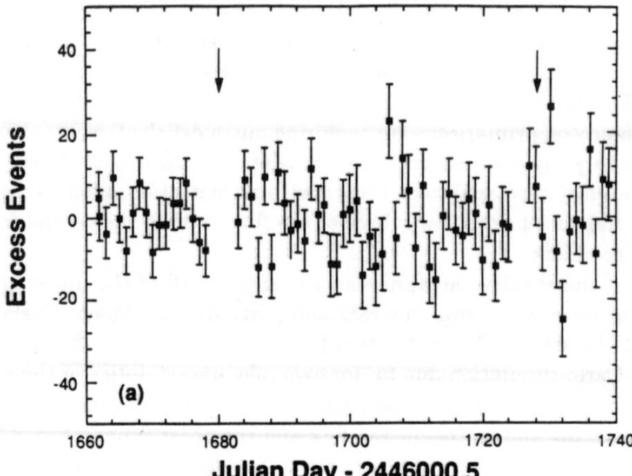

Figure 1: Daily excess from Cygnus X-3.

At low energies (~ 100 MeV), the plane of the galaxy is, except for a few powerful point sources, the dominant source in the sky. The photons at these energies are primarily produced by cosmic rays interacting with the interstellar medium to produce π^0's. This same mechanism must occur at ultra-high energies as well; therefore, the galactic plane is expected to be a continuous source of γ-ray's. The predictions for the fraction of the cosmic ray flux due to these γ-ray's range from 10^{-5} to 10^{-3}. These predictions depend on quantities that are not well measured like the galactic cosmic ray confinement time, the interstellar density, and the galactic cosmic ray flux. Additionally, the cosmic ray flux seems to be anisotropic, with a sidereal modulation observed, at least, at higher energies. Finally, the "right ascension" scan technique for point source searches, as described above, is not well suited to search for many different point sources, especially if they have yet to be identified. It is for these reasons that a technique has been developed to calculate the expected cosmic ray background at every point in the northern sky.

The goal of this analysis is to produce a "background" sky map, that is, the number of events expected from each point in the sky assuming there is no source present there; the analysis has been done on a sample of about 23 million air shower events taken during the first 2 years of operation. This map can then be compared to the observed number of events from every point in the sky in an effort to search for structure. The background map is calculated by first accumulating the distribution of the observed events in local coordinates (zenith and azimuth). For each actual event time, 10 "background events" are chosen randomly from the local coordinate distributions. The celestial coordinates of these events are then used to produce the "background" sky map; these events represent a sample that is as equally likely to have been observed from the cosmic ray background as the actual events, assuming no sources are present and the cosmic rays are isotropic. The advantage of this technique is that the exact running conditions of the experiment are included in the background calculation: the relative exposure to each part of the sky, the trigger conditions

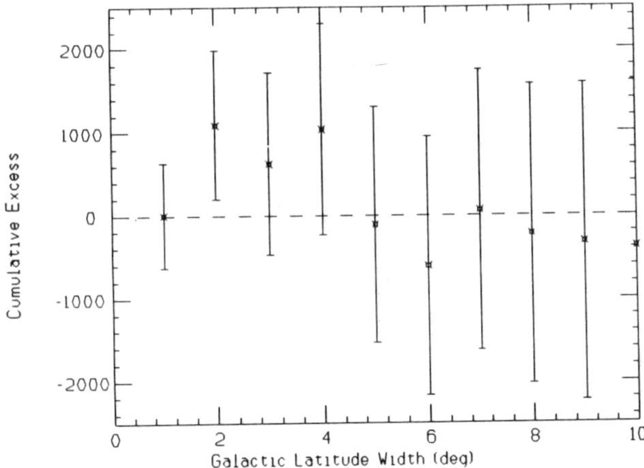

Figure 2: Cumulative excess from the galactic plane as a function of galactic latitude.

on each day, and the trigger rate at each time of day (dependent, for example, upon the local air pressure) are all accounted for. Monte Carlo studies have shown that simulated signals from point sources as well as diffuse sources, such as the galactic plane, are found while effects due to, for example, exposure biases are removed.

The search for diffuse emission from the galactic plane is then straightforward: the observed data is compared with the expected background in galactic coordinates. Figure 2 shows the preliminary result of the excess from the galactic plane. Because the width of the diffuse region at ultra-high energies is unknown, the excess is accumulated as a function of galactic latitude for all galactic longitudes (longitudes from about 30° to about 210° are observable from our site, we do not see the galactic center). Since there is not a significant excess at any latitude, a limit on the fractional excess and the corresponding flux limit (assuming the spectrum is the same as for the cosmic rays) is shown in Figure 3, again as a function of galactic latitude. For an assumed width of ±10° galactic latitude, the 90% CL upper limit on the fractional excess is about 1 part in 1,500 corresponding to a flux limit of 1.1×10^{-11} cm^{-2}s^{-1}sr^{-1}.

Evidence for continuous emission from a list of point sources is easily sought using a bin of 2.4° in declination and 2.4°/cos(δ) in right ascension with this method. A list of potential sources including pulsars, binary x-ray sources, and others, is examined and the excess from each is determined. None of the sources given in Table 1 has a significant excess; therefore, preliminary flux limits, assuming both hard ($\gamma = -1$) and cosmic ray spectral indices, are also given in the Table.

Only slightly more difficult is searching for evidence of emission from "unknown," or at least unspecified, sources. The entire sky from 0° declination to 80° declination is divided into bins of 2.8° in declination by 2.8°/cos(δ) in right ascension and the excess in each bin is calculated. There are about 2,550 such bins in the sky with some regions () excluded because of technical problems with the data there. This procedure is repeated by moving

Source	Limit Frac. Excess	Flux Limit CR Index	Flux Limit Hard Index
CygX3	0.0130	5.59	1.86
HerX1	0.0133	5.68	1.89
Crab	0.0195	8.34	2.78
VirgoA	0.0213	9.13	3.04
K5	0.0195	8.38	2.79
Geminga	0.0240	10.28	3.43
K4	0.0168	7.20	2.40
K1	0.0162	6.95	2.32
4U0115+63	0.0288	12.36	4.12
K6	0.0143	6.15	2.05
M31	0.0202	8.67	2.89
1E2259+58	0.0154	6.58	2.19
2CG095+04	0.0189	8.11	2.70
2CG078+00	0.0130	5.56	1.85
2CG065+00	0.0137	5.85	1.95
PSR1953+29	0.0138	5.91	1.97
PSR1937+21	0.0153	6.57	2.19
PSR1929+10	0.0197	8.43	2.81
401907+09	0.0197	8.43	2.81
PSR0950+08	0.0217	9.31	3.10
2CG135+01	0.0168	7.19	2.40
2CG121+04	0.0190	8.13	2.71
4U0042+32	0.0134	5.76	1.92
4U0316+41	0.0136	5.83	1.94
4U0352+30	0.0142	6.07	2.02
4U0614+09	0.0274	11.76	3.92
4U1837+04	0.0237	10.15	3.38
4U1901+03	0.0329	14.08	4.69
4U1918+15	0.0172	7.36	2.45
4U2321+58	0.0154	6.59	2.20
4U1257+28	0.0142	6.07	2.02
4U1813+50	0.0137	5.86	1.95
4U1957+40	0.0372	15.94	5.31
PSR1957+20	0.0154	6.62	2.21
PSR0355+54	0.0146	6.28	2.09

Table 1: Point source 90% CL limits. Frac. excess is the limit on the fractional excess from the source; flux limits are in units of $10^{-13}\mathrm{cm}^{-2}\mathrm{s}^{-1}$.

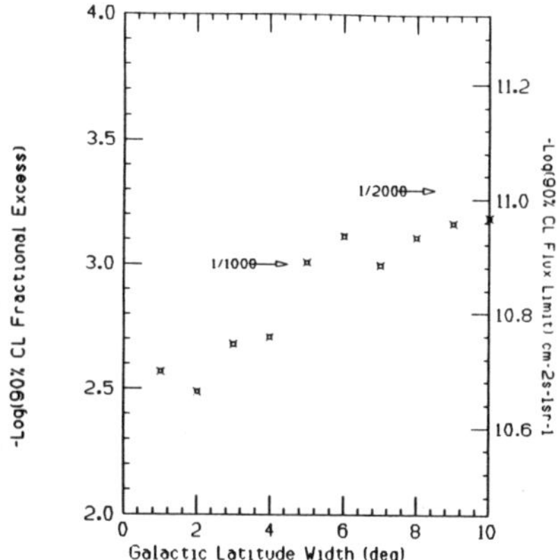

Figure 3: 90% CL limit on the cumulative fractional excess from the galactic plane and the corresponding flux limit.

each bin by 1/2 of a bin width in declination, by 1/2 of a bin width in right ascension, and by 1/2 a bin width in both declination and right ascension, to allow for the possibility that a source is not well centered in a bin. No bin of these four (non-independent) samples has a significant excess after accounting for the total number of bins in the sky. Since a flux limit from each point in the sky is difficult to present, Figure 4 gives corresponding flux limit for each bin, as a function of declination, assuming no excess in the bin of interest.

Although the field of ultra-high energy astronomy is gaining intense worldwide interest and the experiments are continually improving, positive source detections are still rare. There is yet to be an established "standard candle" at these energies, either because no such source exists or because it is too faint to have been detected. Nonetheless, the implications that these studies have had, and will continue to have, for the fields of both astrophysics and particle physics are great enough to warrant the increasing interest in this field for the future.

Many of us would like to thank the MP division at LAMPF for their kind hospitality during all stages of this experiment. This work is supported in part by the Los Alamos National Laboratory, the National Science Foundation, the US Department of Energy, and the Institute of Geophysics and Planetary Physics of the University of California.

* permanent address: The University of Leeds, Leeds, UK.
† on leave from Tata Institute of Fundamental Research, Bombay, India.
‡ now at Argonne National Laboratory, Argonne, IL.
§ present address: National Science Foundation, Washington, DC.

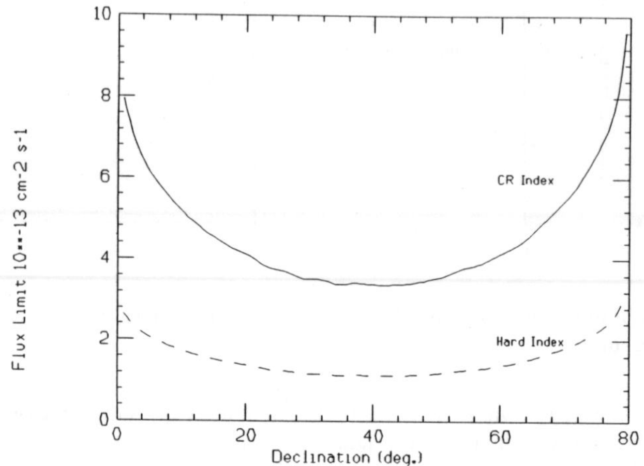

Figure 4: The point source flux limit for each point in the sky assuming no excess at that point.

References

[1] D.E. Alexandreas *et al.*, UCI-HE-GR-90-2, LA-UR-90-426 submitted to Phys. Rev. Lett.; T.J. Haines *et al.*, Phys. Rev. D **41**, 692(1990); T.J. Haines *et al.*, Proc. of the Workshop on the Physics and Experimental Techniques of High-Energy Neutrino and VHE and UHE Gamma-Ray Particle Astrophysics, G. Yodh and D. Wold, eds., Nucl. Phys. B-Proc. Supp. (in press); B.L. Dingus *et al.*, Phys. Rev. Lett. **61**, 1906(1988); B.L. Dingus *et al.*, Phys. Rev. Lett. **60**, 1785(1988); and B.L. Dingus, Ph.D. thesis, University of Maryland Report No. 88-253 and LA-11431-T, 1988(unpublished).

[2] for a review see R.J. Protheroe, Proc. of the 20th Int. Cosmic Ray Conf., V.A. Kozyarivsky *et al.*, eds., (Nauka, Moscow) **8**, 21(1987) and references therein.

[3] M. van der Klis and J.M. Bonnet-Bidaud, Astron. Astro. **214**, 203(1989).

A Search for Evidence of Increased Production of Hadrons by Gamma Rays Above 1 TeV in Balloon-Borne Emulsion Chambers

S.C.Strausz, J.Iwai, T.A.Koss, J.J.Lord, R.J.Wilkes, and J.Woosley
Department of Physics, FM-15
University of Washington, Seattle, WA 98195, USA

Abstract

Gamma rays with energy 1-60 TeV observed in balloon-borne emulsion chambers are examined for evidence of an increased cross section for production of hadrons as suggested by muon-rich air showers from gamma ray sources.

1 Introduction

Several experiments have detected anomalous muon-rich air showers, correlated in arrival direction and phase with galactic point sources such as Cyg X3 and Herc X1 [1] The cosmic rays that produce these showers must be stable, neutral, nearly massless, and have a substantial cross section for atmospheric interactions. Photons are the only known particles that fulfill these requirements, but the observed muon production is up to two orders of magnitude larger than predicted by simulations of gamma ray induced air showers. It has been proposed that the anomalous showers may be due to an increase in the hadronic cross section of gamma rays due to the increased gluonic content of gamma rays at extremely high energies [2]. The present study provides an opportunity to investigate gamma ray interactions at energies above those available from existing accelerators and with detectors that observe the initial gamma ray interactions that are not directly observed by ground-based cosmic ray experiments.

2 The Detector

Emulsion chambers have been flown by stratospheric balloons at altitudes corresponding to an average atmospheric overburden of 5 g/cm^2 as parts of experiments to measure the cosmic ray nuclei and electron spectrum. The emulsion chambers are constructed of multiple layers of nuclear emulsion plates, X-ray films, and lead plates. The analysis consists of the location of high energy showers, the measurement of their energies, the examination of the shower characteristics, the count and positional measurement of tracks, Relative spatial resolution is better than one micron, and energy resolution by conventional calorimetric techniques is accurate to 20% [5].

3   Analysis

Even though there is little air above the chambers cosmic rays occasionally interact in the air and produce families of parallel showers that are well separated and with a negligible probability of secondary interactions in air. The showers from two large families were selected for this analysis: the K family consisting of 31 showers at 67 degrees from vertical with individual energies of 1 to 15 TeV and total photon energy $\Sigma E_\gamma = 130$ TeV observed in a chamber flown by the Galactic Electron Collaboration [3], and the DD family consisting of 17 showers at 57 degrees with individual energies of 1 to 60 TeV and $\Sigma E_\gamma = 310$ TeV observed in a chamber flown by the JACEE Collaboration [4]. The total vertical thickness of 9 radiation lengths and only 0.3 proton interaction lengths (eventhough increased by the slant angle of the nonvertical particles) gives families the appearance of being gamma ray rich. Of the 31 observed showers in the K family, 27 are due to photons (pure electromagnetic shower, no visible incoming charged particle, high starting point) 2 are due to hadrons (deep starting point), 1 is due to a charged hadron with a high starting point, and 1 shower($\#13$) that will be argued to be anomalous. The DD family consisted of 13 photons, 2 electrons, and 2 hadrons.

In the K family, shower $\#13$ has an initial hadronic interaction, a neutral initiator, and a high starting point. Other than statistical fluctuations, the explanations for this are that the initiator of the shower was not neutral, the interaction was not hadronic, the initiator was a projectile fragment neutron, the initiator was a neutral hadron, or there is a increase in the cross section for hadron production by high energy gamma rays.

The initial interaction of shower $\#13$ was not due to an electromagnetic pair production. Hadronic showers differ from electromagnetic showers by having a higher initial multiplicity, a wider spread due to higher transverse momenta, hadronic secondary interactions, and and subcore structuring. Shower $\#13$ has a high initial multiplicity (≈ 35), and a hadronic secondary interaction vertex in emulsion by a charged track.

Shower $\#13$ is not initiated by a charged hadron. Close examination of the emulsion plates above the first appearances of the shower, employing nearby tracks as references, reveals no incoming charged particle track with the expected position and angle.

It is unlikely that shower $\#13$ is initiated by a projectile fragment neutron. If the primary cosmic ray was iron (cosmic ray composition decreases sharply above the iron group nuclei) and all neutrons are fragmented to single particles, the number of neutrons expected to interact in 0.2 interaction lengths is 0.5. Projectile fragments have small deflections from the central event axis, but shower$\#13$ occurs at an angle 100 times wider than that of fragments. Projectile fragments have energies much larger than the produced gamma rays, but shower $\#13$ has energy (6 TeV) less than many of the gamma rays. No projectile fragments that interact deeper in the chamber seen.

There is a small probability that shower #13 is initiated by a neutral hadron. From the average multiplicities of produced particles at CERN UA5 540 GeV $p\bar{p}$ interactions [6], assuming that particle production ratios are similiar at high energies, the ratio of the number of neutral hadrons able reach detector from air vertex to the number of photons is 9%. The 4.6 radiation lengths, in which all photon showers started, is equivalent to 0.20 interaction lengths. Since the Kp is about one half of the pp cross section, the effective number of interaction lengths is decreased to about 0.15. The number of neutral hadrons accompanying 40 photons expected to interact is $N_{interact} = N_{initial}(1 - e^{-x/\Lambda})$ or about 0.50.

The probability for hadron production by gamma rays is small compared with electron pair production. From accelerator studies of 100 GeV gamma ray interactions, the expected cross sections in lead for pair production is 40 b and for hadron production is 25 mb. Accompanied with 40 gamma ray pair production interactions are expected a .025 gamma ray hadron interactions.

In summary, only one apparent gamma ray hadron production event was seen with an calculated neutral hadron background of 0.5 events, and with a predicted gamma ray hadron production of .025 times a proposed increase factor. This suggests (with simple $\sqrt{n}$ statistics) for 1-60 TeV gamma rays that any increase in the hadron production is seen only at the 0.5 sigma level and that the factor of increase is less than 100 at the 2. sigma level. This upper limit on the factor of increase is lower if secondary photons in the showers are considered. Work is in progress to improve this preliminary result.

Work supported by Department of Energy contract DE-AS-06-88ER-40423.

References

[1] M.Samorski & W.Stamm, *18th Int. Cosmic Ray Conf.* **EA1.1-39**, 244 (1983). M.L.Marshak, *et al.*, *Phys. Rev. Lett.* **54**, 2079 (1985). G.Battistoni, *et al.*, *Phys. Lett.* **155b**, 465 (1985). B.L.Dingus, *et al.*, *Phys. Rev. Lett.* **61**, 1906 (1988). M.Teshima, *et al.*, *Phys. Rev. Lett.* **64**, 1628 (1990).

[2] F.Halzen, Proc.of Grande Workshop on the UHE Gamma Ray and Neutrino Int.(Little Rock, AR, May 1989; to be published), MAD/PH/504, July, 1989. M.Drees & F.Halzen, *Phys. Rev. Lett.* **61**, 275 (1988).

[3] T.A.Koss, *et al.*, *17th Int. Cosmic Ray Conf.* **HE3.2-24**, 90 (1981).

[4] T.H.Burnett, *et al.*, *Nucl. Inst. Meth.* **251**, 583 (1986).

[5] J.Nishimura, *et al.*, *Ap. J.* **238**, 394 (1980).

[6] G.J.Alner, *et al.*, *Phys. Rep.* **154**, 247 (1987).

THE HADRONIC STRUCTURE OF THE PHOTON

INA SARCEVIC
Department of Physics,University of Arizona
Tucson, AZ 85721

ABSTRACT: We present results of calculations of the total inelastic photon-air cross sections at ultra-high energies (up to 10^8 GeV in the lab), of relevance to on-going cosmic-ray experiments.[1] We discuss the results in the context of the apparent muon excess of air-showers associated with astrophysical point sources.

1. INTRODUCTION

Recent cosmic-ray data[2] showing muon excesses in air showers generated by neutral, stable particles from point sources (e.g. Cyg X-3, Her X-1, Crab Nebula) hint at new physics at high energies, presently inaccessible with existing colliders. The only candidates for such particles in the Standard Model are photons. Conventionally, a photon-initiated shower is electromagnetic and, therefore muon-poor. The number of muons produced in an electromagnetic cascade is more than an order-of-magnitude smaller than in a hadronic shower. However, if there is a threshold effect for photoproduction at very high energies, the conventional expectations for the muon yield in photon-initiated shower would be altered.[3]

2. ULTRA-HIGH ENERGY PHOTONUCLEAR CROSS SECTIONS

In order to investigate this new threshold effect, we have calculated the total inelastic photon-proton and photon-air cross sections in a QCD-based diffractive model.

The photon-proton cross section is given by

$$\sigma_{\gamma-p} = 2\pi \int bdb[1 - e^{-(P_1(b,s)+P_2(b,s))}] \tag{1}$$

where the term in the brackets is the probability that particle production occurs, either via a hard-(P_1) or via a soft-(P_2)type parton interaction. In the above, P_1 and P_2 are products of the parton cross sections and the overlapping spatial parton densities ($P_1 = A(b)\sigma_{QCD}$ and $P_2 = A(b)\sigma_{soft}$). We determine σ_{soft} and the p_T^{min} parameter in σ_{QCD} from the low energy data.[1] Our results for the photon-proton cross sections at energies $10 \leq \sqrt{s} \leq 10^4$ are presented in Fig. 1. We find that $\sigma_{\gamma-p}$ at $\sqrt{s} > 10^3 GeV$ is one order-of-magnitude larger than the conventional value of $0.114 mb$. The theoretical uncertainties (the value of p_T^{min} and the photon structure

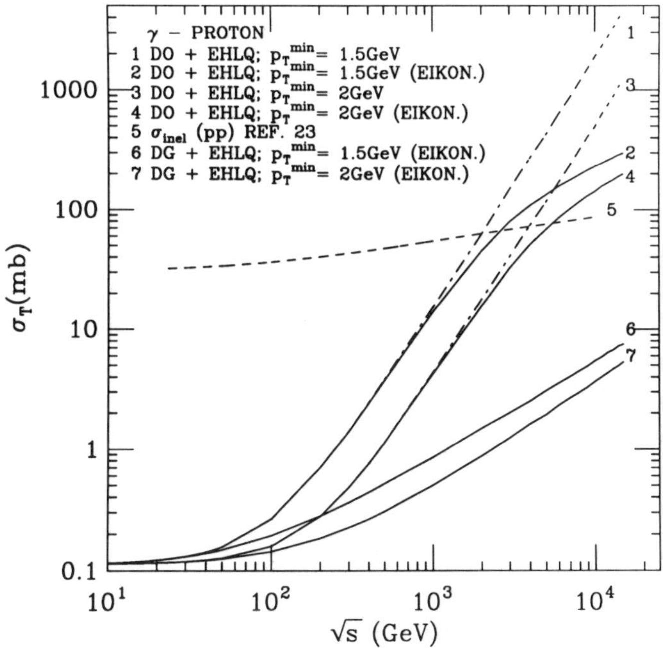

FIG. 1: Jet (dashed curves) and eikonalized (solid curves) inelastic photon-proton cross sections, from Ref. 1.

function at low x) are included in Fig. 1. Here we emphasize that precise photoproduction measurements at Fermilab energies ($E_{lab} \sim 350GeV$) and at HERA energies ($E_{lab} \sim 10^4 GeV$) will reduce these uncertainties substantially.[1]

Next, we calculate the total inelastic photon-air cross section. We use Eq. (1), where P_1 and P_2 now also contain the overlapping spatial densities of the nucleons inside the nuclei. Details of this calculation can be found in Ref. 1. Our results for $\sigma_{\gamma-air}$ at energies $10^4 GeV \leq E_{lab} \leq 10^8 GeV$ are presented on Fig. 2. We note that at ultra-high energies ($E_{lab} \sim 10^8 GeV$) the cross sections are $40 - 500mb$, depending on the choice of the photon structure function. This is to be compared with the conventional value of $1.5mb$. Clearly, the new cross sections will change the conventional expectations for the number of muons produced in the photon-initiated shower.

In order to see this, we calculate the probability (P_a) that the first interaction of the photon is hadron-like and the probability (P_b) that the photon will have at least one hadron-like interaction before reaching $100TeV$ energy. We find P_a to be $14 - 57\%$ and P_b to be about 100% at $10^8 GeV$ energy.[1] Even at much lower energies

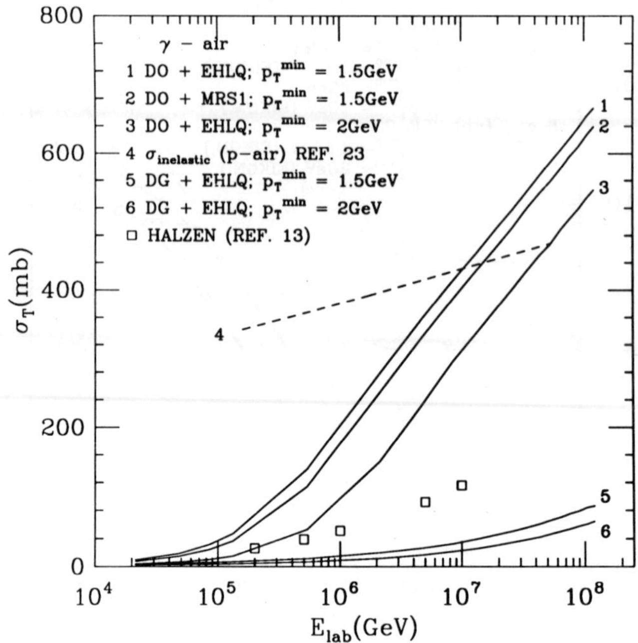

FIG. 2: The total inelastic photon-air cross sections, Ref. 1

($\sim 1TeV$) these probabilities are not negligible. The tendency with increasing energy of the photon to manifest hadronic character is clear and suggestive. We are presently developing a full Monte Carlo shower code which should be able to test whether this mechanism can be the explanation of the "muon puzzle".

ACKNOWLEDGMENTS

The work presented here was done in collaboration with R. Gandhi, A. Burrows, L. Durand and H. Pi and was supported in part by DOE grant DE-F602-85ER40213.

REFERENCES

1. R. Gandhi, I. Sarcevic, A. Burrows, L. Durand, H. Pi, University of Arizona preprint, AZPH-TH/90-3.

2. M. Samonski and W. Stamm, *Ap. J.* **L17**, 268 (1983); B. L. Dingus, et. al. *Phys. Rev. Lett.* **61**, 1906 (1988); Sinha et al., Tata Institute preprint, OG 4 6-23; T. C. Weekes, *Phys. Rep.* **160**, 1 (1988), and reference therein.

3. M. Drees and F. Halzen, *Phys. Rev. Lett.* **61**, 275 (1989).

RECENT RESULTS FROM THE HOMESTAKE CHLORINE SOLAR NEUTRINO EXPERIMENT

R. Davis, Jr., K. Lande, C. K. Lee, P. Wildenhain,
A. Weinberger and T. Daily, University of Pennsylvania,
Philadelphia, PA 19104

B. Cleveland, Los Alamos National Laboratory, Los
Alamos, NM 87545

J. Ullman, Herbert Lehman College of the CUNY, Bronx, NY 10468

ABSTRACT

For the past twenty years the Homestake perchloroethylene
solar neutrino detector has been monitoring the flux of
electron neutrinos reaching the earth from the sun. The
average detection rate is nearly a factor three below solar
model predictions: possible explanations for this are
discussed. There is evidence for a time variation of our
signal with the same period and phase as the solar magnetic
cycle but opposite sign. Fourier analysis shows it to be
significant with high probability, and we have found no
explanation in terms of instrumental or background effects.
Possible explanations for such a time variation are discussed.

1. INTRODUCTION

The chlorine solar neutrino experiment has been operating in
the Homestake gold mine in Lead, S. D. since 1970 with the same
extraction and counting methods. By now over 80 runs have been extracted
and counted, so the average production rate is known with good
precision: the discrepancy between it and the predictions of solar
models is the well-known solar neutrino puzzle.

As early as 1984 it was suggested that our argon production rate
was varying with time in a way related to the solar activity cycle.
Recent results extend our observations over nearly two complete solar
cycles and appear to show an anticorrelation with the solar cycle.

In the following sections of this report we will: describe our
experimental procedure, emphasizing aspects of it that make it unlikely
that observed time variations are an artifact of it or a background
effect (sec.2), compare our observed average neutrino detection rate
with model predictions and discuss some explanations that have been
proposed to explain the discrepancy between them (sec.3), and present
evidence for a statistically significant time variation in our results
and discuss possible explanations of it (sec.4).

2. EXPERIMENTAL PROCEDURE

The Homestake chlorine solar neutrino experiment detects neutrinos by the reaction

$$\nu_e + {}^{37}Cl \rightarrow {}^{37}Ar + e^- \quad \text{(threshold neutrino energy} \quad 814 \text{ MeV)}$$

Enough information is available from nuclear physics experiments to allow the cross-sections for this reaction to the ground and all available excited states to be calculated. This was first done by J. Bahcall in the 1960's, making this experiment possible [1]: since then, some uncertainties have been cleared up by nuclear reaction and decay studies, and now the cross-sections are known with confidence.

Using these cross-sections, argon production rates for neutrinos from the various solar reactions can be calculated for a particular solar model. Results of some of these calculations will be shown later.

The chlorine in this experiment is in the form of 3.7×10^5 liters (610 tons) of perchloroethylene (C_2Cl_4) 4850 feet underground in the Homestake gold mine in Lead, S. D. The underground site is necessary because nuclear reactions initiated by cosmic ray muons can also produce argon 37 in the tank. The rates of these reactions at our depth have been estimated from two independent sets of measurements [2] giving results consistent with each other and theoretical calculations: in our tank they should produce 0.08 ± 0.03 ^{37}Ar atoms per day, about 10% of the observed average signal. Muon flux through the tank is constantly monitored by a large scintillation hodoscope surrounding it and shows the rate and angular distribution expected at that depth.

In the 610 tons of perchloroethylene the average rate of ^{37}Ar production is less than 1 atom per day. If the argon concentration is allowed to build up for 2 half lives of ^{37}Ar (70 days), an average of about 15 atoms of radioactive argon will be extracted from the tank.

At the end of each run, about 0.1 cm^3 of argon carrier, either ^{36}Ar or ^{38}Ar, is mixed in with the perchloroethylene in the tank, to allow monitoring of extraction efficiency. Above the pechloroethylene is very pure helium at an absolute pressure of 1.5 atmospheres. To extract the argon, two powerful pumps circulate the perchloroethylene and draw the helium into it through eductors so that argon in the helium rapidly reaches equilibrium with argon dissolved in the perchloroethylene. At the same time the helium is circulated through a charcoal trap at liquid nitrogen temperature which traps the argon. The trapped argon is then purified by a series of chemical processes including gas chromatography and placed in a small (<0.5 cm^3) proportional counter [3].

For every run, the extraction efficiency estimated from the volume of helium processed is verified by measuring the volume of argon extracted and performing an isotope analysis of it using a mass spectrometer after the counting has been completed. A 20-hour extraction cycle gives an average of over 90% efficiency.

The decay of ^{37}Ar in the counter goes mostly by Auger emission producing ionization in a very small region: this gives a very fast-rising output pulse allowing discrimination of argon decays from background by rise time as well as energy discrimination. The counting is done in a well-shielded location in the mine, with about 40% overall efficiency and a typical background rate of a few counts per month. A run is counted for 8 to 10 ^{37}Ar half lives (200-300 days) to establish the background counting rate precisely. For a typical run, the background rate is much smaller than the initial counting rate.

We are convinced we are counting ^{37}Ar. When the data from many runs are combined on a common energy scale, they show a peak at the ^{37}Ar decay energy with 25% ± 2% resolution and a half life of 35 ± 4.5 days, in agreement with the values obtained from one of our counters filled with ^{37}Ar.

3. AVERAGE NEUTRINO DETECTION RATE

The total number of ^{37}Ar counts, corrected for extraction efficiency, growth of argon in the tank over the run length, and counting efficiency, gives an argon production rate in atoms per day. For comparison with model calculations, this is usually converted into Solar Neutrino Units (SNU): 1 SNU = 10^{-36} neutrino captures per atom per second. For the approximately 80 extractions that have been counted through the end of 1989, the average production rate is

$$0.48 \pm 0.05 \ ^{37}\text{Ar atoms/day or } 2.1 \pm 0.2 \text{ SNU}$$

Theoretical predictions of this rate, and solar energy production reactions contributing to it, are summarized in Table 1. There is little controversy over what these reactions are. The five most important of them, contibuting nearly all the sun's energy production and production of neutrinos observable in this experiment, are shown along with the expected neutrino fluxes and energies associated with them. The branching ratios quoted are from the work of Bahcall and Ulrich [4]. Other groups obtain slightly different results: a comparison with one of these is given in the last two columns, where predictions of rates for our experiment are shown.

Two important features of these reactions should be pointed out:
(1) The proton-proton reaction, which produces the overwhelming majority of the neutrino flux, is one of the chain of reactions that produce the overwhelming majority of the sun's energy. However, these proton-proton (p-p) neutrinos are emitted at low energy, in a continuous spectrum up to 0.43 MeV, below the 0.814-MeV threshold of our experiment. Any solar model that sees hydrogen fusion as the main source of solar energy will predict about the same flux of these neutrinos.
(2) A tiny fraction of the energy production goes through the ^{7}Be - ^{8}B branch, but the neutrinos produced by this branch have higher energies and thus higher cross sections for the reactions used to detect solar neutrinos, so they make an important contribution to the rates of most solar neutrino experiments. In particular, the ^{8}B neutrinos, with energies up to 15 MeV, can induce a superallowed transition to an excited state of ^{37}Ar which has a very large cross section, so most of the signal in the chlorine experiment is produced by them. Their production rate <u>is</u> model dependent. The high Coulomb barrier in the ^{7}Be + P reaction makes its rate vary strongly with the

temperature at the center of the sun: that temperature depends on the rates of energy flow through the radiation and convection zones, which depend on such variables as the assumed element abundances at the center of the sun.

TABLE I

reaction	flux, $cm^{-2}sec^{-1}$	ν energy, MeV	reaction rate in ^{37}Ar, SNU	
			Bahcall and Ulrich [4]	Turck-Chiese et al. [5]
$P+P \rightarrow D+e^{+}+\nu_e$	6.0×10^{10}	0-0.42	0	0
$P+e^{-}+P \rightarrow D+\nu_e$	1.4×10^{8}	1.44	0.2	0.2
^{3}He$+^{4}$He $\rightarrow ^{7}$Be$+$, then either				
$e^{-}+^{7}$Be $\rightarrow ^{7}$Li$+\nu_e$	4.7×10^{9}	.86,.38	1.1	1.1
or $P+^{7}$Be $\rightarrow ^{8}$B$+$, then				
^{8}B $\rightarrow ^{8}$Be$^{*}+e^{+}+\nu_e$	6×10^{6}	0-15	6.1	4.0
CNO cycle: ^{13}N $\rightarrow ^{13}$C$+e^{+}+\nu_e$	6×10^{8}	0-1.2	0.1	--
^{15}O $\rightarrow ^{15}$N$+e^{+}+\nu_e$	5×10^{8}	0-1.7	0.3	--

Note that in the last two columns, both models predict nearly identical rates from all neutrino branches except the ^{8}B branch, and that without the ^{8}B branch, the observed signal would be nearly in agreement with the predicted one. Many nonstandard solar models have been proposed which would decrease the ^{8}B neutrino flux: it is not possible to discuss them adequately in a short presentation.

The discrepancy between observed and predicted neutrino fluxes might also be explained by the physics of the neutrino. An explanation of this kind which has recently attracted much attention is the MSW (Mikheev, Smirnov, Wolfenstein) mechanism [6], by which electron neutrinos undergo a resonant transition to another neutrino state (muon or tauon) in a region inside the sun where the mass density is such that the effective masses of the two neutrino flavors are degenerate, due to their different interactions with matter. The ν_μ or ν_τ will not interact with the chlorine in our experiment. This would require a small mass difference between the various neutrino flavors and a flavor-changing interaction. These phenomena have not manifested themselves in experiments on earth, but would fit naturally into current grand unification theories.

4. TIME VARIATION OF THE OBSERVED NEUTRINO FLUX

It was pointed out as early as 1984 [7] that our observed ^{37}Ar production rate appeared to be varying with time. In fig. 1 we show the

ARGON PRODUCTION RATE

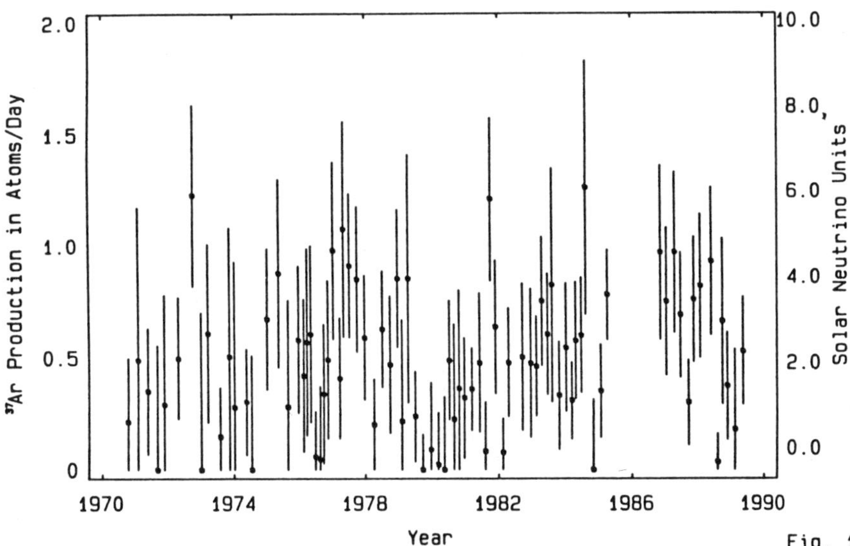

Year

Fig. 1

production rates obtained for every run we have made over the last 20 years, through run 105, extracted in June 1989. Five more runs made since then are still being counted. The gap in the data about 1986 is due to a period of nearly two years when the experiment was not taking data because of the failure of both circulating pumps.

In the data one can find two periods (1977-9 and 1986-8) of about two years when the average production rate of ^{37}Ar was more than twice the overall average rate. These roughly correspond to periods of minimum solar activity, as measured by the number of sunspots and solar flares observed. Following the first of these, in 1980 it remained almost at zero for about a year: this roughly corresponds to a period of maximum solar activity. We are once again passing through a period of maximum solar activity and once again our counting rate has been consistently well below average.

In order to explore the time variation in our signal, we have used Fourier analysis, which had already been applied to our data by Haubold and Gerth [8]. We have been analyzing our data using a date compensated Fourier transform method that is designed to deal with data taken at irregular intervals [9]. The production rate observed at each extraction is weighted by the error assigned to that extraction by maximum likelihood analysis [10]. The computer program used was supplied to us by Dr. Emilia P. Belserene of the Maria Mitchell Observatory.

At the present time we are still working on this analysis so the results presented here are only preliminary. A fundamental period of 9.6 years appears, in good agreement with the rather short length of the

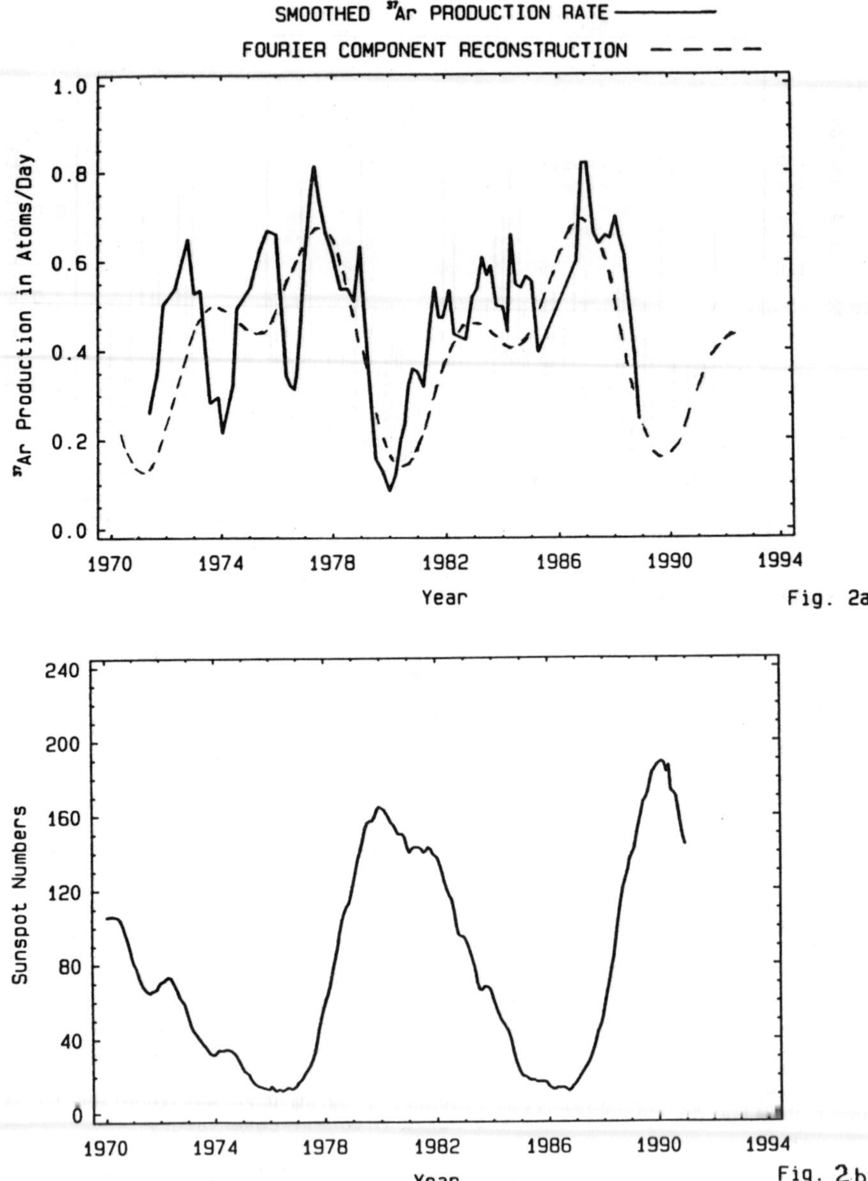

Fig. 2a

Fig. 2b

SMOOTHED SUNSPOT NUMBERS

most recent solar cycle, along with several harmonics of that period.
In fig. 2 we show a sum of the fundamental and harmonics superimposed on
a 5-year running average of our data, compared with sunspot numbers over
the same time period.

To test the significance of this, we have run 1000 Monte Carlo
simulations of our data and analyzed them by the same method. A
constant neutrino flux was assumed, designed to give the correct average
signal. The simulated extractions were given the same exposures,
extraction dates, extraction efficiencies, counter resolutions and
backgrounds, and counting times as the real ones. Based on this
simulation, we estimate that the time series of ^{37}Ar production rates we
observe would arise by chance less than 1% of the time.

This time variation could arise from some background effect that
varies with the solar cycle. Then our signal would have to be mostly
background, increasing the discrepancy between our results and solar
model predictions. We have estimated the ^{37}Ar production rate from many
such possible effects: all fall short of explaining the observed rate by
more than an order of magnitude. Furthermore, if our signal is mostly
background our results are incompatible with those of the Kamiokande
experiment described in a following paper: the solar origin of their
signal is shown by its angular correlation with the sun's position.

If there is really a time variation in the solar neutrino flux,
there are two kinds of explanations for it: either the fusion rates in
the solar core are varying with time, or there is some time variable
absorption or interaction process affecting the neutrinos as they leave
the sun. The first possibilty seems very unlikely. The solar cycle is
generally seen as a manifestation of cyclic variations in the magnetic
fields in the sun's convection zone, far from the core where the
neutrinos are produced. The time scale for variations in the parameters
determining neutrino production in the core is now seen to be 10^7 years,
not 10 years. An explanation of this kind would require drastic changes
in our understanding of both these phenomena.

Explanations involving time-varying neutrino interactions were
proposed soon after the first suggestions that our signal was not
constant. In 1986 Voloshin, Vysotskii and Okun [11] suggested that the
neutrino could have an anomalously large magnetic moment, 10^{-11} μ_B, and
that during periods of solar activity the internal field in the sun
could be as high as several kilogauss extending over distance of 10^9 cm.
Under these conditions, the left-handed ν_e can have their spins flipped
and become right-handed particles, which would not be detectable.

Another proposed class of explanations is similar to the MSW
mechanism in that it involves a resonant matter-induced transition, now
involving a simultaneous spin flip and flavor change brought about by a
transition magnetic moment: the neutrino might still have a very small
or zero static magnetic moment [12]. Whether explanations of this kind
are involved in either the time variation or the overall reduction in
neutrino flux (by MSW effect) will be better known after observations of
the solar neutrino flux are made at different energies, since the
resonant processes are energy-dependent and the spin flip process is
not. We are waiting for the Baksan and GALLEX experiments, sensitive to
the low-energy P-P neutrinos, to give us further clues.

718

ACKNOWLEDGMENTS

We gratefully acknowledge the generosity of the Homestake Mining
Company' in providing us with underground laboratory space and the help
provided by their engineering staff, particularly Mr. Steven Orr, their
Chief Mine Engineer.   Dennis Hothem and Richard Mitchell have provided
enormous help in the care and operation of the detector and laboratory.

This work is supported by the National Science Foundation, the
University of Pennsylvania Research Foundation, and the CUNY Research
Foundation.

REFERENCES

1. Bahcall, J. N., Revs. Mod. Phys. $\underline{50}$, 881 (1978); Bahcall, J.
    N. et. al., Revs. Mod. Phys. $\underline{54}$, 767 (1982)

2. Wolfendale, A. W., Young, E. C. M. and Davis, R., Nature (Physical
    Science) $\underline{238}$, 1301 (1972); Fireman,E. L., Cleveland, B. T.,
    Davis, R. and Rowley, J. K., AIP Conf. Proceedings #126, 22 (1985);
    Zatsepin, G. T., Soviet J. Nucl. Phys. $\underline{33}$, 200 (1981)

3. Operation of the extraction system is described in Davis, R.,
   Harmer, D. F. and Hoffman, K. C., Phys. Rev. Letters $\underline{14}$, 20 (1968)

4. Bahcall, J. N. and Ulrich, R., Revs. Mod. Phys. $\underline{60}$, 297 (1988)

5. Turck-Chieze, S., Cahan, S., Casse, M., and Doom, C., Astrophys.
   J. $\underline{335}$, 415 (1988)

6. Wolfenstein, L., Phys. Rev. $\underline{D17}$, 2369 (1978); Mikheev, S. P.
   and Smirnov, A. Yu., Soviet J. Nucl. Phys. $\underline{42}$, 913 (1985);
   Progr. in Particle and Nuclear Physics $\underline{23}$, 41 (1989)

7. Stozhkov, Yu. I. et. al., Soviet J. Nucl. Phys. $\underline{39}$, 543 (1984);
   Subramanian, A., Siddeshwar, L. and Yvas, P., Astron. Nachr. $\underline{6}$, 363
   (1988)

8. Haubold, H. J. and Gerth, E., Astron. Nachr. $\underline{305}$, 299 (1983)

9. Ferraz-Mello, S., Ap. J. $\underline{86}$, 619 (1981)

10. Cleveland, B. T. , Nucl. Instr. and Meth. $\underline{214}$, 451 (1983)

11. Soviet J. Nucl Phys. $\underline{44}$, 440 (1986): Soviet Phys. JETP $\underline{64}$, 446
     (1986)

12. Lim, C. S. and Marciano, W. J., Phys. Rev. $\underline{D37}$, 1368 (1988);
     Akhmedov, E. Kh., Phys. Lett. $\underline{213b}$, 64 (1988) and Soviet J.
     Nucl. Phys. $\underline{48}$, 382 (1988); Sivaram, C. , 21st Int'l Cosmic Ray
     Conf., Vol. 7, 176 (1990)

The Soviet-American Gallium Experiment at Baksan
A Status Report

A.I. Abazov, D.N. Abdurashitov, O.L. Anosov, S.N. Danshin,
L.A. Eroshkina, E.L. Faizov, V.N. Gavrin, A.V. Kalikhov,
T.V. Knodel, I.I. Knyshenko, V.N. Kornoukhov,
S.A. Mezentseva, I.N. Mirmov, A.I. Ostrinsky, V.V. Petukhov, A.M.
Pshukov, N.Ye. Revzin, A.A. Shikhin, Ye.D. Slyusareva, P.V. Timofeyev,
E.P. Veretenkin, V.M. Vermul, V.E. Yantz,
Yu. Zakharov, G.T. Zatsepin, V.I. Zhandarov
**Inst. for Nuclear Research, Academy of Sciences, USSR
Moscow 1173122, USSR**

T.J. Bowles, B.T. Cleveland, S.R. Elliott, H.A. O'Brien,
D.L. Wark, J.F. Wilkerson
Los Alamos Natl. Laboratory, Los Alamos, NM 87545 USA

R. Davis, Jr., K. Lande
University of Pennsylvania, Philadelphia, PA 19104 USA

M.L. Cherry
Louisiana State University, Baton Rouge, LA 70803 USA

R.T. Kouzes
Princeton University, Princeton, NJ 08544 USA

Abstract

A gallium solar neutrino detector is sensitive to the full range
of the solar neutrino spectrum, including the low-energy neutrinos from
the fundamental proton-proton fusion reaction. If neutrino oscillations
in the solar interior are responsible for the suppressed ^{8}B flux
measured by the Homestake ^{37}Cl experiment and the Kamiokande water
Cherenkov detector, then a comparison of the gallium, chlorine, and
water results may make possible a determination of the neutrino mass
difference and mixing angle. A 30-ton gallium detector is currently
operating in the Baksan laboratory in the Soviet Union, with a ratio of
expected solar signal to measured background (during the first one to
two ^{71}Ge half lives) of approximately one.

1. Introduction

The discrepancy between the solar neutrino capture rate predicted
by standard solar model calculations [1-4] and the ^{37}Ar rate measured by
the chlorine experiment in the Homestake Gold Mine [5,6] (corroborated now
by the Kamiokande II water Cherenkov results [7] has persisted for
eighteen years. Recent calculated values are in the range 6-8 SNU (1
solar neutrino unit $= 10^{-36}$ captures/target atom/sec), while the
measured average $\nu_e + {}^{37}$Cl $\rightarrow {}^{37}$Ar $+ e^-$ production rate is 2.3 +/- 0.3
SNU. In addition, recent analysis of the chlorine experiment suggests

the possibility of time variations correlated with the 11 year solar cycle.[8]

The flux of the high-energy ^{8}B neutrinos (to which the chlorine and water experiments are most sensitive) is critically dependent on the temperature of the sun's core. Numerous nonstandard solar models[9] have therefore been suggested to account for the deficit of high energy solar neutrinos. Unfortunately, these nonstandard models generally have difficulties accounting for other observed features on the sun.

In addition to solar model effects, it has also been suggested that our understanding of the relevant particle physics may be incomplete. Mikheyev and Smirnov [10] and Wolfenstein [11] (and legions of others [12]) have pointed out that matter oscillation effects could enhance neutrino oscillations in the sun and explain the deficit. A neutrino magnetic moment [13-14] and neutrino decays [15] have also been invoked, as have weakly interacting massive particles [16], and extra quarks in nuclei [17].

The plethora of theoretical explanations makes exceedingly clear the need for experimental results capable of discriminating among models. In addition, the key assumption that neutrinos are produced as a result of proton-proton fusion in the sun must be checked directly. So far only radiochemical experiments can provide the low background and low threshold required to study the p-p flux.

Gallium, with a threshold of 0.23 MeV, is so far the only feasible target material capable of detecting the intense flux of the low energy p-p neutrinos via the $\nu_e + {}^{71}\text{Ga} \rightarrow {}^{71}\text{Ge} + e^-$ reaction. [18-21] The current observed energy output at the sun's surface corresponds to a neutrino flux of 6 x 10^{10} cm^{-2}sec^{-1} from the p-p reaction alone, which corresponds to a flux of 70 SNU in the gallium experiment. This conclusion is based largely on energetic grounds, and is insensitive to the details of solar model calculations. This is based on the standard solar model calculations of Bahcall and Ulrich [1], the total counting rate in gallium, including all reactions, is expected to be 132 SNU. Observation of significantly less than 70 SNU with gallium would be very difficult to explain without invoking new neutrino physics.

2. The Baksan Gallium Experiment

Two gallium experiments are currently beginning or about to begin operation: the 30 ton gallium chloride GALLEX detector to be constructed in the Gran Sasso laboratory [22], and the 60 ton gallium metal detector in the Baksan laboratory in the USSR [19].

The Soviet American gallium-germanium neutrino experiment is situated in an underground laboratory specially built in the Baksan Valley of the Northern Caucasus, USSR. The laboratory is 60 m long, 10 m wide, and 12 m high. It is located 3.5 km from the entrance of a horizontal adit driven into the side of Mount Andyrchi, and has an overhead shielding of 4700 mwe.

a) Extraction Procedure

The metal procedure has been worked out independently and fully tested in the USSR [19] and the US [23], and has been proven to work satisfactorily with a full-scale prototype.

The gallium is contained in chemical reactors, each with internal volume 2 m^3 and lined with teflon. The reactors are provided with stirrers and heaters that maintain the temperature just above the gallium melting point. Each reactor holds about 7 tons of gallium.

At the beginning of each run, a known amount of nonradioactive germanium carrier is added to the gallium in the form of a solid Ga-Ge alloy. The reactor contents are stirred so as to thoroughly disperse the carrier throughout the Ga metal. After a suitable exposure interval (3-6 weeks), the Ge carrier and any ^{71}Ge atoms that have been produced by neutrino capture are chemically extracted from the gallium. Details of the chemical extraction procedure have been described in detail elsewhere [28]. Basically, a weak solution of HCl, H_2O, and H_2O_2 is added to the reactors, and stirred together with the gallium. The Ge atoms go into solution and are extracted along with the solution. The solution is then reduced and the germanium is extracted from into H_2O. The Ge is then extracted into CCl_4 and then back-extracted into tritium-free [27] H_2O. The Ge is then synthesized into GeH_4 and mixed with Xe and used to fill small proportional counters.

b) Counting

^{71}Ge decays, with an 11.4 day half life, by electron capture to the ground state of ^{71}Ga. The only way to register such a decay is to detect the low-energy Auger electrons and x-rays produced during electron shell relaxation in the resulting ^{71}Ga atom. These low-energy electrons are detected in a small-volume proportional counter similar to that used in the Cl-Ar experiment. If xenon is mixed in with the germane, then (with a 90%-10% mixture of Xe-GeH_4 and measured detection efficiencies) we observe 41% of the decays in the L peak at 1.2 keV and 41% in the K peak at 10.4 keV.

The standard solar model predicts a production rate of 1.2 atoms of ^{71}Ge per day in 30 tons of Ga. At the end of a 4 week exposure period, and folding in extraction and detection efficiencies, the mean number of detected ^{71}Ge atoms in each run is only 4.0.

The major source of background in the counters is local radioactivity. The counters are therefore made from materials especially selected to have a low radioactive content, and are housed in large passive shields. The counting is conducted deep underground (at a depth of 4700 meters water equivalent) where the muon flux is measured to be only $2.4 +/- 0.3 \times 10^{-9}$ cm^{-2} sec^{-1}. The laboratory is lined with a 6 mm steel shell and 60 cm of low radioactivity concrete. To further reject and to characterize background events, 13 of 19 counting channels currently have an active NaI detector around the proportional counter that is used in coincidence/anticoincidence.

Background radioactivity primarily produces fast electrons in the counter. In contrast to the localized ionization produced by the Auger electrons from ^{71}Ge decay, these fast electrons give an extended ionization signal as they traverse the counter interior. Since the risetime of the induced pulse increases as the radial extent of the ionization distance, it is possible to use pulse shape discrimination to separate the ^{71}Ge decays from the background.

Good rejection of background events is obtained with a counting mixture of 10% GeH_4 and 90% Xe. This gas mixture gives a resolution of 18-21% at 5.9 keV, with a measured total counting efficiency in a 2 keV (FWHM) window around 10.4 keV of 35%. This efficiency includes geometrical effects inside the counter and excludes events whose risetime is outside a 95% acceptance window. The analysis now searches for those events with acceptable values of both pulse height and ADP (or, in the case of the full waveform analysis, for events that can be well fit to a waveform of the expected shape and amplitude), and that occur at times consistent with the 11 day ^{71}Ge decay (i.e., events that occur early in the run, before the counting rate has decreased to a constant background).

c) ^{71}Ge Background

The main source of ^{71}Ge in the reactors other than from solar neutrinos is from protons arising as secondary particles produced by *i)* external neutrons, *ii)* internal radioactivity, and *iii)* cosmic ray muons. These protons can initiate the reaction ^{71}Ga(p,n)^{71}Ge. Extensive work has gone into measurements and calculations of these background channels.[23-26] The sum of all of these backgrounds has been determined to produce less than 0.03 ^{71}Ge atoms/day.

Since these sources of ^{71}Ge background have been made small, the major difficulty for the experiment is the need to remove from the gallium the large quantities of long-lived ^{68}Ge (half life = 271 days) that were produced by cosmic rays while the gallium was on the surface. ^{68}Ge decays only by electron capture, so its decays cannot be differentiated from those of ^{71}Ge. The subsequent decay of ^{68}Ga (half life = 1.14 hours) is by positron emission in 90% of the cases. These ^{68}Ga decays can generally be identified by rise-time analysis of the counter pulse and by detection of a coincidence pulse in the surrounding NaI crystal.

Another type of background that arises only during counting can come from tritium in the counting gas. In order to eliminate this source of counter background, special methods for synthesizing $NaBH_4$ have been developed [26] using starting ingredients selected to have a low tritium content.

3. Present Experiment Status

The experiment began operation in April 1988 when filling of the reactors started, and the experiment now contains 30 tons of gallium in four reactors. Each reactor has undergone at least 20 extractions to remove the germanium isotopes that were produced by cosmic ray interactions while the Ga was on the surface.

The efficiency of extraction of germanium from the reactors is measured at several stages of the extraction procedure. The major uncertainty in these measurements is in the amount of Ge carrier added to the reactors, which is determined to an accuracy of +/- 5%. The overall extraction efficiency from the Ge added to the reactors to the synthesized GeH_4 is approximately 75%. The standard procedure is to conduct three extractions in series within a period of 5 days without adding additional carrier to the reactors.

The total background rate of selected counters filled with 90% Xe, 10% GeH_4 has been measured in the energy interval of 0.7-13.0 keV to be approximately 1.5 events per day. In the region of the Ge K-peak energy, the counter background is 1 event per month.

Counting of the germanium samples began during the extraction of the germanium isotopes produced by cosmic ray interactions. After the gallium had been stored underground for 3 months, the initial activity of the Ge in the 30 tons of Ga was 7700 events per day in the energy region of the Ge K-peak. For recent runs, the activity in the Ge K-peak (35% counting efficiency) is less than 0.2 counts day^{-1}.

Some residual radioactivity is still present,which produces events in the energy range of 1-15 keV. In recent extractions this activity does not appear to be decreasing at the same rate as in the earlier extractions. Measurements of the activity obtained in various isolated steps of the extraction procedure have been conducted, but the statistics available are low. Although the source of this activity has not yet been definitely identified, it clearly comes from either the reactor vessels or the gallium, not from contaminants in the reagents. There is a possibility that there has been some diffusion of long-lived ^{68}Ge from the original dirty gallium into the teflon liners. Also, ^{68}Ge may be bound up in the gallium in another chemical form that is not extracted using the current procedure, although extensive chemical studies suggest this possibility is unlikely.

4. Future Plans

Monthly extractions from the 30 tons of gallium will continue. At the same time, further tests to understand the source of the remaining background will be conducted, and the detector will be expanded to the full 60 tons of gallium. Additional counting channels are also being installed. At the background levels presently achieved, the full 60-ton detector should yield a statistical accuracy of better than 25% after one year of operation (assuming a production rate of 132 SNU).

A calibration of the detector is planned using a ^{51}Cr neutrino source. It is expected that an activity of 0.8 MCi will be obtained by irradiating about 200 g of enriched chromium (86% ^{50}Cr) in Soviet reactor SM-2 (thermal flux = 3.2 x 10^{15} neutrons cm^{-2} sec^{-1}) [24, 27]. Approximately 400 decays from the ^{71}Ge atoms produced by this source are expected to be detected.

5. Acknowledgments

The SAGE collaboration wishes to thank A.E. Chudakov, G.T. Garvey, M.A. Markov, V.A. Matveev, J.M. Moss, V.A. Rubakov, and A.N. Tavkhelidze for their continued interest in our work and for stimulating discussions. We are also grateful to J.K. Rowley and R.W. Stoenner for chemical advice and to E.N. Alekseyev and A.A. Pomansky for their vital help in our work. The U.S. participants wish to acknowledge the support of the U.S. Department of Energy and the National Science Foundation.

6. References

1) J.N. Bahcall and R. Ulrich, Rev. Mod. Phys. **60**, 297 (1988);   B.W. Filippone and D.N. Schramm, Ap. J. **253**, 393 (1982); S. Cahen et al., in *Massive Neutrinos in Astrophysics and Particle Physics*, ed. by O. Fackler and J. Tran Thanh Van, p.83 (1986); B.W. Filippone et al., Phys. Rev. Lett. **50**, 412 (1983); W.A. Fowler, AIP Conf. Proc. No. **96**, 80 (1982);   J.N. Bahcall et al., Rev. Mod. Phys. **54**, 767 (1982).
2) J. Bahcall, Rev. Mod. Phys. **50**, 881 (1978).
3) J.N. Bahcall et al., Ap. J. Letters **292**, L79 (1985).
4) J.N. Bahcall, *Neutrino Astrophysics*, Cambridge Univ. Press, Cambridge (1989).
5) J.K. Rowley et al., in *Solar Neutrinos and Neutrino Astronomy*, ed. by M.L. Cherry et al., AIP Conf. Proc. No. **126**, AIP, NY, p.1 (1985).
6) R. Davis, Jr., 7th Workshop on Grand Unification, (ICOBAN '86), Toyama (1986); R. Davis et al., *Proc. 20th Intl. Cosmic Ray Conf.*, Moscow 4, 328 (1987) and 2nd Intl. Symp. Underground Physics, Baksan (1987); R. Davis et al., *Proc. 21st Intl. Cosmic Ray Conf.*, Adelaide 7, 155 (1990).
7) K.S. Hirata et al., Phys. Rev. Lett. **63**, 16 (1989).
8) R. Davis et al., Proc. 21st Intl. Cosmic Ray Conf., Adelaide 7, 155 (1990); and K. Lande, Highlight talk, 21st ICRC, Adelaide (1990).
9) See articles by Schatzman (p.69), Michaud (p.75), and Roxburgh (p.88) in AIP Conf. Proc. No. **126**, (1985); J. Christensen-Dalsgaard et al., Astron. & Astrophys. 73, 121 (1979), E. Schatzman and A. Maeder, Astron. & Astrophys. **96**, 1 (1981); G. Marx, 2nd Intl. Symp. Underground Physics, Baksan (1987). Also see the discussion by J.N. Bahcall in ref. 4.

10) S.P. Mikheyev and A.Y. Smirnov, Soviet J. Nucl. Phys. **42**, 1441 (1985).

11) L. Wolfenstein, Phys. Rev. **D17**, 2369 (1978) and **D20**,   2634 (1979).

12) H.A. Bethe, Phys. Rev. Lett. **56**, 1305 (1986); S.P. Rosen and J.M. Gelb, Phys. Rev. **D34**, 969 (1986); V. Barger et al., Phys. Rev. **D34**, 980 (1986); A. Dar and A. Mann, Nature **325**, 790 (1987); W. Hampel, Symposium on Weak and Electromagnetic Interactions, Heidelberg (1986); W.C. Haxton, Phys. Rev. Lett. **57**, 1271 (1987); E.W. Kolb et al., Physics Letters **B175**, 478 (1986); P. Langacker et al., Nucl. Physics **B282**, 589 (1987); S.J. Parke, Phys. Rev. Lett. **57**, 1275 (1986); S.J. Parke and T.P. Walker, Phys. Rev. Lett. **57**, 2322 (1986).

13) M.B. Voloshin et al., Sov. J. Nucl. Phys. **44**, 440 (1986); L.B. Okun et al., Sov. J. Nucl. Phys. **44**, 546 (1986).

14) M. Fukugita and T. Yanagita, Phys. Rev. Lett. **58**, 1807 (1987); K.S. Babu and R.N. Mohapatra, Phys. Rev. Lett. **63**, 228 (1989).

15) J.N. Bahcall et al., Phys. Rev. Lett. **28**, 316 (1972).

16) R. Gilliland et al., Ap. J. **306**, 703 (1986); D. Spergel and W. Press, Ap. J. **294**, 663 (1985); J. Faulkner and R.L. Gilliland, Ap. J. **299**, 994 (1985).

17) R.N. Boyd et al., in AIP Conf. Proc. **126**, p. 145 (1985)

18) I.R. Barabanov et al., in AIP Conf. Proc. No. **126**, p.175 (1985); and V.N. Gavrin et al., AIP Conf. Proc. No. **126**, p.185 (1985).

19) V.N. Gavrin et al., in *Proc. Inside the Sun*, Versailles, ed. by G. Berthomieu and M. Cribier, Kluwer Acad. Publ., to be published (1990)

20   G.J. Matthews et al., Phys. Rev. **C32**, 796 (1985).

21) M. Spiro et al., in *Neutrino Masses and Neutrino Astrophysics (Telemark IV)*, ed. by V. Barger et al., p.232 (1987); T. Kirsten, in *Proc. 12th Intl. Conf. on Neutrino Physics and Astrophysics (Neutrino '86)*, Sendai, ed. by T. Kitagaki and Y. Yuta, p.317 (1986); W. Hampel, in AIP Conf. Proc. No. 126, p.126 (1985).

22) J.N. Bahcall et al., Phys. Rev. Lett. **40**, 1351 (1978).

23) I.R. Barabanov et al., Preprint INR AS USSR P-0559 (1987).

24) V.N. Gavrin et al., Preprint INR AS USSR P-0494 (1987).

25) V.N. Gavrin and Yu.I. Zacharov, Preprint INR AS USSR P-0335 (1984).

26) D.P. Alexandrov et al., Proc. Fourth All-Union Meeting on the Chemistry of Hydrides, Dushanbe, USSR, 37 (1987).

27) A.M. Petrosyants, <u>Problems of Atomic Science and Technology</u>, Atomizdat, Moscow (1979); S.M. Fainberg et al., Proc. Third Intl. UN Conference on the Peaceful Uses of Atomic Energy (1969).

28) A.I. Abazov et al., "Measurement of Solar Proton-Proton Fusion Neutrinos with a Soviet-American Gallium Experiment at Baksan: A Status Report", to be published in the Proceedings of the 21st Cosmic Ray Conference, Adelaide, Australia, Jan. 1990.

THE SUDBURY NEUTRINO OBSERVATORY

THE SNO COLLABORATION:
H.C. Evans, G.T. Ewan, H.W. Lee, J.R. Leslie, J.D. MacArthur, H.B. Mak,
A.B. McDonald, W. McLatchie, B.C. Robertson, P. Skensved,
Queen's University,
W.F. Davidson, C.K. Hargrove, H. Mes, D. Sinclair,
National Research Council of Canada,
E.D. Earle, G.M. Milton, E.T.H. Clifford,
Chalk River Nuclear Laboratories,
P. Jagam. J.J. Simpson, **University of Guelph,**
E.D. Hallman, R. Haq, **Laurentian University,**
D. Kessler, B. Hollebone, **Carleton University,**
C. Waltham, R. Schubank, **University of British Columbia,**
G. Buhler, **University of California at Irvine,**
R.T. Kouzes, M.M. Lowry, R.M. Key, **Princeton University,**
E.W. Beier, W. Frati, M. Newcomer, R. Van Berg,
University of Pennsylvania,
T.J. Bowles, B.T. Cleveland, P.J. Doe, S.R. Elliott, M.M. Fowler,
R.G.H. Robertson, T.C. Spencer, D.J. Vieira, J.B. Wilhelmy, J.F.
Wilkerson, J.M. Wouters, **Los Alamos National Laboratory,**
E. Norman, K. Lesko, B. Sur, A. Smith, R. Fulton,
Lawrence Berkeley Laboratory,
N.W. Tanner, N. Jelley, P. Trent, D.L. Wark, J. Barton,
University of Oxford

ABSTRACT

The Sudbury Neutrino Observatory will employ 1000 tonnes of D_2O as
a Cherenkov detector to observe both the charged and neutral
current interactions of solar neutrinos. The expected rates are
an order of magnitude higher than present solar neutrino
experiments. Spectral information will allow a sensitive test for
matter oscillations of neutrinos originating in the sun. This
experiment is now funded and anticipates data acquisition will
commence in mid-1995.

1. INTRODUCTION

The deficit of high energy ^{8}B solar neutrinos, observed by both
the Davis chlorine and the Kamiokande water Cherenkov detectors, is well
established. In order to determine the source of this deficit, it is
essential to measure both the charged and neutral current interactions
of solar neutrinos. In addition, good spectral information with high
statistics would provide the ability to determine the mixing parameters
if neutrino matter oscillation effects, as postulated by Mikheyev,
Smirnov, and Wolfenstein, are the source of the observed deficit. The
SNO collaboration has begun construction of a heavy water Cherenkov
detector which will provide this information.

2. DETECTOR DESIGN

The Sudbury Neutrino Observatory will be located at the Creighton Mine, owned by INCO Limited, near Sudbury, Ontario, Canada. The detector will be constructed in a 22 m diameter barrel shaped cavity to be excavated at a depth of 2270 m (5900 mwe). The detector consists of 1000 tonnes of D_2O contained in an acrylic vessel, which is surrounded by photomultiplier tubes that detect Cherenkov light. The acrylic vessel, constructed from 2.5 cm thick panels, and formed into a sphere of 6 m, is surrounded by at least 4 m of high purity H_2O to shield it from the radioactivity in the rock. Mounted uniformly around the acrylic vessel, at a distance of 2.5 m outside the acrylic are photomultiplier tubes (PMTs) providing 30% photocathode coverage. Each PMT is surrounded by a reflector to increase light collection.

Neutrino interactions in the detector produce either relativistic electrons (charged current interactions) or free neutrons (neutral current interactions). The neutrons thermalize in the water and are subsequently captured, generating gamma rays that produce electrons primarily through Compton scattering. The signals from the PMTs are used to provide the location, energy, and direction of the electron based on the time of arrival of photons.

The expected solar neutrino signal rates of ten to twenty events per day requires an extremely low background environment for the D_2O. Cosmic ray backgrounds at the detector location will consist of only 24 muons per day passing through the acrylic vessel, of which only 0.1 per day will produce detectable spallation events. Activities from naturally occurring radionuclides (^{232}Th, ^{235}U, and ^{238}U and their daughters) require extremely stringent control of the water purity and the materials used in detector construction. Neutrons and gammas originating in the rock are attenuated to a low level by the H_2O shield.

3. SCHEDULE

The detector is expected to begin operation with the acrylic vessel initially filled with H_2O in mid-1995. After approximately six months of operation, the D_2O will be installed and operated for one year to measure the charged current interactions. 2.5 tonnes of NaCl will then be added to the D_2O to increase the efficiency of neutron capture and increase the energy of the emitted gamma ray. The neutral current rate is then determined by subtracting the rate with pure D_2O from the rate with NaCl added. An alternative scheme of using ^{3}He proportional counters or scintillators is also being pursued to allow simultaneous observation of charged and neutral currents. It is anticipated that in excess of 3000 charged and neutral current events will be collected during each year of operation.

Neutrino Masses, Hadronic Axions, and Supernovae

David Seckel

Bartol Research Institute
Univ. of Delaware, Newark, DE 19716

Abstract

Two issues of particle astrophysics concerning supernovae are discussed: a) the possible determination of the mass of the μ or τ neutrino and b) implications for hadronic axions due to the lack of their detection at Kamiokande coincident with the neutrino burst from SN 1987A.

Mass Determination for ν_μ or ν_τ

It is important to particle physicists and cosmologists to know if it is possible to measure a cosmologically interesting mass for the μ or τ neutrino via the time dispersion of a burst of neutrinos from a supernova. This section describes research by Walker, Steigman and Seckel[1] (WSS) which suggests that this will be very difficult. For simplicity, I will assume that ν_τ is heavy and that ν_μ is massless. Both neutrinos would be detected only via neutral current interactions and they should have the same energy spectrum.

The contribution of neutrinos towards a critical density is given by $\Omega_\nu h^2 = m_\nu/91.5\,\text{eV}$, where h is the Hubble constant in units of 100 km/sec/Mpc. A minimal goal for experiment should be $m_\nu = 20$ eV ($\Omega_\nu = .9$, $h = .5$) and since $\Omega_\nu \sim .2$ would still have interesting consequences for galaxy formation, m_ν of a few eV should be a long range target.

With these goals in mind, the time delay for a massive neutrino emitted from a supernova at distance D is

$$\Delta t = D(1 - \beta) = .515 \left(\frac{m_\nu/10\ \text{eV}}{E/10\ \text{MeV}} \right)^2 (D/10\ \text{kpc})\ \text{sec.} \tag{1}$$

This time scale may be compared with the cooling time for the remnant neutron star (~ 10 sec) or to the rise time for the neutrino signal when the thermal emission of τ neutrinos commences ($\sim .1$ sec). Consider the comparison with the cooling time. In order to isolate the events which occur late, it is convenient to define $\tilde{E}$ such that at least 10% of events have time delays $\Delta t > 10$ sec. Then, defining $E_{10} = E/10$ MeV, $\Delta t_{10} = \Delta t/10$ sec, and $D_{10} = D/10$ kpc the measurable mass is

$$m_\nu > 44\tilde{E}_{10} \left(\frac{\Delta t_{10}}{D_{10}} \right)^{1/2}\ \text{eV} \tag{2}$$

Another consideration in measuring m_ν is to have a large enough event sample to perform the measurement. At a minimum it seems reasonable to require a signal S of at least 10 events. It is convenient to scale to the number of events S_{10} that would be detected in a 10 kt detector from a supernova 10 kpc away. The second constraint is then

$$S_{10} \frac{M_{10}}{D_{10}^2} > 10. \tag{3}$$

For electron scattering in a normal water detector WSS estimate $S_{10} = 15$, while for neutral current fission of deuterium[6] they find $S_{10} = 3120$. WSS assumed $T_{\nu_r} = 8$ MeV and $L_{\nu_r} = L_{\bar{\nu}_r} = L_{\bar{\nu}_e}$ with $L_{\bar{\nu}_e}$ chosen to be consistent[4,5] with the SN 1987A results from Kamiokande[2] and IMB[3]. A generic estimate for neutrino detection through nuclear scattering is $S_{10} = 1000$.

A third constraint arises from the need to see the signal above background, $S \gtrsim \sqrt{10B}$. The total background is $B = \Gamma_{10}M_{10}\Delta t$, where Γ_{10} is the background rate for a 10 kt detector. Using Eq. 1 for Δt gives a constraint on the maximum measurable mass

$$m_\nu < 4.4 \left(\frac{\tilde{E}_{10}^2 M_{10} S_{10}^2}{D_{10}^5 \Gamma_{10}} \right)^{1/2} \tag{4}$$

There are many backgrounds that are detector specific, but a common minimal background rate comes from the cosmic supernova background of neutrinos emitted by all previous supernovae. WSS estimate this rate as $\Gamma_{10} \simeq 10^{-11}S_{10}$.

The three constraints, Eqs. 2, 3, and 4 are displayed in Fig. 1 as a function of m_ν and the distance to the supernova. The bold lines are for a generic 1 kt detector with $S_{10} = 1000$, but also shown are constraints for 10 kt normal (H_2O) water and 1 kt heavy (D_2O) water Čerenkov detectors. The signals are calculated assuming νe elastic scattering for the H_2O detector and $\nu D \to pn\nu$ for the D_2O detector. For the H_2O detector the cosmic supernova background was used, but for the D_2O detector the background should be dominated by solar neutrino induced deuterium fission[7].

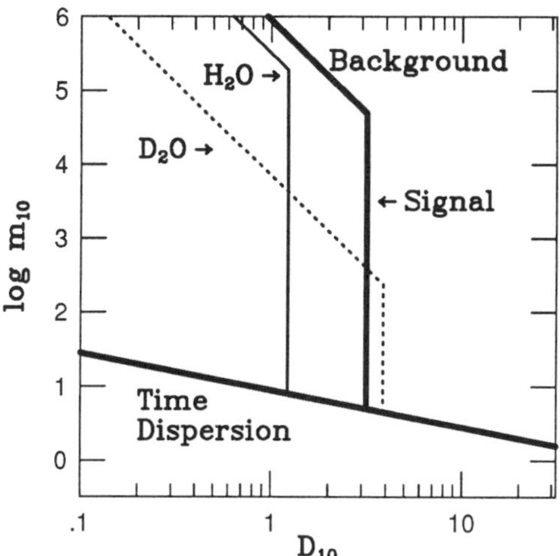

Figure 1: Constraints on Determining m_{ν_r}

Mass measurements are possible only for $D \lesssim 30$ kpc. This range covers the Milky Way, but does not reach to any nearby galaxies. Since the supernova rate in our galaxy is thought to be $.01 - .1$ yr^{-1} this implies a substantial waiting period - testing the patience of scientists and funding agents alike. Also implied in Fig. 1 is that a typical galactic supernova only allows mass measurements down to ~ 50 eV, which is outside the cosmologically interesting range.

To overcome these two problems one would want to detect neutrinos from supernovae in nearby galaxies. Fig. 2 displays the constraint equations again, only here the detector mass required to measure a 10 eV neutrino mass is given as a function of distance. The lines correspond to the same detectors as in Fig. 1. An Andromeda supernova might allow measurement of a 10 eV mass but would be as unlikely as a supernova in our own galaxy, and the measurement would require a 1 Mt detector. To probe the Virgo cluster (1 Mpc)

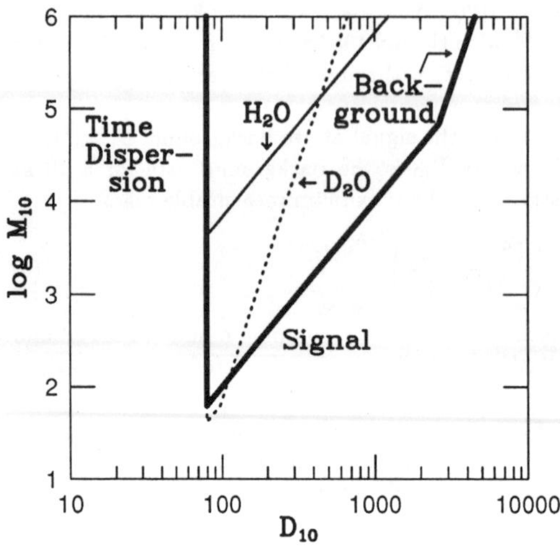

Figure 2: Detector Size Needed to Measure $m_{\nu_\tau} = 10$ eV

would require 1 Gt of detector mass - a cubic kilometer of water. Such a massive engineering project seems unlikely in the near future.

The discussion so far has assumed $\Delta t = 10$ sec, but perhaps one could do better by utilizing the .1 sec rise time of the neutrino signal. Applying Eq. 1 naively gives a factor of 10 improvement in mass sensitivity, but things are not that good. The signal for the $\sim .1$ sec rise consists of missing events. A clear recognition of missing events would require making an energy cut so that all but the highest energy neutrinos are delayed. This implies a higher $\tilde{E}$, which reduces the mass sensitivity by about a factor of 3. In addition, a) one is only using $\sim 20\%$ of the neutrino flux in the first .1 sec; b) there is likely to be a high background rate from ν_μ events if they are indeed massless; and c) neutrinos emitted earlier have higher average energy, further reducing the mass sensitivity. All told, WSS find that a 30 eV neutrino mass may be accessible if $D = 10$ kpc, but 20 eV seems unlikely. Finally, it must be recognized that the behaviour of the neutrino signal at early times is subject to greater modeling uncertainty than is the long term neutrino flux and the overall energy budget.

Based on the arguments given here, it seems unlikely that a definitive measurement of a cosmologically interesting mass for ν_μ or ν_τ may be done using supernova neutrinos.

Hadronic Axions

Many papers[8] have been written concerning axion emission from SN 1987A. Most have concentrated on the limit $g_{aN} \lesssim 10^{-10}$, which arises from requiring that emission of axions from the core[9,10,11] does not disturb the agreement between theory and the observation of $\bar{\nu}_e$ neutrinos at Kamiokande[2] (KII) and IMB[3]. As g_{aN} is increased above 10^{-10} the axion luminosity L_a increases since emissivity increases as g_{aN}^2 - until the probability for axions to be reabsorbed becomes significant. At that point L_a decreases as the surface from which axions may escape moves farther out in the supernova, and the temperature of the emitting surface goes down. Eventually, $L_a < L_\nu$ again and SN 1987A no longer constrains the axion coupling. Turner[10] has used this reasoning to argue that hadronic axions (no coupling to electrons) are allowed in the mass range 2 eV $\lesssim m_a \lesssim 5$ eV.

Recently, however, Engel, Hayes, and Seckel[12] (EHS) pointed out that such axions should have produced an obvious signal at the KII detector. The signal consists of axion absorption by ^{16}O, followed by emission of γ-rays as the nucleus deexcites. Since the signal was not seen hadronic axions seem to be excluded.

The calculation proceeds as follows. 1) The axion mean free path is calculated for the dominant process - axion absorption in nucleon scattering, $NNa \rightarrow NN$. 2) Using this result and a model for the density and temperature structure of the supernova, the temperature and luminosity of the axion flux is determined by comparing to the neutrino flux of a numerical model[5] adjusted to reproduce the observations of SN 1987A. No attempt was made to determine how the axion emission would feedback to change the neutrino emission or affect the structure of the atmosphere. 3) The cross-sections for axion absorption by oxygen, $a^{16}O \rightarrow {}^{16}O^*$ are calculated[13] using a pseudovector form for the axion-nucleon interaction and ^{16}O wave functions from Hayes, Friar and Strottman[14]. 4) The branching ratios for each state to decay in a manner including γ-rays is estimated using data from photoabsorption studies, kinematic thresholds, and a modest amount of shell model based arguments. 5) For a photon of energy E_γ we estimate the efficiency κ_γ with which it would trigger the Kamiokande detector as being equal to that of an electron with energy $E_e = .85E_\gamma$, i.e., $\kappa_\gamma(E) = \kappa_e(.85E)$. 6) The results of 4) and 5) are combined to produce an efficiency ϵ_i for detecting the excitation of state i, and weighting this by the flux of axions an expected signal is produced.

The results of the calculation are shown in Fig. 3 for a model where the isoscalar and isovector axion-nucleon couplings are equal. The vertical line marks where the axion and neutrino luminosities are equal. We take the area to the left of the line, $g_{aN} < 1.2 \times 10^{-6}$, to be excluded by the cooling argument. The solid curve shows the expected number of events as a function of g_{aN}. Five axion events at Kamiokande would present a problem. These would have to be identified with five of their low energy events, but that would make it difficult to understand the remaining neutrino signal at Kamiokande, as well as at IMB[15]. Since a 5 event signal lies in the excluded region EHS conclude that the hadronic axion window is probably closed, although it is not possible to be precise without explicitly including the effects of axion energy transport in a numerical supernova model.

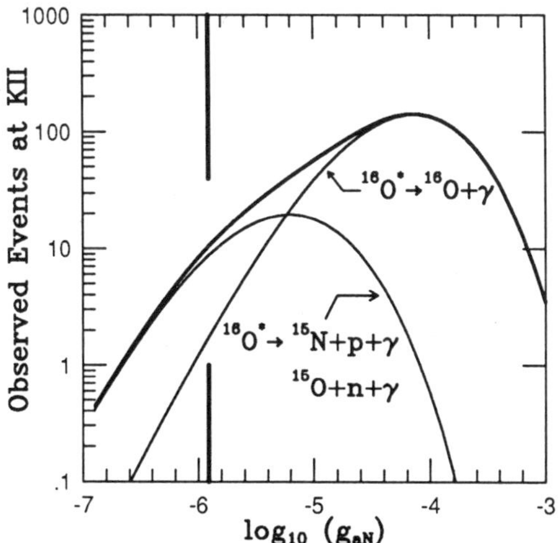

Figure 3: Number of observed axion events expected at Kamiokande

There are two qualitatively different sources for the events, which are shown separately in Fig. 3. If the axion energy is in the $20 - 25$ MeV range there are many states for which absorption is followed by particle emission to a nucleon and a mass 15 nucleus, an event which would be detected with efficiency $\epsilon \sim .15$. The detections may occur when the daughter nucleus is left in an excited state and subsequently photodecays with

$E_\gamma = 5 - 9$ MeV. For axions below the particle emission threshold, there is one interesting state, an isospin zero, $J^\pi = 0^-$, $E = 10.96$ MeV state, which decays exclusively through the emission of two photons to the ground state of ^{16}O. The efficiency for detecting this state is quite high ($\epsilon \sim .7$). For small g_{aN} or high axion temperature the absorption strength in the $20 - 25$ MeV range dominates over the strength into the 10.96 MeV state. However, as g_{aN} increases the emitting surface moves out in the star and the temperature of the emitting surface drops to the point where few axions are energetic enough to excite ^{16}O. Since the particle emission events require higher energy axions, that curve turns over at a smaller value of g_{aN}.

If the isoscalar axion-nucleon coupling is comparable to the isovector coupling then Fig. 3 implies that g_{aN} as large as 9×10^{-4} is excluded. In this case the two supernova arguments would exclude all axion models from those tested by accelerator experiments to those marginally acceptable from cosmological considerations, independent of their coupling to photons or electrons. On the other hand, if the isoscalar coupling is zero, then the signal from the 10.96 MeV state goes away and the lower bound on g_{aN} is correspondingly weaker. The particle emission signal, however, is distributed over many states and is quite insensitive to the isospin breakdown of the axion-nucleon coupling, and so the question of a hadronic window at the limit of the cooling argument depends principly on uncertainties in modeling the supernova.

References

1. T. Walker, G. Steigman, and D. Seckel, *In preparation*

2. K. Hirata *et al.*, *Phys. Rev. Lett.* **58**, 1490 (1987)

3. R.M. Bionta *et al.*, *Phys. Rev. Lett.* **58**, 1494 (1987)

4. L.F. Abbott, A. de Rujula, and T.P. Walker, *Nucl. Phys.* **B299**, 734 (1988)

5. A. Burrows,in *Supernovae*, ed. A.G. Petschek, Springer-Verlag (1988)

6. S. Ying, W. Haxton, and E.M. Henley, *Univ. of Washington Preprint* 40427-6-N9 (1988)

7. G.T. Ewan *et al.*, *Sudbury Neutrino Observatory Proposal* SNO-87-12(1987).

8. For a review, see G. Raffelt, *Stars as Particle Physics Laboratories*, Prepared for *Physics Reports.* (1989)

9. G. Raffelt and D. Seckel, *Phys. Rev. Lett.* **60**, 1793 (1988)

10. M.S. Turner, *Phys. Rev. Lett.* **60**, 1797 (1988)

11. R. Mayle, *et al.*, *Phys. Lett.* **203B**, 188 (1988)

12. J. Engel, D. Seckel, and A. Hayes, *Bartol Preprint* BA-90-11 (1990)

13. J.D. Walecka, in *Muon Physics*, edited by V.W. Hughs and C.S. Wu (Academic, New York. 1975), Vol. 2, p. 113.

14. A.C. Hayes, J.L. Friar, and D. Strottman, *LANL Preprint* (1990)

15. There are no observed axion events at IMB due to the high threshold energy for detecting of Čerenkov light.

DUMAND, a Deep-Ocean Laboratory for Detection of High-Energy Neutrinos

Medford S. Webster, for the DUMAND Collaboration
Department of Physics and Astronomy, Vanderbilt University
Nashville, Tennessee 37235

Abstract

The configuration, sensitivity, and resolution of a detector capable of measuring the high-energy neutrino fluxes from known astronomical sources of high-energy gamma rays are presented. Measurements of the muon flux deep in the ocean with a proto-type detector are presented.

The excellent shielding and the ready availability of an unlimited volume of Cherenkov radiator and interaction medium make the deep ocean an attractive location for a laboratory for the study of high energy neutrino astrophysics and particle physics. The DUMAND (_D_eep _U_nderwater _M_uon _A_nd _N_eutrino _D_etector) collaboration[1] has measured the muon flux in the deep ocean with a prototype detector and has designed and proposed[2] an array which is 100 m in diameter and 250 m high for deployment in the Pacific Ocean. The proposal has been reviewed by the HEPAP subpanel on Major Detectors in Non-Accelerator Particle Physics and by a Technical Cost Review Committee. These reviews have been favorable and DOE has decided to proceed with funding of the project. With the projected funding profile, we expect to deploy the first three strings during 1992 and to complete the array in 1993.

The prototype consisted of seven 16-inch phototubes, each mounted in a glass sphere, and spaced 5m apart vertically. Two calibration modules produced light flashes suitable for monitoring the time and amplitude response of the detectors. Commands were sent to the phototube modules from an electronics module at the bottom of the string which also digitized, with a 5 ns clock, the time of arrival and time over threshold of each pulse from a phototube. The entire array was suspended from a ship. The support cable transmitted electrical power and control signals from the ship to the electronics module, and the digitized and encoded data from the phototubes were sent up to the ship on an optical fiber in the cable. Data were taken at approximately 500 m intervals from 2000 to 4200 m depth. For 5-fold coincidences from the seven detectors, the time of arrival and time over threshold measurements were fitted to the pulses expected from the Cherenkov wavefront of a muon in order to determine the zenith angle and the distance to the line of detectors from the muon. Thus both an angular distribution and a vertical flux were calculated at each depth, and both distributions are in excellent agreement with previous measurements[3]. The vertical flux is shown in Fig. 1.

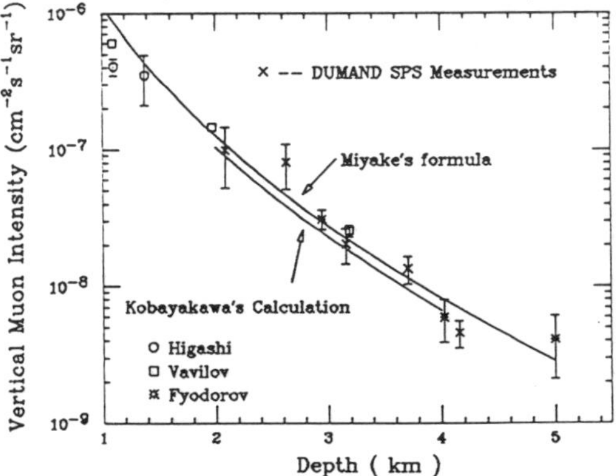

Fig. 1. Comparison of DUMAND vertical flux measurements with previous measurements.

The site for the proposed array is in a subsidence basin at a depth of 4.8 km, 30 km west of Keahole Point, Hawaii. The array consists of nine strings, one at the center and at each of the vertices of an octagon 40m on a side. Each string supports 24 phototubes spaced 10 m apart vertically. A cable from shore supplies electrical power to the array and has an optical fiber for each string. Each string is controlled by an electronics module which digitizes, with a 1 ns clock, the time and charge of each signal received from a phototube on its string and transmits the digitized signal to shore. Muons above about 50 GeV can be reconstructed and will have an rms error less than 1°. For upward muons, the signal is almost entirely due to neutrino interactions beneath the array and the effective area of the array is 20,000 m^2 for upward events. The location deep in the ocean provides essential reduction in the flux of downward muons, so that our signal is dominated by muons produced by neutrino interactions over the entire hemisphere of upward muons and approximately 10° into the downward hemisphere, and in this 2.3π steradian region, the isotropic background from interactions of neutrinos which are due to cosmic ray interactions is 1/3 event per bin of angular resolution size. The number of detected neutrino interactions from one of the more intense point sources of high energy gamma rays, such as Cygnus X-3 or Hercules X-1, is expected to be at least 10 events per year if the gamma rays are from $\pi°$ decay. Thus the significance of such signals will be limited by statistics, not background. The location at 19° from the equator gives us more than 50% duty cycle for all of the sky except for a few degrees around the North Pole.

Footnotes

1. The DUMAND collaboration is:
   Technische Hochschule Aachen, West Germany - P. Bosetti, V. Comminchau, D. Samm, Ch. Ley, H. Geller, K. Hangarter, and C. Wiebusch
   University of Bern, Switzerland - P. K. F. Grieder
   Boston University - S. Dye
   California Institute of Technology - B. Barish and J. Elliott
   University of Hawaii - J. Babson, R. Becker-Szendy, J. G. Learned, S. Matsuno, D. O'Connor, A. Roberts, V. J. Stenger, V. Z. Peterson, and G. Wilkins
   University of Kiel, West Germany - O. C. Allkofer*, P. Koske, M. Preischel, J. Rathlev T. H. Knutz, J. Cohrs W. Greve, and K. Sauerland   *deceased
   Kinki University, Osaka, Japan - T. Kitamura
   University of Kobe, Japan - K. Kobayakawa
   Okayama Technical Institute, Japan - I. Yamamoto
   Scripps Institution of Oceanography - V. Anderson, H. Bradner, M. Sessions, F. Spiess
   Tokohu University, Japan - S. Tanaka, A. Yamaguchi, and T. Hayashino
   Institute for Cosmic Ray Research, University of Tokyo, Japan - K. Mitsui, Y. Ohashi, A. Okada
   Vanderbilt University - J. Clem, C. E. Roos, M. S. Webster
   University of Wisconsin - M. Jaworski and R. March

2. *DUMAND II Proposal to Construct a Deep-Ocean Laboratory for the study of High Energy Neutrino Astrophysics and Particle Physics*, submitted to the particle physics section of the DOE, 1988

3. S. Miyake, *J. Phys. Soc. Japan*, **18**, 1093 (1963); K. Kobayakawa, private communication; S. Higashi *et al.*, **Il Nuovo Cimento 43A**, 334 (1966); V. M. Fyodorov, *Proc. of 19th ICRC La Jolla* **8**, 39 (1985); and Yu. N. Vavilov *et al.*, *Bull. Acad. of Sci. USSR* **34**, 1759 (1970)

HEAVY IONS/LATTICE

Conveners: T. Ludlam & A. Soni

EXPERIMENTAL RESULTS FROM STUDIES
OF NUCLEUS-NUCLEUS COLLISIONS AT HIGH ENERGY

Glenn R. Young
Oak Ridge National Laboratory,* Oak Ridge, Tennessee 37831

ABSTRACT

A short overview is presented of results from the initial accelerator-based studies of collisions of heavy nuclei at high energies. Some comments are made on the global observables studied and their implications for achieving conditions for deconfinement. Finally, information available from certain more specific probes selected for their ability to test whether a deconfined system of quarks and gluons is created in such collisions is presented..

1. INTRODUCTION: MATTER UNDER EXTREME CONDITIONS OF BARYON AND ENERGY DENSITY

It has been realized over the past decade that collisions of energetic heavy nuclei can create extreme conditions of high baryon and energy density over a scale much larger than that given by the QCD scale parameter, Λ_{QCD}. A chance apparently exists to create conditions in the laboratory under which strongly interacting matter undergoes a phase transition from the familiar state of quarks and gluons confined to hadrons to a state in which the quarks and gluons are free to move over a volume more than 100 times that of a proton. Theoretical investigations of QCD using lattice methods suggest that such a phase transition would occur when the energy density of a strongly interacting systems exceeded 2 GeV/fm^3; this should be contrasted with the ground state energy density of a heavy nucleus of 0.15 GeV/fm^3, arising from its rest mass. Scaling from existing experimental information from high-energy $p - p$ and e^+e^- colliders suggests that collisions of heavy nuclei in the range of $\sqrt{s} = 30 - 200$ GeV/nucleon-pair should produce conditions favorable for the creation of such a "deconfined" state, popularly referred to as a quark-gluon plasma. A phase diagram in temperature vs density space is shown in Figure 1, where the dark band represents the region where energy densities of 2 GeV/fm^3 are attained.

An experimentalist considering such collisions would like to find evidence for whether such a deconfinement transition exists and if so what its properties might be and whether a thermalized state can result. One would like to know "where" the transition occurs: what is the critical temperature, or energy density; can it occur in baryon-rich or baryon-free systems; what might its order be? One would then like to know the properties of the matter thus formed: what is its entropy density and baryon density; what is the space-time volume of the system; what proper time is required for it to hadronize; what is its quark flavor content and gluon content; what is its "polarizability," i.e., its response to external perturbation; what is its "new and interesting object" content? A list of likely probes for the matter created in a collision of two heavy nuclei at high energy is given in Table 1. Results of experiments examining each of the three broad classes of probes listed, namely

* Operated by Martin Marietta Energy Systems, Inc., under contract DE-AC05-84OR21400 with the U.S. Department of Energy

global event parameters, indicators of a phase transition, and penetrating probes, are discussed in the remainder of this article.

Heavy ions accelerated to high energies are presently available at the BNL AGS and the CERN SPS. BNL has accelerated ions up to mass 28 (silicon) to energies up to 14.5 GeV/nucleon using its existing Van de Graaff electrostatic accelerator to inject heavy ions into the AGS. The AGS Booster will come on line in 1991 and make ions up to mass 200 (gold) available soon thereafter. The Relativistic Heavy Ion Collider (RHIC) is proposed as a major upgrade utilizing the existing tunnel from the CBA project to provide center of mass energies up to 200 GeV/nucleon-pair for $A = 200$ ions. This collider has been included in the FY 1990 President's budget request to the Congress. CERN has accelerated ions up to mass 32 (sulfur) to energies of 200 GeV/nucleon (6.4 TeV total for sulfur) using a new ECR source plus RFQ to inject the old Linac-I. After Linac-I, the ions pass through in order the PS Booster, PS and SPS and are extracted to the North and West Halls. Upgrade plans for CERN call for construction of a new ion source and Linac capable of furnishing mass 208 (lead) ions to the PS/SPS complex. Long term plans depend upon the fate of the LHC at CERN.

A first round of experiments with these high-energy nuclei began in 1986 and has continued up to the present. There are presently 11 large electronic experiments mounted together with more than two dozen smaller, typically emulsion, experiments. The program involves some 500 physicists and students drawn from nuclear physics and high-energy physics backgrounds. A brief listing of the present experiments, ordered by type of measurement performed and the type of information about the collision thus obtained, is presented in Table 2.

2. GLOBAL EVENT PARAMETERS

One can obtain information on the global distribution of energy and produced particles in such reactions by measuring the transverse energy and number of produced particles as a function of pseudorapidity and azimuth over a large region centered at midrapidity. The energy measurements are accomplished in the usual way via a set of calorimeter towers divided into electromagnetic and hadronic compartments. The multiplicity measurements are performed using large acceptance tracking devices (streamer chambers or TPCs) or large arrays of Iarrocci-type streamer tubes with pad readout. Using a suggestion of Bjorken (Ref. 1), it is possible to relate the measured transverse energy per unit rapidity, $dE_T/d\eta$, to the energy density produced in the reaction, up to an uncertainty arising from an imperfect knowledge of the proper time required for the final state particles to hadronize. If the particle multiplicity were to be a linear function of the transverse energy (or vice versa), as it is found to be, it is also possible to estimate the energy density from the particle multiplicity. It should also be possible to relate the entropy density to the observed particle multiplicity. Furthermore, following a suggestion by van Hove (Ref. 2), a search could be made for evidence of a phase change by examining the mean p_T of the produced particles as a function of energy density, viz. the observed transverse energy or particle multiplicity per unit rapidity, dN_{cp}/dy. The mean p_T, a measure of temperature, would be expected to remain constant as a function of dN_{cp}/dy, which is related to energy content of the system and should increase significantly if a phase transition is passed due to the latent heat of the transition.

It has also proved useful to measure the residual energy that appears in a region within a few tenths of a degree of the incident beam direction but downstream of the target; the devices used to do this are often referred to as Zero Degree Calorimeters.

This is particularly useful in the case of heavy-nucleus-induced reactions as opposed to proton-induced reactions due to the fact that one can relate the observed energy at zero-degrees to the number of nucleons in the incident projectile nucleus that did not participate in the reaction. That is, one can obtain a measurement of the impact parameter of the collision. This occurs because nuclei are loosely bound assemblages of nucleons - those projectile nucleons not coming within a strong absorption radius of a target nucleon will continue along the initial trajectory nearly unperturbed by the violent collisions suffered by their former neighboring nucleons in the projectile that did interact with a target nucleon. This feature has proved useful in selecting nearly head-on collisions of heavy nuclei, which are the collisions expected to exhibit the largest energy density over the largest space-time volume for a given bombarding energy and are thus those expected to offer the best chance for observing formation of a quark-gluon plasma.

An example of the observed spectrum in a zero-degree calorimeter is shown in Figure 2 for $E/A = 200$ GeV ^{32}S collisions on Al, Cu, Ag, and Au nuclei. This experiment had an upper electronic cutoff at 89% of the beam energy; otherwise the "peak" at the upper end would be observed to rise a factor of a thousand above the data shown, since a target equivalent to 0.1% of a nuclear absorption length was used. An increase in the cross section for the lowest energy events is observed for increasing target mass, and therefore increasing target thickness, as the relative probability that the ^{32}S projectile can dive completely into the target increases with target radius and therefore target mass. The fact that no cross section is seen for near-zero energies in the zero-degree calorimeter cannot necessarily be taken as evidence that some fraction of the center of mass energy supplied by the projectile is unavailable for particle production (i.e., that the target does not "stop" all the nucleons of the projectile). Rather, for the acceptance of the particular experiment, a fraction of the created particles are emitted into the acceptance of the zero-degree calorimeter. The distribution of cross section according to the transverse energy produced is shown in Figure 3 for the same reaction and experimental setup. Again the increased thickness of the heavier targets manifests itself in a large extension of the observed plateau. Central collisions of the nuclei lead to the largest observed transverse energies, which exceed 200 GeV in the pseudorapidity interval covered by this experiment. The correlation of energy appearing in the zero-degree calorimeter with that appearing as transverse energy in the calorimeters placed about midrapidity is also shown in that figure. This correlation is observed to be quite strong, showing that either experimental device can be used to select central collisions.

The distribution of transverse energy as a function of pseudorapidity is shown in Figure 4 for the same reactions. Values up to 90 GeV/unit are observed. These correspond to energy densities as high as 2.5 GeV/fm^3 if a hadronization time of 1 fm/c is assumed in applying equation 8 of Ref. 1. The results shown are for $E/A = 200$ GeV fixed target reactions, which corresponds to $\sqrt{s} = 20$ GeV/nucleon. Similar results for $E/A = 60$ GeV/nucleon ($\sqrt{s} = 10$ GeV/nucleon) and $E/A = 14.5$ GeV ($\sqrt{s} = 5$ GeV/nucleon) yield energy density estimates of 1.3 and 0.7 GeV/fm^3 respectively. That is, a roughly linear increase in the observed maximum $dE_T/d\eta$ with $\sqrt{s}$ is observed. No evidence is yet seen in these experiments for the plateau in multiplicity or transverse energy as a function of rapidity that was found at the ISR, Sp$\bar{\text{p}}$S or Tevatron. Such a boost-invariant region is expected to be observed in RHIC experiments.

The distribution of the number of produced particles as a function of pseudo-rapidity is shown in Figure 5 for central collisions of $E/A = 14.5$ GeV ^{28}Si with

740

various targets. The peak in $dN_{cp}/d\eta$ is observed to move from the cm rapidity for nucleon-nucleon collisions for the Al target, for which case the reacting partners have nearly the same mass (to within 4%), to the rapidity expected for a silicon nucleus colliding with the core of a gold nucleus that covers the same transverse area as a silicon nucleus for the case of the Au target. Again, one is seeing that the numbers of participant baryons from the two initial nuclei set the gross features of the observed distributions. The correlation of the observed multiplicity of charged particles produced with the transverse energy produced is shown in Figure 6 for $E/A = 200$ ^{16}O+Au collisions. A narrow correlation is seen, similar to the case for E_T vs zero-degree energy (although the observed slope is of opposite sign, as expected). A similar pattern is seen for other systems and bombarding energies.

The energy appearing at midrapidity and the center-of-mass energy that must be reached before a plateau appears in the rapidity distribution of produced particles are both dependent upon the mean momentum loss, or rapidity downshift of projectile baryons (upshift for target baryons). It is thus of interest to examine the rapidity distribution of observed protons and compare this with that for produced baryons, such as Λ hyperons. This is more instructive in a kinematic regime wherein the projectile and target spectator matter are separated in rapidity. This condition is more nearly attained presently at the SPS. Figure 7 shows the distribution in rapidity of protons and Λ hyperons produced in collisions of $E/A = 200$ GeV ^{32}S with a sulfur target. The protons are seen to exhibit mild peaks 1.5 units away from the projectile and target rapidities, while the Λ hyperons exhibit a plateau covering the central three units (the beam rapidity is 6.06).

A measurement of the mean p_T of produced particles vs dN_{cp}/dy has been presented by several groups. Evidence for a plateau is seen, but not for the later rise expected in the model of Ref. 2 for any of the heavy-nucleus experiments. Such a rise has been seen however in an experiment at the Tevatron (C0, E735) dedicated to measuring properties of the soft particle spectrum. Their result is presented in Figure 8 together with older data at lower $\sqrt{s}$ from UA1. Evidence for a plateau and later rise is present in the C0 results. Observation of this feature with heavy nuclei probably awaits the operation of RHIC.

The above experimental information relates to the *density* achieved for of various quantities. Another compelling reason for using heavy nuclei as reaction partners was to create such conditions over a large volume of space time, of the order of 1000 fm^3. It is possible to study the space-time extent of the system created by observing correlations of identical particles emitted as a function of their relative momentum. The quantum relation between momentum and spatial extent allows an analysis of such correlations to obtain information on the size of the system emitting the particles. Figure 9 shows the measured correlation measured for negative particles (dominantly pions) emitted in reactions of $E/A = 200$ GeV ^{16}O with an Au target. The correlation is presented as a function of relative transverse momentum and is shown for three intervals in rapidity. It is observed that the correlation is markedly narrower for the rapidity interval $2 < y < 3$, which is quite near the rapidity of the nucleon-nucleon center of mass (3.03 in this case) and in fact includes the rapidity calculated for the center of mass of a system composed of the ^{16}O nucleus bombarding the core of the gold nucleus which has the same transverse area as the ^{16}O projectile. Midrapidity of this latter frame was noted above to correspond to the centroid of the observed distribution of produced particles. The transverse and longitudinal radii extracted for the rapidity interval $1 < y < 2$ are 4.3 ± 0.6 fm and 2.6 ± 0.6 fm, respectively, similar to the size of the participant zone in a clean cut geometry. However, the transverse and longitudinal radii extracted for the interval at midrapidity, $2 < y < 3$, are 8.1 ± 1.0 fm and $5.6^{+1.2}_{-0.8}$ fm, respectively, suggestive of an extended system. Such an extended system would be expected were the collision to lead to the formation of a system which required a large proper time to hadronize. Such behavior is expected for a quark-gluon plasma, as it requires a long proper time to expand and cool to the point where hadronization can take place. One must view this result with some caution, however, because the pion correlations will be perturbed strongly by the decay of short-lived resonances. New efforts will concentrate on correlations of K^+ mesons, which should not suffer this problem.

3. INDICATORS OF A PHASE TRANSITION

The expected deconfinement temperature of hadronic matter is in the range of $T = 200$ MeV. This is greater than the mass of the strange quark, meaning that it should be energetically favorable to produce $s\bar{s}$ pairs in such a system. The likelihood of producing such pairs is a strong function of the temperature of the system. This leads to the suggestion that observing a marked increase in strange meson and hyperon production could signal deconfinement and that quantifying this increase could diagnose properties of the system formed (Ref. 3). Straightforward phase space arguments suggest that this signal's discrimination power could be amplified by requiring many strange quarks in a small region of the final state. Experiments have thus been performed to look for hyperons carrying multiple strangeness or look for events that include many hyperons. At present, measurements have been made of kaon abundances and of the production cross sections over a good part of (rapidity-p_T) space of Λ hyperons. First information on multiply-strange hyperons and events with several hyperons is beginning to emerge.

If the region in which the $s\bar{s}$ pairs are produced is baryon rich, then there will be a net excess of u over $\bar{u}$ and d over $\bar{d}$ quarks. In that case it will be easier to form $K^+ = u\bar{s}$ than it will be to form $K^- = \bar{u}s$, and it will be easier to form $\Lambda = uds$ than it will be to form $\bar{\Lambda} = \bar{u}\bar{d}\bar{s}$. Experiments have begun accumulating data on the production ratios of K^+/π^+, K^-/π^-, $\bar{\Lambda}/\Lambda$ and $\bar{\Lambda}/\bar{p}$ over various parts of phase space.

Figure 10 shows results of measurements of K^+, K^-, π^+ and π^- cross sections in collisions of $E/A = 14.5$ GeV ^{28}Si with an Au target. The results are presented as a function of transverse mass for the rapidity range $1.2 < y < 1.5$, which is just behind the nucleon-nucleon midrapidity point. The marked contrast between the K/π ratio for positives vs negatives is apparent. The ratio of the number of $\bar{\Lambda}$'s produced to the number of Λ's produced as a function of p_T for $E/A = 200$ GeV ^{32}S+W reactions is shown in Figure 11. The ratio is less than one as expected from considerations of associated production from the entrance-channel baryons, but it is observed to continue to decrease for increasing p_T. One caveat must be stated. These reactions are in a baryon-rich regime. The production of K^+ is favored due to associated production of hyperons. In addition, the final state interaction $K^- + n \to \Lambda + \pi^-$ is much more likely than $K^+ + \bar{n} \to \bar{\Lambda} + \pi^+$ in the high density hadronic gas formed in the (later stages of the) collision.

If a sufficiently large number of strange quarks are present at hadronization, it might be possible to form multiply strange objects much more than twice as massive as the proton, so-called strangelets (Ref. 4). The postulated H^0 particle is the simplest of these. Under favorable conditions, objects with more than 10 times the mass of a proton and a charge-to-mass ratio of the order of 0.1 to 0.2 might be formed. A first search for these objects has been carried out using the forward spectrometer of E814 at BNL. Any candidates would have fallen into the right hand side of the charge vs mass plot shown in Figure 12; present statistics set a limit of $< 1.2 \times 10^{-4}$ strangelet produced per interaction in $E/A = 14.5$ GeV collisions of ^{28}Si with Au.

4. PENETRATING PROBES

Lepton pairs and photons emitted in reactions of heavy nuclei are expected to carry information about the early stages of the reaction, well before the hadrons observed in the final state are created. It is thought that direct information on the temperature and density of a plasma may be carried in the p_T and number distributions of the emitted real and virtual photons. It is expected that the yield of such photons and lepton pairs will be concentrated in a p_T region characterized by 3 to 7 times the deconfinement temperature. For a deconfinement temperature of 0.2 GeV, direct photon and lepton pair emission would be expected in the p_T range from 0.6 to 1.4 GeV/c. One measurement has been reported to date of the direct photon content of the final state in central and peripheral collisions of $E/A = 200$ GeV ^{16}O with Au. The results are shown in Figure 13 for the two centrality cuts. They are presented as a function of p_T; the points give the inclusive photon yield, normalized to the π^0 yield, while the shaded areas show the yield of photons (again relative to the π^0 yield) which can be accounted for by decays of π^0 and η mesons, which are measured in the same apparatus. Within the errors, no evidence is obtained for a direct photon signal. That is, at the bombarding energies reached at the SPS, there is no requirement to invoke other than familiar hadronic sources to account for the observed photon yield at these low p_T values.

Experiments which detect lepton pairs can also observe the vector mesons. In particular, it is expected that formation of a quark-gluon plasma will have a profound effect on the formation probability for the heavier vector mesons, such as the J/ψ (Ref 5). The formation of a plasma, which consists of color-charged objects, implies the existence of a Debye-radius, beyond which the color charge is screened. For a sufficiently high temperature, this screening radius can fall below the asymptotic separation of a $c - \bar{c}$ pair which is trying to form a J/ψ meson. The formation of J/ψ mesons would thus be "suppressed" relative to its value in $p - p$ collisions or in collisions not leading to plasma formation. The effect should be largest for J/ψ's travelling slowly with respect to the plasma, such as those at low p_T and should be largest for central collisions, if the likelihood of forming a plasma is greatest there. Figure 14 shows the production rate of J/ψ mesons relative to that for Drell-Yan pairs, which should be unaffected by the existence of a plasma, as a function of transverse energy for collisions of $E/A = 200$ GeV ^{16}O with uranium. A significant effect is observed. This suppression is indeed observed to diminish with increasing p_T in the range of $p_T = 0 - 2$ GeV/c. However, again recent theoretical work has shown that reabsorption of the J/ψ's produced in the surrounding dense hadronic matter can mimic the observed effect. Interestingly, calculations reporting an unexpected suppression due to such absorption do however have to assume that matter several times denser than normal nuclear matter is formed.

5. CLOSING REMARKS

The first three years of experimentation with beams of high-energy heavy nuclei have yielded a large number of observations concerning the energy density, space-time volume, and flavor content of the system formed. It has been possible to devise reliable methods for selecting central collisions and to develop experimental techniques for coping with the quite large final-state multiplicities. A number of observations have been made suggesting that a very dense state of excited hadronic matter is formed over a large volume. Future work, particularly at the Relativistic Heavy Ion Collider at BNL should expand these observations to include systems 2-3 times as large and with energy densities an order of magnitude higher than presently attainable. These experiments should be able to address definitively the existence and nature of a quark-gluon plasma.

REFERENCES

For a general review, see the proceedings of the various Quark Matter Conferences, such as Nordkirchen, 1987 in Z. Phys. C **38** (1988) and Lenox, 1988 in Nucl. Phys. **A498** (1989).

1. J. D. Bjorken, Phys. Rev. D **27**, 140 (1983).
2. L. Van Hove, Phys. Lett. **118B**, 138 (1982).
3. R. Hagedorn and J. Rafelski, Phys. Rev. Lett. **97B**, 180 (1980); P. Koch, B. Müller, and J. Rafelski, Phys. Rep. **142**, 453 (1986).
4. S. Chin and A. Kerman, Phys. Rev. Lett. **43**, 1292 (1979); E. Witten, Phys. Rev. D **30**, 272 (1984).
5. T. Matsui and H. Satz, Phys. Lett. **178B**, 416 (1986).

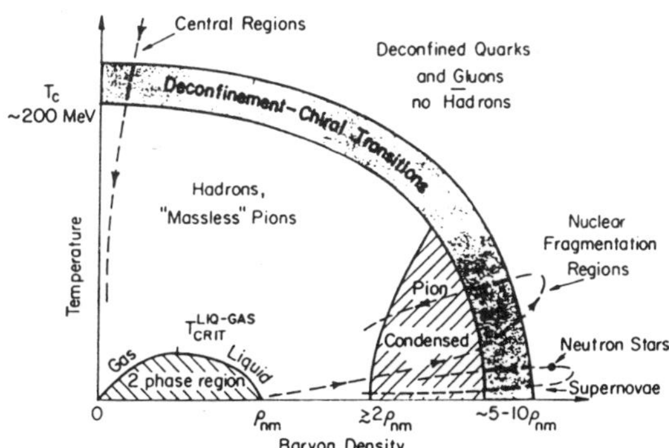

Figure 1

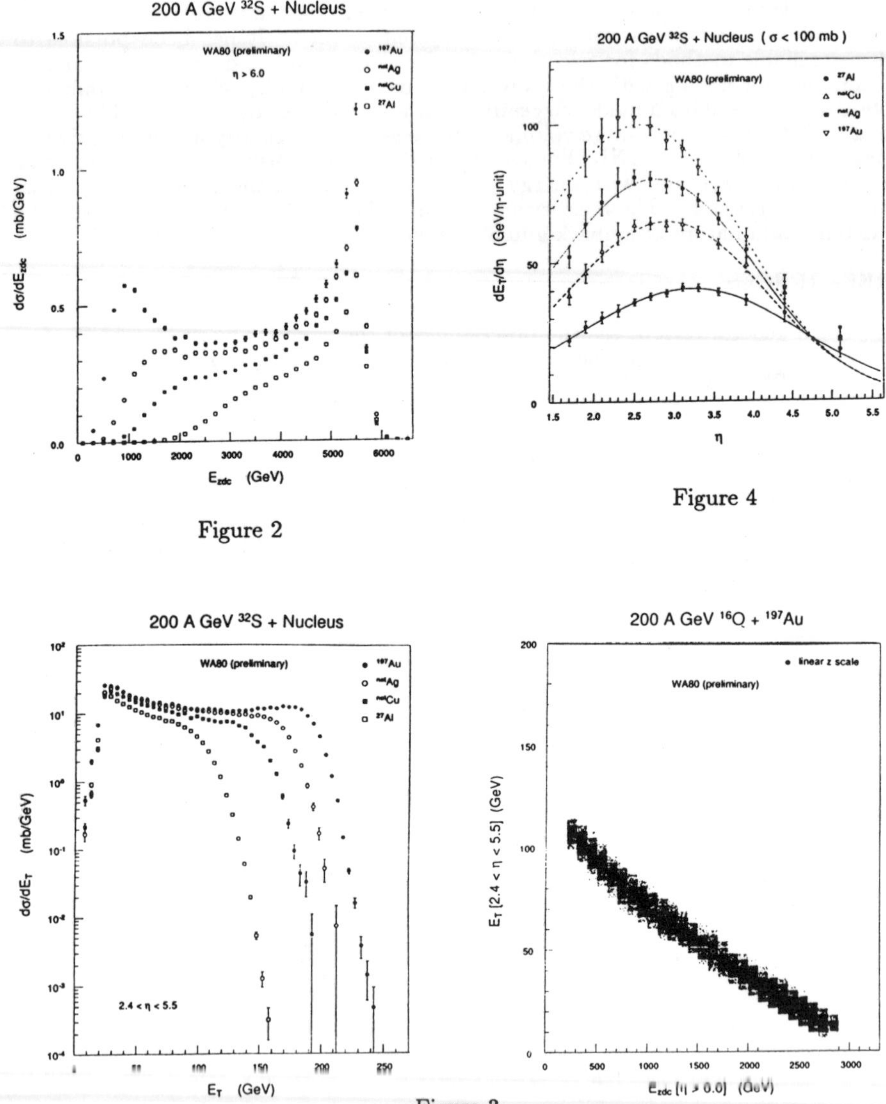

Figure 2

Figure 4

Figure 3

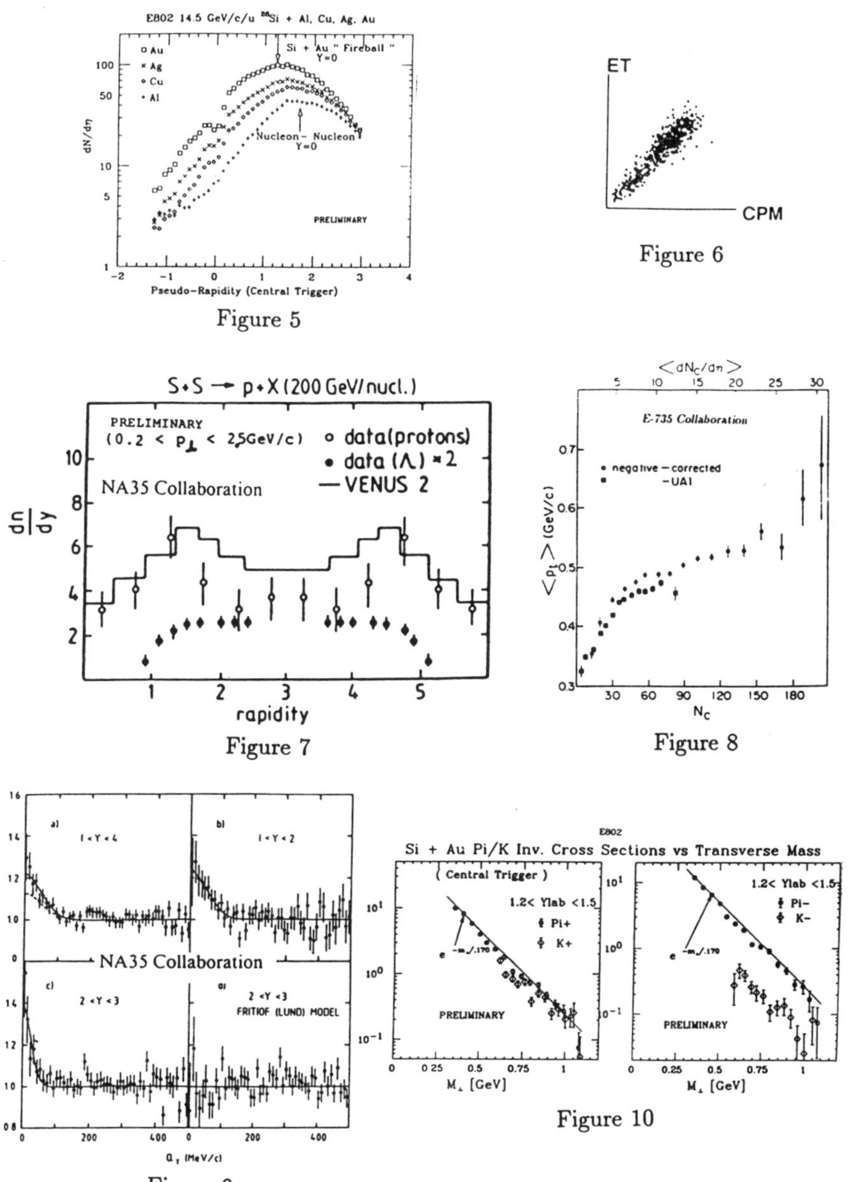

Figure 5

Figure 6

Figure 7

Figure 8

Figure 9

Figure 10

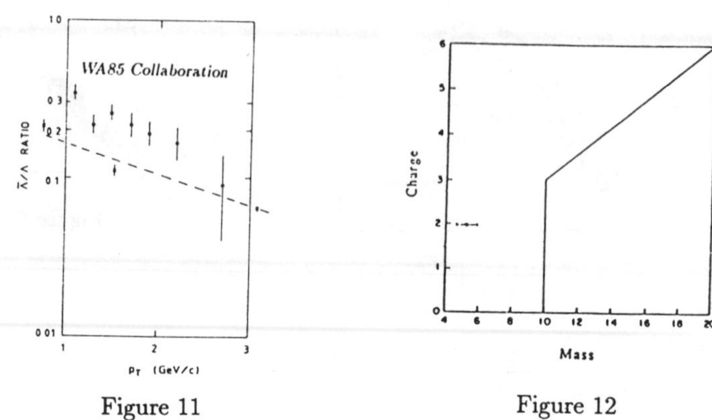

Figure 11

Figure 12

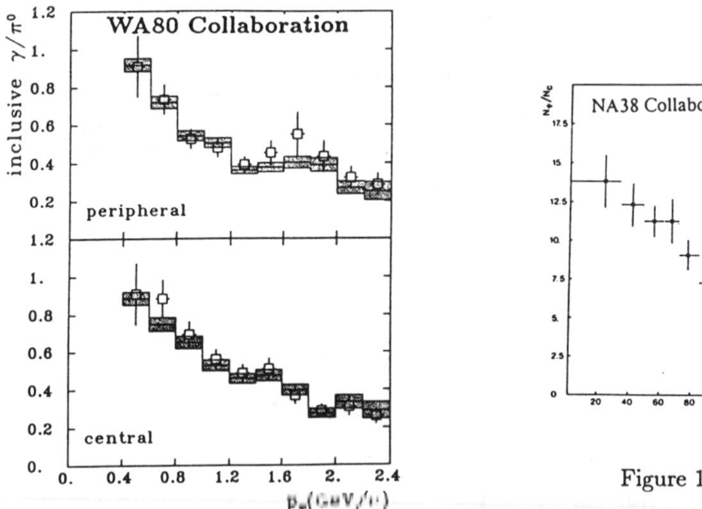

Figure 14

Figure 13

Table 1

SIGNAL	COMMENTS	
Inclusive Particle Spectra Particle Interferometry	Indicators of Temperature, Size and Density	Global Event Parameters
Multi-Particle Correlations in Rapidity; Energy Flow	Long Range Correlations and Macroscopic Fluctuations Characteristic of First-Order Phase Transition	
$J/\psi,\psi'$ Υ,Υ' Suppression in Dilepton Spectra	Color Screening Effects in Deconfined Plasma Suppress Heavy Quark Resonances	Indicators of A Phase Transition
Particle Flavor Ratios	Chemical Equilibrium in Hot Plasma Gives a Large Number of Strange Particles and Enhanced $\bar{\Lambda}/\bar{p}$ Ratio	
Stable Multiquark States	6-Quark and Higher Configurations Readily Assembled in the Plasma	
Direct Photon Production ($m_T = p_T$)	$m_T \leq 50$ MeV: coherent Emission from Local Charge Fluctuations $50 \leq m_T \leq 500$ MeV: Hadronic Decays; Some Coherent Effects $500 \leq m_T \leq 3$ GeV: Direct Emission from Plasma	Penetrating Probes: Direct Information From the Plasma
Lepton Pair Production (Virtual Photon: $m_T^2 = m_{Pair}^2 + p_T^2$)	$m_T \geq 3$ GeV: Approach to Equilibrium; Structure Functions of Quarks and Gluons Change and are Computable in Perturbative QCD	
High-p_T Jets	Measures Propagation of Quarks and Gluons Through Nuclear Matter; Hadronization Properties Reflect the "Real Sea" of Quark-Gluon Plasma	

Table 2. Present High-Energy Heavy-Ion Experiments (circa 1990)

Measurement	Information
Transverse energy spectra E802/859, E814 NA34, NA35, WA80	Estimate of achieved energy density Propagation and collisions of nuclei and nucleons in nuclear matter
Multiplicities and particle spectra E802/859, E810, E814 NA34, NA35, NA36, WA80, WA85 Emulsions C0 (E735, FNAL)	Propagation and collisions of nuclei and nucleons in nuclear matter
Particle identification (All electronic experiments)	Selective information on stopping, hydrodynamic expansion, and signals of quark-gluon plasma
Pion (kaon) interferometry E802/859 NA35, NA44 C0 (E735, FNAL)	Volume/lifetime of system at hadronization
Photon, lepton pair production E855 NA34, NA38, NA45, WA80	Selective probe of deconfined phase
Exotic assemblages (strangelets, antinuclei) E814, E858	Products arising from large phase space density of $s\bar{s}$ pairs and of antiquarks

Resummation of Perturbation Theory in Hot QCD

Eric Braaten
Department of Physics and Astronomy
Northwestern University
Evanston, Illinois 60208

When quarks and gluons have energies small compared to the temperature, resummation of perturbation theory is necessary. The resummation solves the longstanding problem of the gauge dependence of the plasmon damping rate. It also indicates that the production rate of low energy dileptons in a quark gluon plasma is much larger than previous estimates.

QCD, the field theory of strong interactions, predicts that at high temperature, hadronic matter will undergo a phase transition to a quark-gluon plasma. At extremely high temperature T, the running coupling constant $g(T)$ of QCD becomes very small, and many properties of the plasma can be calculated using perturbation theory. At the temperatures that may be accessible in heavy ion collisions, $g(T)$ is not small. Nevertheless, I believe that the most promising approach to understanding the properties of the plasma is to develop a systematic treatment of hot QCD, and then extrapolate down to the temperatures of interest.

In the past year, there has been a significant breakthrough in our ability to calculate the properties of hot QCD using perturbation theory. By hot QCD, I mean hadronic matter at a temperature T much larger than the phase transition temperature and the masses of the relevant quarks (u, d, and s) and also large enough so that $g(T) << 1$. Then the only important energy scales are the well-separated scales T, gT, g^2T, etc. The scale T can be handled by straightforward perturbation theory. The scale g^2T is very complicated; nonperturbative effects, such as the generation of a magnetic screening length, are known to arise at this energy scale. The breakthrough has come about in our understanding of the intermediate scale gT. Naive perturbation theory is not sufficient, but it is possible to treat this scale systematically using a resummation of perturbation theory[1].

The need for a resummation of perturbation theory also arises in zero-temperature field theory. A good example is the cross section for e^+e^- scattering near the Z^0 resonance. In lowest order the cross section diverges at $s = M_Z^2$: $\sigma(s) \propto 1/(s - M_Z^2)^2$. However if $s - M_Z^2$ is of order αM_Z^2, there are loop corrections that are not suppressed by powers of the electromagnetic coupling α. Consider the propagator of the Z^0. At tree level, it is $1/(s - M_Z^2)$, which is of order $1/\alpha M_Z^2$ for $s - M_Z^2$ of order αM_Z^2. The one-loop propagator correction includes a term $iM_Z\Gamma_Z/(s - M_Z^2)^2$, which is also of order $1/\alpha M_Z^2$, because the total width Γ_Z is of order αM_Z. Similarly the two loop propagator correction includes a term $(iM_Z\Gamma_Z)^2/(s - M_Z^2)^3$, which is not suppressed by any powers of α if $s - M_Z^2$ is of order αM_Z^2. A complete calculation requires the resummation of such corrections to all orders of the loop expansion. The resulting expression for the cross section has the familiar form of the relativistic Breit-Wigner resonance: $\sigma(s) \propto 1/((s - M_Z^2)^2 + m_Z^2\Gamma_Z^2)$.

A similar resummation is necessary in hot QCD. For "hard" particles with energies on the order of T, naive perturbation theory is applicable. However for "soft" particles with energies on the order of gT, there are loop corrections that are not suppressed by any powers of the QCD coupling constant g, and a resummation of perturbation theory is therefore necessary. For example, consider the propagator for a massless quark. At tree level, it is simply $1/P\cdot\gamma$. For a soft quark, P is of order gT, so the propagator is of order $1/gT$. But the quark self-energy $\Sigma(P)$ contains a term $\delta\Sigma$ proportional to g^2T^2/E, where E is the energy of the quark. Thus for $E \sim gT$, the term $\delta\Sigma/(P\cdot\gamma)^2$ in the one loop propagator correction is also of order $1/gT$. Similarly there is a two-loop propagator correction $(\delta\Sigma)^2/(P\cdot\gamma)^3$ that is not suppressed by any powers of g compared to tree level. We call the loop corrections that are of the same order in g as the corresponding tree amplitudes "hard thermal loops", because they are thermal corrections that arise from hard loop momenta.

The hard thermal loop corrections to the quark propagator can be resummed to all orders into an effective propagator $1/(P\cdot\gamma - \delta\Sigma)$ for soft quarks. The poles in the effective propagator define the mass shell for quarks propagating through the quark-gluon plasma. The mass shell has two branches whose dispersion relations were first calculated by Klimov and Weldon[2]. The first branch $E = \omega_+(p)$ is a quark with an effective mass $m_q = gT/\sqrt{6}$. The second branch $E = \omega_-(p)$ is a collective mode special to nonzero temperature. We call it the plasmino and label it q_- to distinguish it from the ordinary quark q_+. As $p \to 0$, both dispersion relations approach the effective quark mass m_q while at large p, they both approach the lightcone $\omega = p$. For the plasmino, the minimum of $\omega_-(p)$ occurs not at $p = 0$, but at the nonzero momentum $p = .408m_q$. This has remarkable consequences for the spectrum of soft dileptons emitted by a quark-gluon plasma, as will be discussed later.

The resummation of hard thermal loops in the gluon propagator also produces a nontrivial effective propagator for soft gluons. The mass shell defined by the poles in the effective propagator again has two branches, which were also calculated by Klimov and Weldon[2]. The first branch $E = \omega_t(p)$ is a transversely polarized gluon with effective mass $m_g = gT/3$. The second branch $E = \omega_l(p)$ is a longitudinally polarized gluon, and is called the plasmon. As $p \to 0$, both dispersion relations approach the effective mass $m_g = gT/3$, and at large p, they both approach the lightcone. We denote soft transverse gluons and plasmons by g_t and g_l, respectively.

In hot QCD, the resummation of hard thermal loops in the quark and gluon propagators is not sufficient. There are also hard thermal loops in the one-loop vertex corrections to the 3-gluon, 4-gluon, and quark-gluon vertices.[1] If all momenta entering one of these vertices is soft, it should be replaced by an effective vertex which is the sum of the bare vertex plus the hard thermal loop. There are also hard thermal loops in amplitudes which have no bare counterparts[1]: N-gluon amplitudes for any $N \geq 5$ and the amplitude for a quark-antiquark pair and $N \geq 2$ gluons. Effective vertices must also be introduced for these amplitudes. The complete set of effective propagators and vertices defines a nonlocal effective field theory for soft quarks and gluons with an infinite set of fundamental interactions. It can be thought of as the result of a renormalization group transformation in which all hard quarks and gluons have been integrated out in favor of effective interactions among the soft quarks and gluons.

The resummation of perturbation theory described above has resolved a longstanding puzzle in thermal field theory called the plasmon problem. The problem is to calculate the damping rate of the plasmon, which determines the rate at which fluctuations away from the equilibrium distribution of plasmons damp out as a function of time. There have been a number of independent one-loop calculations of the damping rate, and almost as many different answers. Writing the damping rate as $\gamma = cg^2 T/8\pi$, the calculations have given $c = 1$ in axial and Coulomb gauges, $c \leq -11/4$ in covariant gauges, and $c = -45/4$, $-27/4$, and -11 in three different gauge invariant approaches.[3] The damping rate is determined by the imaginary part of the location of the pole in the gluon propagator and there are general arguments that this must be gauge invariant.[4] The resolution of the puzzle is that one-loop calculations are simply incomplete. There are hard thermal loop corrections that contribute to the damping rate at the same order in g as the one loop diagrams. Only after resumming such diagrams to all orders in the loop expansion can one expect the result to be gauge invariant. The complete set of diagrams that must be resummed has been identified, and the resulting expression for the damping rate has been proven to be gauge invariant[1].

Explicit calculations of the damping rate using this resummation of perturbation theory are in progress.[5] These calculations reveal the physical mechanism for the damping. The damping rate is the difference between the rates for the absorption and production of gluons with zero momentum. I will describe only the production mechanisms, since absorption occurs by the inverse reactions. The mechanism implicit in previous calculations was a $2 \to 1$ scattering process in which a plasmon at rest is produced by the collision of two massless gluons of total energy m_g. This mechanism is clearly inconsistent, because the thermal interactions that generate the mass m_g for the final state gluon have not been included for the incoming gluons. If they are consistently included, this process is kinematically forbidden. The actual mechanisms involve hard quarks and gluons in an essential way. One mechanism is a $2 \to 2$ scattering process involving a plasmon or soft transverse gluon and a hard quark or gluon. The momentum of the soft gluon is absorbed by the hard particle leaving a gluon at rest with energy m_g: $g_l G \to gG$ or $g_t G \to gG$, where G represents a hard gluon (which can equally well be replaced by a hard quark Q or antiquark) and g is the final state gluon with zero momentum. As signified by the lack of a subscript, one can not define transverse or longitudinal polarization states for a gluon at rest. A corollary of this observation is that the damping rates for transverse gluons and plasmons must be equal. The other production mechanism is a $2 \to 3$ scattering processes in which two hard particles scatter with a small amount m_g of the collision energy going into the production of a massive gluon at rest; for example, $GG \to GGg$ or $QG \to QGg$. These processes all contribute to the damping rate at the same order in g as the kinematically forbidden $2 \to 1$ scattering process. Any additional powers of the coupling constant g are compensated by additional power of T coming from integrations over the thermal distributions of the hard particles.

The resummation of perturbation theory described above has also been used to calculate the spectrum of soft dileptons emitted by a quark gluon plasma[6]. A soft dilepton is a lepton pair with total energy on the order of gT. The spectrum of hard dileptons (with energies on the order of T) can be calculated using straightforward perturbation theory. At leading order in g, the production mechanism is quark-antiquark annihilation: $Q\bar{Q} \to \gamma^*$, where Q represents a hard quark and γ^* is a virtual photon. The order g^2 perturbative corrections have also been computed explicitly for the special case of back-to-back dileptons with total momentum $\vec{p} = 0$.[7] For soft dileptons, there are hard thermal loop corrections that contribute at the same order in g as the quark annihilation diagram. These corrections must be resummed to calculate the complete production rate for soft dileptons to leading order in g. For the special case of back-to-back dileptons with $\vec{p} = 0$, this calculation has recently been carried out by Braaten, Pisarski, and Yuan[6]. The final result is the sum of the two curves in the Figure.

The lower curve in the Figure is the partial rate from processes that involve only soft quarks. They include $2 \to 1$ annihilation processes in which a soft quark q_+ or q_- annihilates with a soft antiquark to produce a soft virtual photon. These annihilation processes exhibit thresholds due to the thermal quark mass m_q. For the processes $q_+\bar{q}_+ \to \gamma^*$ and $q_+\bar{q}_- \to \gamma^*$, the threshold is at $E = 2m_q$ as expected. For the plasmino annihilation process $q_-\bar{q}_- \to \gamma^*$, the threshold is lower because the dispersion relation $\omega_-(p)$ has its mimimum away from $p = 0$. It turns on at $E = 1.856\, m_q$, with a sharp peak that is evident in the Figure. The peak is a Van Hove singularity, and is due to a divergence in the density of states that are kinematically able to contribute to this process. There is also a $1 \to 2$ decay process that can produce soft dileptons. It is the decay $q_+ \to q_-\gamma^*$ of a soft quark into a plasmino and a virtual photon. The spectrum from this process terminates at $E = .470m_q$, which is the maximum value of $\omega_+(p) - \omega_-(p)$. There is a Van Hove singularity at this point, which appears as a sharp peak in the Figure. Thus the partial rate from processes involving only soft particles exhibits dramatic structure.

The upper curve in the Figure is the partial rate from processes which involve the scattering of hard particles. The rate from such processes displays little structure except for dramatic growth at low energies. The production mechanisms include $2 \to 2$ scattering processes, such as $q_{\pm}G \to Q\gamma^*$ and $q_{\pm}\bar{Q} \to G\gamma^*$, in which a hard particle absorbs a soft quark and emits a soft virtual photon. The rates from these processes grow like $1/E^2$ at small energy. The remaining production mechanisms are $2 \to 3$ scattering processes, in which two hard particles scatter and in the process radiate a soft virtual photon. Examples are $QG \to GQ\gamma^*$, $Q\bar{Q} \to GG\gamma^*$, and $GG \to Q\bar{Q}\gamma^*$. The rates from these processes grow like $1/E^4$ for small E. While the tree level amplitudes for these scattering processes are of higher order in g than the annihilation process, the additional powers of g are compensated by additional powers of T which arise from integrating over the thermal distributions of the hard particles. Thus they do indeed contribute to the soft dilepton rate at leading order in g. As shown in the Figure, these processes involving hard particles completely overwhelm the dramatic structure in the partial rate from the annihilation and decay of soft quarks. However what is lost in structure is gained in the overall rate. For energies on the order of m_q, the total rate is orders of magnitude larger than the naive prediction from quark-antiquark annihilation, which is shown as a dashed line in the Figure.

In conclusion, progress is being made in our understanding of the high temperature limit of a quark-gluon plasma. The energy scale gT can now be treated systematically using a resummation of perturbation theory. A complete calculation to leading order in g of the production rate of soft dileptons has revealed that the dominant production mechanism involves scattering of hard particles, and that the rate is orders of magnitude larger than naive predictions. The longstanding problem of the plasmon damping rate has been solved, and explicit calculations of that damping rate will be completed soon.

This work was supported in part by the Department of Energy under contract DE-AC02-76-ER022789.

References

[1] R.D. Pisarski, Phys. Rev. Lett. **63**, 1129 (1989); E. Braaten and R.D. Pisarski, Phys. Rev. Lett. **64**, 1338 (1990); Brookhaven preprint BNL-43293, to appear in Nucl. Phys. B; Brookhaven preprint BNL-43882, to appear in Nucl. Phys. B.

[2] O. K. Kalashnikov and V.V. Klimov, Sov. J. Nucl. Phys. **31**, 699 (1980); V.V. Klimov, Sov. J. Nucl. Phys. **33**, 934 (1981); Sov. Phys. JETP **55**, 199 (1982); H.A. Weldon, Phys. Rev. D. **26**, 1394 (1982); Phys. Rev. D. **26**, 1394 (1982).

[3] K. Kajantie and J. Kapusta, Ann. Phys. **160**, 477 (1985); J.C. Parikh, P.J. Siemens, and J.A. Lopez, Paramana **32**, 555 (1986); T. H. Hansson and I. Zahed, Phys. Rev. Lett. **58**, 2397 (1987); U. Heinz, K. Kajantie, and T. Toimela, Phys. Lett. **183B**, 96 (1987); H.–Th. Elze, U. Heinz, K. Kajantie, and T. Toimela, Z. Phys. **37**, 305 (1988); R. Kobes and G. Kunstatter, Phys. Rev. Lett. **61**, 392 (1988); S. Nadkarni, Phys. Rev. Lett. **61**, 396 (1988); M. E. Carrington, T. H. Hansson, H. Yamagishi, and I. Zahed, Ann. Phys. **190**, 373 (1989).

[4] R. Kobes, R. Kunstatter, and A. Rebhan, Winnepeg preprint.

[5] E. Braaten and R.D. Pisarski, Northwestern preprint NUHEP-TH-90-15.

[6] E. Braaten, R. D. Pisarski, and T. C. Yuan, Phys. Rev. Lett. **64**, 2242 (1990).

[7] T. Altherr and P. Aurenche, Annecy preprint LAPP-TH-237/89; Y. Gabellini, T. Grandou, and D. Poizat, Nice preprint NTH 89/1.

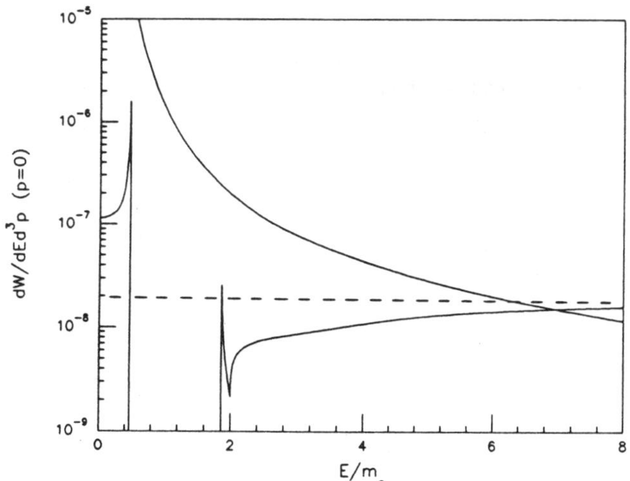

Figure. Contributions to the production rate for soft dileptons dW/dEd^3p with total momentum $\vec{p} = 0$ as a function of E/m_q.

FIREBALL PRODUCTION IN 100 GeV HADRON-NUCLEAR COLLISION

W.D. Walker

Duke University, Durham, N.C.

for E597

ABSTRACT

We have analyzed central 100 GeV hadron nuclear collisions on an event by event basis. We find the production of fireballs which decay into 20 - 30 pions and carry a major fraction of the incoming momentum. We have analyzed a sample of strange particles as well as pions. There seems to be at most a small number of light quark vector bosons produced. Our results seem consistent with a temperature of the order of 150 MeV for the fireball.

INTRODUCTION

We describe here the results of the analysis of experiment E597 done which studied the collisions of 100 GeV/c hadrons with magnesium, silver and gold nuclei. These targets were thin foils placed inside the 30" hydrogen bubble chamber at Fermilab. Downstream of the bubble there were several drift chambers and particle identifiers (CRISIS). These chambers and their results have been described previously.[1,2]

DESCRIPTION OF RESULTS

In the present analysis we look at the results of this exepriment with particular emphasis on the central collisions of the hadrons with nuclei. We do this by requiring an event to have at least three grey protons emerging from the collision. The resulting sample is about 50% of the Ag and Au events with a small fraction of Mg. In these events the out-going shower has traversed 7-10 Fermis of nuclear matter.

From these events there is little evidence of leading particles. There is on the average 1.66 particles emerging with a momentum of greater than 10 GeV/c. The particles seem to be equally distributed between plus and minus charges regardless of the charge of the incident particle.

There are on average 15 pion-like particles and most of these particles produced (presumably pions) have a momentum of less than 10 GeV/c. We have looked at the events on an event by event basis. We characterize the pion production by finding the invariant mass of these produced pions which have a momentum of less than 10 GeV/c. The average invariant mass of these groups of charged pions is 9.4 GeV/c^2 from which we infer an invariant mass of 14.1 GeV.c^2 when π°'s are included. These fireballs have an average rapidity of 1.9 units and carry on average nearly 50 GeV/c in momentum.

We have also analyzed the Λ°'s and K°'s from these reactions. On the basis of the known lifetimes of these particles we have estimated our detection efficiency for these particles. The results of this analysis is that the K°/π° ratio is .10 ± .02. The number of Λ°'s is also quite large. We assume that there are equal numbers of Λ°, Σ°'s, Σ^+ Σ^- produced. If we do this then the number of Λ°, Σ's produced is the order of .5 ± .1 strange baryons per event. This might compare to the order of 4 fast nucleons (p $\geq$ 1.4 GeV/c) per event.

A clear characteristic of these events is that the average P_T is quite dependent on particle type. The average P_T's are as follows.

$$< P_{T\pi} > = .30 \pm .03 \text{ GeV/c}$$
$$< P_{TK} > = .39 \pm .03 \text{ GeV/c}$$
$$\text{fast P} \quad < P_{TP} > = .52 \pm .05 \text{ GeV/c}$$
$$< P_{T\Lambda} > = .50 \pm .05 \text{ GeV/c}$$

We obtain the $< P_T >$ for the protons by taking the difference between the transverse momentum spectra for π^+

and π^- as shown in Figure 1. There are the order of 2 fast protons per central collisions. The fact that there is a fairly strong dependence of $< P_T >$ on particle mass is at least consistent with a temperature which characterizes the processes.

A search has been made for the production of vector bosons. We have looked for the production of ρ^o, ω^o, $K^{*\pm}$. We have set upper limits on the production of these particles which are well below the numbers expected from simulations such as FRITIOF.[3]

We show in Figure 2, 3, 4 the results of our searches for these particles. In the case of the ρ^o and ω^o we are looking for relatively broad bumps in the $\pi^+-\pi^-$ mass spectra. The ω^o gives rise to a bump centered at .470 GeV/c^2 In the case of the ρ^o we should see an increase centered at .77 GeV/c^2. There is a hint of the ρ^o for the peripheral events shown in Figure 2. The more multiple central events show no sign of either the ρ^o or ω^o. It happens that the $\pi^+-\pi^-$ and $\pi^--\pi^-$ mass spectra show nearly the same slope with respect to the invariant masses in the 600-900 MeV/c^2 mass range. We have normalized in the mass region outside the ρ^o region and subtracted. We find an excess of the order of 500 counts in the ρ^o region. This gives us an upper limit of .25 $\pm$ .1 ρ^o's per event for the central events. We find similar upper limits for the number of ω^o's. The same analysis was applied to the peripheral collisions with .15 $\pm$ .05 ρ^o's per event found. In both cases the ρ^o signal seemed to be skewed toward the low mass side. We expected the order of .6 ρ^o's and ω^o's from the FRITIOF simulation program. In Figure 4 we show the invariant mass spectrum of $K^o + \pi^\pm$. As in the previous cases there is no peak at the value expected for the $K^*(890)$. We would expect 150 counts in the region of 890 if there were .5 K^*'s per K^o produced. One might imagine 80 or 90 K^*'s to be present. Thus we can set a limit of .2 to .3 K^*'s per K^o.

CONCLUSIONS

We will comment briefly on the results presented above. One of the important characteristics of this is the large number of fast nucleons knocked on in the process of the fireball production. There is on the average 1 nucleon of 2 GeV/c or more picked up per collision . There are 4 nucleons with momentum between 1.5 and 2.0 GeV/c. It is clear that the fireball has a large cross section for picking up nucleons in its path. The average transverse momenta for the particles that we observe are characteristic of a temperature of the fireball of about 150 MeV. The relatively low upper limits on the number of vector mesons is also consistent with a temperature in the 150 MeV range. The temperature is not obviously related to the multiplicity of the events. The other parameter of importance is the size of the fireball. The average number of "grey" particles (protons and neutrons) which is 16 gives some idea of the size of the fireball. To explain this number of knocked on protons and neutrons we must suppose a diameter of the average fireball is the order of 3 - 4 Fermis as it exits the nucleus. It likely expands even more before it begins hadronization.

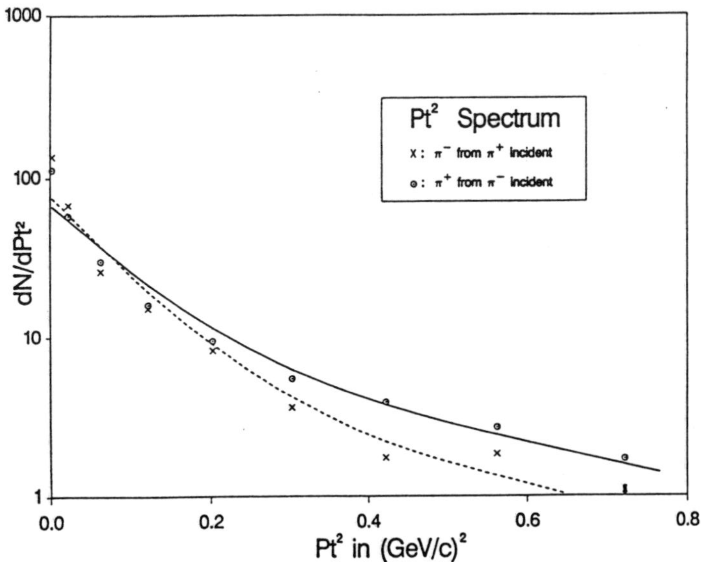

Figure 1: $\dfrac{dN}{dP_T^2}$ spectrum for "π^+" and "π^-".

BIBLIOGRAPHY

1. J. Whitmore et al, Proc. of the Santa Fe Meeting, World Scientific, p. 503 (1984).

2. W.D. Walker, P.C. Bhat in **Multiparticle Production**, edited by P. Carruthers, World Scientific, p. 153 (1988).

3. B. Anderson, G. Gustafson, B. Nilsson-Almquist, Nucl. Phys. B281, p. 289 (1987).

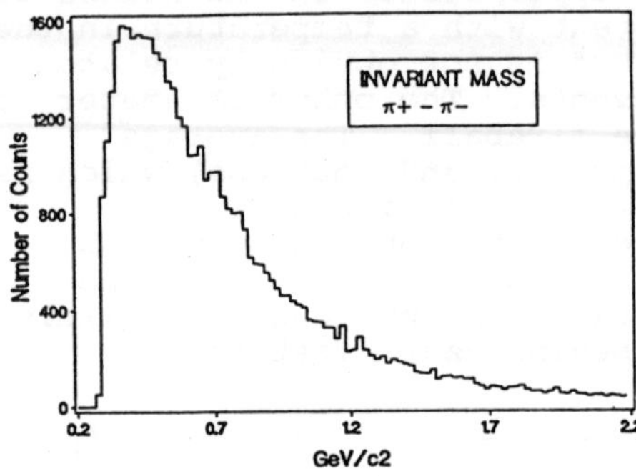

Figure 2: Invariant mass spectrum for $\pi^+ - \pi^-$ from peripheral hadron-nuclear collisions. 3000 events in the sample.

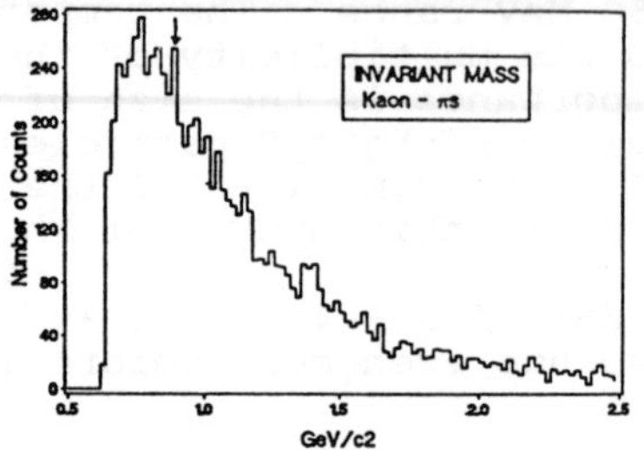

Figure 4: Invariant mass spectrum for $K^0 - \pi^+$ pairs from central hadron-nuclear collisions. 300 K^0's in the sample.

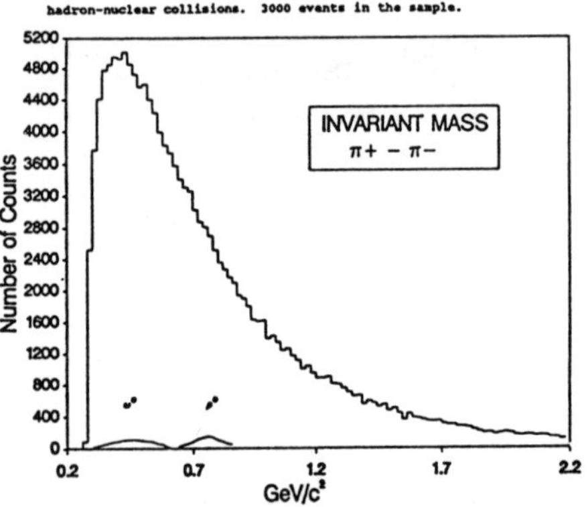

Figure 3: Invariant mass spectrum for $\pi^+ - \pi^-$ from central hadron-nuclear collisions. 2330 events in the sample. We show the number of pairs expected from .5 ρ^0's or ω^0's per event

Results from $p\bar{p}$ Collisions at $\sqrt{s} = 1.8$ TeV for Mass Identified Particle Yields

Tim McMahon
Purdue University
West Lafayette, Indiana 47907

For the FNAL E-735 Collaboration

T. Alexopoulos[6], C. Allen[5], E. W. Anderson[3], H. Areti[2], S. Banerjee[4]
P. D. Beery[4], N.N. Biswas[4], A. Bujak[5], D.D. Carmony[5], T. Carter[1]
P. Cole[5], Y. Choi[5], R. J. De Bonte[5], A. R. Erwin[6], C. Findeisen[6]
A.T. Goshaw[1], L.J. Gutay[5], A.S. Hirsch[5], C. Hojvat[2], V.P. Kenney[4]
C.S. Lindsey[3], J. M. LoSecco[4], T. McMahon[5], A.P. McManus[4]
N. Morgan[5], K. Nelson[6], S.H. Oh[1], J. Piekarz[4], N.T. Porile[5], D. Reeves[2]
R.P. Scharenberg[5], S.R. Stampke[4], B.C. Stringfellow[5], M.A. Thompson[6]
F. Turkot[2], W.D. Walker[1], C.H. Wang[3], and D.K. Wesson[1]

[1] *Department of Physics, Duke University, Durham, North Carolina 27706*
[2] *Fermi National Accelerator Laboratory, Batavia, Illinois 60510*
[3] *Department of Physics, Iowa State University, Ames, Iowa 50011*
[4] *Department of Physics, University of Notre Dame, Notre Dame, Indiana 46556*
[5] *Department of Physics, Purdue University, West Lafayette, Indiana 47907*
[6] *Department of Physics, University of Wisconsin, Madison, Wisconsin 53706*

ABSTRACT

The yields and transverse momentum distributions of pions, kaons, and antiprotons produced in $p\bar{p}$ collisions at $\sqrt{s} = 1.8$ TeV at the Fermilab Tevatron collider have been measured up to $\langle dN_c/d\eta \rangle \approx 20$. The average transverse momentum $\langle p_t \rangle$ versus event multiplicity N_c in the central region for these particles is presented. We also present results of fitting the transverse momentum distributions for each particle with a transverse flow model.

1. Introduction

We report results from the first physics run of experiment E735, which had the primary goal of looking for evidence of Quark-Gluon (QG) Plasma formation in the collisions of protons and anti-protons at $\sqrt{s} = 1.8$ TeV at Fermilab. Conjectures of this phase of strongly interacting matter were made shortly after the formulation of QCD [1]. The experimental search is complicated by the complex nature of detection since the measured particles are typically yielding direct information only on the later stages of the evolution of the collision. With this in mind E735 was designed to measure several expected signatures of QGP formation of which a couple are presented in this article. The experiment has to date recorded the highest energy densities obtained in nucleon-nucleon collisions.

2. Apparatus

The experiment located at the C0 intersection of the Tevatron at Fermilab is a regular lattice $\beta^* = 75$ meter site and for the 1987 physics run acheived a peak luminosity of 1×10^{27} cm^{-2}sec^{-1}. Descriptions of the apparatus appear elsewhere [2]. Here we note the salient points of our detector for the data presented in this article. The measurement of the event charged particle multiplicity, N_c, was provided by a 240 element scintillator hodoscope which had full coverage in phi and extended from -3.25 to +3.25 in pseudo-rapidity η. The transverse momentum, p_t, was measured by a spectrometer situated near 90 degrees to the collision axis with a 0 to 18 degree in phi and -0.36 to +1.0 in pseudo-rapidity acceptance. A TOF array of scintillators located behind the spectrometer drift chambers together with the T0 signal from 18 beam-beam scintillators located 2 meters upstream and downstream from the nominal interaction point gave charged hadron identification up to 1.5 GeV/c momentum.

3. $\langle p_t \rangle$ versus multiplicity

The formation of QG matter is speculated to have a dramatic signature in a temperature T versus entropy density S diagram [3]. Following Van Hove [4] we identify the $\langle p_t \rangle$ with T and the event multiplicity $dN_c/d\eta$ with S. The $\langle p_t \rangle$ is obtained from the invariant distribution $d^2 N_c/dp_t^2 dY$ which is computed by taking the rapidity (Y) distribution to be flat for spectrometer tracks. This was checked for each particle type by plotting dN_c/dY for both low and high values of p_t and N_c over the acceptance of the spectrometer. The invariant distributions for each particle are then fit with the analytic forms $p_0^n/(p_t + p_0)^n$ for π^{+-} , $exp(-bp_t)$ for K^{+-} and $exp(-bm_t)$ for $\bar{p}$ where $m_t = \sqrt{m^2 + p_t^2}$ and were chosen for giving the best χ^2 fit. All data is fit in the interval .3 to 1.5 GeV/c with the parameters from the fit integrated to give the $\langle p_t \rangle$ in the p_t range 0 to 1.5 GeV/c. The π^{+-} fit with p_0 fixed to 1.0 GeV/c gives $n = 8.12 \pm .03$ and $\langle p_t \rangle = 0.362$(stat.$\pm$.001 sys.$\pm$.02) for minimum bias events. The $\langle p_t \rangle$ for K^{+-} is 0.501($\pm$.005$\pm$.05) and the $\langle p_t \rangle$ for $\bar{p}$ is 0.656($\pm$.011$\pm$.033) which is also for minimum bias events. Changes in $\langle p_t \rangle$ from using alternate extrapolation to $p_t = 0.$ methods give approximately 10% variations on the $\langle p_t \rangle$.

The $\langle p_t \rangle$ versus N_c , shown in Fig. 1, is determined by again using the same fitting procedure as above for fixed ranges of N_c. The apparent "plateau" at $N_c > 60$ corresponds to an energy density $\epsilon_0 = 1.5$ to 3.0 GeV/fm^3 [5].

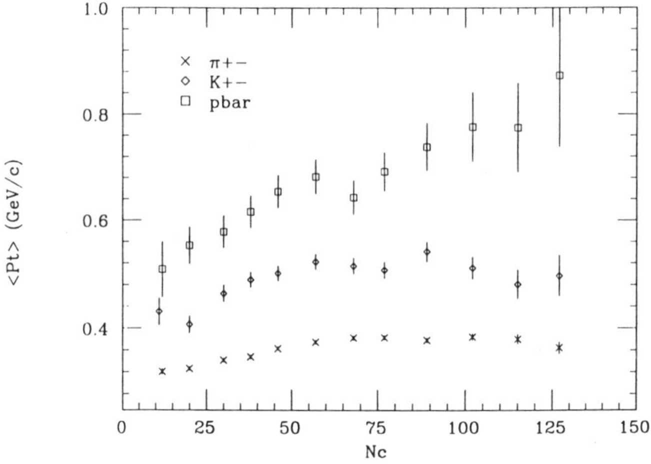

Fig. 1 $\langle p_t \rangle$ versus N_c for π^{+-}, K^{+-} and $\bar{p}$.

4. K/π and $\bar{p}/\pi$ ratios

An enhanced production of strange particles may be of interest in studying QGP formation [6]. Given in Table 1 is the particle ratios of K/π and $\bar{p}/\pi$ as a function of the event multiplicity N_c .

Table 1 K/π and p/π ratio.

N_c	K^{+-}/π^{+-}	$2 \times \bar{p}/\pi^{+-}$		N_c	K^{+-}/π^{+-}	$2 \times \bar{p}/\pi^{+-}$
12	.102(±.009)	.050(±.008)		68	.116(±.005)	.065(±.005)
20	.099(±.006)	.051(±.005)		77	.139(±.006)	.067(±.005)
30	.099(±.005)	.065(±.005)		89	.114(±.006)	.061(±.005)
38	.109(±.005)	.059(±.004)		102	.144(±.008)	.049(±.006)
46	.111(±.005)	.062(±.004)		115	.142(±.012)	.044(±.007)
57	.116(±.005)	.061(±.004)		127	.112(±.013)	.065(±.016)

The Table shows the increase in the production of kaons from low to high N_c relative to pions is approximately 30%. The overall K/π ratio is 11.5(±.1±.6)% and the $\bar{p}/\pi$ ratio is 6.53(±.1±2.5)%. For comparison the CERN UA5 group gives

$K/\pi = 10.0(\pm.8)\%$ at $\sqrt{s} = 900$ GeV [7] and the UA2 experiment gives $\bar{p}/\pi = 6.5(\pm.5)\%$ at $\sqrt{s} = 540$ GeV [8].

5. Transverse flow fits

The p_t spectra for each hadron was fit with a transverse flow model distribution. The model we have used is from Csernai and Barz [9] where the distribution is given to be

$$\frac{1}{N}\frac{dN}{dp_t^2} = \frac{\gamma_t}{2\pi T m^2}\left(m_t \frac{K_1(a)}{K_2(\frac{m}{t})}I_0(b) - p_t v_t \frac{K_0(a)}{K_2(\frac{m}{t})}I_1(b)\right) \tag{1}$$

with $a = \gamma_t m_t/T$, $b = \gamma_t p_t v_t/T$, $m_t = \sqrt{m^2 + p_t^2}$, $\gamma_t = 1/\sqrt{1 - v_t^2}$ and K and I the Bessel functions. This formula is derived from assuming the validity of the Bjorken model [10] and using the Jüttner momentum distribution for an ideal gas of relativistic particles [11] on an expanding envelope of hadrons with transverse velocity v_t.

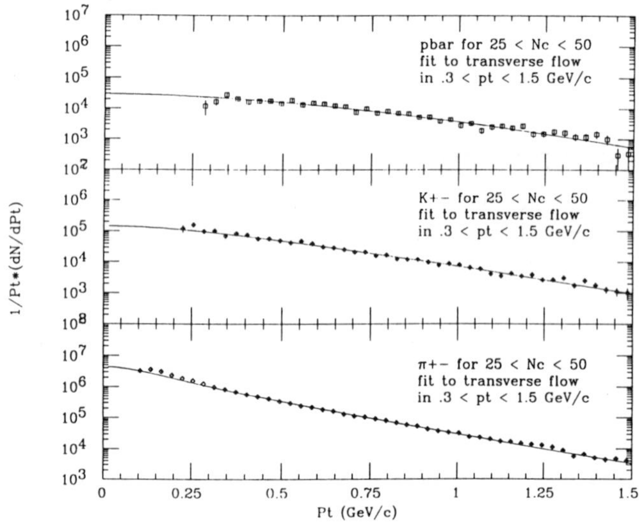

Fig. 2 Fit to $d^2 N_c/dp_t^2 dY$ using transverse flow model.

Fig. 2 shows a fit for each particle in the $N_c = 25$ to 50 range. The fits suggest that the v_t assumes higher values for pions ($v_t/c \approx .7$) and the lowest values for pbar ($v_t/c \approx .1$) and is roughly independent of the multiplicity. We interpret this result as indicating that heavier hadrons are decoupling from earlier stages of the expansion of the collision. The fits also suggest that the temperatures for pions are lower than that for pbars. These results will be given in more detail in an upcoming article.

6. Summary

The data presented can be interpreted as giving the signature of deconfinement. Other interpretations are also possible. The higher statistics data taken during the 1988-1989 collider operation will hopefully distinguish between competing models.

References

[1] J.C. Collins and M.J. Perry, Phys. Rev. Lett. **34**, 1353 (1975).

[2] T. Alexopoulos *et al.*, Phys. Rev. Lett. **60**, 1622 (1988); S. Banerjee *et al.*, Nucl. Inst. and Meth. A **269**, 121 (1988).

[3] L.D. McLerran and B. Svetitsky, Phys. Lett. **98B**, 195 (1981); J. Kuti, J. Polonyi and K. Szlachanyi, Phys. Lett. **98B**, 199 (1981).

[4] L. Van Hove, Phys. Lett. **118B**, 138 (1982).

[5] The energy density ϵ_0 was calculated using Eq. 8 of ref. [10], multiplied by a missing factor of two. Both the radius and lifetime of 1 fm was used.

[6] J. Rafelski and B. Muller, Phys. Rev. Lett. **48**, 1066 (1982).

[7] M. Banner *et al.*, Phys. Lett. **122B**, 322 (1983).

[8] R.E. Ansorge *et al.*, Z. Phys. C **41**, 179 (1988).

[9] Unpublished.

[10] J.D. Bjorken, Phys. Rev. **D27**, 140 (1983).

[11] Relativistic kinetic theory, S.R. de Groot, W.A. van Leeuwen and Ch. G. van Weert (North Holland, 1980).

QCD WITH DYNAMICAL STAGGERED QUARKS[†]

The High Energy Monte Carlo Grand Challenge Collaboration

Khalil M. BITAR[a] Thomas A. DeGRAND[b] R. EDWARDS[a],
Steven GOTTLIEB[c] Urs HELLER[a], A.D. KENNEDY[a],
J.B. KOGUT[d] A. KRASNITZ[c], C. LIU[a], Michael C. OGILVIE[e]
R.L. RENKEN[d], Pietro ROSSI[f], D.K. SINCLAIR[g], R.L. SUGAR[h],
Michael TEPER[i], D. TOUSSAINT[j], K.C. WANG[g]

ABSTRACT

We describe a recent calculation of the hadron spectrum with two flavors of staggered dynamical quarks with a gauge coupling $6/g^2=5.60$ and quark masses of 0.025 and 0.01. The gauge fields were generated using the hybrid algorithm on 12^4 and 16^4 lattices that were doubled or quadrupled to calculate hadron propagators.

1. Introduction

Lattice gauge theory calculations offer the prospect of being able to compute the low lying spectrum of QCD from first principles. However, the calculations involved are much more demanding than was first thought. To perform reliable calculations requires lattices large enough to eliminate finite size effects, lattice spacing small enough to approach the continuum limit and quark masses light enough so that one can extrapolate to the physical pion mass. The Department of Energy has identified lattice QCD as one of its "Grand Challenges." These problems have recently been given much larger blocks of computer time than heretofore available.

Time on the ETA10 supercomputer at SCRI has been devoted to dynamical fermion simulations with weaker coupling and lighter quark mass than previously studied. High statistics and a large volume were also desired. The limited memory of the ETA10 required that studies begin on 12^4 lattices while code was prepared that would run a 16^4 lattice with reasonable efficiency. Initially, DOE had planned to increase the memory on the ETA10; however, the demise of ETA prevented this.

In Sec. 2, the simulation is described. The hadron spectrum is described in Sec. 3, first for staggered quarks and then for Wilson valence quarks. Finally, results are summarized in Sec. 4.

2. Simulation

Gauge configurations are generated using a version of the hybrid-molecular-dynamics algorithm.[1] We use two degenerate flavors of staggered quarks. These represent the light u and d quarks. The effects of the heavier strange quark are not expected to be large for the non-strange spectrum, except perhaps through a rescaling of the lattice spacing. We chose staggered fermions because of their superior chiral properties as compared with Wilson fermions. They also require less storage. (As indicated above, lack of memory was a serious problem.) It is certainly important to do calculations with dynamical Wilson fermions as well. We will have much greater confidence if we can show that results are independent of the way the fermions are placed on the lattice.

We have two main simulations and several auxiliary runs used to test systematic effects. Our main runs are done on 12^4 lattices with quark masses 0.025 and 0.01. In each case, the gauge coupling is given by $6/g^2 = 5.6$. For the heavier mass, we take the molecular dynamics step size equal to 0.02, and for the lighter mass, 0.01. The momenta conjugate to the gauge fields are refreshed every unit of simulation time. In each case, we have run for 5000 units of simulation time after equilibration. We have saved gauge configurations every 10 units of simulation time for later analysis, so on our longer runs we have 500 hadron propagators. To analyze the hadron spectrum, we have gauge fixed to lattice Coulomb gauge using an overrelaxation technique. Then, we have doubled or quadrupled the lattice in the time direction to calculate the fermion and hadron propagators. We have used both point and wall sources for our hadron sources. We find that the wall sources provide a better overlap with the lightest hadrons and that the effective masses from the wall source propagators more quickly reach their long distance limits.

Our auxiliary runs with mass 0.01 were done with increased conjugate gradient accuracy or with larger lattice size. We ran with both $12^3 \times 24$ and 16^4 lattices.

Aside from using non-local source operators, our analysis of hadron propagators is quite conventional. We have analyzed two operators for the pion and two for the rho. For staggered fermions, we use point-like operators at the sink even when we use a non-local source. The full covariance matrix of the hadron propagators is used to determine the masses.[2]

The numerical techniques used in current lattice calculations result in systematic errors and efforts must be made to control those errors. The molecular dynamics integration step size and the conjugate gradient residual must be kept small enough to result in an acceptably small error, but not so small that the computational demands are unduly increased. Due to limitations of space, we will not detail here our selection of these parameters. Other potential sources of error are finite lattice volume, a quark mass larger than its physical value and gauge coupling that must be decreased to approach the continuum limit. A detailed discussion of these effects may be found in Ref. 3.

3. Hadron spectrum

In Fig. 1, we show the simulation time history of the pion propagator at distance 10 from the source. The first 5000 time units are for our long run with mass 0.01. The additional running is done with increased conjugate gradient accuracy, 5×10^{-5} as compared with 5×10^{-4}. The pion propagator in this latter run is smaller; however, the difference may be a fluctuation as the dip here is not that

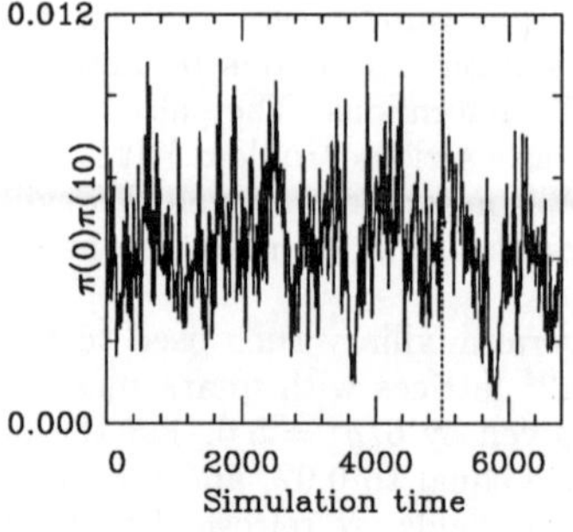

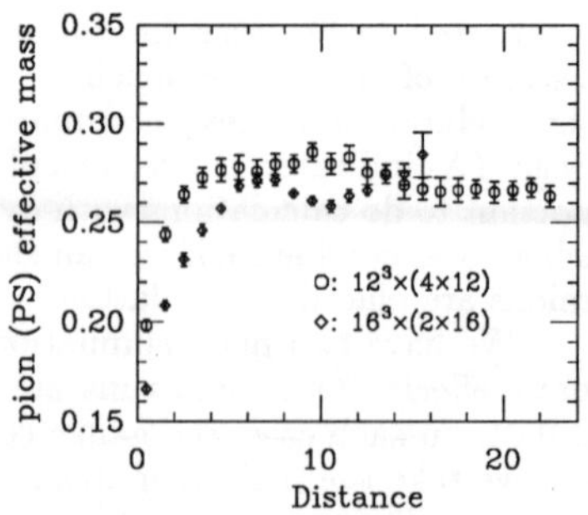

FIG. 1. Time history of the pion propagator a distance 10 from the source on a 12^4 lattice quadrupled in time. The quark mass is 0.01 and $6/g^2 = 5.6$. Beyond time 5000 the conjugate gradient residual is reduced.

FIG. 2. Effective pion mass for quark mass 0.01 on 12^4 and 16^4 lattices quadrupled or doubled in time.

different from the one seen at time 3600.

Averaging over simulation time plots like that in Fig. 1, we may obtain the hadron propagators as a function of the distance from the source. We would then like to determine the particle masses. We do this in two ways. First, we look at two (or four) points of the propagator and fit those points assuming that one (or two) particles determine the propagator. This gives a fit with zero degrees of freedom. Close to the source, there may still be contributions from some heavier particles. We expect a plateau in the mass as the distance from the source increases. If we see the plateau, we are convinced that we have looked at large enough distances to determine the asymptotic falloff of the propagator. A second approach involves fitting the mass over a range of distances, generally from a minimum to the center of the lattice. Once again, we would like to see that as the minimum distance is increased, there is a plateau in the mass. Fitting over a range has the advantage of using more information from the propagator to determine the mass; however, it is very comforting to see the plateau in the simple effective mass approach, and if one cannot find it, some doubt may be cast on more sophisticated procedures. Because of space limitations, we will focus here on effective mass plots only.

Figure 2 contains a pion effective mass plot for two runs, our 12^4 run and the 16^4 run. In each case, the mass is 0.01 and the labels indicate whether we have doubled or quadrupled. The 12^4 does have a plateau at large distance that comes after a dip around 12. For the 16^4 run there is a dip with a minimum at about 10. We believe that these features in the pion effective mass plot are caused by the doubling of the lattice. This results in a systematic uncertainty of approximately 0.01 in the pion mass. Doubling lattices in time is one of the time saving tricks invented in the early days of lattice calculations.[4] In our case, lack of memory was the motivation. We believe that this is the first time that a systematic effect of doubling has been seen. We don't know if it might be as significant in the quenched

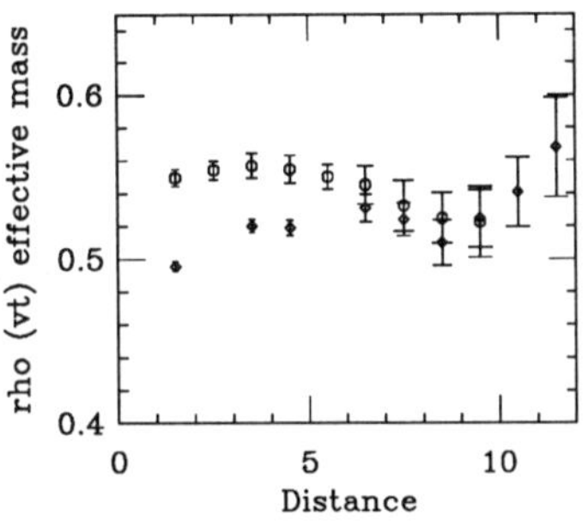

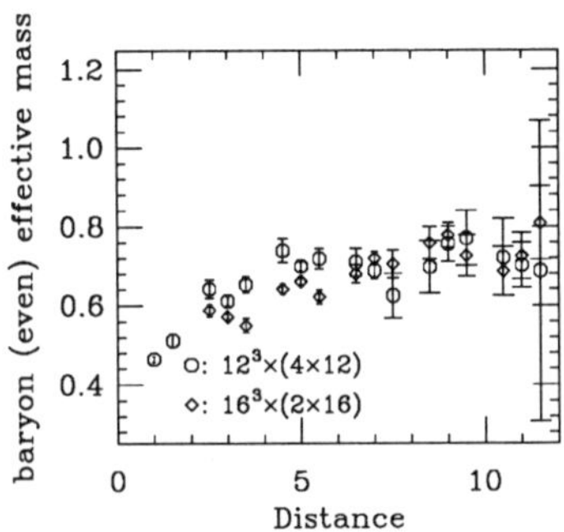

FIG. 3. Same as Fig. 2, but for the rho.

FIG. 4. Same as Fig. 2, but for the nucleon.

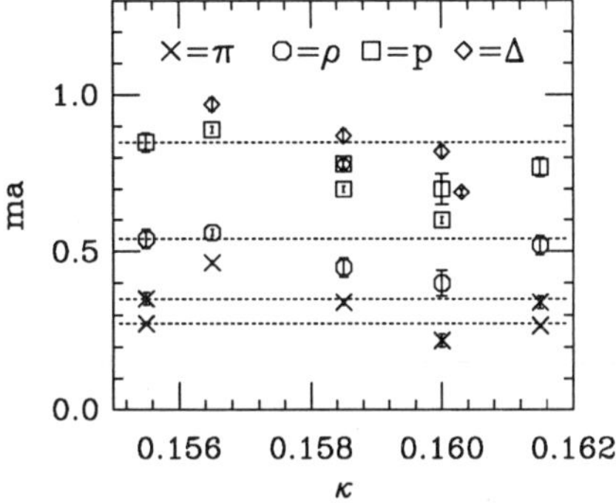

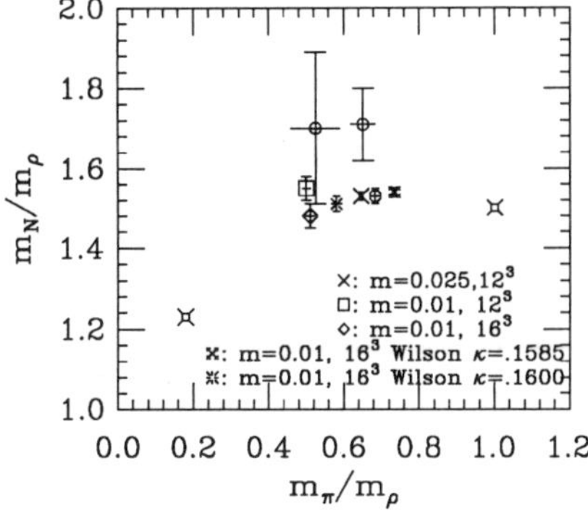

FIG. 5. A summary of the staggered and Wilson masses, with the latter shown as a function of κ and the former shown at the left and right.

FIG. 6. Edinburgh plot of the current calculation. Previous strong coupling results are plotted with an octagon. Physical result and heavy quark limit use the fancy square symbol.

approximation. Given what we have learned here, we certainly will not double lattices again. Figures 3 and 4 contain the effective mass plots for the rho and nucleon.

Although our gauge fields are generated including the effects of dynamical staggered quarks, it is also interesting to calculate the hadron spectrum with Wilson valence quarks. For the staggered spectrum, we take the valence quark mass to be the same as the mass used to generate the gauge configurations. Using Wilson quarks, we have to pick a value for the hopping parameter κ, but it is not clear what value to pick. Just as in a quenched calculation, we pick several κ values and interpolate or extrapolate as necessary. To save space, we will not show individual fits or effective mass plots. Figure 5 contains a summary of all of the Wilson valence quark mass values. At the left edge of the figure are the staggered masses from the 12^4 running. Along the right edge of the graph, we have the staggered masses from the 16^4 run. We see that the nucleon may have a finite size effect. We have two pions because in the staggered formalism there are two propagators that in the continuum represent different pion flavor combinations. The difference between these two masses is a finite lattice spacing effect. The splitting in this case is smaller than with larger lattice spacing.[2] The ratio of masses is 1.21 for $am_q = 0.025$ and was 1.76 for $6/g^2 = 5.4375$ with the same bare quark mass. The dotted lines drawn through the staggered masses are to help find a value of κ for which the hadron masses are the same. For the Wilson spectrum, we have calculations with three values of κ. At $\kappa = 0.1585$ and 0.1600, we show separate values of the nucleon for 12^4 and 16^4 lattices.

We would like to find a κ value at which the staggered and Wilson spectra agree. We find that at $\kappa \approx 0.157$ the nucleon and rho are in very reasonable agreement with the staggered values. However, the Wilson pion is quite a bit heavier than either of the staggered pions. To match the pion mass requires a larger value of κ, but then the nucleon and rho would both be too light.

4. Conclusions

We have carried out a large scale dynamical fermion calculation using weaker gauge coupling than previously attempted. In Fig. 6 we show the "Edinburgh plot" summarizing the mass ratios we have obtained. This plot also includes points from a previous calculation[2] with larger lattice spacing. We see considerable improvement in the nucleon to rho mass ratio; however, this ratio is still far from the physical value. The calculation we have done must be improved to eliminate finite size effects. We plan to continue these efforts using the Connection Machine to be installed at SCRI.

References

[†] Presented by Steven Gottlieb.

[a] SCRI, The Florida State University, Tallahassee, FL 32306-4052, USA.

[b] Physics Department, University of Colorado, Boulder, CO 80309, USA.

[c] Department of Physics, Indiana University, Bloomington, IN 47405, USA.

[d] Department of Physics, University of Illinois, Urbana, IL 61801 USA.

[e] Department of Physics, Washington University, St. Louis, MO 63130, USA.

[f] Department Physics, UC San Diego, La Jolla, CA 92093, USA.

[g] HEP Division, Argonne National Laboratory, Argonne, IL 60439, USA.

[h] Department of Physics, UCSB, Santa Barbara, CA 93106, USA.

[i] Department of Theoretical Physics, Oxford University, Oxford OX1 3NP, UK.

[j] Department of Physics, University of Arizona, Tucson, AZ 85721, USA.

1. S. Duane and J. Kogut, *Phys. Rev. Lett.* **55**, 2774, 1985; S. Gottlieb, W. Liu, D. Toussaint, R. L. Renken and R. L. Sugar, *Phys. Rev. D* **35**, 3531, 1987.
2. Many of the techniques used here are described in S. Gottlieb, W. Liu, D. Toussaint, R. L. Renken and R. L. Sugar, *Phys. Rev. D* **38**, 2245, 1988.
3. K. Bitar *et al.*, Hadron Spectrum in QCD at $6/g^2 = 5.6$, preprint IUHET-183.
4. H. Hamber and G. Parisi, *Phys. Rev. Lett.* **47**, 1792, 1981.

CHIRAL SYMMETRY BREAKING IN QUENCHED QED

SIMON HANDS

Department of Physics, University of Illinois at Urbana-Champaign,
1110 West Green Street, Urbana, IL 61801, U.S.A.

ABSTRACT

The chiral symmetry breaking transition in pure quenched lattice QED and its implications for the theory's continuum limit are discussed. Numerical evidence is presented that it is not described by mean field theory scaling. The role of lattice monopoles in driving the transition is also considered.

There are many ways to introduce the subject of strongly coupled QED. Let me begin with the empirical fact that lattice studies [1] reveal the existence of a continuous phase transition as the electron charge e is raised, between the normal state of (weakly) interacting electrons and photons, and a strongly coupled phase where chiral symmetry is spontaneously broken, as signalled by a non-vanishing condensate $< \bar{\psi}\psi >$. The physical content of this phase is as yet poorly understood, although on general principles we might expect an approximately massless Goldstone pion, together with a rich spectrum of other e^+e^- bound states. This has led to suggestions of a connection between this new phase and anomalous e^+e^- resonances seen in heavy-ion collisions [2]. On a more abstract level, the possibility of an ultraviolet-stable fixed point at the transition has fuelled speculation that the continuum limit of pure QED exists [1,3], in the sense that the cutoff can be removed from the calculation of any physical quantity; this would be the first such example of an interesting non-asymptotically free theory.

I will describe numerical simulations of lattice QED made at Illinois [4]. The lattice action we use is a function of non-compact variables:

$$S_g = \frac{1}{2}\sum_{n\mu\nu}[\theta_\mu(n) + \theta_\nu(n+\hat{\mu}) - \theta_\mu(n+\hat{\nu}) - \theta_\nu(n)]^2 \equiv \frac{1}{2}\sum_{n\mu\nu}\Theta^2_{\mu\nu}(n), \tag{1}$$

where $\theta_\mu(n)$, the lattice version of the vector potential, is a real variable located on the lattice links. In Feynman gauge, S_g is diagonal in momentum space, so it is possible to generate statistically independent configurations at each sweep using a Fourier transformation, and thus eliminate critical slowing down [5]. Of course, the reason we can do this is because S_g just describes free photon modes; in other words when we calculate $< \bar{\psi}\psi >$ in quenched QED we are investigating the properties of fermions in a free field background.

These considerations raise the issue of why we bother with the quenched theory, apart from the obvious reason of its being a warm-up exercise for the full theory. In fact, there are analytic arguments [3] based on a self-consistent calculation of the electron self-energy, to suggest that a UV fixed point arises from the requirement that the pion wavefunction be stable against collapse once the attractive e^+e^- force overcomes the centrifugal barrier. This leads to a renormalisation condition relating the bare fine structure constant α_0 and the cutoff Λ:

$$\frac{\alpha_0}{\alpha_c} = 1 + \left(\frac{\pi}{\ln(\frac{\Lambda}{M})}\right)^2. \tag{2}$$

Here α_c is the critical coupling $\pi/3$ and M is a physical mass scale. This calculation is made using the quenched planar approximation (hence only "rainbow" graphs contribute to the gap equation), and so should be confirmed in a quenched lattice simulation. Of course, in the full theory we expect competition between the collapse mechanism and charge screening from vacuum polarisation [1].

The prediction of the collapse picture which is easiest to test is the scaling of $< \bar{\psi}\psi >$ with $\beta \equiv 1/e^2$ in the strongly coupled phase. It scales as an essential singularity, viz:

$$< \bar{\psi}\psi > \sim \exp\left(\frac{-b}{\sqrt{\beta_c - \beta}}\right), \tag{3}$$

where b, β_c are constants to be determined. There are other equally plausible forms the critical scaling could take, notably that predicted by mean field theory:

$$< \bar{\psi}\psi > \sim (\beta_{MF} - \beta)^{1/2}. \tag{4}$$

Solution of the gap equation away from criticality does indeed yield (4), so we expect this form to be valid deep in the strong coupling region. If it persists right up to the transition, then it will cast doubt on the existence of an interesting continuum limit, since mean field scaling would be consistent with a free field theory at the fixed point [6].

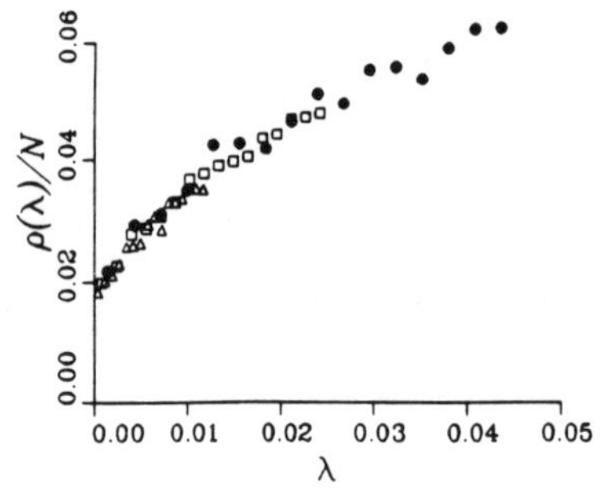

Figure 1

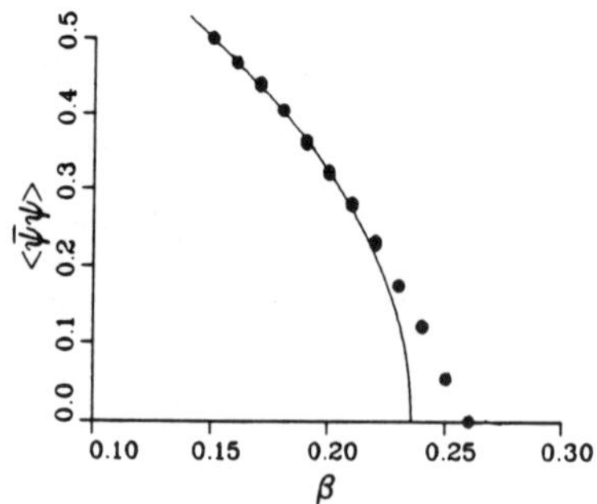

Figure 2

I will initially focus on estimates of $< \bar{\psi}\psi >$ made by calculating the eigenvalues λ_n of the covariant lattice Dirac operator $i\slashed{D}[U]$ [4], where the connection $U_\mu(n)$ is defined as $\exp(ie\theta_\mu(n))$ to ensure gauge invariance. The condensate is then formally given by

$$< \bar{\psi}\psi > = \frac{\pi}{V}\rho(\lambda = 0), \tag{5}$$

where V is the lattice volume and $\rho(\lambda)$ is the spectral density function, which can be estimated via eigenvalue calculations over an ensemble of $\{\theta\}$ configurations. In figure 1 estimates of $\rho(\lambda)$ at $\beta = 0.25$ on lattice sizes 10^4 (filled circles), 12^4 (squares) and 16^4 (triangles) are shown. The intercept at $\lambda = 0$ is clearly non-zero, showing that at this coupling the system exhibits chiral symmetry breaking. The most striking feature is the apparent absence of any systematic change with volume: calculations on larger lattices merely increase the resolution in the vicinity of $\lambda = 0$. This is unexpected, and may either be an artifact of the quenched theory, or a sign that the symmetry breaking is a short-distance phenomenon. Figure 2 plots $< \bar{\psi}\psi >$ as a function of β, showing that the critical coupling β_c lies in the neighbourhood of 0.26. The solid line is a fit of the form (4), showing that there is a small but significant departure from mean field scaling near the transition.

To home in on the critical region , we performed a further study on 16^4 lattices at β values ranging from 0.235 to 0.26 in steps of 0.0025. The resulting $\rho(\lambda)$ are shown in figure 3. For $\beta < 0.25$, the curvature of the plots becomes great enough to make the extraction of $\rho(0)$ using a simple straight-line fit unstable. For this reason I will not attempt a fit of the form (3), except to say that this data puts severe constraints on such a form, with $\beta_c \leq 0.27$ [4]. The numerical evidence for the essential singularity is weaker than first thought [1]. Instead, let me focus on another aspect of the critical behaviour, namely how $< \bar{\psi}\psi >$ scales at the critical coupling as a function of bare electron mass m. Once again on general grounds [7] we expect a form

$$< \bar{\psi}\psi(m) >\sim m^{1/\delta}, \tag{6}$$

where the exponent δ is 3 in mean field theory, and 1 in an asymptotically free theory such as QCD. Figure 4 shows plots of $< \bar{\psi}\psi(m) >$ obtained by inverting the kinetic operator $(\not{D}+m)$, for β values 0.24 (crosses), 0.25 (circles), 0.26 (pluses), 0.27 (triangles) and 0.28 (squares). The dotted lines are power law fits (6). We find the fitted value of δ depends sensitively on the value of β taken as the critical β_c. In particular $\beta_c = 0.25$ yields $\delta = 2.8$, supporting mean field theory, whereas $\beta_c = 0.26$ gives $\delta = 2.0$, which suggests something more exotic. By combining this data with the eigenvalue data of figure 3, we obtain a best fit for β_c of 0.257(1) with a corresponding value of $\delta \simeq 2.2(1)$ [8].

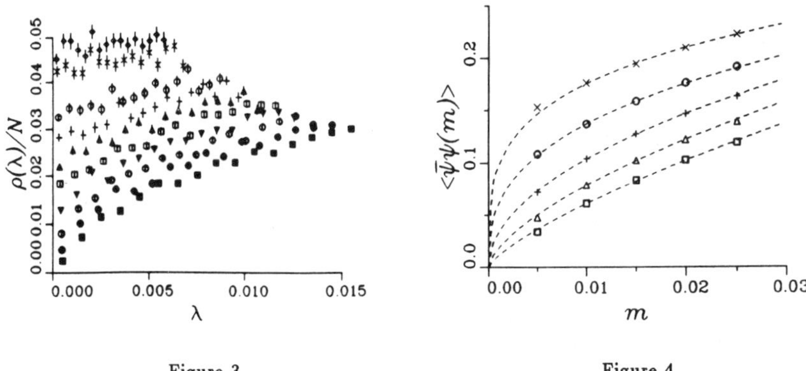

Figure 3   Figure 4

To summarise thus far, we see that in the strong coupling region $< \bar{\psi}\psi >$ scales with the expected mean field form (4) with $\beta_{MF} \simeq 0.24$, but that closer to the transition there is a departure from this behaviour. The critical scaling is difficult to establish, but does not particularly favour the collapse picture (3). Nonetheless, if we assume a power law scaling of the form (6) at criticality, then we find fresh support for non-mean field behaviour.

At this stage, I would like to leave the troublesome issue of the critical scaling, and concentrate on the possibly simpler question of the microscopic mechanism responsible for the transition in the lattice theory. Recall that in the definition of the fermion covariant derivative, the gauge fields only appeared in a compact form, combined with the charge e as $\exp(ie\theta_\mu)$. In other words, the electron fields only have non-trivial gauge transformations under the compact (periodic) group U(1) rather than the non-compact additive group of real numbers R which is the symmetry of (1). Whenever a lattice gauge symmetry has a larger covering group, topological excitations such as magnetic monopoles may be important. Such excitations or defects arise because the fermions are unable to distinguish between flux 0 and 2π through any particular plaquette, although in the latter case we can make the global statement that there is a "Dirac string" present. Magnetic monopoles on the lattice can be identified by the following procedure [9]. First, define an integer $s_{\mu\nu}(n)$, which turns out to be dual to the Dirac sheet, by

$$e\Theta_{\mu\nu}(n) = e\bar{\Theta}_{\mu\nu}(n) + 2\pi s_{\mu\nu}(n), \tag{7}$$

where $e\bar{\Theta}_{\mu\nu} \in (-\pi, \pi]$. Then the monopole current $m_\mu(\tilde{n})$ may be defined on the links of the dual lattice by

$$m_\mu(\tilde{n}) = \varepsilon_{\mu\nu\kappa\lambda}\Delta_\nu s_{\kappa\lambda}(n), \tag{8}$$

with the constraint $\Delta_\mu m_\mu(\tilde{n}) = 0$, implying that the integer-valued monopoles trace out closed loops on the four-dimensional lattice.

It turns out that the most dramatic way to expose the presence of monopoles in non-compact QED is by considering the model as a percolation problem [10]. In other words, neglect the vectorial nature of the m_μ altogether, and simply think of the monopole current elements as bonds connecting (dual) lattice sites into clusters. If monopoles are present in sufficient concentration, then one of the clusters will become dominant and occupy a macroscopic fraction of the lattice sites. In figure 5 I plot results from a 12^4 lattice: the squares show the size of the largest cluster as a fraction of all occupied sites $< n_{max}/n_{tot} >$, and the filled circles show an associated susceptibilty

$$\chi = \left\langle \frac{\sum_n g_n n^2 - n_{max}^2}{n_{tot}} \right\rangle, \tag{9}$$

where g_n is the number of clusters per configuration of size n. There is good evidence for a percolation threshold at $\beta \simeq 0.24$; in other words, monopoles are becoming geometrically dense suspiciously close to the chiral transition.

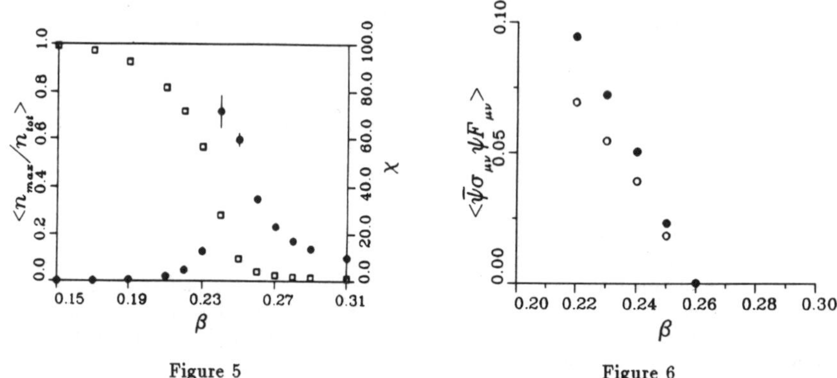

Figure 5   Figure 6

Can we find direct evidence for a causal link? Work on this problem is in progress, but for now let us consider the magnetic condensate defined by $< \bar{\psi}\sigma_{\mu\nu}\psi F_{\mu\nu} >$ ($\sigma_{\mu\nu}F_{\mu\nu} \sim \mu.B$, where μ is the electron's magnetic moment). In terms of lattice fermions, $\bar{\psi}\sigma_{\mu\nu}\psi$ is a two-link operator, so that expectations $< n \mid \sigma_{\mu\nu} \mid n >$ in the basis of eigenfunctions of $i\slashed{D}$ naturally have a large overlap with plaquette operators such as the non-compact $\Theta_{\mu\nu}$ or the (effectively) compact $\bar{\Theta}_{\mu\nu}$. This enables us to estimate the magnetic condensate using the expectation values plus the spectrum density information from previous calculations. Figure 6 shows $< \bar{\psi}\sigma_{\mu\nu}\psi F_{\mu\nu} >$ as a function of β

on a 8^4 system, with $F_{\mu\nu}$ given compact (filled circles) or non-compact (open circles) definitions. As the transition is approached at $\beta_c \simeq 0.26$, the compact signal is enhanced by 25% relative to the non-compact one, despite the fact that at these couplings the two definitions of $F_{\mu\nu}$ differ on only $2\frac{1}{2}\%$ of the plaquettes on the lattice. This would suggest that the most important coherent structures in the fermion eigenfunctions, both in terms of amplitude and phase, are localised on the monopole defects in the $\{\theta\}$ configurations.

If we accept that monopoles are important in strongly coupled lattice QED, then what future directions might be explored? One alternative is to live with the fact that compact variables are present, and incorporate them from the start, in a mixed compact – non-compact gauge action. Since pure compact QED has a first order phase transition [11], this suggests the existence of a tricritical point somewhere in the $\beta_{compact} - \beta_{non-compact}$ plane. Another possibility, at present under active consideration at Illinois [12], is to nullify the effect of monopoles by linearising the fermion – gauge interaction, to make the lattice action more closely resemble the continuum one, and thus perhaps to expose more clearly the true nature of the transition. In either case, it would appear there is still a lot to be learned in this field.

Acknowledgements

This work was supported by NSF grant PHY 87-01775. It gives me great pleasure to thank my collaborators Elbio Dagotto, Sasha Kocić, John Kogut, John Sloan and Roy Wensley.

References

[1] E. Dagotto, A. Kocić and J.B. Kogut, Phys. Rev. Lett. **60**, 772 (1988); Phys. Rev. Lett. **61**, 2416 (1988); Nucl. Phys. **B317**, 253, 271 (1989).

[2] L.S. Celenza, V.K. Mishra, C.M. Shakin and K.F. Liu, Phys. Rev. Lett. **56**, 444 (1986); D. Caldi and A. Chodos, Phys. Rev. **D36**, 2876 (1987); Y. Jack Ng and Y. Kikuchi, *ibid.* p. 2880 (1987).

[3] P.I. Fomin, V.P. Gusynin, V.A. Miransky and Yu.A. Sitenko, Riv. Nuovo Cimento **6**, 1 (1983); V.A. Miransky, Il Nuovo Cimento **90A**, 149 (1985).

[4] S.J. Hands, J.B. Kogut and E. Dagotto, Illinois preprint ILL-(TH)-89-#44, to appear in Nucl. Phys. B.

[5] E. Dagotto, A. Kocić and J.B. Kogut, Illinois preprint ILL-(TH)-89-#34.

[6] G. Schierholz, these proceedings.

[7] A.M. Horowitz, Nucl. Phys. **B**(Proc. Suppl.)**9**, 403 (1989).

[8] E. Dagotto, S.J. Hands, A. Kocić and J.B. Kogut, in preparation.

[9] T.A. DeGrand and D. Toussaint, Phys. Rev. **D22**, 2478 (1980).

[10] S.J. Hands and R.J. Wensley, Phys. Rev. Lett. **63**, 2169 (1989).

[11] E. Dagotto and J. B. Kogut, Nucl. Phys. **B295** [**FS21**], 123 (1988).

[12] S.J. Hands, J.B. Kogut and J. Sloan, in preparation.

A Lattice Simulation of the
Gluon Contribution to the Proton Spin

Jeffrey E. Mandula
Department of Energy
Division of High Energy Physics
Washington, DC 20545

Abstract

The magnitude of the gluon contribution to the proton spin induced by the triangle anomaly diagram in the quark axial current is estimated by means of a lattice QCD simulation of the polarized proton matrix element of the axial current, $<ps|K_\mu|ps>$. The simulation is carried out in the temporal axial gauge, where this quantity is the value of the triangle diagram, and where also K_μ is the gluon canonical spin. The result of the simulation on an ensemble of 204 $6^3\times10$ lattices at $\beta = 5.7$ is that the fractional contribution of the gluon anomaly to the proton spin, $(3\alpha/2\pi)\Delta g$, is less than .05 in magnitude. This is much too small to affect the basic interpretation of the EMC experiment, that the quark spins are responsible for very little, if any of the proton spin.

I. Introduction

The European Muon Collaboration (EMC) has recently published the results and analysis of a measurement of the spin structure of deep inelastic muon scattering from protons.[1] From their measurements of polarized and unpolarized structure functions and SLAC's earlier data[2,3] they conclude

$$\int_0^1 g_1(x)\ dx\ =\ \frac{1}{2}(\ \frac{4}{9}\ \Delta u\ +\ \frac{1}{9}\ \Delta d\ +\ \frac{1}{9}\ \Delta s)\ =\ .126\ \pm\ .010\ \pm\ .015$$

$$(q^2 = -10.7\ \text{GeV})$$

(1)

where the statistical and systematic errors are shown separately. The quoted systematic error of $\pm$ .015 includes an estimate of the uncertainty due to the need to extrapolate the measured value of the structure functions to the regions above x = .7 and below x = .01. The connection of g_1 to the quark spin fractions which are the matrix elements of the quark axial currents follows from operator product expansion analysis.[4]

The significance of the EMC measurement stands out when combined with much earlier measurements of semi-leptonic axial weak decays of neutrons and hyperons. Specifically, neutron β-decay plus isospin symmetry gives[5]

$$\Delta u - \Delta d = g_A = 1.254 \pm .006 \tag{2}$$

while strangeness changing hyperon decay and flavor SU(3) symmetry gives[6]

$$\Delta u + \Delta d - 2 \Delta s = .60 \pm .12 \tag{3}$$

Putting this all together gives

$$\begin{aligned} \Delta u &= +.74 \pm .05 \\ \Delta d &= -.51 \pm .05 \\ \Delta s &= -.19 \pm .07 \end{aligned} \tag{4}$$

and

$$\Delta u + \Delta d + \Delta s = +.04 \pm .16 \tag{5}$$

The errors are dominated by the error on the EMC result.

Eq. (5) is the source of the brouhaha created by the EMC result. The conclusion that only a small fraction, if any, of the proton's spin is carried by the spins of its constituent quarks was completely unanticipated, as was the conclusion that the strange quark's contribution was comparable to that of the down quark.

A very interesting suggestion for understanding (or reinterpreting) the EMC result was made by Efremov and Turyaev,[7] Altarelli and Ross,[8] and Carlitz, Collins, and Mueller.[9] They observed that there is a contribution to each Δq_i associated with the gluonic spin distribution inside the proton. It comes from the triangle diagram, which is the source of the Adler-Bardeen-Jackiw anomaly.[10] The evaluation of this contribution requires a choice of gauge and a careful treatment of high momentum limits. In axial gauges, at least, it comes entirely from infinite loop momenta around the triangle, so that the gluon vertices are effectively contracted to a point. In the (canonical) $A_0=0$ gauge, this gluonic spin contribution to each quark axial current matrix element is

$$2m\, s_\mu\, [\Delta q_i]_{gluonic} = \langle ps\,|\,[j^5{}_\mu]_{Triangle}\,|\,ps\rangle = \frac{\alpha}{2\pi}\,\langle ps\,|\,K_\mu\,|\,ps\rangle = 2m\, s_\mu\, \Delta g \tag{6}$$

The fact that neither the isolation of this gluonic contribution nor even the final formula can be expressed in gauge invariant terms has been criticized and its physical interpretation questioned.[6,11] We regard the meaningful isolation of a gluonic contribution hidden in a quark operator matrix element as at least plausible, if not unambiguously proven. Without entering the controversy, we use the identification in Eq. (6) as the basis for the balance of this talk.

II. A Lattice QCD Calculation of Δg

The main purpose of this article is to describe a lattice gauge simulation of Δg by evaluating the $\langle ps|K_\mu|ps\rangle$ matrix element. We use the standard formulation: Wilson's SU(3) lattice action on a periodic Euclidean lattice to describe the gauge fields and (4-spin $\times$ 3-color component) Wilson fermions to describe the quarks. We work in the quenched approximation, which omits closed quark loop effects. Our calculation uses lattices generated by Bernard, Draper, Hockney, Rushton, and Soni for a different purpose — the calculation of hadronic matrix elements that enter into weak interaction rates.[12] They made available for this purpose an ensemble of 204 $6^3\times10$ lattices with $\beta = 5.7$ together with quark propagators computed through 8 units of Euclidean time both forward and backwards. The quark propagators satisfy open (Neumann) boundary conditions with hopping constant k=.162.

Issues special to the present calculation are (i) the extraction of Minkowski space matrix elements from (dressed) Euclidean space Green's functions; (ii) the formulation of a suitable lattice approximation to the anomalous current; and (iii) the evaluation of polarization asymmetries in a lattice simulation.

(i) *Extraction of On-Shell Matrix Elements*

The coefficient of the leading exponential decay of a Euclidean space Green's function is directly the desired Minkowski space matrix element. This is easily seen by separating the external proton legs (Ψ is a proton operator made of 3 quark operators and all 3-momenta are zero).

$$\langle 0|\Psi(x)\,K_\mu(y)\,\bar\Psi(z)|0\rangle \;=\; \int dx'\;dz'\;S(x\text{-}x')\;\Gamma_\mu(x',y,z')\;S(z'\text{-}z) \tag{7}$$

The leading behavior of the propagator is

$$S(x\text{-}x') \;\to\; P_+\,e^{-m(x\text{-}x')} \qquad (P_+ = \frac{1 + \gamma_0}{2}) \tag{8}$$

and so in the large time limit the Green's function becomes

$$\langle 0|\Psi(x)\,K_\mu(y)\,\bar\Psi(z)|0\rangle \;\to\; e^{-m(x\text{-}z)}\int dx'\;dz'\;e^{mx'\text{-}mz'}\,P_+\,\Gamma_\mu(x',y,z')\,P_+ \tag{9}$$

The coefficient of the overall exponential factor is the truncated vertex evaluated at imaginary Euclidean energy im, which, with the Dirac projector, is just the on-shell Minkowski space matrix element

$$\frac{1}{2m}\;\Sigma\;u_{ps}\;\langle ps|K_\mu|ps'\rangle\;\bar u_{ps'} \tag{10}$$

(ii) *The Lattice Anomalous Current*

For the reasons discussed earlier, we work in the gauge $A_4=0$, which we implement on each lattice by gauge transforming all times like link variables (except at a single reference times outside the quark propagator range) to $U_4=1$. In the continuum, the anomalous current simplifies in this gauge to $K_i = 2\,\epsilon_{ijk}\,\mathrm{Tr}\,A_j\,F_{k4}$. A convenient lattice approximation to F_{k4} is[13]

$$F_{k4}{}^{\mathrm{Lattice}}(n) = \frac{1}{8i}\,[U_k(n)U_k{}^\dagger(n+\widehat{4}) + U_k{}^\dagger(n+\widehat{4}-\widehat{k})U_k(n-\widehat{k})$$
$$+ U_k{}^\dagger(n-\widehat{k})U_k(n-\widehat{4}-\widehat{k}) + U_k(n-\widehat{4})U_k{}^\dagger(n) \tag{11}$$
$$- \text{Hermitean Conjugate}]_{\mathrm{Traceless}}$$

It has the correct continuum limit, transforms simply under the lattice hypercubic symmetry group, and depends under gauge transformation only on the value of the gauge transformation at site n. With this expression for the field strength a convenient lattice approximation to the anomalous current is

$$K_i^{\mathrm{Lattice}}(n) = \mathrm{Im}\,\epsilon_{ijk}\,\mathrm{Tr}\,U_j(n-\widehat{j})\,F_{k4}{}^{\mathrm{Lattice}}(n)\,U_j(n) \tag{12}$$

Like $F_{k4}{}^{\mathrm{Lattice}}(n)$, it has the correct continuum limit and transforms simply under the hypercubic group. Under infinitesimal gauge transformations its change depends on the difference of the values of the gauge transformation at sites $n+\widehat{j}$ and $n-\widehat{j}$ ($j \neq i$). Some of the important properties of the anomalous current that are only recovered in the continuum limit are its response to x_4 independent gauge transformations, the gauge invariance of its divergence and its relation to topology changing gauge transformations.

(iii) *Evaluation of Polarization Asymmetries*

Tracing the Green's function with $i\gamma_5\!\!\!/s$ extracts the difference of the $+s_\mu$ and $-s_\mu$ expectation values. Since the asymptotic behavior of the trace of the propagator is twice the exponential, we may extract Δg using

$$\Delta g = \frac{\mathrm{Tr}\,P_+\,i\gamma_5\!\!\!/s\,\langle 0|\Psi(x_4)\,\vec{s}\cdot\vec{K}_\mu(y_4)\,\bar{\Psi}(z_4)|0\rangle}{(1/2)\,\mathrm{Tr}\,\langle 0|\Psi(x_4)\,\bar{\Psi}(z_4)|0\rangle} \tag{13}$$

This expression is conveniently independent of the normalization of the proton field, which had been tacitly taken to have unit coupling to the proton state in the prior discussion. It is from this formula that we extract the value of Δg in computer simulations. To improve statistics we average over the three principal axis choices for $\vec{s}$.

We "measured" the gluon spin fraction Δg of Eq. (13) using the sample of 204 $6^3\times10$ lattices provided by Bernard, Draper, Hockney, Rushton, and Soni. With the parameters of those lattices, $\beta=5.7$ and $k=.162$, the lattice spacing is about 1.0 GeV^{-1} and the proton mass about 1.4 inverse lattice spacings.

We constructed a color singlet proton field from colored quark fields by projecting on the appropriate spin and symmetry. Of the two acceptable fields we chose

$$\Psi = \epsilon_{ijk} \; i\gamma_\mu\gamma_5 \, \psi_d^i \; \psi_u^{jT} \, S \, i\gamma_\mu\gamma_5 \, \psi_u^k \tag{14}$$

The matrix S is related to charge conjugation. In our representation ($\gamma_i = \sigma_2 \otimes \sigma_i$, $\gamma_4 = \sigma_3 \otimes I$) it is $I \otimes \sigma_2$. To extract Δg using Eq.(13), we should take x_4-y_4 and y_4-z_4 as large as possible, while avoiding the largest values of x_4-z_4, to minimize the sensitivity to boundary conditions. The best compromises are those in the Table. The values of Δg, with forward and backward propagation treated separately, are:

$x_4 - y_4$	$y_4 - z_4$	Forward	Backward
4	3	$+.167 \pm .408$	$-.009 \pm .116$
3	4	$+.307 \pm .222$	$-.010 \pm .076$
3	3	$+.103 \pm .241$	$-.002 \pm .069$

III. Discussion and Conclusions

Several comments are in order regarding the quality of these results. One is that they are all mutually consistent, and consistent with 0. That is, we have not arrived at a (zero) value for Δg, but rather a bound, $|\Delta g| \leq .5$. Another is that the finite lattice spacing and lattice size are serious but probably not fatal limitations. The proton mass is almost 50% too large on these lattices, for example. In the zero spacing and large time limits, the value of Δg gotten from the simulation would be independent of the exact values of x_4-y_4 and y_4-z_4, so long as both were large enough. Within the statistics, there is no evidence that this is failing, and the much closer than statistics self-agreement of the forward and backward results separately indicates that the finite spacing might not be much too coarse. Similarly, quenched approximation results in other calculations, which like this one would not be structurally different in the full theory, are usually reasonable, and have not yet been seen to be qualitatively wrong.

Finally, it is possible to address the limitation that x_4-y_4 and y_4-z_4 are fairly small somewhat quantitatively. By extrapolating the proton propagator back from $\Delta t = 6$ and 7 to $\Delta t = 3$ or 4, we estimate the admixture of higher states still present is about 15% at $\Delta t = 4$ and 40% at $\Delta t = 3$. Thus the error introduced by the relatively short time available to suppress higher states and isolate the proton matrix element is comparable to the other sources of error. The cumulative effect of these uncertainties is roughly to double the bound we have found.

The basic conclusion of the foregoing error considerations is that the gluon spin fraction Δg is apparently less than 1 in magnitude. Since at $\beta = 6/g^2 = 5.7$ the QCD coupling is $\alpha = g^2/(4\pi) \approx .08$, this gives a bound on the correction to the total quark spin fraction of the proton's spin of $(3\alpha/2\pi)|\Delta g| \leq .05$. This is the appropriate measure of the magnitude of the indirect gluonic contribution to the proton spin because it is a renormalization group invariant, in the short distance continuum limit.[9]

The analysis described here was undertaken to evaluate the extent to which the anomalous contribution of gluons to the spin dependent quark structure functions would modify the conclusion of the EMC experiment that the net quark contribution to the proton's spin is very small. In so far as the results using such small lattices can be trusted, the conclusion of this simulation is that the effect is an order of magnitude to small to change the thrust of the EMC conclusion.

Acknowledgements

The author wishes to express his appreciation to the Bernard-Draper-Hockney-Rushton-Soni collaboration for generously making available the lattices that form the basis for this analysis. The author has also greatly benefited from valuable discussions and encouragement to all stages of this work with Robert Carlitz, Robert Jaffe, Marek Karliner, James Labranz, Alfred Mueller, Michael Ogilvie, Amarjit Soni, and Vigdor Teplitz.

References

1. The EMC Collaboration, Nucl. Phys **B328**, 1 (1989)
2. For reviews see J. Drees and H.E. Montgomery, Ann. Rev. Nucl. Sci., **33**, 383 (1983); T. Sloan, G. Smadja, and R. Voss, Phys. Rep. **162**, 45 (1988); M. Diemoz, F. Ferroni, and E. Longo, Phys. Rep. **130**, 295 (1986)
3. SLAC-Yale Collaboration, Phys. Rev. Letts. **37**, 1261 (1976); **41**, 70 (1978); **51**, 1135 (1983)
4. J. Kodaira, Nucl. Phys., **B165**, 129 (1979)
5. J.D. Bjorken, Phys. Rev. **148**, 1467 (1966)
6. R.L. Jaffe and A. Manohar, to be published
7. A.V. Efremov and O.V. Teryaef, JINR report E2-88-278 (1988)
8. G. Altarelli and G.G. Ross, Phys. Letts. **B212**, 391 (1988)
9. R.D. Carlitz, J.C. Collins, and A.H. Mueller, Phys. Letts. **214B**, 229 (1988)
10. J.S. Bell and R. Jackiw, Nuovo Cimento **A51**, 47 (1967); S.L Adler, Phys. Rev. **177**, 2426 (1969)
11. G. Bodwin and J. Qiu, to be published
12. C. Bernard, T. Draper, R. Hockney, A. Rushton, and A. Soni, Phys. Rev. Letts. **55**, 2770 (1985)
13. J.E. Mandula, J. Govaerts, and G. Zweig, Nucl. Phys. **B228**, 109 (1983)

The QCD Phase Transition, with and without Quark Loops*

Norman H. Christ

Department of Physics, Columbia University, New York, NY 10027

ABSTRACT

Recent results regarding the QCD phase transition are reviewed. Scaling and the order of the transition for $N_t = 4$ in pure gauge theory are discussed and the current status of calculations which include the effects of quark loops is described both for two and four quark flavors.

1. INTRODUCTION

The deconfining/chiral phase transition in QCD has long been a very active area in lattice gauge theory for perhaps three reasons. First this is a topic in low energy hadronic physics where predictions can still be made. Second, since so little is known about the properties of hot QCD, even rough numbers are useful for developing the phenomenology of heavy ion collisions. Finally, it is among the easier topics to study in lattice QCD since it requires the calculation of mostly bulk quantities. As we will see, this last reason is beginning to lose it weight as the subject develops — the remaining questions are increasing subtle, requiring large physical volumes and the calculation of correlation lengths. This talk is divided into three sections. The first addresses the phase transition in pure QCD, the second considers QCD with the complete treatment of four flavors of dynamical quarks and the third treats full QCD with two flavors of quarks. There are now a number of fine review articles[1] covering this subject, so in this talk I will concentrate on only fairly new results, referring to these reviews for earlier work.

2. PURE QCD, $N_f = 0$

In this section we will briefly discuss topics probing two limits of pure gauge theory simulations of the deconfining phase transition. The question of scaling requires the limit of very small lattice spacing $a \to 0$ while studies of the order of the phase transition examine the limit of infinite physical volume, $(N_s a)^3 \to \infty$. The limitations of presently available computer resources prevent both limits from being realized at once. At present, studies of the order of the transition are done with a number of sites in the time direction that is quite small, $N_t = 4$, while the number of sites in a spacial direction is much larger, $N_s = 24$ or $N_s = 36$. This corresponds roughly to a lattice spacing $a = 0.25$ Fermi and a volume for the system of 6^3 to 9^3 (Fermi)^3 — a relatively large volume but a lattice spacing far from the continuum limit. In contrast, studies of scaling are done on physically small lattices with spacial extent less than twice the inverse temperature.

2.1 Scaling in pure QCD

This is not an area of very active work but there are two recent results from the Columbia group[2] that might be presented. The most significant tests of scaling are those which study the extent to which dimensionless ratios do not depend on the lattice spacing a. The second investigates the dependence of the physical length scale on the coupling constant g (or the more commonly quoted variable $\beta = 6/g^2$). The latter test is important since a direct comparison with continuum perturbation theory can be made. However, since the length scale depends exponentially on g, there are limits to the values of g that can be explored to systematically improve the validity of

---

* This research supported in part by the U.S. Department of Energy

perturbation theory.

Figure 1a shows a test of perturbative scaling made by comparing the β–dependence of the fraction-deconfined for a $16^3 \times 10$ and a $24^3 \times 16$ lattice. The values of β are converted into GeV by using the perturbative formula for the coupling constant dependence of the lattice spacing in physical units and $\Lambda_{lattice} = 4.5\text{MeV}$. For these relatively small physical volumes, there appears to be considerable mixing of phases as one passes through the transition region and the fraction-deconfined is an adhoc measure of the fraction of the Monte Carlo evolution that is in the deconfined state[3]. The ratio of spacial extents differs from the ratio of the temperatures by 7% so even the finite volume effects coming from the small system size should scale. The agreement between the cross-over region located by the two curves in Figure 1a is quite good and agrees well with the conclusion of earlier work that perturbative scaling is seen for $\beta \geq 6.0$. The different shapes of the two curves may reflect the non-trivial scaling properties of the magnitude of the quantity plotted.

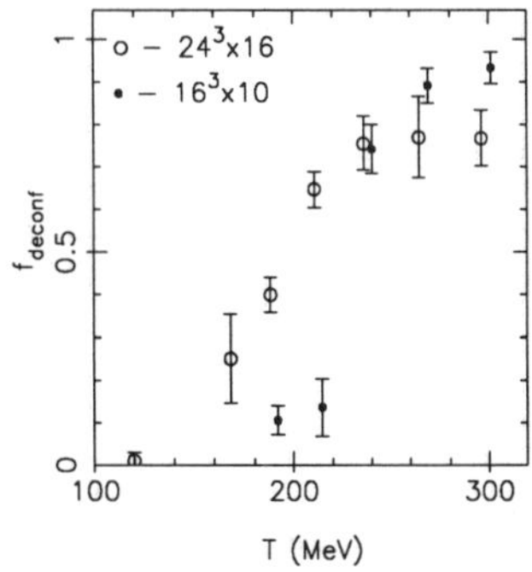

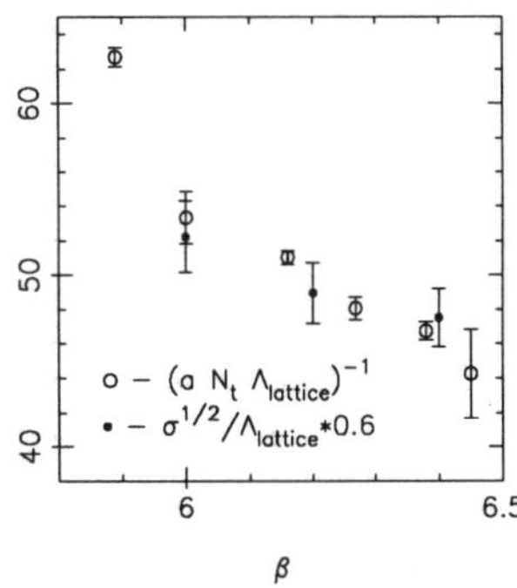

Figure 1. Two comparisons testing the scaling limit $a \to 0$. a) The fraction deconfined for $16^3 \times 10$ and $24^3 \times 16$ lattices is plotted versus temperature. The temperature was varied by varying $\beta = 6/g^2$ through the critical region. b) The critical temperature $(T_c = 1/N_t a)$ and the square root of the string tension $\sqrt{\sigma}$ are plotted versus β. The scale used to plot the string tension points was adjusted to make the two sets of data coincide.

A second demonstration of scaling is shown in Figure 1b. Here a direct comparison is made between the critical temperature and the QCD string tension within a single set of calculations. Both quantities are expressed as a ratio of a quantity with dimensions of mass to a physical scale whose β dependence is given by the 2-loop renormalization group prediction. Here the clearly visible slope of the points indicates a violation of perturbative scaling (seen also in Figure 1a as the $\approx 30\text{MeV}$ offset between the two curves). However, the agreement between the slopes of the T_c- and σ-determined data is clear indication that dimensionless ratios are much more nearly constant. Thus it would appear that ratios of dimensionless quantities computed for $\beta = 6.0$ differ by less than 5% from their continuum limit. Using the experimental value $\sigma = 0.17\text{GeV}^2$ and the ratio of $T_c/\sqrt{\sigma} = 0.6$ suggested by Figure 1a, gives $T_c = 250 \pm 15\text{MeV}$.

2.2 Order of the transition in pure QCD, $N_f = 4$, $N_t = 4$

Early studies of the deconfining phase transition in pure QCD showed evidence for a strong first order transition. Bulk quantities like the expectation values of the Wilson line, the average plaquette and $\bar{\psi}\psi$ showed a large apparent jump as beta was increased. Tunneling from the confined to deconfined phase was identified and Monte Carlo evolutions at the critical point gave distributions of the Wilson line that, when histogramed, showed two peaks corresponding to the presence of two phases. These signals were seen most clearly for $N_t = 2$ and 4 for spacial volumes as large as 8^3.

In the summer of 1988 the APE collaboration[4] and the Columbia[5] group reported results from larger spacial volumes and $N_t = 4$ that questioned some of these results. The APE calculations, done on 8^3 and 12^3 volumes, measured the correlation of smeared Wilson lines with a cold wall source and found correlation lengths that increased by more than a factor of four as $\beta \to \beta_c$. Furthermore, the largest correlation length observed increased as the size of the system was increased. They found their results strongly suggestive of a second order phase transition and inconsistent with a strong first order transition.

The Columbia group measured a number of different quantities on 16^3, 20^3 and 24^3 lattices. Bulk quantities, such the specific entropy showed a very smooth but rapid variation for the 16^3 volume over a very small range of β that was not resolved in the earlier work. As the volume was increased to 24^3 the curve sharpened and tunneling events could be seen quite unambiguously. However, the difference between the entropy of the two phases was about 60% of that seen in the earlier work. The remaining contribution to the earlier, apparently larger discontinuity was simply a very rapid rise of the bulk quantities in the deconfined phase, a rise that did not vary as the volume was increased and was similar to behavior seen in the 3-state Potts model, see Figure 2. We also measured the point-to-point correlation lengths for the 24^3 volume and saw a rapid increase as β_c was approached. However, our values of the correlation lengths were systematically smaller than the APE results and we deliberately did not quote correlation lengths for runs that showed tunneling. We concluded that the transition was clearly first order but with reduced discontinuities and a fairly complex structure.

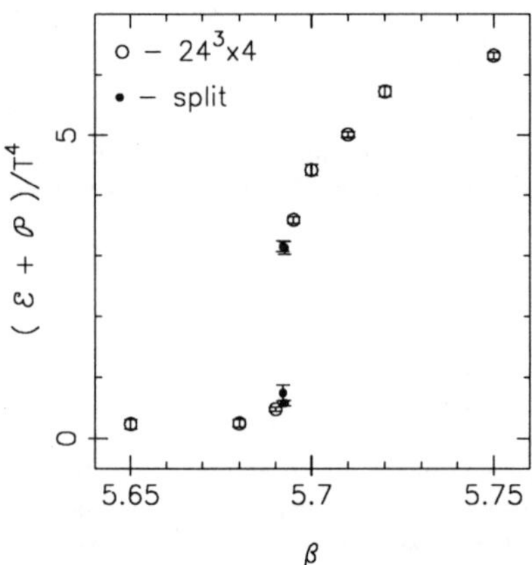

Figure 2. The sum of the specific internal energy and pressure measured in units of T^4 plotted versus β for a $24^3 \times 4$ lattice. The solid points come from runs which showed "flip-flops" between confined and deconfined phases and represent the values from each of the phases, separated by hand[5].

Significant developments were reported at the Capri meeting[6]. The Columbia group had corrected the fitting of the correlation length to include the $1/r$ prefactor expected at finite temperature. This increased our correlation lengths by $\approx$ 20% which never-the-less was not sufficient to resolve the discrepancy with APE. The APE collaboration had extended their calculations to 24^3 volumes and continued to report correlation lengths nearly double those from Columbia. However, data now was available from the Kyoto-Tsukuba group[7] for a variety of volumes up to 36^3. Although for volumes

$\leq 16^3$ their correlation lengths agreed with the APE values, for $\beta = 5.68$ and 5.685, the correlation lengths decreased as the volume was increased from 16^3 to 24^3 just as would be expected if the large spacial correlations were caused by tunneling. The 24^3 results of Columbia and Kyoto-Tsukuba agreed but there was an explicit conflict with the 24^3 APE results.

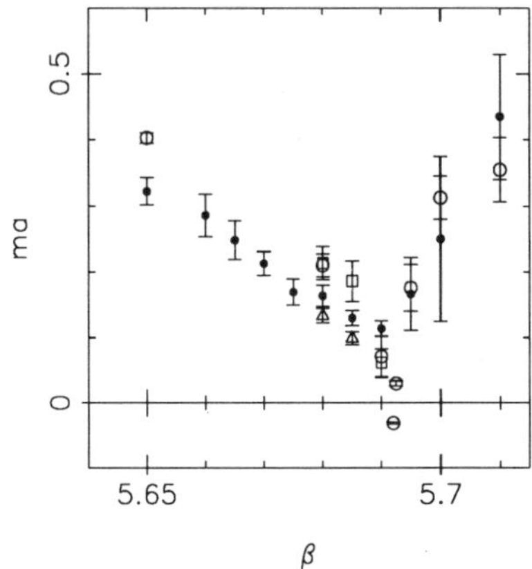

Figure 3. The correlation masses as a function of β. The open circles are Columbia results, all on a 24^3 volume. The closed circles are the original APE values on 8^3 and 12^3 volumes. The open triangles are the more recent APE numbers on 24^3 lattices while the open squares are the 24^3 results of the Kyoto-Tsukuba group.

This situation is displayed in Figure 3. The Columbia and Kyoto-Tsukuba results appear easy to interpret. The points with $\beta \leq 5.685$ on 24^3 volumes may be relatively unaffected by tunneling and represent large but asymptotical finite correlation lengths in the confined phase. Given the clear tunnelings seen at $\beta = 5.692$, it is natural to interpret the long-lived fluctuations seen at $\beta = 5.69$ as caused by partial tunnelings and not to view the much larger correlation lengths found there as a physical result for the confined phase.

However, the much larger 24^3 correlation length of APE represented a serious inconsistency. The most likely cause of the disagreement was the cold wall source used in the APE calculations. The APE collaboration has now repeated their 24^3 calculations at $\beta = 5.685$, evaluating the correlations between two Wilson lines and finds a significantly shorter correlation length in agreement with the Kyoto-Tsukuba point[8]. Thus it appears that the cold wall creates a relatively deep surface layer — making its use in determining asymptotic correlation lengths a bit treacherous. We might conclude that for $N_t = 4$ the pure gauge transition is weakly first order, characterized by quite large fluctuations and significantly smaller discontinuities than previous believed.

3. FULL QCD, $N_f = 4$

For pure QCD the deconfining phase transition is characterized by the spontaneous breaking of the discrete Z_3 symmetry that occurs in the high temperature phase. When coupling between quarks and the gauge fields is introduced this Z_3 symmetry is explicitly broken so that symmetry arguments no longer imply that a phase transition must occur. Of course for sufficiently heavy quarks their effect on the system can be neglected and a first order transition is expected at least for

$N_t = 4$. However, in the other extreme limit, when there are no fewer than two flavors of massless quarks, we expect the spontaneous breaking of chiral symmetry at zero temperature which should be restored at $T \rightarrow \infty$ implying the existance of a phase transition. Again as the quark mass is increased from zero the chiral symmetry is explicitly broken and the existence of a phase transition is no longer guaranteed.

Let us now consider simulations of the complete theory, including all effects of 4 flavors of dynamical quarks described by the Kogut-Susskind formulation beginning with the case $N_t = 4$. The present situation is summarized in Figure 4. There the critical value of beta, β_c, is plotted as a function of the mass of the four degenerate flavors of quarks. It has been recognized for some time that, when viewed as a function of quark mass, four-flavor QCD thermodynamics on coarse lattices divides into three regions: a) A first-order region for large mass which includes the case of pure QCD as $m \rightarrow \infty$; b) A cross-over region of no transition for intermediate values of the quark mass and c) A first-order region for quite small values of m. The phase structure seen in region a) is determined by the Z_3 symmetry of the pure SU(3) while region c) arises from the nearly exact chiral symmetry of the theory with fermions.

Figure 4 shows the current identification[9,10,11] of regions b) and c) for $N_t = 4$. Not much recent attention has been paid to the lower boundary of the pure gauge region a), perhaps the most definitive results available are for $N_t = 2$ where Brown and Hsieh[11] show the transition disappearing in the region $4.00 \leq ma \leq 4.25$. The boundary between regions b) and c) is not precisely known but certainly lies between $ma = 0.05$ and $ma = 0.2$.

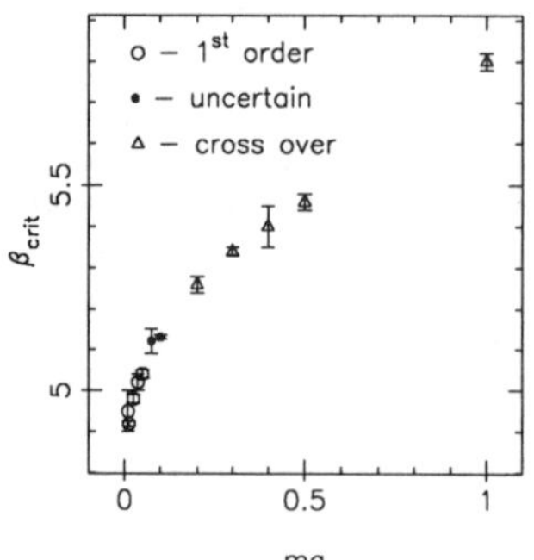
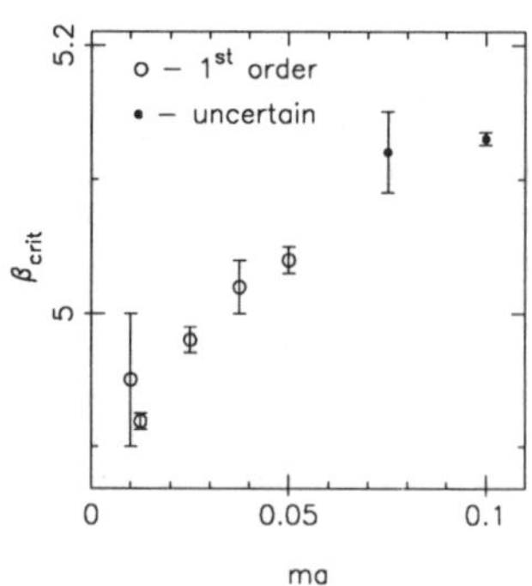

Figure 4. The critical value for β plotted as a function of quark mass for $N_t = 4$ and four flavors from a number of relatively recent calculations[9,10,11]. The graph a) shows all the data while b) has an expanded scale.

The recent series of points for $ma = 0.01$, 0.025, 0.0375 and 0.05 from the Columbia group[10] on $16^3 \times 4$ lattices allows a systematic study of the dependence of the two QCD phases on the quark mass. In Figure 5 we show the variation of $\bar{\psi}\psi$ with quark mass for the confined and deconfined phases for values of β so close to the transition that both phases appear to be stable. The linear vanishing of $\bar{\psi}\psi$ as $m \rightarrow 0$ in the chirally symmetric phase and the m independence seen in the chirally asymmetric phase suggest that these values of m well approximate the chiral limit and that there may be little new found as m is decreased further.

Although very interesting, the above results have been obtained for quite coarse lattices. It is clearly very important to study how the behavior discussed above varies as N_t is increased above 4. There are now interesting results for $N_t = 6[12,10]$ and $N_t = 8[13]$. Figure 6 summarizes the present range of calculations for $N_t = 6$. We will now argue that for $ma = 0.025$ no first-order transition is visible on a 16^3 lattice while a two-state signal can be seen for $ma = 0.01$.

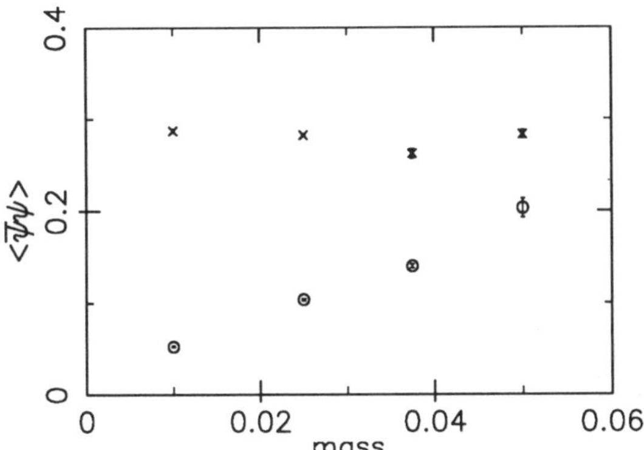

Figure 5. $\bar{\psi}\psi$ is plotted for the confined (crosses) and deconfined (circles) phases as a function of quark mass for $N_f = 4$ on a $16^3 \times 4$ lattice[10]. The values of β used are 4.95, 4.99, 5.02 and 5.04, listed in order of increasing mass.

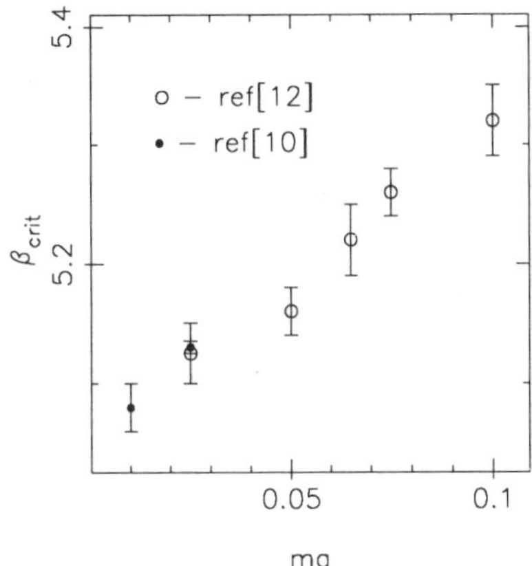

Figure 6. β_c is plotted versus quark mass for $N_t = 6$, four-flavor QCD[12,10].

Evidence for these conclusions is presented in Figure 7. Figure 7a) shows three histograms for $ma = 0.025$ and β in the critical region. No 2-state structure is seen. In contrast, Figure 7b) presents the stable Monte Carlo evolution of an ordered and disordered state both simulated at $\beta = 5.13$ for $ma = 0.01$. Thus, although a first order transition appears to occur for $N_t = 6$, the mass region in which the transition is seen has shrunk much more rapidly than suggested by the naive scale transformation $ma \to 4/6ma$. Although this might be due to the decreasing physical volume as N_t is increased, it raises the possibility that the transition may disappear for all finite mass as the continuum limit is approached.

Finally there are now interesting results[13] for $N_t = 8$ giving evidence that the transition

786

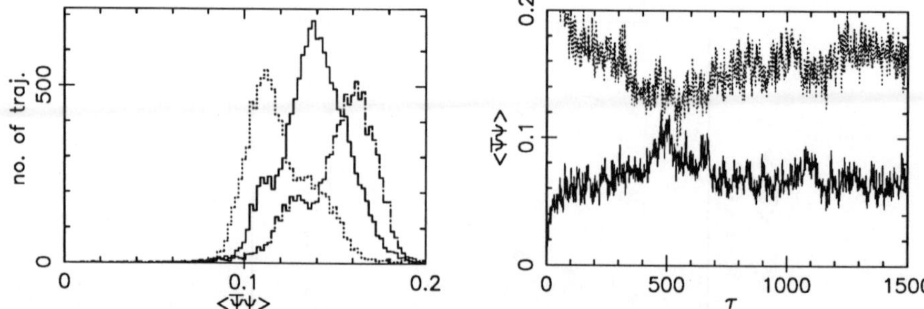

Figure 7. a) Results of a four-flavor simulation on a $16^3 \times 6$ lattice[10] with $ma = 0.025$. The three histograms show the frequency distribution of $\bar\psi\psi$ for the values of $\beta = 5.135$, 5.130 and 5.127 appearing from left to right. Each histogram contains $\approx 6,000$ units of microcanonical time divided equally between an ordered and a disordered start. The absence of a two peak structure suggests there is not a first order transition for this value of the mass. b) In contrast, the two evolutions of $\bar\psi\psi$ shown here for $ma = 0.01$ suggest that a first order transition does occur when the mass is lowered. The upper curve began with a disordered start while the lower curve is an independent run beginning with an ordered start both at $\beta = 5.08$.

remains first order at $ma = 0.01$ when studied on a 16^3 lattice.

4. FULL QCD, $N_f = 2$

Given the large mass of the charmed quark relative to the temperature of the QCD phase transition, simulations with two or three flavors might be expected to be of greater physical interest than those with four degenerate flavors. However, the case of four flavors is frequently treated because the Kogut-Susskind formulation naturally describes four flavors. The standard method for varying the number of Kogut-Susskind flavors is to insert a factor of $N_f/4$ into the force term coming from the fermion action in either the Langevin or molecular dynamics evolution equation. Although this procedure reproduces the N_f-flavor Feynman path integral to all orders of perturbation theory in the continuum limit, it does not correspond to the statistical mechanics of a local four-dimensional theory when the lattice spacing is finite. This prevents the use of the exact hybrid Monte Carlo algorithm now normally used for the $N_f = 4$ case.

The case of two flavors and $N_t = 4$ is addressed in Figure 8 which shows the fermion masses and corresponding values of β_c that have been investigated to date[14,10]. Although many of these studies have reported evidence for a first order transition for sufficiently light quark mass the recent simulations of the Columbia group on typically eight times larger volumes see no evidence for a first order transition for $ma = 0.025$ and 0.01 suggesting that there is in fact no transition for $N_f = 2$ at least for $m \geq 0.01$. Our evidence for that conclusion is shown in Figure 9. There, two evolutions of $\bar\psi\psi$ are shown beginning with ordered and disordered starts for $N_f = 2$, $ma = 0.01$ on a $16^3 \times 4$ lattice. Although there appears initially to be a 2-state signal, it disappears after about 500 time units and both evolutions converge to a value of $\bar\psi\psi \approx 0.018$ which lies midway between the values of 0.10 and 0.33 which are the averages seen at the nearby values of $\beta = 5.275$ and $\beta = 5.25$ respectively. Although no 2-state signal is visible in Figure 9, the very long-scale fluctuations suggest nearly critical behavior, possibly corresponding to the limit $m \to 0$.

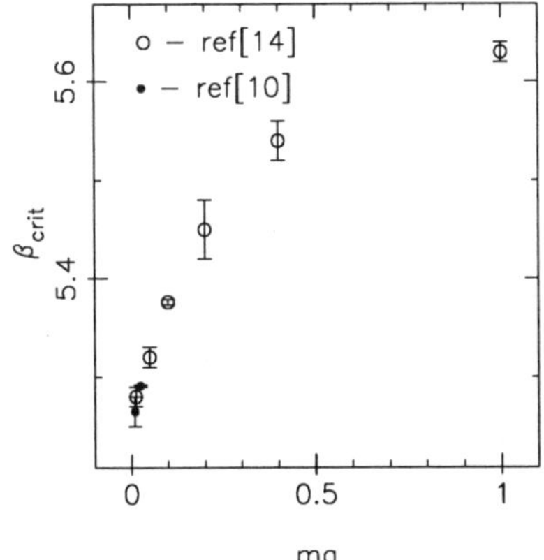

Figure 8. β_c plotted m for two-flavor QCD and $N_t = 4$[14,10].

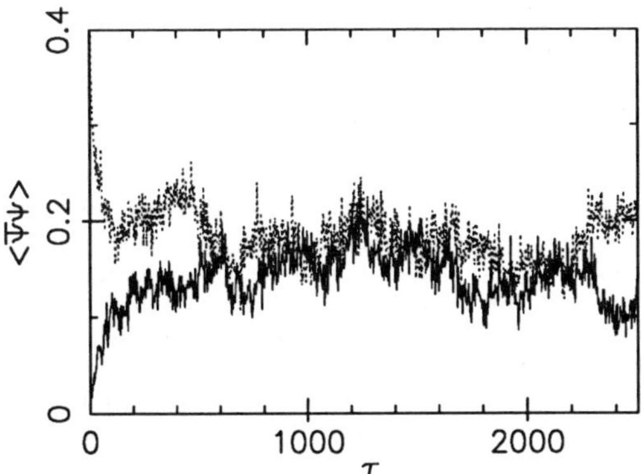

Figure 9. Monte Carlo evolutions of $\bar{\psi}\psi$ starting with an ordered (lower curve) and a disordered (upper curve) start. The calculation was performed on a $16^3 \times 4$ lattice with $N_f = 2$ and $ma = 0.01$ using the R-algorithm of Gottlieb $et\ al.$[15] and a step size of 0.0078.

5. CONCLUSION

The study of the QCD phase transition using Monte Carlo simulations remains one of the most active and interesting in lattice QCD. For the pure QCD case it would appear that the computer resources of one or two years ago were sufficient to investigate either the limit of large physical volume or small lattice spacing but not both. The nature of the transition could be studied with some precision for relatively coarse lattices with $N_t = 4$ while the continuum limit studies were performed on lattices whose spacial dimensions were less than twice $1/T_c$. With the dedicated, multi-gigaflop machines now becoming available it is should be possible to reach a similar point in the study of the complete theory including the effects of dynamical fermions. It may also be practical to give a complete discussion of pure QCD employing both small lattice spacings and large physical volumes.

REFERENCES

[1] M. Fukugita, *Lattice 88*, A. S. Kronfeld and P. B. Mackenzie, eds., Nucl. Phys. B (Proc. Suppl.) 9, 291 (1989); F. Karsch, to appear in *Quark Gluon Plasma*, R. C. Hwa, ed. (1990); A. Ukawa, Nucl. Phys. B (Proc. Suppl.) 10A, 66 (1989); A. Ukawa, *Lattice 89*, E. Marinari *et al.* eds, Nucl. Phys. B (Proc.Suppl.) (1990) to appear.

[2] F. R. Brown, N. H. Christ, Y. F. Deng, M. S. Gao and T. J. Woch, unpublished.

[3] N. H. Christ and A. E. Terrano, Phys. Rev. Lett. **56** 111 (1986).

[4] P. Bacilieri *et al.*, Phys. Rev. Lett. **61** 1545 (1988).

[5] F. Brown *et al.*, Phys. Rev. Lett. **61** 2058 (1988).

[6] G. Parisi, in *Lattice 89*, E. Marinari *et al.* eds, Nucl. Phys. B (Proc. Supp.) (1990), to appear; F. Brown, *ibid*; Fukujita, *et al., ibid.*

[7] M. Fukugita, M. Okawa and A. Ukawa, Phys. Rev. Lett. **63**, 1768 (1989).

[8] E. Marinari, private communication.

[9] S. Duane and J. B. Kogut, Phys. Rev. Lett. **55**, 2774 (1985); J. B. Kogut, Nucl. Phys. **B270**[FS16], 169 (1986); E. V. E. Kovacs *et al.* Phys. Rev. Lett. **58**, 751 (1987); Nucl. Phys. **B290**[FS20], 431 (1987); F. Karsch *et al.* Phys. Lett. **188B**, 353 (1987); S. Gottlieb, *et al.* Phys. Rev. **D35**, 2531, 3972 (1987); R. Gupta, *Field Theory on the Lattice*, A. Billoire *et al.* eds., Nucl. Phys. B (Proc. Suppl.) 4, 462 (1988); S. Gottlieb, *ibid.* 294; D. K. Sinclair, *Lattice 88*, A. S. Kronfeld and P. B. Mackenzie, eds., Nucl. Phys. B (Proc. Suppl.) 9, 331 (1989); S. Gottlieb, *ibid.* 326; R. Gupta, Phys. Rev., **D38**, 1288 (1988); K. M. Bitar, *et al.* SCRI preprint, FSU-SCRI-88-54 (1988); M. Fukugita and A. Ukawa, Phys. Rev. **D38**, 1971 (1988).

[10] F. R. Brown, F. P. Butler, H. Chen, N. H. Christ, Z. H. Dong, W. Schaffer, L. I. Unger, Columbia University preprint, in preparation.

[11] F. R. Brown, P. F. Hsieh, Columbia University preprint, in preparation.

[12] J. B. Kogut, Phys Rev. Lett, **56**, 2557 (1986); J. B. Kogut and D. K. Sinclair Nucl. Phys. **B280**[FS18], 625 (1987); E. V. E. Kovacs *et al.* Phys. Rev. Lett. **58**, 751 (1987); Nucl. Phys. **B290**[FS20], 431 (1987).

[13] R. V. Gavai *et al.* CERN Preprints, CERN- TH.5530/89 and CERN-TH.5663/90.

[14] M. Fukugita *et al.*, Phys. Rev. Lett. **58**, 2525 (1987); S. Gottlieb *et al.*, Phys. Rev. **D35**, 3972 (1987); S. Gottlieb *et al.*, Phys. Rev. Lett. **59**, 1513 (1987); M. Fukugita and A. Ukawa, Phys. Rev. **D38**, 1971 (1988); F. Karsch and H. W. Wyld, Phys. Lett. **213B**, 505 (1988); J. b. Kogut and D. K. Sinclar, Nucl. Phys. **B295**[FS21], 480 (1988); S. Gottlieb *et al.* Phys. Rev. **D40**, 2389 (1989); A. Irbäck *et al.* CERN preprint, CERN-TH-5130/88 (1988).

[15] S. Gottlieb *et al.* Phys. Rev. **D35**, 2531 (1987).

COLOR FLUX DISTRIBUTION IN PURE SU(2)

LATTICE GAUGE THEORY[1]

Richard W. Haymaker, Yingcai Peng, Vandana Singh

Department of Physics and Astronomy, Louisiana State University, Baton Rouge, La 70803, U.S.A

and

Jacek Wosiek

Chair of Computer Science, Jagellonian University, Cracow, Poland

Abstract

The color field distribution around a static $q\bar{q}$ pair is studied in detail for pure SU(2) Lattice Gauge Theory in four dimensions. As a result of large cancellations between electric and magnetic components, the action density dominates the energy density typically by an order of magnitude, at points far from the quarks.

A detailed understanding of the confinement mechanism warrants an exploration of the color field distribution around a static $q\bar{q}$ pair. Early work [1] as well as more recent studies [2], although somewhat limited in scope, have shown hints of interesting features of the flux tube.

Below we present results of a large-scale calculation of the color field distribution in four dimensions. We measure all six components of the color field for $\beta = 2.3$, 2.4 and 2.5, with interquark separations of up to 6 lattice units (in some cases up to 9 lattice units). The highly precise data allow us to measure the energy density in spite of cancellations between various components of the strength tensor.

These computations were performed on the FPS264 array processor at LSU. The lattice was a 17^3 X 20 hypercube with helical boundary conditions. SU(2) configurations were generated by alternating Metropolis sweeps with over-relaxation [3].

We measure the correlation:

$$ f^{\mu\nu}(x) \;=\; \frac{\beta}{a^4}\left(\frac{\langle WP_x^{\mu\nu}\rangle}{\langle W\rangle} - \langle P\rangle\right) \xrightarrow{a\to 0} -\frac{1}{2}\langle(F^{\mu\nu})^2\rangle_{q\bar{q}-vac} \tag{1} $$

where W is the Wilson loop, $P_x^{\mu\nu}$ the plaquette located at x with orientation (μ, ν), and a is the lattice spacing. In the naive continuum limit this is the quantum average of the color field density relative to the vacuum.

Three enhancements (reported elsewhere [4]) are crucial to decrease statistical noise, and consequently the CPU time. As a result of these, statistical errors on the Wilson loops are smaller by an order of magnitude when compared to the calculations of Ref.[5], in spite of the fact that our measurements were fewer in number.

The finite time extent of the Wilson loop implies contamination by excited states in any measurement of the components of $f^{\mu\nu}$. Our data indicate that mainly the $\langle E^2_{parallel}\rangle$ component has a significant T dependance beyond T = 6.

The R (interquark separation) dependence of all components evaluated at the middle of the $q\bar{q}$ axis is shown in Fig. 1. We find that increasing T up to 9a changes only the $\langle E^2_{parallel}\rangle$ contribution. The R dependance is clearly not coulomb-like and is consistent with a constant at large R as expected for a linear confining tube.

---

[1] Presented by V. Singh

Fig. 2 shows the transverse profile of the tube for the $\langle E^2_{parallel}\rangle$ component. Typically, plots of all individual components versus radial distance from the $q\bar{q}$ axis show a sharp fall-off that appears to be consistent with some combination of exponential and gaussian behavior. However the single most interesting result is obtained upon evaluation of the energy density,$(\langle E^2\rangle + \langle B^2\rangle)$. We find that the transverse electric and the transverse magnetic contributions are nearly equal in magnitude (also seen by Sommer for R=4a [6]) and opposite in sign. Color contour plots of the energy density (presented at the conference) show this cancellation. Thus the action density dominates the energy density typically by an order of magnitude in regions not too close to the quarks.

An explicit test of the Michael Sum Rules [7] (which constitute an independent check of our observations) indicates that our data follow the trend predicted by these rules. These will be reported elsewhere [4].

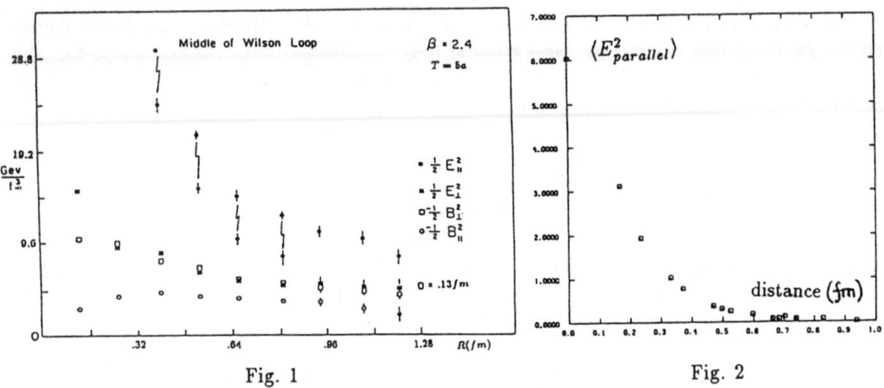

Fig. 1   Fig. 2

Our investigations confirm, within their limits, many of the features of the flux tube expected by the confinement models. Further analysis of our data is underway.

This work was supported in part by the U.S. Department of Energy under Contract No. DE-AS05- 77ER05490 and by the grants of the Polish Agency of Nuclear Energy Nos CPBP-01.09, CPBP-01.03 and RRI-14.

References

1. M. Fukugita and I. Niuya, Phys. Lett.,**132B**, 374(1983); J. Flower and S. Otto, Phys. Lett., **160B**, 128 (1985);

2. R. Sommer, Nucl. Phys. **B291**, 673 (1987); J. Wosiek and R. W. Haymaker, Phys. Rev. Rapid Comm. **D36**, 3297 (1987); J. Wosiek, Nucl. Phys. B (Proc. Suppl.) **4**, 52 (1988); I.H.Jorysz and C.Michael, Nucl. Phys B302 448 (1987)

3. M. Creutz, Phys.Rev. **D36**, 515 (1987). Phys. Lett., **128B**, 418 (1983).

4. Preprint MPI-PAE/PTh 70/89, submitted to Acta Physica Polonica B

5. F. Gutbrod, Z. Phys., **C30**, 585 (1986).

6. R.Sommer Nucl. Phys. **B306** 180 (1988).

7. C.Michael,Nucl. Phys. **B280 [FS18]** 13 (1987).

CONVERGENCE OF LANGEVIN SIMULATION FOR COMPLEX

GAUSSIAN INTEGRALS[1]

Richard W. Haymaker and Yingcai Peng
Department of Physics and Astronomy, Louisiana State University, Baton Rouge, La 70803

Abstract

We studied the convergence of complex Langevin simulations for simple Gaussian integrals analyticaly, and obtained the eigenvalues and eigenfunctions of the corresponding Fokker-Planck equations. This shows the convergence explicitly.

1. Introduction

The application of complex Langevin simulation in Lattice Gauge Theory was first proposed by G.Parisi [1] and J.Klauder [2]. However, it was reported that this method doesn't converge correctly in some cases [3,4]. This motivated us to study the convergence of the complex Langevin simulation.

2. Methods and Results

For simplicity we first consider the one-variable case (real or complex). Suppose one has a physical quantity $O(x)$, it's average can be evaluated using the complex Langevin simulation [5].

$$\langle O(x) \rangle = \frac{\int dx O(x) e^{-S(x)}}{\int dx e^{-S(x)}} = \frac{\int \int dx dy O(x + iy) P_0(x,y)}{\int \int dx dy P_0(x,y)} \tag{1}$$

where the action $S(x)$ is complex with x being real and $P_0(x,y)$ is the real probability distribution of z in complex plane from the complex Langevin simulation.

The real one-variable Langevin equation with real action $S(x)$ is

$$\frac{dx(t)}{dt} = -\frac{\partial S(x)}{\partial x} + \sqrt{2}\eta(t) \tag{2}$$

where the η is a real Gassian random number.

The probability distribution $P(x,t)$ of $x(t)$ generated by this Langevin equation is governed by the corresponding Fokker-Planck equation, if $S(x)$ is real.

$$\frac{\partial}{\partial t} P(x,t) = \frac{\partial^2 P}{\partial x^2} + \frac{\partial}{\partial x}\left(\frac{\partial S}{\partial x} P\right) \tag{3}$$

As the action $S(x)$ becomes complex, eq.(3) no longer represents a probability distribution. The real Langevin eq.(2) has to be extended to the complex one which has the corresponding Fokker-Planck equation.

$$\frac{\partial}{\partial t} P(x,y,t) = \frac{\partial^2 P}{\partial x^2} - \frac{\partial}{\partial x}[v_x P] - \frac{\partial}{\partial y}[v_y P] \tag{4}$$

with $(v_x, v_y) = -[Re(\frac{\partial S(z)}{\partial z}), Im(\frac{\partial S(z)}{\partial z})]_{z=x+iy}$.

One easily see that the general solution of eqs.(3) and (4) can be writen as

$$P(x,t) = \sum_{n=0}^{\infty} e^{-\lambda_n t} P_n(x), \quad \text{or} \quad P(x,y,t) = \sum_{n=0}^{\infty} e^{-\lambda_n t} P_n(x,y)$$

and it converges if $Re(\lambda_n) \geq 0$.

---

[1]Presented by Y. Peng

Usually it's easy to solve eq.(3) analytically, but it's difficult to solve eq.(4) even for some simple action $S(z)$ because the variables are non-separable. We solved the two variable Fokker-Planck eq.(4) with harmonic oscillator action $S(z)=\frac{1}{2}(\sigma_R + i\sigma_I)z^2$. Here we simply list our results [6].

Equations	Actions	Eigenvalues	Eigenfuns.
Eq.(3)	$S(x)=\frac{1}{2}\sigma_R x^2$	$\lambda_n = n\sigma_R$	$P_n(x) \sim H_n(\sqrt{\sigma_R/2}x)$
	$S(x)=\frac{1}{2}(\sigma_R + i\sigma_I)x^2$	$\lambda_n = n(\sigma_R + i\sigma_I)$	$P_n(x) \sim H_n(\sqrt{(\sigma_R + i\sigma_I)/2}x)$
Eq.(4)	$S(z)=\frac{1}{2}(\sigma_R + i\sigma_I)z^2$	$\lambda_{n,m} = n\sigma_R + im\sigma_I$	$P_{n,n}(x,y) \sim H_n(\rho)$ with $\rho = \sqrt{(\sigma_R + i\sigma_I)/2}\times$ $(x + \frac{2\sigma_R - i\sigma_I}{\sigma_I}y)$

where n = 0,1,2,3,......;   m=n,n-2,n-4,......,-n .

The eigenvalue spectrums can be represented in the complex λ- plane by the following figures clearly.

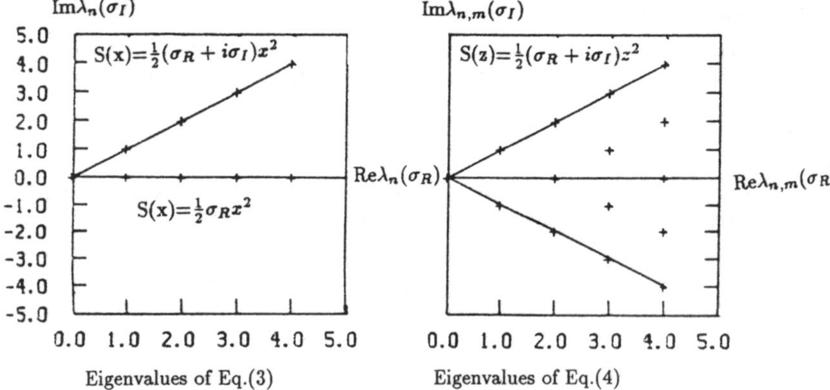

Eigenvalues of Eq.(3)   Eigenvalues of Eq.(4)

where in the above figures the cross sign represents eigenvalues. one can see that $Re(\lambda_n) \geq 0$ for both equations. So the corresponding Langevin simulations converge correctly.

3. Summary, Conclusions, and Acknowledgements

We have shown the convergence of complex Langevin simulation for Gaussian integrals analytically. The results is expected to be applied to the actual problem in LGT. We wish to thank J.Wosiek for many discussions on the complex Langevin method. Supported in part by U.S.DOE contract DE-AS05-77ER05490.

4. References

[1] J.Klauder, Acta Physica Austriaca Suppl.XXV, 251(1983)

[2] G.Parisi, Phys.lett. 131B,393(1983)

[3] J.Flower, S.Otto and S.Callahan, Phys. Rev. D34, 608(1986)

[4] R.W.Haymaker and J.Wosiek, Phys. Rev. D37,969(1988)

[5] J.Ambjorn and S.Yang, Phys. Lett. 165B,140(1985)

[6] R.W.Haymaker and Y.Peng, Phys. Rev. D41,1269(1990)

BEHAVIOR OF HIGHER REPRESENTATION WILSON LINES IN FINITE-TEMPERATURE, $SU(2)$, LATTICE GAUGE THEORY

JOE KISKIS

Department of Physics and Institute for Theoretical Dynamics
University of California
Davis, California 95616

ABSTRACT

The behavior of Wilson lines in representations $J = 1/2$, 1, and 3/2 in the critical region of the deconfined phase of finite-temperature, $SU(2)$, lattice gauge theory is studied. Theoretical considerations based on a flux model and on $(\phi^4)_3$ predict $\langle L_1 \rangle \to$ constant and $\langle L_{3/2} \rangle \propto \langle L_{1/2} \rangle$ as $T \to T_c^+$. For T sufficiently close to T_c, Monte Carlo data favor these predictions over the mean field result $\langle L_J \rangle \propto \langle L_{1/2} \rangle^{2J}$.

1. INTRODUCTION

This talk[1] deals with the behavior of Wilson lines in representations $J = 1/2$, 1, and 3/2 in the critical region of the deconfined phase of finite-temperature, $SU(2)$, lattice gauge theory. In previous work, Damgaard[2] obtained the mean field prediction

$$\langle L_J \rangle \sim \langle L_{1/2} \rangle^{2J} \tag{1.1}$$

and presented some Monte Carlo data from an $8^3 \times 2$ lattice that followed (1.1) down to $4/g^2 \cong 1.93$ $(4/g_c^2 \simeq 1.873)$. Then Redlich and Satz[3] found agreement with (1.1) on an $18^3 \times 4$ lattice down to $4/g^2 \simeq 2.310$ $(4/g_c^2 \simeq 2.296)$.

While these results appear convincing, it is difficult to understand how (1.1) could continue to hold for $T \to T_c^+$. The relation (1.1) is a mean field result. By all measures, much of the data mentioned above is in the critical region where fluctuations are important and mean field theory does not work. If (1.1) were to hold all the way down to T_c, it would mean that the gauge field in the deconfined phase responds to one source in representation J as if it were $2J$ independent fundamental sources. This certainly does not happen in the confined phase where $J = 1/2$ is confined $(\langle L_{1/2} \rangle = 0)$, and $J = 1$ is not confined $(\langle L_1 \rangle \neq 0)$.

The work presented here adds to the previous results in two ways. Section 2 summarizes some theoretical predictions that differ from (1.1). In Section 3, there are Monte Carlo results that favor these predictions over (1.1). The data differ from the previous results in that they are closer to the critical point.

2. THEORY

One approach is based on a percolating flux tube model of the deconfined phase. It leads to

$$\langle L_1 \rangle \to \quad \text{constant} \ > 0 \tag{2.1}$$

and

$$\langle L_{3/2} \rangle \quad \propto \quad \langle L_{1/2} \rangle \tag{2.2}$$

as $T \to T_c^+$. The other approach exploits the connection through universality with $(\phi^4)_3$ field theory and gives the more refined predictions

$$\langle L_1 \rangle = b' + c't^{1-\alpha} \tag{2.3}$$

and

$$\langle L_{3/2} \rangle = (d'' + et)\langle L_{1/2} \rangle . \tag{2.4}$$

(b', c', d'', and e are constants, $t = 4/g^2 - 4/g_c^2$ and $\alpha \simeq 0.12$.)

3. NUMERICAL RESULTS

In this section, Monte Carlo results for $\langle L_1 \rangle$ and $\langle L_{3/2} \rangle$ are described. The simulation used the Wilson action. The lattice size was $15^3 \times 2$ except at $4/g^2 = 1.88$ and 1.879 where $19^3 \times 2$ was used. For the larger values of $4/g^2$, 9,000 sweeps were run, while closer to the critical point, it was 20,000. The first 1,000 were dropped for thermalization.

As expected from the discussion in section 2, $\langle L_1 \rangle \neq 0$ at the critical point. An estimate is $\langle L_1 \rangle_c \simeq .025$. $\langle L_1 \rangle - \langle L_1 \rangle_c$ can be fit reasonably well with a curve proportional to $t^{.88}$. Thus, data for $\langle L_1 \rangle$ are consistent with (2.3). Close to the critical point (1.1) is excluded for $J = 1$.

The $\langle L_{3/2} \rangle$ data are compared with (2.4). $\langle L_{3/2} \rangle/\langle L_{1/2} \rangle$ is fit very well by a straight line, and $d'' \cong 0.018$.

4. CONCLUSION

Previous work showed that the mean-field-motivated relation (1.1) gives a good representation of the data for a surprisingly large range of $4/g^2$. However, theoretical considerations indicate that (1.1) cannot hold very close to the critical point. The relations (2.3) and (2.4) are suggested. These relations are favored by data sufficiently close to the critical point.

ACKNOWLEDGMENTS

It is a pleasure to thank P. Damgaard for stimulating my interest in this problem, and to thank R. Narayanan and P. Vranas for helpful discussions and comments on the manuscript. Support for this research came from the United States Department of Energy, the National Magnetic Fusion Energy Computer Center, and the Institute for Theoretical Dynamics at the University of California at Davis.

REFERENCES

1. A complete description will appear in Phys. Rev. D shortly.

2. P. Damgaard, Phys. Lett. B194, 107 (1987).

3. K. Redlich and H. Satz, Phys. Lett. B213, 191 (1988).

RANDOM WALK, CRITICAL BEHAVIOR
AND FINITE TEMERATURE SU(2) LATTICE GAUGE THEORY

J. KISKIS, R. NARAYANAN and P. VRANAS
Department of Physics, University of California
Davis CA 95616

The SU(2) lattice gauge theory at finite temperature is known to have a second order deconfining phase transition with non-Gaussian exponents. In this talk a qualtitative explanation of the above result will be given by studying the random walk representation of the correlation function of two Wilson lines in the fundamental representation. The details can be found in [1].

Consider an euclidean lattice field theory with a parameter β. The two point function $G(\vec{r})$, between the points 0 and $\vec{r}$, in the random phase can in general be written as [2]

$$G(\vec{r}) = \sum_{w:0 \to \vec{r}} \Omega(w, \beta) = \sum_{l=|\vec{r}|}^{\infty} S_l(\beta) P(l, \vec{r}, \beta). \tag{1}$$

$\Omega(w, \beta)$ is the weight associated with the walk w. S_l is the weighted sum of all walks of length l, and $P(l, \vec{r}, \beta)$ is the probability that a weighted walk of length l starting from 0 ends up at $\vec{r}$. In the large l limit, they obey the following scaling form

$$S_l(\beta) = l^{\gamma_p - 1} e^{-\theta(\beta)l} , \quad P(l, r) = \frac{1}{r^3} \tilde{P}(r/l^{\nu_p}). \tag{2}$$

The existence of a string tension at finite temperature requires that $G(r) \sim e^{-\alpha(\theta)r}$ as $r \to \infty$ where $\alpha(\theta)$ is some function of θ. This, together with (1)and (2) results in $\alpha(\theta(\beta)) \sim [\theta(\beta)]^{\nu_p}$. If the behavior of $\theta(\beta)$ near the critical point β_{cr} is of the form $\theta(\beta) = kt^{\nu_\theta}$, where $t = (|\beta_{cr} - \beta|)/\beta_{cr}$, then $\alpha(\theta(\beta)) \sim t^\nu$, with $\nu = \nu_\theta \nu_p$. Therefore there are two distinct mechanisms by which ν may change. The typical number of steps, $l_0 \sim \theta^{-1}$, in the important part of the sum (1) that represents the two point function, diverges as the critical point is approached. Nonanalytic behavior may enter at this point through the dependence of l_0 on t, viz., $l_0 \sim t^{-\nu_\theta}$. This is due to the interaction of the walk with the background loops and will be referred to as the excluded volume effect hereafter. The second contribution to ν is from $R \sim l^{\nu_p}$ where R is the second moment of the probability distribution in (2). This exponent is the inverse of the dimension of the walk. It gives the distribution of endpoints for a walk of large fixed length. Values of ν_p other than 0.5 reflect non–Markov characteristics of the walk. These interactions of the walk with itself at points separated by many steps are related to nonzero values for the renormalized coupling of the associated field theory.

Some models that illustrate these mechanisms are discussed. First consider the free walk model where the weight of a walk w is given by $\beta^{|w|}$. All walks of the same length are given the same weight. Thus the weight of a walk is independent of its shape and there are no self interactions or interactions with background loops. In fact it is a trivial matter to show that the resulting $G(\vec{r})$ is the free propagator with a mass parameter given by $\sqrt{\beta^{-1} - 6}$ [2], and that leads to Gaussian exponents.

In order to study the effect of self interactions consider a self-avoiding walk. In this case the sum (1) is over self-avoiding walks, and the weight of a walk is still $\beta^{|w|}$. This model is an extreme case of short range repulsion of the walk with itself. Since no loops are introduced there is still no excluded volume effect. This model has been studied in great detail by various techniques [3]. The result is that $\nu_p = 0.592$ and $\nu_\theta = 1$ giving a value of $\nu = 0.592$. The ν_θ exponent did not change from its Gauissian value because no loops were introduced. On the other hand this model has an infinite short range repulsion. This means that the probability that a walk of a given length will end close to its starting point is reduced compared to the free walk. On the other hand the probability of this walk ending at a point far away from the starting point is roughly the same as for the free

796

walk because in this case the majority of the walks are spread out and do not feel the short range repulsion. This effect increases the second moment of the walk and is consistent with an increase in the exponent ν_p.

In order to study the effect of the excluded volume on the universal behavior, consider the O(N) σ-model in the limit N$\to \infty$. For a fixed walk w, the weight $\Omega(w, \beta, N)$ can be expanded in a power series in β/N. The lowest $O((\beta/N)^{|w|})$ contribution accounts for the interactions of the walk with itself. In the N$\to \infty$ limit it can be shown that the coefficient of this leading term is 1 for all walks. Therefore there is no short range repulsion and $\nu_p = 1/2$. But there is an excluded volume effect in this limit and this is easily seen by keeping one more order in the expansion of Ω and seeing that it does not vanish. Further all walks of a fixed length have the same excluded volume effect independent of their shape. This generalizes to any order in β/N so that the weight of a walk w in the $N \to \infty$ limit is $\tilde{z}[(\beta\tilde{z})/N]^{|w|}$ where $\tilde{z}$ accounts for the excluded volume effect to all orders in β/N. The ν_θ exponent can be extracted by considering the two point function $G(0)$ and noting that $G(0) = 1$. The result is $\theta \sim [(\beta/N) - (1.516/6)]^2$ which implies that $\nu_\theta = 2$. Therefore in this limit $\nu = \nu_p\nu_\theta = (1/2) \times 2 = 1$.

These three examples are simple cases that suggest the following conclusions. In the absence of interaction (as in the case of free walks) $\nu_\theta = 1$, $\nu_p = 1/2$, and $\nu = 1/2$. Non–Markov interactions of the walk with itself (as in the example of self avoiding walks) alter the distribution P and ν_p without changing ν_θ. On the other hand if there are no interactions of the walk with itself but there is an effect of excluded volume (as in the example of N$\to \infty$ σ–model) then ν_p has its Gaussian value but ν_θ changes.

The results of the discussion above suggest the effects that may produce nontrivial exponents in a flux tube model of the SU(2), finite temperature, phase transition. Previous work modeled the flux tube as a free random walk and consequently gave $\nu = 1/2$. A more elaborate picture in which the self–interactions and intersections with a background of thermal fluctuating closed loops has the ingredients for changing ν_p, ν_θ, and ν. Toward that end the partition function of the Wilson action for SU(2) lattice gauge theory is considered on a lattice with periodic boundary condition and extent N_t along the time direction and with no space like plaquettes. Under this approximation a random walk representation emerges for the two point function and the weight associated with a walk has an expansion in $u_{1/2}(\beta) = I_2(\beta)/I_1(\beta)$. The coefficient of the lowest order contribution of a given walk w is of the form $\prod_k V_k(w)$. It can be shown that a vertex k that has not been visited at all or has been visited once has $V_k(w)$ equal to one. It is less than one if the vertex has been visited more than once. This shows that there is a short range repulsion. To investigate the excluded volume effect, a straight walk of length $|w|$ is considered and it can be shown that there is a correction to the leading order contribution indicating that there is an excluded volume effect.

The approximation to the finite–temperature gauge theory discussed here gives a picture of the flux tube as a random walk with repulsive interactions, and interactions with background loops. The background loops are thermal fluctuations. The density of the flux that they represent and thus the size of the associated excluded volume effect increases as T $\to$ T$_c$. These physical effects cause changes in ν_θ, ν_p, and ν. Although this understanding improves the physical picture associated with the behavior of the flux tube at the critical point, it is not quantitive enough to give the size of the changes in the ν exponents and the result that the universality class is that of the 3 dimensional Ising model.

REFERENCES:

1. J. Kiskis, R. Narayanan and P. Vranas, UCD-90-? preprint.
2. J. Glimm and A. Jaffe, Quantum Physics, Springer-Verlag, 1987; H. Grosse, Models in statistical physics and quantum field theory, Springer-Verlag, 1988
3. M.N. Barber and B.W. Ninham, Random and restricted walks, Gordon and Breach, Science publishers, 1970; P-G de Gennes, Scaling concepts in polymer physics, Cornell University Press, 1979.

Noncompact Simulations of SU(2)
At Strong Coupling

Kevin Cahill

Department of Physics and Astronomy, University of New Mexico
Albuquerque, New Mexico 87131

Abstract

Wilson loops have been measured on a 12^4 lattice in noncompact simulations of pure SU(2) without gauge fixing at moderate coupling $\beta = 2$ and at strong coupling $\beta = 0.5$. There is no sign of quark confinement, except possibly at $\beta = 0.5$ in the larger loops where the statistics are poor.

In a 1974 paper [1], Wilson showed that in the strong-coupling limit of his compact formulation of lattice gauge theory, static quarks are confined by a linear potential for both abelian and nonabelian gauge groups. A few years later, Creutz [2] and others displayed quark confinement at moderately strong coupling in wilsonian lattice simulations of both abelian and nonabelian gauge theories. Thus it has never been entirely clear whether the confinement seen in these lattice simulations is an artifact of Wilson's formulation or a property of QCD.

The basic variables in Wilson's formulation are elements of a compact gauge group rather than real numbers as in continuum gauge theory. Both abelian and nonabelian theories confine at strong coupling in his formulation because the Wilson action is a function of the product of the group elements of the links around the elementary squares of the lattice. The Wilson action thus possesses multiple minima, some of which correspond to false vacua not present in the continuum theory [3]. These extra vacua contribute importantly to the string tension, as has been shown by Mack and Pietarinen [4] and by Grady [5]. In their simulations of SU(2), they modified Wilson's action gauge invariantly by erecting, in effect, infinite potential barriers between the true vacuum and the false vacua. Mack and Pietarinen found a sharp drop in the string tension; Grady found no string tension at all.

To avoid this artifact of the Wilson action, some physicists have developed Monte Carlo simulation methods [3,6–12] that do not have confinement built in. These methods are called "noncompact" because their basic variables are fields rather than group elements as in Wilson's "compact" method. For U(1) the noncompact method seems accurate for general coupling strengths [9].

In this report I describe the results of measuring Wilson loops on a 12^4 lattice in a noncompact simulation of SU(2) gauge theory without gauge fixing or fermions at strong coupling, $\beta \equiv 4/g^2 = 0.5$, and at moderate coupling, $\beta = 2$. Creutz ratios of large Wilson loops provide a lattice estimate of the $q\bar{q}$-force for heavy quarks. There is no sign of quark confinement at either coupling in the loops that have been measured accurately. In the largest data set at $\beta = 0.5$, however, the force at five lattice spacings is as strong as that at four spacings, which does suggest confinement. But the errors here are very large, and if generous cuts on the data are made to exclude transients and to allow for long periods of thermalization, then the confinement signal fades. The noncompact method used in these simulations has been shown to agree well with perturbation theory at very weak coupling [11,12].

Patrascioiu, Seiler, and Stamatescu [6] performed the first noncompact simulations of SU(2) using a simple discretization of the classical action. They fixed the gauge to avoid the effects of zero modes and saw a force rather like Coulomb's. Seiler, Stamatescu, Wolff, and Zwanziger [7] used a more symmetrical discretization of the classical action with fewer zero modes, but still had to fix the gauge. They also saw a coulombic force.

I have described elsewhere [12] the code that I used in the present simulations. The fields are constant on the links of length a, the lattice spacing, but are interpolated linearly throughout the six transverse plaquettes. In the plaquette with vertices n, $n + e_\mu$, $n + e_\nu$, and $n + e_\mu + e_\nu$, the field is

$$A_\mu^a(x) = (\frac{x_\nu}{a} - n_\nu)A_\mu^a(n + e_\nu) + (n_\nu + 1 - \frac{x_\nu}{a})A_\mu^a(n), \tag{1}$$

and the field strength is given by the continuum formula $F_{\mu\nu}^a(x) = \partial_\nu A_\mu^a(x) - \partial_\mu A_\nu^a(x) + g f_{abc} A_\mu^b(x) A_\nu^c(x)$. The action S is the sum over all plaquettes of the integrals over each plaquette of the squared field strength, $S = \sum_{p_{\mu\nu}} \frac{a^2}{2} \int dx_\mu dx_\nu F_{\mu\nu}^c(x)^2$. The mean-value in the vacuum $|\Omega\rangle$ of a euclidean-time-ordered operator $Q(A)$ may then be approximated by a normalized multiple integral over the $A_\mu^a(n)$'s

$$\langle\Omega|\mathcal{T}Q(A)|\Omega\rangle \approx \frac{\int e^{-S(A)} Q(A) \prod_{\mu,a,n} dA_\mu^a(n)}{\int e^{-S(A)} \prod_{\mu,a,n} dA_\mu^a(n)}, \tag{2}$$

which one may estimate by Creutz's Monte Carlo techniques [2].

The quantity normally used to study confinement in quarkless gauge theories is the Wilson loop, $W(r,t)$, which is the mean-value in the vacuum of the path-and-time-ordered exponential $\mathcal{P}\mathcal{T} e^{-ig \oint A_\mu^a T_a dx_\mu}$ divided by the dimension of the matrices T_a that represent the generators of the gauge group. Although Wilson loops vanish [13] in the exact theory, Creutz ratios [2] of Wilson loops defined as the double difference of their logarithms, $\chi(r,t) \equiv -\ln W(r,t) - \ln W(r-a, t-a) + \ln W(r-a, t) + \ln W(r, t - a)$, are finite; and for large t provide an estimate of a^2 times the force between a quark and an antiquark separated by the distance r.

To measure Wilson loops and their Creutz ratios $\chi(r,t)$, I used a 12^4 periodic lattice and made three independent long runs, each beginning with a cold start, in which all fields were initialized to zero. On the first run, I used Adler's method of overrelaxation [14] with nine values of ω varying in the sawtooth pattern 1.0, 1.1, 1.2, ..., 1.8, 1.9, 1.0, 1.1, etc. This run began with 20,000 thermalizing sweeps at $\beta = 2$ on the last 5000 of which I made 250 measurements. I then switched β from 2.0 to 0.5 and did 1000 thermalizing sweeps. I then did 34,000 sweeps at $\beta = 0.5$ making 1700 measurements. On the second run, I used a heat bath and did 25,000 thermalizing sweeps at $\beta = 2$ on the last 5000 of which I made 250 measurements. I then switched β from 2.0 to 0.5 and did 1000 thermalizing sweeps. I then did 17,000 sweeps at $\beta = 0.5$ making 850 measurements. On the third run, I used a heat bath at $\beta = 0.5$ and did 20,000 thermalizing sweeps. I then did 14,000 sweeps at $\beta = 0.5$ making 800 measurements.

In all three runs, I measured Wilson loops every 20 sweeps, by using Parisi's trick [15] in an implementation that respects the dependencies in the corners of the loops. The values of the Creutz ratios so obtained from 3,350 measurements at $\beta = 0.5$ and 500 at $\beta = 2.0$ are listed in the table. The errors listed in the table are the averages of the asymmetric errors given by the bootstrap method [16]. Binning in groups of 2, 4, and 8 made little difference.

TABLE: *Noncompact Creutz ratios at moderate and strong coupling.*

$\frac{r}{a} \times \frac{t}{a}$	2×2	3×3	4×4	5×5
$\beta = 0.5$	0.2312(2)	0.0356(5)	0.0040(12)	0.0042(36)
$\beta = 2.0$	0.06923(4)	0.01560(6)	0.00374(7)	0.00093(8)

If the static force between heavy quarks is independent of distance, corresponding to a linear confining potential, then the Creutz ratios $\chi(r,t)$ should be independent of r and t at least for large t. There is no sign of confinement in these noncompact simulations at $\beta = 2$. But at $\beta = 0.5$, although the loops fall off at first, the 5×5 ratio is about as big as the 4×4 one. Although the error in $\chi(5a, 5a)$ is huge, these data at $\beta = 0.5$ do suggest that the QCD force between heavy quarks is coulombic at two and three lattice spacings but constant from four to five lattice spacings. While such an interpretation may apply to this data set, if we increase the thermalization on the first two runs by cutting out the first 5000 sweeps instead of the first 1000 after the transition from $\beta = 2.0$ to $\beta = 0$, then the value of the 5×5 Creutz ratio for $\beta = 0.5$ drops from 0.0042(36) to 0.0024(37). Thus the size of that ratio in the uncut data set is more likely to be a transient or a fluctuation than a sign of confinement.

I am grateful to H. Bryant, M. Creutz, G. Kilcup, and D. Topa for useful conversations. This research was supported by the Department of Energy under grant DE-FG04-84ER40166 and was done on a DECstation 3100 computer.

1 K. Wilson, *Phys. Rev. D* **10** (1974) 2445.

2 M. Creutz, *Phys. Rev. D* **21** (1980) 2308, *Phys. Rev. Letters* **45** (1980) 313, and *Quarks, Gluons and Lattices* (Camb. U. Press, 1983).

3 K. Cahill, M. Hebert, and S. Prasad, *Phys. Lett.* **210B** (1988) 198; K. Cahill and S. Prasad, *Phys. Rev. D* **40** (1989) 1274.

4 G. Mack and E. Pietarinen, *Nucl. Phys. B* **205** (1982) 141;

5 M. Grady, *Z. Phys. C* **39**(1988)125.

6 A. Patrascioiu, E. Seiler, and I. Stamatescu, *Phys. Lett.* **107B** (1981) 364.

7 I. Stamatescu, U. Wolff, and D. Zwanziger, *Nucl. Phys. B* **225** [FS9] (1983) 377; E. Seiler, I. Stamatescu, and D. Zwanziger, *Nucl. Phys. B* **239** (1984) 177 and 201.

8 K. Cahill, S. Prasad, and R. Reeder, *Phys. Lett.* **149B** (1984) 377; K. Cahill and R. Reeder, in *Advances in Lattice Gauge Theory* (World Scientific, Singapore, 1985), p. 424.

9 K. Cahill and R. Reeder, *Phys. Lett.* **168B** (1986) 381 and *J. Stat. Phys.* **43** (1986) 1043.

10 K. Cahill, S. Prasad, R. Reeder, and B. Richert, *Phys. Lett.* **181B** (1986) 333.

11 K. Cahill, *Nucl. Phys. B (Proc. Suppl.)* **9** (1989) 529 and *Phys. Lett.* **231B** (1989) 294.

12 K. Cahill, *Comput. Phys.* **4** (1990) 159.

13 K. Cahill and D. Stump, *Phys. Rev. D* **20** (1979) 2096.

14 S. L. Adler, *Nucl. Phys. B (Proc. Suppl.)* **9** (1989) 437.

15 G. Parisi, R. Petronzio, and F. Rapuano, *Phys. Lett.* **128B** (1983) 418.

16 B. Efron, *Ann. Stat.* **7** (1979) 1.

Nonabelian Gauge Theory on a Finite-Element Lattice

Kimball A. Milton
Department of Physics and Astronomy
The University of Oklahoma
Norman, OK 73019

ABSTRACT

We formulate the equations of motion of a nonabelian gauge field coupled to fermions on a finite-element lattice.

A few years ago a new approach to field theory on a Minkowski space-time lattice was initiated, which utilized the linear finite-element method. Advantages of the method include 1) unitary equations of motion, 2) high numerical accuracy, 3) no statistical limitations, and 4) no species doubling for fermions. Earlier we had applied this method to the sine-Gordon model and to Abelian gauge theories. In the last year, we have determined the lattice equations of motion for a nonabelian gauge field coupled to fermions.[1]

A straightforward, iterative procedure, precisely analogous to that in the continuum, allows one to infer, simultaneously, in a self-consistent way, the lattice equations of motion and the gauge transformations. For ease of notation we confine ourselves to $(1+1)$ spacetime dimensions. The Dirac equation is

$$\frac{i\gamma^0}{h}(\psi_{\overline{m},n+1} - \psi_{\overline{m},n}) + \frac{i\gamma^1}{h}(\psi_{m+1,\overline{n}} - \psi_{m,\overline{n}}) + \mu\psi_{\overline{m},\overline{n}} + I_{m,n} + K_{m,n} = 0, \qquad (1)$$

where h is the (square) lattice spacing, the first (second) index represents the spatial (temporal) coordinate, and the overbars represent forward averages: $\psi_{\overline{m},n} = (\psi_{m+1,n} + \psi_{m,n})/2$. Here the interaction terms are given by iterative formulæ summarized by the following difference equations:

$$I_{m,n} + e^{ighC_{m,n}} I_{m-1,n} = \frac{2i\gamma^1}{h}(1 - e^{ighC_{m,n}})\psi_{m,\overline{n}}, \quad C_{m,n} = (A_1)_{\overline{m-1},n}, \; (2a)$$

$$K_{m,n} + e^{ighB_{m,n}} K_{m,n-1} = \frac{2i\gamma^0}{h}(1 - e^{ighB_{m,n}})\psi_{\overline{m},n}, \quad B_{m,n} = (A_0)_{m,\overline{n-1}}, \; (2b)$$

In the process of determining these interaction terms, the gauge transformations for the gauge fields were simultaneously determined to be

$$\delta^{(k)}C_{m,n} = \frac{(igh)^k}{h}\frac{B_k}{k!}[\cdots[\delta\omega_{m,n} - \omega_{m-1,n}, C_{m,n}], \ldots, C_{m,n}], \quad k \neq 1, \qquad (3a)$$

$$\delta^{(1)}C_{m,n} = \frac{ig}{2}[\delta\omega_{m,n} + \delta\omega_{m-1,n}, C_{m,n}], \tag{3b}$$

$$\delta^{(k)}B_{m,n} = \frac{(igh)^k}{h}\frac{\mathcal{B}_k}{k!}[\cdots[\delta\omega_{m,n} - \delta\omega_{m,n-1}, B_{m,n}], \ldots, B_{m,n}], \quad k \neq 1, \tag{3c}$$

$$\delta^{(1)}B_{m,n} = \frac{ig}{2}[\delta\omega_{m,n} + \delta\omega_{m,n-1}, B_{m,n}], \tag{3d}$$

where $\mathcal{B}_k$ is the kth Bernoulli number, and where in (3a) and (3c) k nested commutators are understood. (A slight modification of (3c) and (3d) occurs for the initial time, $n = 0$.)

Once the gauge transformations are found, it is straightforward to determine the field strength $E = F_{01}$. The first two orders are

$$E^{(0)}_{\overline{m},\overline{n}} = \frac{1}{h}(C_1 - C_0 - B_1 + B_0), \tag{4a}$$

$$E^{(1)}_{\overline{m},\overline{n}} = -\frac{ig}{2}([B_1, C_1] + [B_0, C_1] + [B_1, C_0] - [B_0, C_0] + [C_0, C_1] - [B_0, B_1]), \tag{4b}$$

where $B_1 \equiv B_{m+1,n+1}$, $B_0 \equiv B_{m,n+1}$, $C_1 \equiv C_{m+1,n+1}$, and $C_0 \equiv C_{m+1,n}$, while higher order terms involve nested commutators.

The inference of the lattice Yang-Mills equations is identical to that of the Dirac equation. They are $(j^{\mu}_{\overline{m},\overline{n}} = g\overline{\psi}_{\overline{m},\overline{n}}\gamma^{\mu}T\psi_{\overline{m},\overline{n}})$

$$\frac{1}{h}(E_{m+1,\overline{n}} - E_{m,\overline{n}}) + \mathcal{I}_{m,n} = -j^0_{\overline{m},\overline{n}}, \quad \frac{1}{h}(E_{\overline{m},n+1} - E_{\overline{m},n}) + \mathcal{K}_{m,n} = j^1_{\overline{m},\overline{n}}, \tag{5}$$

where the interaction terms satisfy the difference equations

$$\mathcal{I}_{m,n}e^{ighC_{m,n}} + e^{ighC_{m,n}}\mathcal{I}_{m-1,n} = -\frac{2}{h}[e^{ighC_{m,n}}, E_{m,\overline{n}}], \tag{6a}$$

$$\mathcal{K}_{m,n}e^{ighB_{m,n}} + e^{ighB_{m,n}}\mathcal{K}_{m,n-1} = -\frac{2}{h}[e^{ighB_{m,n}}, E_{\overline{m},n}]. \tag{6b}$$

We are now in the process of solving these equations, and the corresponding Abelian versions, to obtain nonperturbative spectral information about gauge theories. This will require the development of new techniques.

ACKNOWLEDGEMENTS

This work was done in collaboration with Ted Grose. I thank the Department of Physics, Ohio State University, for its hospitality during the course of the work reported here. This work was supported in part by the US Department of Energy.

REFERENCES

1. K. A. Milton and T. Grose, *Phys. Rev.* D **41**, 1261 (1990) and earlier references cited therein.

The Status of the Electroweak Interaction on the Lattice

Donald N. Petcher
Supercomputer Computations Research Institute
Florida State University
Tallahassee, Florida 32306-4052 U.S.A.

ABSTRACT

Progress in the formulation of a theory of chiral fermions on the lattice with application to the standard model of the electroweak interaction is reviewed.

1. INTRODUCTION

There has been a quite a bit of progress in the study of chiral fermions on the lattice this past year which I would like to summarise for you in this talk. I can not possibly review all the relevant work in the time available, so I will concentrate on what I consider to be the central aspects. Neither will I spell out all details explicitly, but rather appeal to the literature for further clarification. For a recent and more extensive review, see ref. 1.

Clearly a non-perturbative formulation of chiral fermions is desireable in view of the centrality of chirality in the standard model for electroweak interactions, and because certain phenomena associated with the model are beyond perturbation theory, examples of which are top quark physics and the consequences of triviality of ϕ^4 theory. The major hurdle in the formulation of chiral fermions on the lattice revolves around the well known fermion doubling problem [2,3,4,5]. The most straightforward way to remove the doublers is to add a chiral symmetry breaking term (the 'Wilson' term) [2,6] to the action so that they decouple in the continuum limit by receiving masses of the order of the cutoff. Unfortunately this solution is not available in the standard model because the chiral symmetry is to be gauged, and therefore should not be broken. However, as has been pointed out by Smit [7,8] and Swift [9], a Wilson term could possibly be generated by spontaneous symmetry breaking along with other mass terms, and it may have the desired effect of removing the doublers in the continuum limit. An added complication in such a formulation is that one cannot build such terms with an isolated left-handed chiral fermion without a right-handed partner. Thus a right-handed neutrino must be introduced for every generation in the theory to accompany the corresponding left-handed neutrino.

Neither of these difficulties appears to be insurmountable. The problem of the right-handed neutrino is solved in the form of a theorem due to M. Golterman and myself [11,12] which states that for a class of models, this neutrino decouples en tirely from the theory. Rather convincing evidence from mean field calculations [13], (quenched) Monte Carlo calculations [14,15,16], and more recently, calculations in the $1/d$ expansion [17] indicates that there is a region in the phase diagram where the doublers decouple from the theory in the continuum limit.

2. THE SMIT-SWIFT MODEL

The model is a straightforward discretization of the standard model for electroweak interactions along with a 'Wilson-Yukawa' term with coupling w to generate the Wilson term in the broken phase. Since this term has the Dirac structure of a mass term, it is necessary to introduce a right-handed neutrino for every left-handed one in the theory as mentioned above.

The action for the model is given by [7,9,10,18]

$$S = S_{\text{gauge}} + S_{\text{Higgs}} + S_{\text{fermion}}, \tag{1}$$

where S_{gauge} is the usual plaquette action for the $SU(2) \times U(1)$ gauge degrees of freedom, and the Higgs action is given by

$$
\begin{aligned}
S_{\text{Higgs}} = &-\frac{1}{2}\kappa \sum_{x\mu} \text{tr}[\Phi^\dagger(x)U_{\mu L}(x)\Phi(x+e_\mu)U_{\mu R}^\dagger + \text{h.c}] \\
&+ \lambda \sum_x (\text{tr}[\Phi(x)^2] - 1)^2,
\end{aligned}
\tag{2}
$$

with Φ a 2×2 complex matrix field, whose columns can be related to the usual two-component representation of the Higgs field [7]. U_L and U_R are the left- and right-acting gauge degrees of freedom respectively. Finally the fermion action for the electron-neutrino sector is given by

$$
\begin{aligned}
S_{\text{fermion}} = &\sum_{x\mu} \overline{\psi}\gamma_\mu(\mathcal{D}_\mu + \tilde{\mathcal{D}}_\mu)\psi \\
&+ [\overline{\psi}_L \Phi(Y - \frac{w}{2}\sum_\mu \mathcal{D}_\mu\tilde{\mathcal{D}}_\mu)\psi_R] + [\overline{\psi}_R(Y - \frac{w}{2}\sum_\mu \mathcal{D}_\mu\tilde{\mathcal{D}}_\mu)(\Phi^\dagger\psi_L)]
\end{aligned}
\tag{3}
$$

where $\psi_{L,R} = \begin{pmatrix} \nu_{L,R} \\ e_{L,R} \end{pmatrix}$ is a fermion field containing the electron and neutrino fields, Y is a diagonal matrix of Yukawa couplings with entries y for the neutrino, and y' for the electron, and $\mathcal{D}$ is the lattice covariant derivative given by

$$
\begin{aligned}
\mathcal{D}_\mu\psi(x) &= U_\mu(x)\psi(x+e_\mu) - \psi(x) \\
\tilde{\mathcal{D}}_\mu\psi(x) &= \psi(x) - U_\mu^\dagger(x-e_\mu)\psi(x-e_\mu).
\end{aligned}
\tag{4}
$$

It is important to introduce ν_R here as a gauge singlet, to meet the conditions necessary for its decoupling. Similar fermion action terms can be written for the quarks as well, but we will limit our discussion to the electron-neutrino sector.

Obviously if we substitute $\langle\Phi\rangle = v$ for Φ in eq. (3) as is done in the usual standard model in tree level perturbation theory, the neutrino picks up a mass term with mass matrix Yv and a Wilson term with effective Wilson r parameter wv. Thus a non-zero vacuum expectation value for Φ is expected to generate masses for the fermions just as in the continuum theory.

3. THE NEUTRINO DECOUPLING THEOREM

For $y = 0$, the action given in eq. (3) has a symmetry in the right-handed fermion sector:

$$\nu_R \to \nu_R + \epsilon. \tag{5}$$

This is obvious from the form of the action as presented. Since the right-handed component of the neutrino does not couple to any gauge degree of freedom, the covariant derivative is just a simple derivative on this component. Since the only place that the neutrino field occurs without a derivative acting on it is in the term with coupling y, the symmetry follows. This very simple observation turns out to have important consequences. In fact, the symmetry is enough to insure that the right-handed neutrino decouples entirely from all other particles in the theory in the continuum limit.

To show this, one need only calculate the Ward identitities associated with the shift symmetry given in eq. (5). Making a transformation $\psi_L \to V^\dagger \psi_L$ as the propagating field, the Ward identities (which of course are also the Schwinger-Dyson equations for this field) result first in the following expression for the inverse propagator of the neutrino [11,12]:

$$S_F^{-1}(p) = \sum_\mu \gamma_\mu \left(i f_\mu^L(p) P_L + i \sin p_\mu P_R \right) + y + w \sum_\mu (1 - \cos p_\mu). \tag{6}$$

Except for the arbitrary function $f_\mu(p)$ in the left-handed part of the kinetic term, the inverse propagator is completely specified and has just the form of a free propagator. Thus if $y = 0$, in the continuum limit the mass term vanishes, and the kinetic term of the right-handed part is just that of a free massless neutrino propagator. The left-handed neutrino requires a wave function renormalization.

A second consequence of the Ward identitites is that all one particle irreducible $n-$point functions for $n > 2$ vanish identically (even for finite lattice spacing). Thus the exact propagator for the right-handed neutrino is that of a free propagator in the massless continuum limit, and it couples to no other particle in the theory. Furthermore, the decoupling on the lattice is only broken by the Yukawa and Wilson-Yukawa terms in eq. (6), and hence can be considered a soft breaking similar to chiral symmetry breaking in QCD.

4. THE PHASE DIAGRAM

Now we have a model with some six couplings g,κ,λ,y,y', and w (in the electron-neutrino sector alone) and we must study the phase diagram in this many dimensional coupling space to find out where a continuum limit exists that could have the spectrum of the standard model. Ultimately we will have to find a critical point for which the correlation length associated with each physical state in the spectrum diverges, meaning the mass vanishes in units of the lattice spacing. In contrast, the mass of the doublers should remain finite, thereby decoupling from the physical spectrum. We already know that w cannot be treated perturbatively, because in that case the doubler masses also scale to zero and enter the physical spectrum [22]. This cannot be avoided in the weak coupling regime even with non-perturbative arguments, because of the presumed triviality of Yukawa couplings coupled to scalars. So one thing we know is that for the model to succeed, a scaling region must exist for some rather large value of w (of order unity).

We can simplify our search for a critical region from several facts. First of all, from the presumed triviality of Φ^4 theory, the value of the bare coupling for the Φ^4 interaction shouldn't matter. Hence to simplify matters we can take $\lambda \to \infty$, thereby freezing the Higgs field to unit radius. The radial degree of freedom should of course appear in the spectrum nevertheless, since other operators with the same quantum numbers can create that state. Second, we examine the theory for zero gauge couplings, under similar assumptions to those of Dashen and Neuberger, that these couplings can be treated perturbatively. (Indeed, under this assumption, a similar formula to that of Dashen and Neuberger in the scalar theory can be proven also in the presence of fermions.) Of course the scaling limit for a gauge theory is generally when the bare coupling is taken to zero, so conclusions about the phase diagram with this ansatz should be reliable.

The two Yukawa couplings y and y' are needed for adjusting the masses of the neutrino and the electron respectively. The electron mass is obtained by tuning y' to whatever value is necessary to fit the correct physical electron mass (and similar for Yukawa couplings for quarks). As a consequence of the decoupling theorem above however, the coupling y plays a special role. For zero mass neutrinos, one need only set $y = 0$, and no tuning is required.

Finally we are left with the two couplings κ and w. So we have reduced our investigation to a two dimensional cross section of the phase diagram. In general, the κ-w plane is not completely understood to date, especially in the negative κ region. As it turns out, for slightly different reasons, the fermions decouple from the Higgs field at both extremes in w, so for both $w = 0$ and for $w = \infty$ we have a free fermion theory decoupled from a pure Higgs theory. Along these edges, coming down from large κ, the Higgs theory has a transition from the broken phase to the symmetric phase at some positive κ_c. Further, a transition from the symmetric to an antiferromagnetic phase in the negative κ region also exists at $-\kappa_c$. The latter comes about because if one changes the sign of every scalar field component occupying an 'odd' lattice site, this transforms the action into itself but with negative κ. Thus the ferromagnetic phase is mapped into an antiferromagnetic phase. The interaction term between the bosons and fermions doesn't respect this symmetry so in some sense the fermions work against the Higgs ground state, thus perturbing the long range antiferromagnetic order. However, what actually happens beyond this, and the sorts of states which could be used to characterize the ground state for large negative κ for arbitrary w are not understood.

Using the mean-field approximation [13] combined with an expansion in the Yukawa couplings [23] and quenched Monte Carlo [14,15,16] techniques, a bit more can be said about the interior of this plane of the phase diagram. First it is clear that critical lines continue into the diagram from the four critical points on the boundary for $\pm\kappa_c$ and for $w = 0, \infty$. From mean-field theory (a systematic refinement of this result in the large w region is given by the $1/d$ expansion [17]) one finds that the to first order critical lines have the form $\pm\kappa_c - cw^2$ and $\pm\kappa_c - c'/w^2$ in the small and large w regions respectively where c and c' are constants. For the positive κ transition lines, these continue down into the negative κ region beyond the point where $\kappa = 0$, without intersecting. Thus along the line $\kappa = 0$ the theory goes from the symmetric phase to a broken phase and back to the symmetric phase as w is increased. Results obtained from the hopping parameter expansion [19] probe the symmetric phase of the theory for large w and are in agreement with the above scenario.

5. DECOUPLING OF DOUBLERS

The remaining question concerns the fermion doublers. There are several reasons to believe that if the Wilson-Yukawa coupling is strong enough (say $w \geq 0.4$) the doublers do decouple from the physical spectrum as desired. The first argument comes from mean field theory [13], including corrections from the $1/d$ expansion [17]. Within this scenario which is systematic for large values of $2dw$ (i.e., w of order 1), the function f_μ in eq. (6) takes the form

$$f_\mu(p) = \left(v^2 + \frac{1}{2d}(1 - 2v^2) \right) i \sin p_\mu + O(v^4) + O(\frac{1}{d^2}) \tag{6}$$

to first non-trivial order in $1/d$. With a renormalization constant of $2d$ as the continuum limit is taken ($v \to 0$), the propagator therefore has exactly the same form as a free Wilson propagator. Thus it has only one physical pole , and the remaining (doubler) masses are of the order of the cutoff.

Further evidence comes from (quenched) Monte Carlo calculations, where studies exist both for $SU(2)$ theory [15] and for $U(1)$ theory [16], confirming the $1/d$ results. Thus there is every reason to believe that the fermion doublers actually do decouple.

6. CONCLUSION

We started out by asking whether a version of the standard model can be realized on the lattice with the hope to agree with all experimental constraints. The first test is that the correct field content would be present without additions. So far we have seen that the presence of a right-handed neutrino is not a serious drawback, since it decouples completely from the theory. In addition the fermion doublers apparently decouple from the theory for large enough w. Finally the the gauge vector boson becomes massive in the broken phase, and the coupling κ can be tuned as the phase transition is approached to give this particle a physical mass, so the vector boson is also a part of the spectrum.

There are other issues that one should examine before safely concluding that the standard model is in good shape on the lattice. For example one should make sure Lorentz invariance is restored in the continuum limit. Indeed, in a perturbative expansion in the bare couplings of the theory the vacuum polarization diagram has Lorentz covariance violating terms [10,20,21,24]. As pointed out in ref. 24, these terms can be tuned to zero in principle order by order in perturbation theory by adding new counterterms to the model. In the framework of the present model however, this expansion amounts to an expansion around the point $\kappa = \infty$. Since renormalized perturbation theory presumably occurs around the finite critical value of κ (as in the pure Φ^4 theory [25]) and not for infinite κ, one would expect that Lorentz invariance is restored there without the necessity for tuning. Evidence to this effect has been presented in ref. 26.

Another open question is the role of anomalies in the continuum limit of the model, since the lattice theory is gauge invariant, and therefore anomaly free. Although no severe problems should arise because of this, it points to a puzzle as to how the model relates to the usual perturbative picture.

In conclusion, on the basis of the evidence acquired so far, I think we can conclude with cautious optimism that there is now a realistic candidate for the standard model on the lattice that satisfies every experimental requirement.

ACKNOWLEDGEMENTS

The author is supported by the Florida State University Supercomputer Computations Research Institute which is partially funded by the Department of Energy through Contract No. DE-FC05-85ER250000.

REFERENCES

1. J. Smit, preprint ITFA-89-31, to be published in the proceedings of 1989 Symposium on Lattice Field Theory, Capri, Italy, September, 1989.
2. L. H. Karsten and J. Smit, Nucl. Phys **B183**, 103 (1981).
3. H. B. Nielsen and M. Ninomiya, Nucl. Phys. **B185**, 20 (1981); Nucl.Phys. **B195**, 541 (1982); Nucl. Phys. **B193**, 173 (1981); Phys. Lett. **105B**, 219 (1981).
4. L. H. Karsten, Phys. Lett. **104B**, 315 (1981).
5. D. Friedan, Comm. Math. Phys. **85**, 481 (1982).
6. K. G. Wilson, in *"New Phenomena in Subnuclear Physics"* (edited by A. Zichichi), *Plenum Press, New York* (1977) *(Erice lectures, 1975)*.
7. J. Smit, Acta Physica Polonica **B17**, 531 (1986).
8. L. H. Karsten, in *"Field Theoretical Methods in Particle Physics"* (edited by W. Rühl), *Plenum Press, New York* (1980) *(Kaiserslautern 1979)*.
9. P. D. V. Swift, Phys. Lett. **145B**, 256 (1984).
10. J. Smit, Nucl. Phys. **B** Suppl. 4, 451 (1988).
11. M.F.L. Golterman and D.N. Petcher, Phys. Lett. **225B**, 159 (1989).
12. M.F.L. Golterman and D.N. Petcher, preprint FSU-SCRI-89-124, to be published in the proceedings of 1989 Symposium on Lattice Field Theory, Capri, Italy, September, 1989.
13. J. Smit, Nucl. Phys. **B(Proc. Suppl. 9)**, 579 (1989).
14. A.M. Thornton, Phys. Lett. **221B**, 151 (1989); Phys. Lett. **227B**, 434 (1989).
15. W. Bock, A. K. De, K. Jansen, J. Jersák, T. Neuhaus and J. Smit, Phys. Lett. **B232**, 486 (1989); preprint HLRZ-90-14, Jülich; W. Bock and A. K. De, preprint HLRZ-90-29, Jülich.
16. S. Aoki, I-H. Lee, J. Shigemitsu and R. Shrock, preprint ITP-SB-89-80, RU-89/B_1/55, OSU DOE ER-01545-440.
17. M.F.L. Golterman and D.N. Petcher, preprint preprint UCLA/90/TEP/19.
18. The model proposed in refs. 20 and 21, when all necessary counterterms are taken into account, is equivalent to the Smit-Swift model.
19. S. Aoki, I-H. Lee and S-S. Xue, preprint BNL report.42494 (1989); Phys.Lett. **229B**, 79 ((1989)); S. Aoki, Phys.Rev. **D40**, 2729 ((1989)); I-H. Lee, preprint RU89/B_1/49, to be published in the proceedings of 1989 Symposium on Lattice Field Theory, Capri, Italy, September, 1989.
20. K. Funakubo and T. Kashiwa, Phys. Rev. **D38**, 2602 (1988).
21. S. Aoki, Phys. Rev. **D38**, 2109 (1988); Phys. Rev. Lett. **60**, 618 (1988); Nucl. Phys. **B**, Suppl. 4, 479 (1988), *(Field Theory on the Lattice*, eds. A. Billoire, R. Lacaze, A. Morel, O. Napoly and J. Zinn-Justin, proceedings of the Seillac conference, (1987)).
22. S. J. Hands and D. B. Carpenter, Nucl. Phys. **B266**, 285 (1986).
23. M. A. Stephanov and M. M. Tsypin, Phys. Lett. **B236**, 344 (1990).
24. A. Borrelli, L. Maiani, G. C. Rossi, R. Sisto and M. Testa, preprint I. N. F. N. - Sezione di Roma preprint n.655, 1989.
25. M. Lüscher and P. Weisz, Nucl. Phys. **B290**, 25 (1987); Nucl. Phys. **B295**, 65 (1988); Nucl. Phys. **B318**, 705 (1989).
26. M.F.L. Golterman and D.N. Petcher, preprint preprint FSU-SCRI-89-125, to be published in the proceedings of 1989 Symposium on Lattice Field Theory, Capri, Italy, September, 1989.

THEORY BEYOND THE STANDARD MODEL

Conveners: H. Haber & J. Polchinski

Supersymmetry Breaking in String Theory[*]

LANCE J. DIXON

Stanford Linear Accelerator Center
Stanford University, Stanford, California 94309

ABSTRACT

I briefly review the problems with previous investigations of supersymmetry breaking in string theory — at tree-level, at one-loop, and non-perturbatively. A variant of the original non-perturbative scenario is proposed, in which gaugino condensation takes place in two different strongly-interacting hidden-sector gauge groups. In the new scenario it is possible to generate a large hierarchy of mass scales and to simultaneously stabilize the dilaton at a large expectation value (weak coupling). However, it is still uncertain whether supersymmetry is broken in such a vacuum.

---

[*] Work supported by the Department of Energy, contract DE–AC03–76SF00515.

1. INTRODUCTION

Two of the major uncertainties in extracting low-energy predictions from superstrings involve the choice of vacuum state and the mechanism of supersymmetry breaking. Here I will address the second problem, under the assumption that it can be separated from the first problem. In string theory, many important questions are entangled with supersymmetry (SUSY) breaking, such as:

(a) Can SUSY breaking take place at all?

(b) Can it explain the hierarchy of mass scales M_W/M_{Pl}?

(c) Can it fix the gauge coupling constants to some realistic values that also correspond to a weakly-coupled string?

(d) Can it explain the vanishing of the cosmological constant?

(e) Can it feed into the "observable sector" in a phenomenologically viable way?

I will present a scenario[1] in which the answer to (a), (b) and (c) appears to be yes. First, however, I review earlier work on SUSY breaking in string theory, in order to indicate how problems encountered in that work may be circumvented by the new scenario.

2. REVIEW

Supersymmetry breaking in string theory is made difficult by the existence of massless fields ϕ_i with potentials that are initially flat, $V(\phi_i) \equiv 0$, but that become non-flat in the SUSY breaking process. The non-flat potentials typically have only pathological vacua where fields run off to infinity. The prime example of this phenomenon is the dilaton field ϕ, which couples at zero-momentum to the Euler character of the world-sheet,[2] $\chi = \int d^2\sigma\sqrt{g^{(2)}}R^{(2)}(\sigma)$. Therefore the Polyakov amplitude for a surface with n handles — corresponding to a scattering amplitude at n loops in string perturbation theory — comes with a factor $e^{2(1-n)\phi}$. The effective Lagrangian for ϕ and all the other massless fields has the form

$$\mathcal{L}_{\text{eff}}(\phi, \partial\phi, \ldots) = e^{2\phi}\hat{\mathcal{L}}_{\text{tree}} + \hat{\mathcal{L}}_{1-\text{loop}} + e^{-2\phi}\hat{\mathcal{L}}_{2-\text{loop}} + \ldots, \tag{1}$$

where $\hat{\mathcal{L}}_{\text{tree}}, \hat{\mathcal{L}}_{1-\text{loop}}, \ldots$ depend only on $\partial\phi$ and the other fields. The vacuum expectation value (VEV) of the dilaton is identified[3] as the string coupling constant g_s,

$$e^{2\langle\phi\rangle} \equiv \frac{4\pi}{g_s^2}. \tag{2}$$

A cosmological constant $\Lambda_{n-\text{loop}}$ corresponds to a c-number contribution to $\hat{\mathcal{L}}_{n-\text{loop}}$ in (1), and generates a dilaton potential $V(\phi) \sim e^{-2n\phi}$ (after rescaling $\mathcal{L}_{\text{eff}}$ by $e^{-2\phi}$ to make the dilaton kinetic term in $\hat{\mathcal{L}}_{\text{tree}}$ canonical). This runaway potential leads to an infinite VEV for ϕ, which also means that the string coupling constant vanishes, $g_s = 0$. Certainly such a string vacuum (a free theory) does not describe the real world.

Tree-Level Breaking

The most severe way to break SUSY in string theory is at tree-level, that is, spontaneously. In ten dimensions, the $O(16) \times O(16)$ heterotic string[4] and related models with tachyons[4,5] are vacua with spontaneously broken supersymmetry. They all have $\Lambda_{1-\text{loop}} \neq 0$ (in fact, $\Lambda_{1-\text{loop}}$ is infinite due to tachyons for all except the $O(16) \times O(16)$ vacuum), so the dilaton runs off to infinity. Even had the dilaton VEV been stabilized, the SUSY breaking scale would have been of order the Planck scale M_{Pl}, leaving no hope for low-energy SUSY to explain why M_W/M_{Pl} is so tiny. One could try to stabilize the dilaton by shifting the tree-level vacuum "off-shell" (away from a conformal field theory) in order to cancel tree-level effects against one-loop effects.[6] But here such a cancellation requires the dilaton VEV to be of order one, and so $g_s \sim 1$, a regime where the perturbative analysis is unreliable.

There are also non-supersymmetric compactifications of superstrings to less than ten dimensions, such as "twisted tori"[7] and toroidal compactifications of non-supersymmetric ten-dimensional vacua.[8] The one-loop cosmological constant $\Lambda_{1-\text{loop}}(R_i)$ is now a function of the radii R_i of the tori, which can be tuned to special values R_i^c so that

$$\Lambda_{1-\text{loop}}(R_i^c) = 0. \tag{3}$$

However, the radii also represent massless fields with flat potentials at tree-level in

g_s, and $\Lambda_{1-\text{loop}}(R_i)$ represents a one-loop potential for them, so a stable vacuum at one-loop also requires

$$\partial_{R_i}\Lambda_{1-\text{loop}}|_{R_i^c} = 0, \qquad \text{for all } R_i. \tag{4}$$

Vacua satisfying (3) generally fail to satisfy (4), and so the radii fields run away.[*]

One-Loop Breaking

Henceforth, let the vacuum state be four-dimensional (4d), and let supersymmetry be unbroken at tree-level. Then the effective Lagrangian is supersymmetric, and the dilaton can be organized into a chiral superfield,

$$S(x,\theta) = (e^{2\phi}(x) + ib(x)) + \theta\psi_S(x) + \theta\theta F_S(x). \tag{5}$$

A Peccei-Quinn symmetry for the axion, $b(x) \to b(x) + \text{const.}$, ensures that there is no superpotential for S (no F-term) at any order in string perturbation theory.[9] Generically, 4d vacua have other massless chiral superfields $T_i(x,\theta)$, called moduli, that result from continuous parameters in the compactification (like the torus radii R_i) and that have no superpotential at string tree-level. The same argument of ref. 9 shows that the T_i also acquire no superpotential perturbatively.

Even though F-terms cannot be generated for S, T_i in string perturbation theory, a Fayet-Iliopoulos D-term can be generated at one-loop.[10] This happens whenever there is a $U(1)$ factor in the 4d gauge group with a trace anomaly, $\sum_i Q_i \neq 0$, where the sum is over all left-handed, massless fermions in the spectrum, with $U(1)$ charges Q_i. In field theory the trace anomaly leads to a quadratically divergent D-term for the $U(1)$.[11] In string theory the one-loop D-term is finite and calculable,[12] and results in a non-vanishing two-loop cosmological constant $\Lambda_{2-\text{loop}} \sim D_{1-\text{loop}}^2$.

---

[*] Even if it happened somehow that both (3) and (4) were satisfied at one-loop, the problem would reappear at two-loops, and so on.

As in the tree-level case, $\Lambda_{2-\text{loop}} \neq 0$ means that the dilaton can run off to infinity. This time, however, there is another possibility:[13,10,12] The one-loop D-term can be cancelled against a tree-level D-term induced by giving VEV's to charged, massless scalars A_i with no superpotential, which "happen" to be present in the massless spectrum.[†] The tree-level plus one-loop D-term is

$$D = (\text{Re}S)\sum_i Q_i A_i^\dagger A_i + c\sum_i Q_i, \qquad c > 0. \tag{6}$$

The solution to $D = 0$ with g_s nonzero has $\langle A_i \rangle \sim (\text{Re}S)^{-1/2} \sim g_s$. Note that the shift of VEV's is only $O(g)$ in this case, in contrast to the tree-level case where it was $O(1)$; this makes the shifted vacuum perturbatively reliable here. However, supersymmetry is not broken in this vacuum (at least at one-loop, and most likely to any finite order), the string coupling constant g_s is still not determined, and hence there is still a "dilaton" S' which is a mixture of S and the A_i, with $\langle S' \rangle \sim g_s^{-2}$.

Non-Perturbative Breaking

Non-perturbative SUSY breaking seems to have the best chance of generating a large hierarchy through factors like those occurring in instanton calculations, $\exp(-c/g_s^2)$, if the string coupling g_s can be fixed to a small value. Since there is currently no framework for calculating non-perturbative effects in the full string theory, one has to assume that non-perturbative effects in the low-energy effective field theory dominate over "stringy" non-perturbative effects.

The first attempts to break SUSY non-perturbatively in string theory invoked gaugino condensation in the hidden E_8 in Calabi-Yau compactifications of the heterotic string.[14,15] Up to an overall constant C_G, the gaugino condensate for

---

† I know of no general argument for the existence of such fields, but empirically they are always present. Note that their charge has to be *opposite* in sign to the trace anomaly.

supersymmetric pure Yang-Mills theory with gauge group G follows from renormalization group invariance and dimensional analysis,

$$\langle \lambda^\alpha \lambda_\alpha \rangle \;=\; C_G \Lambda_G^3 \;=\; C_G \left[M \left(\frac{1}{g^2(M)} \right)^{b_1/2b_0^2} \exp\left(\frac{-8\pi^2}{b_0 g^2(M)} \right) \right]^3 \left(1 + O(g^2) \right) . \tag{7}$$

Here Λ_G is a dynamically generated mass scale, approximated by the scale at which the two-loop running coupling blows up, and M will be set to $1/\sqrt{\alpha'}$ for the application to string theory. The renormalization group constants $b_{0,1}$ are defined by

$$\beta(g) \;=\; -\frac{b_0}{(4\pi)^2} g^3 - \frac{b_1}{(4\pi)^4} g^5 + \dots . \tag{8}$$

Now $\lambda^\alpha \lambda_\alpha$ is the *lowest* component of a chiral, composite superfield $W^\alpha W_\alpha$ (W_α is the supersymmetric field strength), so its VEV normally would not break supersymmetry. But in string theory the dependence of the condensate (7) on the gauge coupling translates into dependence on the dilaton superfield S, and in fact it generates (as will be shown below) a superpotential for S which could break supersymmetry,[16,14,15]

$$W(S) \;\sim\; M^3 S \exp\left(-\frac{6\pi}{b_0} S \right) . \tag{9}$$

However, for weak-coupling (large S) the potential

$$V(S, \bar{S}) = e^{K(S,\bar{S})} \left[g^{S\bar{S}} D_S W \bar{D}_{\bar{S}} \bar{W} - 3|W|^2 \right] \tag{10}$$

is monotonic and leads once again to a runaway dilaton. In ref. 15 it was proposed to stabilize the dilaton by an additional c-number term in the superpotential, resulting from a VEV for the antisymmetric tensor field strength $H_{\mu\nu\lambda} = \partial_\mu B_{\nu\lambda} + \dots$ on the internal Calabi-Yau space, $W(S) \to W(S) + c$ with $c = \langle H_{ijk} \rangle$. Unfortunately, c is quantized to be of order 1 in Planck units,[17] so the stabilization takes place at around the Planck scale and for $\langle S \rangle \sim 1$. Thus the semi-classical analysis is unreliable, and in any case it does not predict a large hierarchy. Attempts[18] to stabilize the dilaton through a dependence of the (radiatively-corrected) potential $V(S, T_i)$ on the moduli T_i tend to result in the moduli running away.

In ref. 19 it is argued on general grounds that the results reviewed above are generic, *i.e.* that string theory has *no* acceptable weakly-coupled vacua. However, possible loopholes to the argument are also given. A scenario for a weakly-coupled vacuum will now be presented that relies on loophole #1 of the first ref. 19; specifically it relies on the existence of small (combinations of) discrete, non-dynamical parameters in the theory.

3. A New Scenario

Consider now a variant of the above non-perturbative scenario, in which the hidden sector is more complicated than just a single SUSY pure Yang-Mills theory.[1] Indeed, many 4d string vacua have hidden sectors that are fragmented into the products of several simple Lie algebras, and they may contain charged matter supermultiplets as well. Perhaps the combination of gaugino condensates from the different gauge group factors can stabilize the dilaton VEV at a perturbatively reliable (large) value. For definiteness and simplicity let the hidden sector be SUSY pure Yang-Mills with gauge group $G_1 \times G_2$, where G_1 and G_2 are simple. Then the relative phase between the two condensates $\langle \lambda\lambda \rangle_1$ and $\langle \lambda\lambda \rangle_2$ can adjust to -1 to minimize the dilaton potential. Clearly, in order for this scenario to work the two running couplings $g_1(\mu), g_2(\mu)$ should become strong at roughly the same scale, M_I. Also, $M_I \ll M_{\rm Pl}$ is desired in order to generate a large hierarchy. Thus the couplings should be slightly different at $M_{\rm Pl}$, and should have slightly different β-functions.

Why should $g_1(M_{\rm Pl})$ differ from $g_2(M_{\rm Pl})$? Each tree-level gauge coupling in string theory is given by $g_i = g_s / \sqrt{k_i}$, where the positive integer k_i is the level of the Kac-Moody algebra of world-sheet currents $J^a(z)$ that generate the gauge symmetry G_i.[20] At one-loop, g_i also depends on the spectrum of charged, massive string excitations that have been integrated out to obtain $\mathcal{L}_{\text{eff}}$. The tree-level plus one-loop result is

$$\frac{8\pi^2}{g_i^2(M_{\text{Pl}})} = \frac{8\pi^2 k_i}{g_s^2} + \tfrac{1}{2}\Delta_i \,, \tag{11}$$

where Δ_i is given by an integral over the modular domain for the world-sheet torus, which can be computed for "exactly solvable" 4d string vacua.[21] The identification $4\pi/g_s^2 \equiv \text{Re}S$ means that the S-dependence of holomorphic expressions like the superpotential $W(S)$ are obtained by the replacement

$$\frac{8\pi^2}{g_i^2(M_{\text{Pl}})} \rightarrow 2\pi k_i S + \tfrac{1}{2}\Delta_i + O(S^{-1}). \tag{12}$$

It is also important to know the group-dependence of the constant prefactor C_G in the gaugino condensate (7). One way to determine the constant C_N for $SU(N)$ is to use induction on N.[22] Skipping all the details, the result is $C_N = (C_{N-1})^{1-1/N}$. The large N behavior of this result (which turns out to be the regime of interest) is $C_N \rightarrow$ constant. This behavior can be checked by summing planar diagrams with $g^2 \sim 1/N$ to get $\langle\lambda\lambda\rangle \sim N\, f(g^2 N)$. Using also the renormalization-group-invariant expression (7), with $b_0 = 3N$, $b_1 = 6N^2$ for $SU(N)$, one does get $C_N \approx$ constant.

The dilaton superpotential induced by gaugino condensation is $W(S) = b_0 \langle\lambda\lambda\rangle$. (This equation is related to the trace/axial anomaly equation, $T_\mu^\mu + i\partial \cdot J_A = b_0(F_{\mu\nu}^2 + iF\tilde{F})$, by a supersymmetry transformation; both equations follow from conservation of the super-stress-tensor.) With the replacement (12), the superpotential generated by the hidden sector gauge group $SU(N_1)_{k_1} \times SU(N_2)_{k_2}$ is

$$W_{\text{tot}}(S) = aM^3 S \left[N_1 k_1 e^{-\Delta_1/2N_1} e^{-2\pi\frac{k_1}{N_1}S} - N_2 k_2 e^{-\Delta_2/2N_2} e^{-2\pi\frac{k_2}{N_2}S} \right] , \tag{13}$$

where the numerical constant a has not been computed. Ignoring gravitational corrections, the potential for S is just $V(S,\bar{S}) = F_S^2 = |\partial_S W|^2$ and is minimized (at $V = 0$) by setting $F_S = \partial_S W = 0$, or

$$S = \frac{1}{2\pi}\left(\frac{k_1}{N_1} - \frac{k_2}{N_2}\right)^{-1}\left[\ln\left(\frac{k_1^2}{k_2^2}\right) + \frac{\Delta_2}{2N_2} - \frac{\Delta_1}{2N_1}\right]\left[1 + O\left(\frac{k_1}{N_1} - \frac{k_2}{N_2}\right)\right] . \quad (14)$$

Here SUSY is unbroken in the flat limit (since $F_S = 0$), but should be broken by gravitational corrections. To confirm this expectation, one should properly derive the effective supergravity theory for the dilaton superfield (and possibly other superfields such as the moduli T_i) interacting with a composite chiral superfield that represents the gaugino condensate, following ref. 23; this remains to be done. A 'naive' approach[16] is to simply replace $\lambda\lambda$ by its VEV in the classical supergravity Lagrangian[24] and to leave $W = 0$; using this approach here one would conclude that SUSY is unbroken. However, another approach[15] is to incorporate the effects of the gaugino condensate by substituting the effective superpotential (here, $W_{\text{tot}}(S)$ from eq. (13)) into the supergravity Lagrangian; in this case the result now depends on properties of the other massless chiral superfields (T_i, etc.). If there is only one such field, with a no-scale Kähler potential[25] as suggested by dimensional reduction of tree-level ten-dimensional string theory[26], then one finds (just as in ref. 15) that SUSY is broken, with vanishing cosmological constant in this approximation. The scale of supersymmetry breaking is

$$M_{\text{SUSY}} \sim \langle W_{\text{tot}}\rangle \sim M_I^3/M_{\text{Pl}}^2 , \quad (15)$$

where $M_I^3 = \langle\lambda\lambda\rangle$. It is clearly important to establish which approach (if either) is correct in the present circumstances.

Assume henceforth that SUSY is broken at the scale (15). For supersymmetry to explain the M_W/M_{Pl} hierarchy, M_{SUSY} should be around a TeV, or $M_I \sim 5 \times 10^{13}$ GeV. The desired value of $S = (\alpha_{\text{GUT}}(M_{\text{Pl}}))^{-1}$ depends somewhat on whether there are additional, light fields transforming under the standard model gauge group, but roughly $S \sim 30$ to 40 is desired (assuming level 1 for the standard model Kac-Moody algebras).

To get a sufficiently small M_I and large S, the parameters N_i, k_i have to be "discretely fine-tuned" to make $\frac{k_1}{N_1} - \frac{k_2}{N_2} \ll 1$. (The fine-tuning circumvents the argument of ref. 19 for a strongly-coupled vacuum.) However, in string theory the degree of possible fine-tuning is restricted by the total Virasoro central charge available. Using the Sugawara formula, and taking the standard model to be level 1, gives

$$c_{\text{hidden}} \equiv c_{G_1} + c_{G_2} = \frac{d_{G_1} k_1}{k_1 + C(G_1)} + \frac{d_{G_2} k_2}{k_2 + C(G_2)} \leq 18. \tag{16}$$

This restriction makes it difficult to accommodate higher-level Kac-Moody algebras ($k > 1$) in this scenario. Finally, in the absence of a specific 4d string vacuum, a guess must be made for Δ_1 and Δ_2. Here it is assumed that $\Delta_1/2N_1 = -3/2$, $\Delta_2/2N_2 = +3/2$. Also $a \approx 1$ is assumed.

The "best" results obtained in this scenario so far are for $SU(9)_1 \times SU(10)_1$ ($c_{\text{hidden}} = 17$):

$$S \approx 43, \qquad M_I \sim 5 \times 10^{14} \text{ GeV}. \tag{17}$$

For $SU(8)_1 \times SU(9)_1$ ($c_{\text{hidden}} = 15$) one gets:

$$S \approx 34, \qquad M_I \sim 1 \times 10^{15} \text{ GeV}. \tag{18}$$

And the best result incorporating a level-2 group, $SU(9)_2 \times SU(4)_1$ ($c_{\text{hidden}} = 17.5$), seems to be untenable:

$$S \approx 9, \qquad M_I \sim 9 \times 10^{16} \text{ GeV}. \tag{19}$$

It is interesting that very few choices of $G_1 \times G_2$ (if any!) can give a sufficiently small scale M_I.

Obviously much remains to be done in examining the details of the scenario: Attempts should be made to build actual string models, and to evaluate Δ_i for them; then the gravitino mass can be calculated, as well as the communication of SUSY breaking to the observable sector. There is always the caveat that the above results may be washed out by stronger, "stringy" nonperturbative effects. If one could somehow take the limit $N_1, N_2 \to \infty$ while holding any other non-perturbative effects fixed — and assuming the latter effects to behave like $\exp(-\text{const.}/g_s^2)$ — then the gaugino condensation effects would dominate. Unfortunately no such limit exists. If this problem is neglected, however, a mechanism has been identified for fixing the dilaton VEV at a large value (weak coupling) and perhaps breaking supersymmetry at a phenomenologically interesting mass scale.

REFERENCES

1. L. Dixon, V.S. Kaplunovsky, J. Louis and M. Peskin, to appear.

2. E. Fradkin and A.A. Tseytlin, *Phys. Lett.* **158B** (1985), 316, *Phys. Lett.* **160B** (1985), 64, *Nucl. Phys.* **B261** (1986), 1.

3. E. Witten, *Phys. Lett.* **149B** (1984), 351.

4. L. Dixon and J. Harvey, *Nucl. Phys.* **B274** (1986), 93;
   L. Alvarez-Gaumé, P. Ginsparg, G. Moore and C. Vafa, *Phys. Lett.* **171B** (1986), 155.

5. N. Seiberg and E. Witten, *Nucl. Phys.* **B276** (1986), 272;
   H. Kawai, D.C. Lewellen and S.-H.H. Tye, *Phys. Rev.* **D34** (1986), 3794.

6. W. Fischler and L. Susskind, *Phys. Lett.* **171B** (1986), 383, *Phys. Lett.* **173B** (1986), 262.

7. R. Rohm, *Nucl. Phys.* **B237** (1984), 553.

8. P. Ginsparg and C. Vafa, *Nucl. Phys.* **B289** (1987), 414;
   V.P. Nair, A. Shapere, A. Strominger and F. Wilczek, *Nucl. Phys.* **B287** (1987), 402;
   H. Itoyama and T.R. Taylor, *Phys. Lett.* **186B** (1987), 129.

9. M. Dine and N. Seiberg, *Phys. Rev. Lett.* **57** (1986), 2625.

10. M. Dine, N. Seiberg and E. Witten, *Nucl. Phys.* **B289** (1987), 589.

11. W. Fischler, H.P. Nilles, J. Polchinski, S. Raby and L. Susskind, *Phys. Rev. Lett.* **47** (1981), 757.

12. J. Atick, L. Dixon and A. Sen, *Nucl. Phys.* **B292** (1987), 109;
    M. Dine, I. Ichinose and N. Seiberg, *Nucl. Phys.* **B293** (1987), 253.

13. L. Dixon and V.S. Kaplunovsky, unpublished;
    S. Cecotti, S. Ferrara and L. Girardello, *Nucl. Phys.* **B294** (1987), 537.

14. J.P. Derendinger, L.E. Ibáñez, and H.P. Nilles, *Phys. Lett.* **155B** (1985), 65.

15. M. Dine, R. Rohm, N. Seiberg and E. Witten, *Phys. Lett.* **156B** (1985), 55.

16. S. Ferrara, L. Girardello and H.P. Nilles, *Phys. Lett.* **125B** (1983), 457.

17. R. Rohm and E. Witten, *Ann. Phys.* **170** (1996), 454.

18. G.G. Ross, *Phys. Lett.* **211B** (1988), 315.

19. M. Dine and N. Seiberg, in *Unified String Theories*, eds. M. Green and D. Gross (World Scientific, 1986), *Phys. Lett.* **162B** (1985), 299.

20. P. Ginsparg, *Phys. Lett.* **197B** (1987), 139.

21. V.S. Kaplunovsky, *Nucl. Phys.* **B307** (1988), 145.

22. M.A. Shifman and A.I. Vainshtein, *Nucl. Phys.* **B296** (1988), 445.

23. G. Veneziano and S. Yankielowicz, *Phys. Lett.* **113B** (1982), 231.

24. E. Cremmer, S. Ferrara, L. Girardello, and A. van Proeyen, *Nucl. Phys.* **B212** (1983), 413.

25. E. Cremmer, S. Ferrara, C. Kounnas, and D.V. Nanopoulos, *Phys. Lett.* **133B** (1983), 61.

26. E. Witten, *Phys. Lett.* **155B** (1985), 151.

TOPICS ON SSC*
(Super-String Compactifications)

Fernando Quevedo
Theoretical Division, MS-B285
Los Alamos National Laboratory
Los Alamos, New Mexico 87545
USA

ABSTRACT

A brief overview is presented for recent developments on 4-d strings. In particular, symmetric and asymmetric twisted $N = 2$ coset models and models for which the gauge group comes from a high level Kac-Moody algebra are described and their possible phenomenological implications are discussed.

1. INTRODUCTION

The general motivation for the study of string theory in four-dimensions is to try to find any connection that may exist between string theory and our low-energy world. So far the only approach at hand in this direction is to study the large amount of string vacua, which are given by a subclass of 2-D conformal field theories [1]. In the lack of a principle to select any of these (or other) vacua, the only alternative we have is to look for vacua which have interesting phenomenological perspectives, such as being 4-dimensional, chiral with a gauge group which contains and can be broken to the standard model one, a reasonable number of families (3), etc.

As consistency conditions we have to impose conformal invariance, (which for heterotic strings imply representations of the Virasoro algebra with central charges $c = 9$ and $c = 22$ for right and left movers respectively), modular invariance and unitarity. Also part of the theory should furnish a representation of a Kac-Moody algebra since that is the way gauge symmetries appear in closed strings. The only consistent chiral supersymmetric models have $N = 1$ so we can restrict to exactly this case. This can be translated into the 2-D requirement that the right moving CFT with central charge c=9, should be $N = 2$ supersymmetric.

After imposing the above phenomenological and consistency constraints, we can say that there are two general classes of solutions. One is coming from twisted tensor products of free fermions and bosons. Twisted free bosons can have an interpretation as compactification on geometrical objects denoted orbifolds. The second class of solutions is called coset models [2]. We will first review the space of orbifold vacua and then describe the $N = 2$ coset models for which we obtain a similar understanding. In particular we generalize the idea of asymmetric orbifolds to twisted coset models [3] and find surprisingly that in this approach all of these asymmetric models can be understood as symmetric ones in the presence of discrete torsion. Finally we

---

* Invited talk presented at the APS Division of Particles and Fields Conference, Jan. 3-6, 1990, Houston, Texas

will discuss models for which the gauge group is realized as a high level Kac-Moody algebra and show how they fit in the previous general picture of the space of models [4]. We find a very simple field theoretical interpretation for these models in terms of flat directions in a level-one model. We discuss at the end general constraints on the levels of the standard model groups based on the renormalization group equations and some possible phenomenological properties of these models are discussed.

2.ORBIFOLDS

Being aware of the inmense amount of four-dimensional string models, it is instructive to look for some order and try to have a general view of the space of vacua. Let us describe the space of orbifold vacua first.

The $(c = 22)_L$ left-moving CFT splits into a $(c = 6)_L$ CFT consisting of six (twisted) free bosons and a $(c = 16)_L$ one which can be made out of 32 free fermions or 16 free bosons and which modular invariance restricts to be the Kac-Moody algebra of $E_8 \times E_8$ or $SO(32)$. The $(c = 9)_R$ right-moving CFT consists of 6 bosons together with 6 fermions furnishing a $N = 2$ representation. The six right-moving bosons joint with the six left-moving ones to provide a geometrical interpretation as coordinates of a six-dimensional torus.

Since a torus can be defined as the quotient of R^6 by a lattice Λ, choosing twisted boundary conditions defines an orbifold. This twisting has to correspond to the discrete symmetries of the lattice Λ and it is called the point group P. Let us now describe briefly how to obtain these orbifold vacua.

(i) Choose a twist P and the lattice Λ with P an automorphism of Λ $(P\Lambda = \Lambda)$.

(ii) Embed P on the gauge degrees of freedom by a homomorphism $P \to Q$ where Q acts on the 16-D lattice of say $E_8 \times E_8$.

(iii) For $Z_N \times Z_M$ models, the partition function at one-loop has some arbitrary phases which can not be fixed by all-loops modular invariance, and so each choice of these quantized phases, known as discrete torsion, generates a new model.

(iv) We can consider Wilson lines, that is embed the whole orbifold space group (twists and shifts) on the gauge degrees of freedom by an extra shift on the 16-D lattice.

(v) The point group can be asymmetric i.e. different twists for the left and right movers generating what is known as asymmetric orbifolds $(P_L \neq P_R)$. These constructions have been claimed to provide examples for which there is not geometrical interpretation in terms of compactifications of the 10-D theory.

(vi) Finally for each of the models constructed as above, there can be continuos deformations of them which preserve conformal invariance and thus are valid vacua. In terms of the 4-D effective action they can be found as flat directions to all orders for some massless fields in the theory (moduli).

Following steps (i)-(vi) an extremely large class of models can be generated and most of them can be understood in terms of twisted boundary conditions and background values for the fields in the theory (metric(choice of torii), antisymmetric tensor (discrete torsion), gauge fields (Wilson lines), moduli (flat directions)). This has provided us with a good understanding of the space of orbifold vacua and also allowed us to explicitly construct phenomenologically interesting models, including 3 families and standard model group. We will see next that a similar understanding can be achieved for the coset models.

2.COSET MODELS

There is a simple and general approach to construct consistent 4-D string models due to Gepner [5]. The idea is to split the $(c = 22)_L$ CFT into a direct product of $(c = 9)_L \otimes (c = 13)_L$ and setting $(c = 9)_L \equiv (c = 9)_R$ defining a model with $(2,2)$ worldsheet supersymmetry. The extra $c = 13$ gives $E_8 \times SO(10)$ as required by modular invariance. But now the $c = 9$ left and right theories can be in principle arbitrary and modular invariance is easy to achieve because of the above left-right identification. So we can say that the 4-D theories are parametrized by the different $c = 9$ CFT's. If this theory is made out of free bosons or fermions we reproduce the orbifold models mentioned before. However there is another known way to construct CFT's and that is the coset approach of Goddard-Kent and Olive.

We are thus interested in finding representations of the $N = 2$ superconformal algebra with $c = 9$. Given a representation of a Kac-Moody algebra:

$$J^A(z)J^B(w) \sim \frac{k}{2(z-w)^2}\delta^{AB} + \frac{f^{ABC}J_C(w)}{(z-w)} + ... \tag{1}$$

(f^{ABC} are the structure constants of the group G defining the algebra and k is the level of the algebra), we can construct a representation of the Virasoro algebra (CFT) with stress energy tensor given by $T(z) = \frac{1}{(k+C)}:J^A(z)J^A(z):$. Where C is the Casimir in the adjoint representation. The central charge is given by :

$$c = \frac{k\, dimG}{(k+C)} \tag{2}$$

This procedure was generalized for the coset G/H by GKO for which $T_{G/H} = T_G - T_H$ and $c_{G/H} = c_G - c_H$. The supersymmetric version of this construction was studied by Kazama and Suzuki [2] and found that $N = 1$ G/H is equivalent to an $N = 0$, $(G \times SO(n))/H$ with $n = dimG/H$, and that the $N = 1$ case can be extended to $N = 2$ if the manifold G/H is Kahler (for rankH = rankG). A subclass of these manifolds, known as Hermitian Symmetrric spaces are completely classified. The simplest class of these models is the one based in the coset $SU(2)/U(1)$ which corresponds to the $N = 2$ minimal models with central charge $c = \frac{3k}{k+2}$. In principle this construction looks far richer than the orbifold models because in that case we only had two basic conformal blocks to condstruct the models i.e. free fermions and free bosons. However in the coset models there are infinite series of models that can serve as blocks to make $c = 9$ string vacua. However we have only found of the order of 300 nonequivalent $(2,2)$ models.

Being $(2,2)$ models the gauge group contains always E_6 and the number of generations has been found to be too big (> 12), so none of these models look interesting for phenomenology. Nevertheless, it is known that each of these models has discrete symmetries Z_{k+C} and then the same procedure sketched in the previous chapter to generate models can be followed and again billions of new models can be constructed with different gauge groups and number of families. Of particular interest is to see if it is also possible to construct asymmetric coset models which are consistent. The answer is yes but surprisingly all of them can be understood as symmetric models with nontrivial discrete torsion. This can be very interesting because it suppports the idea that all string vacua are just compactifications of the same theory with different choices of background fields, in precisely the same way as the toroidal compactifications were understood. Many of the orbifold models can be reproduced in this approach, in particular the asymmetric Z_3 orbifold can be constructed and therefore it is equivalent to a symmetric compactification, something that is not possible to see in the orbifold approach.

3.HIGH-LEVEL KAC-MOODY MODELS

We know that the gauge group in closed strings comes from a Kac-Moody algebra (1). The level k has physical significance. The 4-D gauge couplings depend on the level as $g_i = g/\sqrt{k_i}$ where g is the string coupling and the index i labels the different group factors. Also the allowed representations of the gauge group depend on the level. For instance, if the level is one, in the standard model we could only have doublets of $SU(2)$ and triplets of $SU(3)$. Most of the models constructed so far have level one, mostly because the original $E_8 \times E_8$ has level one. There are however higher level models which fit well in the general picture of the space of vacua given in the previous chapters. Rather than presenting explicit string examples we will show a simple field theoretical example which gives the idea of the construction.

Start with a model with symmetry $SU(2) \times SU(2)$. This symmetry can be broken to $SU(2)$ diagonal by a vev of a field transforming as a (2,2) under the group. We can easily check that if there is another (2,2) field in the theory, it will transform under the unbroken $SU(2)$ as a triplet plus a singlet. So if the original symmetry came from a level one model we can see that the new one has to come with at least level two since higher representations are now present. Also if the original gauge coupling was g, the new one will be $g/\sqrt{2}$ reflecting that it is a level two model. In general we can obtain a level n model from products of n factor groups breaking to a common subgroup.

Knowing that there are higher level models it means that there are additional free parameters in the gauge theories coming from the strings. We would like to see if there are some constraints on those parameters. Conformal invariance implies that $c \leq 22$ which from equation (2) means that if dimension of the group exceeds 22, the levels cannot be high ($k \leq 4$ for E_6 and $k \leq 7$ for $SO(10)$). This is not a constraint for the standard model. The contribution to a particle mass comes from the conformal dimension h,

$$h(R) = \frac{C(R)}{k + G} \tag{3}$$

where $C(R)$ is the Casimir of the representation R, for $U(1)$ groups, $h = q^2/k$. So the existence of the right handed electron implies that the $U(1)$ level should be $k \geq 1$. Also the condition for the nonexistence of fractional charged particles can be written in terms of the levels of $SU(3) \times SU(2) \times U(1)$ as $3k_1 + 3k_2 + 4k_3 = 0mod12$

The Weinberg angle at the unification scale is [6]

$$sin^2\theta_W = \frac{k_2}{k_1 + k_2} \tag{4}$$

and evolve according to the renormalization group equations. Knowing the value at the weak scale and also assuming perturbative unification, we can find general constraints on the ratios of the three standard model levels (they cannot differ by more than a factor of 8). Also we find that very few representations are allowed besides the three families, so only one octet of $SU(3)$ and one triplet of $SU(2)$ are allowed, independent of the levels.

Finally let us discuss the possible phenomenological implications from higher level models. The first one is that they open a place for GUT's in string theory, since we need $k \geq 1$ to obtain adjoints. Note that if an $SU(5)$ model is broken by an adjoint (**24**) and to still be conformal invariant this direction should be flat to all orders, then decomposing the **24** under $SU(3) \times SU(2)$ we can see that the $(3,2) + (\bar{3},2)$ are eaten by the supersymmetric Higgs effect but the $(8,1) + (1,3) + (1,1)$ survive at low energies (notice that they are allowed by the renormalization group analysis above.) This seems to be a generic feature of string GUT's. Also higher dimensional representations were proposed in GUT's for the missing partner mechanism to split doublets and triplets and the see-saw mechanism for neutrino masses. They both require $k \geq 5$ so for E_6 the see saw mechanism cannot work.

Another interesting application could be the simultaneous solution of the proton stability and neutrino masses problem by a single field. To illustrate that let us suppose that there is a $U(1)_{B-L}$ at high energies, if it is broken by a field with charge 2, there will be a remaining Z_2 symmetry which still prevents the proton to decay, this same field can couple to the right handed neutrino and give it a large mass as required by the see-saw mechanism. It is crucial that the charge of this field is a multiple of the fundamental one, so higher levels are usually required for that, although in this example only a high $U(1)$ level is actually required. Finally the constraint for integer electric charges sets an strong constraint on the levels of the models, which combined with a reasonable Weinberg angle at the unification scale implies that we should consider only high level models[6]. However nothing prevents the fractional charge particles to be very massive.

REFERENCES

[1]. See for example, B. Schellekens, Superstring Construction, North-Holland (1989).

[2]. P. Goddard, A. Kent and D. Olive, Phys. Lett. B 152B (1985) 88; Y. Kazama and H. Suzuki, Nucl. Phys. B321 (1989) 232.

[3]. A. Font, L.E. Ibáñez and F. Quevedo preprint CERN-TH 5577 (1989) and references therein.

[4]. D. Lewellen, preprint SLAC-PUB-5023 (1989); Font, L.E. Ibáñez and F. Quevedo preprint CERN-TH 5666, LA-UR-90-1089 (1990) and references therein.

[5]. D. Gepner, Nucl. Phys. B296 (1988) 757.

[6]. A.N. Schellekens, preprint CERN-TH 5593/89 (1989).

RECENT PROGRESS IN 4-D STRING
MODEL BUILDING

JOHN S. HAGELIN
Department of Physics
Maharishi International University
Fairfield, Iowa 52556

Abstract

We present a recently revamped three-generation Flipped $SU(5) \times U(1)$ string model with the following properties:

* The complete massless spectrum is derived and shown to be free of all gauge and mixed anomalies apart from a single anomalous $U(1)$.
* The imaginary part of the dilaton supermultiplet is eaten by the anomalous $U(1)$ gauge boson, and the corresponding D-term is cancelled by large v.e.v.'s for singlet fields that break surplus $U(1)$ gauge factors, leaving a supersymmetric vacuum with an $SU(5) \times U(1)$ visible gauge group and an $SO(10) \times SO(6)$ hidden gauge group.
* There are sufficient Higgs multiplets to break the visible gauge symmetry down to the Standard Model in an essentially unique way.
* All trilinear superpotential couplings have been calculated and there are in particular some giving m_t, m_b, $m_\tau \neq 0$.
* A renormalization group analysis shows that $m_t < 190$ GeV and $m_b \simeq 3m_\tau$.
* Spontaneously broken extra U(1) factors give potentially large, generation-dependent D-term contributions to light scalar masses.
* Light Higgs doublets are split automatically from heavy Higgs triplets, leaving no residual dimension-five operators for baryon decay, and the baryon lifetime $\tau_B \sim 2 \times 10^{34\pm2}$y.
* There are no tree-level flavor-changing neutral currents, but $\mu \rightarrow e\gamma$ may occur at a detectable level: $B(\mu \rightarrow e\gamma) \sim 10^{-11}$ to 10^{-14}.

There is a regrettable schism in particle physics today. On the one hand, many theorists are convinced that string is the Theory of Everything (TOE), and strive hard to understand better its formalism without worrying about its physical signatures. On the other hand, experiment is confined dismally within the Standard Model, and many phenomenologists strive hard to constrain its parameters without worrying about its origin. Surprisingly little effort is directed towards deriving the Standard Model from some variant of string theory, or to seeking some other distinctive signature of the TOE. Many would doubtless consider any such effort premature and foolhardy until we understand string better. However, if we wait around for someone to tell us how non-perturbative string effects determine the correct vacuum state, we will have a long wait. Indeed, since we now have many tools–geometric[1] and algebraic,[2] bosonic[3] and fermionic[4,5]–for constructing different string models, surely we should be using them to look for candidate vacua that are phenomenologically realistic.

It may well be that the Standard Model can be derived directly from string without invoking any field-theoretical intermediate scale of gauge symmetry breaking. However, so far all phenomenological string models[6-8] have had four-dimensional groups larger than $SU(3) \times SU(2) \times U(1)$, though often only with extra $U(1)$ factors.[6] We think[8] that if one is to extend the Standard Model gauge group, it is interesting and desirable to embed it in a Grand Unified Theory (GUT), i.e., a simple non-Abelian group containing the $SU(3)$, $SU(2)$ and (maybe) $U(1)$ factors of the Standard Model. Such a framework would combine the physical advantages of GUTs (e.g., slow baryon decay, cosmological baryosynthesis, small neutrino masses, etc.) with the well-known benefits of the string. However, there is an obstacle to such a program that we have emphasized previously[8]: in general, GUTs require Higgs fields in adjoint representations to break the gauge symmetry down to the Standard Model, and these do not exist in string theories with N=1 supersymmetry and/or chiral fermions that have level $K = 1$ Kac-Moody algebras.

The only viable GUT that does not require adjoint Higgses is Flipped $SU(5) \times U(1)$,[8] which also possesses[9] other advantages such as natural Higgs doublet-triplet splitting, a seesaw neutrino mass matrix, no problematic fermion mass relations, and no troublesome $d = 5$ proton decay operators.* We have in the past constructed realistic Flipped $SU(5) \times U(1)$ models in field theory and indicated how they could be derived from manifold compactifications[11] or the fermionic formulation of string theory.[12,13] However, although our efforts using free world-sheet fermions[4,5] encouraged us that we were on the right track, our previous models[12,13] suffered from various technical problems and lacunae, including the following. 1) The spectrum of light states was not correctly stated.[14] 2) Certain desired Yukawa couplings, such as that responsible for the top quark mass, were absent.[15] 3) The hidden sector was not worked out in detail, and hence the absence of gauge anomalies in the effective low-energy field theory could not be checked. 4) The "eating" of the dilaton[16] was not discussed in detail. 5) Quark mass mixing and the weak interaction angles were not completely satisfactory. We have recently revamped[17] our Flipped $SU(5) \times U(1)$ model to remedy the above defects.

The explicit derivation of our four-dimensional string model is presented in ref. 17. It is constructed using the free-fermionic formulation[4] with eight basis vectors of boundary conditions for the world-sheet fermions. The observable gauge group is $SU(5) \times U(1)$, and there is a hidden sector gauge group $SO(10) \times SO(6)$ with vector-like matter representations. In addition, there are four U(1)'s under which both observable sector and hidden sector states transform.

One linear combination of these $U(1)$'s is anomalous, while the three orthogonal combinations are free of all gauge and mixed gravitational anomalies. The anomalous $U(1)_A$ is broken by the Dine-Seiberg-Witten mechanism,[16] in which a potentially large Fayet-Iliopoulos D-term is generated by the v.e.v. of the dilaton field. Such a D-term would, in general, break supersymmetry and destabilize the string vacuum unless, as in our case,[17] there is a direction in the scalar potential $\phi = \sum_i \alpha_i \phi_i$ which is F-flat and also D-flat with respect to the non-anomalous gauge symmetries and in which $\sum_i Q_i^A |\alpha_i|^2 < 0$. If such a direction exists, it will acquire a v.e.v., cancelling the anomalous D-term, restoring supersymmetry[18] and stabi-

---

* Other proposed flipped unification models[10] have not been shown to share all these desirable features.

lizing the vacuum. Since the fields corresponding to such a flat direction typically also carry charges under the non-anomalous gauge symmetries, this mechanism for cancelling the anomalous D-term will in general break some of these symmetries spontaneously.[12,13]

In our string model, at least three and typically all four of the surplus $U(1)$'s are broken by this mechanism. This implies that the superstring vacuum is not really $[SU(5) \times U(1)] \times U(1)^4$, but actually minimal Flipped $SU(5) \times U(1)$. Nevertheless, the extra $U(1)$'s constrain the allowed Yukawa couplings, forbidding in particular trilinear couplings that could give masses to the first two generations of quarks and leptons. This offers a viable framework[17] for understanding the hierarchical pattern of quark and lepton masses, which would be difficult to explain if there were trilinear Higgs couplings of similar magnitude for all three generations.

In addition to three standard Flipped $SU(5) \times U(1)$ matter generations, the light spectrum includes the necessary Higgs representations to break $SU(5) \times U(1)$ down to the Standard Model. In the process, the required Weinberg-Salam Higgs doublets are split automatically from heavy Higgs triplets, leaving no residual dimension-five operators which could lead to rapid baryon decay. All trilinear superpotential couplings have been explicitly calculated,[17] and there are couplings giving masses to the third-generation quarks and leptons t, b, τ. We also recover the successful GUT prediction $m_b = m_\tau$ at $Q^2 = m_{GUT}^2$ (which is not a prediction of minimal $SU(5) \times U(1)$ at the field-theory level), and a renormalization group analysis finds $m_b \simeq 3m_\tau$ and $m_t < 190$ GeV.[17] Plausible assumptions about non-renormalizable couplings yield hierarchical masses for the first- and second-generation quarks and leptons and acceptable CKM-type mixing. A detailed investigation of non-renormalizable terms in the free-fermionic string formulation is underway.

One interesting consequence of the fermion mass terms[17] is that the usual quark-lepton kinship rules are violated. The heaviest quarks and leptons belong to different $SU(5)$ representations. One potentially dramatic consequence is that the baryon decay modes $p \to \mu^+\pi^0$ or $\bar{\nu}\pi^0$, $n \to \mu^+\pi^-$ or $\bar{\nu}\pi^0$ could dominate over $p \to e^+\pi^0$, $n \to e^+\pi^-$.[17] The baryon lifetime should be similar to the estimate of ref. 9, namely $2 \times 10^{34\pm2}$y. In addition, the lepton flavor-changing decay $\mu \to e\gamma$ may be greatly enhanced[19] relative to conventional supersymmetric GUTs. In $SU(5)$, for example, $\mu \to e\gamma$ results from generation-changing slepton masses $\delta\tilde{m}_{\mu e}^2$ induced by renormalization effects between m_P and m_{GUT}.[20] These have the form $\delta\tilde{m}_{\mu e}^2 \propto (K_{32}^*\lambda_t^2 K_{31})\ln(m_{GUT}/m_P)$, where the combination $K_{32}^*K_{31}$ of Cabibbo-Kobayashi-Maskawa mixing angles is known experimentally to be $< 10^{-3}$, so that $B(\mu \to e\gamma) \sim 10^{-14}$ to 10^{-17} in conventional supersymmetric $SU(5)$. However, in our Flipped $SU(5) \times U(1)$ string model,[17] since the usual quark-lepton kinship rules break down, one of $K_{32,31}$ is expected to be $O(1)$, while the other is supressed by a single Cabibbo-like factor $\sin\Phi$. The branching ratio is therefore enhanced relative to $SU(5)$ by the ratio $(\sin\Phi/10^{-3})^2 \sim 10^3$, giving $B(\mu \to e\gamma) \sim 10^{-11}$ to 10^{-14} in the range accessible to the MEGA experiment.[19]

One final point of phenomenological importance is the D-term contributions to scalar masses (Higgs, squarks and sleptons) induced by spontaneous breaking of $SU(5) \times U(1)$ and of the extra $U(1)$ gauge factors. It was recently shown[21] that these D-terms leave their distinctive thumbprint on the low-energy scalar spectrum, providing a critical test of the presence of gauge structure beyond the Standard Model which extends to arbitrarily high mass scales. The scalar spectrum thereby affords a viable means of discriminating among various competing string models. In this respect our $SU(5) \times U(1)$ model is particularly distinctive, since the spontaneously broken extra $U(1)$'s give D-term contributions that are generation-dependent, causing significant departures from the usual expectation of universal squark and slepton masses.[21]

In this talk we have described an improved version[17] of our three-generation Flipped $SU(5) \times U(1)$ string model[12,13] in which there are non-vanishing trilinear superpotential terms giving $m_t, m_b, m_\tau \neq 0$. We recover the successful GUT relation $m_b \simeq 3m_\tau$ and find that $m_t < 190$ GeV after renormalization. The low-energy spectrum is completely free of gauge and mixed gravitational anomalies, and all surplus $U(1)$ gauge factors are automatically broken by the Fayet-Iliopoulos D-term generated by the Dine-Seiberg-Witten[16] mechanism. Baryon decay is automatically suppressed so that $\tau_B \simeq 2 \times 10^{34\pm2}$y,[9] with the likelihood of unusual decay modes such as $\bar{\nu}\pi$ or $\mu^+\pi$ dominating. There are no tree-level FCNC's, but slepton mixing is likely to induce observable $\mu \to e\gamma$ decay at the one-loop level. Plausible hypotheses about non-renormalizable superpotential terms yield the minimal Flipped $SU(5) \times U(1)$ matter spectrum with non-zero light fermion masses and CKM flavor mixing. We believe we have identified a pheomenologically viable string vacuum.

This work was supported in part by the National Science Foundation under grant no. PHY-8907816.

REFERENCES

1. P. Candelas, G.T. Horowitz, A. Strominger and E. Witten, *Nucl. Phys.* **B258** (1985) 46; L. Dixon, J.A. Harvey, C. Vafa and E. Witten, *Nucl. Phys.* **B261** (1985) 678 and **B274** (1986) 285.

2. D. Gepner, *Nucl. Phys.* **B290** (1987) 10 and **B296** (1987) 757; *Phys. Lett.* **199B** (1987) 380.

3. K.S. Narain, *Phys. Lett.* **169B** (1986) 41; W. Lerche, D. Lüst and A.N. Schellekens, *Nucl. Phys.* **B287** (1987) 477.

4. I. Antoniadis, C. Bachas, C. Kounnas and P. Windey, *Phys. Lett.* **171B** (1986) 51; I. Antoniadis, C. Bachas and C. Kounnas, *Nucl. Phys.* **B289** (1987) 87; I. Antoniadis and C. Bachas, *Nucl. Phys.* **B298** (1988) 586.

5. H. Kawai, D.C. Lewellen and S.-H.H. Tye, *Phys. Rev. Lett.* **57** (1986) 1832; *Phys. Rev.* **D34** (1986) 3794; *Nucl. Phys.* **B288** (1987) 1; R. Bluhm, L. Dolan and P. Goddard, *Nucl. Phys.* **B309** (1988) 330.

6. L.E. Ibañez, J.E. Kim, H.P. Nilles and F. Quevedo, *Phys. Lett.* **191B** (1987) 282; L.E. Ibañez, J. Mas, H.P. Nilles and F. Quevedo, *Nucl. Phys.* **B301** (1988) 157; A. Font, L.E. Ibañez, H.P. Nilles and F. Quevedo, *Phys. Lett.* **210B** (1988) 101; J.A. Casas, E.K. Katehou and C. Muñoz, *Nucl. Phys.* **B317** (1989) 171; J.A. Casas and C. Muñoz, *Phys. Lett.* **209B** (1988) 214; **212B** (1988) 343; **214B** (1988) 63; D. Bailin, A. Love and S. Thomas, *Phys. Lett.* **194B** (1987) 385; *Nucl. Phys.* **B298** (1988) 75.

7. B. Greene, K.H. Kirklin, P.J. Miron and G.G. Ross, *Phys. Lett.* **180B** (1986) 69; *Nucl. Phys.* **B278** (1986) 668 and **B292** (1987) 606;

R. Arnowitt and P. Nath, *Phys. Rev.* **D39** (1989) 2006; *Phys. Rev. Lett.* **60** (1988) 1817 and **62** (1989) 1437, 2225; Texas A&M University Preprints CTP-TAMU-74/88 (1988) and 36/89 (1989).

8. I. Antoniadis, J. Ellis, J.S. Hagelin and D.V. Nanopoulos, *Phys. Lett.* **194B** (1987) 231.

9. J. Ellis, J.S. Hagelin, S. Kelley and D.V. Nanopoulos, *Nucl. Phys.* **B311** (1988) 1.

10. K. Tamvakis, *Phys. Lett.* **201B** (1988) 95;
J. Rizos and K. Tamvakis, *Phys. Lett.* **212B** (1988) 176;
T.W. Kephart and T.C. Yuan, Vanderbilt University Preprint VAND-TH-89-1 (1989);
C. Panagiotakopoulos, Bartol Preprint BA-89-20 (1989).

11. B. Campbell, J. Ellis, J.S. Hagelin, D.V. Nanopoulos and R. Ticciati, *Phys. Lett.* **198B** (1987) 200.

12. I. Antoniadis, J. Ellis, J.S. Hagelin and D.V. Nanopoulos, *Phys. Lett.* **205B** (1988) 459.

13. I. Antoniadis, J. Ellis, J.S. Hagelin and D.V. Nanopoulos, *Phys. Lett.* **208B** (1988) 209.

14. H. Dreiner and J. Lopez; G. Cleaver, Private communications.

15. D. Bailin, D.C. Dunbar and A. Love, *Phys. Lett.* **219B** (1989) 76.

16. M. Dine, N. Seiberg and E. Witten, *Nucl. Phys.* **B289** (1987) 585.

17. I. Antoniadis, John Ellis, J.S. Hagelin and D.V. Nanopoulos, *Phys. Lett.* **231B** (1989) 65.

18. S. Cecotti, S. Ferrara and M. Villasante, *Int. J. Mod. Phys.* **A2** (1987) 1839.

19. J.S. Hagelin and S. Kelley, Maharishi International University preprint no. MIU-THP-89/34, submitted to Nucl. Phys. B.

20. L.J. Hall, V.A. Kostelecky and S. Raby, *Nucl. Phys.* **B267** (1986) 415.

21. J.S. Hagelin and S. Kelley, Maharishi International University preprint no. MIU-THP-89/51, to appear in Nucl. Phys. B.

LOW ENERGY PREDICTIONS FROM CALABI-YAU
HETEROTIC STRING THEORY[†]

R. ARNOWITT[*]
Center for Theoretical Physics, Department of Physics
Texas A&M University, College Station, TX 77843-4242

PRAN NATH
Department of Physics, Northeastern University
Boston, MA 02115

ABSTRACT

Low energy predictions are given of three generation Calabi-Yau manifolds where E_6 breaks to $[SU(3)]^3$ at compactification, and $[SU(3)]^3$ subsequently breaks to the Standard Model at an intermediate scale. It is shown that if the intermediate scale breaking is triggered by a SUSY soft breaking mass and preserves matter parity, then there always exist four $SU(3) \times SU(2) \times U(1)$ neutral light (electroweak mass) states that couple to Higgs and leptons. These give rise to neutrinos of mass O $(10^{-6}$ eV), O $(1$ eV), and O $(1$ eV) and charged leptons where $m_e/m_\tau =$ O (10^{-3}). Neutrino oscillations also occur with $\nu_{e,\mu} \leftrightarrow \nu_\tau \gg \nu_e \leftrightarrow \nu_\mu$.

1. INTRODUCTION

We consider here properties of the heterotic string[1] compactified on a three generation Calabi-Yau manifold. At compactification, the theory contains an E_6 internal symmetry with massless particles in the 27 and $\overline{27}$ representations. Two such manifolds have been well studied: the $CP^3 \times CP^3/Z_3$ manifold of Tian and Yau[2] and the $CP^3 \times CP^2/Z_3 \times Z_3'$ of Schimmrigk[3]. A unique feature of string theory is that in principle it determines all Yukawa couplings. In practice, however, Yukawas can be calculated only for the simplest symmetric manifolds and then only partially. The non-renormalizable $(27 \times \overline{27})^n$ couplings (which are needed for low energy predictions) are still mostly inaccessible. One can, however, learn a great deal about these models by imposing the simple requirement that the theory reduce to the Standard Model at low energy. (Models that do not do this are, of course, phenomenologically uninteresting.)

For non-simply connected manifolds, flux breaking of E_6 at the compactification scale M_c can occur. If one Z_3 is non-trivially embedded in E_6, then the only flux breakings that preserve the Standard Model are $E_6 \rightarrow [SU(3)]^3$; $SU(6) \times U(1)$; E_6. For the ith 27 generation, the $[SU(3)]^3$ content is

$$27_i = L_i(1,3,\bar{3}) \oplus Q_i(3,\bar{3},1) \oplus Q_i^c(\bar{3},1,3) \tag{1.1}$$

where

$$L = [\ell = (\nu,e)\ ;e^c\ ;H\ ;H'\ ;\nu^c\ ;N] \tag{1.2}$$

$$Q = [q^a = (u^a,d^a)\ ;H_3 \equiv D^a]\ ;Q^c = [u_a^c,d_a^c\ ;H_3' \equiv D_a^c]\ . \tag{1.3}$$

Here ℓ, H, H', q are the lepton, Higgs and quark $SU(2)_L$ doublets, e^c, u_a^c, d_a^c are the lepton and quark conjugate singlets, D^a, D_a^c are $SU(5)$ color triplets of 5 and $\bar{5}$ representations, and ν^c and

---

[†] Research supported in part under NSF Grants Nos. PHY-8907887 and PHY-8706878 and the Texas Advanced Research Program under Grant No. 3043.

[*] Conference Speaker

N are $SU(5)$ singlets. Further dynamical symmetry breaking at an intermediate scale M_I, which preserves the Standard Model, can occur only if the VEVs of N_i and ν_i^c are non-zero. For the $SU(6) \times U(1)$ and E_6 cases, this would leave a residual $SU(5)$ symmetry with no apparent way of breaking it further. But $< N_i >, < \nu_i^c > \neq 0$ does break $[SU(3)]^3$ to the Standard Model. Thus there is a natural way of arriving at the Standard Model[4]:

$$E_6 \xrightarrow[M_c]{\text{flux breaking}} [SU(3)]^3 \xrightarrow[M_I]{< N_i >,\, < \nu_i^c >} SU(3)_C \times SU(2)_L \times U(1)_Y \ . \tag{1.4}$$

The VEV formation at M_I is dynamical and in general requires a $(\text{mass})^2$ to turn negative. The only mass present for the massless 27 and $\overline{27}$ modes is the soft breaking SUSY masses which are of size $m \lesssim O$ (1 TeV). These $(\text{mass})^2$ are, in fact, expected to turn negative at[5] $M_I \gtrsim 10^{16}$ GeV. Then the non-renormalizable $(27\ \overline{27})^n$ interactions will produce $< N_i >$, and $< \nu_i^c >$ VEV growth[4] with $< N_i >, < \nu_i^c > \gtrsim 10^{15}$ GeV provided[6] $n \gtrsim 3, 4$.

One can always make a transformation in generation space such that only $< N_1 >$ is non-zero and all other $< N_i >$ vanish. If also $< \nu_1^c > = 0$, one can similarly arrange the generation labels so that only $< \nu_2^c >$ is non-zero. The vanishing of $< \nu_1^c >$ occurs naturally if N_i and ν_i^c are in different symmetry classes. In fact, this automatically happens if the theory is "matter parity", M_2, invariant. Here M_2 is defined by[7,4] $M_2 = U_z C$ where U_z flips the sign of all $SU(2)_{L,R}$ doublets and C acts only on the generation index and obeys $C^2 = 1$. One has that $U_z N_i = N_i$, $U_z \nu_i^c = -\nu_i^c$ and so M_2 invariance requires $i = 1$ to be a C even state (e.g. $C N_1 = N_1$) and $i = 2$ to be C odd (e.g. $C N_2 = -N_2$).

2. NEW LOW ENERGY STATES

We now prove the following theorem[8]:

Consider any Calabi-Yau string model where

   (i) E_6 breaks to $[SU(3)]^3 \equiv SU(3)_C \times SU(3)_L \times SU(3)_R$ at M_c.

   (ii) $[SU(3)]^3$ breaks to $SU(3)_C \times SU(2)_L \times U(1)_Y$ at intermediate scale M_I triggered by a mass $m \lesssim O$ (1 TeV) whose square turns negative.

   (iii) The VEVs at M_I obey $\sum_i < N_i >< \nu_i^c > = 0$.

Then there are always four new "exotic" light [mass $\lesssim O$ (1 TeV)] chiral multiplets which are $SU(3) \times SU(2) \times U(1)$ neutral. In the basis where only $< N_1 >$ and $< \nu_2^c >$ are non-zero, these states are

$$n_2 = \cos \theta\, N_2 + \sin \theta\, \bar{\nu}_1^c \ ; \quad \bar{n}_2 = \cos \theta\, \bar{N}_2 + \sin \theta\, \nu_1^c$$
$$n_1 = (N_1 + \bar{N}_1)/\sqrt{2} \ ; \quad \hat{\nu}_2^c = (\nu_2^c + \bar{\nu}_2^c)/\sqrt{2} \tag{2.1}$$

where $\tan \theta = < \nu_2^c > / < N_1 >$.

We briefly sketch the proof of the theorem. When $[SU(3)]^3$ breaks to $SU(3) \times SU(2) \times U(1)$ at M_I, 12 vector bosons become massive members of 12 SUSY massive vector multiplets. These multiplets include 12 massive Dirac states (and 12 scalar bosons). The Dirac states are $\psi = \chi_L + P_R \lambda$ where λ are the gaugino partners of the massive vector bosons and χ_L are the Weyl spinor partners of the Goldstone fields. In the following we adopt the short hand notation of $\psi = (\lambda, \chi_L)$.

To find the massive Dirac fields, one looks at the gaugino interactions which produce mass. For $SU(3)_L$ one has

$$\mathcal{L}_L^{\text{mass}} = -i\sqrt{2} g_L \bar{\lambda}_L^\alpha [(T^\alpha)_{\ell'}^\ell \chi_{irL}^{\ell'} < (\phi_i^\dagger)_r^\ell > - \bar{\chi}_{ir}^{\ell'} (T^\alpha)_\ell^{\ell'} < (\bar{\phi}_i^\dagger)_r^\ell >] + h.c. \tag{2.2}$$

where $\ell, r = 1,2,3$ are the $SU(3)_{L,R}$ triplet indices and (χ_{iL}, ϕ_i) are chiral multiplets of the 27_i generation. (Note that $N_i \equiv (\phi_i)_3^3$ and $\nu_i^c \equiv (\phi_i)_2^3$.) Similar mass terms arise from $SU(3)_R$ interactions. If only $< N_1 >$ and $< \nu_2^c >$ are non-zero, one easily finds the following 12 massive Dirac fields:

$$(\lambda_L^{(+)} ; c\ell_1 + sH_2') ; (\lambda_L^{(-)} ; c\bar{\ell}_1 + s\bar{H}_2')$$

$$(\lambda_{1-}^R ; e_2^c) ; (-\lambda_{1+}^R; \bar{e}_2^c) ; (\lambda_{4-}^R ; e_1^c) ; (-\lambda_{4+}^r ; e_1^c) \qquad (2.3a)$$

$$(\lambda_{6-}^R ; c\nu_1^c - s\bar{N}_2) ; (\lambda_{6+}^R ; c\bar{\nu}_1^c - sN_2)$$

$$(\hat{g}_R \lambda_8^R - \hat{g}_L \lambda_8^L) ; (N_1 - \bar{N}_1)/\sqrt{2} ; \frac{1}{2}\hat{g}_R(\sqrt{3}\lambda_3^R - \lambda_8^R) - \hat{g}_L \lambda_8^L ; (\nu_2^c - \bar{\nu}_2^c/\sqrt{2}) \qquad (2.3b)$$

where $\lambda_L^{(+)} = \pm(\lambda_{4\pm}^L, \lambda_{6\pm}^L)$ are doublets, $\lambda_{4\pm} = (\lambda_4 \pm i\lambda_5)/\sqrt{2}$, etc. and $c = \cos\theta$, $s = \sin\theta$, $\tan\theta \equiv < \nu_2^c > / < N_1 >$ and $\hat{g}_{R,L} = g_{R,L}/\sqrt{g_R^2 + g_L^2}$. We see that the four combinations of fields stated in Eq. (2.1) are just the ones orthogonal to the fields entering in Eq. (2.3b). Consequently, the modes of Eq. (2.1) receive no mass from the gauge interactions. One may check that they also receive no mass from the $(27)^3$ and $(\overline{27})^3$ Yukawa interactions. The non-renormalizable $(27\,\overline{27})^n$ interactions (which produce the intermediate scale breaking) give these modes a mass $m \lesssim O\,(1\ \text{TeV})$ where m is the mass that triggers the intermediate scale breaking.

Thus the four modes of Eq. (2.1) remain light i.e. of electroweak size mass.

3. NEW PHYSICS AT LOW ENERGY

The existence of the four new states of Eq. (2.1) depends on relatively weak assumptions, and thus their presence represents a signal for a fairly wide class of Calabi-Yau string models. To see what these signals are, we need the interactions these new particles have with the Standard Model particles. To do this we now become a little more specific about the properties of the manifolds being considered.

We assume now explicitly that M_2 is conserved and write the generation indices as $i = (n, r)$ where $n = C$-even and $r = C$-odd. Below M_I, the mass matrix can be divided into M_2-even ($M^{(e)}$) and M_2-odd ($M^{(0)}$) parts, i.e. $\phi_a M_{ab}^{(e)} \phi_b'$ and $\xi_a M_{ab}^{(0)} \xi_b'$ where $\phi_a \equiv (N_n, \bar{H}_n', \bar{\ell}_r)$, $\phi_b' \equiv (H_n', \bar{H}_n, \ell_r)$ and $\xi_a \equiv (\lambda_L^{(-)}, \ell_n, \bar{H}_r, H_r')$, $\xi_b' \equiv (\lambda_L^{(+)}, \bar{\ell}_n, H_r, \bar{H}_r')$. For three generation manifolds, the index theorem guarantees three light lepton generations ℓ_p, $p = 1,2,3$. We assume here that these lie in the ξ_a states, which is the case for the Tian-Yau and Gepner models. (One may easily consider more general possibilities.) Thus the mass diagonal states of $M_{ab}^{(0)}$ may be written as $\xi_a = \ell_p U_{pa}^\dagger + \eta_\alpha U_{\alpha a}^\dagger$ where $(U_{pa}^\dagger, U_{\alpha a}^\dagger)$ is the unitary matrix used to diagonalize $M_{ab}^{(0)}$ and η_α, ξ_b' are the superheavy [O $(10^{15}$ GeV)] remaining M_2 odd states. Any phenomenologically acceptable model also must have one pair of light Higgs doublets H and H' to complete the breaking of $SU(2) \times U(1)$ at the electroweak scale. With the above assumptions, one may show that these Higgs lie in the M_2 even sector[9] ϕ_a and ϕ_b' i.e. one may write $\phi_a = V_{aH} H + V_{a\alpha} \chi_\alpha$ and $\phi_b' = V_{bH'}' H' + V_{b\alpha}' \chi_\alpha'$ where V and V' diagonalize $M_{ab}^{(e)}$ and χ_α, χ_α' are superheavy.

One may now integrate out all the superheavy fields in the $(27)^3$ and $(\overline{27})^3$ Yukawa interactions to obtain an effective low energy superpotential. One finds[10]

$$W_{\text{eff}} = \left\{ \lambda^{(d)}_{pp'} H' e^c_p \ell_p + \lambda^{(u)}_{pp'} H q_p u^c_{p'} + \lambda^{(d)}_{pp'} H' q_p d^c_{p'} \right\}$$
$$+ [\lambda_p H \ell_p n_2 + \bar{\lambda}_p H \ell_p \bar{n}_2) + (m_1 n_2 \bar{n}_2 + \tfrac{1}{2} m_2 n_2 n_2 + \tfrac{1}{2} m_3 \bar{n}_2 \bar{n}_2)$$
$$+ (m_4 n_1 \hat{\nu}^c_2 + \tfrac{1}{2} m_5 n_1 n_1 + \tfrac{1}{2} m_6 \hat{\nu}^c_2 \hat{\nu}^c_2)] + W_{\text{seesaw}} \ . \tag{3.1}$$

Here $\lambda^{(\ell)}_{pp'}$, etc. and λ_p, $\bar{\lambda}_p$ are related to the $(27)^3$, $(\overline{27})^3$ couplings and the unitary transformations needed to diagonalize $M^{(e)}$ etc.[10] and the masses m_i are $\lesssim O$ (1 TeV). In addition, the $\lambda^{(-)}_L$ gaugino interaction gives rise to the low energy interaction

$$\mathcal{L}_{\text{gaugino}} = g_L \ell_p \gamma^0 [\lambda^{(g)}_{pp'} e^c_{p'} H^\dagger + \frac{1}{\sqrt{2}}(s n_1 - c \hat{\nu}^c_2) \ell^\dagger_{p=1} + H(\lambda n^\dagger_2 + \bar{\lambda} \bar{n}^\dagger_2)] U^\dagger_{p\lambda(-)} + h.c. \tag{3.2}$$

where $s = \sin\theta$, $c = \cos\theta$, and the fields with "$\dagger$" are the scalar components of the corresponding chiral multiplets.

We see that the terms in the brace of Eq. (3.3) is just the SUSY Standard Model superpotential. Eq. (3.4) and the remainder of Eq. (3.3) represents the new physics predicted by the string theory.

4. NEUTRINO MASSES

One of the important features of the new physics interactions is the prediction of neutrino masses arising below the electroweak scale. These come from two sources:

(i) W_{seesaw}

These terms arise from couplings of the form $\nu_p H \chi_\alpha$ where χ_α is one of the superheavy fields below the M_I scale with mass term $\tfrac{1}{2} M \chi_\alpha \chi_\alpha$, $M = O\ (M_I)$. Eliminating the superheavy χ_α, one finds below the electroweak scale where $< H > \neq 0$, a Majorana mass term $W_{\text{mass}}(\text{seesaw}) = \tfrac{1}{2}\nu_p \mu_{pp'} \nu_{p'}$ where $\mu_{pp'} \approx m^2_\ell/M_I$ with m_ℓ a charged lepton mass[4]. Thus $\mu_{pp'} \approx 10^{-6}$ eV for $m_\ell = m_\tau$.

(ii) $\lambda_p H \ell_p n_2 + \bar{\lambda}_p H \ell_p \bar{n}_2$ Interactions

These lead to coupling between ν_p and the new fields n_2, $\bar{n}_2$ and a mass term $W_{\text{mass}}(n_2 - \nu_p) = \mu_p \nu_p n_2 + \bar{\mu}_p \nu_p \bar{n}_2$ where $\mu_p = \lambda_p < H >$ and $\bar{\mu}_p = \bar{\lambda}_p < H >$.

Thus Eqs. (3.3), (3.4) leads to a 5×5 mass matrix for ν_p ($p = 1, 2, 3$), n_2 and $\bar{n}_2$. Since $\mu_{pp'} \ll \mu_p$, $\bar{\mu}_p \ll m_i$, the five eigenvalues have the following size:

$$m_{\nu_1} = O(\mu_{pp'}) \approx 10^{-6} \text{ eV} \ ; \quad m_{\nu_2}, m_{\nu_3} = O(\frac{\mu^2_p}{m_i} ; \frac{\bar{\mu}^2_p}{m_i}) \tag{4.1}$$

and $m_{4,5} = O\ (m_i) \lesssim 1$ TeV. Thus two seesaws operate: the conventional one for $\mu_{pp'} \approx m^2_\ell/M_I$, and a new one of μ^2_p/m_i. Since experimentally, $m_\nu \lesssim 10$ eV, one finds that $\mu_p = \lambda_p < H >$ and $\bar{\mu}_p = \bar{\lambda}_p < H >$ must be $\lesssim 1$ MeV. This in turn puts constraints $(27)^3$ couplings λ^3_{ijk} and $\epsilon \equiv < \nu^c_2 > / < N_1 >$ since $\lambda_p \approx \lambda^3_{12r} \equiv \delta^2$ and $\bar{\lambda}_p \approx \epsilon^3$. One finds[10]

$$\epsilon \lesssim 0.03 - 0.05 \ ; \quad \delta^2 \approx \epsilon^3 \tag{4.2}$$

Thus λ^3_{12r} must be very small or else neutrino masses will violate the present experimental bounds.

5. CHARGED LEPTON MASS HIERARCHY

The smallness of λ^3_{12r} can be related to the charged lepton mass hierarchy i.e. the smallness of m_e/m_τ. Thus the charged lepton mass matrix, $e^c_p M^{(\ell)}_{pp'} e_{p'}$, $(p, p' = 1, 2, 3$ are the three light generations) has the form[10] $M^{(\ell)}_{p1} \equiv \sigma_p$, $M^{(\ell)}_{p2} \equiv a_p$, $M^{(\ell)}_{p3} \equiv b_p$ where $\sigma_p \approx \delta^2/\epsilon$ and is quite small, while $a_p, b_p = O\ (1)$. One may calculate the eigenvalues of $M^{(\ell)}_{pp'}$. One finds $m_e/m_\tau \cong (\sigma/a)(1 + r^2)^{-\frac{1}{2}}$ where $a \cdot b \equiv a^2 r \cos \alpha$ and $r \equiv b/a$. Thus from Eq. (4.3) one finds $m_e/m_\tau \approx \epsilon^2 \approx 10^{-3}$ and the smallness of m_e/m_τ is due to the smallness of δ. Note, however, that δ cannot vanish as this would leave the electron massless i.e. δ cannot be much less than the bounds of Eq. (4.3), and hence one expects neutrinos of mass O (1 eV). For the muon one finds[10] $m_\mu/m_\tau = (r \sin \alpha)/(1 + r^2)$ showing correctly that $m_\mu < m_\tau$. In fact, the correct experimental lepton mass ratios occur with the choices $r \simeq 3$ and $\sin \alpha \simeq 0.2$.

6. CONCLUSIONS

We have examined here the consequences of three generation Calabi-Yau compactifications of the heterotic string in which contraints have been imposed so that the theory reduce to the Standard Model at low energy. In particular we assumed that (i) E_6 breaks to $[SU(3)]^3$ at M_c by flux breaking, (ii) Matter parity invariance holds (to stabilize the proton) and (iii) $[SU(3)]^3$ breaks to $SU(3) \times SU(2) \times U(1)$ at M_I triggered by SUSY breaking masses. Then one finds that there always exists four new $SU(3) \times SU(2) \times U(1)$ neutral chiral multiplets with electroweak size masses (in addition to the usual three light generations of quarks and leptons). We note that Calabi-Yau manifolds that obey (i) and (ii) exist, and calculations for special manifolds shows that (iii) does indeed occur for $M_I \gtrsim 10^{16}$ GeV.

The four new states interact with Higgs and leptons giving rise to neutrino masses, one very light [O (10^{-6} eV)] and two others constrained by the size of $\epsilon = < \nu^c_2 > /\ < N_1 >$ and $\delta^2 = \lambda^3_{12r}$. Experimental bounds on m_ν then require $\epsilon \lesssim 0.05$ and $\delta^2 \approx \epsilon^3$ to be very small. The smallness of δ^2 and ϵ gives a natural explanation of the lepton mass hierarchy i.e. one finds $m_e/m_\tau = O(\delta^2/\epsilon) \approx 10^{-3}$. The neutrino mass growth also gives rise to neutrino oscillations[10] with the unique signal that $\nu_{e,\mu} \leftrightarrow \nu_\tau$ is considerably larger than $\nu_e \leftrightarrow \nu_\mu$. These oscillations may be accessible to experiment.

REFERENCES

1. D. Gross, J. Harvey, E. Martinec and R. Rohm, Phys. Rev. Lett. **55** 502 (1985); Nucl. Phys. **B258** 253 (1985).
2. G. Tian and S. T. Yau, Proc. of Argonne Symposium, Anomalies, Geometry and Topology, ed. A. White (World Scientific, Singapore, 1985).
3. R. Schimmrigk, Phys. Lett. **193B** 75 (1987); D. Gepner, Nucl. Phys. **B296** 757 (1988).
4. B. Greene, K. H. Kirklin, P. J. Miron and G. G. Ross, Phys. Lett. **180B** 69 (1986); Nucl. Phys. **B278** 667 (1986); **B292** 606 (1987).
5. F. del Aguila and C. D. Coughlan, Phys. Lett. **215B** 93 (1988).
6. P. Nath and R. Arnowitt, Phys. Rev. **D39** 2006 (1989); R. Arnowitt and P. Nath; Phys. Rev. **D40** 191 (1989).
7. M. C. Bento, L. Hall and G. G. Ross, Nucl. Phys. **B292** 400 (1987).
8. P. Nath and R. Arnowitt, NUB#2991-CTP-TAMU-10/90.
9. P. Nath and R. Arnowitt, NUB#2993-CTP-TAMU-12/90; R. Arnowitt and P. Nath, Proc. Int. Europhysics Conf., Madrid (1989).
10. R. Arnowitt and P. Nath, CTP-TAMU-11/90-NUB#2992.

OPEN SUPERSTRINGS AS THE THEORY OF EVERYTHING?*

Z. Bern
Theoretical Division, MS-B285
Los Alamos National Laboratory
Los Alamos, New Mexico 87545
USA

D.C. Dunbar
D.A.M.T.P.
Liverpool University
Liverpool, UK

ABSTRACT

We discuss the first examples of one-loop finite four-dimensional superstrings. These examples can be either space-time supersymmetric or not depending on the details of the models.

1. INTRODUCTION

One of the most important features of all string theories is that they provide a framework for a consistent theory of quantum gravity unified with all other forces [1]. The recent interest in string theories began with the observation of anomaly cancellation in the Green-Schwarz SO(32) ten-dimensional open superstring [2]. Since then open strings have been neglected as potential theories of nature, although there is no a priori reason for this. Although there has been great progress in constructing four-dimensional heterotic closed string theories [3-6], there has been no analogous progress for four-dimensional open superstrings. Part of the reason for this is sociological and part is technical. The basic technical difficulty with lower-dimensional open strings in contrast to closed strings is that there is no simple symmetry principle analogous to modular invariance. This makes the construction of lower-dimensional open superstrings less mechanical than the corresponding closed constructions constructions, but is not a fundamental obstruction. Indeed, we have succeeded in constructing the first examples of sensible four-dimensional open superstrings.

The main motivation for studying open string theories as compared to closed string theories is the hope that some of the important unsolved problems in string theories such as the dilaton problem [7], and the fact that there seems to be no fundamental way to choose between the large number of consistent lower-dimensional string models [3-6], may have a simpler resolution in open string theory. The construction of open string models is considerably more constrained [8,9,10] so that the number of viable open string vacua seems to be considerably less than that of closed strings alone. Also since open string field theory is much simpler than closed string

---

* Talk presented by Z.B.

field theory, a string field theory understanding of why a particular vacuum is chosen over others would be much simpler for the open string case.

2. EXAMPLE OF CONSTRUCTION

For simplicity, we only focus on the simplest example of a four-dimensional superstring, although we have developed a formalism for constructing a variety of open superstrings. For our example we take the world sheet fermion's contribution to the partition function on the torus to be

$$Z^{\text{torus}} = \frac{1}{2}(Z_{W_0}^{W_0} + Z_{\mathbf{0}}^{W_0}) + \frac{1}{2}(Z_{W_0}^{\mathbf{0}} - Z_{\mathbf{0}}^{\mathbf{0}})\,, \tag{1}$$

where we are following the notation of refs. [4,8]. The upper vectors specify the space boundary conditions and the lower vectors specify the time boundary conditions of the world sheet fermions on a torus. In this example, $W_0 = ((1/2)^{10}|(1/2)^{10})$ and $\mathbf{0} = (0^{10}|0^{10})$ respectively correspond to Neveu-Schwarz and Ramond boundary conditions on all twenty left- and right-mover complex world-sheet fermions. The complete model consists of tensoring the world-sheet fermionic contributions with the bosonic contributions and integrating over the modular parameter as described in, for example, refs. [1,4].

The consistency of this closed model follows from its modular invariance properties which can be easily checked using the transformations which may be found, for example, in ref. [4]. Furthermore, the model can be truncated to a type I unoriented closed model [8], as follows from the left-right symmetry of the model.

Our construction of open strings from a given closed string [8] is done through the use of the boundary and crosscap states [11]. The set of states of the closed string model which may couple to a boundary or crosscap state are the left-right-symmetric (LRS) ones, that is states which have the same left- and right-mover content. Thus, as discussed in refs. [8,12] the cylinder with two boundary states is

$$\begin{aligned}
Z^{BB}(\tau') &= \text{tr}_{\text{LRS}}[q^{\hat{H}^{\text{left}}_{W_0}} \bar{q}^{\hat{H}^{\text{right}}_{W_0}} \hat{P}(W_0)] \ + \text{tr}_{\text{LRS}}[q^{\hat{H}^{\text{left}}_{\mathbf{0}}} \bar{q}^{\hat{H}^{\text{right}}_{\mathbf{0}}} \hat{P}(\mathbf{0})] \\
&= \text{tr}\,[|q|^{2\hat{H}^{\text{left}}_{W_0}}] \\
&\equiv F^{U_0}_{U_0}(\tau')\,,
\end{aligned} \tag{2}$$

where $\tau' = \ln|q|/i\pi$ and $U_0 = ((1/2)^{10})$ is a Neveu-Schwarz boundary condition vector of half the length of W_0, P is a GSO projector and $\hat{H}_{W_0}$ is the closed string hamiltonian for the world sheet fermions in the Neveu-Schwarz sector. The GSO projectors collapse because they are resricted to LRS states so that

$$\begin{aligned}
\hat{P}(W_0)\Big|_{\text{LRS}} &= \frac{1}{2}(1 + e^{2\pi i W_0 \cdot \hat{N}_{W_0}})\Big|_{\text{LRS}} = 1 \\
\hat{P}(\mathbf{0})\Big|_{\text{LRS}} &= \frac{1}{2}(1 - e^{2\pi i W_0 \cdot \hat{N}_{\mathbf{0}}})\Big|_{\text{LRS}} = 0\,.
\end{aligned} \tag{3}$$

For LRS states, $N = (N_l|N_l)$ and so $\exp(2\pi i W_0 \cdot N)$ is unity for such states. By Jacobi transforming this, we obtain the annulus contribution to the open string partition function

$$Z^{\mathrm{ann}}(\tau) = F_{U_0}^{U_0}(\tau) = \mathrm{tr}\,[e^{2\pi i \tau \hat{H}_{U_0}^{\mathrm{open}}}] \equiv \mathrm{tr}\,[w^{\hat{H}_{U_0}^{\mathrm{open}}}]\,, \tag{4}$$

where $\tau = -1/\tau'$ and $\hat{H}_{U_0}^{\mathrm{open}}$ is the open string hamiltonian in the U_0 sector.

Given the annulus contribution, we can then construct the möbius contribution by requiring a physically sensible projection between the annulus and möbius contributions. This is equivalent to requiring the action of the twist operator [1,8] $\hat{\Omega}$ on open string states to satisfy $\hat{\Omega}^2 = 1$ when acting on a state. Since the open string U_0 sector, which does not have a GSO projector, contains states at both integer and half integer mass levels, the naive twist operator $\hat{\Omega} \sim e^{\pi i \hat{H}_{U_0}}$ is not sensible. In the presence of a GSO projector which would ensure that this twist operator is well defined, consistent open string models exist only in $D = 2, 6, 10$ as described in ref. [8,9] and not in $D = 4$.

The solution to this difficulty is actually quite simple: Use a GSO projector to separate the states into integer and half-integer mass levels before applying the twist operator and associate different phases with the two parts of the twist operator so as to ensure $\hat{\Omega}^2 = 1$. The explicit form for the twist operator is then

$$\hat{\Omega} = e^{\pi i d/24}\left[\eta_1\left(\frac{1+(-1)^{\hat{N}_{U_0}}}{2}\right)e^{\pi i \hat{H}_{U_0}^{\mathrm{open}}} - i\eta_2\left(\frac{1-(-1)^{\hat{N}_{U_0}}}{2}\right)e^{\pi i \hat{H}_{U_0}^{\mathrm{open}}}\right]\,, \tag{5}$$

where $\hat{N}$ is the usual fermion number operator and $d = 10$ for a four-dimensional model. The phase $e^{\pi i d/24}$ absorbs the phase in $e^{\pi i \hat{H}_{U_0}^{\mathrm{open}}}$ due to the zero-point energy of the hamiltonian $\hat{H}_{U_0}^{\mathrm{open}}$. For $\eta_1 = \eta_2$ this twist operator is equivalent to the one given by Clavelli and Shapiro many years ago [13]. The choice of the $\eta_i = \pm 1$ determines the Chan-Paton gauge group representation [14] of the various mass levels. (See ref. [1] for details.) (We are allowing for an independent choice of the Chan-Paton gauge group representation at integer and half-integer mass levels because these lie in different representations of the two-dimensional Kac-Moody symmetry.) The Chan-Paton representation of the massless gauge bosons corresponds to the adjoint representation of the gauge group, so for $\eta_2 = +1, -1$, which controls the representation of the massless states, the gauge group will respectively be a Sp or SO group.

With this choice of twist operator the möbius contribution is

$$Z^{\mathrm{mob}} = e^{i\pi d/24}\left(\frac{1}{2}\eta_1\mathrm{tr}\,[w^{\hat{H}_{U_0}^{\mathrm{open}}}(1+(-1)^{\hat{N}_{U_0}})\hat{\Omega}] - \frac{i}{2}\eta_2\mathrm{tr}\,[w^{\hat{H}_{U_0}^{\mathrm{open}}}(1-(-1)^{\hat{N}_{U_0}})\hat{\Omega}]\right)$$

$$= e^{i\pi d/24}\left(\frac{1}{2}\eta_1\left(F_{U_0}^{U_0}(\tau+1/2) + F_0^{U_0}(\tau+1/2)\right)\right.$$

$$\left. - \frac{i}{2}\eta_2\left(F_{U_0}^{U_0}(\tau+1/2) - F_0^{U_0}(\tau+1/2)\right)\right)\,. \tag{6}$$

The Jacobi transformation properties of the möbius contributions to the partition function are given by

$$F_V^U(\tau + 1/2) = e^{4\pi i (U - U_0) \cdot (V - U_0)} e^{2\pi i (V \cdot V - d/6)} e^{\pi i (U \cdot U - d/6)} F_{\overline{V-U}}^U(\tau'/4 + 1/2) \, , \quad (7)$$

where U and V consist of Neveu-Schwarz and Ramond open string boundary conditions and the overbar on the boundary condition vector indicated that it should be evaluated mod 1. By Jacobi transforming the möbius partition function (6) (after the usual rescaling [2] $\tau'/4 \to \tau'$) we obtain the cylinder with one boundary and one crosscap ($D = 4, d = 10$)

$$Z^{BC} = e^{i\pi d/24} \left(\frac{1}{2}\eta_2 \left(F_{U_0}^{U_0}(\tau' + 1/2) + F_{\mathbf{0}}^{U_0}(\tau' + 1/2) \right) \right.$$
$$\left. - \frac{i}{2}\eta_1 \left(F_{U_0}^{U_0}(\tau' + 1/2) - F_{\mathbf{0}}^{U_0}(\tau' + 1/2) \right) \right) , \quad (8)$$

which contains the same closed string states as the cylinder with two boundary states (2).

The potential divergences of the amplitudes in eqs. (2) and (8) are determined by the leading and next to leading terms in the Taylor expansion, in $e^{2\pi i \tau'}$, corresponding to the tachyon and massless scalars propagating into the vacuum. In the Jacobi trasformed möbius contribution (8), the relative normalization of the tachyon singularity is $2^2\eta_2$ while for the massless scalar it is $2^2\eta_1 M$, where we are including the factor $2^{D/2}$ arising from the bosonic [1,8]. Similarly for the Jacobi transformed annulus (2), the relative normalizations of these singularities is M for gauge groups $Sp(M)$ or $SO(M)$. Thus, for cancellation of both tachyon and massless scalar divergences

$$M + \eta_2 2^2 = 0 \, , \qquad\qquad M + \eta_1 2^2 = 0 \, , \qquad (9)$$

so the gauge group is $SO(4)$.

Further models can be constructed by the addition of extra boundary condition basis vectors. In ref. [12] examples of open-closed string models with $N = 4, 2, 1$ space-time supergravity were presented.

3. CONCLUSIONS

Much work remains to be done before open strings are as well developed as the heterotic string for phenomenological purposes. For example, the question of chirality remains a difficulty with lower-dimensional open superstring constructions, although some progress has been made on this question within the context of the orbifold approach to string theory [15]. Additionally, there is the question of higher loop amplitudes [16] as well as the potential of using non-trivial projections on the Chan-Paton factors [17,10,15]. However, we have explicitly shown that it is possible to construct a variety of four-dimensional one-loop finite open superstring theories and are hopeful that such constructions will eventually lead to phenomenologically viable models.

ACKNOWLEGEMENTS

This work was supported in part by S.E.R.C. and in part by the US Department of Energy.

REFERENCES

1. J.H. Schwarz, Phys. Rep. 89 (1982) 223; M.B. Green, J.H. Schwarz, and E. Witten, Superstring Theory (Cambridge University Press) (1987).

2. M.B. Green and J.H. Schwarz, Phys. Lett. 151B (1985) 21.

3. L. Dixon, J. Harvey, C. Vafa, and E. Witten, Nucl. Phys. B261 (1985) 678; Nucl. Phys. B274 (1986) 285; K.S. Narain, Phys. Lett. 169B (1986) 41; K.S. Narain, M.H. Sarmadi and C. Vafa, Nucl. Phys. B288 (1987) 551; W. Lerche, D. Lüst and A.N. Schellekens, Nucl. Phys. B287 (1987) 477.

4. H. Kawai, D.C. Lewellen and S.-H.H. Tye, Phys. Rev. Lett. 57 1832 (1986); Nucl. Phys. B288 (1987) 1.

5. I. Antoniadis, C.P. Bachas and C. Kounnas, Nucl. Phys. B289 (1987) 87.

6. D. Gepner, Nucl. Phys. B296 (1988) 757.

7. Z. Bern and D.C. Dunbar, preprint LA-UR-90-1121, LTH-90-251.

8. Z. Bern and D.C. Dunbar, Phys. Lett. 203B (1988) 109; D.C. Dunbar, Nucl. Phys. B319 (1989) 72; Z. Bern and D.C. Dunbar, Nucl. Phys. B319 (1989) 104.

9. L. Clavelli, P.H. Cox, B. Harms and A. Stern, University of Alabama preprint UAHEP-882; L. Clavelli, P.H. Cox and B. Harms, Phys. Rev. Lett. 61 (1988) 787; L. Clavelli, P.H. Cox, P. Elmfors and B. Harms, preprint UAHEP-891.

10. N. Ishibashi and T. Onogi, Nucl. Phys. B318 (1989) 239; N. Ishibashi, Mod. Phys. Lett. A4 (1989) 251.

11. M. Ademollo et al, Nucl. Phys. B94 (1975) 221; C.G. Callan, C. Lovelace, C.R. Nappi and S.A. Yost, Nucl. Phys. B293 (1987) 83; J. Polchinski and Y. Cai, Nucl. Phys. B296 (1988) 91.

12. Z. Bern and D.C. Dunbar, Phys. Rev. Lett. 64 (1990) 124; preprint LA-UR-89-3653, LTH-89-246, to appear in Phys. Lett. B; preprint LA-UR-89-3893, LTH-89-249, to appear in Int. J. Mod. Phys.

13. L. Clavelli and J.A. Shapiro, Nucl. Phys. B57 (1973) 490.

14. J.E. Paton and Chan Hang-Mo, Nucl. Phys. B10 (1969) 519.

15. J.H. McCown, Princeton University Ph.D. Thesis (1989).

16. M. Bianchi and A. Sagnotti, Phys. Lett. 211B (1988) 407; preprint ROM2F-89/15.

17. J.A. Harvey and J.A. Minahan, Phys. Lett. 188B (1987) 44.

FERMIONS and UNIFICATION in WEYL SPACETIME[1]

David Hochberg[2]
Department of Physics and Astronomy, Vanderbilt University
Nashville, TN 37235

Abstract

Including spinors in Weyl's geometry leads to a unique, nonminimal, coupling between fermions and the Weyl vector. In view of this fact, we reconsider Weyl geometry as a framework for the unification of Einstein gravitation with abelian vector interactions and briefly comment on some possible applications.

1. Introduction

The introduction of new degrees of freedom beyond those which are empirically established, is a common feature shared by all proposals aiming at a synthesis of nature's elementary interactions. Thus for example, theories in higher spatial dimensions (Kaluza-Klein), with higher internal symmetry groups (grand unification), and those mixing fields of different spins (supersymmetry), to name only a few, all introduce additional degrees of freedom in one way or another. Another feature these and related schemes all share is the assumption that spacetime is Riemannian, but one might consider generalizing the spacetime geometry itself as a way to unify gravitation with other types of interactions. Indeed, such a proposal was conceived by Weyl long ago in a first attempt to bring electromagnetism and gravitation into a common geometric framework [1]. In this talk we discuss recent work incorporating matter (fermions) in Weyl space and comment on possible physical identifications for the Weyl vector [2].

2. Spinors in Weyl Spacetime

Weyl geometry is characterized by two elementary fields: the metric and the Weyl vector, which is the gauge field of local scale or Weyl transformations. Under such a transformation, the metric and Weyl vector transform as

$$g'_{\mu\nu} = e^{w(g)\phi(x)}g_{\mu\nu} \qquad \text{and} \qquad W'_\mu = W_\mu - \partial_\mu\phi(x),$$

where $\phi(x)$ is a real-valued function (the local calibration field) and $w(g)$, the Weyl weight of the metric, is a real number which we now set equal to one (any other choice is equally good). This transformation extends to any field T in Weyl spacetime: $T'(x) = e^{w(T)\phi(x)}T(x)$, where $w(T)$ is the corresponding Weyl weight, and is real. The tensor calculus for Weyl geometry is developed in complete analogy with the Riemannian case and the construction of covariant derivatives, connection and curvature tensor is straightforward [3].

1 This work was done in collaboration with Günter Plunien

2 Research supported in part by DOE grant No. DE-FGO5-85ER40226

To incorporate spinors in Weyl space requires the existence of a globally defined system of tetrads, e_a^μ, where $g_{\mu\nu} = \eta_{ab} e_\mu^a e_\nu^b$ and $e_\mu^a = \eta^{ab} g_{\mu\nu} e_b^\nu$. Interactions between fermions and W_μ is best addressed at first in the context of elementary 2-spinors, denoted here as ξ^A and $\phi_{\dot{B}}$, respectively. The Weyl transformation of these fields is completely characterized by their respective weights: $w(e_a), w(\xi^A)$ and $w(\phi_{\dot{B}})$. The seperate weights appear to be free, arbitrary parameters, but important constraints are imposed by the dynamics. Indeed, let

$$A = \int d^4x \sqrt{-g}\,\mathcal{L},$$

be any Weyl-invariant action containing tetrads, Weyl vector and 2-spinors. Invariance requires $w(\sqrt{-g}\mathcal{L}) = 0$, and hence $w(\mathcal{L}) = -2$. This constraint leads to a vanishing minimal coupling [2]. To see how this comes about, consider the kinetic terms for the spinors:

$$\mathcal{L}_{kin} = \frac{i}{\sqrt{2}}(\xi^{\dot{B}} \sigma^\mu_{A\dot{B}} D_\mu \xi^A + \phi_A \sigma^{\mu A\dot{B}} D_\mu \phi_{\dot{B}}) + hc.,$$

where D is the spinor covariant derivative in Weyl space [3]. Applying the above constraint to $\mathcal{L}_{kin}$ leads to the relations $w(\xi^A) = -1 - \frac{1}{2}w(e_a)$ and $w(\phi_{\dot{B}}) = -\frac{1}{2} - \frac{1}{2}w(e_a)$ [2,3]. Furthermore, all the W-dependence in the covariant derivatives assembles into a single factor given by

$$\sigma^\mu_{A\dot{B}} D_\mu \xi^A |_W = [w(\xi^A) + 1 + \frac{1}{2}w(e_a)]\sigma^\mu_{A\dot{B}} W_\mu \xi^A,$$

which vanishes identically, with a similar result holding for the ϕ sector. Thus, spinors do not couple to the Weyl vector through the covariant derivative, i.e., there is no minimal coupling. The effective spinor covariant derivative reduces to $D_\mu \xi^A \to \partial_\mu \xi^A + \Gamma^A_{B\mu} \xi^B$, where the spinor connection depends only on the pure Christoffel part of the Weyl-space connection. These conclusions carry over immediately to Dirac spinors

$$\Psi = \begin{pmatrix} \phi_A \\ \xi^{\dot{A}} \end{pmatrix}.$$

Demanding that Ψ itself have a well-defined Weyl transformation immediately fixes the weights of all the spinors and tetrads at once: $w(\Psi) = w(\xi) = w(\phi) = -\frac{3}{4}$ and $w(e_a) = -\frac{1}{2}$. The covariant derivative for Ψ is given by $D_\mu = (\partial_\mu + \Gamma_\mu)$, where Γ_μ is just the spin connection for Riemannian space.

3. Geometric Couplings

Despite the above conclusions, potentially important couplings of a manifestly geometric character exist which provide the means by which matter can interact with the Weyl vector. By geometric, we simply mean that fermions couple to some function of the Weyl space curvature tensor. To find the explicit form of a local coupling term, we consider bilinears of Dirac spinors and functions f of the Weyl curvature $\sim \bar{\Psi} f(\bar{R}^{\lambda}_{\mu\nu\kappa})\Psi$, where

$$\bar{R}^{\lambda}_{\mu\nu\kappa} = 2(\partial_{[\kappa}\bar{\Gamma}^{\lambda}_{|\mu|\nu]} + \bar{\Gamma}^{\eta}_{\mu[\nu}\bar{\Gamma}^{\lambda}_{\kappa]\eta}), \text{ and } \bar{\Gamma}^{\alpha}_{\mu\nu} = \{^{\alpha}_{\mu\nu}\} + \frac{1}{2}(\delta^{\alpha}_{\mu}W_{\nu} + \delta^{\alpha}_{\nu}W_{\mu} - g_{\mu\nu}W^{\alpha}),$$

are the curvature tensor and connection of Weyl space, respectively. The function f must be a coordinate scalar and a spinor tensor and in addition, must satisfy the constraint $w(\bar{\Psi} f \Psi) = -2$. As discussed in [2], there is only one function satisfying these conditions and which at the same time, is directly interpretable as a spinor operator: $f = \bar{R}^{1/2}$, where $\bar{R} = \delta^{\nu}_{\lambda}g^{\mu\kappa}\bar{R}^{\lambda}_{\mu\nu\kappa}$. This comes about through linearization: we seek an operator $\mathcal{R}$ such that $\mathcal{R} \times \mathcal{R} = \bar{R} = R + 3/2W_{\mu}W^{\mu}$ (Here, R is the scalar of Riemannian curvature, and a total divergence term has been dropped since it vanishes by virtue of the equations of motion). With the ansatz $\mathcal{R} = A^{\mu}(x)W_{\mu} + B(x)R^{1/2}$, we find the coefficient functions must obey the nonintrivial algebra $\{A^{\mu}, A^{\nu}\} = 3g^{\mu\nu}$, $\{A^{\mu}, B\} = 0$, and $B^2 = 1$. The explicit solution is given by $A^{\mu}(x) = \sqrt{3/2}\gamma^{a}e^{\mu}_{a}(x)$ and $B(x) = det(e_{\mu})\gamma_5$, where $[\gamma^{a}, \gamma^{b}] = 2\eta^{ab}$ are the flat-space Dirac matrices, and γ_5 is the curved space pseudoscalar operator [2]. Assembling these results for the kinetic and interaction terms for a Dirac spinor in Weyl space leads to the following lagrangian for matter:

$$\mathcal{L}_{matter} = \bar{\Psi}(i\gamma^{\mu}D_{\mu} - m)\Psi + \lambda\bar{\Psi}(\sqrt{3/2}\gamma^{\mu}W_{\mu} + e\gamma_5 R^{1/2})\Psi,$$

where λ is a real, dimensionless coupling parameter, and m is a mass term with $w(m) = -\frac{1}{2}$.

4. The Full Action and Gauge Fixing

To complete the action, we need to specify the Lagrangian for the metric and Weyl field. We take

$$\mathcal{L}_{(g,W)} = \bar{R}^2 + aW_{\mu\nu}W^{\mu\nu} + \Lambda^2,$$

where $W_{\mu\nu} = (1/2)\bar{R}^{\alpha}_{\alpha\mu\nu}$ is the Weyl field strength, a is a dimensionless parameter and Λ is a Weyl space cosmological constant with $w(\Lambda) = -1$. We now exploit the gauge invariance of the action to pick a gauge wherein Einstein's equations for the metric are manifest: this is accomplished by taking $\bar{R} = \Lambda = constant \neq 0$

with $\phi(x) = ln(\bar{R}/\Lambda)$. The variation of the gauge-fixed action leads to the set of equations:

$$G_{\mu\nu} = (a/\Lambda)\Omega_{\mu\nu} - (1/2\Lambda)T_{\mu\nu}, \quad \text{and} \quad \nabla_\mu W^{\mu\nu} = (3\Lambda/2a)W^\nu + (\lambda/4a)\bar{\Psi}\gamma^\nu\Psi,$$

and

$$[i\gamma^\mu D_\mu - m + \lambda(\sqrt{3/2}\gamma^\mu W_\mu + e\gamma_5 R^{1/2})]\Psi = 0,$$

for the metric, Weyl field and Dirac spinor, respectively. $G_{\mu\nu}$ is the (Riemannian) Einstein tensor, and $\Omega_{\mu\nu}, T_{\mu\nu}$ denote the stress tensors for W_μ and Ψ, respectively. At this point we see that Λ is no longer arbitrary, but must be related to Newton's constant in order that we recover the Newtonian limit: $\Lambda = (8\pi G_N)^{-1}$. This identification completely fixes the gauge.

5. The Pauli and Schrödinger Equations for Matter

To gain further insight into the nature of the interaction between matter and Weyl's vector implied by the geometric coupling term, it is instructive to take the nonrelativistic limit of the above Dirac-type equation. For this purpose we adopt the method of Foldy and Wouthuysen, which can be applied to spinors in curved backgrounds. We have performed the nonrelativistic reduction for the class of diagonal metrics; this presents no loss of generality and tremendously simplifies the algebra. Details are presented in Ref [2]. The Pauli equation we obtain is given by (χ refers to the upper components of the Dirac spinor)

$$i\sqrt{g^{00}}\frac{\partial\chi}{\partial t} = H_{Pauli}\chi,$$

where

$$H_{Pauli} = (1/2m)(-i\vec{\nabla} + \lambda\vec{W})^2 - \lambda\sqrt{g^{00}}W_0 + (\lambda/2m)\vec{\sigma}\cdot(\vec{\nabla}\times\vec{W})$$
$$+ (1/2m)(-\lambda^2 R - i\lambda\vec{\sigma}\cdot\vec{\nabla}R^{1/2}).$$

The first three terms in H_{Pauli} bear a striking resemblance to the kinetic, potential and spin-orbit terms for a charged particle in an electromagnetic field. The last two terms reflect the explicit coupling of the Pauli spinor to scalar curvature. The associated Schrödinger Hamiltonian is easily obtained and is given by

$$H_{Schrodinger} = (1/2m)(-i\vec{\nabla} + \lambda\vec{W})^2 - \lambda\sqrt{g^{00}}W_0(x) - (\lambda^2/2m)R.$$

The potential is completely geometric in origin, and although we started with a geometric interaction, the Weyl field appears in both the Pauli and Schrödinger operators *as if* it had been introduced there via the minimal coupling prescription. Furthermore, the precise couplings between the wave functions and the scalar curvature are a direct consequence of the original geometric interaction term.

6. Summary and Discussion

We have reconsidered Weyl's geometry of spacetime based on the existence of a unique, nonminimal coupling between fermions and the Weyl vector. This interaction can be linearized and thus be interpreted as a spinor operator. The Weyl field then appears in the Dirac equation, as well as in the nonrelativistic Pauli and Schrödinger equations, as if it had been introduced there by the standard minimal prescription. In addition, this linearized coupling leads to direct couplings between spinors and the gradient of scalar curvature at the level of the Pauli equation, and makes the origin of the curvature-dependent potential in the Schrödinger hamiltonian manifest. We make use of the gauge invariance of the full theory and fix a gauge wherein the contact to Einstein's metric theory is easily made. This fixes the Weyl space cosmological constant to be proportional to Newton's constant and implies a massive Weyl vector (Proca field). Aside from the bare fermion mass m, this theory contains two adjustable parameters: $a = a(m_W)$ and λ, where $m_W \sim \Lambda/a$ is the Proca mass, a sets the proportionality between the Weyl and physical vector field $(\sqrt{a}W_\mu = V_\mu)$, and λ is the coupling to matter.

To get some feeling for possible candidates for the force mediated by Weyl's field, consider electromagnetism and 'fifth' forces. Empirical bounds on a finite photon mass are on the order of $m_\gamma < 10^{-33} MeV$. To satisfy this bound requires $\sqrt{a} \sim 10^{54}$ for the proporionality between the Weyl field (geometry) and the electromagnetic potential, A_μ At the same time, one identifies $\lambda^2/\sqrt{a}$ with the fine structure constant. In the case of vectorial fifth forces, current experimental limits for the mass and coupling strength are $m_V \sim 10^{-15} MeV$ and $g_V \sim 10^{-20}$, respectively. Identifying W_μ as this carrier requires $\sqrt{a} \sim 10^{37}$. Thinking of the Weyl vector more in terms of a phenomenological effective interaction (i.e., if the Weyl vector has a composite structure), one could then get tight bounds on its mass and coupling, for example, to electrons and positrons, from constraints on radiative corrections to high precision QED processes, like the $g - 2$ measurement and Lamb shift.

Finally, one would like to know how this theory (or some variant thereof) stands in relation to renormalizability. In this regard, provided there exists a Weyl invariant regularization prescription, then the number of possible counterterms would be <u>finite</u>: $\mathcal{L}_{counterterm} = b_1 \bar{R}_{(\mu\nu)}\bar{R}^{(\mu\nu)} + \bar{R}_{\mu\nu\alpha\beta}(b_2 \bar{R}^{\mu\nu\alpha\beta} + b_3 \bar{R}^{\alpha\beta\mu\nu} + b_4 \bar{R}^{\mu\alpha\nu\beta})$, where the b_i are real dimensionless constants.

References

1. H. Weyl, Sitzung. d. Preuss. Akad. d. Wiss. (1918) 465.

2. D. Hochberg and G. Plunien, Vanderbilt University preprint VAND-TH-89-13 (Dec. 1989).

3. J. Audretsch, Phys. Rev. **D27** (1983) 2872.

Gravity from Multiloop Calculations
in Superstring Theory

Gerardo Cristofano[1,3], Marco Fabbrichesi[1,2] and Kaj Roland[2]

[1] *NORDITA, Blegdamsvej 17, DK-2100 Copenhagen Ø, Denmark.*
[2] *The Niels Bohr Institute, Blegdamsvej 17, DK-2100 Copenhagen Ø, Denmark.*
[3] *Dipartimento di Scienze Fisiche, Università di Napoli and INFN, Sezione di Napoli, I-80125 Naples, Italy.*

Abstract

We explore quantum gravity by considering the elastic scattering of two on-shell string states in the $s \to \infty$, $t \to 0$ limit. The amplitude is computed using the Covariant Loop Calculus which turns out to provide a powerful method of systematically exploring such a régime. By a resummation of the perturbation series we compute the deflection angle up to the second order in Newton's constant.

1. Introduction

The major attraction of superstrings is their unique role as consistent, finite [1] quantum theories containing gravity. They allow for the first time a well-defined investigation of gravitational multiloop interactions. However, as it is well known, the low-energy limit of the type II superstring is a theory of supergravity in ten space-time dimensions [2]. Therefore, in any realistic theory we have to compactify down to four dimensions and, in the process, break supersymmetry and get rid of the unwanted massless states (such as the dilaton and the gravitino). In spite of much work this program is nowhere near finished.

In the foreseeable future, therefore, the simplest way of dealing with these problems is by circumventing them, i.e. to study string scattering in situations where –for kinematical reasons– pure gravity dominates all other interactions.

Since gravitational interactions couple to energy, the Planckian regime of very large center-of-mass energies $\sqrt{s}$ and infinitesimal transferred momentum $\sqrt{-t}$ (corresponding to large impact parameter b) is exactly such a regime. As long as gravity dominates, the details of the compactification are irrelevant and we may compactify down to four space-time dimensions in a very simple-minded way on tori of equal radii. Since we are essentially in the classical limit, we expect to recover general relativity, modified by small quantum and string-like corrections. Since gravitational tree amplitudes violate unitarity at high energies [3] we expect this to happen through a resummation of the entire perturbation series.

A program of calculations to connect string theory directly to general relativity along these lines has been initiated by the pioneering work of Amati, Ciafaloni and Veneziano [4].

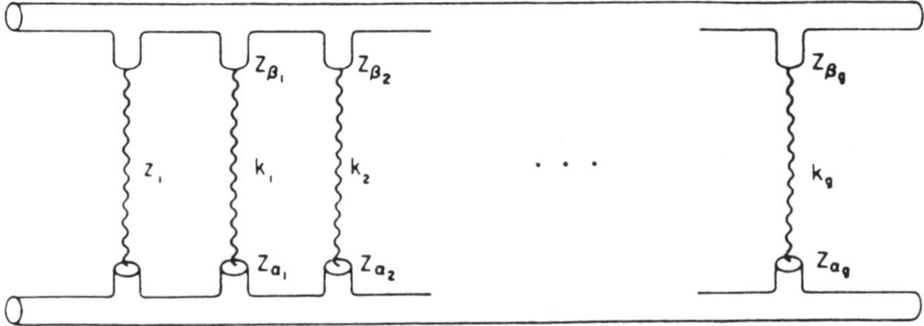

Figure 1: The leading diagram in the two-string scattering amplitude at the g-loop order (the wiggling lines representing the exchange of the leading trajectory)

Whereas their calculations are based on the Regge-Gribov technique [5], our approach [6], like that of ref.[7], is to compute the string amplitudes directly from superstring first principles. To do so, we use the so-called Covariant Loop Calculus —developed concurrently by several groups [8]. This formalism gives an explicit expression for the vertex of three arbitrary string states, and then proceeds to any tree diagram and any g-loop order by sewing these vertices together. This sewing procedure may be viewed simply as a step-by-step recipe for computing the Polyakov path integral on a punctured super Riemann surface with arbitrary boundary conditions [9]. The super-Riemann surface thus constructed is described in terms of the super-Schottky parametrization [8].

Going to large impact parameters we stretch any leg of the string world sheet that connects the two states being scattered. For this reason, one is not interested in all of moduli space, but only in the contributions coming from particular corners corresponding to such a geometrical pinching. This is very fortunate, since the full integration region (i.e. the fundamental domain of the modular group) is unknown in the Schottky parametrization.

When sewing together two pieces of world sheet, the length of the connecting leg is given by minus the logarithm of some modular parameter. So, to pinch a given tube, one simply considers the limit where the corresponding parameter is taken to zero.

2. Resummation of the Loop Expansion

The simplest exercise to undertake is to consider at each loop order only the dominant contribution to the amplitude, the one most singular in t and containing the greatest number of powers of s. At g-loop order this contribution corresponds to the ladder diagram (fig.1) in which the two incoming string states proceed parallel and exchange ($g+1$ times at the g loop order) the leading trajectory, that is the (bosonic) trajectory to which the graviton belongs. (In the $t \to 0$ limit this exchange is completely dominated by the long-range gravitational interaction).

The amplitude obtained from the Covariant Loop Calculus is of course in the form of an integral over the $3g+1$ bosonic and $2g+2$ fermionic moduli of a 4-punctured Super Riemann surface (see ref.[10] for a complete discussion). In the pinching limit considered, it is relatively straightforward to single out the kinematical piece dominant as $s \to \infty$, integrate out the Grassmannian moduli and sum over the spin structures.

This leaves only the integrations over the phase and the absolute value of each of the $3g + 1$ bosonic moduli. These moduli can be grouped into two sets: First there are $2g$, defining the position of insertion of the $g+1$ exchanged tubes on the two fast legs. By varying these parameters we include —in field theory terms— not just the ladder diagram of fig.1 but also all "cross-ladders". Then there are $g + 1$ pinched moduli defining the length of the $g + 1$ exchanged tubes.

All phases are trivially integrated in the $t \to 0$ limit. Of the absolute values, the first $2g$ are dealt with by a stationary phase approximation brought about by the limit $s \to i\infty$. The absolute values of the $g + 1$ pinched moduli we integrate between zero and some small positive number, the dependence on which drops out in the end, and these integrals may be identified as the Schwinger parametrization of the exchanged legs of the corresponding field theory diagram [10], and thus may be replaced by $D - 2$ dimensional integrations over the transversal momenta $\mathbf{q}_1, \ldots, \mathbf{q}_{g+1}$ exchanged in the legs.

By an appropriate choice of the overall numerical coefficient N_g appearing in the topological expansion at g-loop order, the amplitude can be written as

$$a_{g-loop}(s,t) \;=\; \left(\frac{i}{2s}\right)^g \epsilon_1 \cdot \epsilon_2 \, \epsilon_3 \cdot \epsilon_4$$

$$\times \int \frac{d^{D-2}\mathbf{q}_1 \cdots d^{D-2}\mathbf{q}_{g+1}}{(g+1)! \, (2\pi)^{g(D-2)}} \delta^{D-2}(\mathbf{q} - \sum_{\mu=1}^{g+1} \mathbf{q}_\mu) \prod_{\mu=1}^{g+1} a_{tree}(s, \mathbf{q}_\mu) \tag{1}$$

Here ϵ_i, $i = 1, \ldots, 4$ denotes the polarizations of the external states, g_D is the string coupling constant in D space-time dimensions and

$$a_{tree}(s, \mathbf{q}_\mu) = \frac{1}{2}\alpha' g_D^2 \frac{s^2}{\mathbf{q}_\mu^2} \tag{2}$$

is the tree amplitude.

Eq.(1) agrees with ref.[4] and is an explicit proof that the leading part of the amplitude factorizes in the product of $g + 1$ tree amplitudes. In this form it can be diagonalized by going into the impact-parameter space where the series is readily summed into the exponential form of ref. [4]

$$a(s,t) = \epsilon_1 \cdot \epsilon_2 \, \epsilon_3 \cdot \epsilon_4 \, 4s \int d^{D-2}\mathbf{b} \exp\left(i\mathbf{q} \cdot \mathbf{b}\right) \left[\frac{\exp[2i\delta(s,\mathbf{b})] - 1}{2i}\right] , \tag{3}$$

where the so-called eikonal phase δ is given by

$$\delta(s, \mathbf{b}) = \int \frac{d^{D-2}\mathbf{q}}{(2\pi)^{D-2}} \frac{a_{tree}(s,t)}{4s} \tag{4}$$

It is a remarkable feature that the requirement of exponentiation, needed to restore unitarity, allows one to fix the value of the numerical constant N_g at any loop order [10].

3. Calculation of the Deflection Angle

The eqs.(3,4) above were derived for the case of arbitrary string states. This gives us the means to study, among the possible phenomenological consequences of string theory, the deflection angle experienced to this order by a massless test particle (the graviton) in the rest-frame of a state of mass M.

The respective geodesic traced out by the graviton and the massive scalar can be calculated by a stationary phase in the integral (3). From this one computes [4],[10] the deflection to be (for $D = 4$):

$$\Delta\varphi = 4M \frac{g^2 \alpha'}{16\pi\hbar} \frac{1}{b},$$ (5)

where g is the string coupling for $D = 4$ and α' is the square of the string length scale λ_s. In order for this angle to be in agreement with the well-known result obtained in the long-distance (linear) approximation of general relativity, the string coupling constant must be related to Newton's in the following way:

$$g^2 = \frac{16\pi G_N \hbar}{\alpha'}.$$ (6)

This identification is analogous to the one in quantum electrodynamics that relates the coupling constant to the charge of the electron by requiring the Born cross section to go into the classical one in the long distance limit.

Now, the eikonal phase δ, which to leading order is just given by eq.(4), has corrections coming from string diagrams subleading in powers of s and/or less singular in t than the ladder diagram. The corrections involved are essentially of a threefold nature: 1) Corrections of $o(\lambda_s/b)$, corresponding to stringy effects becoming more important at shorter distances. 2) Corrections of $o(G_N \sqrt{s}/b)$, that is, classical non-linear corrections. 3) Quantum corrections. This obviously allows for a wide variety of physical investigations, depending on the relative importance of the three types of corrections.

If we want to compare with general relativity, we should look for the first non-linear correction, while still steering clear of the stringy and quantum regimes. Such a correction is of the second order in Newton's constant, and hence, by eq.(6), of the fourth order in the string coupling constant. It therefore has to originate at 1-loop order. As discussed in detail in ref.[11] the relevant piece may be identified and —assuming exponentiation to hold also for this subleading contribution— one arrives at the following expression for the deflection angle:

$$\Delta\varphi_{\text{STRING}} = \frac{4G_N M}{b} + \frac{\pi}{2}\left(\frac{G_N M}{b}\right)^2.$$ (7)

We can compare it to the post-linear approximation [12] to general relativity that gives, to the same order,

$$\Delta\varphi_{\text{GR}} = \frac{4G_N M}{b} + 15\frac{\pi}{4}\left(\frac{G_N M}{b}\right)^2,$$ (8)

There is obviously a discrepancy between the two second-order coefficients. A careful investigation reveals that the string result contains comparable contributions from all the massless states of the low-energy theory ($N = 8$ supergravity using our toroidal compactification), and therefore, at this order, it is not surprising to find disagreement with general relativity. To obtain agreement, we

852

would have to face again the problems of supersymmetry breaking and of giving a mass to the unwanted massless states.

4. Summary

In summary, the Planckian regime allows one to extract several interesting informations connecting general relativity and string theory in a manner that is independent of compactification details. In particular, resummation of the loop contributions leading in powers of s lead to a determination of the absolute normalization of all terms in the topological expansion of string perturbation theory and to a relation between Newton's constant and the string coupling constant. Extending the analysis to subleading diagrams, however, care should be taken; for, even though the Planckian regime is indeed dominated by gravity, it does not necessarily imply the automatic decoupling of all superpartners but the graviton. Rather, one should be careful and select only those processes where such a decoupling is present. The collision of two gravitons, instead of a graviton and a massive state, is a promising candidate for such a process [13].

References

[1] S. Mandelstam, Talk given at the Workshop on String Theories (Santa Barbara, 1986).

[2] See, e.g., M.B. Green, J.H. Schwarz and E. Witten, *Superstring Theory* (Cambridge University Press, 1987).

[3] M. Soldate, *Phys. Lett.* **B186** (1987) 321 and references cited therein.

[4] D. Amati, M. Ciafaloni and G. Veneziano, *Phys. Lett.* **B197** (1987) 129; *Int. J. Mod. Phys.* **A3** (1988) 1615.

[5] See, e.g., P.D.B. Collins, *Regge Theory and High Energy Physics* (Cambridge University Press, 1977) and references cited therein.

[6] G. Cristofano, M. Fabbrichesi and K. Roland, *Phys. Lett.* **B236** (1990) 159.

[7] E. Gava, R. Iengo and C.-J. Zhu, *Nucl. Phys.* **B323** (1989) 585

[8] P. Di Vecchia, M. Frau, K. Hornfeck, A. Lerda and S. Sciuto, *Phys. Lett.* **B211** (1988) 301, and references cited therein; J.L. Petersen, J.R. Sidenius and A.K. Tollsten, *Nucl. Phys.* **B317** (1989) 109 and G. Cristofano, R. Musto, F. Nicodemi and R. Pettorino, *Phys. Lett.* **B217** (1989) 59.

[9] J.R. Sidenius, *Reggeons and the String Path Integral*, in Perspectives in String Theory, Copenhagen, October 1987 (P.Di Vecchia and J.L. Petersen eds., World Scientific, 1988)

[10] G.Cristofano, M. Fabbrichesi and K. Roland, *Eikonal Resummation and Topological Expansion Coefficients in Superstring Theory*, preprint NORDITA-90/13 P.

[11] R. Iengo and K. Lechner, *Schwarzschild-like Corrections to Gravity from Superstring at One Loop*, preprint SISSA 98EP (1989)and G. Cristofano, M. Fabbrichesi and K. Roland, *Quantum Correcctions and Non Linearity in Gravity. A Superstring 1-Loop Calculation*, preprint NORDITA-90/20 P.

[12] For a review, see: K. Westpfahl, *Fortschr. Phys.* **33** (1985) 417.

[13] D. Amati, M. Ciafaloni and G. Veneziano, *Higher Order Gravitational Deflection and Soft Bremsstrahlung in Planckian Energy Superstring Collisions*, preprint CERN-TH.5636/90

QUANTUM MAPS FOR DEFORMED ALGEBRAS

C.K. ZACHOS

High Energy Physics Division *, Argonne National Laboratory, Argonne, IL 60439
zachos@anlhep

We find explicit functionals that map SU(2) algebra generators to those of several quantum deformations of that algebra. We indicate how any such quantized algebra can be mapped to any other, and how representations of any such algebra can be expressed as simple functions of SU(2) representations. The representation theory and its comultiplication rules are thus systematized and streamlined by direct reference to their SU(2) correspondents. We speculate on quantum deformations of the Virasoro algebra.

I will outline and sample work I have done in collaboration with T. Curtright[1] on Quantized Universal Enveloping algebras (QUE-algebras; also see [2-4]) pioneered by Drinfeld and Jimbo[5]. These remarkable mathematical structures (distinctive Hopf algebras) are emerging in several disparate physics contexts[5-7] which they link together: in solutions of the Yang-Baxter equation for integrable models; in spin chains, and in vertex operators of rational conformal field theory; in string noncommutative geometry, Wilson lines of the 3-d Chern-Simons action, and knot theory. They are called "quantum deformations", since they generally depend on one or more parameters which yield a conventional Lie algebra in their "classical" limit to special values, analogous to the $\hbar \to 0$ limit of the correspondence principle or the well-known Wigner-Inönü contractions.

We assign more explicit meaning to the term "deformation", by providing simple invertible functionals of the generators of SU(2) which satisfy each QUE-algebra, respectively. These functionals then directly deform SU(2) continuously and reversibly (except for special values of the deformation parameter) into each QUE-algebraic structure, and thereby connect all these structures. Substituting any representation of SU(2) into these functionals consequently produces the corresponding representation of these QUE-algebras. Thus their rich and important representation theory[8-9] is codified and systematized, and its comultiplication rules are directly referred to their classical correspondents.

For specificity, consider the SU(2) algebra:

$$[j_0, j_+] = j_+ \qquad [j_+, j_-] = j_0 \qquad [j_-, j_0] = j_- \ . \tag{1}$$

*Work supported by the U.S.Department of Energy, Division of High Energy Physics, Contract W-31-109-ENG-38. Talk at the DPF90 Meeting of the APS, Rice U., January 4, 1990.

854

The Casimir invariant is

$$C \equiv 2j_+j_- + j_0(j_0 - 1) = 2j_-j_+ + j_0(j_0 + 1) \equiv j(j + 1) \,, \tag{2}$$

where the operator j is straightforward to solve for in terms of the operator C. Consequently, $-2j_+j_- = (j_0 + j)(j_0 - 1 - j)$. Note the symmetry $j_+ \leftrightarrow j_-$, $\quad j_0 \leftrightarrow -j_0$. Further note $j_+ f(j_0) = f(j_0 - 1)j_+$.

Also for specificity, consider Drinfeld–Jimbo's quantum deformation[5] :

$$[J_0, J_+] = J_+ \qquad\qquad [J_+, J_-] = \frac{1}{2}\,[2J_0]_q \qquad\qquad [J_-, J_0] = J_- \,, \tag{3}$$

where the Chebyshev polynomial of the 2nd kind $[x]_q \equiv (q^x - q^{-x})/(q - q^{-1})$ is called the "q-deformation"[10] of x. The r.h.s. of the middle eq.(3) transforms to $[J_0]_{q^2}$ provided the generators $J_\pm$ absorb a factor $\sqrt{2/(q + 1/q)}$ in their normalization. The invariant of this algebra is

$$C_q \equiv 2J_+J_- + [J_0]_q[J_0 - 1]_q = 2J_-J_+ + [J_0]_q[J_0 + 1]_q \equiv [j]_q[j + 1]_q \,. \tag{4}$$

Since this is not a closing Lie algebra, its group structure (exponentiation) is beset with subtlety.

The classical limit $q \to 1$ yields SU(2). Can we reverse this starting from SU(2), i.e. can we find functionals of the generators (1) that satisfy the "quommutation" relations (3)? The answer is yes, in analogy to the well-known "group expansions"[11]. Recall the elementary physics example of inflating a 2-sphere to a plane: $\mathbf{S}^2 \sim \mathrm{SO}(3)/\mathrm{SO}(2) \to \mathbf{R}^2 \sim \mathrm{E}(2)/\mathrm{SO}(2)$,

$$[j_x, j_y] = j_z \qquad\qquad [j_y, j_z] = j_x \equiv rP_y \qquad\qquad [j_z, j_x] = j_y \equiv -rP_x \,, \tag{5}$$

by sending the radius $r \to \infty$, which contracts the SU(2) algebra to

$$[P_x, P_y] = 0 \qquad\qquad [j_z, P_x] = P_y \qquad\qquad [j_z, P_y] = -P_x \,. \tag{6}$$

E.g. this is how the Lorentz group is introduced via the de Sitter group. But conversely, SO(3) may be reconstituted out of the *expansion*[11] of the Euclidean operators. Since $P_x^2 + P_y^2$ commutes with all three E(2) generators, it may be handled as a constant. Defining

$$j_x' \equiv \frac{[j_z^2, P_x]}{2i\sqrt{P_x^2 + P_y^2}} = \frac{j_zP_y + P_yj_z}{2i\sqrt{P_x^2 + P_y^2}} \qquad\qquad j_y' \equiv \frac{[j_z^2, P_y]}{2i\sqrt{P_x^2 + P_y^2}} = \frac{-j_zP_x - P_xj_z}{2i\sqrt{P_x^2 + P_y^2}}, \tag{7}$$

it follows directly that they obey the original SO(3) commutation relations:

$$[j_x', j_y'] = j_z \qquad\qquad [j_y', j_z] = j_x' \qquad\qquad [j_z, j_x'] = j_y' \,. \tag{8}$$

In the same spirit, we find deformation functionals Q which convert the SU(2) generators g to operators $G = Q(g)$ which obey the quommutations (3). To start with, we take for simplicity q real (or, in an exceptional case below, a phase) which allows hermitean conjugation $\mathcal{O}_- = \mathcal{O}_+^\dagger$ for the generator operators under discussion:

$$J_0 = Q_0(j_0) = j_0 \qquad J_+ = Q_+(g) = \sqrt{\frac{[j_0 + j]_q [j_0 - 1 - j]_q}{(j_0 + j)(j_0 - 1 - j)}} \; j_+ \qquad J_- = Q_+^\dagger(g). \quad (9)$$

The maps $Q_\pm$ are functionals of all three SU(2) generators j_0, j_+, j_-, since they depend on the operator j. Note that $C_q = [j]_q [j+1]_q$. These deformation maps are readily invertible, except for special cases when the parameters are roots of unity, which can be analyzed as limits of the above.

Substituting specific matrix representations in the above connection formulas produces the corresponding representations of the QUE-algebra in a direct, mechanical fashion. For example, the ($j = 1/2$, Pauli-matrices) **2** representation of SU(2) :

$$j_0 = \frac{1}{2}\begin{pmatrix} 1 & 0 \\ 0 & -1 \end{pmatrix} \qquad\qquad j_+ = \frac{1}{\sqrt{2}}\begin{pmatrix} 0 & 1 \\ 0 & 0 \end{pmatrix} \qquad\qquad (10)$$

maps to $J_0 = j_0$, $\quad J_+ = j_+$. The **3**:

$$j_0 = \begin{pmatrix} 1 & 0 & 0 \\ 0 & 0 & 0 \\ 0 & 0 & -1 \end{pmatrix} \qquad\qquad j_+ = \begin{pmatrix} 0 & 1 & 0 \\ 0 & 0 & 1 \\ 0 & 0 & 0 \end{pmatrix} \qquad\qquad (11)$$

maps to $J_0 = j_0$, $J_+ = \sqrt{(q + 1/q)/2}\, j_+ = \sqrt{[2]_q/2}\, j_+$. The **4**:

$$j_0 = \begin{pmatrix} 3/2 & 0 & 0 & 0 \\ 0 & 1/2 & 0 & 0 \\ 0 & 0 & -1/2 & 0 \\ 0 & 0 & 0 & -3/2 \end{pmatrix} \qquad j_+ = \begin{pmatrix} 0 & \sqrt{3/2} & 0 & 0 \\ 0 & 0 & \sqrt{2} & 0 \\ 0 & 0 & 0 & \sqrt{3/2} \\ 0 & 0 & 0 & 0 \end{pmatrix} \quad (12)$$

maps to

$$J_0 = j_0 \qquad\qquad J_+ = \begin{pmatrix} 0 & \sqrt{[3]_q/2} & 0 & 0 \\ 0 & 0 & [2]_q/\sqrt{2} & 0 \\ 0 & 0 & 0 & \sqrt{[3]_q/2} \\ 0 & 0 & 0 & 0 \end{pmatrix}, \qquad (13)$$

and so forth.

Now observe that $[3]_q = 0$ for $q = \exp(2\pi i/3)$, and hence the **4** representation J_+ now has only one nontrivial entry and $J_+^2 = 0$; the middle commutator in (3) breaks up, so the representation reduces: $\mathbf{4} = \mathbf{1} \oplus \mathbf{2} \oplus \mathbf{1}$. This reduction obtains for roots of unity with period smaller than the dimensionality of the representation[9].

In the more general nonhermitean case, use instead $J_0 = j_0$ and

$$J_+ = \left(\frac{[j_0 + j]_q}{j_0 + j}\right)^{1/2+\lambda}\left(\frac{[j_0 - 1 - j]_q}{j_0 - 1 - j}\right)^{1/2-\lambda} j_+, \quad J_- = j_-\left(\frac{[j_0 + j]_q}{j_0 + j}\right)^{1/2-\lambda}\left(\frac{[j_0 - 1 - j]_q}{j_0 - 1 - j}\right)^{1/2+\lambda},$$

$$(14)$$

for arbitrary λ. Our papers similarly treat SU(1,1), and further popular deformations of SU(2), such as Witten's deformation[6]

$$[W_0, W_+]_r \equiv rW_0W_+ - \frac{1}{r}W_+W_0 = W_+ \qquad [W_+, W_-]_{1/r^2} = W_0 \qquad [W_-, W_0]_r = W_-, \quad (15)$$

which interpolates between SU(2) for $r = 1$ and SU(1,1) for $r = -1$; or the 2-parameter generalization of Fairlie[3],

$$[I_0, I_+]_r = I_+ \qquad [I_+, I_-]_{1/s} = I_0 \qquad [I_-, I_0]_r = I_- , \qquad (16)$$

(reducing to the previous one upon $s \to r^2$), and several others. These deformation functionals and their inverses serve to transmute any QUE-algebra to any other.

Our method leading to these involves inserting natural Ansätze such as $J_0 = F(j_0)$, $J_+ = H(j_0)j_+$, $J_- = j_-H(j_0)$, into the proper quommutation relations and solving the resulting functional equations for F and H. It extends to graded algebras[2] and SU(N) algebras[4].

Having referred the representation theory of the QUE-algebras to the representation theory of SU(2), it follows that the composition laws for representations are likewise linked. In the addition of angular momenta, two parallel operators tensor-multiply to an operator satisfying the same SU(2) commutation relations; this operator is a reducible representation of SU(2), the reduction effected by the Clebsch-Gordan operator C:

$$\Delta(g) = \mathbb{1} \otimes g + g \otimes \mathbb{1} = C(g_1 \oplus g_2 \oplus g_3 \oplus ...)C^{-1} . \qquad (17)$$

It is evident that the invertible map Q from SU(2) generators g to SU(2)$_q$ generators $G = Q(g)$ induces a tensor coproduct of G's

$$Q(\Delta(g)) = Q(\mathbb{1} \otimes Q^{-1}(G) + Q^{-1}(G) \otimes \mathbb{1}) \qquad (18)$$

which obeys SU(2)$_q$ quommutations, since its argument obeys SU(2). Any similarity transformation $U^{-1}QU$ on the above coproduct will also produce an isomorphic induced tensor coproduct (comultiplication). In particular, the same Clebsch operator C will automatically also reduce the coproduct (18): $C^{-1}Q(\Delta(g))C = G_1 \oplus G_2 \oplus G_3\oplus$.

Recall, however, the standard[5] comultiplication rule for (3)

$$\Delta_q(J_0) = J_0 \otimes \mathbb{1} + \mathbb{1} \otimes J_0 \qquad\qquad \Delta_q(J_\pm) = J_\pm \otimes q^{J_0} + q^{-J_0} \otimes J_\pm , \qquad (19)$$

which appears different than the induced coproduct (18). It is reduced to a direct sum by the unitary q-Clebsch operators[12] C_q, and consequently

$$Q(\Delta(g)) = CC_q^{-1} \Delta_q(G) C_q C^{-1}. \qquad (20)$$

Thus our induced comultiplication is related to the well-known one by a similarity transformation $U_q = C_qC^{-1}$, the unitary operator that converts C to C_q .

As an aside, application of these techniques and our older construction[10], leads to the q-generalization of the Virasoro operators: $Z_m = x^{-m} (r^{2x\partial} - 1)/(r - 1/r)$, which satisfy a deformation of the centerless Virasoro algebra

$$[Z_m, Z_n]_{r^{n-m}} = [m - n]_r Z_{m+n} . \qquad (21)$$

In the limit $r \to 1$, Z_m yields the standard Virasoro realization $x^{1-m}\partial$. However, in the absence of comultiplication rules, this structure may not yet be claimed to constitute a bonafide Hopf algebra—may it serve as an inspiration for one?

To sum up, the techniques introduced here connect all QUE-deformations among themselves and refer their representation theory and coproducts to those of the underlying classical Lie algebra.

REFERENCES

1. T. Curtright and C. Zachos, Argonne preprint ANL-HEP-PR-89-105, October 1989, to appear in Phys.Lett. B; T. Curtright, G. Ghandour and C. Zachos, Argonne preprint ANL-HEP-PR-90-08, March 1990.

2. T. Curtright, Miami preprint TH/1/89, to appear in *Physics and Geometry*, L.-L. Chau and W. Nahm, eds., (Plenum); T. Curtright and G. Ghandour, Miami preprint TH/8/89.

3. D. Fairlie, IAS preprint, IASSNS-HEP-89/61, to appear in J.Phys.A, and private communications.

4. A. Polychronakos Florida preprint, UFTP-89-23, November 1989.

5. V. Drinfeld, Sov.Math.Dokl. **32** (1985) 254; M. Jimbo, Lett.Math.Phys. **10** (1985) 63; **11** (1986) 247; Commun.Math.Phys. **102** (1986) 537.

6. E. Witten, Institute for Advanced Study preprint, IASSNS-HEP-89/32.

7. L. Faddeev, N. Reshetikhin, and L. Takhtajan, Steklov preprint LOMI-E-14-87; M. Jimbo, Int.J.Mod.Phys. **A4** (1989) 3759; S. Majid, Int.J.Mod.Phys. **A5** (1990) 1.

8. M. Rosso, Comm.Math.Phys. **117** (1988) 581.

9. P. Roche and D. Arnaudon, Lett.Math.Phys. **17** (1989) 295; L. Alvarez-Gaumé, C. Gomez, and G. Sierra, Phys.Lett. **220B** (1989) 142; Nucl.Phys. **B319** (1989) 155; V. Pasquier and H. Saleur, Nucl.phys. B, in press.

10. G. Andrews, *q-Series*, AMS Regional Conference Series in Mathematics, **66**, Providence, 1985; R. Askey and J. Wilson, SIAM.J.Math.Anal. **10** (1979) 1008.

11. R. Gilmore, *Lie Groups, Lie Algebras, and Some of Their Applications*, Wiley, New York, 1974.

12. A. Kirillov and N. Reshetikhin, LOMI preprint (1988); V. Pasquier, Com.Math.Phys. **118** (1988) 355; M. Nomura, J.Math.Phys. **30** (1989) 2397; J.Phys.Soc.Jap. **58** (1989) 2694; ibid. **59** (1990) No2; Bo-Yu Hou, Bo-Yuan Hou, and Z-Q. Ma, Beijing preprints BIHEP-TH-89-7;8; H. Ruegg, Geneva preprint UGVA-DPT 1989/08-625; C-P. Sun and H-C. Fu, J. Phys. **A22** (1989) L983; L. Biedenharn, Duke preprint Print-89-0864.

13. D. Fairlie, J. Nuyts, and C. Zachos, Phys.Lett. **202B**, 320 (1988).

QUANTUM WORMHOLES IN LORENTZIAN SIGNATURE

Matt Visser
Physics Department
Washington University
St. Louis, Missouri 63130–4899

ABSTRACT

This note presents a minisuperspace analysis of a simple class of wormholes that are based on surgically modified Reissner–Nordstrøm spacetimes. By solving the Wheeler–de Witt equation for Einstein gravity on this minisuperspace the quantum mechanical wave function of the wormhole is obtained in closed form.

1.   INTRODUCTION

The past two years have seen a marked resurgence of interest in wormholes defined on Lorentzian signature spacetimes [1–5]. A major result of these investigations is that violations of the weak energy hypothesis occur at the throats of such wormholes. Experimentally, it is known that quantum mechanics violates the weak, strong, and dominant energy hypotheses [1–5]. This suggests that quantum mechanics may be the key to interesting wormhole physics in Lorentzian spacetime, and it is this topic that shall be addressed in this note.

2.   THE EINSTEIN–HILBERT ACTION IN MINI–SUPERSPACE

To construct the class of wormholes of interest, consider two copies of the Reissner–Nordstrøm geometry. Use Schwarzschild co-ordinates, as use of the maximally extended solutions proves unproductive. Remove from each copy identical four–dimensional regions of the form $\{(r,\theta,\phi,t)|r < a(\tau)\}$. One is left with two geodesically incomplete manifolds with boundaries given by the timelike hypersurfaces $\{(r,\theta,\phi,t)|r = a(\tau)\}$. Identify these two hypersurfaces. The resulting spacetime is geodesically complete. It possesses two asymptotically flat regions connected by a wormhole. The throat of the wormhole is at $r = a(\tau)$. Because this manifold is piecewise Reissner–Nordstrøm, the Ricci scalar is everywhere zero, except at the throat itself. At the throat the Riemann curvature tensor is proportional to a delta function. The Ricci tensor at the junction can be calculated in terms of the second fundamental form

$$\mathcal{K}^i{}_j = \tfrac{1}{2}\, g^{ik} \cdot \left.\frac{\partial g_{kj}}{\partial \eta}\right|_{\eta=0}.$$

Here η denotes the normal coordinate to the throat, while τ denotes proper time along the throat. The Ricci tensor is [4]:

$$R^\mu{}_\nu(x) = R^{(RN)\mu}{}_\nu(x) - 2 \cdot \begin{bmatrix} \mathcal{K}^i{}_j(x) & 0 \\ 0 & \mathcal{K}(x) \end{bmatrix} \cdot \delta(\eta).$$

The Einstein–Hilbert action reduces to

$$S = \frac{c^3}{16\pi G} \int_{\mathcal{M}} \sqrt{g_4}\, R = \frac{-c^3}{4\pi G} \int_{\partial\Omega} \sqrt{g_3}\, \mathcal{K}.$$

Henceforth adopt "geometrodynamic units" so that $G \equiv 1$, $c \equiv 1$, and $\hbar \equiv L_P^2 \equiv M_P^2$, where L_P and M_P are the Planck length and Planck mass respectively. The two non–trivial components of the second fundamental form may conveniently be extracted from reference 4.

$$\mathcal{K}^{\theta}{}_{\theta} \equiv \mathcal{K}^{\phi}{}_{\phi} = \frac{1}{a} \cdot \sqrt{1 - \frac{2M}{a} + \frac{Q^2}{a^2} + \dot{a}^2};$$

$$\mathcal{K}^{\tau}{}_{\tau} = \frac{\ddot{a} + \frac{M}{a^2} - \frac{Q^2}{a^3}}{\sqrt{1 - \frac{2M}{a} + \frac{Q^2}{a^2} + \dot{a}^2}} = \frac{d}{d\tau} \mathrm{Sinh}^{-1}\left(\dot{a}/\sqrt{1 - \frac{2M}{a} + \frac{Q^2}{a^2}}\right) + \frac{1}{a^2}\left(M - \frac{Q^2}{a}\right)\frac{dt}{d\tau}.$$

Since $\sqrt{g_3}\, d^3x \mapsto 4\pi\, a^2\, d\tau$, an integration by parts leads to

$$S_g = 2 \int \left\{ a\,\dot{a}\,\mathrm{Sinh}^{-1}\left(\dot{a}/\sqrt{1 - \frac{2M}{a} + \frac{Q^2}{a^2}}\right) - a\sqrt{1 - \frac{2M}{a} + \frac{Q^2}{a^2} + \dot{a}^2}\right\} d\tau - \int \left(M - \frac{Q^2}{a}\right) dt.$$

It should be noted that this calculation is a direct analogue of the minisuperspace techniques more commonly used in quantum cosmology. The electromagnetic contribution to the action is

$$S_{em}^{(1)} = \frac{-1}{16\pi} \int F^2 = \frac{-1}{16\pi} \int 4\pi r^2 \frac{Q^2}{r^4} dr dt = \frac{Q^2}{2} \int_a^\infty \frac{dr}{r^2} dt = -\frac{Q^2}{2} \int \frac{dt}{a(t)} = S_{em}^{(2)}.$$

A convenient technical trick at this stage is to cut off the Reissner–Nordstrøm geometry at some large fixed radius ρ. Then attach this truncated Reissner-Nordstrøm geometry to a piece of Minkowski space. This procedure contributes to the Einstein–Hilbert action an amount

$$S_{\rho}^{(1)} = \frac{1}{2}\int [\mathcal{K}]\rho^2 dt = \frac{1}{2}\int M\,dt + o(1/\rho) = S_{\rho}^{(2)}.$$

Here $[\mathcal{K}]$ denotes the discontinuity in the second fundamental form at $r = \rho$. Now let $\rho \to \infty$ and combine terms to obtain

$$\begin{aligned}
S_{eff} &= S_g + S_{em}^{(1)} + S_{em}^{(2)} + S_{\infty}^{(1)} + S_{\infty}^{(2)}\\
&= 2\int\left\{a\,\dot{a}\,\mathrm{Sinh}^{-1}\left(\dot{a}/\sqrt{1 - \frac{2M}{a} + \frac{Q^2}{a^2}}\right) - a\sqrt{1 - \frac{2M}{a} + \frac{Q^2}{a^2} + \dot{a}^2}\right\} d\tau.
\end{aligned}$$

3.   CLASSICAL BEHAVIOUR OF THE WORMHOLE

In order to understand better the classical behaviour of this wormhole, we shall add some matter to the model. We choose pressureless dust that is confined to lie on the throat of the wormhole. Then $m = 4\pi\sigma a^2$, the mass of the dust shell, is a constant of the motion. The matter Lagrangian simply reduces to $L_m = -m$, and the total Lagrangian describing the matter plus gravity system is

$$L_{tot} = 2\left\{a\,\dot{a}\,\mathrm{Sinh}^{-1}\left(\dot{a}/\sqrt{1 - \frac{2M}{a} + \frac{Q^2}{a^2}}\right) - a\sqrt{1 - \frac{2M}{a} + \frac{Q^2}{a^2} + \dot{a}^2}\right\} - m.$$

The classical Wheeler–de Witt Hamiltonian is now easily extracted; the conjugate momentum to a is

$$p \equiv \frac{\partial L}{\partial \dot{a}} = 2a\,\mathrm{Sinh}^{-1}\left(\dot{a}/\sqrt{1 - \frac{2M}{a} + \frac{Q^2}{a^2}}\right).$$

This relation may be easily inverted to yield $\dot{a} = \sqrt{1 - \frac{2M}{a} + \frac{Q^2}{a^2}} \cdot \mathrm{Sinh}(p/2a)$ so that the Wheeler-de Witt Hamiltonian is

$$H_{tot}(p, a) \equiv p\dot{a} - L_{tot} = 2a \cdot \sqrt{1 - \frac{2M}{a} + \frac{Q^2}{a^2}} \cdot \mathrm{Cosh}\left(p/2a\right) + m = 2a \cdot \sqrt{1 - \frac{2M}{a} + \frac{Q^2}{a^2} + \dot{a}^2} + m.$$

The classical dynamics of the wormhole is now obtained by setting $H_{tot} = 0$. The fact that the Hamiltonian is zero is a standard consequence of the reparameterization invariance of the theory.

4. QUANTIZATION

With the classical Wheeler-de Witt Hamiltonian now in hand, quantization is straightforward. The only remarkable aspect of the analysis is that in some cases closed form exact expressions are obtained. Canonical quantization proceeds via the usual replacement $p \mapsto -i\hbar \frac{\partial}{\partial a}$. The resulting quantum Hamiltonian has a factor ordering ambiguity. This factor ordering ambiguity may be removed in a natural (though not unique!) way by demanding Hermiticity:

$$\hat{H}_{tot} = \sqrt[4]{1 - \frac{2M}{a} + \frac{Q^2}{a^2}} \cdot 2a \, \mathrm{Cos}\left(\frac{L_P^2}{2}\frac{1}{a}\frac{\partial}{\partial a}\right) \cdot \sqrt[4]{1 - \frac{2M}{a} + \frac{Q^2}{a^2}} + m.$$

This Hamiltonian acts on $L^2[0,\infty)$, the space of square integrable functions on $[0,\infty)$. The wave function of the wormhole is determined in the usual fashion by the Wheeler-de Witt equation $\hat{H}_{tot}\psi(a) = 0$, which may be rewritten as $\hat{H}_{eff} \, \psi = -m \, \psi$. Thus m may be interpreted as an eigenvalue of the effective Hamiltonian associated with L_{eff}. The mass of the dust shell is therefore quantized in this formalism. For nonzero values of m exact solutions have proved elusive, and one must resort to WKB techniques [6]. For the special eigenvalue of $m = 0$ exact solutions have been obtained:

$$\psi_n(a) = \frac{\exp\left[-(n + \frac{1}{2})\pi(a/L_P)^2\right]}{\sqrt[4]{1 - \frac{2M}{a} + \frac{Q^2}{a^2}}}.$$

Here n is integer valued quantum number describing the internal state of the wormhole. Note that the dynamics thus implies that $\psi(0) = 0$. The expectation value of the wormhole radius is $\langle \psi_n|a|\psi_n \rangle \approx L_P$.

5. DISCUSSION

Perhaps the most damaging criticism that can be made concerning this calculation is that it is performed in minisuperspace instead of using Wheeler's full superspace. It is quite possible that the brutal truncation from an infinite number of degrees of freedom $g_{ij}(\vec{x},t)$ down to one degree of freedom $a(\tau)$ has also brutally truncated the real physics. There is no satisfactory answer to this objection, except to point out that given our current calculational abilities (or rather, lack of them) we simply have no choice if we are to be able to calculate anything. However, although the application is unique, the minisuperspace technology employed is a standard quantum gravity technique.

Finally, I should mention some other potentially serious problems: (1) Though the factor ordering choice made in $\hat{H}_{tot}$ is in some sense "natural", it is by no means unique. (2) Since the expected wormhole radius is of order the Planck length, it is far from clear that the Einstein–Hilbert action is an appropriate description for gravity. If R^2 terms are present (and we expect this on rather general grounds) the analysis of this note fails. In particular the "boundary layer" formalism is no longer appropriate since this would now involve squares of delta functions.

I wish to acknowledge fruitful discussions with Steve Blau, Edward Farhi, David Garfinkle, Henry Kandrup, Ian Redmount, and Wai-Mo Suen.

6. REFERENCES

[1] M.S. Morris and K.S. Thorne, Am. J. Phys. 56 (1988) 395.
[2] M.S. Morris, K.S. Thorne, and U. Yurtsever, Phys. Rev. Lett. 61 (1988) 1446.
[3] M. Visser, Phys. Rev. D39 (1989) 3182.
[4] M. Visser, Nucl. Phys. B328 (1989) 203.
[5] M. Visser, Phys. Rev. D41 (1990), 1116.
[6] I.H. Redmount, W.M. Suen, and M. Visser, in preparation.

RADIATIVE GENERATION OF QUARK AND LEPTON MASSES

Ernest Ma
Department of Physics
University of California
Riverside, CA   92521

ABSTRACT

I review some of the basic ideas and mechanisms of radiative mass generation in gauge models, and comment briefly on the present status and future outlook.

INTRODUCTION

This talk is a brief review of some ideas and mechanisms of radiative mass generation that others and I have been exploring in the last year or two.[1] There are two underlying assumptions: (1) We deal only with renormalizable spontaneously broken gauge theories, and (2) there should be no mass scale greater than a few TeV. Four types of mass are then possible: (1) A gauge-invariant mass, (2) a spontaneously generated mass, i.e., $m = fv$ where v is the vacuum expectation value of a neutral scalar field, (3) a see-saw mass from the matrix.

$$\underline{M} = \begin{bmatrix} 0 & m_1 \\ m_2 & m_3 \end{bmatrix} \tag{1}$$

and (4) a radiative mass. Fermion masses are zero at tree level either because (A) there is some extra symmetry, or because (B) the particle content is restricted, or both. In many cases, (B) actually implies (A). For example, in the standard SU(2) X U(1) electroweak gauge model, the absence of ν_R (and a scalar triplet which can couple to $\nu_L\nu_L$) implies $m_\nu = 0$ __and__ the conservation of lepton number.

RADIATIVE MECHANISMS

Fermion mass terms are radiatively generated if (A) is softly or spontaneously broken, or (B) does not result in an exact extra symmetry. For example, in the standard model, if we add just __one__ ν_R and allow it a (gauge-invariant) Majorana mass, then lepton number is not conserved and one linear combination of ν_e, ν_μ, and ν_τ, acquires a see-saw mass with $m_1 = m_2 = fv$ in Eq. (1). The other two orthogonal combinations are massless at tree level but obtain radiative masses through double W-boson exchange,[2] as shown in Fig. 1.

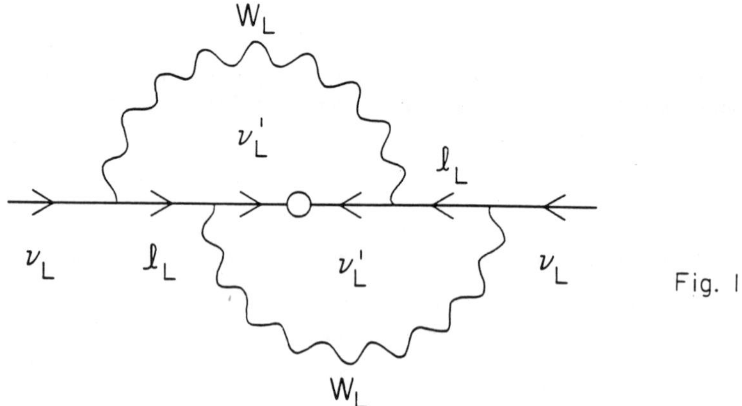

Fig. 1

This is called Mechanism (5) in Ref. 1 and is the only one which does not require extra Higgs bosons or additional symmetry.

There are five other basic mechanisms as discussed below.

(1) One-loop scalar exchange with one fermion mass insertion as shown in Fig. 2.

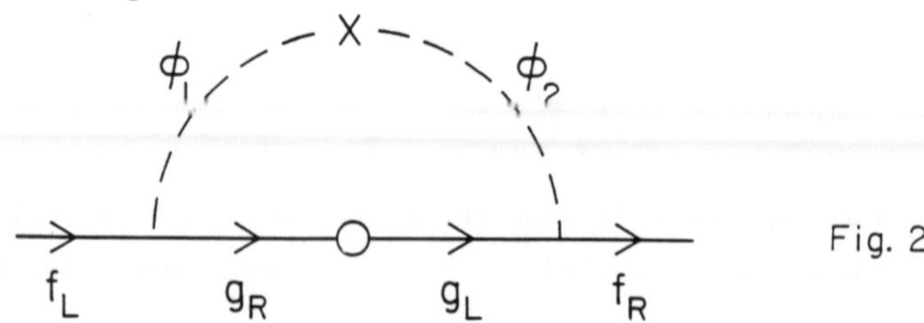

Fig. 2

Here ϕ_1 (ϕ_2) must not couple to f_R (f_L), but ϕ_1 and ϕ_2 must mix as the result of symmetry breaking such as $Z_4 \to Z_2$ in Ref. 3 or $SU(2)_L$ X $SU(2)_R$ X $U(1) \to U(1)$ in Ref. 4.

(2) One-loop scalar exchange with three fermion mass insertions as shown in Fig. 3.

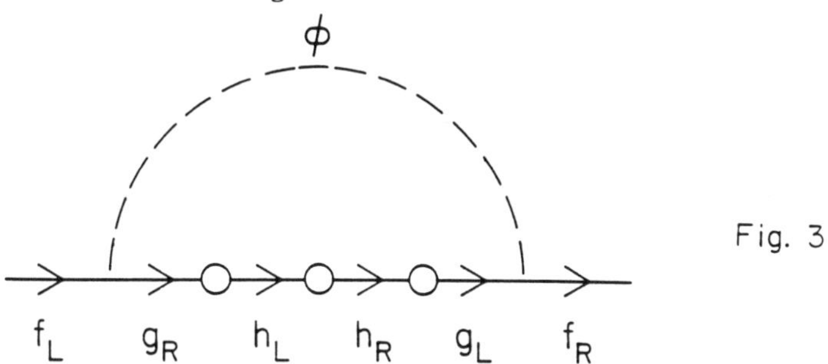

Fig. 3

The first model of this type was that of Ref. 5. Note that g in Fig.3 must not have a tree-level mass by itself, hence the fermion mass matrix spanning g and h is of the see-saw form as given by Eq. (1). I should remark also that in the above two diagrams, a line labeled by $g_{R,L}$ may be replaced with one labeled by $g'_{L,R}$ with the arrow reversed.

(3) One-loop induced scalar expectation value with five fermion mass insertions as shown in Fig. 4.

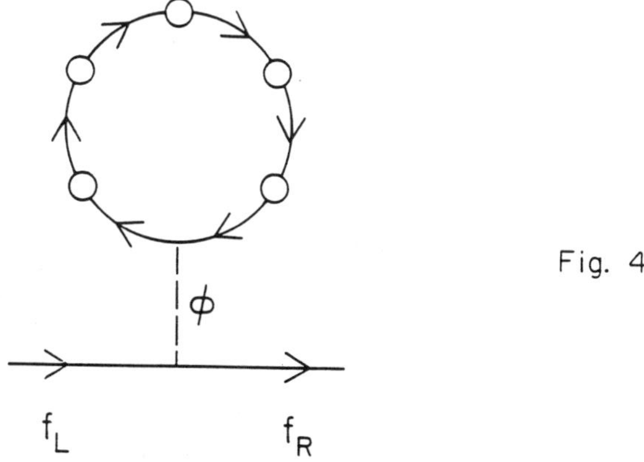

Fig. 4

The first model of this type was that of Ref. 6. Note that the mass matrix spanning the three fermions occurring in the loop is of the double see-saw form.

(4) Two-loop scalar exchange with cubic scalar coupling as shown in Fig. 5.

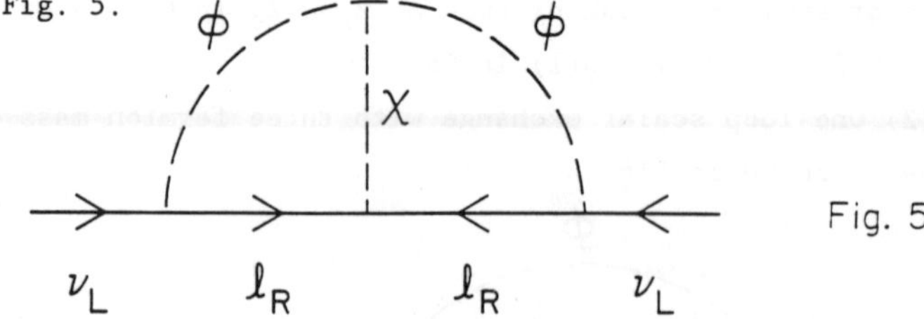

This has been used to obtain very small Majorana neutrino masses.[7] Note that two fermion mass insertions or one scalar mixing is required to render this diagram finite. The latter has the interesting feature that the radiative fermion mass is <u>not</u> proportional to some other, heavier fermion mass.

(6) Two-loop exchange of W_L and W_R bosons as shown in Fig. 6.

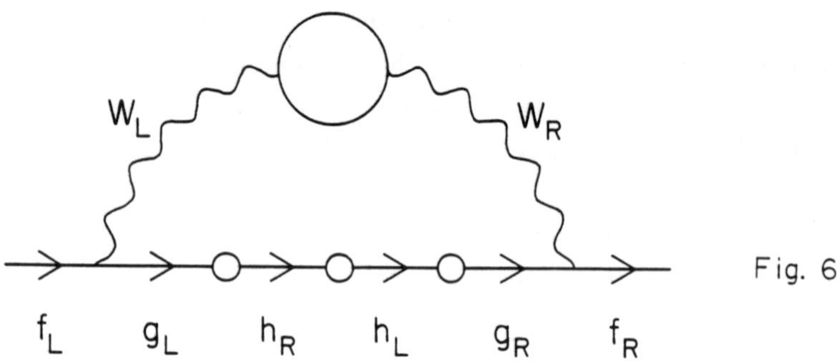

This has been used to obtain very small Dirac neutrino masses.[8] The gauge group used is $SU(2)_L \times SU(2)_R \times U(1)$ and the scalar and fermion sectors are arranged so that W_L and W_R do not mix at tree level, but do so in one loop.

COMMENTS AND OUTLOOK

Many models with radiative masses contain new couplings among known particles, resulting in possible serious phenomenological constraints. For example in the model of Ref. 9, doubly charged scalar bosons couple to both e and μ, hence the process $\mu \to eee$ exists at tree level and fine tuning is needed to suppress it below the present experimental upper bound of 10^{-13} on its branching fraction.

A more subtle effect concerns the process $\mu \to e\gamma$ with a branching fraction known to be less than 5×10^{-11}. In any model which does not forbid μ-e transitions explicitly, the radiative mass matrix spanning e and μ is in general not diagonal. This by itself is of course not a problem because we can always redefine e and μ to be the physical mass eigenstates. However, the resulting magnetic-dipole-moment matrix is then not automatically diagonal and $\mu \to e\gamma$ will occur at a certain level. We estimate[10] the typical rate to be some three orders of magnitude above the experimental bound.

In contrast to the fermion mass hierarchy, the hierarchy of the various entries of the 3 x 3 changed-current mixing matrix is derived radiatively in only a few models.[11] As such models are supposed to eliminate all small parameters at tree level, this problem must not be overlooked. If new physics at or just above the electroweak mass scale is indeed responsible for the enormous range of observed fermion masses, then many new particles are expected at forthcoming high-energy accelerators with many clear experimental signatures.

ACKNOWLEDGEMENT

This work was supported in part by the U. S. Department of Energy under Contract No. DE-AT03-87ER40327.

REFERENCES

1. Recent developments are briefly reviewed by K. S. Babu and E. Ma, Mod. Phys. Lett. A **4**, 1975 (1989).

2. K. S. Babu and E. Ma, Phys. Rev. Lett. **61**, 674 (1988); Phys. Lett. B **228**, 508 (1989); S. T. Petcov and S. T. Toshev, Phys. Lett. **143B**, 175 (1984).

3. E. Ma, Phys. Rev. Lett. **62**, 1228 (1989).

4. E. Ma, Phys. Rev. Lett. **63**, 1042 (1989).

5. B. S. Balakrishna, Phys. Rev. Lett. **60**, 1602 (1988)

6. D. Chang and R. N. Mohapatra, Phys. Rev. Lett. **58**, 1600 (1987).

7. K. S. Babu, Phys. Lett. B **203**, 132 (1988).

8. K. S. Babu and X. G. He, Mod. Phys. Lett. A **4**, 61 (1989).

9. B. S. Balakrishna and R. N. Mohapatra, Phys. Lett. B **216**, 349 (1989).

10. E. Ma, D. Ng, and G. G. Wong, Z Phys. C (in press).

11. B. S. Balakrishna, A. L. Kagan, and R. N. Mohapatra, Phys. Lett. B **205**, 345 (1988); A. L. Kagan, Phys. Rev. D **40**, 173 (1989); E. Ma, in preparation.

Gauge hierarchy and long range forces[†]

Palash B. Pal
Institute of Theoretical Science
University of Oregon, Eugene, OR 97403
Wai–Yee Keung
Department of Physics
University of Illinois at Chicago, IL 60680
Darwin Chang
Department of Physics and Astronomy
Northwestern University, Evanston, IL 60208

With the aid of simple examples, we show how a long range attractive force can arise in a gauge theory with a hierarchy. The force is due to the exchange of a Higgs boson whose mass and matter couplings are both naturally suppressed by the hierarchical mass ratio. Such bosons appear if there is an accidental global symmetry in the low-energy renormalizable Lagrangian after the high energy symmetry breaking.

Introduction • To discuss Physics beyond the Standard model, quite often we need large mass scales. For example, grand unified theories have the unification scale $M_{\mathrm{GUT}} \gtrsim 10^{15}$ GeV; any unification including the gravitational interaction involves the Planck scale, $M_{\mathrm{Planck}} \simeq 10^{19}$ GeV. Since the scale of the standard model is given by $M_W \simeq 10^2$ GeV, this means that the models have a small parameter, defined by

$$\epsilon \equiv M_W / \Lambda_G \lesssim 10^{-13}, \qquad (1)$$

where Λ_G is a high mass scale whose magnitude is left undefined at this point.

The basic question that we ask is the following. In models with such a hierarchy, are there scalars whose masses are suppressed by some powers of the parameter ϵ compared to the weak scale? If such particles exist, they mediate long range forces between fermions. We show that it is indeed possible to construct such models.

The motivation of this undertaking is obvious. There are a lot of experiments going on to find out any deviation from Newtonian gravity. They might be able to test the models we propose. At present, these experiments indicate no unequivocal evidence for any deviation from Newtonian gravity. If we denote the two-body potential energy by

$$W(r) = -\frac{G_N m_1 m_2}{r} \left(1 + \alpha_5 e^{-r/\lambda} \right), \qquad (2)$$

we can summarize the various bounds [1] on the parameter λ roughly as in Table 1. For the rest of this talk, we will roughly take $|\alpha_5| \lesssim 10^{-4}$.

If the force is mediated by a scalar of mass m_ω whose coupling with fermions is

---

[†]Talk given by Palash B. Pal at the DPF90 meeting held at Houston, Texas, January 3-6, 1990.

Table 1: Rough upper bounds on the strength of long ranges forces with various ranges.

λ (in cm)	Rough upper limit on $|\alpha_5|$
1	10^{-4}
10^2	10^{-3}
10^5	10^{-6}
10^{10}	10^{-8}

g_5, the strength of the interaction is given by $(g_5^2/r)e^{-m_\omega r}$. From Eq. (2), we get $g_5^2 = G_N m_1 m_2 \alpha_5$. Since $G_N m_1 m_2 \simeq 10^{-38}$ for two nucleons, we get $g_5 \lesssim 10^{-20}$. In view of this, we look for models where the scalar coupling is also suppressed by the hierarchy ϵ.

There have been many attempts to construct models of long range forces. These involve higher dimensional Kaluza-Klein type theories [2], or superstring models [3], or non-perturbative effects [4]. We take a more modest approach involving pseudo-Goldstone bosons (PGB). The idea here is simple which can be realized in the context of four dimensional perturbative field theory. At the Lagrangian level, there is an approximate global symmetry or *pseudosymmetry*, in the sense that there are terms which explicitly break this symmetry, albeit by a small amount. If this pseudosymmetry is then broken spontaneously, we would obtain not a true Goldstone boson, but rather a PGB whose mass is governed by the explicit breaking terms. People have constructed models of this type where the small explicit breaking terms are put in by hand [5]. It is here that our idea [6]

seems to be much elegant than the previous work in this field.

Accidental pseudosymmetries and PGB • The basic approach is as follows. At high energy, there is no global symmetry at all. There is a gauge symmetry G which breaks at the high scale. However, a discrete subgroup D of G remains unbroken. In the low energy effective Lagrangian $\mathcal{L}_{\text{eff}}$, this D-invariance automatically implies a larger, continuous symmetry Ω on the renormalizable terms. However, Ω is not a symmetry of the high energy theory, so $\mathcal{L}_{\text{eff}}$ must contain some terms that break Ω. We will show that in a class of models, such Ω-breaking terms are the non-renormalizable terms of $\mathcal{L}_{\text{eff}}$, i.e., they are suppressed by factors like Λ_G^{-n} for some $n > 0$. If then Ω is broken spontaneously at low energy, we obtain a PGB whose mass is suppressed by inverse powers of Λ_G.

Let us illustrate this idea with a toy model involving no fermions. Consider a gauge theory with a symmetry $U(1)_G$. There are three Higgs fields S, σ and s, with charges as follows:

$$Q_G(S) = 5, \ Q_G(\sigma) = -3, \ Q_G(s) = 1. \quad (3)$$

The Higgs potential can be written as

$$V = V_0 + V', \quad (4)$$

where V_0 contain terms like S^*S or $(S^*S)^2$ or $(S^*S)(s^*s)$ which are invariant under independent phase rotations of S, σ and s. So, if these were the only terms present, the potential would have had a $[U(1)]^3$ symmetry. This is not the case because of the other terms in V:

$$V' = hS\sigma^2 s + f\sigma s^3 + kS\sigma s^{*2} + \text{h.c.}. \quad (5)$$

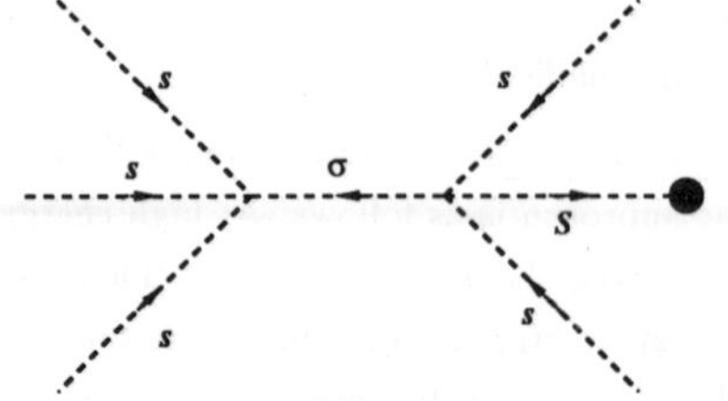

Figure 1: The s^5 coupling generated when S develops a VEV at high energy.

At a high scale Λ_G, if S develops VEV V_S, this VEV breaks the gauge symmetry $U(1)$. However, since S has 5 units of charge, the Lagrangian after symmetry breaking is still invariant under a discrete Z_5 symmetry under which

$$S \to S, \ \sigma \to e^{-6\pi i/5}\sigma, \ s \to e^{2\pi i/5}s. \qquad (6)$$

To write down the low energy Lagrangian, we first have to know which particles are heavy and which are light. Let us assume that both S and σ are heavy, with mass of order Λ_G, whereas the mass of s is made much lighter by the usual fine tuning which is necessary to produce the hierarchy. In that case, the low energy effective Lagrangian, $\mathcal{L}_{\text{eff}}$, contains only the s field. Terms consistent with the symmetry of Eq. (6) have the form $s^{5n_1} s^{*5n_2} (s^*s)^{n_3}$ for different integers n_1, n_2 and n_3. As far as renormalizable terms are concerned, the only possibilities are (s^*s) and $(s^*s)^2$. But both these terms are invariant under the larger, continuous global symmetry $U(1)_\Omega$ under which $s \to \exp(i\theta)s$. Thus, if this were the end of the story, at a low scale when s develops a VEV v, we would have obtained a Goldstone boson.

However, this is not the end of the story. Certainly, $\mathcal{L}_{\text{eff}}$ contains non-renormalizable terms. In Fig. 1, we show how the lowest such term is generated. Recalling the couplings from Eq. (5), it is trivial to see that the interaction is given by

$$\frac{fkV_S}{M_\sigma^2} s^5 \simeq \frac{fk}{\Lambda_G} s^5, \qquad (7)$$

where in the last step, we used the fact that without any extra fine-tuning, $M_\sigma \sim \Lambda_G$. This term is not invariant of the accidental pseudo-symmetry $U(1)_\Omega$ mentioned before. Therefore, when s develops VEV v, we would not obtain a true Goldstone boson but a PGB ω whose mass is given by

$$m_\omega^2 \sim \frac{fk}{\Lambda_G}v^3 = fkv^2\epsilon \qquad (8)$$

where ϵ is the hierarchy parameter whose presence makes the mass small.

We can construct similar models in which the discrete symmetry after the high energy symmetry breaking is Z_N for any other N. In that case, the PGB mass would be given by

$$m_\omega^2 \sim \beta_0 v^2 \epsilon^{N-4}, \qquad (9)$$

where β_0 is some combination of coupling constants of the model. Thus, for example, if $N = 6$, $v \sim M_W$ and $\epsilon \sim 10^{-17}$, we get $m_\omega \sim 10^{-7}$ eV for $\beta_0 \sim 10^{-2}$. This corresponds to a range of about 1 meter.

Model with fermions • Now that we have seen how PGB's can arise in a gauge theory with hierarchy, let us discuss how they might couple with fermions. We adopt a simple model based on the gauge group $SU(2)_L \times U(1)_Y \times U(1)_G$. All particles present in the standard

$SU(2)_L \times U(1)_Y$ model would be assumed to be singlets of the $U(1)_G$ group. On the other hand, the particles having $U(1)_G$ charges as given in Eq. (3), which are assumed to be singlets of the standard electroweak group.

It must be emphasized at this point that we also need CP-violation in the theory. In absence of this element, the PGB is a pseudoscalar (i.e., CP negative), as we show later. In the non-relativistic limit, the exchange of this particle between fermions produces spin-dependent forces, which are hard to detect.

Of course, CP violation occurs in the standard model through the Kobayashi-Maskawa mechanism. But these effects are suppressed by ratios like m_q/M_W, where m_q is some quark mass. In order to make room for an alternative source of CP violation, we introduce two particles which have the quantum numbers of the s and call them s_1 and s_2. We can always define them in a way that $\langle s_2 \rangle = 0$. Neither s_1 nor s_2 has any Yukawa coupling with the known fermions in the unbroken theory. They can couple to fermions only so far as they can mix with the $SU(2)_L$ doublet ϕ.

Consider only the renormalizable terms at first. They have the $U(1)_\Omega$ symmetry. We can use it to ensure that $\langle s_1 \rangle$ is real. In that case, the Goldstone mode is given by $\omega = \operatorname{Im} s_1$. In the Higgs potential, there are CP violating interactions like

$$\sum_{i,j} C_{ij}\, \phi^\dagger \phi\, s_i^\dagger s_j + \text{h.c.} \qquad (10)$$

With the VEV mentioned above, this produces mixing between $\operatorname{Re}\phi_0$ and $\operatorname{Im} s_2$ which we will denote by $\sin\Theta_{CPX}$. However, no

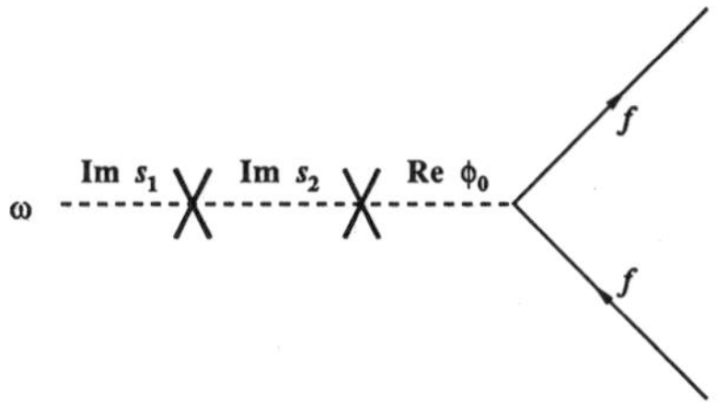

Figure 2: The scalar coupling of the PGB.

mixing is generated between $\operatorname{Re}\phi_0$ and $\operatorname{Im} s_1$. The PGB, therefore, does not have scalar coupling with fermions at this level.

Now take the non-renormalizable terms into account. There are terms like

$$\frac{1}{\Lambda_G^{N-4}} \sum_{n=0}^{N} \beta_n s_1^{N-n} s_2^n + \text{h.c.} \qquad (11)$$

in the low-energy effective Lagrangian, where β_n are combinations of coupling constants of the basic theory. After $\operatorname{Re} s_1$ develops a VEV, the $n = 1$ term of the above expression produces mixing between $\operatorname{Im} s_2$ and $\operatorname{Im} s_1$, whose magnitude is of order $\beta_1(v/\Lambda_G)^{N-4}$ In conjunction with the mixing between $\operatorname{Re}\phi_0$ and $\operatorname{Im} s_2$ produced by the renormalizable terms, this now gives a scalar coupling of the PGB, which is schematically shown in Fig. 2. The magnitude of the coupling can easily be read off from the figure to be

$$g_5 \simeq \beta_1 \left(v/\Lambda_G\right)^{N-4} \sin\Theta_{CPX} \frac{m_f}{v_W}, \qquad (12)$$

where $v_W = \langle\phi_0\rangle$. For $N = 5$ and $\epsilon \simeq 10^{-17}$, it is easy to obtain $g_5 \simeq 10^{-20}$.

Conclusions and outlook • We showed how, in models involving gauge hierarchy, one might obtain Higgs bosons whose mass and scalar coupling are both suppressed by the hierarchy parameter ϵ. Since we do not impose any ad hoc global symmetry by hand but rather obtain a pseudosymmetry as a consequence of the high energy gauge symmetry breaking, it seems plausible that this kind of model can be generalized to realistic grand unified models.

There is one problem of the simple version presented here. Comparing Eq. (9) and Eq. (12), we see that the ratio m_ω^2/g_5 does not depend on the hierarchy parameter. In fact, we get

$$\frac{m_\omega^2}{g_5} \simeq \frac{\beta_0}{\beta_1 \sin \Theta_{CPX}} \cdot \frac{v_W}{m_f} \cdot v^2. \qquad (13)$$

If $m_\omega \sim 10^{-5}$ eV corresponding to a range of about a centimeter and g_5 is close to the experimental limit for that range, we obtain $v \lesssim 10^{-4}$ GeV. This is far below the weak scale and might create a new hierarchy. On the other hand, if $v \sim v_W$, our simple version predicts forces much weaker than the present experimental bound. However, this problem concerns the specific model rather than our main idea of the presence of PGBs, and the situations can be different in more realistic models.

References

[1] See, e.g., B R Heckel *et al*, *Phys. Rev. Lett.* **63**, 2705 (1989).

[2] I Bars & M Visser, *Phys. Rev. Lett.* **57**, 25 (1986);
S M Barr & R N Mohapatra, *Phys. Rev. Lett.* **57**, 3129 (1986).

[3] G Lazarides, C Panagiotakopoulos & Q Shafi, *Phys. Rev. Lett.* **56**, 432 (1986).

[4] D Chang, R N Mohapatra & S Nussinov, *Phys. Rev. Lett.* **55**, 2835 (1985).

[5] A Halprin, M Barnhill and S M Barr, *Phys. Rev.* **D39**, 1467 (1989);
C T Hill and G G Ross, *Phys. Lett.* **203B**, 125 (1988) & *Nucl. Phys.* **B311**, 253 (1988/89).

[6] D Chang, W Y Keung and P B Pal, *Phys. Rev.* **D**, to be published.

Updating the Bounds on Charged Higgs Bosons

J.L. Hewett
Physics Department, University of Wisconsin
Madison, WI 53706

Abstract

We summarize the constraints placed on the parameters in the charged Higgs sector in two-Higgs-doublet models from a recent analysis of possible charged Higgs contributions to both low-energy processes and high energy hadron collider data.

1. Introduction

At present there exists no experimental information on the nature of the Higgs sector of the Standard Model (SM). The possibility of an enlarged Higgs sector beyond the minimal one-doublet version of the SM [1] is thus consistent with data. The simplest extensions are models with two Higgs doublets (requiring the existence of charged Higgs bosons) which are naturally present in many scenarios beyond the SM, including supersymmetry (SUSY), Peccei-Quinn models, and E_6 superstring-inspired theories. Due to the lack of direct experimental knowledge, it is prudent to combine all information on the indirect effects of charged Higgs bosons in order to place constraints on this sector. In this talk I summarize recent calculations [2] of the possible charged Higgs contributions to low-energy processes, such as B_d^0-$\bar{B}_d^0$ and D^0-$\bar{D}^0$ mixing, Δm_K, ϵ, and ϵ'/ϵ, as well as the influence of charged Higgs bosons on the top-quark search and missing E_T measurements at $\bar{p}p$ colliders. The results from this combined analysis are then used to restrict the parameters in the charged Higgs sector and determine the region which is consistent with all data.

There are four two-Higgs-doublet models which naturally avoid tree-level flavor changing neutral currents which can be induced by Higgs exchange. A summary of which type of fermion couples to which doublet, ϕ_1 or ϕ_2 , in Models I-IV is given in Table I. Each doublet obtains a vacuum expectation value (VEV) v_i subject only to the constraint that $v_1^2 + v_2^2 = v^2$, where v is the usual VEV of the SM. The generic charged Higgs coupling to fermions in Models I-IV (assuming massless neutrinos) is given by

$$\mathcal{L} = \frac{g}{2\sqrt{2}\, M_W} H^\pm \tag{1}$$
$$\times \left\{ V_{ij} m_{u_i} A_u \bar{u}_i (1 - \gamma_5) d_j + V_{ij} m_{d_j} A_d \bar{u}_i (1 + \gamma_5) d_j + m_\ell A_\ell \bar{\nu}(1 + \gamma_5)\ell \right\} + \text{h.c.}$$

where the values of the coefficients A_f for all four models are given in Table I; g is the usual $SU(2)_L$ coupling constant, V_{ij} is the relevant element of the Kobayashi Maskawa (KM) matrix, and $\tan\beta \equiv v_2/v_1$ is the ratio of VEVs. The value of $\tan\beta$ determines the relative strength between the hadronic and leptonic $H^\pm$ couplings. Model II is that which is present in SUSY and E_6 theories. Note that charged Higgs contributions to processes which depend only on the quark sector are the same in Models I and IV and in Models II and III.

The mass $m_{H^\pm}$ and the ratio v_2/v_1 are *a priori* free parameters in a general two-doublet model. The current experimental limits [3] on the $H^\pm$ mass are $m_{H^\pm} > 19$ GeV from PETRA with ALEPH at LEP extending this bound to $m_{H^\pm} > 35.4$ GeV with $B(H^\pm \to cs) = 100\%$ and $m_{H^\pm} > 43.0$ GeV with $B(H^\pm \to \tau\nu) = 100\%$. In minimal SUSY models $m_{H^\pm}$ is restricted to lie above M_W [4], while the bound is less stringent in E_6 superstring models giving $m_{H^\pm} > 53$ GeV [5]. Recent data from LEP on searches for the SM Higgs boson can also be used [6] to severely constrain $m_{H\pm}$ and $\tan\beta$ in SUSY models. However, there are no theoretical restrictions on the value of $m_{H\pm}$ and $\tan\beta$ in general two-doublet models. Clearly, any constraints on the charged Higgs parameters from low-energy and hadron collider data will also be strongly dependent on the value of the top quark mass.

TABLE I. Summary of which doublet, ϕ_1 or ϕ_2, gives mass to each type of fermion and the values of the coefficients A_f in the charged Higgs coupling to fermions in Eq. (1) for models I–IV.

Models	I		II		III		IV	
	VEV	A_f	VEV	A_f	VEV	A_f	VEV	A_f
$\begin{pmatrix} u \\ d \end{pmatrix}$	2	$\cot\beta$	2	$\cot\beta$	2	$\cot\beta$	2	$\cot\beta$
	2	$-\cot\beta$	1	$\tan\beta$	1	$\tan\beta$	2	$-\cot\beta$
$\begin{pmatrix} \nu \\ l \end{pmatrix}$	2	$-\cot\beta$	1	$\tan\beta$	2	$-\cot\beta$	1	$\tan\beta$

Semi-quantitative restrictions on $m_{H\pm}$ and $\tan\beta$ can be set by requiring that the $H^\pm$ width not be too large and that the $\bar{t}bH$ coupling remain perturbative. It is clear from Eq. (1) that the width and $\bar{t}bH$ coupling both grow rapidly with increasing m_t and hence perturbation theory may become endangered for some values of the parameters. If we demand that the theory remain perturbative, a bound on $\tan\beta$ as a function of m_t is obtained. If we require $\Gamma_{H\pm}/m_{H\pm} < 1/2$ and the $\bar{t}bH$ coupling to be smaller than the QCD coupling $g_s^2 = 4\pi\alpha_s(M_W^2)$ as definitions of the perturbative region, the bound $\tan\beta \gtrsim m_t/(600 \text{ GeV})$ is obtained. This restriction applies in Models I-IV. In Models II and III there are corresponding upper limits on $\tan\beta$ arising from the $\bar{t}bH$ coupling proportional to $m_b V_{tb} \tan\beta$ which are given by $\cot\beta \gtrsim m_b/(600 \text{ GeV})$.

2. Constraints from Low-Energy Data

Charged Higgs bosons may contribute to a variety of low-energy processes via $H^\pm$ exchange in internal loops. Consistency with data places overall restrictions on the mass $m_{H\pm}$ and coupling parameter $\tan\beta$. Results are presented from a combined analysis [2] of these contributions which takes the large freedom in the allowed ranges of all the parameters into account.

A Monte Carlo analysis is employed in order to probe the freedom in the KM sector. 10^6 sets of the four parameters (3 mixing angles $\theta_{1,2,3}$ and 1 phase δ) present in the 3 generation KM mixing matrix are generated subject to the restriction that the KM elements lie within their allowed ranges [7] as determined from charged current experiments and unitarity considerations. 10^6 lotteries are sufficient to insure that the parameter space is well sampled and converges at the boundaries of the allowed ranges for the KM elements. Consistency (at the $\pm 2\sigma$ level) with the recent [8] ARGUS and CLEO result $|V_{ub}/V_{cb}| = 0.10 \pm 0.03$ is also required.

For a particular value of m_t, $m_{H\pm}$ and $\tan\beta$ are varied in the ranges $m_{H\pm} = 25 - 500$ GeV and $\tan\beta = 0.1 - 10.0$ and $B_d^0 - \bar{B}_d^0$ and $D^0 - \bar{D}^0$ mixing, ϵ, ΔM_K, and ϵ'/ϵ are calculated for each set of KM parameters which survive the above constraints on V_{ij}. In these calculations, reasonable ranges [2] for the values of the meson lifetimes, decay constants, and bag factors are explored. The requirement is then made that each of the above calculated quantities be consistent with current experimental measurements [7] as follows

$$x_{B_d} = 0.70 \pm 0.13,$$

$$\Delta M_D < 1.3 \times 10^{-13} \text{ GeV},$$

$$\epsilon = (2.259 \pm 0.018) \times 10^{-3},$$

(2)

$$\Delta M_K(\text{new physics}) < (3.521 \pm 0.014) \times 10^{-15}\text{ GeV}, \qquad\qquad (2\ cont.)$$

$$|\epsilon'/\epsilon| < 5.0 \times 10^{-3}.$$

The resulting bounds from this data are essentially equivalent in Models I and II. The branching fraction $B(b \to s\gamma)$ is also calculated in Models I and II for the parameters which have survived all of the above constraints. Demanding consistency with the inferred [9] bound $B(b \to s\gamma) < 4.0 \times 10^{-3}$ resulting from measured limits [8] on the decay $B \to K^*\gamma$, additional constraints on $m_{H^\pm}$ and $\tan\beta$ are obtained in both models. Further constraints can be imposed by requiring ϵ'/ϵ to lie within $\pm 2\sigma$ of the published CERN NA31 data [10] of $\epsilon'/\epsilon = (3.3 \pm 1.1) \times 10^{-3}$.

The results from this analysis are presented in Fig. 1 which shows the allowed regions in the $m_{H^\pm}$, $\tan\beta$ plane for $m_t = 90$, and 150 GeV. The solid curve corresponds to the combined restrictions from the data in Eq. (2) the short-dashed (dash-dotted) curve depicts the constraints from the decay $b \to s\gamma$ in Model II (I), the long-dashed curve represents the limits from the NA31 data on ϵ'/ϵ, and the dotted curve illustrates the perturbative bound $\tan\beta \gtrsim m_t/(600\text{ GeV})$. The excluded regions lie to the left of the curves. Clearly, the constraints from these processes become stronger as the top quark becomes heavier. It is interesting to note that the perturbative bound alone provides a rough estimate of these combined restrictions.

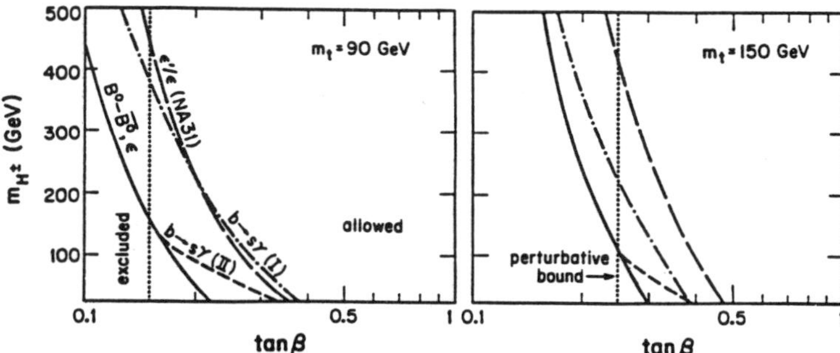

Fig. 1. Summary of restrictions from low-energy data on the charged Higgs mass $m_{H^\pm}$ and $\tan\beta$ for $m_t = 90$ and 150 GeV. Regions to the left of the boundaries are excluded.

3. Constraints from $\bar{p}p$ Collider Data

Top quark searches at high energy hadron colliders are based on the SM semileptonic decay $t \to b\ell\nu$ with $\ell = e$ or μ. These searches have come up empty-handed [11] and place lower bounds on the mass m_t in the SM. These bounds may be evaded if the SM decays are sufficiently suppressed; in two-Higgs-doublet models, such suppression [12] can be provided by competition with non-standard top decays via a real or virtual charged Higgs boson, $t \to bH^+$. Hence for values of m_t excluded in the SM (*i.e.*, for $m_t < 77$ GeV, as given by the CDF experiment at the Tevatron collider [11]) the non-observation of top signals may be used to restrict the parameters in the charged Higgs sector. This places the approximate constraint $m_{H^\pm} < m_t - m_b$, since top decays to real H^+ are very competitive with virtual W^+ modes, whereas decays via virtual H^+ are generally not competitive. However, the present analysis is based on exact calculations of the competing H^+ and W^+ channels and the results deviate from this approximate bound on $m_{H^\pm}$.

For a given sign of lepton, say ℓ^+, the $\bar{t}t$ production signal arises mainly from $t \to bW^{*+} \to b\ell^+\nu$ with $\bar{t}$ decaying either via W^{*-} or via H^-. The experimental acceptance for the $W^{*+}H^-$ modes are found to be similar to (and in fact slightly exceed) that for $W^{*+}W^{*-}$ modes. Hence, the effective bound that can be placed on $B(t \to bW^*)$ is

$$B(t \to bW^*) \leq \left[\frac{\text{experimental } \sigma_{\bar{t}t} \text{ upper limit}}{\text{theory } \sigma_{\bar{t}t} \text{ lower limit}} \right] \equiv R(m_t) \,. \tag{3}$$

This branching fraction is calculated as in [12] for both real and virtual Higgs bosons, including the small contributions from $t \to sW$, sH and $H \to us$, ub. The result is insensitive to the precise values of the KM matrix elements, since the dominant channels have matrix elements $\simeq 1$.

The CDF experiment has also searched for events with large missing transverse energy ($\not{E}_T$) and jets, since this can be a signal for production of SUSY particles or other new physics. It can in particular also be a signal for charged Higgs boson production because of the undetected neutrinos in $H^+ \to \bar{\tau}\nu \to \bar{\nu}\nu q\bar{q}'$ (or $\bar{\nu}\nu\nu\bar{\ell}$) decay. Assuming that $H^\pm$ are produced from top quark decays (the most copious mechanism), upper limits on the frequency of events with large $\not{E}_T$ will imply constraints on the two–Higgs–doublet model parameters for any given m_t.

For $\not{E}_T \geq 60$ GeV, with an integrated luminosity of 4.6 pb^{-1}, CDF observed 34 events; the expected background from standard model sources (excluding unknown top decays) is 38 ± 17 events [11]. Treating the background as having a gaussian probability distribution the bound

$$\sigma(\text{nonstandard physics}: \not{E}_T > 60 : \text{CDF cuts}) \leq 4.2\,\text{pb} \tag{4}$$

is derived [2] at 90% CL. To evaluate the constraints from this $\not{E}_T$ bound, $H^\pm$ production and decay are calculated by Monte Carlo methods [12] and consistency is required.

Figure 2 shows the effects of the constraints (3) and (4) in the $m_{H^\pm}$, $\tan\beta$ plane for Model II with $m_t = 50$, 60, 70 GeV. In this figure, the regions which lie on the same side of the curves as the cross-hatching are excluded. The top search bound excludes the upper part of the figure, while the $\not{E}_T$ limits exclude most of the remaining parameter space for larger $\tan\beta$. Also displayed in the figure are the constraints obtained from low-energy data and from the ALEPH experiment [3]. It is clear that only a small sliver of the $m_{H^\pm}$, $\tan\beta$ plane survives. In Model I an even smaller

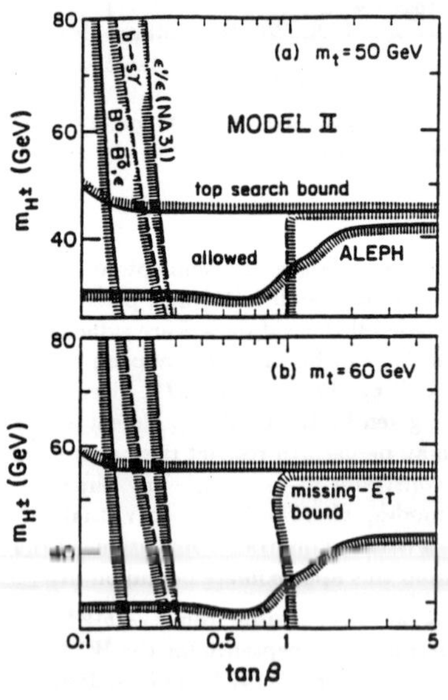

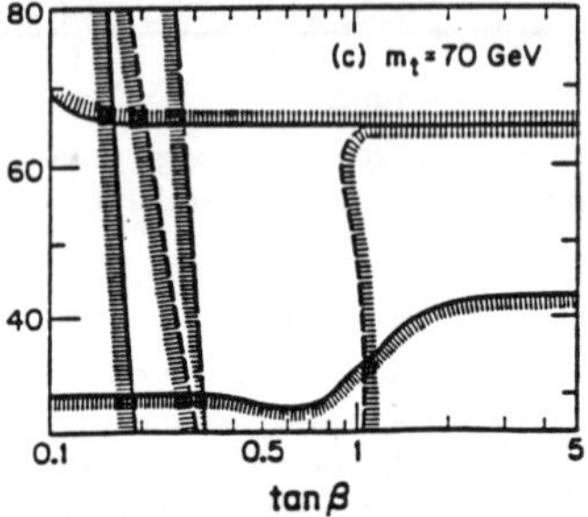

Fig. 2. The overall allowed region in the $m_{H^\pm}$, $\tan\beta$ plane for (a) $m_t = 50$, (b) $m_t = 60$, (c) $m_t = 70$ GeV in model II. The thin solid curve indicates the bounds placed from the CDF SM top-quark search; the thin dashed curve from the CDF missing-E_T events; the thick solid curve from B^0-$\bar{B}^0$ and D^0-$\bar{D}^0$ mixing, ΔM_K, ε, and $|\varepsilon'/\varepsilon| < 6.0 \times 10^{-3}$; the thick short-dashed curve from $b \to s\gamma$; and the thick long-dashed curve from ε'/ε (NA31). The excluded regions lie on the cross-hatched side of the curves.

fraction [2] of the parameter space is allowed by the data, with the entire region being excluded for $m_t < 70$ GeV.

4. Summary

A comprehensive analysis of the bounds that low-energy and $\bar{p}p$ collider data place on the parameters of the charged Higgs sector is reviewed. As can be seen from Fig. 1, the compilation of restrictions on $m_{H^\pm}$ and $\tan\beta$ from low-energy data excludes the regions of small $\tan\beta$ and light $m_{H^\pm}$. When these constraints are combined with those from hadron colliders, Model I is ruled out for $m_t < 70$ GeV, and only a tiny fraction of the parameter space remains in Model II. One of the most important results of this analysis is that the CDF bounds on the top-quark mass can probably not be evaded via the possible decay $t \to bH^+$. Hence the CDF bound of $m_t > 77$ GeV most likely holds in two-Higgs-doublet models as well as in the SM.

Charged Higgs contributions to ϵ'/ϵ and rare decays can also be evaluated with better accuracy. It is found [2] that the resulting values of ϵ'/ϵ can be negative or positive, and can agree with either the CERN NA31 or Fermilab E731 results, depending on the values of the parameters. The large enhancements of rare decay modes, such as $K_L \to \mu\mu$ or $K^+ \to \pi^+\nu\bar{\nu}$, only occur [2] for the ranges of $\tan\beta$ and $m_{H^\pm}$ which have been ruled out from the low-energy data. These rare decay modes and ϵ'/ϵ tend to their SM predictions as $\tan\beta$ becomes large.

The effects of two-Higgs-doublet models are wide and varied. The present study determines the overall regions of the $m_{H^\pm}$, $\tan\beta$ parameter space which are consistent with present data and thus helps to pinpoint the best signatures for the possible discovery of charged Higgs bosons.

Acknowledgements

I am indebted to my colleagues V. Barger and R. Phillips. This research was supported in part by the U.S. Department of Energy under contract DE-AC02-76ER00881.

References

1. For a recent review of the minimal Higgs sector and its possible extensions see J. F. Gunion, H. E. Haber, G. L. Kane, and S. Dawson, *The Higgs Hunter's Guide*, UCD-89-4 (1989).

2. This talk is based on the work of V. Barger, J.L. Hewett, and R.J.N. Phillips, Univ. of Wisconsin Report MAD/PH/530 (*Phys. Rev. D*, in press). See also the related work of A. Buras, P. Krawczyk, M.E. Lautenbacker, and C. Salazar, Max Planck Institute Report MPI-PAE/PTh 52/89 (1989); J. F. Gunion and B. Grzadkowski, UCD-89-30 (1989).

3. H. J. Behrends *et al.*, CELLO Collaboration, Phys. Lett. **193B**, 376 (1987); D. Decamp *et al.*, (ALEPH Collaboration), CERN Report CERN-EP/90-34 (1990).

4. J.F. Gunion and H. E. Haber, Nucl. Phys. **B272**, 1 (1986); *ibid.* **B278**, 449 (1986).

5. V. Barger and K. Whisnant, Int. J. Mod. Phys. **A3**, 1907 (1988). For a recent review see, J. L. Hewett and T. G. Rizzo, Phys. Rep. **183**, 193 (1989).

6. T.G. Rizzo, Phys. Lett. **B237**, 88 (1990).

7. Particle Data Group, Phys. Lett. **B239**, 1 (1990).

8. M.V. Danilov (ARGUS Collaboration) talk presented at the *14th Int. Symp. on Lepton and Photon Interactions*, Stanford, CA, 1989; D.L. Kreinick (CLEO Collaboration) *ibid.*

9. We use the recent estimate of the ratio $\Gamma(B \to K^*\gamma)/\Gamma(b \to s\gamma) = 6\%$ from N. G. Deshpande, P. Lo, and J. Trampetic, Z. Phys. **C40**, 369 (1988).

10. H. Burkhardt *et al.*, (NA31 Collaboration), Phys. Lett. **206B**, 169 (1988).

11. F. Abe *et al.*, (CDF Collaboration), Phys. Rev. Lett. **64**, 142 (1990); L. Nodulman, plenary talk given at *Int. Europhysics Conf. on High Energy Physics*, Madrid, Spain, 1989.

12. V. Barger and R.J.N. Phillips, Phys. Lett. **201B**, 553 (1988); Phys. Rev. **D41**, 894 (1990).

Probing the Three Vector Boson Vertex in Future Collider Experiments

U. Baur
Physics Department, University of Wisconsin
Madison, WI 53706

Abstract

We summarize the prospects for testing the $WW\gamma$ and WWZ couplings in future collider experiments. LEP II will be able to probe these couplings with $10 - 70\%$ accuracy in $e^+e^- \to W^+W^-$. Although cross sections are small, the $WW\gamma$ vertex can be measured at HERA in $ep \to eW^\pm X$ with $30 - 50\%$ accuracy after a few years of running. At the Tevatron, for an integrated luminosity of 100 pb^{-1}, the $WW\gamma$ and WWZ couplings can be determined with a precision of $25 - 40\%$ in $W^\pm\gamma$ and $W^\pm Z$ production.

1. Introduction

During the past decade the standard elctroweak theory (SM) has proven extremely successful in describing experimental results on electroweak interactions. In spite of this success one has to bear in mind that so far only the fermion-vector boson interactions have been scrutinized, while the scalar sector of the theory and the predictions of the SM for the vector boson self interactions have not yet been tested experimentally. These couplings are a manifestation of the non abelian gauge symmetry upon which the standard electroweak theory is based. The observation of the vector boson self interactions is a crucial test of the theory.

In Section 2 we briefly review the structure of the most general WWV ($V = \gamma$, Z) vertex. In Section 3 we summarize the prospects for measuring the $WW\gamma$ and WWZ couplings in future experiments at LEP II, HERA and the Tevatron. Our results are summarized in Section 4.

2. Three Vector Boson Couplings

In order to find out how well future collider experiments can test the three vector boson couplings one must employ a general WWV ($V = \gamma$, Z) vertex. The most general WWV coupling consistent with Lorentz invariance may be parametrized in terms of seven form factors, g_1^V, κ^V, λ^V, g_4^V, g_5^V, $\tilde{\kappa}^V$ and $\tilde{\lambda}^V$, if we exclude terms proportional to $\partial_\mu W^\mu$ and $\partial_\mu Z^\mu$ [1]. These terms vanish if the vector bosons are coupled to massless fermions, which, to a good approximation, is the case in the processes we are going to consider. g_4^V violates CP and C, g_5^V violates C and P but is CP even, while $\tilde{\kappa}^V$ and $\tilde{\lambda}^V$ are P odd and CP violating. Within the SM they all vanish at tree level and so does λ^V. The only nonzero couplings within the SM are

$$g_1^V = 1, \qquad \kappa^V = 1, \qquad V = \gamma, Z. \tag{2.1}$$

It is well known that tree level unitarity, *e.g.* for the process $e^+e^- \to W^+W^-$, uniquely restricts the $WW\gamma$ couplings to their (SM) gauge theory values at asymptotically high energies [2]. This implies that any deviation $a = g_1^V - 1, \ldots, \tilde{\lambda}^V$ from the SM expectation results in a contribution to the helicity amplitudes which rapidly grows with energy and therefore has to be described by a form factor $a(q^2, \bar{q}^2, q_V^2)$ which vanishes when one of the arguments, the square of the four momentum of one the W-bosons, q^2 or $\bar{q}^2$, or the square of the four momentum of V, q_V^2, becomes large. Electromagnetic gauge invariance puts an additional constraint on $g_1^\gamma - 1$, g_4^γ and g_5^γ: for onshell photons these anomalous couplings are forbidden and the corresponding form factors must hence all be of $O(q_\gamma^2/\Lambda^2)$, where Λ represents the scale at which new physics becomes important in the weak boson sector [1].

3. Signals for Anomalous $WW\gamma$ and WWZ Couplings

LEP II

$WW\gamma$ and WWZ couplings can be studied in W pair production at LEP II ($\sqrt{s} \approx 190$ GeV). At the tree level $e^+e^- \to W^+W^-$ proceeds via s-channel photon and Z exchange, and via t-channel neutrino exchange. The interference between the graphs and in particular the gauge cancellations between them make this reaction sensitive to deviations of the three vector boson vertices from the SM. Since LEP II will produce W pairs close to threshold, its sensitivity to anomalous $WW\gamma$ and WWZ couplings is limited because the full strength of the gauge theory cancellations can only be exploited at much higher energies. Moreover, both the WWZ and the $WW\gamma$ vertex contribute roughly equally to W pair production. Hence only a linear combination of the two will effectively be measured at LEP II, unless longitudinal or transverse beam polarization is available [3]. As a result experiments at LEP II can measure anomalous $WW\gamma$ and WWZ couplings only with an accuracy of $\pm 0.1 \ldots \pm 0.7$.

HERA

An independent measurement of those couplings from other colliders would thus be a valuable input for the analysis at LEP II. A possibility for such a measurement is offered by single W production at HERA ($\sqrt{s} \approx 314$ GeV). By the time LEP II will go into operation HERA should have accumulated data over several years. Charged weak bosons can either be produced together with neutrinos or with electrons in ep collisions. Because the cross section for νW production is much smaller than the one for eW production [4] (about a factor 100 at HERA energies) we shall concentrate on the process $ep \to eWX$.

The dominant contributions to $ep \to eWX$ arise from photon exchange graphs due to the t-channel pole induced by the photon propagator. Z-exchange diagrams are strongly suppressed by the large mass of the Z-boson. Hence, $ep \to eWX$ is quite insensitive to anomalous WWZ couplings, and subsequently we shall only consider anomalous contributions to the $WW\gamma$ vertex. Since the exchanged photon is almost real, $g_1^\gamma - 1$, g_4^γ and g_5^γ must effectively vanish, due to electromagnetic gauge invariance. As a result, single W production in ep collisions only probes the anomalous $WW\gamma$ couplings κ^γ, λ^γ, $\tilde{\kappa}^\gamma$ and $\tilde{\lambda}^\gamma$. Since the energy region covered by HERA is smaller than what is typically expected for the form factor scale Λ ($\Lambda \gtrsim O(1 \text{ TeV})$) we also may assume the form factors $a = \kappa^\gamma - 1, \ldots, \tilde{\lambda}^\gamma$ to be approximately constant.

The signal we are investigating consists of a jet produced by the quark struck inside the proton, an onshell W, and possibly the original beam electron. Since the

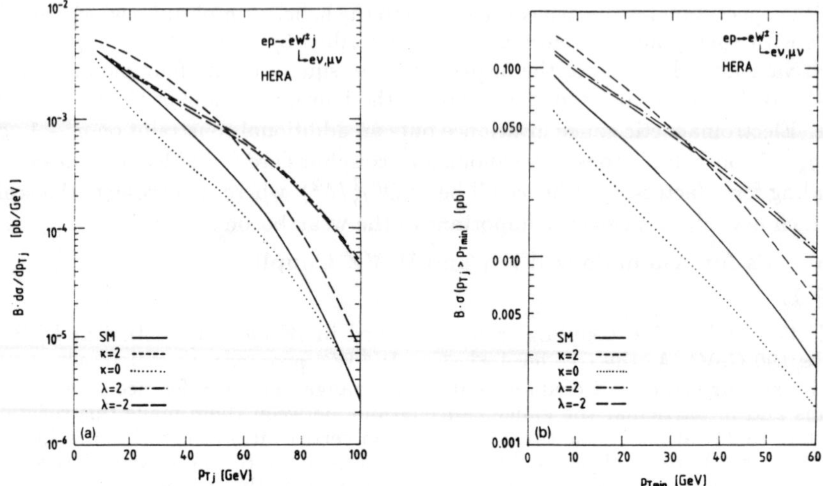

Fig. 1: *(a) Transverse momentum spectrum of the jet and (b) the total cross section $B \cdot \sigma(p_{Tj} > p_{T\min})$ versus $p_{T\min}$ for $ep \to eW^{\pm}jX$ at HERA. From Ref. 6.*

dominant contributions to the cross section arise from the exchange of an almost real photon, the scattered electron will in general be lost in the beam pipe. The W may either decay leptonically or hadronically. The hadronic decay modes of the W will be difficult to observe at ep colliders due to the QCD two- and three-jet background [5]. By considering leptonic W decay modes only, about 20% of all W decays are still observable (we neglect the decay $W \to \tau\nu$ in the following). The signal we are considering thus is $e^- p \to \ell^{\pm} + \text{jet} + \not{p}_T (+e^-)$ with $\ell^{\pm} = e^{\pm}, \mu^{\pm}$.

Our analysis [6] of the process $ep \to eW^{\pm}X$ is based on the calculation of helicity amplitudes for the complete process $eq \to eW^{\pm}q'$, $W^{\pm} \to \ell^{\pm}\nu$. The transverse momentum distribution of the jet, $B \cdot d\sigma/dp_{Tj}$ ($B = \mathrm{Br}(W \to e\nu,\ \mu\nu)$), turns out to be quite sensitive to anomalous $WW\gamma$ couplings. Fig. 1a shows the p_{Tj} spectrum above 5 GeV for the SM and various anomalous values of $\kappa \equiv \kappa^{\gamma}$ and $\lambda \equiv \lambda^{\gamma}$.

At large p_T the jet transverse momentum spectrum directly reflects the different high energy behaviour of the various anomalous contributions to the helicity amplitudes. Terms proportional to λ grow much faster than κ terms with the center of mass energy $\sqrt{\hat{s}}$ and the p_{Tj} spectrum for anomalous values of λ is thus harder than the one for a non-standard κ. At small values of p_{Tj}, however, anomalous values of κ produce larger deviations from the SM than equal values of λ do.

In Fig. 1b we show the integrated cross section $B \cdot \sigma$ above a mimimum transverse momentum cutoff $p_{T\min}$ for the jet. From there it is apparent that a rather large anomalous value of λ is necessary in order to produce a measurable effect in $ep \to eWX$ at HERA. Consequently κ can be measured at HERA more precisely than λ. A quantitative analysis shows that for an integrated luminosity of 10^3 pb^{-1}, κ can

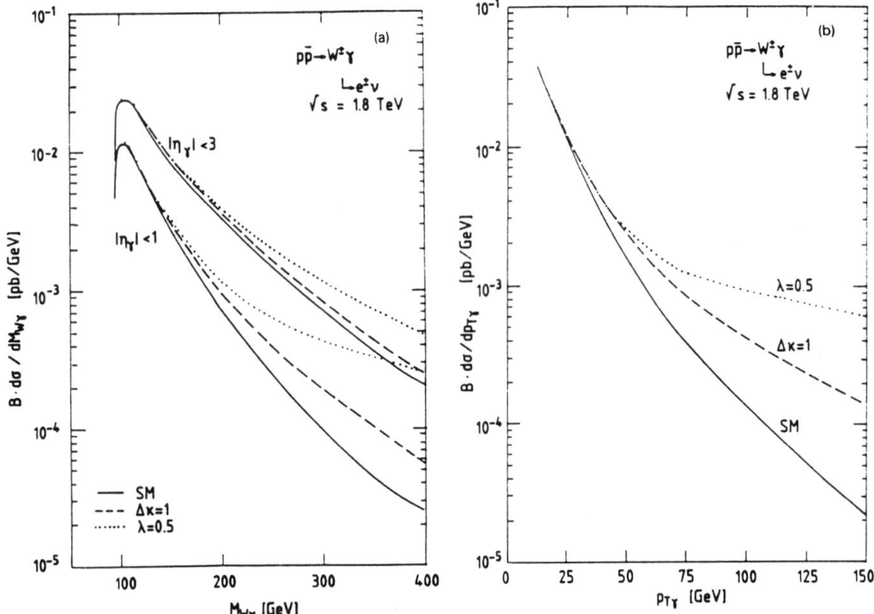

Fig. 2: *(a) W-photon invariant mass spectrum and (b) the transverse momentum spectrum of the photon in $p\bar{p} \to W^{\pm}\gamma$; $W^{\pm} \to e^{\pm}\nu$ at the Tevatron. From Ref. 8.*

be determined with 30% (50%) accuracy at the 69% (90%) confidence level (CL), while λ and $\tilde{\lambda}$ can only be limited to $|\tilde{\lambda}|$, $|\lambda| < 0.9$ (1.2) [6]. $|\tilde{\kappa}|$ is constrained to be smaller than $O(10^{-3})$ by the electric dipole moment of the neutron [6] and no visible effects are possible in collider experiments.

Tevatron

The $WW\gamma$ vertex can also be tested via $W \to f\bar{f}'\gamma$ and $W\gamma$ production in future experiments at the Tevatron. Since the photon is real in both processes one again probes only κ^{γ}, λ^{γ}, $\tilde{\kappa}^{\gamma}$ and $\tilde{\lambda}^{\gamma}$. The hadronic decays of the W will be diffcult to observe due to the QCD $jj\gamma$ background [7]. We therefore focus on final states which contain either an electron or a positron and consider the signal $p\bar{p} \to e^{\pm} + \gamma + p\!\!\!/_T$ [8].

As mentioned above, the anomalous contributions to the helicity amplitudes grow rapidly with $\hat{s}$. In radiative W decays $\hat{s} \approx M_W^2$ whereas much larger values of $\hat{s}$ are available in $p\bar{p} \to W\gamma$. W-photon production is therefore much more sensitive to anomalous contributions to the $WW\gamma$ vertex than $W \to e\nu\gamma$ [8].

A typical signal for anomalous couplings will be a broad increase in the W-photon invariant mass spectrum at large values of $M_{W\gamma} = \sqrt{\hat{s}}$. The resulting effect on $B \cdot d\sigma/dM_{W\gamma}$, $B = \mathrm{Br}\,(W \to e\nu)$, is shown in Fig. 2a. The increase of the cross section at large $M_{W\gamma}$ is much more pronounced for photon pseudorapidities $|\eta_\gamma| < 1$ than for $|\eta_\gamma| < 3$. This is due to the fact that anomalous couplings only contribute via s-channel W-exchange, and hence only to the $J = 1$ partial wave, when fermion masses are neglected. The anomalous contributions are, therefore, almost isotropic in the center of mass frame.

The population of the small rapidity region increases the average photon transverse momentum of events produced at a fixed value of the W photon invariant mass. The p_T distribution of the photon, $d\sigma/dp_{T\gamma}$, should be particularly sensitive to anomalous couplings. This fact is visible in Fig. 2b where the $p_{T\gamma}$ spectrum is plotted. A quantitative analysis [8] shows that at the Tevatron with $\int \mathcal{L}\,dt = 100$ pb^{-1} limits of $|\Delta\kappa| < 0.8\,(1.2)$ and $|\lambda|$, $|\tilde\lambda| < 0.40\,(0.25)$ can be obtained at the 69% (90%) CL. From the present 4.7 pb^{-1} of data sensitivities of $|\Delta\kappa| \approx 3$ and $|\tilde\lambda|$, $|\lambda| \approx 1.2$ can already be reached. $\Delta\kappa^Z$ and λ^Z can be measured with similar precision in $p\bar{p} \to W^\pm Z$ [9].

4. Discussion and Conclusions

Combined, the reactions $p\bar{p} \to W\gamma$ at the Tevatron and $ep \to eWX$ at HERA will yield a precise direct measurement of the $WW\gamma$ vertex before W pair production can be studied at LEP II. In W^+W^- production it is actually a linear combination of $WW\gamma$ and WWZ couplings which will be measured. In view of our present poor knowledge of the anomalous three vector boson couplings from low energy data [10], the direct measurement of the $WW\gamma$ couplings via $p\bar{p} \to W\gamma$ at the Tevatron and $ep \to eWX$ at HERA will constitute major progress. In the mid 1990's, these values will also be helpful in disentangling $WW\gamma$ and WWZ anomalous couplings in W pair production at LEP II.

Acknowledgements

This research was supported in part by the University of Wisconsin Research Committee with funds granted by the Wisconsin Alumni Research Foundation, and in part by the U. S. Department of Energy under contract DE-AC02-76ER00881.

References

1. K. Hagiwara *et al.* , Nucl. Phys. **B282**, 253 (1987).

2. J. M. Cornwall, D. N. Levin and G. Tiktopoulos, Phys. Rev. Lett. **30**, 1268 (1973); Phys. Rev. **D10**, 1145 (1974); C. H. Llewellyn Smith, Phys. Lett. **46B**, 233 (1973); S. D. Joglekar, Ann. Phys. **83**, 427 (1974).

3. D. Zeppenfeld, Phys. Lett. **183B**, 380 (1987).

4. E. Gabrielli, Mod. Phys. Lett. **A1**, 465 (1986).

5. H. Baer, J. Ohnemus and D. Zeppenfeld, Z. Phys. **C43**, 675 (1989).

6. U. Baur and D. Zeppenfeld, Nucl. Phys. **B325**, 253 (1989).

7. V. Barger *et al.* , Phys. Lett. **232B**, 371 (1989).

8. U. Baur and E. L. Berger, Phys. Rev. **D41**, 1476 (1990).

9. K. Hagiwara, J. Woodside and D. Zeppenfeld, MAD/PH/521 preprint (1989).

10. G. Kane, J. Vidal and C. P. Yuan, Phys. Rev. **D39**, 2617 (1989).

PHENOMENOLOGY OF EXTRA GENERATIONS OF HIGGS PARTICLES

MARC SHER

Physics Department,
College of William and Mary,
Williamsburg, VA 23815

ABSTRACT

We study the possibility that the Higgs sector of the minimal supersymmetric model consists of three generations in the same manner that the fermion sector does. Including only extra doublets and assuming that tree-level flavor changing neutral currents are naturally eliminated by some symmetry, we find that the lightest of the new scalars is absolutely stable, and that the second lightest is only slightly more massive. The coupling of these scalars to the Z is independent of any new parameters, and if light enough they could easily be discovered if a search is made for the somewhat unusual signature. Other phenomenological implications of these additional scalars is also discussed.

One of the attractive features of supersymmetric models is that they treat scalar bosons and fermions in the same framework, and allow for the possibility of placing the Higgs boson(s) of the standard model in the same representation as the fermions. Yet, in spite of the fact that the fermion sector of the standard model is mysteriously replicated twice, the possible replication of the Higgs sector of the minimal model has been largely ignored. In the popular "superstring-inspired" E_6 models[1], the Higgs bosons are, in fact, placed in the same representation as the fermions, and thus the minimal Higgs sector is replicated twice. However, even here the phenomenology of these additional Higgs bosons has attracted relatively little attention.[2,3].

In this work, we study the properties of a possible second and third generation of Higgs bosons in the context of a minimal supersymmetric model. We find, by making the rather mild assumption that tree-level flavor changing neutral currents vanish naturally (i.e. they are removed by some symmetry), that the properties of the additional generations of Higgs bosons are rather remarkable. We find that the lightest of these particles is absolutely stable and that the next lightest is probably only a GeV or so more massive. If under 45 GeV in mass, both of these particles are detectable at SLC and LEP, if a search is made for the somewhat unusual signature. This work was done in collaboration with Kim Griest, at U.C. Berkeley.[4,5]

The Higgs structure of the minimal supersymmetric model[6] consists of two doublets of opposite hypercharge. If there are several pairs of doublets, we will show that, under our assumptions, a basis can be chosen in which only one pair acquires a vacuum expectation value (VEV). The other pairs do not acquire VEVs and so do not, strictly speaking, participate in spontaneous symmetry breaking via the Higgs mechanism. For this reason we will refer to these particles as "pseudo-Higgs" bosons.

We extend the structure of the minimal supersymmetric model to include several pairs of doublets, H_i and $\overline{H}_i$. The most general gauge-invariant superpotential is given by

$$W = \mu_{ij} H_i \overline{H}_j + f_{ijk} Q_i U_j H_k + g_{ijk} Q_i D_j \overline{H}_k + h_{ijk} L_i E_j \overline{H}_k$$

where our phase convention will be chosen so that $H_i \overline{H}_j = \varepsilon_{ab}(H_i)_a(\overline{H}_j)_b$. As usual, Q, U, D, L and E are the quark and lepton superfields.

A potential problem with extending the Higgs sector of the standard model is the existence of flavor changing neutral currents (FCNC). As shown by Glashow and Weinberg[7] and by Paschos[8], the only way to naturally eliminate such FCNC is to couple all quarks of a given charge to a single Higgs multiplet. We will explicitly assume that there are no tree-level FCNC (in either the quark or lepton sector), and that the vanishing of FCNC occurs due to some symmetry (discrete, continuous, local or global). A basis in which only H_3 and $\overline{H}_3$ couple to fermions can then be chosen. (It should be noted that one could, in principle, couple the other Higgs fields to fermions in a flavor-diagonal manner, however our assumption that a symmetry eliminates FCNC will generally cause Cabbibo mixing to disappear in such models.) In this case, H_3 and $\overline{H}_3$ must have different quantum numbers under the symmetry than the other H_i and $\overline{H}_i$ (in order for the latter's couplings to fermions to vanish while the former's do not). The detailed nature of the symmetry is irrelevant. Note that we are now following the conventional notation in which the third generation Higgs fields are the ordinary fields while the first and second generation fields are the pseudo-Higgs fields.

The most general soft supersymmetry breaking terms are mass terms for the scalar fields plus terms proportional to superpotential terms. They are given by

$$m^2{}_{ij} H_i{}^\dagger H_j + \overline{m}^2{}_{ij} \overline{H}_i{}^\dagger \overline{H}_j - B\mu_{ij} (H_i \overline{H}_j + h.c.).$$

Since H_3 and $\overline{H}_3$ have different quantum numbers under the symmetry which eliminates tree-level FCNC, $m^2{}_{i3} = m^2{}_{3i} = \overline{m}^2{}_{i3} = \overline{m}^2{}_{3i} = 0$ for i=1,2; i.e. the mass terms will not mix the third generation of Higgs fields with the other two. What about the m_{i3} and m_{3i}

terms? Because of the above symmetry, if a m_{33} exists, then the m_{i3} and m_{3i} terms do not, and vice versa. It is not hard to see[4,5] that if a m_{33} term is absent, then the resulting potential will have a global U(1) symmetry under which the third generation fields transform non-trivially. This leads to an unacceptable axion. Thus, the m_{33} term must be present, and so the m_{3i} and m_{i3} terms are not. We see that none of the quadratic terms mix the third generation fields with the first two. Since the quartic terms are supersymmetric, they also will not mix the third generation fields with the first two. Thus, when the third generation fields get VEVs (as they must to give the fermions mass), the pseudo-Higgs fields will not be forced to get VEVs.

In fact, the pseudo-Higgs fields will not acquire VEVs in general. In the minimal model derived from minimal supergravity, the mass-squared parameters are equal at the GUT scale. In this model, the mass-squared parameters of the pseudo-Higgs fields have identical beta-functions (since none of them couple to fermions), and thus these parameters will be equal at all scales. It is not hard to see[4,5], looking at the most general potential, that they will thus not acquire VEVs at all. Thus, there are no terms in the Lagrangian which can mix the ordinary Higgs bosons with the pseudo-Higgs bosons, and the latter do not get VEVs. It is then easy to see that the Lagrangian has a symmetry in which each pseudo-Higgs field changes sign (the symmetry is actually a global U(1) symmetry). As a result, the lightest pseudo-Higgs is absolutely stable. We emphasize that this is entirely due to our assumption that some symmetry eliminates all tree-level FCNC.

One can now calculate the mass matrices of the pseudo-Higgs fields (which decouple from the ordinary Higgs mass matrices). There are four new Higgs doublets, corresponding to eight neutral fields. The mass matrices divide up into two 4 x 4 matrices; each of which is composed of essentially arbitrary parameters. However, the two matrices turn out to be identical except for the signs of the even-odd elements[4,5]. As a result, the eigenvalues are identical. This degeneracy will be split by radiative corrections, and we conclude that the second lightest pseudo-Higgs is only a few percent (0.2 to 4 GeV) heavier than the lightest. From the charged pseudo-Higgs mass matrix, which is degenerate in the limit of the SU(2) gauge coupling vanishing, we find that the lightest charged pseudo-Higgs is typically O(30) GeV heavier than the lightest pseudo-Higgs.

Thus, given the assumptions that the minimal supersymmetric model is extended to include several doublets and that some symmetry eliminates tree-level flavor changing neutral currents, we find that the lightest of the additional scalars, which we call S, is absolutely stable, the second lightest, which we call P, is only a GeV or so heavier, and the lightest charged scalar, which we call C is about 30 GeV heavier. How could these scalars be detected? They do not couple to fermions at all, and (since they have no VEVs) there is no scalar-vector-vector coupling, so they cannot be produced via Z-brehmstrahlung. The only interaction which is of phenomenological relevance is the scalar-scalar-vector

884

coupling.

The signature for S and P production is very clean if their mass is below 45 GeV. The Z-boson will decay into an S and a P. The S is absolutely stable and disappears. The P decays into an S, which disappears, and a virtual Z, which goes into a lepton pair. The signature is then missing energy and a low energy fermion (usually muon or electron) pair. Especially intriguing is the fact that the P lifetime is very long, due to the mass degeneracy, and is typically between 10^{-13} and 10^{-7} seconds, thus the low energy fermion pair might not point back to the vertex. The signature is thus very distinct, and would be easily detectable *if* the branching ratio is large enough.

Generally, when one introduces new fields into the standard model, the couplings of these fields have various mixing angles (which are usually arbitrary). A remarkable property of this model is that the coupling of the scalars S and P to the Z-boson turns out to be independent of all mixing angles. This is because the mass matrices (of the neutral scalars) are identical except for the sign of the even-odd elements; this means that the unitary matrices which diagonalize the mass matrices are identical except for the sign of the even-odd elements. The Z-S-P coupling can then easily be shown to be proportional to the sum of the squares of the elements of a row of one of these matrices; this is unity, and so all mixing angles in this coupling cancel. The branching ratio is then completely determined, and is given by

$$\Gamma (Z \longrightarrow SP) \;\; = \;\; (1/2) \, (1 - 4m^2_S/ m^2_Z)^{3/2} \; \Gamma (Z \longrightarrow \nu\bar{\nu})$$

where we have taken the S and P masses to be degenerate (which is a very good approximation). If they are light, then this is 3% of all Z decays, and could be easily detected *if* the unusual signature is looked for. In fact, this process will enable detection of the S and P at future electron-positron colliders up to the kinematic limit of the machine.

What about the charged pseudo-Higgs, C^+? It can also be produced in Z-decays, although it is not likely to be light enough. It decays into an S (or a P) and a virtual W. As a result, it would appear to be identical to a heavy lepton with an associated neutrino which is massive (and about 30 GeV lighter). One significant difference would be in the angular distribution, which would have the usual $\sin^2 \theta$ distribution of scalars. Heavy lepton searches should then be sensitive to this particle.

At hadron colliders, the above mechanism for S and P detection would not be practical, because of the difficulty in picking out a low energy fermion pair from the background. The only plausible mechanism for detecting pseudo-Higgs particles at a hadron collider would appear to be through Drell-Yan production of a C^+C^- pair, each of which decays into an S or P and a virtual W. Here, again, heavy lepton searches would be relevant. To our knowledge, no detailed analysis of heavy lepton production at the SSC has been done when the associated neutrino is so heavy that decays into a real W are forbidden. Such an analysis would be necessary to examine the possibilities of pseudo-Higgs detection at hadron colliders.

We have ignored the supersymmetric partners of the pseudo-Higgs bosons. These could be stable in one of two ways: the lightest could be the lightest supersymmetric particle or the lightest could be lighter than the lightest pseudo-Higgs boson. Note that there is no mixing of these fermions with the regular Higgsinos or gauginos. Detection of these partners is very similar to detection of regular pseudo-Higgs bosons; there are some factor of two differences is cross-sections, and different angular distributions due to the spin. There is one significant difference, however. In this case, the mass matrices of the charged pseudo-Higgsinos is also degenerate with the neutral pseudo-Higgsinos at tree-level. As a result, the supersymmetric partner of the C^+ is not much heavier than its neutral counterpart. In this case, it would appear to be a heavy lepton with a neutrino only a few hundred MeV lighter. The lifetime could be quite long (somewhere between the muon and tau lifetimes), and it might leave a detectable track. Analysis of heavy leptons with nearly degenerate neutrinos can be quite complicated, but would be necessary to look for psuedo-Higgsinos if they are lighter than the pseudo-Higgs bosons.

To conclude, we have made two assumptions: the Higgs sector of the minimal supersymmetric model is extended to include several generations, and there is a symmetry which eliminates tree-level FCNC. We conclude that the lightest of the extra Higgs bosons is absolutely stable, the second lightest is only about a GeV heavier, and the lightest charged scalar is only about 30 GeV heavier. The former two can be detected in Z decays at an electron-positron collider *if* the unusual signature is looked for. The charged scalar would appear to be a heavy lepton with an associated heavy neutrino, with a decay distribution characteristic of scalars.

REFERENCES

1. M. Dine, V. Kaplunovsky, M. Mangano, C. Nappi and S. Seiberg, Nucl. Phys. B259 (1985) 519; J. Breit, B. Ovrut and G. Segre, Phys. Lett. 158B (1985) 33.
2. J. Ellis, D.V. Nanopoulos, S. Petcov and F. Zwirner, Nucl. Phys. B283 (1987) 93.
3. M. Drees, Int. J. Mod. Phys. A4 (1989) 3635.
4. K. Griest and M. Sher, Phys. Rev. Lett. 64 (1990) 135.
5. K. Griest and M. Sher, U.C. Berkeley preprint CfPA-TH-90-002, 1990.
6. See J. F. Gunion and H.E. Haber, Nucl. Phys. B272 (1986) 1; B278 (1986) 449.
7. S. Glashow and S. Weinberg, Phys. Rev. D15 (1977) 1958.
8. E. A. Paschos, Phys. Rev. D15 (1977) 1966.

RARE K AND B DECAYS: BEYOND THE STANDARD MODEL

George Wei-Shu Hou
Paul Scherrer Institute
CH-5232 Villigen PSI, Switzerland

ABSTRACT

Theoretical expectations for beyond the Standard Model effects in rare K and B decays are briefly reviewed.

1. INTRODUCTION

At the outset I should confess the fundamental presentation difficulty in covering the (two) vast subject(s) completely. There are just two many Mode(l)s.
Modes:

$K \to \mu\bar{e},\ \pi\mu\bar{e}.$ $b \to s\gamma,\ sg.$
$\ \pi\nu\bar{\nu},\ \pi\ell^+\ell^-$ (and ϵ'). $s\ell^+\ell^-,\ s\nu\bar{\nu}.$
$\ \pi H^0,\ \pi a,\ \pi f,\ \pi+$ majorons, *etc.* $\mu\bar{e} + X;\ h^0 + X,$ *etc.*

Models, i.e. BSM; 1-$1\frac{1}{2}$ steps (according to conveners) beyond the Standard Model:

* gauge sector   left-right symmetry (LRS), horizontal, ...
* fermion sector   4th family (SM4), vector-like, exotics, ...
* scalar sector   extra doublets (2HDM, 3HDM), triplets, singlets, ...
* New symmetry   SUSY, ...
* New structure   ETC, WTC, compositeness, ...

Crossing the two lists, it is clear that each entry of this vast "matrix" has to be covered in only a few seconds time. An elegant exposition is not possible, and I will only attempt at a glossary of things. References will be made only to recent theoretical work. Others can be easily traced by looking up many excellent earlier reviews. Experimental references are suppressed, but many of the modes discussed here are covered in this Conference.

"Rare", usually flavor-changing neutral current (FCNC), decays become interesting when lifetime is prolonged. Just as in real life, in such cases, one (or the particle) is more likely to "decease by weird diseases". It is these "weird diseases" that allow rare K and B decays (usually $s \to d$ and $b \to s,\ d$) to probe "weird" stuff, namely, BSM effects. The decay mechanisms are sometimes contrived. It is useful to remember, however, that the early puzzle of absence of $K \to \pi\nu\bar{\nu}$, $K_L \to \mu\bar{\mu}$ vs. $K \to \pi\ell\bar{\nu}_\ell$ was resolved by the G.I.M. mechanism, which was a crucial step that lead towards the present day SM. The moral is that rare (FCNC) decay studies may still help us in choosing from the many competing BSM's, the Standard Model of Tomorrow.

It should be noted that BSM physics may already be lurking in *deviations*, or *relations* between SM parameters (e.g. fermion masses and mixing), rather than the derived quantities that we will discuss. For instance, the major uncertainties in KM aside from the Higgs sector are: V_{ub}, V_{cb}, V_{td}, V_{ts}, V_{cs}, V_{tb} and m_t. One may eventually find that the 3×3 KM matrix is not unitary, or one may verify, say, the Fritzsch type mass-mixing relations, e.g. $V_{cb} \cong \sqrt{m_s/m_b} - e^{i\phi}\sqrt{m_c/m_t}$, which would imply the existence of some underlying BSM physics. The top mass cannot be too heavy if the relation above is to hold, since the smallness of V_{cb} is achieved by a fine tuned cancellation. Indirect and direct evidence ($m_t > 89$ GeV, CDF) indicate a rather heavy top, so this seems to be ruled out already. However, since all the known quarks (u, d, s, c, b) are much lighter than M_W, having $m_t > M_W$, especially if $m_t > v.e.v.$ (weak symmetry breaking scale), would make the top quark look rather odd. Could this in itself be an indication for BSM? Chris Hill gives us an interesting account[1] in his "Two Steps Beyond SM" talk on minimal dynamical symmetry breaking, with the view that an ultra heavy top could be an indication that the $v.e.v.$ is due to the condensation of $t\bar{t}$, while the Higgs boson is a $t\bar{t}$ bound state. Another possibility I would like to mention is perhaps replacing[2] the Fritzsch relation above by $V_{cb} \cong \sqrt{m_c/m_t}$. This can be achieved if one assumes that when the u-quark mass matrix is of the Fritzsch form, the b quark mass is already diagonal. In this way the fine tuning in both the phase and m_t is evaded, replaced by having an ultra heavy top. The 10 mass and mixing parameters of the quark sector become only 6, plus a CP phase which is very close to $\pi/2$. The weak part of this ansatz is that m_t tends to be too heavy, perhaps in the domain where the effects of m_t may be nonperturbative. In any case, m_t cannot be brought below 200 GeV, but this is precisely the domain where the effect of loop induced FCNC is the largest in SM. One can therefore state the punchline: perhaps the long B lifetime was the first indication of BSM effects ($\tau_B \propto m_t!$), and if such is the case, FCNC K and B decays that I discuss here could be at the maximum allowed by SM, even if no other BSM effect is found. Remember, however, this is just an ansatz.

Let me now turn to our main subjects.

2. RARE K DECAYS

In general, long distance (LD), i.e. SM effects dominate K decays with π and/or γ and/or $\ell\bar{\ell}'$ in final state (e.g. $K_L \to \mu^+\mu^-$ comes mostly via $K_L \to \gamma\gamma \to \mu^+\mu^-$),

$$\text{except,} \quad \ell \neq \ell': \qquad K \to \mu\bar{e},\, \pi\mu\bar{e},\, \text{etc.}$$
$$\ell,\, \ell' = \nu: \qquad K \to \pi\nu\bar{\nu} \text{ (and } K \to \pi + \text{"nothing")}.$$
$$\text{CP violation:} \quad K_L \to \pi^0\ell^+\ell^-,\, \pi^0\nu\bar{\nu};\, \epsilon',\, \text{etc.}$$

These processes could be dominated by short distance (SD) effects. Kaons are produced abundantly, and experimental techniques have been developed for decades. Very impressive BR limits have already been achieved with promise for further improvements.

2.1 Lepton Number Violation: $K_L \to \mu\bar{e},\, \pi\mu\bar{e}$

Present experimental limit is $\mathcal{O}(10^{-10})$, which will be improved to $\mathcal{O}(10^{-11}\text{-}10^{-12})$ in near future. SM gives 0. These modes probe ultra-high energy scales, even beyond 70 TeV, although perhaps diluted by some "angles". Candidate "theories" are: horizontal, compositeness, ETC, *etc.* Experiments on these modes *must* continue, and it would be exciting if such decays were found.

2.2 $K^\pm \to \pi^\pm\nu\bar{\nu}$ ($K^\pm \to \pi^\pm + $"nothing")

From Δm_K, Δm_B and ϵ, one infers that the BR for this mode is $(0.5\text{-}8) \times 10^{-10}$ within SM. Present experimental limit is $\mathcal{O}(10^{-8})$, which will be down to 10^{-9} soon, eventually reaching 10^{-10} level in a couple of years. So, the sooner it's found (e.g. 10^{-9} or more), the more likely an indication for new physics. Likewise, if it is not found at the 10^{-10} level, it may also indicate new physics. Candidate BSM's:

* **SM4:** This is truly BSM now that ν_4 (perhaps call it ν_σ) has to be heavy. New flavor distinguishing free parameters $V_{t'd}$, $V_{t's}$ as well as $m_{t'}$ allow evasion of 3×3 KM unitarity, i.e. evade the constraints above. Thus, BR up to $10^{-9}\text{-}10^{-8}$ is possible.

* **SUSY:**

888

- Direct: $K^+ \to \pi^+ \tilde{\gamma}\tilde{\gamma}$ would need very light $\tilde{\gamma}$. This is unlikely but $\mathcal{O}(10^{-9})$ is of course possible.
- Indirect: Quark (q) and squark ($\tilde{q}$) mass matrices cannot be simultaneously diagonalized at low energy. That is, running down from some high (SUGRA) scale, effective $\tilde{g}d_i\tilde{d}_j$ ($i \neq j$) couplings are generated. Note that this involves the strong coupling constant α_s. Unfortunately, Δm_K with SUSY particles in box diagrams are very restrictive, and SUSY effects are unimportant for $K \to \pi\nu\bar{\nu}$. Taking ϵ_K, ϵ' into account would make matters only worse. Likewise, extra Higgs boson (i.e. H^+) effects suffer a similar fate.

* **other mode(l)s:**
  - $K \to \pi H^0$, H^0 ultra-light. This has become unlikely with the stringent mass limits from ALEPH. Perhaps it is still possible with non-standard Higgs bosons, e.g. 2HDM.
  - $K \to \pi a$. Unlikely. In general, the invisible axion leads to invisible BR.
  - $K \to \pi f$. Familon decay modes are possible, but of $\mathcal{O}(SM)$.
  - $K^+ \to \pi^+ m^0 m^{0\prime}$. Majorons emerge if lepton number is spontaneous violated. This mode may exist at the one neutrino family level[3]. The missing $M_{m^0 m^{0\prime}}$ mass spectrum may be useful.

In conclusion, $K^\pm \to \pi^\pm \nu\bar{\nu}$ is an extremely exciting mode. If found above 10^{-9} level, BSM physics is implied. Even if found at the 10^{-10} order, detailed study may disentangle submerged BSM effects.

2.3 $K_L \to \pi^0 e^+ e^-$ (or $\mu^+\mu^-$)

This has charged lepton final states, so according to our rule, LD effects are expected to dominate. However, at lowest order one would need CP violation, so SD effects may also be important.

LD: $K_L(K_2) \to \pi^0\gamma\gamma \to \pi^0 e^+e^-$; CP even but $\mathcal{O}(\alpha^2)$.

SD: $K_L(K_2) \overset{\epsilon}{\to} K_1 \to \pi^0\gamma^* \to \pi^0 e^+e^-$;   indirect CP odd (via mixing).

$K_L(K_2) \overset{CP\ odd}{\longrightarrow} \pi^0(\gamma^*, Z^*) \to \pi^0 e^+e^-$;   direct CP odd (in decay).

The direct CP violating contribution is sensitive to heavy top effects. All three of the above channels could be comparable in amplitude[4], but things are not yet settled. Theoretically the current prevailing expectation is at the 10^{-10}-10^{-11} level, which hopefully would not change. Experimentally, at present the limit is at the 10^{-8} level, with hopeful improvement down to 10^{-11}-10^{-12} level in the future. If so, there could be a three order of magnitude range for discovering BSM effects. Candidates: BSM CP violating currents; light scalars that go into e^+e^- (unlikely due to LEP limits); extra Higgs or other weird stuff. Given that the discussions have centered around the SM contributions, more work could be done on BSM effects in $K_L \to \pi^0 e^+e^-$.

2.4 $K_L \to \pi^0 \nu\bar{\nu}$ and ϵ'

The mode, which is analogous to $K_L \to \pi^0 \ell^+\ell^-$, is very nice theoretically, with neutrinos in final state but no CP conserving two photon contributions, thus coming purely from SD effects. The question is: How does one detect such a mode?

If the E731 value bears out, $\epsilon'/\epsilon \approx 0$ once again. This may be another indirect indication[5] that m_t is very heavy in SM. It should also constrain BSM effects. However, one suffers from rather punishing uncertainties in LD effects.

2.5 Rare K Summary and Comments

- Rare K decay studies are quite special and worthy since very small BR's can be reached experimentally.

- $K_L \to \mu\bar{e}$, $\pi\mu\bar{e}$ searches are absolutely necessary. But perhaps theorists should wait and see the new results before constructing more models.
- $K^{\pm} \to \pi^{\pm}\nu\bar{\nu}$ is very interesting and could prove to be exciting. If found at $\mathcal{O}(10^{-9})$ or higher, the most likely BSM candidate is SM4, which now demands a heavy 4th neutrino. Below that, detailed study of missing mass spectrum may reveal effects of familons, non-standard Higgs, majorons, *etc.*
- $K_L \to \pi^0 e^+ e^-$ holds an exciting future as well. The theoretical aspects within SM should be settled. Experimentally, one needs to push down to $\mathcal{O}(10^{-11}\text{-}10^{-12})$ to be able to rule out or discover BSM effects. Whether this is experimentally doable is not yet clear, but theorists should also do some work on BSM effects in this mode.

It is clear from our discussions above that there remains a lot of Experimental/Theoretical work to be done, after all these years. The future continues to look rather bright. The kaon system is truly a miracle of Nature.

3. RARE B DECAYS

As compared to the K system, the B system may be even more promising:

- $|V_{cb}| \ll |V_{us}|$, hence the B lifetime is even more prolonged than the K lifetime.
- The b quark is a 3rd generation quark, thus it is more sensitive to extra heavy flavor or other loop effects involving heavy particles.
- $m_b \gg \Lambda_{\text{QCD}} \approx$ hadronic scale. Hence, long distance effects are much less important, and various processes such as $B \to K^*\gamma$ which has no counterpart in the K system (due to phase space limitations) may occur.
- As a result of the above, SM loop effects are already rather rich and sizeable.
- Charm counting (error bars $\approx 15\%$), or the present "discrepancy" between $BR(B \to e\nu + X) = 10\%$ (experiment) vs. 11% (theory) in principle allow some unseen B decay mode(s) to exist even at the 10% level.

There is a basic problem: For (semi-)inclusive decays, the theory is well founded but experimentation is quite diffcult, especially for the rare decays that we will discuss; exclusive two body hadronic final states are experimentally the cleanest, but theory suffers from a high level of uncertainty, the more so if the final states are light mesons. In between, the B meson will have multi-particle exclusive final states, where experiment is more diffcult, but the theoretical situation seems hopeless.

The last three years or so we have witnessed rather intense theoretical activities on rare B decays, which is gradually subsiding, except on the topic of CP violation. One has reached a stage where it is rather hard to make drastic new improvements on either SM or BSM physics, and one is waiting for new experimental results. On the experimental side, there has been tremendous efforts by ARGUS, CLEO and others, but compared to K physics, B physics is still at its "infancy". In the past year or two, efforts have intensified to realize one or more "B Factories". Definition: a facility that provides 10^7-10^8 $B\bar{B}$'s per year, *together with* a perfect detector that can realize all the B physics potentials. The Holy Grail is of course CP violation, but rare decays should not and will not be missed along the way.

Physics is rich in the B system, but it would demand a lot of hard work, patience, and perhaps some ingenuity, just like the efforts on K physics in the past 30 years.

3.1 B_d-$\bar{B}_d$ Mixing: Surprise from ARGUS in 1987

In SM, this is accounted for easily if m_t is large. It came as a surprise in 1987 due to the prevailing expectations for m_t at that time. Now we know that m_t is rather heavy, which in itself could be an indication for BSM physics. What kind of BSM physics could also account for large B mixing, without invoking a heavy top?

* **SM4:** This is easily achieved, given that even SM3 would do the job if m_t is heavy

enough. Of course, given m_t and $m_{t'}$ this should lead to constraints on V_{td}, $V_{t'd}$. An intriguing indication for SM4 would be if B_s-$\bar{B}_s$ mixing turns out to be nonmaximal. If experiments (at B Factories) find $x_s < 3$-5, then BSM physics is implied. SM4 would be the most natural candidate theory, at least the least contrived one.

 * **Extra Higgs bosons?** H^+ effects could easily account for Δm_B. Note that m_t may well evade the CDF limit if top decays via $t \to bH^+$.

 * **SUSY**: possible, but its effects are weaker than SM in minimal model.

One could name many other models, but in any case, B_d-$\bar{B}_d$ mixing provides a rather strong constraint on various BSM's.

3.2 $b \to u$

This standard decay mode is of order 1%, so it is also "rare". The importance of measuring this transition rate is that it provides us a measurement of V_{ub}, which is crucial for CP violation in SM, and therefore crucial to see whether SM works, and whether the 3×3 KM matrix is unitary .

A special kind of $b \to u$ transition comes in the form of $B^+ \to \tau^+ \nu_\tau$. The decay rate is proportional to $V_{ub}^2 f_B^2$, with both parameters being of great interest. Given the present bound on V_{ub}, the BR should be less than 10^{-4}, unless f_B turns out to be much larger than expected. Experimentally, this mode is a must, and one may take advantage of the one-charged-prong dominance of τ decays.

3.3 $B \to \mu\bar{e} + X$, $\mu\bar{\tau} + X$ or $e\bar{\tau} + X$

$B \to \mu\bar{e}$ itself suffers from either helicity or coupling constant suppression (depending on whether vector or scalar coupling), which invariably leads to a suppression factor of $(m_\mu/m_B)^n$, with n a positive integer. These lepton number violating decay modes are independent of $K \to \mu\bar{e}$, $\pi\mu\bar{e}$. They are routinely searched for but existing limits are far from restrictive. Just like their K counter part, these experiments are definitely needed, but further theoretical speculations should probably wait.

3.4 $b \to s\gamma$, $sg^{(*)}$

SM effects for these modes are interesting enough already, and we digress to give some discussion. Effective $bs\gamma$, bsg couplings are of the form (ignoring m_s),

$$F_1 \left(q^2\gamma_\mu - q_\mu \not{q}\right)L + F_2\, i\sigma_{\mu\nu}q_\nu m_b R,$$

due to current conservation. We shall use calculations for the "charge" form factor F_1 and "dipole" form factor F_2 that are done in the vanishing external mass (but keeping exact internal mass dependence) approximation. That is, one expands to leading nonvanishing order in m_b and q, where q is the vector boson momentum. We single out these modes not only because of their importance, but also the theoretical subtleties involved. In SM, heavy internal quark (t) effects are damped by powers of $1/(m_t^2 - M_W^2)$ for both F_1 and F_2. However, for light internal quarks with mass m_i, $F_{1i}^{\gamma,\ g} \propto \log(m_i^2/M_W^2)$, while $F_{2i} \propto m_i^2/M_W^2$. Thus, the F_1 form factor shows an unusual sensitivity in the infrared, i.e. semi-long-range effects (longer range than $1/m_b$ but short range compared to $1/\Lambda_{\rm QCD}$), but the F_2 form factor is completely insensitive to low mass intermediate states. This can be traced physically to the fact that QCD allows rescattering of color octet "on-shell" light $i\bar{i}$ states into light quark final states via a virtual gluon (analogously for photon), but if an extra (quark level) spin flip is required, as is the case for the F_2 contribution, one suffers from a suppression by powers of m_i. This subtelty, after activating the G.I.M. mechanism by combining the contributions of internal u, c, and t quarks, leads to the result $F_2^{\gamma,\ g} \ll F_1^{\gamma,\ g}$ at the one-electroweak-loop level. Interestingly, QCD corrections bring in the IR sensitive logarithm at the α_s order and greatly *enhances* the effective F_2^γ form factor, but for the gluonic coupling the coefficient is small and works destructively against the already small lowest order result. Thus, QCD effects tend to *suppress* F_2^g. Note that for on-shell $b \to s\gamma$, sg, the F_1 contribution vanishes due to the additional q^2 factor in the coupling, and only the F_2 term contributes.

The upshot is that in SM, $b \to s\gamma$ is at the 10^{-4} level, while $b \to sg$ is very suppressed, but $b \to sg^* \to sq\bar{q}$ is at the 1% level. The $b \to sgg$ mode is suppressed by cancellations between $b \to sg^* \to sgg$ and other channels, as a consequence[6] of an extension of Low's low energy theorem. Another mode, $b \to s\gamma\gamma$, could[7] also be of interest, leading to $B_s \to \gamma\gamma$ at 10^{-7} level, but could be higher with BSM effects. Exclusive modes are experimentally more tangible, but they suffer from hadronic form factor and matrix element suppressions, e.g. $\Gamma(B \to K^*\gamma)/\Gamma(b \to s\gamma)$ is "predicted" to be anywhere between 5%-40%, while $\Gamma(B \to K\pi)/\Gamma(b \to sq\bar{q})$ is $\mathcal{O}(10^{-3})$.

The subtelty described above has interesting implications. Since $F_2^{\gamma,\ g}$ (but not $F_1^{\gamma,\ g}$) is suppressed, it is rather sensitive to BSM effects, as we shall see below.

* **SM4**: For $b \to sq\bar{q}$, since F_1 is insensitive to heavy m_Q effects, t and t' are effectively degenerate, and BR remains of order SM. For $b \to s\gamma$, F_2^γ is slightly more sensitive to m_Q, and some enhancements are possible. Since there is only one single amplitude for this mode, a more interesting possibility is when the t and t' contributions are destructive, resulting in a BR far below the SM result.

* **SUSY**: Within minimal SUSY, if $m_{\tilde{q}}$ and $m_{\tilde{g}}$ are light, sizeable enhancements are possible. The impact is on the F_2 form factor. As stated above, this paramter is very suppressed in SM, so it is a sensitive probe to certain kinds of effects. SUSY is one of them and $b \to s\gamma$, sg both tend to get enhanced. In fact, they may well be at the current limits, which can be turned around to constrain $m_{\tilde{q}}$-$m_{\tilde{g}}$, and is competitive w.r.t. current collider limits. However, as CDF improves on SUSY limits, the effects would be reduced down to the SM level. Furthermore, if one has minimal SUSY *plus* radiative SUSY breaking, then SUSY effects are necessarily at the SM level. The dominant BSM effect turns out to be due to H^+, which exists with minimal SUSY.

* H^+ **effects**: Almost all extensions of SM contain extra Higgs bosons, and almost invariably they contain physical charged Higgs bosons. The minimal extension to 2HDM (two doublets) serves to illustrate the possible effects. To have Natural Flavor Conservation, each fermion charge type can couple to only one doublet. In 2HDM, the single physical H^+ boson has the coupling

$$(g/\sqrt{2}M_W)\ \bar{u}_i\,(\xi\, m_{u_i} V_{ij} L - \xi' V_{ij} m_{d_j} R) d_j\ H^+,$$

where $\xi = v_2/v_1$ is the ratio of *v.e.v.* of the two doublets, and $\xi' = -1/\xi$ (Model I; motivated by PQ symmetry, minimal SUSY...), ξ (Model II; why not?). Model I is more appealing but Model II usually leads to stronger effects.

The H^+ contribution to Δm_K, Δm_B is proportional to $\xi^4\, m_t^2/M_W^2\, f(m_t^2/M_W^2)$, where $f(x)$ is a mild function of x. Thus, Δm_B constrains $\xi^4 m_t^2$ strongly, but weaker on m_H^2/m_t^2. Now, the induced bsZ coupling with H^+ running around the loop is $\delta_H G_Z \propto \xi^2\, m_t^2/M_W^2\, g(m_t^2/m_H^2)$, where $g(x)$ is again a weakly varying function. Thus, Δm_B severely constrains the H^+ induced bsZ coupling. Similarly, for the "charge radius" form factor, $\delta_H F_1 \propto \xi^2\, h\, f_1(h)$ (where $h \equiv m_t^2/m_H^2$), but it does not contain the IR sensitive log as the SM contribution. Thus, H^+ effects via the F_1 form factor are not pronounced. However, $\delta_H F_2 \propto [\xi^2 f_2(h) + \xi\xi' f_2'(h)]\, h$, which distinguishes between Model I and II! This comes about because of the spin-flip nature of the F_2 form factor. To get the required m_b when doing the calculation, one invokes either the Dirac equation on external b momentum, i.e. $\not{p}_b \to m_b$, and is ξ'-independent, or via a direct Yukawa coupling to b, and is therefore ξ'-dependent. The F_2 coupling not only distinguishes between the two types of 2HDM, but because the SM effect is very subdued here, the H^+ effects could turn out to be very pronounced, even with the constraints from Δm_B. In general, generous enhancements are possible[8, 9] for both $b \to s\gamma$ and $b \to sg$, especially for the less motivated Model II. In Model II, H^+ effect may be destructive[8] against SM contribution for $b \to s\gamma$ mode, while for the remaining combination of modes and models, the general effect is constructive.

I emphasize that $b \to s\gamma$ with H^+ enhancement is still consistent with existing data. When, say, $B \to K^*\gamma$ is finally observed, it will take a while before we can have a trustworthy hadronization model. On the otherhand, from the 1% "discrepancy" between theory and experiment regarding the inclusive BR for semi-leptonic B decays, we stated that a 10% unknown decay channel may well exist. Could this be $b \to sg$, either from SUSY enhancements or from H^+ effects? They could easily lead to $BR(b \to sg) \approx 10\%$. This may not be inconsistent with the existing exclusive limits on say $B \to K\pi$. One expects $B \to K\pi$ to come about mostly through $b \to sg^* \to sq\bar{q}$, which is *not* enhanced since it follows from the F_1 coupling. Given that $b \to sg$ probably has a rather energetic, on-shell gluon ($E_g \cong p_g \approx 2.5$ GeV), i.e. "jet-like", it probably hadronizes less frequently into two body final states. Unfortunately, no theoretical calculation method is available, given that the 4-quark operator that one can construct from the F_2 coupling would be non-local. Thus, there may well be another great surprise waiting to be discovered in the B system, coming as a very "strange", "light-like", "penguin". It would be a challenge to rule this out.

3.5 $b \to s\ell^+\ell^-$, $s\nu\bar{\nu}$

These modes are analogous to the $K \to \pi\nu\bar{\nu}$ mode. In SM3, with $m_t \sim 200$ GeV, the $\ell^+\ell^-$ mode is at the 10^{-5} level semi-inclusively per charged lepton species, while for the $\nu\bar{\nu}$ mode it is at the 10^{-4} level summing over three light neutrinos. These modes are very sensitive to heavy quark effects. Possible candidate BSM's are:

* **SM4:** One can gain through both KM (now 4×4) or very heavy $m_{t'}$. Without pushing $V_{t's}$ to rather large values, one could still gain an order of magnitude in BR. The photonic, Z and box contributions to $b \to s\ell^+\ell^-$ never mutually cancels, but for $b \to s\nu\bar{\nu}$, only Z and box diagram contribute, and one may have large cancellation between t and t' effects, leading to suppression in rate. Of course, once B_s mixing is measured, one should be able to predict these decay rates better .

* **SUSY:** These effects are constrained by Δm_B to be less than the SM effect.

* H^+ **effect:** $b \to s\nu\bar{\nu}$ is hardly affected, since it is highly constrained by Δm_B. However, fed by the F_2 coupling through the virtual photon ($\gamma^* \to \ell^+\ell^-$) contribution, $b \to s\ell^+\ell^-$ can be enhanced by up to a factor of 5 in Model II, but for Model I it hardly changes if $m_t > M_W$.

The $b \to s\nu\bar{\nu}$ mode arises purely through short distance effects, and is theoretically clean, just like $K \to \pi\nu\bar{\nu}$. However, its experimental detection would be very challenging. Although it contains strangeness in the final state, it does not have the advantage of one-charged-prong dominance of τ decays in $B^+ \to \tau^+\nu_\tau$.

3.6 CP Violation

A major driving force behind B Factories is the potential richness of CP violating phenomena in the B system. CP violation in a sense is a particular kind of "rare decay" mechanism. But it is such a vast field that it deserves its own detailed discussion, which is outside the scope of this talk. The phenomenology is rich enough within SM, but, CP violation effects in B system may well be the first place where BSM effects show up.

3.7 $b \to sh^0$

The ALEPH limit makes standard Higgs mode unlikely, but perhaps it ·can be evaded by enlarging the Higgs sector. Whether this is possible remains to be seen.

4. HOPEFUL FUTURE: FROM RARE TO WELL DONE

K and B physics is already so rich, as we have seen. Although the likelihood has diminished due to neutrino counting results at the Z peak, 4th generation fermions may well exists. They actually become more exciting since the 4th neutrino has to be heavy. There would be an extra bonus in that the b' quark may have the unusual property of FCNC $b' \to b$ decay dominance. Please see my other talk[10] for SM4 properties of b'. The relevant point here is that since the b' lifetime could be so prolonged, if it exists, not only it may have extraordinary SM(4) phenomena, it may provide us with a great view of all kinds of "weird" physics as well.

5. CONCLUSIONS AND OUTLOOK

$\Diamond$ K **Physics:**

- Quite mature, with very impressive limits that will continue to improve.
- $K^\pm \to \pi^\pm \nu \bar{\nu}$ and $K_L \to \pi^0 e^+ e^-$ are best bets for BSM lookout, and could be seen in a few years. ϵ'/ϵ would be harder to disentangle.

$\heartsuit$ B **Physics:**

- Theory
  - S.D. part "settled".
  - For exclusive modes, severe difficulty in calculating hadronic matrix elements.
- Experiment
  - Far from mature yet.
  - B Factories wanted: 10^7-10^8 $B\bar{B}$ per year, plus "perfect" detector.
- $b \to s\gamma$, sg sensitive to SUSY or H^+ effects, which could be substantial.
- $B \to K^*\gamma$ would probably be the first FCNC B decay mode to be discovered.
- CP violation studies itself a great way to look for BSM.
- Imaginative BSM work still lacking.

$\spadesuit$ B' **Physics?**

If b' exists, may give us the biggest boost yet in uncovering BSM physics via "Rare" Decays.

REFERENCES

1. Talk by C. Hill, this proceedings.
2. X.G. He and W.S. Hou, Phys. Rev. **D41** (1990) 1517.
3. S. Bertolini and A. Santamaria, Nucl. Phys. **B315** (1989) 558.
4. J. Flynn and L. Randall, Phys. Lett. **B216** (1989) 221; Nucl. Phys. **B326** (1989) 31. C. Dib, I. Dunietz and F.J. Gilman, Phys. Lett. **B218** (1989) 487; Phys. Rev. **D39** (1989) 2639.
5. J. Flynn and L. Randall, Phys. Lett. **B224** (1989) 221; G. Buchalla, A.J. Buras and M.K. Harlander, MPI-PAE/PTh-63/89; C.S. Kim, J.L. Rosner and C.P. Yuan, EFI-89-47.
6. J. Liu and Y.P. Yao, UM-TH-89-19; H. Simma and D. Wyler, private communication.
7. G.L. Lin, J. Liu and Y.P. Yao, Phys. Rev. Lett. **64** (1990) 1498.
8. W.S. Hou and R.S. Willey, Phys. Lett. **B202** (1988) 591; Nucl. Phys. **B326** (1989) 54.
9. B. Grinstein and M.B. Wise, Phys. Lett. **B201** (1988) 274; T.D. Nguyen and G.C. Joshi, Phys. Rev. **D37** (1988) 3220.
10. See my talk on b' decays, this proceedings.

DYNAMICAL GENERATION OF FERMION MASS

Thomas Appelquist
Department of Physics, Yale University
New Haven, Ct. 06511

ABSTRACT

Some recent developments in dynamical electroweak symmetry breaking are reviewed. Particular attention is paid to the problem of generating large fermion masses and large splittings among these masses.

The past few years have seen a modest revival of interest in dynamical electroweak symmetry breaking, in particular in technicolor theories where the breaking is driven by an asymptotically free gauge interaction. This interest has been stimulated partly by some new observations [1,2,3] about the dynamics of these theories and the natural scale of particle masses that can be generated. The theories contain ingredients that can enhance the high momentum components, giving much larger quark and lepton masses than had been naively expected while leaving the weak scale itself relatively unaffected. It may even be possible to achieve realistic mass levels -- including the current lower bound on the t-quark mass, while keeping flavor changing neutral currents adequately suppressed.

None of this work has yet led to a compelling model or to any deep insight into the strange pattern of quark and lepton masses. However, I think it has opened up some new possibilities and it has provided some new perspectives on dynamical symmetry breaking. In this talk, I will review some of these developments and then highlight a few of the problems that future work in this area will have to confront.

In technicolor theories, the Higgs sector is taken to be a set of massless fermions interacting via an asymptotically free vector-like gauge theory with confinement scale of order several hundred GeV. With two or more fermions, the global chiral symmetry will be at least $SU(2)_L \times SU(2)_R$, where L and R label the chiralities of the technifermions. Described in this way, technicolor is simply a scaled-up version of QCD. Since the QCD interactions among the nearly massless u and d quarks provides the only known example in particle physics of a spontaneously broken $SU(2)_L \times SU(2)_R$ symmetry, technicolor is a very conservative theory -- perhaps the most conservative theory-- of the Higgs sector. As in the QCD prototype, the fact that the coupling only becomes strong near the confinement scale provides a natural explanation of why chiral symmetry breaking takes place there and not at much higher energies. As in QCD, an effective, low energy nonlinear sigma model can be employed, and it breaks down only at the highest energy $(O(2\pi F))$ that it can. Strong interactions develop at these energies and then decrease in strength at higher energies as the asymptotic freedom sets in.

While technicolor may be a more natural theory of gauge boson mass generation than the seemingly contrived and technically unnatural scalar-field interactions of the minimal standard model, it begins to seem less natural when the problem of fermion mass generation is confronted. The fermion masses must also arise when the $SU(2)_L \times SU(2)_R$ symmetry breaks spontaneously, and the only way to implement that is to introduce new (ETC) interactions at some higher scale Λ (or perhaps a set of higher scales), connecting technifermions to the quarks and leptons.

These new interactions can be described at lower energies in terms of effective higher dimension operators. The dominant contribution at low energies should arise from dimension six, four-fermion operators. In technicolor theories, the effect of these operators (ETC interactions) was considered some time ago [4]. They were treated as perturbations to the dominant technicolor gauge interactions, giving rise to quark and lepton masses, pseudo-Goldstone boson masses, and typically to flavor-changing neutral currents among the quarks and leptons. Without some GIM-type mechanism, they must be suppressed by taking the ETC scale Λ to be at least several hundred TeV, and this then leads to a potential problem for technicolor theories: Quark and lepton masses are given by $m(q,l) \sim (g^2/\Lambda^2)$ $\langle\bar\psi\psi\rangle$ where g^2 is a dimensionless coupling strength and $\langle\bar\psi\psi\rangle$ is the technicolor condensate, naively expected to be of order Λ_c^3. These masses then turn out to be no more than a fraction of an MeV.

Another embarrassing fact is that some of the pseudo Goldstone bosons generically present in these theories seem to have masses that are too small. They are the ones that do not get mass from QCD interactions, but instead must rely on the four-fermion (ETC) interactions. These masses are then also proportional to $\langle\bar\psi\psi\rangle$ and turn out naively to be no larger than a few GeV. The electrically charged pseudos in this mass range should probably have been seen by now.

The revival of interest in technicolor theories is partly due to the observation [1,2] that some of them contain within themselves a partial solution to these problems. The large number of technifermions that typically populate these theories can slow the running of the coupling while maintaining the asymptotic freedom. The technicolor condensate, and therefore the quark, lepton and pseudo masses, are quite sensitive to this modest enhancement of the high momentum components. In many candidate technicolor theories ("walking theories"), the slowing of the running is sufficient to raise the masses by up to nearly two orders of magnitude [5]. Even this effect, however, does not seem adequate to produce a realistic quark and lepton mass spectrum.

Another enhancement effect, however, has also been suggested and studied recently. The ETC interactions, perhaps in the form of four-fermion operators, can also produce enhancement. If they are strong enough, they cannot necessarily be treated as perturbations to the gauge interactions. Instead, they should be included in the gap equation that governs the breaking and they can dramatically affect the mass scales that emerge. That four-fermion interactions can drive dynamical breaking has of course been known since the classic paper of Nambu and Jona-Lasinio [6]. If they begin to dominate the dynamics in the present framework, however, the entire low energy expansion could be breaking down and even higher dimension operators could then play a role. I will conclude this talk by describing some recent work that goes beyond the four-fermion approximation.

To begin, flavor symmetry violation will be neglected. The technifermions together with their technicolor and four-fermion interactions will then respect a global $SU(N_f) \times SU(N_f)$ symmetry. Spontaneous chiral symmetry breaking will lead to a residual $SU(N_f)_{L+R}$ symmetry with a technicolor singlet fermion mass determined by an appropriate gap equation for the dynamical technifermion mass $\Sigma(p)$. With the gauge sector treated in dressed ladder approximation [7], the gap equation will take the form

$$\Sigma(p) = \frac{1}{4} \int_0^{\Lambda^2} \frac{k^2 dk^2}{M^2(k,p)} \frac{\alpha(M)}{\alpha_c} \frac{\Sigma(k)}{k^2 + \Sigma^2(k)} + 2 \frac{g^2}{\Lambda^2} [\frac{1}{4\pi^2} \int_0^{\Lambda^2} k^2 dk^2 \frac{\Sigma(k)}{k^2 + \Sigma^2(k)}] \quad (1)$$

The first term is the technicolor contribution in Landau gauge. $M(k,p)$ is the maximum of k and p, and $\alpha_c = \pi/3C_2$ is the critical gauge coupling required for spontaneous breaking in the absence of other interactions. C_2 is the quadratic Casimir of the technifermion representation. The second term comes from the chirally invariant four-technifermion interaction. Its coupling strength is defined to be g^2/Λ^2, and it is assumed that the physics giving rise to this interaction damps rapidly beyond Λ. It is also assumed that the four-fermion interaction is attractive in the technicolor-singlet channel. Factors arising from the closed loop sum over N_f and technicolor have been absorbed into g^2. The quantity in square brackets, when multiplied by the dimensionality of the tecnifermion representation, is the technifermion condensate.

How do $\Sigma(p)$, F, and the condensate $\langle \bar{\psi}\psi \rangle$ depend on the four fermion strength $\lambda \equiv g^2/2\pi^2$? For very small λ, the physics will be dominated by the technicolor interaction. $\Sigma(p)$ will be of order Λ_c for $p \sim \Lambda_c$. For $p >> \Lambda_c$, it will fall with increasing p at a rate roughly between $1/p$ and $1/p^2$ depending on the rate of running of the coupling. The integral expression giving F in terms of $\Sigma(p)$ [8] converges rapidly for any rate of fall of $\Sigma(p)$ in the above range, and leads to a value of F on the order of Λ_c. The condensate depends sensitively on the running and can vary from being roughly of order Λ_c^3 (ordinary running) up to nearly of order $\Lambda_c^2\Lambda$ (very slow running).

Suppose next that λ is very large, corresponding to the existence of new strong interactions among the technifermions at energies of order Λ. If the gauge interaction is neglected, chiral symmetry breaking will take place for $\lambda \equiv g^2/2\pi^2 > 1$. If λ is well above 1, the gauge interaction can be treated as a perturbation and $\Sigma(p)$ will turn out to be of order Λ ($>> \Lambda_c$) and essentially independent of p. In this limit, $F = O(\Lambda)$ and $<\bar{\psi}\psi> = O(\Lambda^3)$. This of course is an unphysical limit from the point of view of technicolor. For the reasons described above, a heirarchy must be maintained between F and Λ. It is also a limit in which the low energy expansion has broken down since the physics is taking place dominantly at the cutoff, and the above estimates are not reliable in detail.

It is the range of intermediate values of λ that is of interest for technicolor theories. We have studied the evolution of F and $<\bar{\psi}\psi>$ throughout this range, both analytically and numerically. [1,9] For most of the range, $\Sigma(0) << \Lambda$, and an analytic treatment is made possible because the integral equation can be linearized for $p,k >> \Sigma(0)$. The analytic treatment can then be supplemented numerically for all λ. The numerical results for a typical case, an SU(4) gauge theory with $\Lambda/\Lambda_c > 10^3$ and with the number of technifermions varied to adjust the rate of running of the gauge coupling, are displayed in Table I. It is of course only the orders of magnitude of these results and their sensitivity to λ that are important.

For essentially any rate of running, the four-fermion term makes a relatively small contribution to $\Sigma(0)$ for most of the range of λ between 0 and 1. For normal running ((b-δb)$\alpha_c \sim 1$, where b and δb are the gauge and fermion contributions to the beta function $\beta(\alpha) = -(b-\delta b)\alpha^2$), it becomes comparable to the technicolor contribution only for $\lambda \cong$ 0.93. For somewhat slower running ((b-δb) $\alpha_c \sim 0.4$), as might be expected in many technicolor theories, it becomes comparable for $\lambda \cong 0.85$. For very slow running ((b-δb)$\alpha_c \sim 0.1$), it becomes comparable for $\lambda \cong 0.7$.

For any λ below the above values, $\Sigma(0)$ and F remain essentially unchanged and of order Λ_c. The condensate, however, increases substantially throughout each range. The increase from the $\lambda = 0$ value is at first modest, but it is at least three orders of magnitude near the end of each range. In the intermediate running case, the condensate is $O(\Lambda_c^3)$ in the absence of the four-fermion interaction, corresponding to a quark or lepton mass on the order of one Mev. The condensate then increases by nearly an order of magnitude by the time λ reaches 0.7 and by more than another two orders of magnitude in the range between

0.7 and 0.85. At the end of the range, a quark or lepton mass exceeding 1 GeV is generated. In the case of very slow running, the additional effect of the four-fermion interaction can produce a dramatically enhanced condensate within the above range. Near the end, quark and lepton masses in the range of 5 - 10 Gev can be generated.

For any rate of running, as λ is increased beyond the ranges described above, some dramatic effects set in. The changes are very rapid and by the time λ exceeds unity, the theory becomes unphysical for technicolor purposes. Consider the intermediate running case $(b-\delta b)\alpha_c \sim 0.4$. As λ increases from 0.85 to 0.92, F grows from being somewhat less than Λ_c to being several times larger. Since F is the fixed weak scale of the theory, this means that the technicolor confinement scale drops below the weak scale. The resulting light physics would consist of relatively narrow techni-glueballs which could be produced at existing or soon-to-exist accelerators. With $\lambda = 0.92$, for example, the technicolor confinement scale drops to roughly 20 GeV, compared to F = 90 GeV, $\Sigma(0) = 120$ GeV, and the condensate is large enough to produce a quark or lepton mass greater than 50 GeV. While, the low energy expansion still appears to remain valid for λ in this range, the results are becoming more sensitive to physics at Λ. It will be important to explore this sensitivity in more detail.

A natural question is whether there is an expected range of values of λ in technicolor theories. It is clear from Table I that substantial condensate enhancement, due partly to the four-fermion interaction, occurs for a reasonably large range of λ. An enhancement large enough to produce a t quark mass and light techniglueballs, on the other hand, emerges only for a narrower range of λ values nearer to unity. If the four-fermion interactions are due to ETC boson exchange, g will be the ETC boson coupling. It would then not be unreasonable to speculate that $(\pi/2)\lambda = g^2/4\pi$ should be on the order of $\alpha(\Lambda)$, the technicolor coupling at the ETC scale. This would be expected if these interactions become unified at Λ. In the SU(4) examples of Table I, $\alpha(\Lambda)$ ranges between 0.2 and 0.35, somewhat less than the "interesting" range of values of $(\pi/2)\lambda$. On the other hand, λ contains group theory factors depending on the unified theory and therefore not present in the technicolor coupling $\alpha(\Lambda)$. Because the unified group is larger than the technicolor group, these factors might naturally be expected to be larger than unity.

While the appearance of fermion masses up into the GeV range is welcome, the problem of producing a realistic quark and lepton mass spectrum remains. In technicolor theories, the mysteries of the large hierarchies in these spectra get shifted into the effective four-fermion interactions among the technifermions and the quarks and leptons. It should be clear from the above discussion, however, that this mapping can be far from linear. In particular, a small breaking of weak isospin or flavor symmetry in the four-technifermion couplings can be considerably magnified in the quark and lepton mass spectrum. This breaking could be due, for example, to group factors arising from the spontaneous breaking of some ETC theory at scale Λ.

We [10] have studied the weak isospin breaking effect in more detail in a simple two-coupling example constructed from the SU(4) theory. With four-technifermion couplings λ_U and λ_D (for the up and down members of the weak doublets), a variety of hierarchies will appear depending on the relative size of λ_U and λ_D. If they are large enough, a small difference between them can lead to a large ratio between the up condensates and down condensates, and in turn to a large splitting between the masses of up fermions and down fermions. An important point is that large quark and lepton mass hierarchies can develop in this framework without a hierarchy of ETC scales, but instead from modest coupling differences at a single ETC scale.

The large amount of weak isospin breaking in the quark and lepton mass spectrum, however, poses a special problem for technicolor theories. Not only must a sufficiently large t-b quark mass splitting be generated, but this must be done in a way that doesn't overly infect the experimentally measured ρ parameter. The fact that $\rho - 1 \equiv M_w{}^2/M_z{}^2 \cos^2 q_W - 1 < 10^{-2}$ [11], means that very little weak isospin violation can be allowed in ρ. The analysis of Ref. 10 includes a crude estimate of the technicolor contributions to $\delta\rho$. The estimate uses the fact that $\rho = F_\pm^2/F_0^2$ where $F_\pm$ and F_0 are the decay constants of the charged and neutral Goldstone bosons. Sum rules analogous to that of Ref. 8 are then employed to estimate $F_\pm$ and F_0. Although these estimates are not much better than order of magnitude, because of the presence of strong technicolor interactions, they give uncomfortably large values. With λ_U and λ_D arranged to give the t-b splitting, it appears difficult to keep $\delta\rho$ below a few percent. It will be important to refine the ρ parameter estimates in the near future.

Finally, it is worth remarking that small amounts of weak isospin breaking will be induced by the electroweak interactions. The magnification effect I have described could then mean that these small increments to the four-fermion couplings could produce large weak isospin splitting in the quark and lepton mass spectrum. QCD effects could be magnified in a similar way, leading to important contributions to quark-lepton mass differences. There will be analogous contributions to the masses of possible pseudo-Goldstone bosons. In general, the new interactions at scale Λ, if they are strong enough, can add to and magnify the contributions of standard model gauge interactions to the mass parameters of the standard model and its extensions.

During the past few months, Opher Shapira and I have been exploring the QCD splittings, and at the same time going beyond the four-fermion approximation for the ETC interactions. In a recent paper [12], we assumed that the ETC interactions were due to the exchange of the massive spin-one bosons that are generated when an ETC group breaks to the smaller TC group. A conventional family structure was assumed for the technifermions, and the ETC group was taken to act sideways, commuting with the standard model interactions. Large fermion mass enhancement then depends on whether the ETC gauge coupling $\alpha(M)$ at the breaking scale M is near the critical coupling necessary for the full ETC theory to break the global chiral symmetries. If it is, then, as in the four-fermion example, the dynamical technifermion mass $\Sigma(p)$ can fall more slowly than $1/p$ for $p < M$, and give rise to very large quark and lepton masses. In the examples explored, if M is taken to be roughly $50\ \Lambda_C$, fermion masses above $100\ \text{GeV}$ be generated without excessive fine tuning. This of course is too low a scale for the suppression of flavor-changing neutral currents. Therefore the generation of mass for the lighter particles, where the FCNC constraints are severe, may have to rely on higher ETC scales. Within this framework, Shapira and I then computed the mass splitting effects of the QCD interactions. With the ETC coupling $\alpha(M)$ adjusted at about the 10% level, the additional effect of color (roughly 1/10 the strength of the ETC interactions at the breaking scale M) can split quark and lepton masses by up to two orders of magnitude. Whether this can play a role in explaining, for example, the mass splitting between the t quark and τ lepton remains to be seen.

TABLE CAPTION

The results of a numerical investigation of chiral symmetry breaking, in an SU(4) theory with $N_f = 8, 16, 20$ fermion flavors, are tabulated for various values of the four-fermion coupling $\lambda \equiv g^2/2\pi^2$. The gauge and fermion contributions to the β-function are given by $b = 44/6\pi$ and $\delta b = N_f/3\pi$ respectively. Since there are $N_f/2$ weak doublets, $F = 250\ \text{GeV}/(N_f/2)^{1/2}$. The confinement scale is Λ_c and Λ is the (ETC) cutoff. With $\Lambda \cong 2500\ \text{TeV}$, $\Lambda/g > \Lambda/(2\pi^2)^{1/2} \cong 600\ \text{TeV}$. The dynamical technifermion mass is $\Sigma(0)$ and the size of $\Sigma(\Lambda)$ is a measure of the rate of fall of $\Sigma(p)$. To normalize F relative to $\Sigma(0)$ and Λ_c, we have used the relation appearing in Ref. ??, which is accurate for the high momentum components. We note that this expression underestimates f_π in QCD by roughly a factor of two. In the numerical solution of Eq. 1, confinement dynamics is approximated crudely by allowing $\alpha(p)$ to plateau at $\alpha(\Lambda_c) = 2.5\ \alpha_c$ for $p \leq \Lambda_c$. For $\Lambda_c < p < 2\Sigma(0)$, $\alpha(p)$ evolves according to $\beta(\alpha) = -b\alpha^2$, and for $p > 2\Sigma(0)$ it evolves according to $\beta(\alpha) = -(b-\delta b)\ \alpha^2$. The output numbers, in particular the dependence on the rate of running and on λ, are high momentum phenomena and not terribly sensitive to the low momentum approximation for $\alpha(p)$. For values of λ small enough so that Λ_c and $\Sigma(0)$ remain essentially unchanged relative to Λ, $\alpha(\Lambda)$ remains constant. For larger values of λ, our procedure leads to a small decrease in $\alpha(\Lambda)$. This, however, doesn't play an important role in the results for fermion masses or for the F/Λ_c hierarchy.

TABLE I

$N_f = 8$ Case. (F = 125 GeV, $b\alpha_c = 1.304$, $\delta b\alpha_c = 0.474$)

$\lambda \equiv g^2/2\pi^2$	$\Sigma(0)/\Lambda_c$	$\Sigma(\Lambda)/\Lambda_c$	F/Λ_c	$\langle\bar\psi\psi\rangle/\Lambda^2\Lambda_c$	Λ_c	$m_f/\lambda_F f$
0.00	0.48	8.2×10^{-9}	0.13	2.3×10^{-8}	970 GeV	0.4 MeV
0.50	0.48	2.7×10^{-7}	0.13	5.0×10^{-8}	970 GeV	1.0 MeV
0.70	0.48	6.9×10^{-7}	0.13	9.6×10^{-8}	970 GeV	1.8 MeV
0.80	0.48	1.4×10^{-6}	0.13	1.8×10^{-7}	970 GeV	3.4 MeV
0.90	0.48	9.5×10^{-6}	0.13	1.0×10^{-6}	970 GeV	20 MeV
0.91	0.48	1.9×10^{-5}	0.13	2.0×10^{-6}	970 GeV	39 MeV
0.92	0.48	3.6×10^{-4}	0.13	3.9×10^{-5}	960 GeV	730 MeV
0.93	0.84	6.3×10^{-2}	0.26	6.6×10^{-3}	480 GeV	62 GeV
0.94	1.6	0.30	0.77	3.2×10^{-2}	160 GeV	100 GeV

$N_f = 16$ Case. (F = 88 GeV, $b\alpha_c = 1.304$, $\delta b\alpha_c = 0.948$)

$\lambda \equiv g^2/2\pi^2$	$\Sigma(0)/\Lambda_c$	$\Sigma(\Lambda)/\Lambda_c$	F/Λ_c	$\langle\bar\psi\psi\rangle/\Lambda^2\Lambda_c$	Λ_c	$m_f/\lambda_F f$
0.00	0.55	3.3×10^{-8}	0.15	4.6×10^{-8}	590 GeV	0.5 MeV
0.50	0.55	6.4×10^{-7}	0.15	1.1×10^{-7}	590 GeV	1 MeV
0.70	0.55	2.1×10^{-6}	0.15	2.8×10^{-7}	590 GeV	3 MeV
0.80	0.55	8.8×10^{-6}	0.15	1.0×10^{-6}	590 GeV	12 MeV
0.82	0.55	1.9×10-5	0.15	2.2×10^{-6}	590 GeV	25 MeV
0.84	0.57	1.0×10^{-3}	0.15	1.2×10^{-4}	570 GeV	1.3 GeV
0.86	0.78	1.9×10^{-2}	0.23	2.1×10^{-3}	390 GeV	17 GeV
0.88	1.1	8.2×10^{-2}	0.42	8.9×10^{-3}	210 GeV	37 GeV
0.90	1.9	0.31	1.0	3.3×10^{-2}	85 GeV	55 GeV
0.91	2.8	0.62	1.8	6.6×10^{-2}	49 GeV	64 GeV
0.92	4.6	1.3	3.5	0.14	26 GeV	72 GeV

$N_f = 20$ Case. (F = 79 GeV, $b\alpha_c = 1.304$, $\delta b\alpha_c = 1.185$)

$\lambda \equiv g^2/2\pi^2$	$\Sigma(0)/\Lambda_c$	$\Sigma(\Lambda)/\Lambda_c$	F/Λ_c	$\langle\bar\psi\psi\rangle/\Lambda^2\Lambda_c$	Λ_c	$m_f/\lambda_F f$
0.00	0.63	4.9×10^{-7}	0.17	3.4×10^{-7}	470 GeV	3 MeV
0.50	0.63	9.6×10^{-6}	0.17	1.5×10^{-6}	460 GeV	14 MeV
0.55	0.63	1.6×10^{-5}	0.17	2.3×10^{-6}	460 GeV	21 MeV
0.60	0.63	3.6×10^{-5}	0.17	4.9×10^{-6}	460 GeV	45 MeV
0.65	0.65	3.1×10^{-4}	0.18	3.9×10^{-5}	450 GeV	350 MeV
0.70	0.75	3.2×10^{-3}	0.21	4.0×10^{-4}	370 GeV	3 GeV
0.75	0.95	1.6×10^{-2}	0.30	1.9×10^{-3}	270 GeV	10 GeV
0.80	1.3	6.8×10^{-2}	0.54	7.7×10^{-3}	150 GeV	22 GeV
0.90	8.4	2.6	7.0	0.27	11 GeV	61 GeV

REFERENCES

[1] Bob Holdom, Phys. Lett. **B150**, 301 (1985); T. Appelquist, D. Karabali, and L.C.R. Wijewardhana, Phys. Rev. Lett. **57**, 957 (1986); T. Appelquist and L.C.R. Wijewardhana, Phys. Rev. **D35**, 774 (1987); **D36**, 568 (1987).

[2] T. Appelquist, M. Einhorn, T. Takeuchi, and L.C.R. Wijewardhana, Phys. Lett. B **220**, 223 (1989).

[3] For related discussions see V. Miranksy and K. Yamawaki, Mod. Phys. Lett. **A4**, 129 (1989) and V. Miransky, m. Tanabashi and K. Yamawaki, Mod. Phys. Lett. **A4**, 1043 (1989).

[4] S. Dimopoulos and L. Susskind, Nucl. Phys. **B155**, 237 (1979); E. Eichten and K. Lane, Phys. Lett. **90B**, 125 (1980). For a recent view of technicolor see R. Kaul, Rev. Mod. Phys. **55**, 449 (1983).

[5] T. Appelquist, D. Carrier, L.C.R.Wijewardhana, and W. Zheng, Phys. Rev. Lett. **60**, 1114 (1987); R. Casalbuoni, S. De Curtis, and R. Gatto, Nucl. Phys. **B319**, 367 (1989); S. King and D. Ross, Phys. Lett. **B228**, 363 (1989).

[6] Y. Nambu and G. Jona-Lasinio, Phys. Rev. **122**, 345 (1961).

[7] T. Appelquist, K. Lane, and U. Mahanta, Phys. Rev. Lett. **61**, 1553 (1988); U. Mahanta, Phys. Rev. Lett. **62**, 2349 91989).

[8] Pagels and Stokar, Phys. Rev. **D20**, 2947 (1979). The integral isconvergent if $\Sigma(p)$ falls with p as any power. If $\Sigma(p)$ falls logarithmically or does not fall at all as is the Nambu-Jona-Lasinio case, $F \sim \log \Lambda$. Thus as $\lambda \to 1$, F will increase relative to $\Sigma(0)$.

[9] T. Takeuchi, Phys. Rev. **D40**, 2697 (1989).

[10] T. Appelquist, T. Takeuchi, M.B. Einhorn and L.C.R. Wijewardhana, Phys. Lett. **B232**, 211 (1989).

[11] P. Langacker, University of Pennsylvania preprint, to be published in the 1990 Review of Particle Properties.

[12] T. Appelquist and O. Shapira, Yale preprint, YCTP-P5-90, May 1990, to be published in Physics Letters B.

[13] B. Holdom, Phys. Rev. Lett. **60**, 1233 (1988).

Minimal Dynamical Symmetry Breaking
of the Electroweak Interactions and m_{top}

Christopher T. Hill

Fermi National Accelerator Laboratory

P. O. Box 500, Batavia, Illinois 60510

Abstract

We review the recent idea of a mechanism for breaking the electroweak interactions which relies upon the formation of condensates involving the conventional quarks and leptons. In particular, such a scheme would indicate that the top quark is heavy, greater than or of order 200 GeV, and gives further predictions for the Higgs boson mass. It may be imbedded either into a GUT setting using supersymmetry or applied to a fourth generation with new strong TEV scale flavor–interactions.

1. Theoretical Implications of a Heavy Top Quark

We now know from CDF that $m_{top} > 89$ GeV. From the perspective of the theoretical structure of the Standard Model a large top quark mass, of order the weak scale ~ 175 GeV, implies a strongly coupled theory with a Higgs–Yukawa coupling constant of top, $g_{top} \sim 1$. That is, a heavy top quark is strongly coupled to the *agent or dynamics which breaks the electroweak interactions*. A large m_{top}, moreover, implies difficulties for conventional extended technicolor, and even walking technicolor for very large m_{top} requires fine–tuning (see *e.g.*, [1]). This, in turn, suggests that the top quark might, itself, play a fundamental role in the breaking of electroweak symmetries [2 – 5] by acting as a "techniquark," as a consequence of some new interaction.

Let us first consider the behavior of the parameters of the Standard Model Lagrangian as we evolve upwards in energy scale. Those associated with $d = 4$ terms satisfy the conventional renormalization group equations and evolve logarithmically with scale. For example, the coupling constants g_1, g_2 and g_3 evolve according to the conventional β–functions. g_2 and g_3 are asymptotically free, thus tending toward zero for large energy scales, while g_1 has a Landau singularity at very high energies. On the other hand, the $d = 2$ Higgs boson mass term has an additive contribution and evolves between scale μ and Λ as $m_H^2 \rightarrow m_H^2 + c(\Lambda^2 - \mu^2)$ with c determined from vector boson and fermion loops.

In the behavior of Higgs–Yukawa couplings, which are associated with the terms leading to fermion masses, there occurs a special trajectory which we shall refer to as the "Pendleton–Ross trajectory" [6]. This trajectory defines a particular $g_{PR}(\mu)$ such that if $g_{top} < g_{PR}(\mu)$ $(g_{top} > g_{PR}(\mu))$ then g_t is asymptotically free (asymptotically diverges at some scale). This is equivalent to the asymptotic smoothness criteria discussed by Kubo et $al.$ [6]. To one loop precision one obtains this trajectory by combining the RG equations for g_3 and for g_{top} and demanding the combined β–function vanishes:

$$16\pi^2 \frac{d}{dt}\ln(g_{top}/g_3) = \frac{9}{2}g_{top}^2 - (8 - b_0)g_3^2 = 0 \tag{1}$$

The physical top quark mass associated with the Pendleton–Ross trajectory is the solution to $m_{PR} = g_{PR}(m_{PR}) \times (175\ GeV)$ and Marciano [5] has recently given a precise estimate of $m_{PR} \approx 98$ GeV for $N_g = 3$ [4]. Remarkably, we are experimentally on the verge of crossing the Pendleton–Ross trajectory and possibly observing a second coupling constant in the Standard Model, g_{top}, that has a Landau singularity!

If $m_{top} > m_{PR}$ and if the Standard Model is a valid effective Lagrangian up to a given large energy scale Λ then the top quark mass is bounded from above by the so–called "triviality bound" [7]. Moreover, this corresponds to an infrared "quasi–fixed point" in the sense that over a large range of initial values of at Λ, g_{top} is swept to a universal physical value at low energies [8]. Thus, on purely probabilistic grounds one might expect $m_{top} \approx 230$ GeV for $\Lambda \approx 10^{15}$ GeV. Similiar results have been obtained for Higgs bosons in the Standard Model and in multi–Higgs boson generalizations [8]. We will see below that *this fixed point actually corresponds to a dynamically broken Standard Model by a top condensate and a composite Higgs boson composed of $\bar{t}t$*.

2. Dynamical Symmetry Breaking

We [4] have straightforwardly implemented a BCS or Nambu-Jona-Lasinio mechanism in which a new fundamental interaction associated with a high energy scale, Λ, is used to trigger the formation of a low energy condensate, $\langle \bar{t}t \rangle$. The bootstrapping of the symmetry breaking mechanism to the top quark introduces no *fundamental* Higgs scalar bosons (though a dynamical 0^+ boundstate emerges) and, by virtue of its economy, leads to new predictions which are in principle testable, or which constrain or rule out the mechanism altogether. In particular, we are able to derive predictions for m_{top} and m_{Higgs} in this scheme. The usual Cabibbo–Kobayashi–Maskawa structure and fermion mass spectrum is readily accomodated, but *bona fide predictions* of mixing angles and light quark masses are not derivable until one specifies the dynamics at the scale Λ more precisely. The usual single–Higgs–doublet Standard Model emerges as the low energy effective Lagrangian, but with new constraints that lead to nontrivial predictions.

If we consider, for discussion, the approximation in which all quarks and leptons other than the top quark are massless we may then define the theory at the scale Λ to be:

$$L = L_{kinetic} + G(\bar{\Psi}_L^{ia} t_{Ra})(\bar{t}_R^b \Psi_{Lib}) \tag{2}$$

Here $\Psi_L = (t, b)_L$ and i runs over $SU(2)_L$ indices, (a, b) run over color indices. $L_{kinetic}$ contains the usual gauge invariant fermion and gauge boson kinetic terms.

We have first considered a solution based upon the effects of the fermionic determinant alone, *i.e.*, a fermion bubble approximation. This is equivalent to a large–N_{color} expansion in the limit in which the QCD coupling constant is set to zero, and it captures nonperturbative features of the theory from the point of view of a small–coupling constant expansion.

We demand a solution to the gap equation for the induced top quark mass:

$$m_t = -\frac{1}{2}G \langle \bar{t}t \rangle = 2GN_c m_t \frac{i}{(2\pi)^4} \int d^4l \, (l^2 - m_t^2)^{-1} \tag{3}$$

or:

$$1 = \frac{GN_c}{8\pi^2} \left(\Lambda^2 - m_t^2 \ln(\Lambda^2/m_t^2) \right). \tag{4}$$

which has solutions for sufficiently strong coupling, $G \geq G_c = 8\pi^2/N_c\Lambda^2$ where G_c is the "critical" coupling constant. We regard G and Λ as fundamental parameters of the theory and we solve for m_t. Normally, for very large Λ, perhaps of order the GUT scale 10^{15} GeV, we would expect the solution of this equation to produce a large mass, $m_t \sim \Lambda$ in the broken symmetry phase. We see that a solution for $m_t \sim M_W$ for such large Λ constitutes a fine–tuning problem in that $G^{-1} - G_c^{-1}$ must then be very small. This is, indeed, the usual fine–tuning or gauge hierarchy problem of the Standard Model. The gap equation contains a quadratic divergence, corresponding to the usual Higgs mass quadratic divergence in the Standard Model. *However, the fine–tuning problem will be isolated in the gap equation, i.e., once we tune G to admit the desirable solution we need cancel no other quadratic divergences in other amplitudes.*

If we now consider the sum of leading large-N_c scalar channel fermion bubbles generated by the interaction eq.(2) we find:

$$\Gamma_s(p^2) = \frac{1}{2N_c}\left[(p^2 - 4m_t^2)(4\pi)^{-2}\int_0^1 dx \log\left\{\Lambda^2/(m_t^2 - x(1-x)p^2)\right\}\right]^{-1} \quad (5)$$

Γ_s is the propagator for a dynamically generated 0^+ boundstate, a scalar composite particle composed of $\bar{t}t$. In particular, owing to the pole at $p^2 = 4m_t^2$, we see that the theory predicts the boundstate mass of $2m_t$. This is a standard result for the Nambu–Jona-Lasinio model. We emphasize that this boundstate is the physical, observable, low energy Higgs boson. The prediction holds here only to leading order in $1/N_c$ in the absence of gauge boson corrections. We can also infer from eq.(5) that this particle is described by a field with a wave–function renormalization constant, Z_H, given by:

$$Z_H = \frac{N_c}{8\pi^2}\int_0^1 dx \log\left\{\Lambda^2/(m_t^2 - x(1-x)p^2)\right\} \quad (6)$$

This is *a relativistic boundstate*, and normal intuition from nonrelativistic potential models does not apply. In fact, the compositeness of this state is reflected by the behavior of Z_H:

$$Z_H \to 0 \qquad \text{as} \qquad -p^2 = \mu^2 \to \Lambda^2. \quad (7)$$

In fact, the essential point is embodied in eq.(7) and this allows us to give a more precise determination of the top mass upon considering the full renormalization group behavior of the complete theory.

Since this mechanism is indeed a dynamical breaking of the continuous $SU(2) \times U(1)$ symmetry, it implies the existence of Goldstone modes. Moreover, the symmetry breaking transforms as $I = \frac{1}{2}$ and will produce the same spectrum of Goldstone bosons as in the Standard Model Higgs–sector. Of course, we have a dynamical Higgs–mechanism and the gauge bosons acquire masses by "eating" the dynamically generated Goldstone poles. We obtain a second prediction of the theory in the form of a relation between the W boson mass and the top quark mass as follows.

Consider now the inverse propagator of the gauge bosons. We rescale fields to bring the gauge coupling constants into the gauge boson kinetic terms, *i.e.*, we write the kinetic terms in the form $(-1/4g^2)(F_{\mu\nu})^2$. It is useful to write the induced inverse W boson propagator in the form:

$$\frac{1}{g_2^2} D_{\mu\nu}^W(p)^{-1} = (p_\mu p_\nu/p^2 - g_{\mu\nu}) \left[\frac{1}{\bar{g}_2^2(p^2)} p^2 - \bar{f}^2(p^2) \right] . \tag{8}$$

The W boson mass is the solution to the the mass–shell condition:

$$M_W^2 = p^2 = \bar{g}_2^2(p^2)\bar{f}^2(p^2) \tag{9}$$

while the Fermi constant is the zero–momentum expression:

$$\frac{G_F}{\sqrt{2}} = \frac{1}{8\bar{f}^2(0)} \tag{10}$$

In the bubble approximation we find:

$$\begin{aligned}
\frac{1}{\bar{g}_2^2(p^2)} &= \frac{1}{g_2^2} + N_c(4\pi)^{-2} \int_0^1 dx\, 2x(1-x) \\
&\quad \times \log\left\{ \Lambda^2/(xm_b^2 + (1-x)m_t^2 - x(1-x)p^2) \right\}
\end{aligned} \tag{11}$$

and:

$$\begin{aligned}
\bar{f}^2(p^2) &= N_c(4\pi)^{-2} \int_0^1 dx\, (xm_b^2 + (1-x)m_t^2) \\
&\quad \times \log\left\{ \Lambda^2/(xm_b^2 + (1-x)m_t^2 - x(1-x)p^2) \right\}
\end{aligned} \tag{12}$$

A quantitative prediction for m_t in terms of G_F results when eq.(10) is combined with eq.(12):

$$\begin{aligned}
\bar{f}^2(0) = \frac{1}{4\sqrt{2}G_F} &\approx N_c(4\pi)^{-2} \int_0^1 (1-x)m_t^2 \log\left\{ \Lambda^2/((1-x)m_t^2) \right\} \\
&\approx \frac{1}{2} N_c(4\pi)^{-2} m_t^2 \log\{ \Lambda^2/m_t^2 \}
\end{aligned} \tag{13}$$

For example, with $\Lambda = 10^{15}$ GeV one finds $m_t \approx 165$ GeV.

To what extent is this an accurate prediction for m_t? For one, it is valid only in leading order of $1/N_c$ with $g_3 = 0$. This result, moreover, neglects the full dynamical effects of gauge bosons and the composite Higgs boson, which should be included in the renormalization group running below the scale Λ. We note that this result is substantially less than the full RG–improved Standard Model result as described below.

Analogous results are obtained for the neutral gauge boson masses, but they contain no additional information beyond that described here, a consequence of the conventional $I = \frac{1}{2}$ breaking mode. Moreover, the usual ρ parameter relationship for m_t emerges.

3. Effective Lagrangian

The dynamically generated scalar boundstates are described by the following effective Lagrangian:

$$\begin{aligned} L \;=\;& L_{kinetic} + (\bar{\Psi}_L t_R H + h.c.) \\ &+ Z_H |D_\mu H|^2 - m_H^2 H^\dagger H - \frac{\lambda_0}{2}(H^\dagger H)^2 \end{aligned} \tag{14}$$

We include here the gauge invariant kinetic terms of the Higgs doublet and its induced quartic interaction coming from top quark loops, as well as the wave–function normalization constant, Z_H.

In the present case, however, the Higgs field is dynamical with a vanishing wave–function renormalization constant at the scale $\mu \sim \Lambda$. That is, we have the following conditions at Λ (in terms of the unconventional normalization):

$$Z_H \propto N_c \ln(\Lambda/\mu) \to 0|_{\mu \to \Lambda}. \tag{15}$$

$$\lambda_0 \propto N_c \ln(\Lambda/\mu) \to 0|_{\mu \to \Lambda} \tag{16}$$

Conventionally one normalizes the kinetic terms of a field theory at any scale, μ, with a condition that the kinetic terms have free–field theory normalization. That is, we may exercise our freedom of rescaling the various fields, H, Ψ_L, t_R, etc., to define the coefficient of $|D_\mu H|^2$ to be unity. In the present case $H \to H/\sqrt{Z_H}$. The physical coupling constants, such as top quark Higgs–Yukawa coupling, $\bar{g}_t$, and the quartic Higgs coupling, $\bar{\lambda}$, are

912

then:

$$\bar{g}_t = \frac{1}{\sqrt{Z_H}}; \qquad \bar{\lambda} = \frac{1}{Z_H^2}\lambda_0 \tag{17}$$

It is clear from eqs.(17) that as $\mu \to \Lambda$ then $\bar{g}_t$ and $\bar{\lambda}$ diverge, while $\bar{g}_t^2/\bar{\lambda} \to$ *constant*.

To obtain a renormalization group improvement over the large–N_c Nambu-Jona-Lasinio model we may utilize these boundary conditions on $\bar{g}_t$ and $\bar{\lambda}$ and the full β-functions (neglecting light quark masses and mixings) of the Standard Model. To one–loop order we have:

$$16\pi^2 \frac{d\bar{g}_t}{dt} = (\frac{9}{2}\bar{g}_t{}^2 - 8\bar{g}_3{}^2 - \frac{9}{4}\bar{g}_2{}^2 - \frac{17}{12}\bar{g}_1{}^2)\bar{g}_t \tag{18}$$

and, for the gauge couplings:

$$16\pi^2 \frac{d\bar{g}_i}{dt} = -c_i\, \bar{g}_i{}^3 \tag{19}$$

with

$$c_1 = -\frac{1}{6} - \frac{20}{9}N_g; \quad c_2 = \frac{43}{6} - \frac{4}{3}N_g; \quad c_3 = 11 - \frac{4}{3}N_g \tag{20}$$

where N_g is the number of generations and $t = \ln\mu$.

The precise value of the top quark mass is determined by running $\bar{g}_t(\mu^2)$ from very high values at a given compositeness scale Λ down to the mass–shell condition $\bar{g}_t(m_t^2)v/\sqrt{2} = m_t$. The nonlinearity of eq.(18) focuses a wide range of initial values into a small range of final low energy results. For an estimate one can assume that the gauge couplings are constant, which indicates why the solutions are attracted toward the effective low energy fixed–point [8]:

$$\bar{g}_t{}^2(\mu^2) \approx \frac{16}{9}\, \bar{g}_3{}^2(\mu^2) \tag{21}$$

The action of the effective fixed–point makes the top quark mass prediction very insensitive to the initial high values of the coupling constant close to Λ. The uncertainties of higher orders can be viewed as an uncertainty in the precise position of Λ, and the fixed point behavior implies that m_t is determined up to $O(\ln\ln\Lambda/m_t)$ sensitivity to Λ. In Table I we give the resulting physical m_{top} obtained by a numerical solution of the renormalization group equations as a function of Λ.

The Higgs boson mass will likewise be determined by the evolution of $\bar{\lambda}$ given by:

$$16\pi^2 \frac{d\bar{\lambda}}{dt} = 12(\bar{\lambda}^2 + (\bar{g}_t^2 - A)\bar{\lambda} + B - \bar{g}_t^4) \tag{22}$$

where:

$$A = \frac{1}{4}\bar{g}_1^2 + \frac{3}{4}\bar{g}_2^2; \quad B = \frac{1}{16}\bar{g}_1^4 + \frac{1}{8}\bar{g}_1^2\bar{g}_2^2 + \frac{3}{16}\bar{g}_2^4 \tag{23}$$

The resulting prediction of the full Standard Model analysis is a top quark mass that might be considered large in comparison to certain published upper limits. Indeed, it has been claimed that the ρ parameter limit implies $m_t \lesssim 180$ to 200 GeV [9], and this is the most stringent quoted limit. However, without doing sufficient justice to any one of them, a number of recent experimental results have central values that are tantalizingly suggestive of a very heavy top quark (e.g., the E-731 measurement of ϵ'/ϵ; recent UA-2 W-mass determination; the LEP combined results on $\Gamma_{Z\to hadrons}$).

We feel that it it is premature to reject theoretical predictions of a very heavy top quark, up to at least ~ 250 GeV, based upon the present status of the precision measurements to date. Note that, by incorporating the data with our prediction, we favor $\Lambda \gtrsim 10^{11}$ GeV. The phenomenological predictions for $\sin^2\theta_W$ (Marciano–Sirlin definition) and M_W are summarized by the following equations:

$$\sin^2\theta_W = 0.215 \pm 0.002 + 0.00017(230 - m_t), \tag{24}$$

$$M_W = 80.73 \pm 0.15 + 0.009(m_t - 230). \tag{25}$$

If, ultimately, the theoretical top quark mass prediction proves to be too high then it is still possible, albeit possibly less compelling, to maintain this mechanism by assuming that the gap equation is saturated by a fourth generation. The top quark then plays no important role itself in the symmetry breaking of the Standard Model and should have a mass between current lower bounds, but presumably much less than the predictions for the masses of the fourth generation. In Table II we include the corresponding predictions for the masses of a degenerate fourth doublet. The resulting modified predictions for the Higgs mass as well as the corresponding errors are also shown.

We note that the two–Higgs boson generalization of this scheme has recently been analyzed in detail by Suzuki and Luty [10]

$\Lambda[GeV]$	10^{19}	10^{17}	10^{15}	10^{13}	10^{11}	10^{10}	10^9	10^8	10^7	10^6	10^5	10^4
$m_t^{phys}[GeV]$	218	223	229	237	248	255	264	277	293	318	360	455
pert.	± 2	± 3	± 3	± 3	± 5	± 6	± 7	± 9	± 12	± 16	± 25	± 45
$m_H^{phys}[GeV]$	239	246	256	268	285	296	310	329	354	391	455	605
pert.	± 3	± 3	± 4	± 5	± 8	± 9	± 11	± 15	± 21	± 32	± 56	± 142

Table I: The predictions for the physical top–quark and Higgs boson mass for different scales Λ. One loop β–functions are used with $g_1^2(M_Z) = 0.127 \pm 0.009$, $g_2^2(M_Z) = 0.446 \pm 0.020$, $\alpha_S(M_Z) = 0.115 \pm 0.015$ and $M_Z = 91.17$ GeV as input. The numbers are obtained for the central value of these input data and requiring the on-shell condition $\bar{m}(m) = m$. Variation of the gauge couplings within their errors results to a very good approximation in a change of ± 6 GeV for the top mass and ± 4 GeV for the Higgs mass. The rows labeled 'pert.' show the change in the result if we change the couplings at the cutoff to unity instead of infinity, as a measure of the errors induced by using perturbation theory.

$\Lambda[GeV]$	10^{19}	10^{17}	10^{15}	10^{13}	10^{11}	10^{10}	10^9	10^8	10^7	10^6	10^5	10^4
$m_4^{phys}[GeV]$	199	202	206	212	220	226	233	243	257	277	312	388
pert.	± 1	± 2	± 2	± 2	± 3	± 4	± 5	± 7	± 10	± 14	± 22	± 39
$m_H^{phys}[GeV]$	235	241	248	258	272	282	294	310	333	365	423	553
pert.	± 1	± 2	± 2	± 3	± 4	± 6	± 7	± 10	± 15	± 22	± 39	± 99

Table II: Predictions for a degenerate fourth–generation quark doublet with the same input data as in Table I. The top–quark and the fourth–generation leptons are assumed to be much lighter than this quark doublet. The variation of the gauge couplings results in a change of ± 7 GeV for the quark masses and ± 5 GeV for the Higgs mass.

4. Naturalness and Other Issues

One might object to this scheme on the basis of naturalness and the fine–tuning that is implicit in demanding a solution to eq.(4) in the limit $m_t << \Lambda$. Of course, all known physical quantum field theories have a naturalness problem in association with the cosmological constant, and whatever mechanism controls this problem commutes with many successful predictions. Perhaps such a mechanism is operant for the quadratic divergences that plague theories with scalars. However, we should investigate whether there exist natural generalizations of the above mechanism and what kinds of natural theories might exist.

(i) Supersymmetric extensions of the model described above exist and the most straightforward has been studied by Bardeen, Love and Clark [11]. One imagines an effective supersymmetric four–fermion containing interaction to exist on scale $\mu < \Lambda$ and supersymmetry is broken softly on a scale Δ. Here the quadratic divergence of the gap equation is essentially replaced by the SUSY soft–breaking scale Δ. Thus, if $\Delta \sim m_t$ and $G \sim 1/\Delta^2$ there is no large hierarchy. One generates a low energy effective Lagrangian which now contains the two Higgs bosons as demanded by supersymmetry and chirality. One of these (the one associated with top) is now composite with analogous compositeness conditions as in eq.(7). The renormalization group improvement is thus similar to the preceding case the net result for $\Delta \sim 100$ GeV, $\Lambda \sim 10^{19}$ GeV is $m_t \approx 200$ GeV.

There is, however, a problem with schemes like this. In particular, solutions to the gap equation require $G \sim 1/\Delta^2$ while the four–fermion effective Lagrangian is viewed as valid up to scales $\mu \sim \Lambda$. This implies that G is unacceptably large on scales $\mu >> \Delta$ and thus there would be unitarity violations on scales large compared to Δ but small compared to Λ. While the fermion bubble sum implies that a partial unitarization has been performed in some channels, there would presumably be large violations in more complicated processes. This defect is also problematic for any "walking" version of this scenario in which a new force is invoked to provide a large anomalous dimension to the four–fermion operator thus reducing the degree of (quadratic) divergence of the gap equation loop integral. We thus feel the following line is more promising.

(ii) Perhaps the simplest solution to the naturalness problem is to consider

theories in which $\Lambda \sim 1$ TeV. Then the top can probably no longer be upheld as the condensate since we see that m_t becomes ~ 500 GeV and unacceptably large. However, a fourth generation may be workable. We emphasize that such has not been ruled out by neutrino counting at LEP and in fact, it is very reasonable in such a scheme to consider the see–saw mechanism to be operant at the electroweak scale. In this case a remarkable thing happens: light neutrinos go down to their experimental limits a heavy neutrino goes up to the electroweak scale [12]!

Such a theory is a novelty in terms of its dynamics, being a "Strong Broken Horizontal Gauge Symmetry." We have experience with the weak broken symmetries of the standard model and the strong confining gauge force of QCD, but it is unusual (albeit perfectly reasonable) to ponder a force that is, itself, broken yet sufficiently strong to drive the formation of chiral condensates. In fact, some preliminary work [13] suggests that there may be some dynamical possibilities for engineering a natural hierarchy, $m_t/\Lambda << 1$, though $\sim 10^{-13}$ would seem unlikely. Thus, a fourth generation with $\Lambda \sim TeV$ seems an intriguing possibility and from Table II we expect $m_{quarks} \sim 500$ GeV. The details of the lepton sector are under investigation [12,13].

5. Conclusions and Discussion

Our principal conclusions are as follows:

(1) The gauged Nambu–Jona-Lasinio mechanism within the framework of the Standard Model, dynamically broken by a strongly coupled top quark which forms a condensate $\langle \bar{t}t \rangle$, may be implemented in the fermion loop approximation (or large–N_c with vanishing g_3). It yields primitive relationships between M_W and m_t and the cut–off Λ, and the Higgs mass is determined as $m_H = 2m_t$. The latter relationship has been emphasized by Nambu [2].

(2) We infer that $Z_H \to 0$ as $\mu \to \Lambda$ is the general compositeness condition for the Higgs boson of the full theory. The conventional normalization, $Z_H = 1$, implies, equivalently, that g_t and λ diverge as $\mu \to \Lambda$. This constraint, in turn, implies that the low energy values of these coupling constants are controlled by the renormalization group infra–red fixed–points. Consequently, the low energy results are insensitive to the detailed behavior of g_t and λ as $\mu \to \Lambda$.

(3) We give our results for the full Standard Model as functions of the scale of new physics (or Higgs boson compositeness scale), Λ. Our favored results for m_t are the lower values, as constrained by phenomenological considerations, hence $m_{top} \approx 230$ GeV and $m_{Higgs} \approx 260$ GeV with $\Lambda \approx 10^{15}$ GeV. The mechanism may be adapted to a fourth generation, or a multiple dynamical Higgs scheme, though our primary impetus is in the connection with the top quark since the lower mass limits on the top quark are suggestive of a strongly coupled system.

As our discussion has indicated, the compositeness of the auxiliary Higgs field leads to predictions for the top quark and Higgs masses which are equivalent to effective fixed–point arguments. We have in this mechanism a *raison d'etre* for the single–Higgs doublet Standard Model with a heavy top quark.

References

1. T. Appelquist, in these proceedings, and refs. therein.

2. Y. Nambu, "BCS Mechansim, Quasi-Supersymmetry, and Fermion Mass Matrix," Talk presented at the Kasimirz Conference, EFI 88-39 (July 1988); "Quasi-Supersymmetry, Bootstrap Symmetry Breaking, and Fermion Masses," EFI 88-62 (August 1988); a version of this work appears in "*1988 International Workshop on New Trends in Strong Coupling Gauge Theories*," Nagoya, Japan, ed. Bando, Muta and Yamawaki; "Bootstrap Symmetry Breaking in Electroweak Unification," EFI Preprint, 89–08 (1989).

3. V. A. Miransky, M. Tanabashi, K. Yamawaki, *Mod. Phys. Lett.* **A4**, 1043 (1989); *Phys. Lett.* **221B** 177 (1989).

4. W. A. Bardeen, C. T. Hill, M. Lindner, *Phys. Rev.* **D41**, 1647 (1990).

5. W. J. Marciano, *Phys. Rev. Lett.* **62**, 2793 (1989);
   related works are: P. Kaus, S. Meshkov, *Mod. Phys. Lett.* **A3**, 1251 (1988); *ibid.* (erratum), **4**, 603 (1989);
   J. D. Bjorken, *Annals of Physics*, **24**, 174 (1963);
   T. Eguchi, *Phys. Rev.* **D14**, 2755 (1976);
   F. Cooper, G. Guralnik, N. Snyderman, *Phys. Rev. Lett.* **40**, 1620, (1978).

6. B. Pendleton and G. G. Ross *Phys. Lett.* **98B**, 291 (1981);
J. Kubo, K. Sibold and W. Zimmermann, *Phys. Lett.* **B220**, 191 (1989);and *Nucl. Phys.* **B259**, 331 (1985).

7. L. Maiani, G. Parisi and R. Petronzio, *Nucl. Phys.* **B136**, 115 (1978);
N. Cabbibo et al., *Nucl. Phys.* **B158**, 295 (1979);
R. Dashen and H. Neuberger, *Phys. Rev. Lett.* **50**, 1847 (1983);
D. J. E. Callaway, *Nucl. Phys.* **B233**, 189 (1984);
M. Lindner, *Zeit. Phys.* **31C**, 295 (1986).

8. C. T. Hill, *Phys. Rev.* **D24**, 691 (1981);
C. T. Hill, C. N. Leung, S. Rao, *Nucl. Phys.* **B262**, 517 (1985).

9. U. Amaldi et al., *Phys. Rev.* **D36**, 1385 (1987).

10. M. Suzuki, U. C. Berkeley preprint UCB-PTH-89/37 (1989) and LBL preprint LBL-27968 (1989);
M. Luty, U. Chicago preprint EFI-89-66 (1990) to appear in *Phys. Rev. D*.

11. T. E. Clark, S. T. Love, W. A. Bardeen, Purdue University preprint PURD-TH-89-11 (1989).

12. C. T. Hill, E. A. Paschos, Fermilab preprint FERMILAB-PUB-89/228 (1989), to appear in *Phys. Letters B*; C. T. Hill, M. Luty, E. A. Paschos, in preparation.

13. C. T. Hill, work in progress.

SUPERSYMMETRY AT e^+e^- AND $p\bar{p}$ COLLIDERS[*]

HOWARD A. BAER

Physics Department, Florida State University, Tallahassee, FL 32306 USA

ABSTRACT

Searches for non-standard decay modes of the Z^0 boson at LEP and SLC have resulted in model-dependent supersymmetric particle mass limits of $m \gtrsim M_Z/2$. We extract model-independent mass limits on sparticles from constraints on the total Z width and peak hadronic cross-section. We also point out what remains to be done at LEP in regards to the search for supersymmetry: this mainly entails searching for fractional number of neutrinos, and missing energy events. At $p\bar{p}$ colliders, the search for supersymmetry at the Tevatron is reviewed. Cascade decays of squarks and gluinos can be best searched for via usual searches for missing energy plus jet events, but also by searching for isolated same-sign dileptons, and events containing real Z bosons plus jets plus missing transverse energy.

1.   SUPERSYMMETRY FROM $e^+e^- \to Z^0$

The search for supersymmetry at the four LEP experiments at CERN and at Mark II at SLC has produced mass limits[1] on various charged sparticles which are typically half the beam energy, *i.e.* $m(\text{sparticle}) \gtrsim M_Z/2$. These mass limits depend in detail on various assumptions about the supersymmetric particle couplings and decay modes. For instance, if $Z \to \tilde{W}_-\bar{\tilde{W}}_-$ and the light chargino decays to the lightest neutralino via $\tilde{W}_- \to q\bar{Q}\tilde{Z}_1$ and $m_{\tilde{Z}_1} \simeq m_{\tilde{W}_-}$, then the visible decay products may be very soft, and the event may not be recognized as of supersymmetric origin.

Model independent bounds[2] can be obtained from using information on the total Z width Γ_Z and the measured number of neutrinos N_ν, which is obtained from measurements of the peak hadronic cross-section σ_h^0. The combination of data from the four LEP experiments[1] results in an upper limit $\Gamma_Z < 2.588$ GeV at 90% CL. The total Z width in the Standard Model has been recently computed[3]

---

[*] Research supported in part by the U.S. Department of Energy

by Bardin *et al.* to be 2.468 GeV$< \Gamma_Z <$ 2.514 GeV. Hence, the contribution of non-standard physics to Γ_Z should be bounded by

$$\Delta\Gamma < 0.120 \text{ GeV} \quad (90\%\text{CL}). \tag{1}$$

We may now use the couplings as given by the minimal supersymmetric model to put limits on sparticle masses for sparticles which couple to the Z^0 boson. We find the following significant mass limits:

$$m_{\tilde{d}_L} \geq 32.8 \text{ GeV} \quad (90\%\text{CL}) \tag{2}$$

$$m_{\tilde{u}_L} \geq 27.7 \text{ GeV} \quad (90\%\text{CL}) \tag{3}$$

$$m_{\tilde{\nu}} \geq 28.3 \text{ GeV} \quad (90\%\text{CL}) \tag{4}$$

where the sneutrino mass limit assumes three degenerate sneutrino flavors and the squark limit two degenerate squark flavors. Bounds on other sfermions are not as significant due to smaller couplings, but will get better as more data accumulates.

We can place bounds on masses of charginos and neutralinos also, using $\Delta\Gamma_Z$. The gauge/higgsino mass matrices depend in the minimal model on three parameters: the gluino mass $m_{\tilde{g}}$, the Higgsino mixing mass $2m_1$, and the ratio v'/v of vacuum expectation values of the two Higgs doublets. By scanning all allowed parameters, we find the mass of the lightest chargino $\tilde{W}_-$ and third lightest neutralino $\tilde{Z}_3$ must satisfy

$$m_{\tilde{W}_-} \geq 37 \text{ GeV} \quad (90\%\text{CL}) \tag{5}$$

$$m_{\tilde{Z}_3} \geq 57 \text{ GeV} \quad (90\%\text{CL}) \tag{6}$$

or else the bound (1) is violated. (Bounds on $m_{\tilde{Z}_{1,2}}$ are not possible.) As the bound (1) on contributions of non-standard physics to the total Z width becomes more precise, the sparticle mass limits should approach $M_Z/2$. These limits are completely independent of sparticle decay modes, and hence are considerably less model dependent than similar bounds from exclusive searches.

Restrictions on the Z invisible width can be obtained from measurements of the peak hadronic cross-section σ_h^0 at the Z pole. Such measurements have been used to restrict the total number of neutrino generations to be $N_\nu < 3.32$ at the 90% CL. This bound assumes no non-standard hadronic Z decays, and translates to a bound on the non-standard, non-hadronic Z width

$$\Delta\Gamma_{nh}^{NSM} < 53 \text{ MeV} \quad (90\%\text{CL}). \tag{7}$$

We may now put a limit on the sneutrino mass assuming no sneutrino hadronic decays:

$$m_{\tilde{\nu}} \geq 23.1 \text{ GeV} \quad (90\%\text{CL}) \tag{8}$$

where (8) holds for just one species of sneutrino. The Kamiokande experiment[4] excludes the region 3 GeV $< m_{\tilde{\nu}} <$ 25 GeV from non-observation of high energy neutrinos due to sneutrino annihilation in the sun. Also, lack of observation of $\tilde{\nu}$'s in double-beta decay search experiments[5] excludes the region $12-20$ GeV$< m_{\tilde{\nu}} <$ 1.0 TeV. Combining these astrophysical bounds with (8) excludes the sneutrino as the lightest supersymmetric particle (LSP).

A further bound on the parameter space of the minimal Higgs sector in supersymmetry can be obtained using the result from LEP that $m_H > 24$ GeV. This bound excludes points in the m_{H^+} vs. v'/v plane, and restricts $v'/v < 0.76$.

Is there anything left to be done in the search for supersymmetry by e^+e^- machines operating at the Z pole? Yes! Fig. 1 shows the SUSY parameter plane $m_{\tilde{g}}$ vs. $2m_1$ for $v'/v = 0.5$. The regions excluded by the bound (1) are shown by the dotted lines. More precise measurements of the Z total width can search even for the invisible decay $Z \to \tilde{Z}_1\tilde{Z}_1$. We have plotted the region where $\Delta\Gamma_Z > 10$ MeV as the long-dashes lines in Fig. 1. Also, visible decays such as $Z \to \tilde{Z}_1\tilde{Z}_2$ can be searched for as missing energy events. The region of parameter space where there are extra Z visible decays with $\Delta\Gamma_Z > 0.1$ MeV is shown by the short-dashed lines in Fig. 1. These regions roughly deliniate the parameter space left to be explored in e^+e^- collisions on the Z.

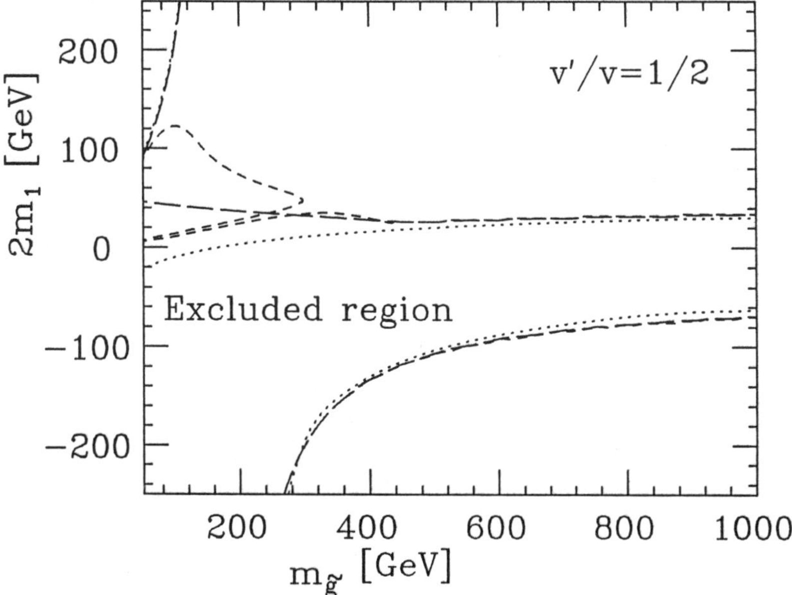

Fig. 1. Regions of SUSY parameter space excluded by constraint (1) (*dotted-lines*), and open to future exploration at LEP I via the total (visible) Z width being greater than 10 MeV (0.1 MeV), *long-dashed* (*short-dashed*).

2.   SUPERSYMMETRY AT $p\bar{p}$ COLLIDERS

The best limits on supersymmetric particles at hadron colliders come from searching for the processes $p\bar{p} \to \tilde{g}\tilde{g}, \tilde{q}\tilde{q}$ and $\tilde{g}\tilde{q}$ at the CDF and UA2 experiments, where limits[6] of $m_{\tilde{g}}, m_{\tilde{q}} \gtrsim 70 - 80$ GeV have been obtained. These limits were obtained by assuming

$$\tilde{g} \to q\bar{q}\tilde{Z}_1 \ (m_{\tilde{g}} < m_{\tilde{q}}) \text{ or } \tilde{g} \to \tilde{q}q \ (m_{\tilde{q}} < m_{\tilde{g}}) \tag{9}$$

$$\tilde{q} \to q\tilde{Z}_1 \ (m_{\tilde{q}} < m_{\tilde{g}}) \text{ or } \tilde{q} \to \tilde{g}q \ (m_{\tilde{g}} < m_{\tilde{q}}). \tag{10}$$

The signature is jets plus missing p_T ($\not{p}_T$), where the missing energy is due to escaping neutralinos $\tilde{Z}_1$, here assumed to be the LSP. The assumptions (9) and (10) were generally valid for squark and gluino searches at the $Sp\bar{p}S$ collider when masses $\lesssim 50$ GeV were being probed.

However, as the sparticle mass search regions increase, the minimal SUSY model predicts many more decay modes opening up[7] to all kinematically accessible gaugino/higgsino states, and usually these decay modes are dominant. For instance, we may also have

$$\tilde{g} \to q\bar{q}\tilde{Z}_i \text{ and } \tilde{g} \to q\bar{q}\tilde{W}_j \tag{11}$$

where $i = 1 - 4$ and $j = -+$, and $m_{\tilde{Z}_{i-1}} < m_{\tilde{Z}_i}$ and $m_{\tilde{W}_-} < m_{\tilde{W}_+}$, for the four neutralinos and two charginos. Furthermore, the radiative decays

$$\tilde{g} \to g\tilde{Z}_i \tag{12}$$

may also take place through top sfamily loops. Left-squarks may decay via

$$\tilde{q}_L \to q\tilde{Z}_i \text{ and } \tilde{q}_L \to q\tilde{W}_j \tag{13}$$

while right-squarks may decay via

$$\tilde{q}_R \to q\tilde{Z}_i \tag{14}$$

only, since they do not couple to charginos. The charginos and neutralinos will decay via two-body modes to vector-bosons $W^\pm$ and Z^0 or to Higgs bosons, if the decays are kinematically allowed; otherwise they decay via three-body modes to lighter -ino states, where the decay chain proceeds until the LSP (assumed to be $\tilde{Z}_1$) is reached, which escapes as missing energy. Hence, supersymmetry events from squarks and gluinos contain many jets, leptons and missing transverse energy.

The $\not{p}_T$ in SUSY cascade events is generally less[7] than that expected assuming the direct decays to the LSP of (9) and (10). This $\not{p}_T$ degradation leads to diminished mass bounds of typically $10 - 30$ GeV from those quoted assuming the direct decays. Also, the jet multiplicity is increased, due to further decays to quarks farther down the cascade decay chain.

An interesting way to search for $p\bar{p} \to \tilde{g}\tilde{g}$ is to look for same-sign hard isolated dileptons[7,8] from the case where each gluino decays via $\tilde{g} \to \tilde{W}_- q\bar{Q}$, and

$\tilde{W}_- \to l\nu\tilde{Z}_1$. These events can also come from $\tilde{g}\tilde{q}$ and $\tilde{q}\tilde{q}$ production. For $m_{\tilde{q}} = m_{\tilde{g}}$ up to 100 GeV, cross-sections can range up to 0.3 pb at the Tevatron collider, after cuts to suppress backgrounds from heavy flavor production. These events occur in a wide region of parameter space, and backgrounds appear to be controllable.

Another SUSY signal[7] can occur at the Tevatron collider when the Higgsino mixing parameter $2m_1$ is small: -20 to 40 GeV. In this case, the $\tilde{Z}_4$ mass is frequently lighter than $m_{\tilde{g}}$ and $m_{\tilde{q}}$. Then $\tilde{g}$ and $\tilde{q}$ may decay to $\tilde{Z}_4$ via tree or loop decays. The $\tilde{Z}_4$ then decays to $Z^0\tilde{Z}_1$, resulting 6% of the time in a leptonic Z^0 event, with associated jets plus $\displaystyle{\not}p_T$. This signal can range up to 1 pb for $m_{\tilde{g}} \sim m_{\tilde{q}} \sim 130$ GeV.

Backgrounds occur due to Z^0+jets, $b\bar{b}Z^0$ and $c\bar{c}Z^0$ production, Z^0Z^0+ jets, and $t\bar{t}$ production. However, requiring at least two jets and $\displaystyle{\not}p_T \gtrsim 25$ GeV should allow the signal to stand firmly above background over much of the parameter space where the signal is large.

ACKNOWLEDGEMENTS

I thank M. Drees, X. Tata, and J. Woodside for collaborations on these various projects, and T. Xerxes for reading the manuscript.

REFERENCES

1) See MARK II, ALEPH, DELPHI, L3 and OPAL talks, this proceedings.

2) H. Baer, M. Drees and X. Tata, FSU-HEP-891104 (1989) (Physical Review D- in press).

3) D. Bardin, *et al.*, CERN preprint TH-5468 (1989).

4) K. S. Hirata *et al.*, in *Proc. of the Second Workshop on the Elementary Particle Picture of the Universe*, KEK , Feb.4-6, (1988), p.73.

5) D. Caldwell *et al.*, Phys. Rev. Lett. <u>61</u>, 510 (1988);
   S. P. Ahlen *et al.*, Phys. Lett. <u>195B</u>, 603 (1987).

6) CDF Collaboration, F. Abe *et al.*, Phys. Rev. Lett. 62, 1825 (1989);
   UA2 Collaboration, J. Alitti *et al.*, CERN-EP/89-151 (1989).

7) H. Baer, X. Tata and J. Woodside, Phys. Rev. Lett. <u>63</u>, 352 (1989);
   Phys. Rev. <u>D41</u>, 906 (1990); FSU-HEP-900215 (1990).

8) M. Barnett, J. Gunion and H. Haber, UCD-88-30 (1988).

B+L Violation at High Energy

Larry McLerran

Theoretical Physics Institute
School of Physics and Astronomy
University of Minnesota
116 Church St.
Minneapolis, Mn. 55455

Abstract

Recent work which suggests that baryon plus lepton number might be violated in high energy scattering is discussed. The status of B+L violation at high temperatures is reviewed.

1   Introduction

In this talk, I will discuss the following issues:

1. Baryon number violation at finite temperature, and the reliability of computations of the rate.

2. The breakdown of perturbation theory in electroweak theory for B+L violating processes, and the possibility that B+L violating cross sections might be large in the collision of high energy particles.

I will begin by briefly reviewing the fact that B+L is not conserved in electroweak theory due to an anomaly in the conservation of the baryon number current. Although the rate for this process is small at low energy and zero temperature, I will review considerations which argue that the rate is large for high temperature. I discuss some apparent paradoxes which arise from having a large rate at high temperature. The resolution of this paradox requires that there be large matrix elements for processes which involve a large number of particles.

The implication of the finite temperature result is that there might be processes which are large at high energy corresponding to multi-particle production amplitudes for B+L violation. This implication is discussed, as are recent instanton estimates of high such amplitudes. I argue that perturbation theory breaks down in this sector of the theory. A supersymmetric Yang-Mills example is presented where the rate for B+L changing processes can be shown to be large.

2 Baryon Number is Not Conserved in Electroweak Theory

The equations of motion for electroweak theory naively conserve baryon number. This should be true since we would like the rate to be very small, and since the scale of electroweak physics is a mere 100 GeV, and if there was explicit violation of baryon number, it should be observably large.

When the equations of motion of electroweak theory are quantized, baryon number is no longer conserved. This is a consequence of the Adler-Bardeen anomaly.[1]–[3]. Baryon number is violated as

$$\partial_\mu J_B^\mu = \frac{g^2}{32\pi^2} tr F^{\mu\nu} F_{\mu\nu}^d = \partial_\mu K_{CS}^\mu \tag{1}$$

In this equation, $F^{\mu\nu}$ is the field strength of the electroweak W and Z fields,

$$F_{\mu\nu}^d = \frac{1}{2}\epsilon_{\mu\nu\lambda\sigma} F^{\lambda\sigma} \tag{2}$$

The product $F F^d$ can be written as the divergence of the Chern-Simons current where

$$K_{CS}^\mu = \frac{1}{16\pi^2}\epsilon_{\mu\nu\lambda\sigma}\, Tr(F^{\nu\lambda} A^\sigma - \frac{2}{3}A^\nu A^\lambda A^\sigma) \tag{3}$$

The fields A are the W and Z fields in a matrix basis

$$A = \tau^a \cdot A^a \tag{4}$$

Therefore the Adler-Bardeen anomaly requires that the change of baryon number be equal to the change in Chern-Simons number,

$$\Delta N_B = \Delta Q_{CS} = \int_{t_1}^{t_2} dt \int d^3x\ \partial_\mu K_{CS}^\mu \ . \tag{5}$$

The change in Chern-Simons charge is a gauge invariant quantity (but the the Chern-Simons charge density is not).

It is not at all obvious that the anomaly generates baryon number violation at a sufficiently small rate that it would have escaped detection. We will later see that this is the case, where we shall argue that in weak coupling, the rate of change of topological charge at zero temperature is of order $e^{-4\pi/\alpha_W} \sim 10^{-170}$ and can be ignored.

The existence of a topological charge Q_{CS} is a consequence of the underlying vacuum structure of electroweak theory. As a function of ΔQ_{CS}, the minimal energy or effective potential energy is multiply periodic. This is shown in Fig. 1.

With such a multiply periodic structure, it is straightforward to gain some physical insight into the nature of the baryon non-conserving anomaly.[4] Consider Figures 2a -

2c. On the left hand side of each figure, there is a plot of the effective potential. On the right hand side, there is a plot of the fermi levels. In Fig. 2a, we imagine being at the minimum of the effective potential labeled by x which corresponds to $\Delta Q_{CS} = 0$. In this case, the fermi levels are assumed to be unoccupied for energies greater than zero. For all energies less than zero, the fermi levels are occupied. Particles are interpreted to be occupied positive energy states and anti-particles unoccupied negative energy states.

The fermi level picture is a consequence of quantizing the Dirac equation. What we will show here is that the picture of multiply periodic vacuum structure, together with fermi levels for the Dirac equation leads to the anomaly.

Before proceeding further, we should mention that the minima at integer values of ΔQ_{CS} are all gauge transforms of one another. Therefore the spectra of the Dirac equation corresponding to these different minima must be identical. As we shall soon see, if we continuously deform the fields from one minimum to another, the occupation numbers of positive energy states is not invariant. Transitions from one minimum to another may therefore produce particles.

In Fig. 2b, we have deformed the W and Z fields from a minimum to a maximum separating two minima. In the process of this variation, the fermi levels move. As shown by Manton[5] (for not too strong a scalar coupling λ [6]) the configuration which sits at the top of the barrier corresponds to a level crossing of a fermion at zero energy.

The configuration of W, Z and Higgs fields at the top of the barrier between the two minima is called the sphaleron. It was first constructed for electroweak theory by Klinkhammer and Manton.[7] The sphaleron is a solution of the static classical equations of motion. Since it corresponds to a top of an energy barrier, it is classically unstable under time dependent perturbations. By making small fluctuations around the sphaleron, it is also possible to construct the shape of the potential in the neighborhood of the top of the barrier. The sphaleron has Chern-Simons charge of 1/2.

In Fig 2c, the W, Z and Higgs fields have been deformed to the minimum to the right of the original minimum. The fields are now gauge transforms of the fields near the original minimum, and therefore the spectrum of fermi levels is the same. The only difference is that in the deformation, a filled energy level which originally had negative energy has gained positive energy. A particle has been produced.

This adiabatic level crossing is a general feature of electroweak theory. The relation between it and the original computation of the anomaly is not obvious. The original computation required a careful use of ultraviolet regulators. In this level crossing picture, it is necessary to know that such a regulation scheme exists, since we have ignored what is happening at the bottom of the fermi sea.

The non-conservation of baryon number is related to transitions between degenerate minima of the effective potential with differing topological charge. At zero temperature, the only way such transitions can take place is by quantum mechanical tunneling, as shown in Fig. 3. This tunneling process was first computed by t' Hooft[8]. In the computation of the rate, a classical solution of the equations of motion is used which is called an instanton,[9] which corresponds to the classical solutions used in the WKB method of ordinary quantum mechanics. The process computed by t' Hooft is therefore often referred to as instanton mediated.

The result of t' Hooft's computation is that the rate is

$$Rate \sim e^{-4\pi/\alpha_w} \sim 10^{-170} \tag{6}$$

In this formula, there are no dimensional factors. Remember that even if we measure the space time volume of the universe in terms of the W boson compton wavelength, it is only

$$Vt_{universe} \sim 10^{168} \tag{7}$$

Since there are in fact very few protons per Compton volume of the W boson, the rate given by t' Hooft is so small that never in the history of universe was it even marginally likely that an instanton generated a baryon number change. Nevertheless, the computation does prove that the rate of baryon number change in electroweak theory is non-zero.

The process generated by instantons is visualized in Fig. 4. This amplitude violates both baryon number and lepton number by 3 units. The sum of baryon number plus lepton number is not conserved, but the difference $B + L$ is conserved. This is generally true in electroweak theory. There is a corresponding anomaly for the lepton current which is identical with that of the baryon current. Note also, that each generation of lepton and baryon is produced in the process. The electromagnetic and color charges are also conserved.

Even if such a process as shown in Fig. 4 existed at a large rate in electroweak theory, it would be very difficult to detect experimentally. Most proton decay experiments are designed to measure only single proton decays. Requiring a three body decay in a nucleus would surely reduce the rate due to the small probability that three nucleons overlap. In addition, requiring that only light mass quarks are produced would involve strong suppression due to weak mixing angles. Finally, there would in general be much missing energy in the event due to the large probability of emitting a neutrino.

3 The Rate of Baryon Number Non-conservation at Finite Temperature

At finite temperature, an entirely new method of generating baryon number change arises. There is some probability that there is a thermal fluctuation with an energy higher than the height of the barrier between two degenerate minima. In this case, the fluctuation can classically traverse the maximum and enter the new minimum, as shown in Fig. 5. The probability that the thermal fluctuation has an energy higher than the top of the barrier is[7],[10]–[11]

$$Rate \sim e^{-E_{sph}/T} \tag{8}$$

As discussed above, if we know the sphaleron solution of the classical equations of motion with topological charge $Q_{sph} = 1/2 \, \Delta Q_{cs}$, then we know the height of the barrier E_{sph}. The energy of the sphaleron comes from a classical solution, and classical solutions correspond to the limit of very many quanta $\langle N \rangle \sim 1/\alpha_w$, so that the energy is of order

$$E_{sph} \sim M_W/\alpha_W \sim 1 - 10 \; TeV \tag{9}$$

The typical energy per quanta is of order M_W, and therefore the size of the sphaleron is of order

$$R_{sph} \sim 1/M_W \tag{10}$$

We can visualize the process of making a classical transition over the top of the barrier by the diagram of Fig. 6. The diagram is identical to that of the instanton process, and the conservation laws associated with the event are identical. The sphaleron process may be thought of as analogous to an instanton process, except that we imagine the sphaleron process taking place in real time at finite temperature.

The above estimate of the probability that there is a thermal fluctuation with an energy larger than the height of the barrier clearly gives an overestimate of the rate. We must also know the probability that the thermal fluctuation is in some sense near the top of the barrier. For example, if we are lost in a forest in the high Himalayas, we may be much higher than a mountain pass in the Rockies, but we may never find our way to it. Put another way, the function space in which we imagine our thermal fluctuation is multi-dimensional and complicated. It is not obvious what weight a thermal fluctuation has for inducing a baryon number changing transition.

There is a reliable method for computing thermal transition processes if the coupling is weak.[12]–[13]. We must first ask if there is a range of temperatures where weak coupling methods are reliable. First we must recall that at high temperature, the mass of the W boson becomes temperature dependent $M_W \rightarrow M_W(T)$.[14] At some temperature, the masses of the W and Z bosons approach zero, and the broken symmetry of electroweak theory is restored.

In general, it is possible to show that perturbation theory breaks down in electroweak theory when $T > M_W(T)/\alpha_w$.[15]–[16] This is because infrared integrations over thermal propagators become so singular that the factors of alpha generated at each order of the loop expansion are canceled. In addition, the use of the Langer-Affleck formalism for computing thermal transitions over the barrier is only valid in the high temperature limit, $T >> M_W(T)$. This latter case is not much of a restriction since only in this limit are thermal transitions larger than quantum tunneling, that is $e^{-M_W/T} \geq e^{-4\pi/\alpha_W}$. We therefore conclude that there is a region of temperatures where the rate of baryon number changing transitions may be reliably computed in electroweak theory,

$$M_W(T) << T << M_W(T)/\alpha_W \tag{11}$$

In the above range of temperatures, the rate of baryon number change has been computed in electroweak theory for $\lambda/g^2 \sim 1$ where λ is the four Higgs coupling strength.[17]-[18] The rate is large compared to the expansion rate of the universe in cosmology,

$$\Gamma_{\Delta B}/\Gamma_{universe} \geq 10^9 \ for \ T \sim 100 \ GeV \tag{12}$$

The rate has also been recently computed for arbitrary values of λ/g^2, with the result that there is some suppression for values not close to one, but with essentially the same conclusions as was the case for $\lambda/g^2 \sim 1$ [19]-[20].

For temperature $T \geq M_W(T)/\alpha_W$, it is not possible to do reliable weak coupling estimates. It is however possible to make estimates based on scaling arguments or approximate classical solutions.[17],[21] The rate here is also many orders of magnitude larger than the expansion rate of the universe.

4 The Problem with Instantons

The sphaleron estimates of the rate of baryon number change have been criticized by several authors.[22]-[23] The paper by Ellis et. al. points out several puzzling apparent paradoxes between instanton computations and those of sphalerons, and we will address these apparent puzzles here. The resolution of these apparent paradoxes sheds some interesting light on the mechanism of sphaleron transitions.[17]. We will not discuss the second set of criticisms due to Cohen et. al., as this would go beyond the scope of this talk, except to say that this paper has, in my opinion, at least two essential mistakes which invalidate the conclusions, as has been recently pointed out in the work by Dine et. al., and by Mottola et. al.[24]-[25]

The first set of apparent problems with the sphaleron computation is that the sphaleron is a large classical object. It has $\langle N \rangle \sim 1/\alpha_W$ quanta in it and is large compared to its Compton wavelength. It should be hard to make such a fat, fluffy object.

In fact, when the method of Langer-Affleck is applied to the computation of the rate of sphaleron processes, one estimates the probability of this production. There is a suppression, roughly proportional to α_W^4, which results. There are some large numerical factors however which enhance the rate. When all the factors are combined together, the overall suppression is about 10^{-5}. This suppression is not enough to make the rate irrelevant cosmologically since the rate of expansion of the universe is slow compared to the natural time scale set by thermal processes $T/M_{planck} \sim 10^{-16}$.

It is easy to get confused by this issue nevertheless. If we think in terms of exclusive probabilities, the probability to have only a sphaleron in a box the size of a sphaleron compared to that to have ordinary thermal fluctuations is very small $P \sim e^{-T^3/M_W^3}$ The question we are asking is however an inclusive one. We ask what is the probability that there is a sphaleron transition within the volume of the sphaleron plus all thermal excitations within the same volume. The sphaleron is weakly interacting, and the probability that a highly excited Z or W boson interacts with it during its lifetime is small, that is, the sphaleron is transparent to the high energy modes and one can compute using inclusive probabilities. This probability is not a priori small.[24],[26]

A more difficult instanton argument to evade is provided by the following "theorem". Consider the baryon number changing expectation value $\langle qqql \rangle$ where q is a quark field and l is a lepton field. We will consider for purposes of this argument a theory with only one generation of quarks and leptons. We will consider this expectation value at finite temperature. This expectation value gives the amplitude for a baryon number changing process.

We can estimate this expectation value using the path integral,

$$\langle qqql \rangle = \frac{\int qqql\, e^{-S}}{\int e^{-S}} \tag{13}$$

This amplitude may be computed in Euclidean space, and later analytically continued to Minkowski space. The numerator of the above ratio of path integrals is nonzero only if

$$\int d^4x\; FF^d \neq 0 \tag{14}$$

Now we also have

$$F^2 \geq FF^d \tag{15}$$

and the Higgs part of the action is positive definite. Therefore

$$S \geq S_{inst} \tag{16}$$

and

$$\langle qqql \rangle \leq e^{-2\pi/\alpha_W} \tag{17}$$

The conclusion seems to be that baryon number processes are exponentially suppressed in contradiction with the sphaleron arguments.

How can this argument be wrong? One possibility is that instanton computations may break down in a weak coupling expansion for $E \geq M_W/\alpha_W$, and that the estimate above is simply incorrect.

Even if the above argument is correct, it does not prove that all baryon number changing processes are suppressed.[26] Recall that the sphaleron is composed of many quanta. We would expect at high temperature that the dominant matrix elements involve of order $1/\alpha_W$ quanta. If these matrix elements are Poisson distributed in the number of participating quanta

$$P_N = e^{-\langle N \rangle}\, \frac{\langle N \rangle^N}{N!} \tag{18}$$

so that while the total probability is of order one, the probability of a few prong event is exponentially suppressed

$$P_1 \sim e^{-\langle N \rangle} \tag{19}$$

If the above situation is true, then it must also be true that if instanton methods are used to compute these multi-particle processes, the weak coupling expansion around the instanton will break down, since instanton processes always have a prefactor of $e^{-2\pi/\alpha_W}$.

As we will see in a later section, this is what seems to happen when instantons are employed to estimate the rate of baryon number changing processes.[27]–[29] In these calculations, instanton methods are used to estimate matrix elements for processes which have baryon number change in the presence of a large number of quanta. If taken at face value, these calculations would imply that the rate is large. These calculations are however plagued by the breakdown of weak coupling methods, and conclusions are still in dispute.

We can also see in an explicit quantum mechanics example that a large suppression associated with instantons does not imply a similar suppression for sphaleron processes.[26] Consider the pendulum shown in Fig. 7. In a gravitational field, the potential for the pendulum is periodic as is shown in Fig. 8. The instantons of this theory are classical solutions which in Euclidean space interpolate between $\theta = 0$ and $\theta = 2n\pi$ in a time $1/T$. The sphaleron is the classical solution which sits on top of the barrier, $\theta = 2(n + 1/2)\pi$.

We can easily compute both the sphaleron and instanton contributions at high temperature, where the potential can be ignored. The sphaleron contribution is $e^{-E_{sp}/T} \sim 1$. The instanton classical solution is

$$\theta = 2\pi t T \tag{20}$$

The Euclidean action is

$$S \sim \int_0^\beta dt \frac{1}{2}\dot{\theta}^2 \sim 2\pi^2 T \tag{21}$$

Therefore at high temperatures, the instanton contribution is greatly suppressed, $e^{-2\pi^2 T}$

In the case of the simple pendulum at high temperature, we of course know the correct answer. The highly excited pendulum will roll over many times. The sphaleron therefore predicts qualitatively correct results where the instanton naively gives incorrect intuition.

5 Instantons and High Energy Scattering

In Fig. 9, the amplitude for a baryon number changing process in the presence of a large number of scalars and gauge bosons is shown. Let us for definiteness assume that the number of scalars and gauge bosons is N. By crossing symmetry, this amplitude describes a large number of particles going to a large number of particles, and as well two particles going to a large number. The difference between these amplitudes is that in the first case, we can have a transition where all of the particles have a low energy, $E \sim M_W$, and where all invariant subenergies are small. In the latter case, the invariant subenergies of all particles connected by interactions with one of the particles in the initial state has a large invariant subenergy.

By the considerations of the previous sections, the amplitude for a large number of particles to go into a large number of particles with baryon number change must become large if the total energy is of order $E \sim E_{sp}$. This does not necessarily imply that the amplitude for a few particles to go into a large number is large. Because of the large invariant subenergies, the amplitude might be cut off by a variety of form factor like effects. In addition, at fixed energy, the phase space available per particle in the final state is less by about a factor of two, and since there are a large number of produced particles, the magnitude of the relative suppression$(1/2)^{2N}$ can be big.

It is interesting that when the amplitude for few particle collisions to make large numbers of particles plus baryon and lepton number change are computed, one finds that naively the total cross section becomes large.[27]–[29]. This can be seen to be a consequence of the pointlike nature of instanton induced amplitudes. To leading order in the classical computation, the instanton induced amplitude for multiple scalar emission, ignoring the contributions arising from fermion zero modes, is

$$\int [d\rho] e^{-2\pi/\alpha_W} e^{-2\pi^2 \rho^2 v^2} \phi(p_1,\rho)....\phi(p_N,\rho) \tag{22}$$

The integration over ρ is over the scale size of the instanton. Upon going to mass shell and amputating the poles, we find that in the classical limit the amplitude is point like. We find

$$A_N \sim N! v^{-N} e^{-2\pi/\alpha_W} \tag{23}$$

If we compute the total cross section, we must square the amplitude, divide by $N!$, and integrate over final state phase space. The integration over phase space gives a factor of $(E/N)^{2N}$. We find

$$\sigma \sim e^{-4\pi/\alpha_W} e^{CE^2/v^2} \tag{24}$$

The total cross section shows a truly remarkable growth with energy. It becomes of order one when $E \sim M_W/\alpha_W$, and when the total multiplicity is of order $1/\alpha_W$.

Unfortunately, the naive instanton computation breaks down well before the energy is high enough so that the total cross section becomes of order one. It is possible to study this model non-perturbatively in some approximations, [29] and the result above holds. It is also possible to compute non-perturbatively in a supersymmetric model, and again the cross section becomes of order 1.[30]. For the realistic electroweak case however as yet uncomputable corrections to the formula for the total cross section make definite conclusions hard to establish, and there appear to be large corrections due to form factors.[31]–[32].

While it is not yet established that the cross section for baryon number violating processes do in fact become large in electroweak theory, it is amusing to speculate on the phenomenological consequences if such behavior does happen. We expect that unitarity breaks down at some scale of order $E \sim M_W/\alpha_W$. In this case, the weak interactions become strong. The total cross section is expected to be $\sigma \sim (1/E_{sp})^2 \sim \alpha_W^2/M_W^2$, a cross section typical of off resonance weak interaction processes. In baryon number changing processes, the total multiplicity of jets is large, $N \sim 1/\alpha_W$, and the typical transverse momenta of jets is expected to be $p_t \sim M_W$. The events would be quite spectacular, and common at high energy.

6 Acknowledgments

I gratefully acknowledge my colleagues at Minnesota, L. Carson, Xu Li, A. Vainshtein, M. Voloshin, R. Wang with whom much of the work discussed here was carried out. I also grateful acknowledge many useful conversations with V. Zhakharov on the breakdown of perturbation theory around instantons.

References

[1] S. Adler, *Phys. Rev.* **177**, 2426 (1969).

[2] J. S. Bell and R. Jackiw, *Nuovo Cimento* **51**, 47 (1969).

[3] W. Bardeen, *Phys. Rev.* **184**, 1841 (1969).

[4] J. Ambjorn, J. Greensite, and C. Peterson *Nuc. Phys.* **B221**, 381 (1983).

[5] N. Manton, *Phys. Rev.* **D28**, 2019 (1983).

[6] L. Yaffe, University of Washington Preprint (1989).

[7] F. Klinkhammer and N. Manton, *Phys. Rev.* **D30**, 2212 (1984).

[8] G. 't Hooft, *Phys. Rev. Lett.* **37**, 8 (1976).

[9] A. Belavin et. al., *Phys. Lett.* **59B**, 85 (1975).

[10] S. Dimopoulos and L. Susskind, *Phys. Rev.* **D18**, 4500 (1975).

[11] V. Kuzmin, V. Rubakov and M. Shaposhnikov, *Phys. Lett.* **155B**, 36 (1985).

[12] J. Langer, *Ann. Phys.* **41**, 108 (1967); **54**, 258 (1969).

[13] I. Affleck, *Phys. Rev. Lett.* **46**, 388 (1981).

[14] D. A. Kirzhnits and A. Linde, *Ann. Phys.* **101**, 195 (1976).

[15] A. D. Linde, *Phys. Lett.* **96B**, 289 (1980).

[16] D. Gross, R. Pisarski and L. Yaffe, *Rev. Mod. Phys.* **53**, 43 (1981).

[17] P. Arnold and L. McLerran, *Phys. Rev.* **D36**, 581 (1987).

[18] A. Ringwald, *Phys. Lett.* **201B**, 510 (1988).

[19] L. Carson and L. McLerran, *Phys. Rev.* **D41**, 647 (1990)

[20] L. Carson, Xu Li, L. McLerran and R. T. Wang, University of Minnesota Preprint TPI-MINN-90/13-T

[21] S. Khlebnikov and M. Shaposhnikov, *Nuc. Phys.* bf B308, 885 (1988)

[22] J. Ellis, R. Flores, S. Rudaz, and D. Seckel, *Phys. Lett.* **194B**, 241 (1987).

[23] A. Cohen, S. Dugan and A. Manohar, *Phys. Lett.* **222B**, 91 (1989).

[24] M. Dine, O. Lechtenfeld, B. Sakita, W. Fischler and J. Polchinski, City College Preprint CCNY-HEP-89/18

[25] J. Cline, E. Mottola and S. Raby, in preparation.

[26] P. Arnold and L. McLerran, *Phys. Rev.* **D37**, 1020 (1988).

[27] A. Ringwald, *Nuc. Phys.* **B330**, 1 (1990).

[28] O. Espinosa, California Institute of Technology preprint CALT-68-1586.

[29] L. McLerran, A. Vainshtein and M. Voloshin, University of Minnesota Preprint TPI-MINN-89/36-T

[30] L. McLerran, A. Vainshtein and M. Voloshin, University of Minnesota Preprint TPI-MINN-90/8-T

[31] V. Zakharov, University of Minnesota Preprint TPI-MINN-90/7.

[32] P. Arnold and M. Mattis, Los Alamos National Lab Preprint.

Figures:

Fig. 1 The minimal energy, or effective potential, as a function of ΔQ_{CS}. The effective potential is multiply periodic.

Fig. 2a The effective potential and the fermi levels for W, Z and Higgs fields near the minimum with zero topological charge. 2b The W, Z and Higgs fields have been deformed to near the top of the energy barrier. The fermi levels move, and an occupied state crosses zero energy. 2c The W, Z, and Higgs fields are now near the next minima and the topological charge has changed by one unit. An occupied level has crossed zero energy and is now at the position of the first unoccupied level near the previous minimum.

Fig. 3 Tunneling through the effective potential from one minimum to another.

Fig. 4 The amplitude for baryon number change generated by instantons.

Fig. 5 Transitions across the maximum between two degenerate minima of the effective potential for a thermal fluctuation.

Fig. 6 A sphaleron induced baryon number violating process.

Fig. 7 A simple pendulum.

Fig. 8 The multiply periodic potential of the simple pendulum in a gravitational field.

Fig. 9 A baryon number changing process in the presence of the emission of a large number of scalar and vector gauge quanta.

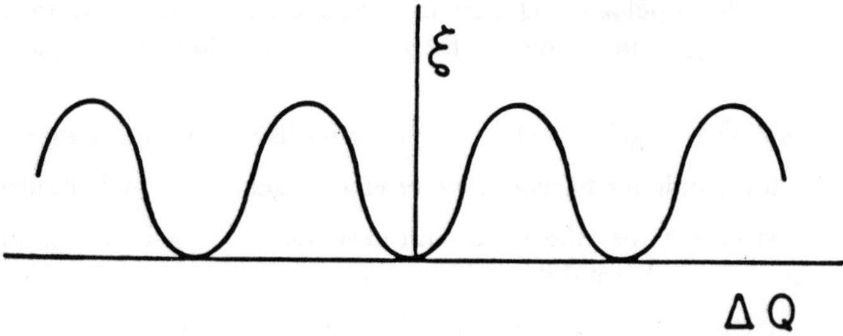

Figure 1

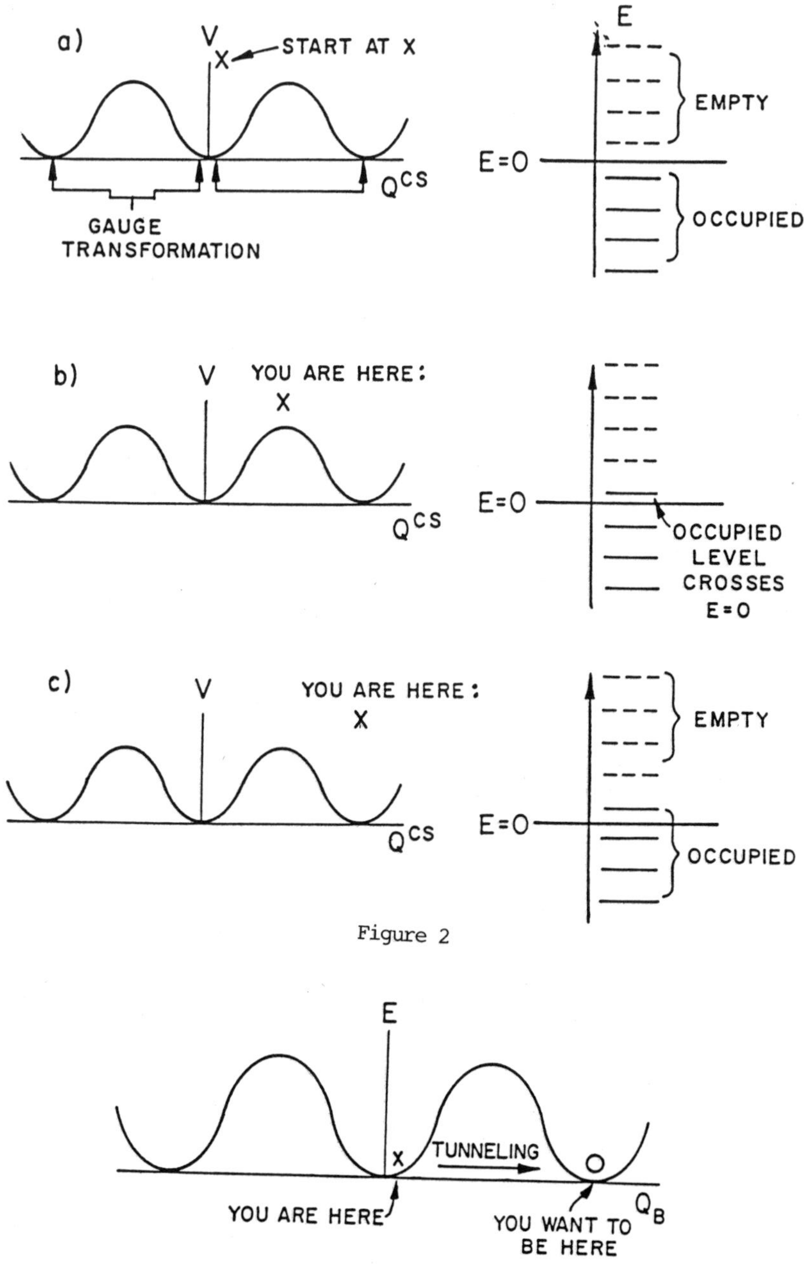

Figure 2

Figure 3

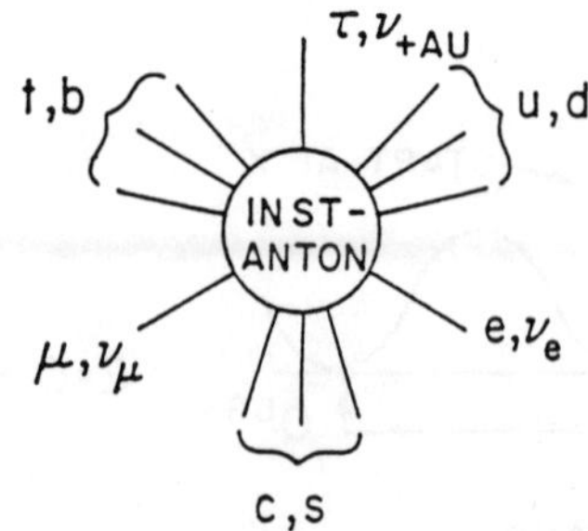

Figure 4

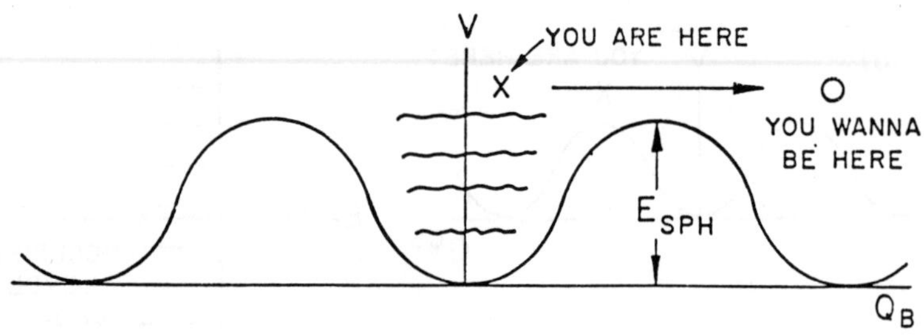

Figure 5

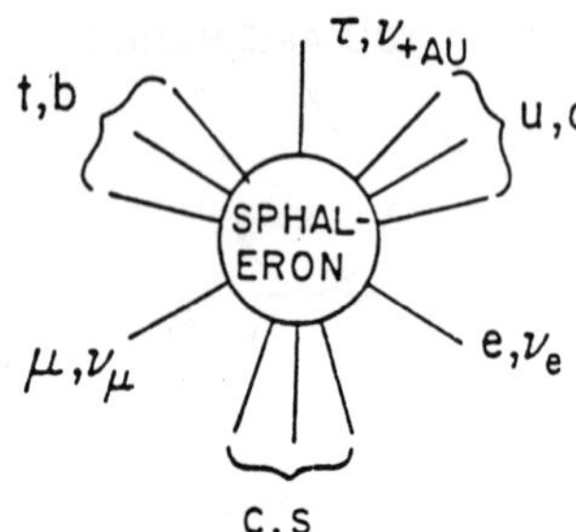

Figure 6

Figure 7

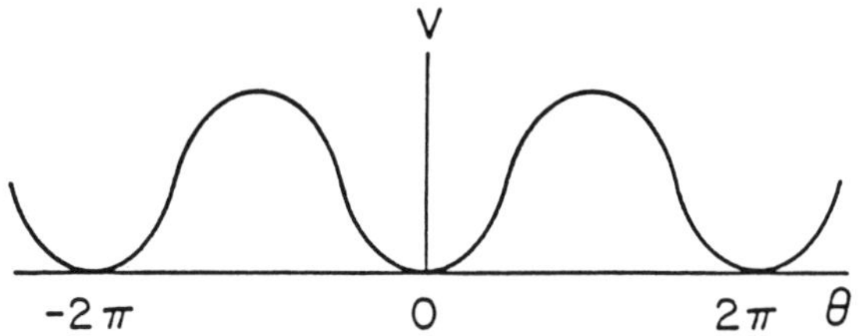

Figure 8

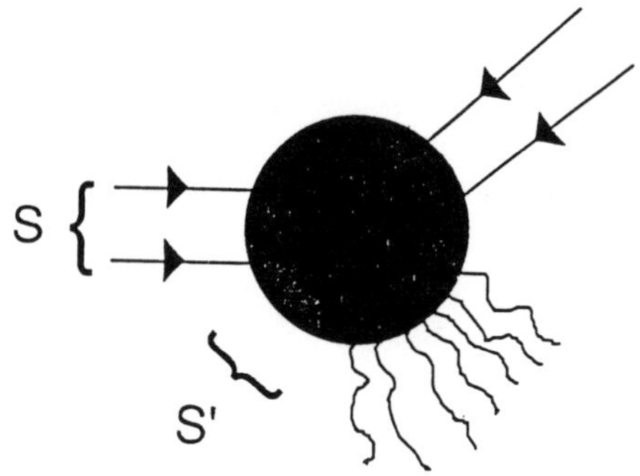

Figure 9

ACCELERATORS AND INSTRUMENTATION

Conveners: R. Huson & M. Month

LARGE HADRON COLLIDERS

S.G. PEGGS
Fermi National Accelerator Laboratory
Batavia, Illinois 60510

ABSTRACT

A review of some of the accelerator physics issues which arise in the design of the contemporary large hadron colliders discusses the following topics: Survey of Hadron Colliders, Primary Constraints, Magnet Style, and Dipole Field Errors.

1. INTRODUCTION

The first two high energy proton accelerators to be brought into storage ring operation, the ISR and the SPS, were built using conventional iron dominated magnets. In order to reduce the power consumption, and to achieve fields above the saturation limit of iron, contemporary proton storage rings use superconducting magnets, whether they exist (TEV), are under construction (HERA, SSC), or are in the design phase (LHC, RHIC, UNK). Another significant trend is that the number of circulating bunches which must be stored in order to achieve useful luminosities increases dramatically, as energies rise and colliders get longer. There are about 15,000 bunches in the SSC. A "large hadron collider" is defined, for the purposes of this discussion, as using superconducting magnets, and storing very many bunches per beam. Both of these assumptions have fundamental consequences. For example, the magnetic field quality is significantly worse in superconducting magnets than in conventional magnets.

The discussion which follows concentrates on selected issues which dominate large hadron collider design. Some important issues (such as collective effects) are ignored, and some (such as the beam-beam interaction) are merely sketched, in order to retain pedagogical clarity within the space available.

2. SURVEY OF HADRON COLLIDERS

The list of the worlds' large hadron colliders shown in Table 1 reveals a healthy diversity of particle species, and accelerator parameters, reflecting the spectrum of high energy physics goals that they seek to address. Only CERN's Super Proton Synchrotron (SPS), the oldest of the machines, and DESY's HERA-e, which is not a hadron collider, do not use superconducting (SC) technology. These colliders are included in the table for the sake of completeness. It is significant that the only two colliders to use antiparticles (antiprotons), the SPS and Fermilab's Tevatron, are also the only two colliders currently in operation. As is discussed in more detail below, the designers of future colliders avoid the difficulties of producing antiprotons, at the cost of complicating the magnet design, in order to achieve higher luminosities.

Asymmetric beams of electrons and protons will collide in HERA, at collisions points where the beams in two very different rings, HERA-e and HERA-p, will cross.

CERN's Large Hadron Collider, the LHC, lies in the LEP tunnel, and thus has a well defined circumference, and a somewhat constrained geometry.

Collider name	Particle species	Magnet type	B field (Tesla)	Energy (TeV)	Circumference (kilometers)	Finish date
(HERA-e	e	conventional	.18	.030	6.3	built)
HERA-p	p	SC	4.5	.82	6.3	1991
LHC	p-p	SC	8.0 ?	7.0 ?	27.3	1997
RHIC	ions	SC	3.5	Au: 0.1u	3.8	1996
SPS	p-$\bar{\text{p}}$	conventional	1.8	.45	6.9	built
SSC	p-p	SC	6.6	20.0	86.3	1997
Tevatron	p-$\bar{\text{p}}$	SC	4.3	.95	6.3	built
UNK	p-p	SC	5.0	3.0	21.0	1997

Table 1   General parameters of contemporary large hadron colliders

Brookhaven's Relativistic Heavy Ion Collider, RHIC, is perhaps the most exotic of the colliders in the list. It will collides a range of species of ions, with a momentum per atomic mass number of 100 GeV/c, in the case of gold.

The Superconducting Super Collider, SSC, is under construction just south of Dallas, Texas, and has the highest energy and the largest circumference of all the colliders quoted.

UNK is an accelerator presently under construction in the Soviet Union, at the Serpukhov laboratory near Moscow.

The nominal time scale for these projects, show in the finish dates, suggests that most will come to fruition towards the close of the twentieth century. Only HERA will become active in the first half of the 1990's, to stand in the limelight as the worlds most modern hadron collider for at least five years. Accelerator physicists will learn a great deal from the commissioning and operation of HERA, just as a great deal has already been learnt from the Tevatron, the worlds first superconducting hadron collider.

3. PRIMARY CONSTRAINTS

Suppose that the design goal of a hadron collider is to reach a given luminosity at a given storage energy, with a given normalized emittance. The luminosity for head on collisions at a single interaction point is

$$L = \frac{N_B^2}{S_B} \frac{c}{4\pi\sigma^{*2}} \qquad 1$$

where c is the speed of light, σ^* is the root mean square transverse beam size (assuming the same Gaussian distribution in both dimensions), N_B is the bunch population, and S_B

is the longitudinal spacing between bunches in one beam. The bunch spacing must be reduced as far as possible, in order to make large luminosities with reasonable bunch populations which will not, for example, violate beam-beam dynamical limits. Another reason to reduce S_B at fixed luminosity is in order to reduce the average number of events per crossing,

$$<n> \; = \; L \, S_B \, \frac{\Sigma_{inel}}{c} \qquad\qquad 2$$

to a level which can be tolerated by the experiments. Some experiments may only be able to interpret one event per interaction. The total proton-proton inelastic cross section Σ_{inel} is expected to be about 90 millibarns at SSC energies.

As the bunch spacing is decreased at fixed luminosity, the total beam current $I \sim N_B/S_B$ increases, since $L \sim I^2 S_B$. This has unfortunate consequences if synchrotron radiation is significant, since the total radiated power per ring is

$$P \; = \; \frac{Z_0}{3} \, \frac{e^2 c^2 \gamma^4}{C\rho} \, N_T \qquad\qquad 3$$

where Z_0 is the impedance of free space, e is the electronic charge, γ is the usual relativistic factor, C is the ring circumference, ρ is the bending radius, and N_T is the total population in one beam. This heat load must be removed by the cryogenic refrigeration system cooling the vacuum pipe, with significant cost implications. Another limit to short bunch spacing is the ability of the experiments to reset their electronics in the time available between bunch crossings.

Short bunch spacing means that there are typically tens of locations in an interaction region, separated longitudinally by $S_B/2$, where parasitic collisions may occur. This mandates the introduction of a crossing angle at the principal collision point, since the experimenters want collisions at only one place. Worse, without a crossing angle the cumulative beam-beam interaction effects are usually intolerable, since the tune shift at each head on collision is

$$\Delta\nu_{HO} \; = \; \xi \; \equiv \; -\frac{N_B r_p}{4\pi\varepsilon} \qquad\qquad 4$$

where ξ is the beam-beam tune shift parameter, r_p is the classical proton radius, and ε is the invariant emittance. When the crossing angle α is large compared to σ^*/β^*, the angular size of the beam at the interaction point, then in addition to the single head on tune shift given by (4), there is also a long range tune shift, due to a total of n_c collisions per interaction region, which is well approximated by

$$\Delta\nu_{LR} \; = \; n_c \, \xi \, \frac{2}{[\, \alpha \, / \, (\sigma^*/\beta^*) \,]^2} \qquad\qquad 5$$

The head on tune shift is proportional to the bunch population N_B, while the long range shift depends on the total population N_T, so that varying the bunch separation trades one off against the other.

When the crossing angle is large compared to σ^*/σ_S, the aspect ratio of the beam, a significant loss of luminosity results, because the longitudinal tails of the bunches do not collide. The ratio of the true luminosity to the head on luminosity is given by

$$\frac{L}{L_0} = [\, 1 + (\alpha\, \frac{\sigma_S}{2\sigma^*})^2\,]^{-\frac{1}{2}} \qquad\qquad 6$$

Attempts to increase the true luminosity by continuously reducing β^* (and hence σ^*) eventually run into a conflict between the need for α to be large to control the long range tune shift, and the need for α to be small in order to achieve a reasonable fraction of the available head on luminosity. Angles much smaller than the longitudinal aspect ratio are also preferred for beam-beam reasons, because odd resonances and satellite resonances are then partially suppressed.

4. MAGNET STYLE

If large numbers of nominal intensity bunches in each counter-rotating beam are allowed to pass through each other head on, at locations all around a collider, the cumulative beam-beam effect is disastrous. Left unsolved, this problem limits the attainable luminosity, since either the number of bunches or their intensity must be reduced. Two solutions are possible - separate the orbits of the two beams in a single vacuum pipe, or separate them into two vacuum pipes. When the counter-rotating beams are separated into two different magnet bores, long range beam-beam collisions only occur near the interaction points. Their effect is then of only secondary importance, except when collider performance is pushed to its ultimate limits.

Separation of orbits inside a single magnet bore has the advantage of requiring fewer magnets - or, at least, requiring simpler magnets, since a single dipole magnet may contain two bores. However, the long-range beam-beam interactions which are still present, although greatly reduced in strength, still limit the attainable luminosity. Hence, this solution has not been adopted for any of the large hadron colliders in currently in design or construction. It has been adopted as an upgrade strategy in both of the existing hadron colliders, the SPS and the Tevatron, the only two contemporary colliders colliding particles with antiparticles. The residual limitation on the number of bunches is not of great concern in these machines, since the limited numbers of antiprotons available are anyway most efficiently employed in a relatively small number of bunches. Table 2 shows very clearly the correlation between a single bore and a small number of bunches. The apparent exception to this rule is RHIC, which uses a relatively small number of bunches to concentrate the limited number of heavy ions available.

Assuming from here on the (now conventional) design choice of separating two beams of identical species into two bores, the question of choosing one or two magnets still remains. In a typical "two-in-one" dipole magnet, two bores in a single magnet are horizontally separated in the mid-plane of the magnet, so that the flux passing up through one bore goes down through the other. The shared flux means that slightly higher fields are reached more easily, but also means that the two beams are magnetically coupled. This coupling is a potential source for serious operational difficulties, since not only will the two rings share magnetic errors, but also the corrections applied to the two rings tend not be orthogonal. However, the total heat leak per meter is less than for two separate magnets, and the total cross-sectional space occupied in the tunnel is less. It is because of the compactness of this solution that it has been adopted for the LHC, which must share a tunnel with the LEP. Table 2 shows that, in contrast, the other comparable colliders (RHIC, SSC, and UNK) all use two independent magnets.

Name	Species	Number of				Comment
		BUNCHES	BORES	MAGS	CRYOS	
HERA	e-p	210	2	1+1	1	one ring normal
LHC	p-p	10,000 ?	2	1	1	space constraint
RHIC	ions	Au: 57	2	2	2	horizontal separation
SPS	p-$\bar{\text{p}}$	12	1	1	-	horizontal pretzels
SSC	p-p	15,000	2	2	2	vertical separation
TeV	p-$\bar{\text{p}}$	36	1	1	1	helical orbits
UNK	p-p	12,000	2	2	2	horizontal separation

Table 2   Magnet styles in contemporary hadron colliders

Independent "one-in-one" magnets have the virtue of simplicity of design, and of independent operation. They may be placed side by side, or one on top of the other. The magnetic errors of each ring are independent, as are the misalignment errors, and the beam separation may be increased at will, above a minimum value. One ring may be operated independently of the other, perhaps even at a different energy. If two magnets are chosen, there is still the question of using one or two cryostats - two magnets could be enclosed in a single cryostat. However, as Table 2 shows, in practice each magnet is assigned its own cryostat.

A vertical separation of two one-in-one rings has the advantage of simpler beam injection and abort, easier magnet installation and change procedures, and more efficient use of tunnel floor space. Horizontal separation has the slight advantage of making dispersion control easier at the collision points, but makes it difficult to match the two ring circumferences when the IRs are "clustered" with an even number of IRs in each cluster.

None of the arguments presented here conclusively favor one or other magnet construction style, or separation orientation.

5. DIPOLE FIELD ERRORS

The two dimensional field in any magnet can be specified by a multipole expansion according to

$$B_y + i\,B_x = B_0 \sum_{n=0}^{\infty} (b_n + i a_n)\,(x + iy)^n \qquad\qquad 7$$

For example, a perfect dipole has $b_0=1$, and a perfect skew quadrupole has $a_1=1$, with all other terms zero. This notation also lends itself naturally to a description of magnetic field errors in dipoles, which are the dominant sources of optical errors in large hadron colliders. The normal and skew coefficients b_n and a_n can be derived almost directly from Fourier analysis of the voltage on a rotating coil inside a magnet.

Conventional "iron dominated" magnets are limited in the field they can attain to the saturation field of their laminations, about 2 Tesla. If the laminations do not saturate - so that their relative permeability is very much larger than one - the magnetic field is constrained to be normal to the surface of the pole tips. This simple boundary condition, and the precision with which laminations can be stamped and aligned, results in the high field quality usually associated with iron dominated magnets. The boundary conditions in superconducting "current dominated" magnets is, by contrast, relatively awkward. It is easy to show that a thin circular current shell, with a density proportional to $\cos(\theta)$, where θ is the azimuthal angle, results in a perfect dipole field. However, in a realistic "cosine theta" magnet the distribution of conductors is not thin, and it is not trivial to design even an idealized coil which will suppress several low order systematic multipole fields. In operation, the conductor braids must not slip relative to each other or their mechanical supports, despite the large forces to which they are subjected, otherwise the magnet will quench. Their cross-sectional size is subject to fluctuation, and they are harder than laminations to place accurately. Even if all these practical problems could be overcome, the fundamental phenomenon of "persistent currents" in the filaments, discussed below, causes fields which are both hysteretic and time dependent.

Random contributions to the normal and skew multipole coefficients come from conductor placement errors, and from the spread in the distribution of filament sizes and critical currents. Measurements on many different superconducting dipoles of various bore diameters d have lead to an empirical scaling of the coefficients,

$$\langle a_n^2 \rangle^{\frac{1}{2}}, \quad \langle b_n^2 \rangle^{\frac{1}{2}} \quad \sim \quad d^{-n-\frac{1}{2}} \qquad\qquad 8$$

which is also in good agreement with theoretical models of construction errors. The physical meaning of this scaling is expressed as a rule of thumb which ignores the $-1/2$ term in the exponent: the fraction of the bore containing good field quality is independent of the bore size. Another rule of thumb is that random field errors are not significant contributors to betatron tune shifts with amplitude, but are dominant contributors to resonance effects, such as the distortion of phase space orbits away from linear behavior.

Systematic contributions to even order normal multipole coefficients, b_{2n}, come from persistent current magnetization in the superconducting wires, from saturation in the iron in the cryostat, and from systematic conductor placement errors. The strongest of these sources is the persistent current effect. Here, each individual superconducting filament partially excludes magnetic flux by rearranging its internal current distribution, so that more current flows down one side of the filament than the other. This nonuniform current distribution depends on the history of the filament - persistent current multipole fields exhibit hysteresis. Less flux can be excluded at higher fields, so that the problem is most severe at the injection energy of a collider. One way to alleviate the situation is to use the smallest possible filament diameter. Other ways are to raise the injection energy (and field), and to enlarge the bore of the magnet.

The worst property of systematic errors is their ability to cause large tune shift variations with particle betatron amplitude A_β, and with δ, the momentum offset. The horizontal tune shift due to an error b_n is

$$\Delta \nu_n \;=\; b_n < \beta \cos(\phi) \, [\eta\delta + A_\beta \, (\beta/\beta_{max})^{1/2} \cos(\phi)]^n > \qquad\qquad 9$$

where two independent averages are taken, over the phase of a betatron oscillation ϕ, and over the beta and dispersion values. In a "smooth" approximation, where β and η take on their average cell values, this reduces to

$$\Delta \nu_n \;=\; b_n \beta \sum_i 4^{-i-1} \, C^n_{2i+1} \, C^{2i+2}_{i+1} \, A_\beta^{2i} \, (\eta\delta)^{n-2i-1} \qquad\qquad 10$$

where

$$C^i_j \;\equiv\; \frac{i}{j! \, (i-j)!} \qquad\qquad 11$$

These simple expressions demonstrate that the effects of systematic dipole errors are easier to calculate than those of random errors, whose analysis relies on computer simulation.

Operationally, systematic errors are much easier to diagnose and correct than random errors. This is fortunate, because the strong persistent current multipole fields not only exhibit hysteresis, but also decay with time. The flux pinning centers, which are responsible at a quantum mechanical level for maintaining the current distribution, are subject to diffusion. These pinning centers are a feature of Type II superconductors, which permit the penetration of a flux quantum by energetically favoring its passage through a small island of normal material, surrounded by a sea of superconductivity. When the dipole field is held constant, during an injection phase typically lasting tens of minutes, the persistent current sextupole component (for example) drifts according to the logarithm of the time. However, as soon as the main field starts to rise, at the beginning of the energy ramp, the sextupole component quickly snaps back almost to its original value. This rapid change in the nonlinear behavior of hadron colliders is a major operational impediment, which has still to be completely understood and controlled. A special feedback circuit is being created at the Tevatron for this purpose.

It is good to end on this challenging note. Large hadron colliders are by no means as boring and "conventional" as some more daring accelerator physicists would have us believe. They still present important and intriguing problems whose proper analysis requires the development of radically new techniques.

STATUS OF THE RELATIVISTIC HEAVY ION COLLIDER

S.Y. LEE*

Accelerator Development Department
Brookhaven National Laboratory
Upton, Long Island, NY 11973

ABSTRACT

Accelerator Physics issues, such as the dynamical aperture, the beam lifetime and the current–intensity limitation are carefully studied for the Relativistic Heavy Ion Collider at Brookhaven National Laboratory. The single layer superconducting magnets, of 8 cm coil inner diameter, satisfying the beam stability requirements have also been successfully tested. The proposal has generated wide spread interest in the particle and nuclear physics.

1. INTRODUCTION

In 1983, the DOE/NSF Nuclear Science Advisory Committee declared, "the construction of relativistic heavy ion collider facility expeditiously" as the long range plan for nuclear science. Brookhaven National Laboratory submitted to the proposed design of the Relativistic Heavy Ion Collider (RHIC) in May 1986 with a detailed construction plan and cost estimate. In May 1987, the RHIC proposal was subjected by DOE to a full scale technical, cost and management review, and was followed by an independent cost estimate review. The project was declared ready to proceed to the construction phase.

Beginning in fiscal year 1987, R&D funding has been allocated to the project by DOE totaling about 15M$ through FY 1989. During this period, a series of full size superconducting RHIC arc dipole magnets have been built and tested. Industrial manufacturing studies have been carried out for dipole and quadrupole magnets, and the design has been tailored to make them suitable for mass production. There are 8 dipoles and 2 quadrupoles tested. These R&D magnets serves important manufacturing protocol for the industrial manufacturing procedures.

The project generates wide spread excitement and interest in the nuclear and particle physics community. Several workshops and schools and International Conferences on the experiments, detectors and physics for RHIC have been organized in LBL, BNL, etc. in USA and Universities in European countries. These studies generate a specified requirement on the properties of the beam at the interaction point, and energy range for the discovery potential. It is generally agreed, based on theoretical estimate, that 100 GeV/u per beam for Au ion should induce phase transition for the quark gluon plasma. The collider should cover the entire energy range, e.g. 1.5 GeV/u to 100 GeV/u. The luminosity should be $> 10^{25}/\mathrm{cm}^2\mathrm{sec}$ and the diamond length σ_I should be smaller than 20 cm. The collider should have the capability of colliding unequal species such as p and Au ions.

Based on these considerations, the RHIC conceptual design has evolved through several workshops to assure machine performance. In this report, we shall review the basic machine design and performance.

---

* Work performed under the auspices of the U.S. Department of Energy.

2. RHIC INJECTION SOURCE

The AGS–Booster, to be completed in 1990, should be able to deliver the heavy ion, e.g. Au, from the Tandem to the energy for the AGS injection. Table 1 lists the parameters expected from the Booster. In the lower part of the table,[1] we list also the beam parameters assumed for the RHIC and the expected design luminosity. We note that the expected performance should be better than what we have assumed. Since the Booster can deliver more intensity than that assumed for the RHIC design parameter, we can expect better performance in RHIC.

Table 1. Number of Heavy Ions per Bunch in the Booster

Species	^{16}O	^{28}Si	^{63}Cu	^{127}I	^{197}Au	Units
Booster						
Harmonic, h	2	2	2	3	3	
f_{rf} @ injection	350	318	232	264	213	kHz
f_{rf} @ top energy	2.68	2.60	2.08	2.23	1.66	MHz
Top kinetic energy	1249	998	376	144	72	MeV/u
Ions/Bunch@S.C. limit	38	24	19	11	10	$\times 10^9$
Ions/Bunch	8.32	7.06	8.7	5.64	4.12	$\times 10^9$
Normalized emittance	6.0	5.4	3.9	3.0	2.3	π mm·mrad
Bunch area	0.07	0.07	0.07	0.07	0.07	ev·sec/u
Stripping efficy. @ S_B	100	100	100	40	50	%
Charge State after S_B	8	14	29	53	77	
AGS						
Ions/bunch after S_B	8.32	7.06	8.7	2.26	2.0	$\times 10^9$
Norm. emitt. after S_B	6.1	5.5	4.1	3.6	4.0	π mm·mrad
RHIC						
Ions/bunch	8.3	5.6	2.7	1.5	1.0	$\times 10^9$
Normalized emittance	10	10	10	10	10	π mm·mrad
Bunch area	0.3	0.3	0.3	0.3	0.3	ev·sec
Injection γ	15.6	15.6	14.4	13.0	12.2	
Top Energy γ	135.3	135.3	124.5	112.9	108.4	
Luminosity ($\beta^* = 2$m)	98	44	9.5	2.7	1.1	$\times 10^{27}$cm^{-2}sec^{-1}

3. RHIC COLLIDER

For 100 GeV/u Au ion, the magnetic field required is 3.45 Tesla in RHIC dipole. The circumference of the RHIC tunnel is 4.75 times that of the AGS circumference, ($C_{AGS} = 807.12$ m), the natural choice for the RHIC rf system would be $4.75 \times n \times h_{AGS}$, where $h_{AGS} = 12$. It is thus natural to choose 57 bunches for bunch to bucket transfer.

The collider is composed of two identical non–circular rings in a common horizontal plane, oriented to intersect with one another at 6 crossing points. Each ring consists of three inner arcs and three outer arcs and six insertions joining the inner and outer arcs. Each arc is composed of 12 FODO cells. The insertion has nine quadrupoles and four dipoles (two for dispersion suppression and two for the beam crossing) on each side of the crossing point. Figure 1 shows the general layout of the collider ring, which conforms well with the existing tunnel geometry.

The arc consists of 12 FODO cells. Each cell (with length 29.622 m) is composed of two quadrupoles and two dipoles. The total dipole bending angle per cell is 77.701 mrad. The phase advance is $90°$. The maximum betatron amplitude function and dispersion function are 50 m and 1.52 m respectively.

The insertion is composed of: (1) dispersion suppression section, (2) low-β matching sections, and (3) beam crossing section. Figure 2 shows the schematic layout of the half insertion. By exciting BC1 and BC2 independently, the collision of unequal species can be achieved as shown in Figure 3.

The important feature of the RHIC lattice is that the half integer stopband are minimized by the achromatic arcs and cancellation between nearby insertions. With four family of sextupoles, the chromatic aberration of the RHIC lattice can be minimized. The β^* value can be dynamically tuned from $\beta^* = 0.5$m to $\beta^* = 10$m without changing the phase advance across the insertion. The transition energy and the betatron tune are:

$$\gamma_T = 24.6$$

$$Q_x = 28.826$$

$$Q_y = 28.822$$

where the betatron tunes Q_x, Q_y can be adjusted within ± 1.

4. DYNAMICAL APERTURE

The dynamical aperture issues for the superconducting magnets arises from the multipoles generated by the systematic effect of the magnetization and/or saturation effect and the random coil placement errors. Using the Herrera model, which was developed through the experience of the CBA and Tevatron magnets, one can estimate the multipole components of the RHIC magnets. The dynamical aperture calculation based on these multipole field is shown in Figure 4. Note here that the aperture limitation is due mainly to the aperture limitation of the high β quadrupoles and the beam crossing dipoles. Note here that the dynamical aperture is taken to be the smallest of 10 random seeds for these random error in 400 turns tracking. Long term survival tracking is yet to be studied.

The dynamical aperture limitation and the long term stability must be related to the nonlinear resonances. To minimize the resonance effect, normal and skew quadrupole correction systems, six families of sextupole, and octupole and decapole correction system are used in the RHIC lattice. The closed orbit can be corrected to less than 1 mm.

Besides the transverse dynamical aperture, the longitudinal dynamical aperture is also important to RHIC. Note that the small Z/A ratio for the heavy ions, the injection γ into RHIC range from 12 to 15 for heavy ion species (see Table 1). Thus these heavy ion beams should be accelerated through the transition energy in RHIC. At the transition energy region, the problems related to the longitudinal beam dynamic arise:

(1) Due to small synchrotron frequency, the Landau damping needed for microwave instability disappears. The microwave instability has been observed in many high intensity accelerators in the transition energy region. The instability can be controlled by increasing the phase space area and the bunch distribution.

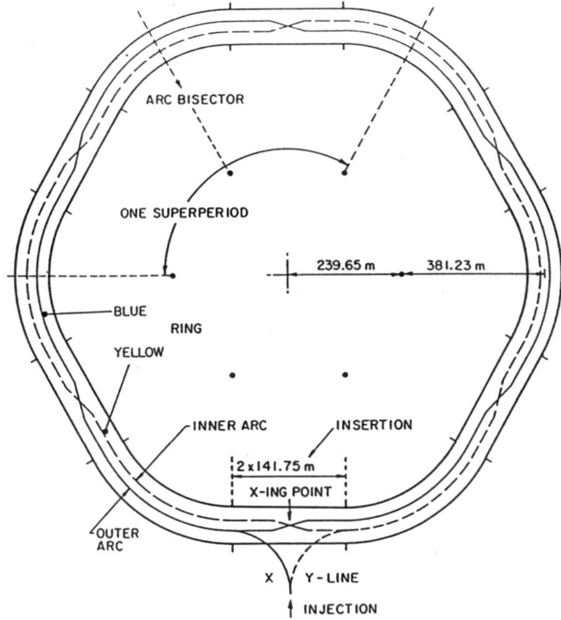

Fig. 1. Layout of the Collider.

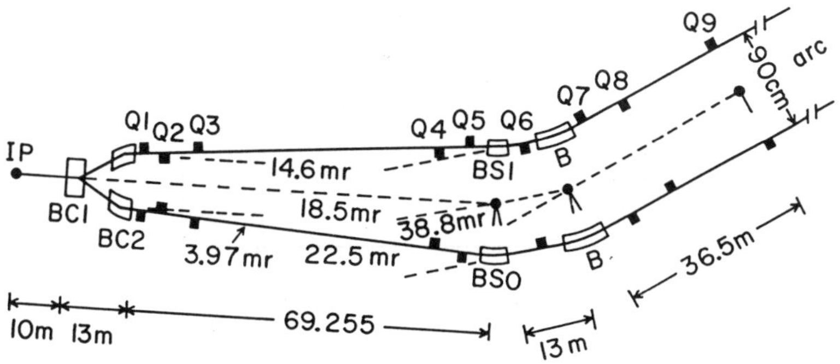

Fig. 2. Expanded layout of half–insertion.

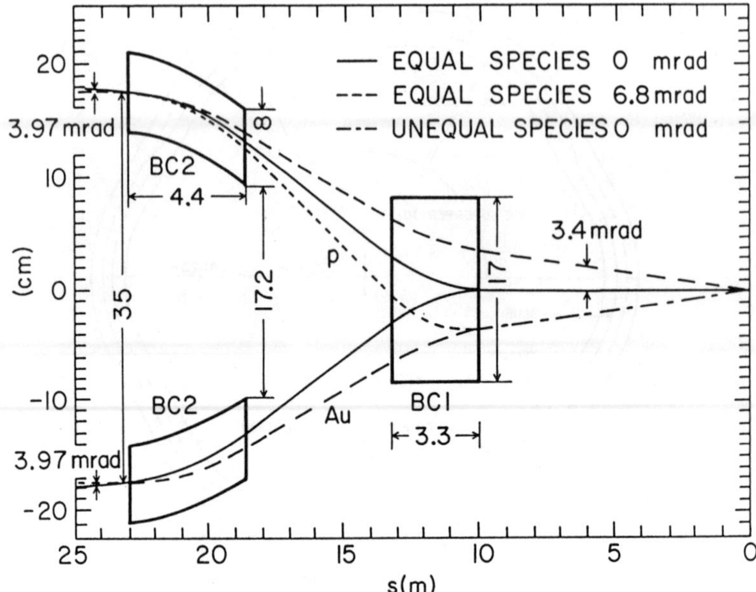

Fig. 3. Beam crossing geometry.

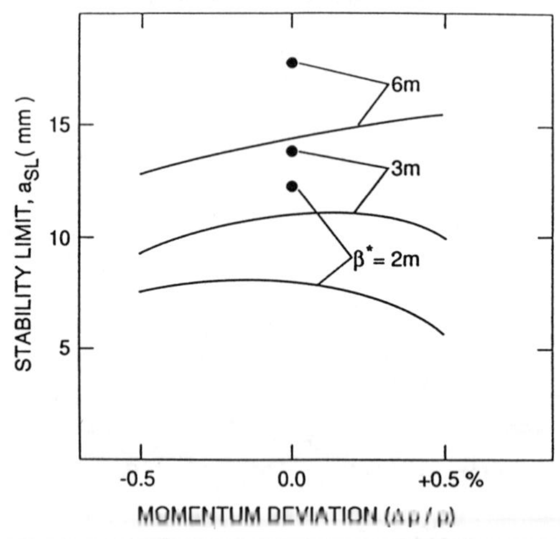

Fig. 4. Variation of the dynamical aperture with $\Delta p/p$ and β^* (Tracking program PATRICIA). The physical aperture limits of the error–free machine are marked by dots.

(2) The nonlinear synchrotron motion: At the transition energy region, the synchrotron phase equation can be expressed as

$$\dot\varphi = h\Omega_0(\eta_0 + \eta_1\delta + \ldots)\delta$$

where h and Ω_0 are respectively the harmonic number, and the revolution frequency, $\delta = \Delta P/P$ is the momentum error, $\varphi = -h(\theta - \theta_0)$ is the relative rf phase of the particle, and

$$\eta_0 = \frac{1}{\gamma_T^2} - \frac{1}{\gamma^2}$$

$$\eta_1 = \frac{3\beta^2}{2\gamma^2} + \frac{\alpha_1}{2\gamma_T^2} \simeq \frac{2}{\gamma_T^2}$$

where βc is the velocity of the particle and $\alpha_1 = -2\gamma_T^{-1}(d\gamma_T/d\delta)$. Note that when $\eta_0 = 0$, at the transition energy region, the second order term $\eta_1\delta$ becomes important. Assuming a linear acceleration, $\gamma = \gamma_T + \dot\gamma t$, we obtain $\eta_0 \simeq -\frac{2\dot\gamma t}{\gamma_T^3}$.

For a bunch with momentum spread $\hat\delta$, the time when the nonlinear term dominates is

$$\tau_{NL} = \frac{\eta_1\hat\delta}{2\dot\gamma}\gamma_T^3 \simeq \frac{\gamma_T}{\dot\gamma}\hat\delta .$$

The nonlinear time is inversely proportional to $\dot\gamma$ and increases with the $\hat\delta$ or the square root of the phase space area. During the nonlinear time at the transition energy region, part of the bunch can experience a de–focusing force in the longitudinal phase space. The bunch area can grow exponentially. The growth in phase space area is especially important for RHIC, where the accelerator rate $\dot\gamma$ is small.

(3) Beam induced rf voltage across the high–frequency rf system. Since the 160 MHz ($h = 2052$) rf system is needed at the storage time for the 20 cm diamond length for collision, these rf stations should be paraphased to minimize their effect on the beam during the acceleration. Assuming a 1% error in paraphasing the high frequency rf system, we expect a remnant rf voltage of $\pm$ 10 Kvolts at $h = 2052$. This is equivalent to the microwave source at a fixed frequency. The effect is important only at the transition energy region. The tolerance for RHIC is $\pm$ 25 Kvolts.

The solution to the transition energy problem is to jump the γ_T by 0.6 within 60 msec. The task of γ_T jump can be accomplished by two families of quadrupole correctors in the arc by changing 0.8 Tesla in 60 msec of the integrated quadrupole field.

5. LIFETIME LIMITATION

For the heavy ion colliders, the lifetime limitation is determined by the following processes: Nuclear reactions and Coulombic interaction between colliding bunches, Coulombic interaction between particles within the bunch and the beam gas scattering. The Coulombic interaction between particles within the bunch, the so called Intrabeam Coulomb Scattering (IBS), plays the major role in the lifetime limitation. Due to the intrabeam Coulomb scattering, the beam size increase diffusively. The aperture of the magnets is determined by ensuring enough linear aperture for a minimum 10 hr collider operation. Besides the intrabeam scattering, the Beam–Beam Bremsstrahlung and the Beam–Beam Coulomb dissociation are important also for high Z collision, such as Au on Au or Pb on Pb, etc. Table 2 summarizes the initial lifetime and the reaction rate for various operation modes.

Table 2. Initial Reaction Rate $\lambda = -I^{-1}\,dI/dt$ and Total Half Life of Particle Beams for Head–On Collisions

	Beam–gas nuclear reaction λ_1	Beam–beam nuclear reaction λ_2		Beam–beam Coulomb dissociation λ_3	Beam–beam Bremsstrahlung electron pair production λ_4	Initial Half Life
Beam @ 10^{-10} Torr	A on A	p on A		A on A	A on A	A on A
	$\times 10^{-3}$/h	$\times 10^{-3}$/h	$\times 10^{-3}$/h	$\times 10^{-3}$/h	$\times 10^{-3}$/h	h
p	0.15	2.4	2.4	–	–	272
O	0.37	5.5	32.5	–	–	118
Si	0.54	5.4	42.5	–	–	117
Cu	0.76	4.1	62.3	0.55	0.12	125
I	1.08	1.1	29.2	4.6	4.6	61
Au	1.37	2.8	102.3	20.7	31.6	12

6. THE BEAM CURRENT LIMITATION

A high–brightness beam in the storage ring can offer higher luminosity for the collider. Yet high–brightness is also prone to various collective instabilities:

(1) The beam–beam tune shift for the round beam is given by

$$\Delta\nu_{b-b} = \frac{3r_0}{2}\frac{Q^2}{A}\frac{N_B}{\epsilon_N}$$

where the classical proton radius $r_0 = 1.535 \times 10^{-18}$m and ϵ_N is the normalized emittance for 95% of the beam. Thus the beam–beam tune shift is proportional to the two dimensional brightness. Similarly, the Laslett space charge tune shift at the injection is proportional to the four dimensional brightness, i.e. the bunching factor is also important. For RHIC, the space charge at injection is less than 0.02 for all ion species and the beam–beam tune shift is less than 0.004 per crossing. The total beam–beam tune shift is also less than 0.02.

(2) Microwave instability: (a) The threshold impedance for the microwave instability is given by

$$|Z/n| \leq 2\pi\,\frac{E|\eta|}{eI_p}\left(\frac{\sigma_E}{E}\right)^2\frac{A}{A^2}$$

$$|Z_\perp| \leq 2\pi\nu\,\frac{J_0 L}{eI_p^e}\frac{A}{Q^2}\left[(n-\nu)\eta+\nu'\right]\left(\frac{\sigma_E}{E}\right)$$

Table 3 summarizes the single bunch instability.

Table 3. Tune Shift and the Threshold Impedance

Species	p	O	Si	Cu	I	Au
$\epsilon_N[\pi\text{mm·mrad}]$	20	10	10	10	10	10
$N_B[\times 10^9]$	100	8.3	5.6	2.7	1.5	1.0
$\Delta\nu_{BB}[\times 10^{-3}]$	3.7	2.4	2.9	2.6	2.4	2.3
$\sigma_E/E[\times 10^{-4}]$	0.8	1.2	1.2	1.3	1.5	1.5
$|Z/n|(\Omega)$	2.0	9.1	7.7	9.4	11	12

(3) Coupled–Bunch Instability: Using ZAP code, the coupled–bunch instability has been estimated with the impedance due to space charge, resistive wall, broad band, and higher order parasitic cavity modes. In the absence of Landau damping, the transverse instability growth rate is expected to be 43 sec^{-1} for the stainless steel vacuum chamber or to be 1.3 sec^{-1} for the copper–coated chamber (25 μm thickness). Similarly, the longitudinal bunch to bunch instability is expected to have a growth time of 60 msec due to the parasitic modes of h=342 cavities. Special care is required to control the spacing between the parasitic modes. Wide–band damper systems have been designed to cure the collective motion.

7. STATUS AND OUTLOOK

The single layer $\cos\theta$ magnet has been successfully designed to achieve the quench field of 4.5 Tesla, which is considerably larger than the operation field of 3.45 T. Eight dipole and two quadrupole magnets have been successfully built and tested. All magnets achieve the quench field beyond the operation field without training. The multipoles of each magnet have also been measured. The geometric components of the sextupole and decapole are yet to be understood. The random multipoles are within the prediction of the Herrera model. Besides these arc dipoles and quadrupoles, the beam crossing dipoles BC1 and BC2 are important R&D items.

Since the intrabeam Coulomb scattering is important for the lifetime of the heavy ion beam, a stochastic cooling system capable of canceling the growth due to the intrabeam Coulomb scattering shall enhance the luminosity more than 4 times. The stochastic cooling can also eliminate the costly high rf voltage, which is needed to contain the bunch within the bucket area.

RHIC is in the President budget of FY 1991 for a total cost of 397 M\$ including three detector facilities. The cryogenic system of helium refrigerator has been fabricated, installed and tested in 1986. The load capacity is about two times the estimated heat loads for RHIC. A total of six detector area are available. The machine, to be completed in 1997, shall offer challenging and exciting physics opportunity for the new and unexplored kingdom of the Quark–Gluon plasma. Similarly, the luminosity for the proton–proton collision mode can achieve $2 \times 10^{32}\text{cm}^{-2}\text{sec}^{-1}$ at 250 GeV/beam (for unpolarized proton as well as polarized protons).

REFERENCES

Conceptual Design of the Relativistic Heavy Ion Collider, BNL 51932 (1986); BNL 52195 (1989).

APPROACHING ULTIMATE LUMINOSITY AT THE SSC

Alex Chao

Superconducting Super Collider Laboratory[*]
Accelerator Division
2550 Beckleymeade Avenue
Dallas, Texas 75237

ABSTRACT

The SSC is designed[1,2] to operate conservatively at a luminosity of 10^{33} $cm^{-2}\,s^{-1}$ over essentially its entire energy range, up to the peak energy of 20 TeV per beam. This luminosity is dictated by the envisioned detector limitations. This note attempts to explore the possible scenarios to approach an ultimate SSC luminosity from the accelerator point of view. We find and reconfirm earlier studies[3,4] that the ultimate luminosity limit is much beyond the 10^{33} $cm^{-2}\,s^{-1}$ level.

The nominal operation of the SSC, as presently conceived, requires two counter-rotating proton beams, each consisting of a train of about 17000 beam bunches. Each bunch contains 7×10^9 protons and adjacent bunches are spaced by 5 meters. To avoid spurious bunch collisions, the two beams cross at an angle, which is nominally 75 μrad, but can be varied in the range up to 150 μrad.

The design model assumes 4 interaction regions, 2 with high-luminosity interaction points (IP) and 2 with medium-luminosity IP's. For a typical detector that can handle 1 event per crossing, or an event rate of about 6×10^7 events per second, the corresponding luminosity is given by the event rate divided by the pp total cross section, which is about 60 mb. This reproduces the nominal SSC luminosity of 10^{33} $cm^{-2}\,s^{-1}$. In fact, the luminosity is

$$L = N^2 c\gamma/4\pi\epsilon_N\beta^*S \; , \tag{1}$$

where N = number of protons per bunch, S = bunch spacing, β^* = beta-function at the high-luminosity IP, ϵ_N = normalized rms emittance. With $N = 7 \times 10^9$, $\gamma = 21000$ (20 TeV), $\epsilon_N = 1\,\pi$ mm-mrad, $\beta^* = 0.5$ m, S = 5 m, the luminosity is 10^{33} $cm^{-2}\,s^{-1}$.

---

[*]Operated by the Universities Research Association, Inc., for the U.S. Department of Energy under Contract No. DE-AC02-89ER40486.

This nominal luminosity is dictated by the physics requirements for the SSC and by considerations of practical limitations to the event rates that can be handled by typical detectors expected to be employed at the SSC. For example, the speed of electronic devices prevents the bunch spacing to be much less than 5 meters, while the event resolution prevents the number of events per bunch crossing to be approximately less than 1. However, these limitations are process related and it is useful to explore the ultimate limits to the SSC luminosity as far as the accelerator—not the detector—is concerned. It is possible that specialized experiments can be designed to exploit higher luminosity when very rare, but distinctive processes are to be studied. Indeed, one finds that this ultimate luminosity limit is much beyond the 10^{33} cm^{-2} s^{-1} level, as we will discuss below.

The question is how to increase the luminosity, given in Eq. (1), by varying N, ε_N, β^*, or S without violating known limits of accelerator physics. Because of its very high energy, synchrotron radiation plays an important role in the performance of the SSC. At the highest energies, synchrotron radiation, which is absorbed on the vacuum chamber walls at liquid helium temperatures, will limit the total beam current and, hence, the luminosity. At lower energies, it is expected that the highly nonlinear fields of one beam acting on particles of the other beam will ultimately limit the luminosity—the beam-beam limit. These considerations are specified by three quantities: the synchrotron radiation power per unit length, P/C, the head-on beam-beam tune shift per IP, $\Delta\nu_{HO}$, and the long range beam-beam tune shift, $\Delta\nu_{LR}$:

$$P/C = Z_oNe^2c^2\gamma^4/3CS\rho \tag{2}$$

$$\Delta\nu_{HO} = Nr_p/4\pi\varepsilon_N \tag{3}$$

$$\Delta\nu_{LR} = Nr_pL_{LR}/\pi\gamma\beta^*\alpha^2S \ , \tag{4}$$

where Z_0 = 377 ohms, C = circumference, ρ = bending radius, r_p = classical proton radius = 1.535×10^{-18} m, α = crossing angle, L_{LR} = length over which two beams interact with long range beam-beam interaction. In this study, we ignore the collective instability limits, assuming they are taken care of by other means.

In the attempt to find the ultimate luminosity limit in the SSC, we fix the following quantities:

$$\beta^* = 0.35 \text{ m}, \ \alpha = 150 \text{ μrad}, \ L_{LR} = 170 \text{ m}, \ \rho = 10 \text{ km}, \ C = 85 \text{ km}, \ S = 5 \text{ m}.$$

The value of β^* assumes a higher peak IR triplet quadrupole gradient than that of the present design. The crossing angle has been opened up to the envisioned maximum. It is possible to consider variation of the bunch spacing S.[5] It turns out that, due to a scaling property, the same ultimate luminosity is obtained, although different combination of the other parameters will be used.

We consider a beam-beam limit imposed by the condition that the total tune shift induced by the beam-beam interactions be less than 0.02, which is about the value that can fit into a tune space without crossing low order resonances. In other words, we consider

$$\Delta v_{tot} = 4\,\Delta v_{HO} + 2\Delta v_{LR} < 0.02 \ . \tag{5}$$

The factors 4 and 2 in Eq. (5) follow from the fact that we assumed 2 high-luminosity-IP's and 2 medium-luminosity IP's, and the fact that Δv_{HO} is the same for high- and medium-luminosity IP's, while Δv_{LR} is significant only for high-luminosity IP's.

It turns out that, at low energies when condition (5) is dominating, the best luminosity is achieved when the head-on and the long-range components equally share the 0.02 allowance, i.e.,

$$4\,\Delta v_{HO} = 0.01 \tag{6a}$$

and

$$2\Delta v_{LR} = 0.01 \ . \tag{6b}$$

At high energies, we must consider the limit imposed by synchrotron radiation power per unit length

$$P/C < 0.4\,\text{W/m} \ . \tag{7}$$

In this energy range, the beam-beam limit is mainly imposed by the head-on component (6a); the long-range component will stay below the limit (6b). The present SSC design allows 0.1 W/m synchrotron radiation power. By strengthening cryogenic pumping and introducing more pumping locations, a higher synchrotron radiation does of 0.4 W/m may be tolerable.

Figures 1(a)-(f) show the energy dependence of the various parameters (note bunch spacing has been assumed fixed at 5 m). The transition between low- and high-energy behaviors occurs at 17.8 TeV. Below and above the transition energy, we have

	Below	Above
N	$\sim E$	$\sim E^{-4}$
ε_N	$\sim E$	$\sim E^{-4}$
P/C	$\sim E^5$	$\sim$ const
Δv_{HO}	$\sim$ const	$\sim$ const
Δv_{LR}	$\sim$ const	$\sim E^{-5}$
L	$\sim E^2$	$\sim E^{-3}$

The peak luminosity occurs at the transition energy and has the value of $2.4 \times 10^{34}\ \text{cm}^{-2}\,\text{s}^{-1}$. Luminosity at 20 TeV is $1.7 \times 10^{34}\ \text{cm}^{-2}\,\text{s}^{-1}$ as compared with the nominal value of $10^{33}\ \text{cm}^{-2}\,\text{s}^{-1}$.

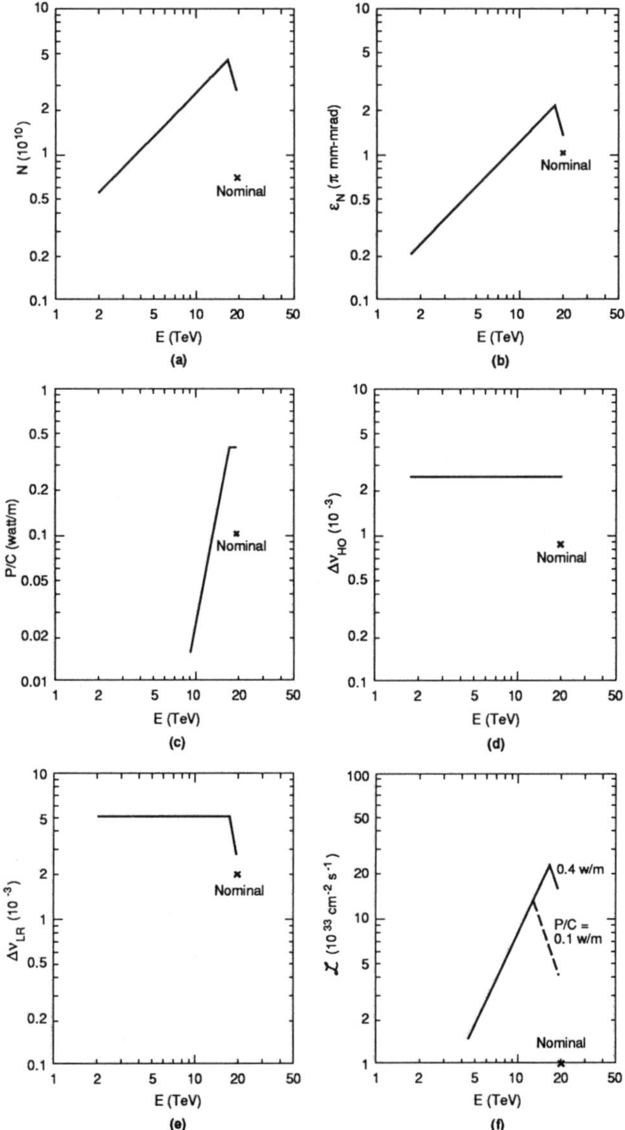

Figure 1. Energy dependence of various parameters. (a) number of protons per bunch, (b) normalized emittance, (c) synchrotron radiation power, (d) head-on tune shift, (e) long-range tune shift, and (f) luminosity. The nominal operation at 20 TeV are marked by crosses. In (f), a dotted curve representing 0.1 W/m limit has been added for reference.

It may be useful to mention a few additional considerations which seem to be potential for further increasing the ultimate luminosity of the SSC. One consideration concerns the beam-beam limit. Another concerns the crossing angle. Still one more concerns a different operating temperature of the superconducting magnets. These considerations are more speculative than the scenario discussed so far and have to be considered correspondingly.

It is conceivable that the beam-beam limit is less stringent than that assumed in Eq. (5). This is because the limit is likely to be imposed on the tune spread instead of the tune shift. In case of head-on collisions, the tune spread is equal to the tune shift, but in case of long-range collisions, tune spread is related to tune shift approximately by

$$\Delta v_{LR,\,spread} = \Delta v_{LR} \,(3/2)\,(n^2\,\epsilon_N / \alpha^2\,\gamma\,\beta^*) \, . \tag{8}$$

Consider a particle in the 6-sigma tail in its betatron amplitude, n=6, the nominal SSC has $\Delta v_{LR,\,spread}$ smaller than Δv_{LR} by about a factor of 4. If we consider the beam-beam limit to be Eq. (5) except for replacing Δv_{LR} by $\Delta v_{LR,\,spread}$, the ultimate luminosity in the lower energy regime will increase.

It turns out that in this case, the optimal sharing of the tune spread between the head-on and the long-range components are different from that of Eq. (6). The optimum occurs when

$$4\,\Delta v_{HO} = 0.015 \tag{9a}$$

and

$$2\Delta v_{LR,\,spread} = 0.005 \, . \tag{9b}$$

In this case, the luminosity in the low energy regime is about 50% higher than shown in Fig. 1(f). This possibility is shown in Fig. 2 as the dashed curve.

The ultimate luminosity depends sensitively on the crossing angle in the low energy regime due to the long range beam-beam effect. The 150 μrad crossing angle for the SSC comes from the long interaction region quadrupoles needed for a 20 TeV beam. If for some reason it becomes important to maximize the luminosity at a lower energy, it is possible to redesign the interaction region using shorter quadrupoles to allow a larger crossing angle. For example, doubling the crossing angle to 300 μrad would increase the ultimate luminosity in the low energy regime by a factor of 2 and 4, assuming the long-range limit due to Δv_{LR} and $\Delta v_{LR,\,spread}$, respectively. The dot-dash curve in Fig. 2 shows an increase of a factor of 4 below 10 TeV.

It is presumably possible to use the cryogenic power, not to absorb the synchrotron radiation in a luminosity upgrade, but to lower the operating temperature of the superconducting magnets. This would significantly complicate the magnet design, but assuming the magnet issues are all resolved, one could imagine lowering the temperature from the nominal 4.35 to 3 degree Kelvin, which translates into a field increase from the nominal 6.6 to perhaps 7.7 Tesla, allowing an operation at 23 TeV per beam. The corresponding luminosity curve would look perhaps like that shown as the dotted curve in Fig. 2.

One should also note that the considerations below for further increasing the ultimate luminosity are not meant to be exhaustive. A vacuum pipe liner would substantially reduce the impact of synchrotron radiation. Alternating the beam crossings horizontally and vertically would reduce the long-range beam-beam effects. Allowing for fewer IP's will make the luminosity at each IP's higher. Crossing the beam with a much larger angle could be possible if one uses 2-in-1 design of interaction region quadrupoles[6] Use of pre-splitters that separate the beam would help long-range encounters.[7]

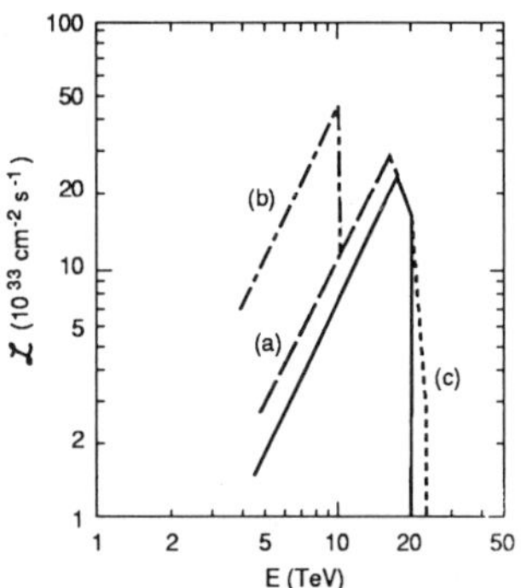

Figure 2. Three speculations that might lead to higher ultimate luminosities than shown in Figure 1(f), which is reproduced here as the solid curve. Curves (a): using $\Delta\nu_{LR, \text{spread}}$ instead of $\Delta\nu_{LR}$. Curve (b): with 300 µrad crossing angle below 10 TeV. Curve (c): operating the magnets at 3 degree Kelvin.

References
1. SSC Conceptual Design Report, March, 1986.
2. SSC Site-Specific Conceptual Design, 1990.
3. R. Diebold, Snowmass Proceedings on Physics of the SSC, 1986, p. 585.
4. D. Bintinger, SSC-60, 1988.
5. R. Schwitters, private communications, 1989.
6. R. Palmer, private communication, 1989.
7. R. Diebold, private communication, 1989.

RESEARCH AT SLAC TOWARDS
A 0.5 TeV LINEAR COLLIDER*

RONALD D. RUTH

Stanford Linear Accelerator Center
Stanford University, Stanford, California 94309

1. INTRODUCTION

The purpose of this paper is to review the ongoing research at SLAC towards a next generation linear collider (NLC). The energy of the collider is taken to be 0.5 TeV in the CM with a view towards upgrading to 1.0 TeV. The luminosity is in the range of 10^{33} to 10^{34} cm^{-2}sec^{-1}. The energy is achieved by acceleration with a gradient of about a factor of five higher than SLC, which yields a linear collider approximately twice as long as SLC. The detailed trade-off between length and acceleration will be based on total cost. A very broad optimum occurs when the total linear costs equal the total cost of RF power.

The luminosity of the linear collider is obtained primarily in two ways. First, the cross-sectional area of the beam is decreased primarily by decreasing the vertical size. This creates a flat beam and is useful for controlling beamstrahlung. Secondly, several bunches ($\sim$10) are accelerated on each RF fill in order to more efficiently extract energy from the RF structure. This effectively increases the repetition rate by an order of magnitude.

An overall layout and brief table of important parameters are shown in Fig. 1 and Table 1, respectively. In the next several sections, we trace the beam through the collider to review the research program at SLAC. More details of ongoing work at SLAC and throughout the world can be found in Refs. 1-5.

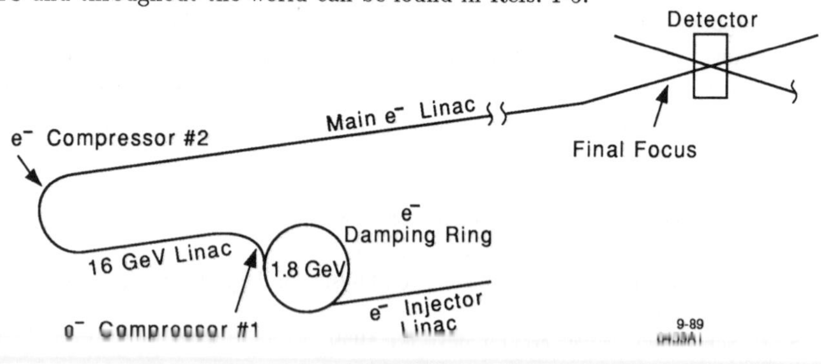

Fig. 1.   Schematic of a Next Linear Collider.

* Work supported by the Department of Energy, contract DE–AC03–76SF00515.

Table 1. Parameters for a 0.5 TeV NLC.

Parameter	Value
CM energy	0.5 TeV
Luminosity 10^{33}	3.9 cm^{-2} sec^{-1}
RF frequency	11.4 GHz
Repetition rate	360.0 Hz
Acceleration gradient	93.0 MV/m
Number of bunches	10
Particles/bunch (at IP)	1.6×10^{10}
Length	7.2 Km
β_y^*	0.08 mm
Crossing angle (no crab)	4.8 mrad
Crab crossing angle	30–100 mrad
σ_y	3.1 nm
σ_x/σ_y	180
σ_z	70.0 μm
Disruption D_y	10.0
Luminosity enhancement H	1.5
Beamstrahlung δ	6%

2. DAMPING RINGS

In Refs. 6 and 7, T. Raubenheimer *et al.* discuss many of the basic design considerations for the damping ring. The basic parameters are shown in Table 2 where they are compared to those of the SLC. The key differences are the decrease of the horizontal emittance by an order of magnitude, the increase of the repetition rate, and the requirement of $\epsilon_x/\epsilon_y = 100$. Although asymmetrical emittances have been measured in the SLC damping ring, they are not required for SLC operation.

The desired repetition rate is obtained by having many batches of bunches in the ring. Each batch of 10 bunches is extracted on one kicker pulse and accelerated on one RF fill in the linac. The remaining batches are left in the ring to continue damping, while an additional batch is injected to replace the extracted one.

The basic layout of a possible damping ring is shown in Fig. 2. Notice that there are several insertions which contain wigglers. In order to obtain the high repetition rate, it may be necessary to decrease the damping time by the addition of wigglers in straight sections. However, if the desired repetition rate is decreased by a factor of two, it is probably not necessary to include wigglers to decrease the damping time.

Table 2. Basic parameters of NLC and SLC damping rings.

	NLC	SLC
Energy, E_0	1.8 GeV	1.15 Gev
Length, L	155.1 m	35.3 m
RF frequency, f_{RF}	1.4 GHz	714 MHz
Repetition rate	360 Hz	120 Hz
Emittance, $\gamma\epsilon_x$	3.0 μmrad	36 μmrad
Emittance, $\gamma\epsilon_y$	30 nmrad	500 nmrad
Damping time, τ_x	2.50 ms	3.4 ms
Damping time, τ_y	3.98 ms	3.4 ms
Energy spread, σ_ϵ	0.0010	0.0007
Bunch length, σ_z	5.2 mm	5 mm[8]
Vertical size	3 μm	20 μm
Horizontal size	30 μm	100 μm

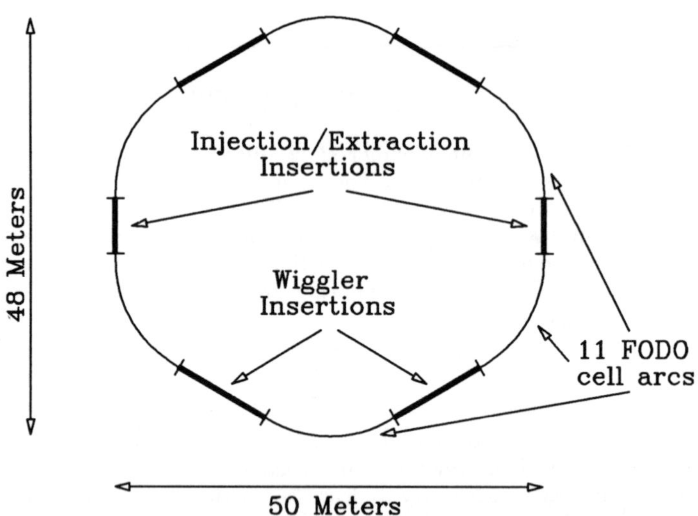

Fig. 2. Schematic of an NLC damping ring.

3. BUNCH COMPRESSION AND PRE-ACCELERATION

In order to obtain the very short bunches necessary for the linac, it is necessary to perform at least two bunch compressions after the damping ring. Designs for such bunch compression are presented in Refs. 9 and 10. A bunch length of about 50 μm in the linac puts a constraint on the longitudinal emittance of the damping ring. In addition, during the bunch compressions, it is necessary to keep the energy spread small to avoid the dilution of the transverse emittance. If we assume that we can transport 1% energy spread without diluting either transverse emittance, then at least two bunch compressions are needed. For example, if we consider a 1.8 GeV damping ring with an energy spread of $\Delta E/E = 10^{-3}$ and a bunch length of 5 mm, the two compressions are shown in Table 3. The first one decreases the bunch length by an order of magnitude. This is followed by a pre-acceleration section to decrease the relative energy spread in the beam by about an order of magnitude. One must avoid an increase of energy spread due to the cosine of the RF wave (and also due to beam loading). If this pre-acceleration is done at the present SLAC frequency and if the bunch intensity is in the range $1 - 2 \times 10^{10}$, then the additional energy spread induced is quite small. Neglecting this small increase, the next bunch compression happens around 18 GeV and serves to reduce the bunch length to about 50 μm. This is suitable for injection into the high frequency, high gradient structure.

Table 3.   Bunch compression.

E	$\Delta E/E$	σ_z	Compress $\rightarrow$	$\Delta E/E$	σ_z
1.8 GeV	10^{-3}	5 mm	Compress $\rightarrow$	10^{-2}	0.5 mm
[pre-acceleration at long wavelength, $\lambda = 10.5$ cm]					
18 GeV	10^{-3}	0.5 mm	Compress $\rightarrow$	10^{-2}	50 μm

Reference 10 presents several designs in which the bend angle for the final compressor is 180° as shown in Fig. 1. This low energy bend allows easy upgrades in energy, and also makes it possible to do direct feedback to compensate jitter from the damping ring kicker magnet.

4. LINAC

The linac is envisioned to be similar to the SLAC disk-loaded structure with a frequency of four times the present SLAC frequency. The irises in the design are somewhat larger (relative to the wavelength) to reduce transverse wakefields. The structure may have other modifications to damp long-range transverse wakefields. This would be driven by a power source capable of about 220 MW/m in order to obtain an accelerating field of 100 MV/m.

The remainder of this section is divided into two subsections. In the first subsection we discuss structures, while the second deals with RF power sources.

4.1 STRUCTURES

Since the acceleration gradients being considered range from 100 MV/m to 200 MV/m, the first question that arises is RF breakdown. This question is treated in Refs. 11 and 12. In these papers, G. Loew and J. Wang present results from many experiments at various frequencies. If the scaling laws thus obtained are extrapolated to 11.4, the breakdown-limited surface fields obtained are 660 MV/m. To convert this to an effective accelerating gradient, a reduction factor of 2.5 is typically used, which yields an accelerating gradient above 200 MeV/m. However, the measurements also indicated significant "dark currents" generated by captured field-emitted electrons. The question of the effects of dark current on loading and beam dynamics is not yet resolved and needs further study.

As mentioned in the Introduction, in order to make efficient use of the RF power and to achieve high luminosity, it seems essential to accelerate a train of bunches with each fill of the RF structure. This leads to two problems: (1) the energy of the bunches in the train must be controlled and (2) the transverse stability of the bunch train must be ensured. Both of these problems are helped greatly by damping higher modes (both transverse and longitudinal) in the RF structure. In Ref. 13, R. Palmer describes a technique of using slotted irises coupled to radial waveguides to damp these modes: Q's as low as 10–20 have been measured in model structures. This encouraging evidence has led to a development program at SLAC and KEK to do more detailed studies of slotted structures. Recently at SLAC, Q's as low as 8 have been measured in low-power models at 2.8 GHz.[14,15] The beam dynamics consequences of damping the higher modes are explored in Section 6.

4.2 RF POWER SOURCES

Before discussing results on power sources, it is useful to contrast and compare two basic approaches, RF pulse compression and magnetic pulse compression.

4.2.1 RF Pulse Compression and Conventional Klystrons

In Fig. 3(a), you see illustrated the basic principle of RF pulse compression. A long modulator pulse is converted by a high-power, "semi-conventional" klystron or some other power source into RF power with the same pulse width. This RF pulse is then compressed by slicing the pulse using phase shifts and 3 db hybrids and re-routing the portions through delay lines, so that they add up at the end to a high peak power but for a small pulse width. This scheme was invented by D. Farkas at SLAC and is presently under experimental investigation.[16] With a factor of 6 in pulse compression, a 100 MW klystron could power about 3 m of structure to achieve an accelerating gradient of 100 MV/m.

In Ref. 17, P. Wilson describes RF pulse compression in some detail including estimates of efficiencies. An experimental test at SLAC of a low-loss, low-power system has been completed which yielded a factor of 3.2 power gain.[18] Recently, a high power RF pulse compressor, designed to yield a power gain of 6, has been completed at SLAC. This has been tested at low power and has achieved a gain

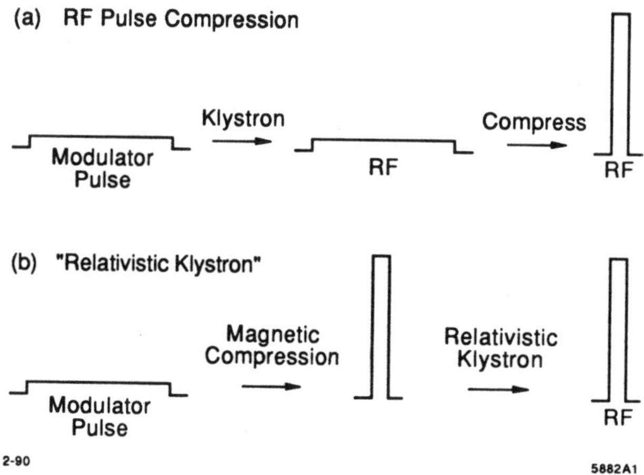

Fig. 3a. Illustration of RF pulse compression. 3b. Illustration of the relativistic klystron with magnetic compression.

of 5.5.[19] This will be powered with a high-power 11.4 GHz klystron designed to yield 100 MW peak power. An initial prototype of this klystron has achieved 65 MW in short pulses and 25 MW in long pulses.[20]

4.2.2 Magnetic Pulse Compression and the Relativistic Klystron

In Fig. 3(b), you see the principle of magnetic pulse compression and the relativistic klystron illustrated. In this case, the pulse compression happens *before* the creation of RF. This technique makes use of the pulsed power work done at LLNL in which magnetic compressors are used to drive induction linacs to produce multi-MeV e^- beams with kiloampere currents for pulses of about 50 nsec. These e^- beams contain gigawatts of power. The object, then, is to bunch the beam at the RF frequency and then to extract a significant fraction of this power. This can be done either by velocity modulation or by dispersive magnetic "chicanes." After bunching, the beam is passed by an RF extraction cavity, which extracts RF power from the beam.

Experiments on the relativistic klystron are described in Ref. 21. The best power output achieved to date is 330 MW.[20] Although higher acceleration gradients have been achieved, the best breakdown-free acceleration gradient in this experiment is 84 MV/m with 80 MW of RF power input into a 30 cm long accelerator structure.

4.2.3 Other RF Sources

It is also possible to consider other sources driven by magnetic pulse compressors which directly produce short, high-power RF pulses. One example is a cross-field amplifier (CFA). This device has the geometry of a magnitron but is configured as an amplifier rather than an oscillator. SLAC has completed the construction of a CFA designed to produce 100 MW. This initial prototype has achieved about 10 MW peak output power. Although it is a large extrapolation from existing sources, it holds the promise of being less expensive than an equivalent power klystron.

Another interesting possible RF source is the cluster klystron. In Ref. 22, R. Palmer and R. Miller describe a multiple beam array of "klystrinos" which when coupled together can give impressive results. By dividing a single beam into many beams shielded from each other, the problems of space charge are effectively eliminated. This source could be used as a driver for RF pulse compression. Alternatively, with the addition of a grid and an oil-filled transmission line for energy storage, the device could directly produce short RF pulses. Thus far, there has been no experimentation; but calculations and cost estimates are encouraging.

To conclude this section, if high-power tests of RF pulse compression show positive results, and if the 100 MW klystron achieves its design power, this combination could provide a power source for a Next Linear Collider.

5. FINAL FOCUS

The final focus, as described in the parameters in Table 1, is a flat beam final focus with a crossing angle. The purpose of the flat beam is to increase the luminosity while controlling beamstrahlung and disruption. The crossing angle is to allow different size apertures for the incoming and outgoing beam. Another invention, "crab-wise crossing", discussed in Ref. 23, allows a much larger crossing angle than the diagonal angle of the bunch. As discussed in Ref. 23 and in Ref. 24, this type of geometry may now be essential due to the production of e^+e^- pairs by beamstrahlung photons in the field of the bunches.

5.1 FINAL FOCUS OPTICS AND TOLERANCES

The first job in the final focus is to demagnify the beam to provide a small spot for collision. A design for such a system is presented in Ref. 25 by K. Oide. This is a flat beam final focus which achieves the parameters shown in Table 1 for vertical and horizontal beam size. The vertical size is limited by a fundamental constraint, the "Oide limit", due to the synchrotron radiation in the final doublet coupled to the chromatic effect of a quadrupole. The quadrupole gradients necessary are very high, and in Oide's design are obtained by conventional iron magnets with 1 mm pole-to-pole distance. Tolerances are very tight in such a final focus. The most restrictive vibration tolerance is on the final doublet which must be stable pulse-to-pulse to about 1 nm.

Since vibration of the final doublet is the most serious problem, it is considered in some detail in Ref. 26. In this paper, it is shown that passive vibration isolation seems to be more than adequate to handle the vibrations above 10 Hz at the high frequency end. For low frequencies, an interferometric feedback system can be used to control motion down to about 1 μm. Beam steering feedback can then be used to control slow variations in the 1 nm to 1 μm region.

5.2 BEAM-BEAM EFFECTS

When a small bunch of electrons collides with a small bunch of positrons, the fields of one bunch focus the other causing disruption. Since the opposing particles are strongly bent, they also emit radiation called beamstrahlung. These are the two basic beam-beam effects. The disruption enhances the luminosity by a small amount, while the beamstrahlung causes significant energy loss during collision and increases the effective momentum spread for physics. These issues are discussed in detail in Ref. 24.

In addition, there are several other important effects which should be mentioned here. If the beams are offset relative to each other, a kink instability develops. For small disruption, this effect actually causes the luminosity to be less sensitive to offsets because the beams attract each other and collide anyway. There is also a multibunch kink instability, which is more serious since it can cause the trailing bunches to miss each other entirely. This places restrictions on the product of the vertical and horizontal disruption per bunch.

As mentioned earlier in the Introduction, it has been discovered that beamstrahlung photons pair-produce in the coherent field of the bunch.[27-30] The corresponding incoherent process has been known for some time, but its importance has only just been realized.[31] The problem is that low energy e^+e^- pairs are produced in an extremely strong field, which then deflects the charge of the appropriate sign while confining the other.

These stray particles can lead to background problems, which must be addressed by detailed interaction point design. In Ref. 23, it is suggested that crab-crossing combined with large crossing angles and solenoidal fields would allow one to channel these electrons out through a large exit hole to a beam dump. Further studies of this option have indicated that crossing angles in the range 30–100 mrad would be necessary to avoid particle impact in the detector.[32]

The measurement of the final spot size is an extremely important, but as yet unsolved, problem. From SLC experience, it is probably possible to use beam-beam effects to minimize spot sizes. However, for the initial tune-up of the final focus, a single-beam method is almost essential. Initial studies of this problem use the energy and angular distribution of ions from a gas jet which is intercepted by the beam.[33] However, there is clearly much more work needed in this area.

6. MULTIBUNCH EFFECTS

As mentioned earlier, in order to efficiently extract energy from the RF to obtain high luminosity, it is essential to have many bunches per RF fill. This, however, leads to transverse beam breakup. The invention of damped structures discussed in Section 5.1 helps but may not completely solve the problem for the linac. It may also be necessary to tune the frequency of the first dipole mode of the accelerating structure.[34] In Ref. 35 the problem of multibunching is traced all the way through the linear collider, subsystem by subsystem. Damped accelerating cavities are required for the main linac and the damping rings, while other systems can get by with very strong focusing. Thus, from the transverse point of view, stability seems possible.

In addition, it is necessary to control the energy spread from bunch to bunch very precisely ($\Delta E/E \lesssim 10^{-3}$). This can be accomplished by injecting the bunches before the RF structure is full to match the extraction of energy by the bunches to the incoming energy as the structure fills. This leads to tight tolerances on phase and amplitude of the RF, as well as tight control of the pulse-to-pulse number of particles in a batch of bunches.[36] However, the benefits of multibunching seem to far outweigh any difficulties they impose due to the order of magnitude increase in luminosity.

7. OUTLOOK

During the past few years, there has been tremendous progress towards a next generation linear collider. We now have a much clearer picture of how to obtain both the energy and luminosity required. An important development is the increased interest in a linear collider with 0.5 TeV in the CM which would be upgradable to 1.0 TeV with additional power sources or length. We will probably see the development of a power source and structure during the next couple of years. This would yield the energy of the collider; what about the luminosity?

Designs of damping rings, bunch compressors and final focus systems will continue. Studies of BNS damping in the linac and emittance dilution will continue both experimentally with the SLC and theoretically for the next generation high-frequency linac. However, to really understand tolerances, new measurement techniques, and final focus optics, it is probably essential to build a scale model final focus at SLC energy. This is being planned at SLAC (Final Focus Test Beam) and is being supported as a collaborative effort of SLAC, INP, KEK, Orsay, and DESY.

One key aspect of all linear collider design is background control. This problem is complicated by pair production during the bunch crossing; however, detailed studies of interaction point design are underway at SLAC, KEK, and INP.

To conclude, it looks like we are on the path towards a next generation linear collider, and with proper funding of R&D over the next few years we may see a detailed conceptual design in the mid-990's.

REFERENCES

1. *Proceedings of the Workshop on Physics of Linear Colliders*, Capri, Italy, June 1988.

2. *Proceedings of the Summer Study on High Energy Physics in the 1990's*, Snowmass, Colorado, July 1988, World Scientific, Singapore, 1989.

3. *Proceedings of the International Workshop on Next Generation Linear Colliders*, SLAC, Stanford, CA, Dec. 1988, SLAC-Report-335.

4. *Linear Collider Working Group Reports From Snowmass '88*, Ed. R. D. Ruth, SLAC-Report-334.

5. *Proceedings of the 1990 Workshop on Next Generation Linear Colliders*, KEK, Tsukuba, Japan, March 1990.

6. T. O. Raubenheimer, L. Z. Rivkin and R. D. Ruth, *Damping Ring Designs for a TeV Linear Collider*, SLAC–PUB–4808 (1988) and in Refs. 2 and 4.

7. T. O. Raubenheimer, *et al.*, *A Damping Ring Design for Future Linear Colliders*, Proc. of 1989 IEEE Part. Acc. Conf., p. 1316, and in SLAC–PUB–4912 (1989).

8. L. Z. Rivkin *et al.*, *Bunch Lengthening in the SLC Damping Ring*, SLAC–PUB–4645 (1988).

9. S. A. Kheifets, R. D. Ruth, J. J. Murray and T. H. Fieguth, *Bunch Compression for the TLC. Preliminary Design*, SLAC–PUB–4802 (1988) and in Refs. 2 and 4.

10. S. A. Kheifets, R. D. Ruth and T. H. Fieguth, *Bunch Compression for the TLC*, Proc. of Int. Conf. on High Energy Acc. Tsukuba, Japan, 1989 and in SLAC–PUB–5034 (1989).

11. G. A. Loew and J. W. Wang, *RF Breakdown Studies in Room Temperature Electron Linac Structures*, SLAC–PUB–4647 (1988) and in Refs. 2 and 4.

12. G. A. Loew and J. W. Wang, *Field Emission and RF Breakdown in Copper Linac Structures*, Proc. of the 14th Int. Conf. on High Energy Acc., Tsukuba, Japan (1989) and in SLAC–PUB–5059.

13. R. B. Palmer, *Damped Accelerator Cavities*, SLAC–PUB–4542 (1988) and in Refs. 2 and 4.

14. H. Deruyter, *et al.*, *Damped Accelerator Structures*, Proceedings of the 2nd European Part. Acc. Conf., June, 1990.

15. N. M. Kroll and D. U. L. Yu, *Computer Determination of the External Q and Resonant Frequency of Waveguide Loaded Cavities*, SLAC–PUB–5171, to be published.

16. Z. D. Farkas, *Binary Peak Power Multiplier and its Application to Linear Accelerator Design*, IEEE Transcripts on Microwave Theory and Techniques, **MTT–34**, No. 10 (1986) 1036, and also SLAC–PUB–3694.

17. P. B. Wilson, *RF Pulse Compression and Alternative RF Sources*, SLAC–PUB–4803 (1988) and in Refs. 2 and 4.

18. Z. D. Farkas, G. Spalek, and P. B. Wilson, *RF Pulse Compression Experiment at SLAC*, Proc. of IEEE Part. Acc. Conf. 1989 and in SLAC–PUB–4911 (1989).

19. T. L. Lavine, *et al.*, *Binary RF Pulse Compression Experiment at SLAC*, Proceedings of the 2nd European Part. Acc. Conf., June 1990, and in SLAC–PUB–5277

20. M. A. Allen, *et al.*, *RF Power Sources for Linear Colliders*, Proceedings of the 2nd European Part. Acc. Conf., June 1990, and in SLAC–PUB–5274.

21. M. A. Allen *et al.*, *High Gradient Electron Accelerator Powered by a Relativistic Klystron*, Phys. Rev. Lett. **63** (1989) 2472.

22. R. B. Palmer and R. Miller, *A Cluster Klystron*, SLAC–PUB–4706 (1988) and in Refs. 2 and 4.

23. R. B. Palmer, *Energy Scaling, Crab Crossing and the Pair Problem*, SLAC–PUB–4707 (1988) and in Refs. 2 and 4.

24. P. Chen, *Disruption, Beamstrahlung, and Beamstrahlung Pair Creation*, SLAC–PUB–4822 (1988) and in Refs. 2 and 4.

25. K. Oide, *Final Focus System for TLC*, SLAC–PUB–4806 (1988) and in Refs. 2 and 4.

26. W. W. Ash, *Final Focus Supports for a TeV Linear Collider*, SLAC–PUB–4782 (1988) and in Refs. 2 and 4.

27. R. Blankenbecler, S. D. Drell, and N. Kroll, *Pair Production From Photon Pulse Collisions*, Phys. Rev. D, **40**, 2462 (1989) and SLAC–PUB–4954.

28. P. Chen and V. I. Telnov, *Coherent Pair Creation in Linear Colliders*, Phys. Rev. Lett., **63**, 1796 (1989) and SLAC–PUB–4923.

29. M. Jacob and T. T. Wu, *Pair Production in Bunch Crossing* , Phys. Lett. B, **221**, 203 (1989).

30. V. N. Baier, V. M. Katkov, and V. M. Strakhovenko, *Proc. 14th International Conf. on High Energy Particle Accelerators*, Tsukuba, Japan, 1989.

31. M. S. Zolotarev, E. A. Kuraev and V. G. Serbo, *Estimates of Electromagnetic Background Processes for the VLEPP Project,* Inst. Yadernoi Fiziki, Preprint 81-63, 1981; English Translation SLAC TRANS-0227, 1987.

32. J. Irwin, Private communication.

33. J. Buon, *Proceedings of the 14th International Conference on High Energy Accelerators*, Tsukuba, Japan, 1989.

34. K. A. Thompson and R. D. Ruth, *Controlling Transverse Multibunch Instabilities in Linacs of High Energy Linear Colliders*, Phys. Rev. D, **41**, 964 (1990), and in SLAC–PUB–4801 (1989).

35. K. A. Thompson and R. D. Ruth, *Multibunch Instabilities in Subsystems of 0.5 and 1.0 TeV Linear Colliders*, SLAC–PUB–4800 (1988) and in Refs. 2 and 4.

36. R. D. Ruth, *Multibunch Energy Compensation*, SLAC–PUB–4541 (1989) and in Ref. 1.

The Brookhaven Accelerator Test Facility

K. Batchelor, I. Ben-Zvi, R.C. Fernow, J. Fischer, A.S. Fisher, J. Gallardo,
Xie Jialin, H.G. Kirk, Z. Parsa, R.B. Palmer, T. Rao, J. Rogers, J. Sheehan,
T.Y.F. Tsang, S. Ulc, A. Van Steenbergen, M. Woodle, R.S. Zhang
Brookhaven National Laboratory
Upton, NY 11973

K.T. McDonald   D.P. Russell
Princeton University
Princeton, NJ 08544

Z.Y. Jiang
SUNY, Stony Brook
Stony Brook, NY 11794

C. Pellegrini   X.J. Wang
UCLA
Los Angeles, CA 90024

Presented by D.P. Russell

Abstract

The Accelerator Test Facility (ATF), presently under construction at
Brookhaven National Laboratory, is described. It consists of a 50-MeV
electron beam synchronizable to a high-peak-power CO_2 laser. The interaction of electrons with the laser field will be probed, with some emphasis
on exploring laser-based acceleration techniques.

I. Introduction

The present generation of charged-particle accelerators, based on RF-cavities,
is rapidly approaching the limits set by size and technology. Significant increases
in beam energy and quality will likely require new acceleration techniques. One
possibility is to make use of coherent electromagnetic radiation of shorter wavelength
than radio-frequency waves. Optical and infrared lasers are natural candidates for
supplying such accelerating fields.

The Accelerator Test Facility is being constructed by the Center for Accelerator
Physics at Brookhaven to provide electron and laser beams of sufficient quality for

the study of their interaction: A major theme is to be the demonstration of certain laser-based acceleration schemes. Central to the construction of the ATF is the production and transport of a high-brightness (low-emittance) electron beam, an important issue for present and future accelerators. The ATF is to be a user facility at which experiments can be performed in several different beamlines.

II. The Facility

The Accelerator Test Facility consists of the following components:

1. 50-MeV high-brightness electron beam
    a. 4.5-MeV high-brightness photocathode RF gun
    b. 50- to 100-MeV S-band linac
2. Synchronizable laser system
    a. Nd:YAG laser with 6-picosecond pulse length
    b. CO_2 laser with 6-psec pulse length and 20-GW peak power.

A schematic diagram of the ATF is shown in Figure 1.

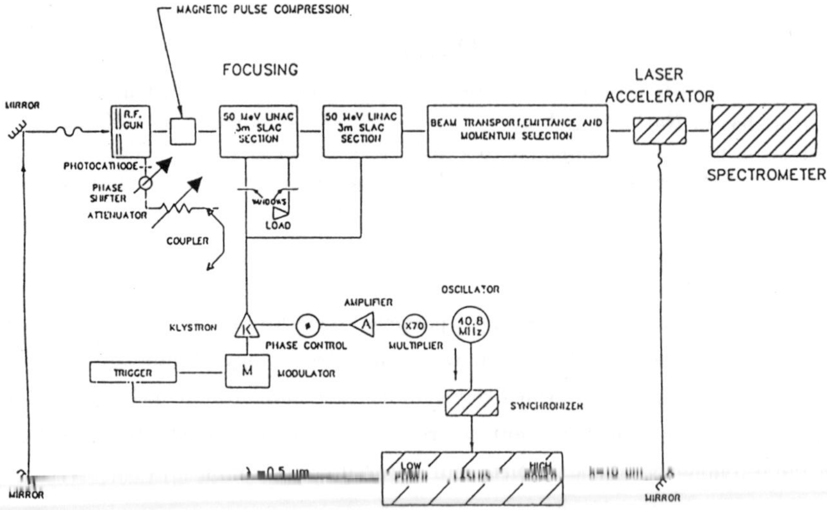

Figure 1. Schematic diagram of the ATF.

The RF gun[1] is shown in Fig. 2. It is based on a design[2] in which a photocathode is incorporated into the end wall of an RF cavity. The Yttrium photocathode is illuminated by a YAG laser pulse which is synchronized with the RF system. The transverse and longitudinal phase spaces of the resulting electron pulse can then be altered by varying the laser spot size and duration. A small invariant transverse emittance may be achieved in this manner. The gun was first operated during the Summer of 1989, during which time an emittance of about 8π mm-mrad was measured.

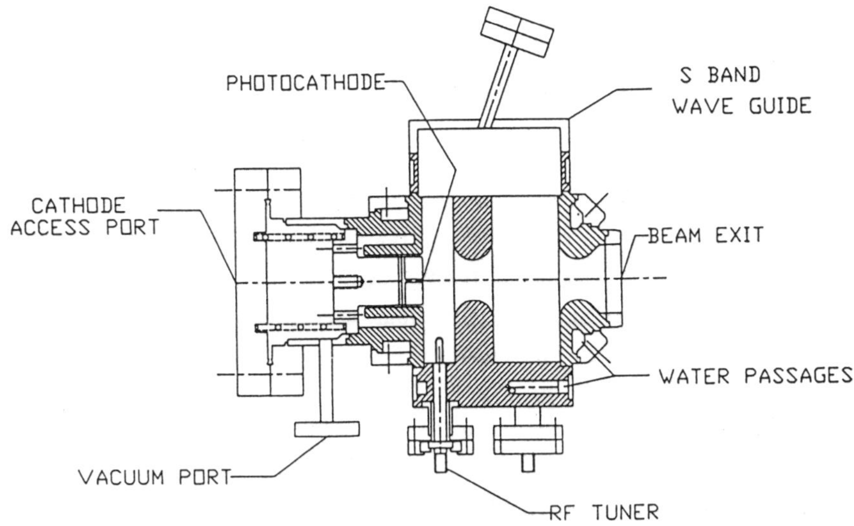

Figure 2. The RF gun of the Brookhaven Accelerator Test Facility.

The line transporting the beam from the gun for injection into the linac, as well as the post-linac transport to the experimental stations must be designed and constructed so as to minimize emittance growth. This is a major challenge, and the initial running phase at the ATF is concentrated on diagnosing the beam quality at various points along the beamline. A beam-profile monitor[3] is used to measure transverse emittances, and in conjunction with an RF kicker and dipole bending magnet, should allow diagnosis of the longitudinal phase space as well.

Another challenge is the synchronization of the electron beam with the laser beams on the picosecond scale. The RF system and YAG laser are driven by a common master oscillator. The pulse from the mode-locked YAG is shortened to 6 picoseconds by fiber-optic chirping[4] and then drives the photocathode. Part of this pulse also trims the CO_2 laser pulse to 6 picoseconds for use with experiments.

Experimental Program

The following is a partial list of experiments to be run at the ATF.[5]

- Smith-Purcell radiation.
- Laser-grating accelerator experiment.
- Inverse Čerenkov accelerator experiment.
- Visible free-electron laser.
- Nonlinear Compton scattering experiment.

Beam through the linac is expected by Spring, 1990. The first experiments are planned for Fall, 1990.

References

1. K. Batchelor *et al.*, in Proceedings of the European Particle Accelerator Conference, 1988; K.T. McDonald, IEEE Trans. Elec. Dev. **35**, 2052 (1988).
2. J.S. Fraser and R.L. Sheffield, IEEE Jour. Quant. Elect. **23**, 1489 (1987).
3. D.P. Russell and K.T. McDonald, in Proceedings of the 1989 IEEE Particle Accelerator Conference, p. 1510.
4. T. Shimada *et al.*, contributed to the Conference on Lasers and Électrooptics, 1989.
5. Proposals presented at the ATF User's Meeting, Brookhaven National Laboratory, November, 1989.

High Luminosity, Electron-Positron Colliders as Strangeness, Charm, and Beauty Factories[+]

William A. Barletta[*]

Department of Physics, UCLA,
and
Lawrence Livermore National Laboratory

ABSTRACT

High luminosity electron-positron colliders operating at the mass of the ϕ meson (1.02 GeV) can produce copious $K\overline{K}^o$ pairs from a single quantum state. Temporal correlations in the decays of the K's provide a measure of the direct CP violating amplitude and also allow a high precision test of CPT invariance. A low energy collider with high luminosity can serve as a beam physics testbed to evaluate novel approaches to collider design that may be necessary for B factories to attain luminosities $\geq 10^{34}\,cm^{-2}s^{-1}$.

I. INTRODUCTION

Colliding beam accelerators have become the mainstay in advancing the high energy frontier of elementary particle physics. Beyond the upgrade of LEP to 100 GeV per beam, the extension of e^+-e^- colliders to still higher energies will take the form of linear accelerators producing sub-micron beams. In contrast, the study of rare processes such as CP violation in the B meson system, or the high precision measurement of CP violation parameters in the K system, or high precision tests of CPT invariance are likely to be performed with a new generation of e^+-e^- storage ring colliders operating at extremely high luminosity (flavor factories). The high collision rates and high beam currents implied by the combination of high luminosity and low-to-moderate beam energy introduce a complex of accelerator design difficulties.

The lowest energy flavor factory operates at the formation energy of the ϕ meson (1.02 GeV). Several groups throughout the world are actively considering the design of a ϕ factory with a luminsoty $>10^{32}\,cm^{-2}s^{-1}$. One group (UCLA) is aiming at a configuration sufficiently flexible to permit the exploration of novel approaches to realizing very high luminosity.

I.A. Physics motivation

The $K\overline{K}^o$ system has been the most sensitive "laboratory" for tests of discrete symmetries, having provided conclusive evidence[1] of CP violation in the two π decay modes of the K_L. Twenty five years later, the physical origin of the violation is still not understood. Present experimental evidence for CP violation can be described in terms of a single, complex mixing parameter, ε, that expresses the mass eigenstates, K_L and K_S, in terms of the CP eigenstates, K_1 and K_2. Direct (not mass-matrix) sources of CP violation imply a second complex mixing parameter, $\varepsilon' \leq O(\varepsilon^2)$, that leads to differences in the amplitudes for K_L decay into $\pi^+\pi^-$ and $\pi^o\pi^o$. Recent measurements at CERN[2] find that

[+] Invited paper, Meeting of the American Physical Society Div. of Particles and Fields, Jan, 1990
[*] Work partially performed under the auspices of LLNL under DOE contract W-7405-eng-48.

$\varepsilon'/\varepsilon \approx 3 \times 10^{-3}$; in contrast, the E731 group at Fermilab reports that ε'/ε is consistent with zero to within an accuracy of 10^{-3}. This conflict suggests the utility of measuring CP violation and of CPT invariance directly in a system with different systematics and one in which assumptions about the relative phase of the mixing parameters are unneccessary.

An especially sensitive probe[3] of CP physics is provided by measuring the decays of the neutral kaons produced in the process

$$e^+ e^- \to \phi\ (1020) \to K_L\ K_S. \qquad (1)$$

The quantum numbers of the $\phi(1020)$ meson ($J^{PC} = 1^{--}$) ensure that only $K_L\ K_S$ are produced in the final state even when CP (or CPT) is violated in the kaon system. By measuring asymmetries in the distribution of times of the decays $K_L \to \pi^+\pi^-$ (at time t_1) and $K_S \to \pi^0\pi^0$ (at t_2) near $(t_2 - t_1) = 0$ and by measuring the branching ratio of the decay into all charged versus all neutral pions, one determines directly both $\mathrm{Im}(\varepsilon'/\varepsilon)$ and $\mathrm{Re}(\varepsilon'/\varepsilon)$ respectively.

Knowing the cross section for ϕ production ($\approx 2.4\ \mu$b) and the branching ratio for the decay of the ϕ into $K_L\ K_S$ (34%), one estimates the range of luminosity of an e^+-e^- collider neccessary to obtain $10^{10} - 10^{11}$ clean ϕ events per year to be 2×10^{32} to 2×10^{33} cm^{-2}s^{-1}. This range can yield an effective measurement of ε'/ε of order 10^{-4}. By measuring the ratio of decay into two pion channels versus that into semi-leptonic channels, one can determine the relative mass difference between the K and $\overline{K}^0$ to better than 10^{-20}. The sensitivity of the experiment is limited by the background channel, $\phi \to K\overline{K}^0\gamma$, which may have a branching ratio[4] as large as 10^{-6}.

Other physics accessible in this luminosity range include study of CP violation in decays of the K$^+$ and K$^-$, search for CP violation in K^0_L and $K^0_S \to \gamma\gamma$ (to 10^{-6} in branching ratio) and in K^0_L and $K^0_S \to 3\pi^0$ (to 10^{-8} in branching ratio). Measurements of the branching ratio at low energy of $e^+e^- \to \omega/\rho$ are important for calculations of $(g\text{-}2)_\mu$.

I.B. Detector considerations

Typically detector requirements will tightly constrain the design of flavor factories. In a ϕ factory one need not record each ϕ formation event; with proper triggering of the detector only 10^4 to 10^5 per 10^{10} ϕ events need actually be recorded for kinematic reconstruction. For these events one must measure the ϕ formation and decay at a rate of 10 to 100 Hz with good particle identification, efficiency, vertex measurement, and background rejection over a range of ≈ 10 K_S lifetimes (10cm). Thus the detector will need a uniform B field, tracking with high spatial resolution, minimal material in the tracking volume to minimize multiple Coulomb scattering of the pions, precise energy determination plus some directional information for the γ's. Indentification of the μ, π, e is essential. Fortunately, the decay of the ϕ into K$^+$ and K$^-$ allows for a self-calibration of detector efficiencies and resolutions. This question is presently under study by the UCLA group.

Avoiding K_L regeneration requires that the beam pipe must be large (>6 cm) in the interaction region (IP), permitting injection of undamped positron beams. In contrast, detectors for B factories demand a small beam pipe ≈ 2 cm for adequate vertex resolution, implying that positron bunches must be damped to their equilibrium emittance in a separate ring prior to injection into the collider. As the c.m. energy of the collider is raised from 1 GeV to 10.6 GeV (B-factory), the difficulties of masking the detector from the synchrotron radiation from the beams will increase considerably may preclude some design options.

II. LUMINOSITY CHALLENGE OF FLAVOR FACTORIES

Several approaches are available for the design of the flavor factory; these are a) a true linear collider, b) a linac-on-ring collider, c) single or multiple storage rings, d) a beam from a normal or superconducting linac colliding with a bunch in a storage ring by-pass (quasi-linear collider). In all the cases except that of a single storage ring one can choose either equal or unequal beam energies.

The luminosity of a collider is given by the well known relation

$$L = \frac{N_- N_+ f_c H(D)}{4\pi\, \sigma_x \sigma_y} \, , \tag{2}$$

where f_c is the collision frequency, N_- and N_+ are the charges in the electron and positron bunches, $H(D)$ is the disruption enhancement due to the pinch effect, and σ_x and σ_y are the horizontal and vertical beam sizes respectively. The disruption parameter, D, is related to the bunch length, σ_z, and transverse size by

$$D = \frac{r_e N\, \sigma_z}{\gamma\, \sigma_t^{\,2}} \, , \tag{3}$$

where r_e is the classical electron radius. For linear collider, the interaction area $\pi\sigma_x\sigma_y$ should be $<10^{-8}$ cm (optimistic for beam energies < 1GeV). The maximum practical N_- is $\approx 10^{11}$. If $H(D)$ is 4 (an optimistic assumption), a luminosity of 10^{33} cm^{-2}s^{-1} implies that positrons must be supplied to the collision point at a rate, $f_c N_+ \approx 10^{14}$ s$^{-1} \approx 2 \times 10^4$ nA. Based on the typical performance of positron targets at many laboratories, one can expect to generate, capture, cool and transport $\approx 5 \times 10^{-3}$ e$^+$ / e$^-$ /GeV or ≈ 20 nA of positrons per kW of electrons incident on the positron production target. A luminosity of 10^{33} cm^{-2}s^{-1}, therefore, requires $\geq$ 1MW on the target and fast damping rings. For comparison the present SLC target is a 30 kW design. One concludes that true linear colliders are impractical for beam energies < 10 GeV. This consideration drives one to a storage ring based scenario in which the positrons can be recovered and reused after each collision.

An immediate modification of the linear collider scheme is to pass an electron beam from a linac through the positron bunches contained in the ring. In this scheme the disruption (i.e., equivalent tune shift) of the positron beam must be kept sufficiently small, although the electrons could be strongly disrupted. Collision frequencies are limited to the kilohertz range with room temperature linacs making high luminosity extremely difficult. In contrast, use of a superconducting linac allows megahertz collision rates. As the electron bunch is always newly formed, disruption compensation techniques may permit higher tune shifts (≈ 0.1) in linac-ring colliders than in storage ring colliders, especially if the damping time of the ring is reduced to $\approx$ 1ms by the use of superconducting dipoles or high field wigglers. A superconducting linac-on-ring ϕ factory with $L = 3 \times 10^{32}$ cm^{-2}s^{-1} has been proposed[5] by Amaldi and Coignet. However, the consistency of a practical lattice with the high collision rates, small normalized emittance and short bunch length in their design is unclear. Moreover, the use of a superconducting linac raises the cost of such a ϕ

factory to well over 100 M\$. A B-factory using this approach has also been suggested by Amaldi and Coignet and is under study at CEBAF.

The megahertz collision rates achievable with superconducting linacs can also be realized with both beams contained in storage rings. If the beams have equal energies the electrons and positrons can share a common ring. The linear beam-beam tune shift, ξ, in a storage ring collider is typically limited to ≈ 0.5 where

$$\xi_\iota = \frac{r_e}{2\pi} \frac{N \beta_\iota^*}{\gamma \sigma_\iota (\sigma_x + \sigma_y)}, \quad (\iota = x, y) \tag{4}$$

where σ_i is related to the geometrical emittance, ε, and the β^* at the interaction point by

$$\sigma_\iota = \sqrt{\varepsilon_\iota \beta_\iota^*}. \tag{5}$$

A commonly used, simplifying assumption is that of "optimal coupling"; i.e.,

$$\xi_x = \xi_y \Rightarrow \frac{\beta_x^*}{\sigma_x} = \frac{\beta_y^*}{\sigma_y}. \tag{6}$$

In this case, for round beams the tune shift and disruption are related by

$$D_{ring} = 4\pi \, \xi \left(\frac{\sigma_z}{\beta^*} \right). \tag{7}$$

Generally, one assumes that the beam-beam tune shift will limit the allowable value of N; in that case, one has the scaling law

$$L = 2.17 \times 10^{34} \, cm^{-2} s^{-1} \, \xi \, (1 + r) \left[\frac{E\,(GeV)\,I\,(A)}{\beta_y^*\,(cm)} \right], \tag{8}$$

where $r = \sigma_y / \sigma_x$. For an asymmetric double ring collider in which the beams may have unequal energy, the proposed condition of energy transparency[6] (equal beam sizes, equal tune shifts, equal damping decrements for both beams) implies the equality of the quantity in brackets for both low and high energy beams.

The figure plots the luminosity scaling vs physics requirements for a symmetric (Φ) and quasi-linear (QLΦ) ϕ factory, a τ-charm factory (τCF), and for symmetric (SBF) and asymmetric B factories (ABF). The curve labelled medium risk corresponds to a collider with round beams (r=1), a tune shift of 0.05, a current of 1 Amp, and a β_y^* of 4 cm. Even this combination represents a difficult design, especially in the case of a B factory where difficulties of masking the detector from synchrotron radiation are severe and may preclude round beams colliding head-on.

Synchrotron radiation effects aside, at the lower end of the desirable luminosity range both the ϕ and B factories represent comparable technical risk for a storage ring collider. With respect to beam dynamics increasing the luminosity of an energy symmetric ϕ factory to $>10^{33}$ cm^{-2}s^{-1} is similar in difficulty to designing an asymmetric B factory of 10^{34} cm^{-2}s^{-1}; the risk will be high unless an innovative design approach is adopted.

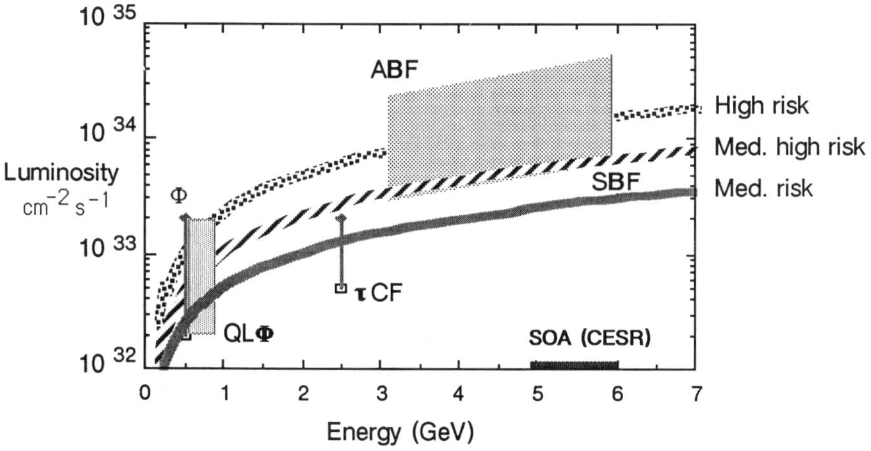

III. UCLA φ FACTORY DESIGN

The UCLA group has been studying the design of a φ factory in two phases, with the following goals : 1) Start operation at a minimum luminosity ($> 10^{32}$ cm^{-2}s^{-1}) in the shortest possible time and at the minimum cost; 2) Subsequently increase the luminosity to $\approx 10^{33}$ cm^{-2} s^{-1}, with the option of unequal e$^-$ and e$^+$ energies. In phase 1 the collider would consist of a single, compact storage ring holding one bunch of electrons and one bunch of positrons. The characteristics of the storage ring are given in the table.

Circumference	12.9 m	Z/n	0.7 Ω	Emittance	4.5 mm-mrad
Beam Energy	510 MeV	Mom. compaction	0.12	Coupling	0.2
Beam Current	$\approx$ 2 A per species	Charge/bunch	100 nC	Loss.turn	14.1 kV
Bending Field	4 T	$\xi_x = \xi_y$	0.05	RF freq.	348 MHz
Luminosity	2.9×10^{32} cm^{-2}s^{-1}	ν_s	0.007	σ_z	2.7 cm

The compact ring with superconducting dipoles has a large collision rate, $\approx$23 MHz, producing high luminosity with a few positrons. In such a ring the synchrotron radiation damping time becomes very short ($\approx$2 ms), increasing beam stability, lifetime and the allowable beam-beam tune shift. The luminosity limit for this configuration comes from the beam-beam bremsstrahlung and Touschek lifetimes, from the maximum current that can be stored without excessive bunch lengthening, and from the beam-beam interaction. The technical approach is similar to that adopted by INP, Novosibirsk, albeit with a different topology.

The design of the cold ring is a modification of the SXLS ring being developed at Brookhaven National Laboratory. The SXLS design has been modified in four significant ways. a) The 4 T dipole configuration is divided into six 40 cm segments with a field index n $\approx$1 as compared with two long ($\approx$ 3 m) segments planned for SXLS. b) As injection is at full energy, the superconducting dipoles can operate at constant field. These two differences should reduce both the technical risk and the cost of the magnets. c) The straight sections are lengthened to allow easier injection and extraction. d) The number of quadrupoles is increased to allow operation in both high and low equilibrium emittance conditions and to allow variation of the momentum compaction from (10^{-5} to >0.1)

A promising lattice[7] for the ring employs a triple bend achromat configuration.

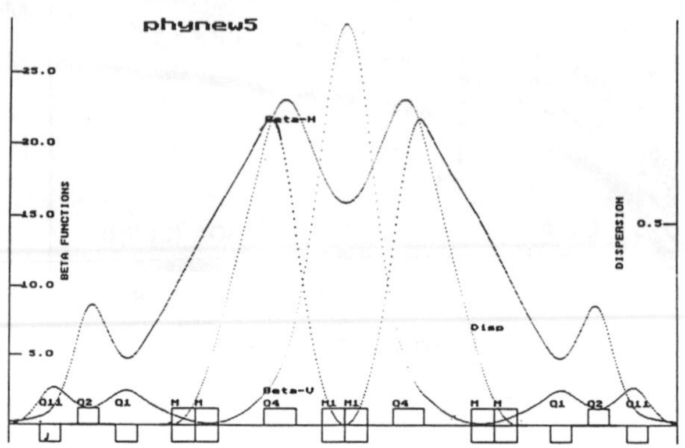

Placing a quadrupole doublet approximately 50 cm from the interaction point reduces the vertical chromaticity and the maximum β_y. To correct the residual chromaticity of this design (-12 and -5) while maintaining adequate dynamic aperture, one may use strong, modified sextupoles of the type proposed by Cornacchia and Halbach[8]. The consistency of this design with the high precision detectors needed for the ϕ factory must still be studied.

The injector[9] uses a full energy, travelling wave S-band linac with a low power (300 W) positron source to fill the ring in $\approx$ 60 s.. For the ϕ factory (unlike a B factory) the beam pipe must be large (several K_S^0 path lengths) near the interaction point to avoid regeneration. Even at injection the aperture will be 5 times the size of an undamped positron pulse. The quantum lifetime is related to the damping time τ_x by

$$\tau_q = \tau_x \frac{\exp\left(\wp^2/2\right)}{\wp^2} .$$

$$(9)$$

As the $\tau_q \gg \tau_x$ for $\wp = 5$, most of the originally injected pulse is should survive the damping process. In contrast, the small beam pipe at the interaction region of a B-factory plus the lower equilibrium emittance associated with higher energy operation imply the a B-factory will require a damping ring for the positrons.

IV. INCREASING THE LUMINOSITY OF FUTURE B AND ϕ FACTORIES

To increase the luminosity and reduce the cost of future colliders one must increase the beam density in 6-D phase space. One can adopt a number of design strategies[10]:

a) Increase the beam current by increasing the number of bunches or the bunch current. In the latter case one must lower the longitudinal impedance of the ring to avoid bunch lengthening instabilities. Unfortunately increasing current also increases the difficulty of handling synchrotron radiation and increases the cost of rf-power.

b) Reduce the degree of asymmetry of the collisions. This choice may restrict the possible particle physics experiments and/or can make detector design more difficult.

c) Try to raise ξ above 0.05. This choice risks instabilities. The systematics of the beam-beam tune shift limit need careful experimental study with existing rings.

d) Reduce β^* by decreasing the bunch length ($L \propto \sigma_z$). To avoid synchro-betatron resonances β^* should be greater than the bunch length. One can decrease the bunch length by raising the rf-frequency or voltage or by reducing the momentum compaction, α, of the ring to a value near zero. At UCLA an on-going study[11] of nearly isochronous rings is exploring the latter option. For stable motion in longitudinal phase space (Ψ, $\Delta p/p$)

$$\alpha_1 \approx \alpha_2 \left(\frac{\Delta p}{p}\right),$$

(10)

where the momentum compaction is defined as

$$\alpha \equiv \frac{\left(\dfrac{\Delta \Psi}{2\pi}\right)}{\left(\dfrac{\Delta p}{p}\right)} \approx \alpha_1 + \alpha_2 \left(\frac{\Delta p}{p}\right).$$

(11)

For large damping decrements ($10^{-3} - 10^{-2}$), the minimum value of α_1 required for stability is reduced even further. With $\Delta p/p \approx 10^{-3}$ (common in many rings), one should be able to reduce α to $<10^{-4}$, two orders of magnitude lower than is typical. Because $\sigma_z \propto \alpha^{0.5}$, σ_z and β^* in a nearly isochronous ring could be reduced ten-fold to ≈ 1mm. The consequence is that the current (and operating cost) of a very high luminosity B factory could be reduced by an order of magnitude. The proposed UCLA ring would allow one to test this approach.

e) Strongly disrupt and discard the electrons ($D_- \gg 1$). As $L \propto \sigma_t^{-2}$, by allowing the electron bunch length to be much shorter than the positron bunch length, the elctrons could be focussed to a very small spot. If $D_+ < 1$, the electrons should form a stable pinch channeled by the positrons. In the quasi-linear collider, the storage ring serves as a positron accumulator/recovery ring that can be continually "topped-off". The collisions take place in a by-pass with the electrons being supplied by a bright, high repetition rate, high gradient linac. By moving the interaction point to a by-pass outside the storage ring, one may reduce the coupling of the ring dynamics and the beam-beam effects. In the by-pass the positron bunch must be compressed to permit a small value of β^* at the interaction point and then stretched before re-insertion in the ring. A B-factory using this approach has been proposed by Cline and Pellegrini.[12] If the damping ring can operate isochronously, the buncher and debuncher could be eliminated from the by-pass. The technical difficulties of quasi-linear collider include frequent beam manipulation without degradation of emittance, the development of high frequency extraction and injection systems, and the development of very bright, high repetition rate electron beam sources. These issues could also be tested in the second phase of the proposed UCLA ϕ factory.

V. CONCLUSIONS

Pushing the luminosity of future flavor factories to the limits that full exploration of CP violation in the K and B systems may demand will require electron-positron colliders of novel design. Several approaches compared in the Table 2, could be evaluated experimentally with a compact storage ring collider that is proposed for building on the UCLA campus. This collider will also serve as a ϕ factory with sufficient luminosity to explore CP violation and test CPT invariance to a sensitivity beyond that of previous experiments.

Table 2. Comparison of collider schemes

	Linear	Linac-Ring	Storage ring
IP	outside	inside	inside
$\beta^*(-)$	< 1 mm	< 5 mm	<5 cm
$\beta^*(+)$	< 1 mm	<5 cm	<5 cm
N^-	~10 nC	~1 nC	~ 100 nC
N^+	~10 nC	~ 100 nC	~ 100 nC
ε^-_n	very low	very low	medium
ε^+_n	low	medium	medium
σ_t	(< 1 µm)	(~ 10 µm)	(~100 µm)
f_c	~1 - 10 kHz	~ 10 MHz	~10 MHz
D^+D^-	>> 100	~ 10	< 1
H(D)	5 - 10	1 - 2	1
Source issues	Low emittance e^-, e^+	lifetime, refill	lifetime
Dynamics issues	ε^+_n preservation large D effects	ξ limits beam-beam effects	ξ limits beam-beam
Structure issues	S.C. cavities high gradients	Z/n, I	Z/n, I

The author thanks C. Buchanan, D. Cline, C. Pellegrini, and his other colleagues in the UCLA ϕ factory group for many suggestions in preparing this summary paper.

References:
1. H. Christianson, J. W. Cronin, V. L. Fitch, R. Turlay, Phys. Rev. Lett. 13,138 (1964)
2) CERN experiment No. NA-31, H. Wahl spokesman, also R. H. Bernstein et al., Phys. Rev. Lett. 54, 1631 (1985) and K. Black et al., Phys. Rev. Lett. 54, 1628 (1985)
3) I. Dunietz, J. Hausner, J. L. Rosner, Phys. Rev. D, 35, 2166 (1987)
4) B. Nussinov and T. N. Trong, "The Decay $\phi \to K^0 K^0 \gamma$ and its Possible Effects on Future Kaon Factories", UCLA report, UCLA/89/TEP/11
5) U. Amaldi and G. Coignet, (1989)
6) Feasibility Study for an Asymmetric B-Factory Based on PEP, S. Chattophadyay and M. Zisman, ed. LBL-PUB-5244 and SLAC-352, Oct. 1989
7) F. Agahmir et al., The Conceptual Design of the UCLA Phi Factory and Partical Beam Physics Complex , UCLA report, CAA0053-UCLA-2/90
8) M. Cornacchia and K. Halbach, Private communication (1989)
9) W. A. Barletta and C. Pellegrini, CAA0048-UCLA-11/89 (1989)
10) W. A. Barletta, D. B. Cline, and C. Pellegrini, High Luminosity B $\overline{\text{B}}$ Factories using Novel Techniques, CAA00050-UCLA-12/89 and Proc. of European Part. Accel. Conference, Nice, 1990
11) D. Rubin and C. Pellegrini, Proc. of International Workshop on B Factories and Related Physics, Blois, 1989
12) D. B. Cline and C. Pellegrini, to be published in Nuc. Inst. and Meth., (1990)

Physics at a Tau Charm Factory†

Joseph M. Izen*
*University of Illinois at Urbana-Champaign,
Urbana, Illinois 61801*

ABSTRACT

A high luminosity e^+e^- collider is proposed to explore decays of charmed D and D_s mesons, τ leptons, and charmonia. A brief description of the accelerator and detector design is given, followed by a discussion of selected experiments in D and τ physics.

Electron-positron colliders operating at $\sqrt{s}$ from 3 to 4.5 GeV are ideally suited to studying D, D_s, τ and J/Ψ particles. Kinematics constraints and straightforward particle identification help to reconstruct exclusive final states, and below-threshold data can be used to measure backgrounds from other processes. A proposed new collider[1] with a design luminosity of $1\times10^{33}\,\mathrm{cm^{-2}s^{-1}}$ at $\sqrt{s} = 4.4\,\mathrm{GeV}$ represents a 100-1000 increase over existing facilities. The accelerator is a double storage ring with a dedicated injector. It will operate with a total beam current of 498 mA, 21 bunches per beam and 52.4 ns between bunch crossings. It is designed for high reliability, and it is expected to deliver beams to the experiment with a duty factor of 5000 hours per year.

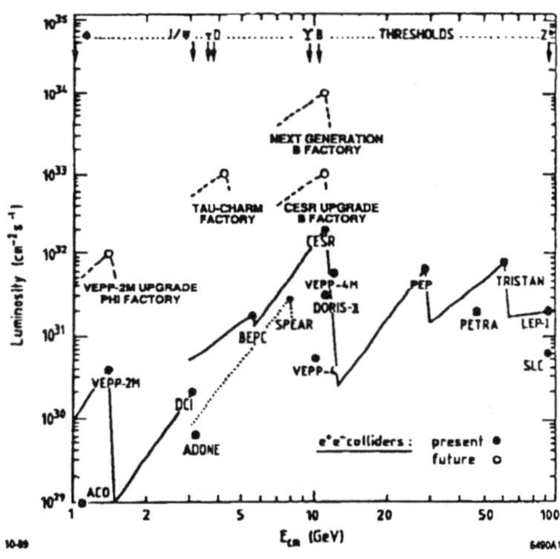

Fig 1. The luminosity of present and future colliders.

The Tau Charm Factory detector differs from detectors of the Mark III/BES generation several respects. CsI crystals are use for electromagnetic calorimetry to maximize photon

† This work was supported in part by the U.S. Department of Energy, under contract DE-AC02-76ER01195

* Representing the Tau Charm Collaboration: CERN, Massachusetts Institute of Technology, SLAC, University of Cincinnati, University of California at Santa Cruz, University of Illinois at Urbana-Champaign, University of Oregon, University of Washington

efficiency for low energy photons. It is 90% efficient at 10 MeV with an energy resolution of better than $\sqrt{E}$ [GeV]. A double layer of time of flight counters covers both the barrel and the end cap. Many analyses ($D_s \to \tau \nu$, τ tagging, D^0-$\overline{D^0}$ mixing with a pair of semileptonic decays...) rely on a missing energy signature. A hadron veto calorimeter rejects a potential K_L background.

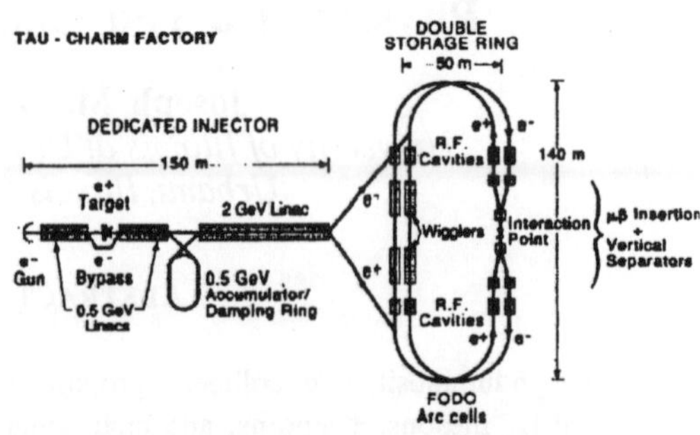

Fig. 2. Layout of the Tau Charm Factory Accelerator

The calorimeter doubles as a muon identifier. The range measurement that is obtained by sampling every 1.25 cm through the iron results in better 90% muon efficiency with less than 1% pion feedthrough for momenta above 350 MeV. Continuous monitoring of the detector's performance is essential since many branching fraction measurements will have miniscule statistical errors. Our goal is to reduce the systematic errors to the 1% level by interleaving higher energy running with short J/Ψ runs. The J/Ψ is an ideal test beam since its simple decays (J/$\Psi \to K_s K_L \eta$, J/$\Psi \to \rho \pi \to \pi^+ \pi^- \pi^0$, J/$\Psi \to K^* K$, J/$\Psi \to e^+ e^- \gamma$, J/$\Psi \to \mu^+ \mu^- \gamma$) can be used for calibrating the detector.

DECAYS OF D^0, D^+, AND D_S MESONS

The study of charmed mesons produced near threshold offers several essential advantages over mesons produced in the high energy continuum or secondary B meson

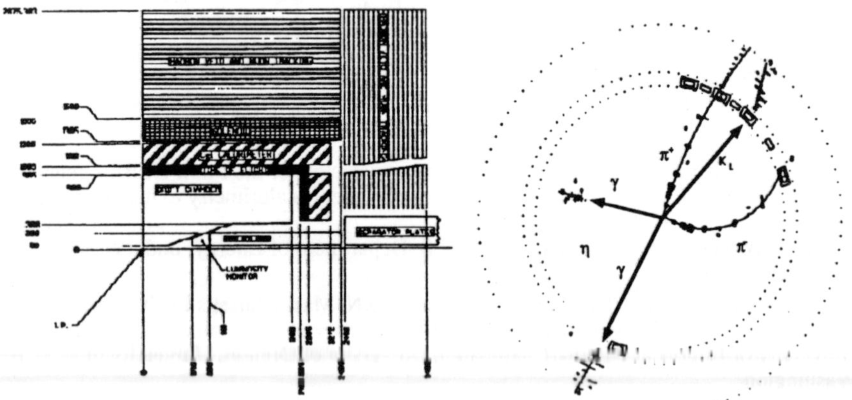

Fig. 3. Side view of the Tau Charm Detector

Fig. 4. J/$\Psi \to K_s K_L \eta$, $\eta \to \gamma\gamma$, $K_s \to \pi^+\pi^-$. This decay may be used as a K_L test beam.

decay. Meson pairs are produced exclusively near threshold, insuring a low combinatoric background and offering kinematic constraints for further background rejection. The quantum statistics of the production and decay processes can be used to distinguish the rare D^0-$\bar{D}^0$ mixing signal. The charm production cross section is measured, and it is ten times larger than at $\sqrt{s} = 10.6$ GeV. The number of single tags per year using well established decay modes is given in Table 1.

HADRONIC D^0, D^+, AND D_s DECAYS [2]

Much progress has been made in understanding hadronic D^0 and D^+ decay by observing the complicated pattern of Cabibbo allowed and singly suppressed channels. Weak annihilation, W exchange, interference, QCD, and final state interactions all complicate the simple spectator decay picture. This study will be extended to doubly-Cabibbo suppressed decays for the first time at a Tau Charm Factory. Doubly suppressed D^+ decays are particularly interesting because, and unlike allowed decays, there is neither interference nor the mixing complications of the D^0 system. Several hundred events per year will be reconstructed with backgrounds reduced to the 10% level.

Experimentally, we have catalogued $85 \pm 15\%$ of the D^0 and D^+ decays with an absolute scale error of 7% and 12% respectively. With a thousand fold increase in data, these measurements would be limited only by systematics at the 1% level. In contrast, D_s are poorly known. Probably less than 50% of the decays are accounted for, and no absolute scale has yet been set. In the next several years, we anticipate that BES will determine the absolute scale of D_s decays at the 15% level. However, the Tau Charm Factory CsI calorimeter is needed to untangle the large fraction of the decays with one or more neutrals (π^0 and η). A year of Tau Charm factory running will reduce the scale error for D_s decays to below 3%, and ultimately to a systematic 1%.level.

$D^0 \rightarrow$	BR	# Detected/yr
$K^-\pi^+$	0.042	2.7×10^6
$\bar{K}^0\pi^0$	0.020	2.1×10^5
$\bar{K}^0\pi^+\pi^-$	0.064	9.4×10^5
$K^-\pi^+\pi^0$	0.130	3.0×10^6
$K^-\pi^+\pi^+\pi^-$	0.091	3.3×10^6
$\bar{K}^0\phi$	0.010	3.1×10^4
$\pi^+\pi^-$	0.002	1.6×10^5
K^+K^-	0.005	2.5×10^5
$K^-\pi^+\pi^0\pi^0$	0.149	1.2×10^6
Total		1.2×10^7
$D^+ \rightarrow$		
$\bar{K}^0\pi^+$	.0320	4.1×10^5
$K^-\pi^+\pi^+$	0.091	3.6×10^6
$\bar{K}^0\pi^+\pi^0$	0.130	5.9×10^5
$\bar{K}^0\pi^+\pi^-\pi^+$	0.066	3.5×10^5
$K^+K^-\pi^+$	0.011	3.5×10^5
$\bar{K}^0 K$	0.010	1.1×10^5
$K^-\pi^+\pi^+\pi^-\pi^+$	0.007	4.2×10^4
Total		5.5×10^6
$D_s \rightarrow$		
$\phi\pi$	0.030	1.1×10^5
$S^*\pi$	0.009	1.0×10^5
$\bar{K}^0 K^+$	0.030	1.7×10^5
$\eta\pi^+$	0.080	3.4×10^5
$K^{0*}K^+$	0.030	1.1×10^5
Total		8.3×10^5

Table 1. Number of D^0, D^+, and D_s single tags per year

Fig. 5. Weak annihilation.

LEPTONIC D^+, AND D_s DECAY [3]

The leptonic decays of the D^+ and D_s are of great theoretical interest, but are as yet unmeasured. The decay occurs through the W annihilation diagram. The weak decay constant f_D may be understood as a measure of the overlap of the c and d(s) quark wave function as described by QCD. A precise measurement of the leptonic branching fraction leads directly to a determination of f_D :

$$B(D^+ \rightarrow \mu^+ \nu) = \frac{G_F^2}{8\pi} f_D^2 \tau_D M_D m_\mu^2 |V_{cd}|^2 \left(1 - \frac{m_\mu^2}{M_D^2}\right)^2$$

Much work has gone toward the prediction of f_D and f_{Ds}., but theoretical uncertainties are still quite large. Lattice calculations offer most exciting prospects. Precision measurements of both f_D and f_{Ds} will serve as important benchmarks in the development of Lattice QCD because a similar measurement for the B^+ system is unattainable; we will have to trust the lattice calculations of f_B. These measurements take on an added importance in second order weak processes such as B^0-$\overline{B^0}$ mixing where f_B enters under the vacuum insertion approximation:

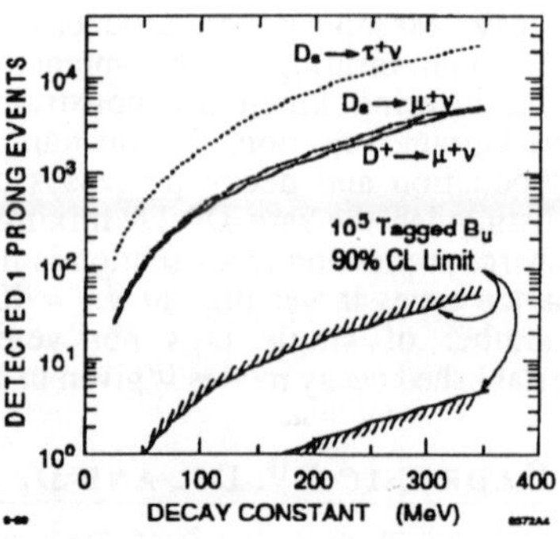

Fig. 6. Number of detected leptonic decays

$$\langle B^0 | J_\mu J^\mu | \overline{B^0} \rangle = |\langle B^0 | J_\mu | 0 \rangle|^2 B_{B_d} = \frac{4}{3} M_{B_d} f_{B_d}^2 B_{B_d}.$$

The signature for $D^+ \rightarrow \mu^+ \nu$ and $D_s \rightarrow \mu^+ \nu$ events is a tagged event containing one additional muon and zero missing mass. The leptonic decay, $D_s \rightarrow \tau^+ \nu$ with $\tau^+ \rightarrow l^+ \nu \nu$ is also detectable although the missing mass constraint is lost. Instead, we will have to rely on the hermeticity of the detector and the low photon threshold of the CsI calorimeter to assure that neutrinos are the only undetected neutrals. The hadron veto calorimeter which is 95% efficient for K_L momenta above 350 MeV, is important to reject the Cabibbo suppressed semileptonic decay, $D_s \rightarrow K_L e^+ \nu$. Fig.7 shows signal and background distributions for a year's running at a Tau charm factory. The decay constants will be measured to within 3% with a year's running.

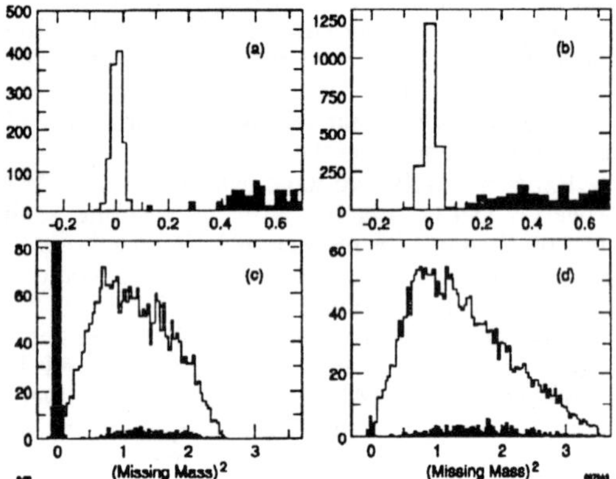

Fig. 7. Missing mass for leptonic decays. Shaded areas are backgrounds. a) $D^+ \rightarrow \mu^+ \nu$, b) $D_s \rightarrow \mu^+ \nu$, c) $D_s \rightarrow \tau^+ \nu$, $\tau^+ \rightarrow e^+ \nu \nu$, d) $D_s \rightarrow \tau^+ \nu$, $\tau^+ \rightarrow \mu^+ \nu \nu$

SEMILEPTONIC D^0, D^+, AND D_S DECAYS [4]

Precision measurements of KM matrix elements exist for V_{ud} and V_{us} with 0.1% and 1.0% errors respectively; V_{cs} and V_{cd} are only know to 20%. Imposing unitarity and three generations reduces the errors in the charm sector, but this precludes a test of the standard model. Direct measurement of D semileptonic branching fractions will reduce these errors to ~1%. The decay rate for D_{l3} decays is given by

$$\Gamma(D^0 \rightarrow K^- e^+ \nu) = (G_F^2 M_c^5 / 192\pi) \left| V_{cs} \right|^2 \int \left| f_+(t) \right|^2 p^3 dt$$

where $f_+(t)$ is the vector form factor (Terms with the $f_-(t)$ form factor are suppressed by order $(m_{lepton} / M_D)^2$.) The t dependence of the form factor is usually parametrized with a single pole at the D_s^* mass. The Tau Charm Factory will explicitly measure the t dependence with great sensitivity. If the radial excitation of the D_s^* couples with 1% of the strength of the D_s^*, it will be detected. Form factor values at zero four momentum transfer, $f_+(0)$ are calculated by the same lattice QCD techniques used to predict the leptonic decay constants. Therefore the nine D_{l3} $f_+(0)$ form factors independently serve as additional benchmarks for the lattice calculations.

The D_{l4} decays are governed by a vector form factor $V(t)$, and three axial vector form factors, $A_0(t)$, $A_1(t)$, $A_2(t)$. The third, $A_2(t)$ is suppressed by order $(m_{lepton} / M_D)^2$.). The relative strengths of the D_{l4} form factors may be determined by a fit to the K^* and "W" helicity angles and the angle between the K^* and "W" decay planes. It is crucial to eliminate backgrounds since the their angular dependence is difficult to determine. Backgrounds typically are cause by undetected low energy photons. Fig. 8 demonstrates the dramatic difference in background rejection afforded by the CsI calorimeter. The measurement of every D_{l3} and D_{l4} channel is only possible at threshold. Photoproduction and high energy continuum measurements of some of the easier channels are possible even though the kinematic constraint has been lost, but the presence of fragmentation particles precludes the measurement of any channel with a π^0 or η.

The Tau Charm Factory can also provide information on states not permitted by spectator diagrams such as $D \rightarrow g\, g\, l^+ \nu_l$ and $D \rightarrow (glueball) l^+ \nu_l$. The couplings to the η', the θ

$D^0 \rightarrow$	BR	$N_{X e \nu_e}$	$N_{X \mu \nu_\mu}$	CKM
$K^- \ell^+ \nu$	0.034	2.9×10^5	2.2×10^5	V_{cs}
$\pi^- \ell^+ \nu$	0.004	3.7×10^4	3.0×10^4	V_{cd}
$K^{*-} \ell^+ \nu$	0.06	1.5×10^5	1.2×10^5	V_{cs}
$\rho^- \ell^+ \nu$	0.004	1.6×10^4	1.3×10^4	V_{cd}
$D^+ \rightarrow$				
$K^0 \ell^+ \nu$	0.07	1.1×10^5	8.6×10^4	V_{cs}
$\pi^0 \ell^+ \nu$	0.004	1.4×10^4	1.1×10^4	V_{cd}
$\eta \ell^+ \nu$	0.0015	3.3×10^3	2.6×10^3	V_{cd}
$\eta' \ell^+ \nu$	0.0005	9.2×10^2	6.2×10^2	V_{cd}
$\bar{K}^{*0} \ell^+ \nu$	0.05	2.0×10^5	1.5×10^5	V_{cs}
$\rho^0 \ell^+ \nu$	0.0025	1.3×10^4	1.0×10^4	V_{cd}
$\omega \ell^+ \nu$	0.0025	5.5×10^3	4.0×10^3	V_{cd}
$D_S \rightarrow$				
$\eta \ell^+ \nu$	0.02	6.7×10^3	5.1×10^3	V_{cs}
$\eta' \ell^+ \nu$	0.006	1.5×10^3	8.5×10^2	V_{cs}
$K^0 \ell^+ \nu$	0.002	4.7×10^2	3.6×10^2	V_{cd}
$\phi \ell^+ \nu$	0.034	4.4×10^3	3.2×10^3	V_{cs}
$K^{*0} \ell^+ \nu$	0.0013	4.5×10^2	3.4×10^2	V_{cd}

Table 4. Number of semileptonic decays reconstructed per year

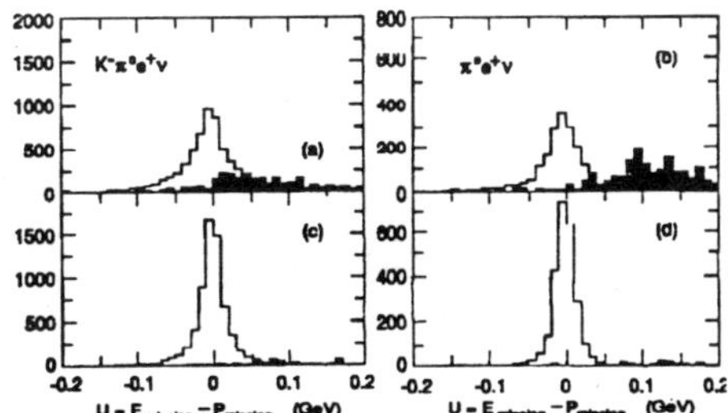

Fig. 8. Semileptonic channels with a π^0
Backgrounds are shaded.
a) and c) Mark III/BES style calorimeter
b) and d) CsI crystal calorimeter

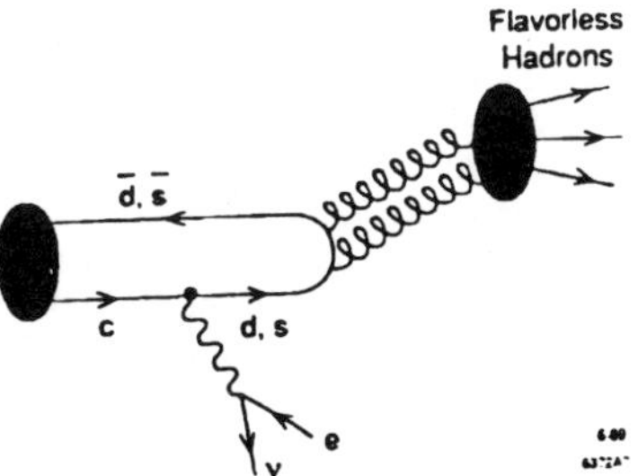

Fig. 9. Semileptonic decay $D \rightarrow gg l^+ \nu$

and the iota in semileptonic decay may provide new insights into their gluonic makeup. Branching fractions as small as 10^{-3} will produce tens of detected, background free events.

D^0-$\overline{D}^0$ MIXING [5]

In the Standard Model, D^0-$\overline{D}^0$ mixing is a second order weak process which can occur either through the box diagrams, or through long range effects. In an experiment that doesn't explicitly observe the time evolution of the decay, the mixing parameter r_D (the ratio of mixed to unmixed events) is obtained from the mass matrix:

$$r_D = \frac{\left(\frac{\Delta M}{\Gamma}\right)^2 + \left(\frac{\Delta\Gamma}{2\Gamma}\right)^2}{2}.$$

The GIM mechanism suppresses box diagram contributions to $r_D \leq 10^{-6}$ but long range contributions may be as large as $r_D \sim 10^{-4} - 10^{-5}$.

Mixing may be measure experimentally when both mesons decay semileptonically, or by using quantum statistics to distinguish doubly Cabibbo suppressed decays (DCSD) from mixing[6] Bose statistics forbids the D^0 mesons in a relative p-wave from decaying to the same final state as shown in Table 3. If they are in a relative s-wave, both mixing and interference occur. By observing this pattern, $\Delta M/M$ and $\Delta\Gamma/\Gamma$ may be observed separately.

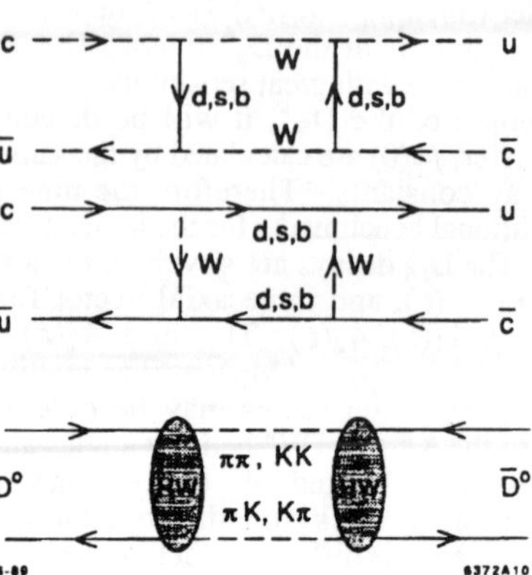

Fig. 10. Weak mixing: a) box diagrams b) through intermediate final states

$e^+e^- \to$	No Mixing	Mixing	Signature
$D^0\overline{D}^0$	0	r_D	$\dfrac{(K^\mp\pi^\pm)(K^\mp\pi^\pm)}{(K^+\pi^-)(K^-\pi^+)}$
$D^0\overline{D}^0\gamma$	$4\tan^4\theta_c\lvert\beta\rvert^2$	$3r_D + 8\left(\frac{\Delta\Gamma}{2\Gamma}\right)\tan^4\theta_c\beta + 4\tan^4\theta_c\lvert\beta\rvert^2$	
$D^0\overline{D}^0\pi^0$	0	r_D	
$D^0\overline{D}^0$	0	r_D	$\dfrac{(K^\mp l^\pm\nu)(K^\mp l^\pm\nu)}{(K^+l^-\nu)(K^-l^+\nu)}$
$D^0\overline{D}^0\gamma$	0	r_D	
$D^0\overline{D}^0\pi^0$	0	r_D	

Table 3. The pattern of mixing and doubly Cabibbo suppressed decays.
Bigi defines $\beta = \dfrac{1}{\tan^2\theta_c} \dfrac{T(D^0 \to K^-\pi)}{T(D^0 \to K\pi)}$

This analysis may be carried out during D_s running at either $\sqrt{s}$ = 4.04 or 4.28 GeV. Instrumental backgrounds due to double time of flight misidentifications are easily removed due to the excellent resolution of the two layer system (σ_{TOF} = 120 ps). Reconstructed samples of greater than 10^5 events per year permit the measurement of mixing for $r_D \sim 10^{-4}$ and the detection of mixing for $r_D \sim 10^{-5}$.

BRANCHING FRACTIONS OF THE TAU LEPTON [7]

The Tau Charm Factory is ideally suited to measure τ branching fractions. The difficulties of performing these measurments with high energy continuum data is demonstrated by the "one prong" problem. The sum of measured one prong branching branching fractions and theoretical limits for unmeasured decays adds to 0.804 ± 0.018, but the measured topological branching fraction is 0.860 ± 0.003.[8] The best energy for measuring τ branching fractions is just above threshold; there is no contamination from charm or bottom production, and any residual hadronic contamination may measured by running below threshold. The τ production cross section 2 Mev above threshold $\sqrt{s} = 3.57$ GeV) is 0.5 nb because of an enhancment due to radiative corrections.[9] Events may be single tagged by 1) a monochromatic pion plus missing energy, 2) a monochromatic kaon plus missing energy, or 3) an identified lepton plus missing energy. Notice the perfect kinematic separation due of the pions and kaons,and the easy range of mometa for lepton identification in Fig. 11. In a year of running at threshold, 2.5×10^6 singly-tagged τ leptons will be collected with measured backgrounds less than 10^{-3}. Statistical precisions of 0.15% for $\tau \rightarrow e\nu\bar{\nu}$, 0.15% for $\tau \rightarrow \pi\nu\bar{\nu}$, 0.8% for $\tau \rightarrow \pi\nu$, and 0.8% for $\tau \rightarrow K\nu$ for a year's running will permit a check of radiative corrections which are a 1% effect.

THE MASS OF THE TAU NEUTRINO [10]

The Standard Model does not predict the masses of any fermions, including the neutrino. In extensions of the Stardard Model, one might expect a mass hierarchy; one model predicts[11] that ratio of neutrino masses equals the ratio of lepton masses of the corresponding flavors. If so, then ν_τ offers our best opportunity to measure a non-zero neutrino mass; a ν_τ mass of 1 MeV corresponds to a ν_e

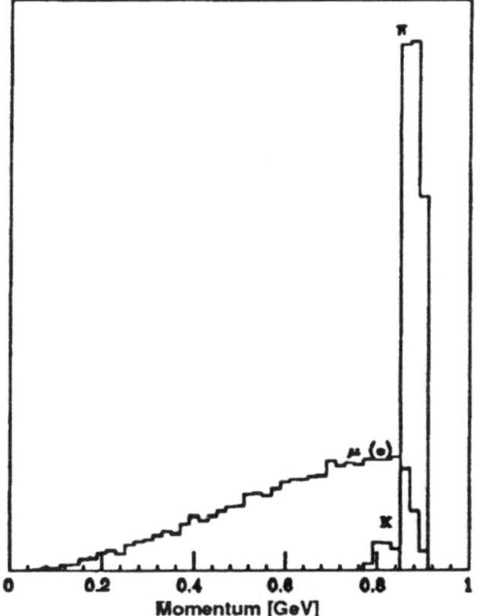

Fig 11.   One prong momenta at $\sqrt{s} = 3.57\,\text{GeV}$.

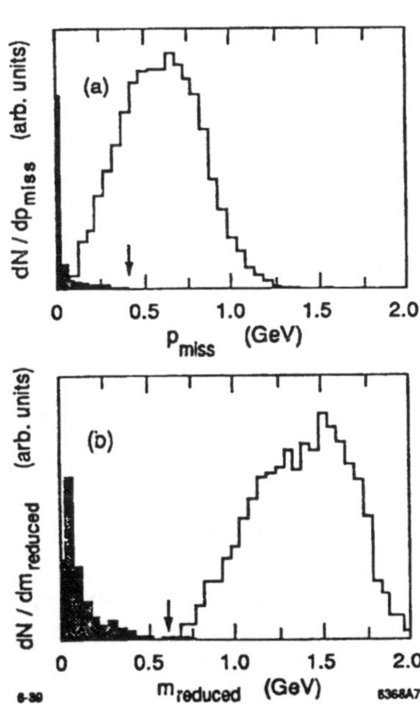

Fig 12.   Selection cuts for"1 vs 5" signal and hadronic background (shaded)
a)   missing momentum
b) $M_{(\text{lepton + missing}}$ momentum)

of 0.08 eV. Current limits are m_{ν_e} < 18 eV[12], m_{ν_μ} < 285 KeV[13], m_{ν_τ} < 35 MeV.[14]

The ν_τ mass is best measured in events with the "1 vs 5" topology corresponding to the decays: $\tau^+ \to e^+ \nu \bar{\nu}, \tau^- \to \pi^- \pi^+ \pi^- \pi^+ \pi^- \nu$. The mass is determined by the endpoint of the five pion mass spectrum. The analysis works both above and below threshold so data from charm runs may be used. Events with the correct topology and no photons are selected. A missing momentum cut ($p_{missing}$ > 400 MeV) rejects hadronic backgrounds (Fig. xx) and a cut on the invariant mass of the lepton + missing momentum system ($m_{reduced}$ > 600 MeV) rejects leptons from $K \to \mu\nu$ and $\pi \to \mu\nu$ (Fig 12). Endpoint spectra for several values of ν_τ mass are shown in Fig. 13. In two years of running, 1400 events with $M_{5\pi}$ > 1.75 GeV will be collected leading to a 2 MeV limit.

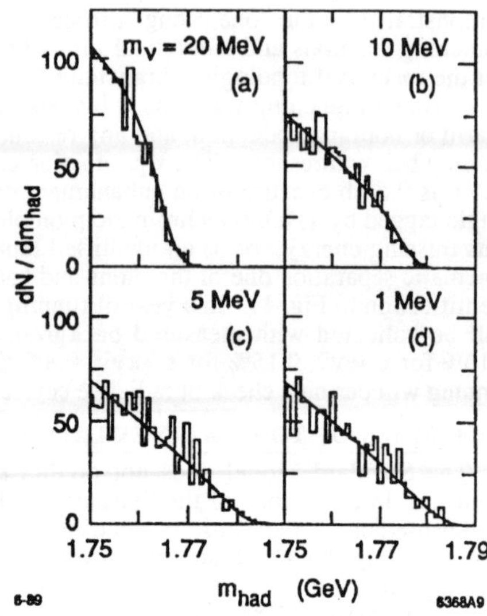

Fig. 13. Endpoint spectra of five-pion mass for different values of ν_τ mass

1 Tau-Charm Factory Design, SLAC-PUB-5180, 1990

2 R.H. Schindler, Proceedings of the Tau Charm Factory Workshop, SLAC Report 343 (1989)

3 P..Kim, Proceedings of the Tau Charm Factory Workshop, SLAC Report 343 (1989)

4 J.M. Izen, Proceedings of the Tau Charm Factory Workshop, SLAC Report 343 (1989)

5 G. Gladding, Proceedings of the Tau Charm Factory Workshop, SLAC Report 343 (1989)

6 I.I.Y.Bigi, Proc. of the Sixteenth SLAC Summer Institute (1988) 30.

7 A. Seiden, Proceedings of the Tau Charm Factory Workshop, SLAC Report 343 (1989)

8 F.J. Gilman and J.M. Rhie, Phys. Rev. **D31**, (1985) 1066
F.J. Gilman, Phys., Rev. **D35**, 3541 (1987)
For a recent review see:
K.G. Hayes, Proceedings of the Tau Charm Factory Workshop, SLAC Report 343 (1989)

9 M. Voloshin, Talk given at he Tau Charm Factory Workshop, SLAC

10 J.J. Gomez-Cardenas, Proceedings of the Tau Charm Factory Workshop, SLAC Report 343 (1989)

11 R.N. Mohapatra and G. Senjanovic, Phys Rev. Lett **44**, (1980) 912.

12 M. Frischi, *et al.*, Phys. Lett. **173B**, (1986) 485.

13 R. Abela, *et al.*, Phys Lett. **146B**, (1984) 431.

14 H. Alberecht, *et al.*, Phys Lett. **202B**, (1988) 149.

SSC PHYSICS AND DETECTORS

Conveners: D. Cassel & T. Dombeck

Physics Goals and Signatures at the SSC

Robert N. Cahn

Lawrence Berkeley Laboratory
University of California
Berkeley, California 94720

Abstract

The physics goals of the SSC are presented and the capabilities of
the SSC to achieve them are assessed. New gauge bosons, electroweak
symmetry breaking, supersymmetry, and quark substructure are the
primary targets for particle physics research and the SSC is the most
effective means to find them.

The greatest achievement of the Standard Model is that it delineates
sharply the known from the unknown. We know there are three generations
of quarks and leptons, and that they interact through gauge interactions —
the strong gluonic interactions and the electroweak interactions. Beyond this
we known almost nothing. We confront a Periodic Table of Particles that
is as inscrutable as the one that faced Mendele'ev. The patterns are clear
but the reasons for them are, for now, beyond our understanding. The goal
of the SSC is no less than to provide the experimental discoveries that will
make the Standard Model part of a larger structure in which its arbitrariness
will be reduced or eliminated.

If the list of fermions has been completed, at least at the present level of
structure, the list of gauge bosons seems patently incomplete. The elegant
solution of Georgi and Glashow that all forces are derived simply from $SU(5)$
has fallen into disfavor because of the negative results of the proton-decay
experiments. Nonetheless there are compelling aesthetic reasons for expect-
ing further unification. The quarks and leptons have electric charges that are
simply related to each other and this suggests that they are somehow unified into larger families containing both colored and uncolored particles. We
then expect that there are additional gauge bosons that transform one into

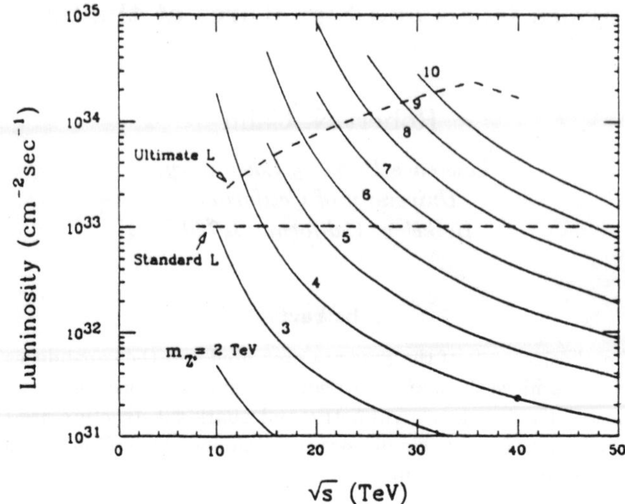

Figure 1: The Schwitters plot for Z' production[2]. The contours show the luminosity required to produce 100 Z's in a year of SSC running (10^7 s). The couplings of the Z' are assumed to be the same as those of the known Z.

the other. Moreover, the expansion of the number of gauge bosons generally will lead to additional neutral gauge bosons similar to the Z. Models based on left-right symmetry and on string-inspired E_6 theories provide explicit examples.

The production of new Zs has been extensively discussed. Of course the basic reference is EHLQ [1]. Recently an ad hoc committee was convened at the SSC to reanalyze the physics objectives and it generated "Schwitters plots," which show the luminosities required to achieve particular goals as a function of the cms machine energy[2]. Figure 1 is the Schwitters plot for producing 100 Z', with curves showing fixed values of the Z' mass. Also shown is the design luminosity, 10^{33}cm^{-2}s^{-1}, and an estimate for the peak luminosity that might eventually be achieved at the SSC. The present design has a reach of 8 TeV with these definitions.

The most immediate problem in particle physics is to understand electroweak symmetry breaking. The standard model is based on a symmetry, $SU(2) \times U(1)$, that is not apparent in the particle spectrum. If it were, the W and Z would be massless like the photon, and indeed all the fermions would be massless, too. The symmetry is broken because everywhere in spacetime, there is a nonzero quantity that points out a direction in the symmetry space, just as an external magnetic field breaks the rotational symmetry of an atom and splits degenerate levels. What we do not know is the nature of that quantity. In the simplest version of the Standard Model there is a funda-

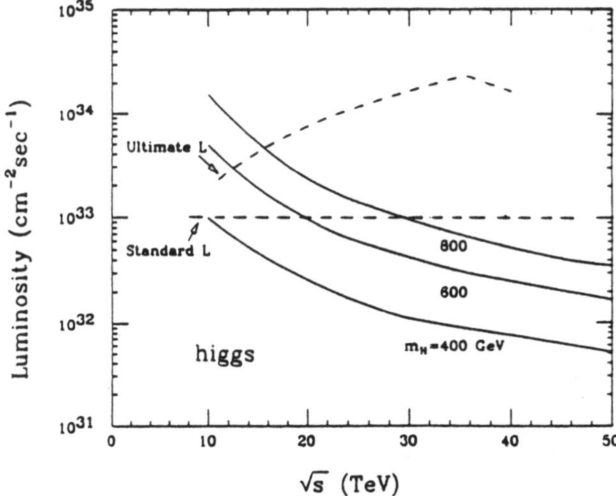

Figure 2: The Schwitters plot for Higgs production[2]. The contours show the luminosity required to produce 20 Higgs decays in the gold-plated decay channel in a year of SSC running (10^7 s).

mental scalar field that chooses a nonzero value in the vacuum state. The fluctuations away from this value represent a dynamical quantity, the Higgs boson.

If there is a conventional Higgs boson and its mass is between 180 GeV and 800 GeV, it will be directly accessible at the SSC. It will be produced by gluon-gluon fusion, through a t-quark loop, and by WW and ZZ fusion. The real problem is to identify the Higgs bosons that are produced. A Higgs boson heavier than twice the mass of the Z will decay primarily to WW and ZZ. Of these decays, the most identifiable are the "gold-plated" ones, in which the Zs both decay into e^+e^- or $\mu^+\mu^-$. Taking into account the 1/3 branching ratio for $H \to ZZ$, the total branching ratio for the gold-plated events is just $(1/3) \times (0.066)^2 = 14 \times 10^{-4}$. Now each picobarn of cross section generates 10^4 events in an SSC year and the cross section for Higgs production varies from more than 100 pb to just 1 pb or so for a very heavy Higgs boson. This leaves precious few events. However, each such event will be very impressive and the only serious background will be from actual ZZ pair production by $q\bar{q}$ annihilation. The Schwitters plot for gold-plated Higgs events is shown in Figure 2.

The second-best signature is the silver-plated event, in which one Z decays to charged leptons (not τ) and one decays to neutrinos. These give a large missing transverse energy recoiling against an identified Z. The branching ratio for these is 88×10^{-4}. The mass of the Higgs cannot be directly deduced from a single event, but measurement of the transverse momentum spectrum of the Zs in such events would lead to an approximate value of the mass.

These searches can be extended to lower and higher masses. For masses between about 130 GeV and 180 GeV, one looks for $H \rightarrow ZZ^*$, with both Zs decaying into charged leptons. For masses above 800 GeV, one still uses the ZZ channel, but in addition one may look for WW with the Ws decaying leptonically. One solution to the electroweak symmetry breaking problem is to have strongly interaction Ws and Zs, with possible resonances in the 1 - 2 TeV region. Even the like-sign WW channel may show effects.

While the simple Higgs model does provide an adequate mechanism for electroweak symmetry breaking, it is not altogether attractive. Elementary scalar fields are anathema to theorists except in one special circumstance: when they are paired with fermionic fields in a supersymmetric theory. In a supersymmetric theory, for every known particle there is a superpartner. While it may seem extravagant to postulate the existence of as many new, and so far unobserved, particles as there are known particles, the payoff is that quadratic divergences are tamed. To many theorists this seems a real bargain. In the presence of quadratic divergences we cannot explain why particles like the Higgs boson do not have masses of 10^{15} GeV.

In any event, supersymmetry is such a beautiful idea that it cannot be ignored. All we have to do is find the fermionic partners of the photon, gluon, W, Z, etc. and the scalar partners of the electron, muon, quarks, etc. Of these superpartners one will be the lightest. In most supersymmetric theories the superpartners must be pair produced. As a result, the lightest superpartner is absolutely stable. Provided it is neutral, it acts just like a neutrino. It follows that a key to finding supersymmetry is to look for missing transverse energy. Such searches are of course underway at the Tevatron, where the exploration extends to masses of roughly 100 GeV. At the SSC we will be able to reach beyond 1.5 TeV as shown in Figure 3. The searches are complicated by the intricate decay chains that do not produce the lightest supersymmetric particle at the outset, but only after emission of quarks, W's etc. The characteristic signal is missing transverse momentum on the order of 200 GeV.

Supersymmetry has Higgs bosons, too. In fact it must have lots of Higgs bosons, both neutral and charged. If there is supersymmetry, there will be a marvelous spectroscopy to explore. A more subtle approach to electroweak symmetry breaking is to do without scalar fields altogether. In technicolor

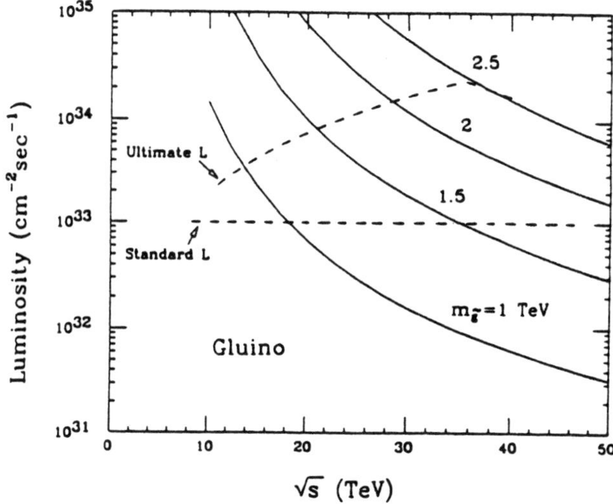

Figure 3: The Schwitters plot for supersymmetry[2]. The contours show the luminosity required to produce 10,000 gluinos in a year of SSC running (10^7 s).

there are new fermions — technifermions — that interact through a new technicolor interaction. A condensate forms, that is, a vacuum expectation value for the operators $U\overline{U}$ and $D\overline{D}$, where U and D are the techniquarks. This vacuum expectation value plays the role of the vacuum expectation value of the Higgs field - but there is no elementary Higgs field. The technipions are eaten like charged Higgs bosons, but there is a technirho and a techniomega to find. In grander technicolor models there are myriad new particles. Technicolor will turn the desert into a forest or a jungle.

Mendeleév could not possibly have understood the chart he proposed. For him the atom was structureless. Without underlying structure there could be no explanation of the rows and columns. It would not be surprising then if we found that the masses and repetitions of the quark and lepton families could not be understood without learning about their substructure. We already know that quarks and leptons are structureless down to a very small distance. At the SSC we will be able to extend the search substantially. The technique that will be used is to look for an excess of two-jet events with large transverse momentum. QCD predicts that the cross section to produce a pair of jets each with transverse momentum greater than $p_{\perp 0}$ is roughly $\alpha_s^2/p_{\perp 0}^2$. On the other hand, if there are composite interactions with a scale Λ, that is a size $1/\Lambda$, then there will be two-jet events due to these interactions with a cross section $1/\Lambda^2$. Thus the composite interactions will be comparable to the QCD interactions when $p_{\perp 0} \approx \alpha_s\Lambda$.

If we say that discovering composite interactions means finding twice as

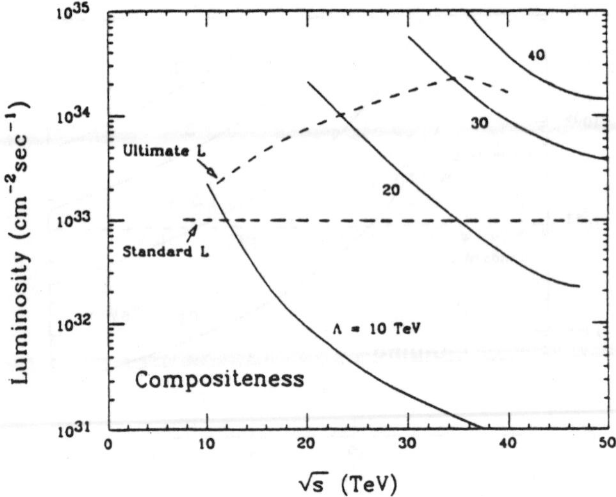

Figure 4: The Schwitters plot for compositeness[2]. The contours show the luminosity required to find evidence for compositeness at a scale Λ. Adequate evidence is defined to be twice as many events in a 100 GeV bin as expected from QCD alone, with a minimum of 50 events in the bin.

many events in a 100 GeV bin as QCD would predict (and at least 50 events in the bin), then the SSC will probe to over 20 TeV in Λ, as shown in Figure 4. The present limit is about 1 TeV.

The t-quark will be prominent in SSC physics. If it is not discovered at the Tevatron, it will surely be found quickly at the SSC. The clearest signature will come from the semileptonic decays of Ws generated by $t \rightarrow bW$. More detailed studies will require find the hadronically decaying W from one of the t-quarks. Since t-quarks will be copiously pair-produced, they will be come an enormous source of WW events, making even more problematical the $H \rightarrow WW$ process.

All this is a theorist's delight. But it is an enormous challenge for the experimenter. There is a clear need to measure charged leptons well. Moreover, both the Higgs search and the supersymmetry search require reliable measurements of missing transverse energy, and thus good hermeticity. The compositeness search requires calorimetry out to very large energies. The search for new Zs puts a premium not just on electrons, whose energies can be accurately measured, but also on muons since their charges can be determined out to very high transverse momenta. This will allow a measurement of the forward-backward asymmetry for the new Z. Simply put, an ideal detector must do everything. Each detector will make its own compromises, one emphasizing one feature, another something else. It is essential of course that the detectors complement each other.

The SSC will not be the only physics at the beginning of the 21st century. We can look forward to a B-factory and ultimately a linear e^+e^- collider. But the SSC will truly be the vanguard machine. If we are to make real progress on fundamental questions like what are the basic particles and forces, I believe it will come only from discovering those very particles — the new Zs, the technirho, the squark, or whatever — and not from inferring their existence in precision measurements. Precision measurements have a special place in the history of particle physics and the discoveries of parity violation and CP violation rank as premier achievements. But the heart of particle physics is spectroscopy and that is what the SSC is about.

References

[1] E. Eichten et al., *Rev. Mod. Phys.* **56**, 579 (1984), *ibid.* **58**, 1065E (1986).

[2] Report of the Ad Hoc Committee on SSC Physics, SSC-250, December 1989.

EMPACT
An Optimized SSC Detector

John P. Rutherfoord

Department of Physics
University of Arizona, Tucson, AZ 85721, USA

ABSTRACT

The EMPACT detector concept is a contender for a first-round SSC experiment looking at the physics of hard collisions. We describe the unique approach, list the basic parameters, and indicate some of the flexibility in the design.

1. INTRODUCTION

EMPACT is an SSC detector concept which addresses the premier physics goals for which the SSC is being built, i.e. the source of electro-weak symmetry breaking, and the search for new gauge interactions, new fermion families, and/or structure within leptons, partons, and the gauge bosons. EMPACT is an acronym for Electrons, Muons, and Partons with Air Core Toroids[1]. The air core toroids give high resolution muon momentum measurements and the low mass (and Z) of the superconducting toroids and cryostat walls induce little of the electromagnetic showers which accompany very high energy muons exiting denser materials thereby confusing the tracking. With no central magnetic field, the EMPACT calorimetry can be optimized for uniform coverage, for superior energy and angle resolution over a wide range of η, and for hermeticity. The calorimeter will include a fine grained electromagnetic section followed by a deep hadronic section. Inside the calorimeter is a combined tracking and TRD system to identify event topologies and to improve electron identification. And there is room for a vertex detector. This detector concept provides superior coverage in η with uniform technology yielding balanced measurements of electrons, muons, photons, jets, and missing neutrinos, i.e. the energy and angle resolution are more comparable for the different particle species than with other approaches. The EMPACT concept is quite flexible and will allow ready upgrades to focus on unanticipated physics phenomena and to higher luminosities following accelerator improvements.

In section 2 the general philosophy of the EMPACT approach is presented. The physics goals are enumerated in section 3. Next, the major elements of the EMPACT detector are described starting from the outside and progressing inward towards the interaction region. So in subsequent sections we describe the muon system with the air core toroids, the hadronic and electromagnetic calorimeters, and the combined tracker and TRD system. Then we step back to get a view of the whole detector to see how the parts fit together to make an optimized device.

2. THE EMPACT CONCEPT

SSC detector concepts have been discussed for many years now and reported at several SSC detector workshops[2,3,4]. The conventional approach, based on past and existing hadron collider experiments, has advantages and inadequacies. Muons are measured with massive iron toroids which provide poor momentum resolution compared with the superior energy measurement of electrons in almost any calorimeter. One must accept many compromises to accommodate a central magnetic field and the magnetic flux return. There are dead regions in the calorimeter for the coil, cryostat, services, and mechanical supports. There must be a change in technology and/or performance at $\eta \approx 1.5$ where magnetic field coverage ends. It is difficult to achieve optimal calorimetry and hermeticity. A magnetic field also spreads the energy in a jet over a large area of the calorimeter. At lower luminosities this just requires some work in the event reconstruction to extract the jet parameters but at higher luminosities there are unavoidable backgrounds which will further degrade the measurement of jet parameters when the jet is smeared out from the naturally rather narrow cone. The optimization question the EMPACT group asked was "Will the freedom allowed by foregoing an internal magnetic field more than make up for the lost capabilities?" We believe the answer is yes. The freedom won allows significantly improved performance of the other detector elements at lower cost. A detector with no internal magnetic field provides the opportunity to optimize calorimetry and tracking. It is more practical to consider photomultipliers and/or transformers for read-out. The detector can have better hermeticity and uniformity of response in η. Tracking in simplified. Since the tracking is not needed to measure momentum in a limited magnetic field, position resolutions of 1 mm are adequate compared to 200 μm or better. And there are no loopers. These advantages are likely to allow the EMPACT detector more flexibility in responding to changes in direction which might be dictated by unanticipated physics hints seen in early SSC running and to perform well in the higher luminosities in the 10^{34} cm^{-2} sec^{-1} range that the SSC will eventually provide.

3. PHYSICS GOALS

The EMPACT detector will concentrate on high p_T physics with thresholds of order 20 GeV. The source of electro-weak symmetry breaking is very likely to be found within the energy and luminosity window available to the SSC and this detector will be optimized for most of the expected signatures. So the list of physics benchmarks considered include 1) Higgs bosons of various masses, types, and decay modes, 2) new gauge bosons, supersymmetry, technicolor, etc., 3) heavy quarks and leptons, and 4) compositeness. These topics were covered very nicely in the review paper by Robert Cahn in these proceedings[5].

In order to detect the signatures for all of these possibilities and to respond to unanticipated results as well, a detector must cover a wide range in η with good energy and angle resolution for electrons, muons, jets, photons, and missing E_T.

4. THE MUON SYSTEM

Rather than a > 30 kiloton solid iron magnet for muon identification and momentum measurement, EMPACT is exploring the use of superconducting, air-core toroids with a total weight of about 2 kilotons and power usage of about 1 Megawatt. While new to particle physics[6,7], there is much experience with superconducting toroids in the magnetic fusion community[8]. A quarter section along the beamline of the toroid arrangement is shown in Fig. 1. The magnetic field is normal to the page and falls off as R^{-1}. Field integrals are indicated for rays near 90° and near $\eta=3$. Along the ray near 90° the present design has one chamber station of 16 planes which lies between the calorimeter and the toroids (see Fig. 2, 7, and 8) while two more of 24 and 12 planes respectively surround the toroids with lever arm of more than 3 m. The total number of muon read-out channels is about 500,000. Including chamber position resolution of a modest 200 μm per plane and full multiple scattering in all the material, the muon momentum resolution for a wide range of muon momenta, as shown in Fig. 3, is really quite good. The modest thickness of material and the low atomic number means that the outer muon chambers will be relatively clean of the electromagnetic showers that would otherwise lead to unacceptable tracking efficiencies. The muon chamber design includes a precise optical alignment scheme to maintain the survey constants to better than 50 μm. The calorimeters are deep enough to reduce the hadron punch-though background to negligible levels. Since the only significant material the muons penetrate is the calorimeter, which is active, the not infrequent catastrophic energy loss will be measured.

A great deal of engineering design of the Toroids has been carried out by a group from Grumman with help from the fusion community from Princeton and Oak Ridge.

5. CALORIMETRY

Without interference from the systems necessary to produce a central magnetic field, (coil, cryostat, support structures, services, and return yoke) the very important calorimetric measurement of electrons, photons, and hadrons can be optimized for position and energy resolution and for particle identification. Partial electron ID comes from electromagnetic shower shape, both longitudinal and transverse. Transverse segmentation also allows accurate Jet-Jet invariant mass measurements while longitudinal segmentation helps determine those rare events with missing E_T and cases of catastrophic muon energy loss. Magnetic analysis would be useful only for charged particles at lower values of p_T and over a limited range of η. The sign of the charge of electrons is the only really important feature that is lost. With uniform calorimetry over a wide η range and a minimum of dead material before and within the sensitive volume, superior hermeticity is possible.

The calorimeter group in the EMPACT collaboration is working to optimize the segmentation in $\Delta\eta$, $\Delta\phi$, and in depth, the total depth of the calorimeter to minimize punch-through and missing E_T resolution, for e/π "compensation" for optimal jet energy resolution, radi-

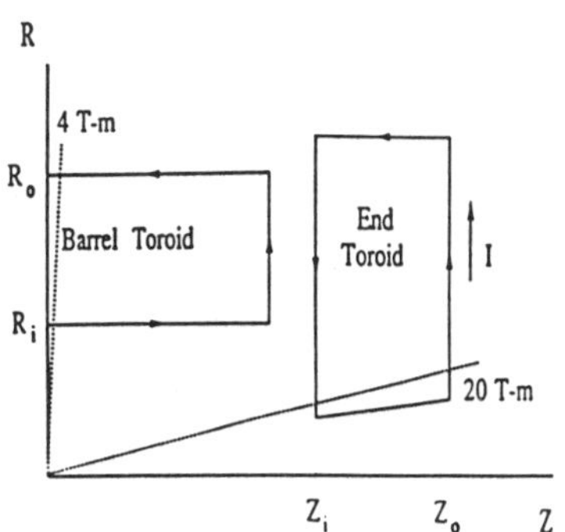

Fig. 1

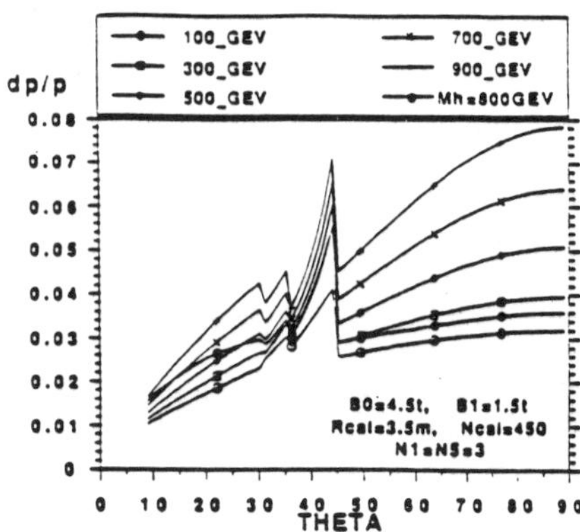

Fig. 3

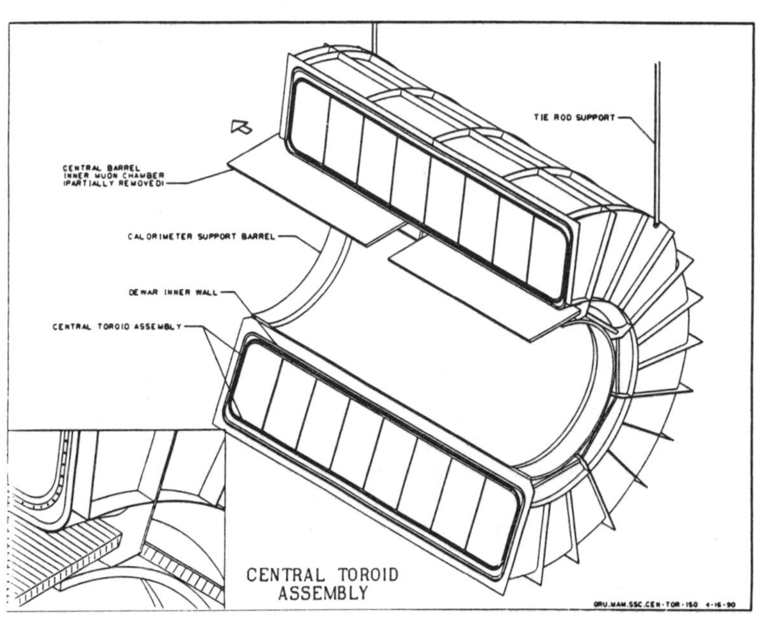

Fig. 2

ation hardness, read-out time, stability and uniformity of calibration, structural integrity, etc. Several technologies are being considered. Liquid argon sampling ionization calorimetry is very attractive for realizing all the goals listed above, some better than others. An e/π ratio below 1.05 is deemed adequate for the required hadronic energy resolution. There are various tricks that can speed the charge read-out to the point where unavoidable physics noise sources dominate. Fig. 4 shows a recent LAC design using depleted uranium absorber. EM and hadronic towers are segmented at $\Delta\eta \times \Delta\phi = 0.025 \times 0.025$ and 0.05×0.05 respectively with 4 depth read-outs in each out to $\eta = 3$. Forward hadronic calorimeters (see Fig. 7 and 8) cover $3 < \eta < 5.5$ with $\Delta\eta \times \Delta\phi = 0.2 \times 0.2$. Special care has gone into engineering the transition between the barrel and endcap calorimeter cryostats near $\eta = 1.5$ to reduce the dead material, particularly near shower maximum.

A "spaghetti" calorimetry design has also been developed for EMPACT including a variation with a silicon/Tungsten EM section in front. The various calorimeter options have been engineered in detail by a group from Martin Marietta.

6. THE INNER TRACKER/TRD SYSTEM

In a non-magnetic environment the EMPACT tracker/TRD system requirements are to 1) determine the primary vertex position, 2) separate multiple vertices, 3) determine event topologies and multiplicities, 4) provide track linkages from the vertex to the calorimeter and, for muons, through the muon chambers, 5) provide less than 10% radiation length of material, and 6) augment electron/photon ID. The TRD's, in combination with the calorimeter shower shape, cut down on the charged hadron backgrounds in the electron candidates[9]. Requiring a track near electromagnetic deposition reduces the background from photons, mostly from π^0 decay. Position resolution of ≈ 1 mm is sufficient. The dE/dx measurement in the TRD's also reduces the background from gamma conversions before the tracker since the e^+e^- look like a single charged track with at least twice minimum ionization.

A schematic of the tracker/TRD system is shown in Fig. 5. The system is either hexagonal or octagonal. In the octagonal scheme there are about 384,000 azimuthal straws of 4 mm diameter each[10]. The inner straws start at 35 cm from the beam line and the outer straws end at 77 cm from the beam line. Two planes of straws follow each 1 cm thick slab of 6% density HD polyethylene foam. The gas is a 50/50 mixture of Xenon and CO_2 giving a maximum drift time of 40 ns to the 50 μm anode wire[9]. Admixtures to speed the drift velocity are under study. Each wire feeds a 2-bit ADC. Minimum ionization fires the first level, twice minimum ionization (such as an e^+e^- from a low energy gamma) triggers the second, transition radiation sporadically fires the third, and heavily ionizing particles consistently fire the third or fourth.

Monte Carlo studies show that the system works remarkably well, allowing electrons to be easily picked out from the large backgrounds. In Fig. 6 are depicted two octants of the

Fig. 4

Azimuthal straws
Radiators
Calorimeter
y = 1.5
1000
924
770
350
300
Pad Chambers
Pad Chambers
y = 3
High Rate Module
193
2500
1000
1470
1926
2346

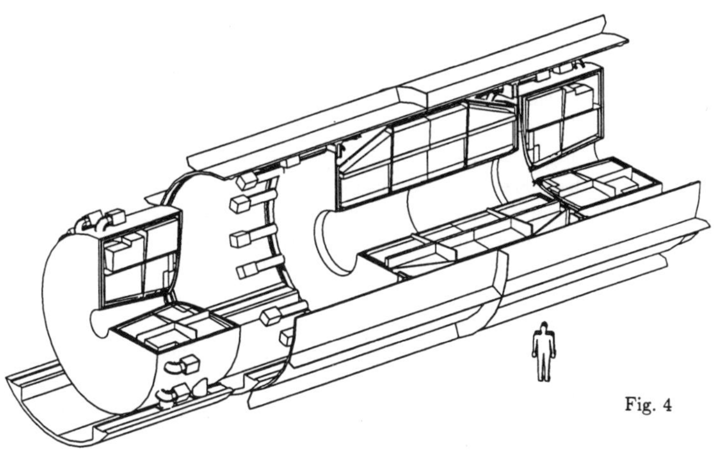

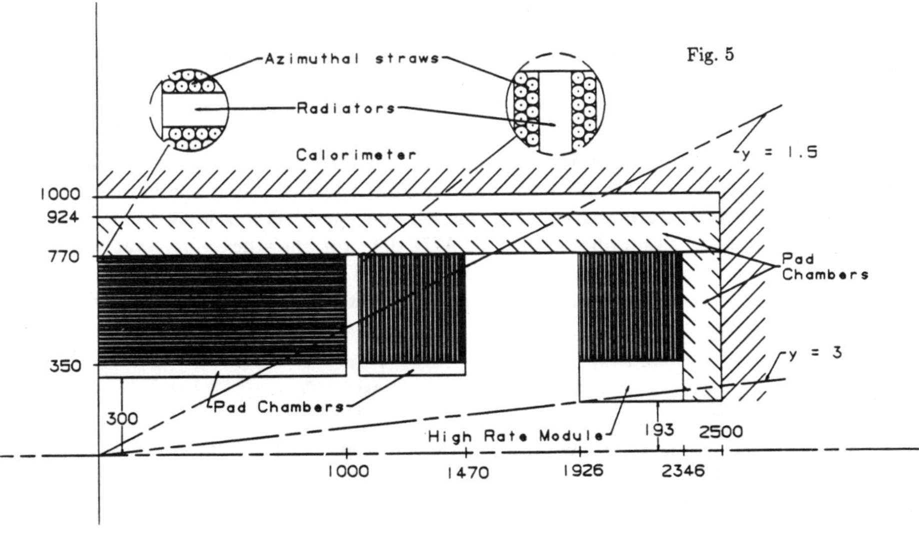

Fig. 5

tracker/TRD system with the ADC level indicated by the thickness of the blob representing each "hit". In 24 layers of radiator, there will be 8 or more detected TRD X-ray hits at least 90% of the time for electrons above 1 GeV. From inner to outer the mean occupancy per straw per crossing at 10^{33} cm^{-2} sec^{-1} is 0.016 to 0.007. The resolving time of a straw is three to four crossings.

7. THE EMPACT DETECTOR

As can be seen from Fig. 7 and 8 the technology of each subsystem is quite uniform as a function of η and there are few weak areas in the coverage so little inefficiency in the acceptance. Particularly noticeable is the complete muon coverage. Due to the light weight of the toroids and, hence the total detector (roughly 6.5 kilotons), the whole detector can be suspended from overhead support frames via small diameter rods which can be placed conveniently to avoid undue interruption of the muon coverage. Careful inspection of the figures shows evidence of the engineering of all the utilities and structural members. Much simulation effort has gone into optimizing the design described in this paper including studies of the effects of this necessary dead material. However there are many areas where improvement is still possible so this design will surely evolve as work continues.

8. CONCLUSIONS

Many years of summer studies indicate the importance of electrons, photons, muons, and jets in extracting physics from the SSC. Complete solid angle coverage with uniform technology provides a manifestly superior attack on the hard subprocesses in hadron collisions that are the premier physics goals. Calorimetry will be the prime tool for measurement of electrons, photons, jets, and missing E_T. The EMPACT approach is to compromise the calorimetry as little as possible with ancillary detector subsystems. Superconducting, air-core toroids provide muon momentum measurements more comparable to the precision of the calorimetric measurements of electrons, photons, and jets providing a balanced emphasis on leading physics probes. The EMPACT detector has a flexibility that will allow it to more readily change in response to unanticipated results from the first round of experiments and to evolve to the higher luminosities that the accelerator will likely provide in the years to come.

9. REFERENCES

1. Michael Marx, "EMPACT—An Alternative Approach to a High P_T SSC Experiment", SSC-219 (May 1989).

2. "Proceedings of the 1984 Summer Study on the Design and Utilisation of the Superconducting Super Collider", ed. Rene Donaldson and Jorge G. Morfin, Snowmass (1984).

3. "Proceedings of the 1986 Summer Study on the Physics of the Superconducting Super

Collider", ed. Rene Donaldson and Jay Marx, Snowmass (1986).

4. "Proceedings of the Workshop on Experiments, Detectors, and Experimental Areas for the Supercollider", ed. Rene Donaldson and M.G.D. Gilchriese, Berkeley (1987).

5. R.Cahn, These proceedings.

6. B.G.Pope, L.Rosenson, and T.M.Taylor, "Toroidal Magnets", Proceedings of the 1978 ISABELLE Summer Workshop, BNL 50885. B.G.Pope *et al.*, "A General Purpose Toroidal Detector", Proceedings of the 1981 ISABELLE Summer Workshop, BNL 51443 (1981).

7. L.W.Jones, "Air Core Superconducting Toroids as Magnets around Interior Points", Proceedings of the Workshop on Experiments, Detectors, and Experimental Areas for the Supercollider", ed. Rene Donaldson and M.D.G. Gilchriese, Berkeley (1987).

8. "The IEA Large Coil Task", Fusion Engineering and Design", 7,1-232 (1988).

9. B.Dolgoshein, Proceedings of the ECFA Instrumentation Meeting, Barcelona, Spain (1989); B.Dolgoshein *et al.*, CERN-EP/89-161 (1989).

10. A.Tomasch *et al.*, Nucl. Instrum. Meth. **A241**, 265 (1985).

Central TRD/Tracker Hits

Fig. 6

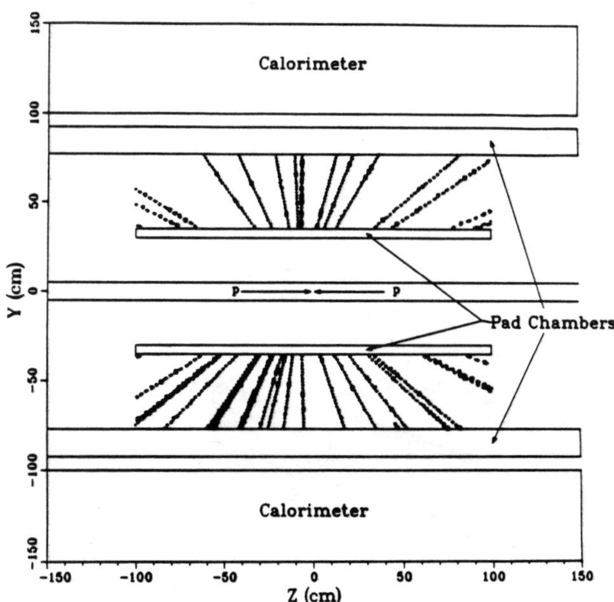

Fig. 7

ETA=1.12
1856
ETA=0.64
MUON CHAMBERS
MUON CHAMBERS
350
1306
CENTRAL TOROID
END TOROID
ETA=2.5
830
ETA=3.0
350
BARREL CAL
ENDCAP CAL
100
FWD CAL
307
QUADRAPOLE (REF.)
CENTRAL TRACKER
700
1492
265
2000

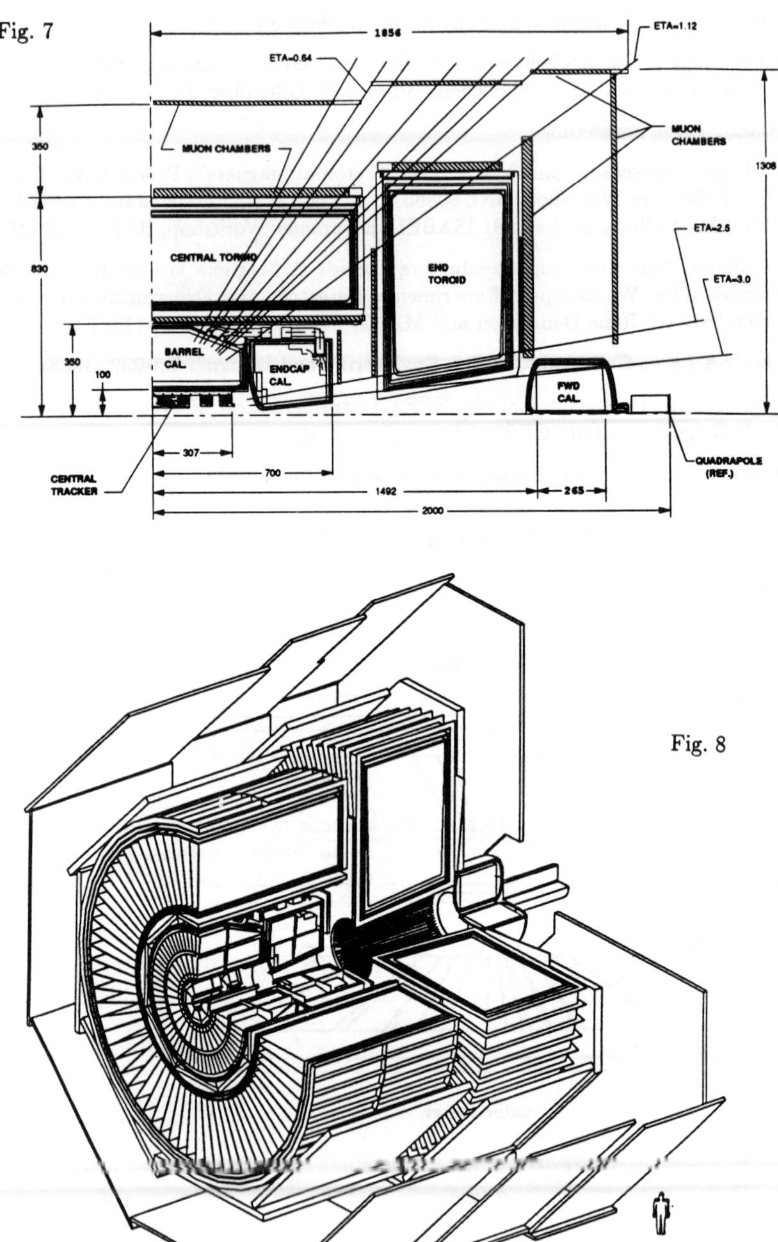

Fig. 8

SOLENOIDAL DETECTOR COLLABORATION FOR THE SSC

L. E. PRICE

Argonne National Laboratory

Argonne, IL 60439

ABSTRACT

Several groups who have been studying the possiblity of a general-purpose solenoidal detector for the SSC have recently joined forces to form the Solenoidal Detector Collaboration. The goal of the collaboration is to build a detector that will be ready when the SSC turns on and which is sensitive to as broad a range of new physics phenomena as possible. While the overall parameters of the detector are determined in a preliminary way, many competing technologies are under investigation for each subsystem. The result of this competition should be the most capable detector possible.

1. GOALS OF COLLABORATION

A collaboration has recently been formed to design a solenoidal detector for the SSC. The collaboration has been drawn from a number of groups which have separately been studying conceptual designs of solenoidal detectors. Approaches have differed and will have to be debated within the new collaboration, but there is broad agreement on designing a detector aimed at sensitivity to as wide a spectrum of possible new physics processes as possible.

Many possible new physics signatures have been suggested, and one goal of the Solenoid Detector Collaboration is to be able to observe essentially all of those that actually exist. Primary on the list are the two missing pieces of the otherwise highly successful "standard model" of elementary particle physics. The first missing element is the top quark, whose existence is all but certain, and whose mass lies somewhere above the 80 GeV mass limit set by CDF. The top quark may well still be discovered at the FNAL Tevatron as it undergoes upgrades during the next decade. If it has not yet been discovered, its discovery will be a major task fo the SSC. Even if it is discovered before the SSC turns on, rates will be low at current machines, and it will fall to the SSC to determine its detailed properties. Since a primary channel for discovery of the top quark is $e^{\pm}\mu^{\mp}$ combinations, this search puts a premium on lepton detection.

The other standard-model element whose discovery will probably await the SSC is the Higgs scalar or other mechanism of electroweak symmetry breaking. At high masses (200-800 GeV), the Higgs can best be observed by decays to two Z^0 bosons, each of which decays in turn to 2 leptons. At the highest masses, it may be necessary to observe $2l2q$ or $2\ell2\nu$ final states. At masses below the $2W$ threshold, the Higgs can be detected either through 4ℓ final states (via ZZ^*) or through decays to $b\bar{b}$ or $\gamma\gamma$. The latter two possibilities require excellent vertex identification and excellent electromagnetic calorimeter resolution respectively. Otherwise, discovery of the Higgs scalar again puts a premium on lepton detection.

Other commonly suggested processes include alternatives to the Higgs scalar such as strong WW scattering (requiring good lepton detection and missing-E_t) and supersymmetric particles (missing-E_t resolution, calorimeter hermeticity) and other possible phenomena such as new gauge bosons (lepton resolution) and quark compositeness (calorimeter linearity).

We are designing a detector to be sensitive to each of these processes that have been proposed as possibilities at SSC energies. But we are also designing a detector that will be as sensitive as possible to the unpredicted new physics phenomena that past experience teaches will be waiting in this new energy region.

2. ORGANIZATION OF THE COLLABORATION

The Solenoidal Detector Collaboration (SDC—at the time of this talk a name had not been officially chosen, but this has subsequently become the collaboration's choice) was formed in late 1989 from groups centered at ANL, FNAL, KEK, and LBL, each of which had spent several months studying approaches to solenoidal detectors for the SSC. An international Interim Steering Committee was set up to draft bylaws for the collaboration. These bylaws were currently before the collaboration at the time of this talk and were to be voted on by January 12. The bylaws were approved, thereby bringing the collaboration into full existence.

The collaboration bylaws call for an Institutional Board with a member from each institution. The Institutional Board deals with general policy issues, including addition of new collaboration members. An Executive Board is defined, consisting of seventeen members elected by the collaboration, which together with the Spokesperson provides the scientific direction of the experiment. Seven of the seventeen members of the Executive Board are from outside the US. The Executive Board appoints the Spokesperson and three Deputy Spokespersons, and approves other major appointments and decisions. A technical organization consisting of a Technical Manager and Technical Board provides the technical base for the experiment. The initial election of the Executive Board took place on January 30, after this talk was given. The Executive Board subsequently chose George Trilling as Spokesperson of the collaboration for the initial period through approval of the proposal.

In January, 1990, the founding membership of SDC consisted of about 270 physicists at 41 institutions in the US, 100 physicists from 14 institutions in Japan, and 30 physicists from 5 institutions in Europe and the Soviet Union.

3. INITIAL PARAMETERS OF THE DETECTOR

Several conceptual detectors were developed by the study groups before SDC was formed. Their broad topology was quite similar, as shown by the sampling in Fig. 1. Each emphasized tracking in a magnetic field provided by a superconducting solenoid inside of the calorimeter. In broad outline, there was consensus on an initial list of parameters as shown in Table 1 (taken from the Expression of Interest (EoI) submitted by the collaboration).

Approximate dimensions have been agreed on as shown in the sketches of Fig. 2. While many technology choices remain to be made, the choice that most affects the overall layout of the detector is the magnet geometry. Fig. 2a is drawn with a "short" magnet that terminates inside the calorimeter, to allow as hermetic a calorimeter as possible. Fig. 2b shows a "long " magnet which penetrates the calorimeter in order to terminate in the iron flux return. The long magnet geometry emphasizes uniform magnetic field, while the short magnet geometry emphasizes hermetic calorimetry. A possible compromise is to use the short geometry with enough iron in the endcap calorimeter to provide a uniform magnetic field in the tracking volume. In order to be as up-to-date as possible, the drawings of Fig. 2 have also been taken from the EoI.

4. CHOICE OF SUBSYSTEM TECHNOLOGIES AND R&D EFFORTS

Although the broad outline of the detector has been sketched by the SDC collaboration, much work remains. Not only must the detailed engineering be carried out, but the basic technology of each subsystem must be chosen. Primarily because of the great technical challenge to SSC detectors from the high interaction rates, short time between bunch crossings, and high radiation levels, none of the developed technologies can be guaranteed to meet all requirements, particularly in tracking and calorimetry. Thus extensive R&D is under way to improve and demonstrate the capabilities of several technologies and there is not general agreement on which is most likely to be satisfactory after the R&D is complete.

Technologies under consideration for the tracking system include silicon detectors, arranged both in pixels and in strips, "straw" drift tubes, and scintillating fibers. It is highly desirable to have silicon detectors at small radius, both as a vertex detector for heavy quark studies, and also for their intrinsically high spatial resolution. The strip arrangement is a relatively mature technology, but will need to be extended to unprecedented numbers and density. The principle concern is about the radiation levels that will be encountered. Pixel-oriented silicon devices are still under development, but should offer substantial advantages at the smallest radii both in superior pattern recognition and in improved radiation tolerance. At larger radii, starting at about 50 cm, straw tubes are the appropriate evolution of the drift chambers that have been used for central tracking in 4π detectors. The cathode surrounding each wire provides isolation and permits higher counting rates. R&D is emphasizing small straw radii appropriate to the high track fluxes. but even with straw diameters of about 4 mm, occupancy is high. Scintillating fibers of about 0.5 mm diameter have been proposed to alleviate the occupancy problem, and also

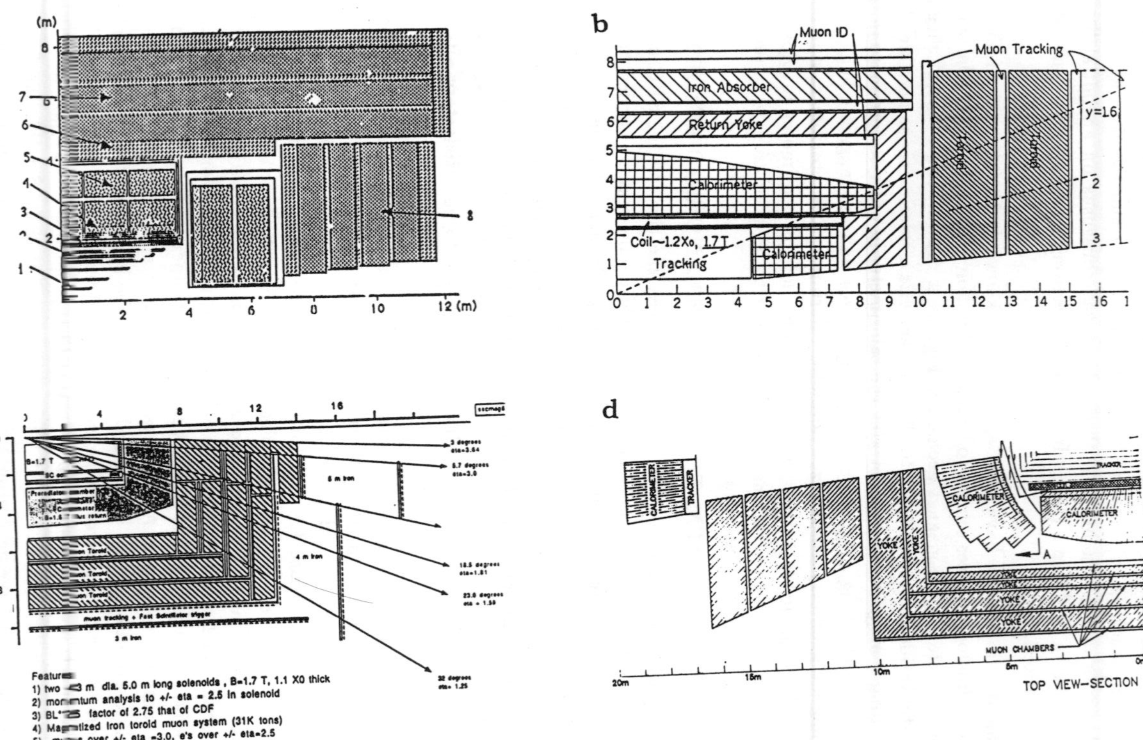

Fig. 1. Examples of detector drawings from pre-SDC study groups. a) Air Core Solenoid detector, b) Hermetic Calorimeter Detector, c) Fast Tracking detector, and d) Fiber and Scintillator Technology.

Table 1. Description and design goals of the SDC detector as presented in the Expression of Interest.

	Central $\lvert\eta\rvert \lesssim 1.5$	Intermediate $1.5 \lesssim \lvert\eta\rvert \lesssim 3.0$	Forward $3.0 \lesssim \lvert\eta\rvert$
Inner tracking:			
Magnetic field	2.0 T	2.0 T	No
Radius	1.85 m	1.85 m	
Combined $\delta p_t/p_t^2$ at 1 TeV/c	0.2 (TeV/c)$^{-1}$	1.0 (TeV/c)$^{-1}$ [a]	
Calorimeter:			
Inner radius or z	2.2 m	4.7 m	17 m[b]
Depth	$\gtrsim 9\lambda$	$\gtrsim 11\lambda$	$\gtrsim 14\lambda$
Segmentation (Had)	0.05–0.10[c]	0.05–0.10[c]	10 cm $\times$ 10 cm
Resolution (Had) $\delta E/E$	$< 0.7/\sqrt{E} \oplus 0.04$[d,e]	$< 0.7/\sqrt{E} \oplus 0.04$	$< 1.0/\sqrt{E} \oplus 0.05$
Resolution (EM) $\delta E/E$	$< 0.25/\sqrt{E} \oplus 0.02$	$< 0.25/\sqrt{E} \oplus 0.02$	
Electron ID	Yes	Yes	None
Muon system:			
Total absorber	$\geq 14\lambda$	$\geq 14\lambda$	—
Combined $\delta p_t/p_t^2$ at 1 TeV/c (with central tracker and toroids)	$\lesssim 0.18(\text{TeV}/c)^{-1}$	$\lesssim 0.25(\text{TeV}/c)^{-1}$ [a]	—

[a] At $\lvert\eta\rvert = 2.5$.
[b] For version shown in Fig. 3.
[c] $\Delta\eta = \Delta\phi$.
[d] E is in GeV unless otherwise specified.
[e] Here and elsewhere, $\oplus$ indicates addition in quadrature.

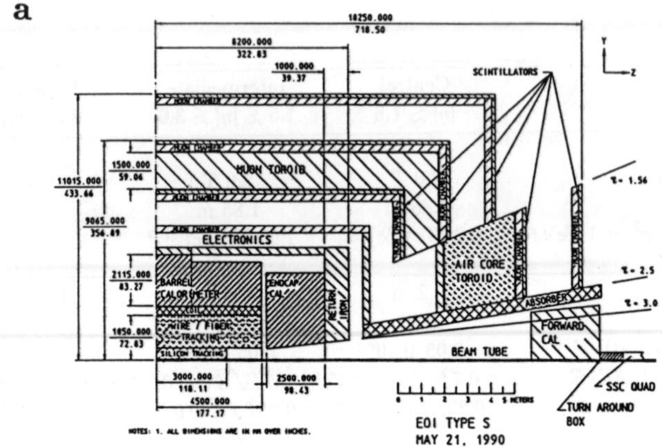

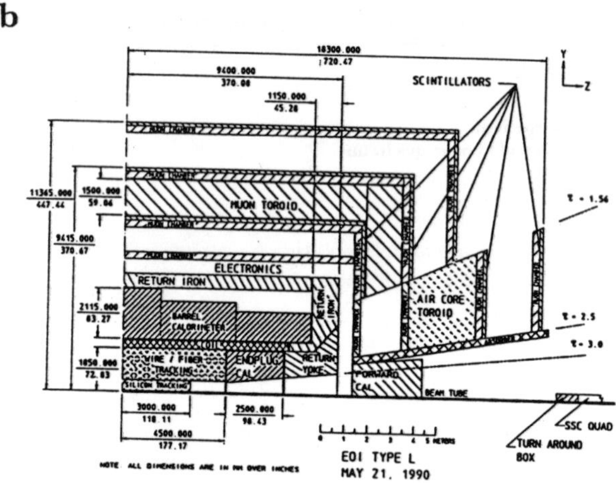

Fig. 2. Initial conceptual drawings of GDC. a) shows the short magnet geometry with the calorimeter surrounding the magnet coil and b) shows the long magnet geometry with the coil penetrating the calorimeter to terminate in the iron flux return.

to provide fast signals appropriate to the 16 ns time between beam crossings at the SSC. Scintillator is traditionally subject to radiation damage, and extensive R&D is underway by several scintillator companies to provide more radiation-hard scintillators. Readout devices are also a subject of much R&D, with interest in both avalanche photodiodes and in new solid-state photomultipliers[].

As noted above, the major choice that must be made on the magnet is the long or short geometry relative to the iron flux return. Detailed engineering is still needed, but conceptual designs indicate that superconducting magnet technology can provide about 2T fields with material of about a radiation length before the calorimeter.

Both liquid ionization and scintillator technologies are under development for calorimetry. Liquid ionization has the advantage of radiation hardness and can easily be configured for the desired projective geometry. Liquid argon is a mature technology, but may be too slow for use at SSC. In addition, the necesary cryostat requires additional dead material both in front of the calorimeter and in boundaries between major segments of the calorimeter. Warm liquids such as TMP and TMS have been investigated in recent years. They offer an improvement in speed and the possibility of better compensation compared to liquid argon. They are still in development, however, and must be proven on an appropriate scale before they can be adopted for use at SSC. Scintillators have a long history as readout media for calorimeters. As with their use in tracking, more radiation hardness is needed, but there seems good probability that the necessary radiation hardness will be forthcoming from current R&D. They also offer advantages in speed of readout. Traditional readout with plates of wavelength-shifting (WLS) plastic limits the geometric possibilities, although Zeus has demonstrated that this technology can be used effectively. Alternative readout systems under development include WLS fibers imbedded in the scintillator plates, giving excellent light collection and allowing flexibility in optical paths to the photomultiplier or other optical transducer. The final possibility is to make the scintillator out of fibers and couple them directly to photomultipliers.

In the central region, the momentum resolution for muons will come from the central tracking system. A system of iron toroids will provide filtering to identify muons behind the calorimeter. For $|\eta| \geq 1.5$, muons are no longer tracked through the full radius of the magnetic field. In this region, air core superconducting toroids are under investigation. Their magnetic field would be chosen to provide comparable resolution in p_t for muons in the range $0 < |\eta| < 2.5$.

Much of the work of moving from the EoI to the proposal will consist of making these choices of technologies, in addition to the detailed engineering of the chosen subsystems and the integration of subsystems into a total detector. All of the candidate technologies are being investigated through the SSC Laboratory's Major Subsystem R&D program. The results of these Subsystem projects will provide the input for deciding between the competing technologies. It will fall to the technical organization of the collaboration to evaluate the merits of each technology and make recommendations to the Spokesperson and Executive Board.

5. CONCLUSION

A strong collaboration has been formed to build a capable, general purpose detector for the SSC. Tracking in a solenoidal magnetic field surrounded by calorimetry and muon identification has been chosen as the approach that will be sensitive to as broad a range of physics as possible. Although much detailed engineering remains to be done, the central task of preparing the proposal consists of completing the R&D underway on several subsystems and making the optimum choice of technologies.

A SCINTILLATING FIBER PRESHOWER DETECTOR FOR TAGGING ELECTRONS WITH SYNCHROTRON RADIATION PHOTONS

P. Mélèse and R. Rusack

Physics Department, Rockefeller University, New York, NY 10021

P. Cushman and V. Singh

Physics Department, Yale University, New Haven, CT 06511

Presented by P. Mélèse

Abstract

We are constructing, and will soon test, a preshower detector that is sensitive to synchrotron radiation photons. High energy electrons (25 to 200 GeV) traversing a magnetic field of the magnitude visualized for SSC detectors will produce synchrotron radiation photons that can be detected by pair conversion. By sampling the early development of showers with fine spacial segmentation, both transversely and in depth (radiation lengths), the synchrotron radiation can be detected. Our prototype detector uses ribbons composed of 0.5 mm scintillating fibers sandwiched between 1/4 radiation length sheets of lead. A simulation of such a detector clearly shows the synchrotron radiation signal. This synchrotron signal could be used to positively identify electrons in the SSC environment.

1.) Introduction

Lepton identification will be crucial to the success of proposed LHC and SSC detectors for Higgs and exotic particle searches. With a high field solenoid magnet, the high energy electrons will radiate synchrotron photons that can be absorbed and detected in a lead/scintillating fiber detector. This synchrotron radiation signal can uniquely identify electrons.

Some characteristics of the synchrotron radiation signal are:

1.) The total energy emitted by synchrotron radiation $I_{SR} = 1.27E^2B^2L$ [KeV], where E is the e^- energy [GeV], B is the magnetic field [Tesla], and L is the path length within the magnetic field [m];

2.) The average synchrotron photon energy $\overline{E_\gamma} = \hbar\omega_c/4$ [KeV], where the critical frequency ω_c is given by $\hbar\omega_c = 1.33E^2B$ [KeV];

3.) The average number of synchrotron photons $\overline{N_\gamma} = 3.8BL$; and

4.) The transverse extent $\delta_{SR} \propto BL^2/E$ of the synchrotron radiation signal is equivalent to the e^-'s deflection within the magnet.

Table 1 assumes a 2 Tesla, 2 m radius solenoid magnet to illustrate the magnitudes of the synchrotron radiation signal for electrons energies between 50 and 250 GeV.

E_e [GeV]	I_{SR} [MeV]	$\overline{E_\gamma}$ [MeV]	δ_{SR} [mm]	$\overline{N_\gamma}$
50	25	1.7	15.0	14
100	102	6.8	7.5	14
150	229	15.0	5.0	14
200	406	26.5	3.7	14
250	635	41.5	3.0	14

Table 1. Synchrotron Parameters for $B = 2$ T and $L = 2$ m.

In addition to lepton identification, precise electron and photon shower localization will be necessary within the high multiplicity environment of the proposed hadron colliders. In particular, the $H^0 \rightarrow \gamma\gamma$ search, will require good γ to π^0 separation. The preshower detector described below will have excellent shower resolution both transversly and longitudinally (in radiation lengths) and should answer the question of how well electromagnetic showers can be isolated.

2.) Apparatus

This detector prototype (Fig. 1) is designed with ribbons of scintillating fibers to achieve the excellent transverse position resolution needed to separate the synchrotron photons from the spreading electron shower.

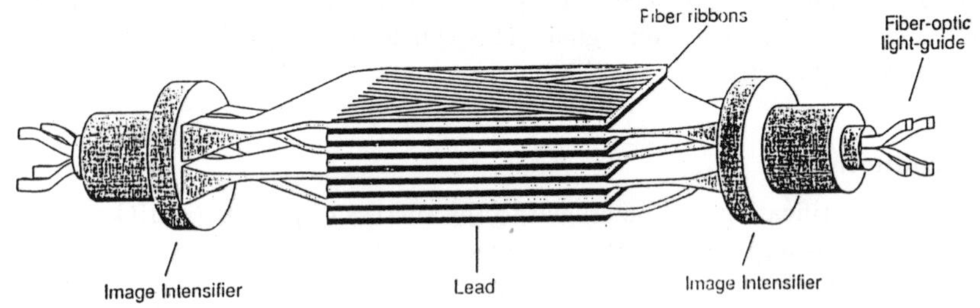

Figure (1)

Yale/Rockefeller Prototype Imaging Preradiator.

The sampling every 1/4 radiation lengths, is sufficient for photon conversion and yet will not completely absorb the relatively low energy (2-50 MeV) synchrotron photons. There are 8 sampling layers in all, spanning 2 radiation lengths.

Each sampling layers is made up of three fiber ribbons, one vertical, one at -15°, and on at +15° to the vertical. Each ribbon is made by winding the 500 μ round scintillating fiber[1] onto a 1 m circumference drum containing groves with a 550 μ pitch. One layer consists of 200 windings of fiber (11 cm wide). A second layer is wound on top of the first, laying within the groves left by the first layer of round fibers. The two layers are glued together to form a ribbon of two, 200 fiber layers offset by 1/2 fiber diameter. The active area of the detector is a diamond 11 cm wide horizontally.

The light from each fiber is detected using an Image Intensifier/Micro-channel Plate/CCD readout chain modelled after experiment UA2[2]. One end of each fiber in the first four sampling layers, approximately 5000 fibers, are glued into a flat plate in 4 grids, one for each sampling layer, and this plate butts up against the front face of the Image Intensifier. The light is demagnified within the Image Intensifier (so that the 500 μ fiber diameter becomes 100 μ), amplified within the Micro-channel Plate, and then the output image is split by 4 fiber optic light guides onto 4 CCD's. There is an identical readout chain for the last four sampling layers. The light from the other end of the fibers is detected by 8 photomultiplier tubes, such that each photomultiplier detects the total amount of light in one sampling layer.

The Pb/scintillating fiber detector is mounted within a light-tight box that stands behind a 2 Tesla analyzing magnet.

3.) Simulation and Status

The detector described above has been Monte Carlo-simulated for a 5 Tesla, 1 m magnet. To extract the synchrotron radiation signal, the electron shower is excluded by cutting the signal within 1.5 mm of the electron shower's centroid. The remaining signal is projected onto the y-axis, where $y = 0$ is the bend plane of the magnet. Figure 2 shows the energy deposited [KeV] versus y [mm] at various Pb depths within the detector for 200 GeV and 50 GeV incident electrons.

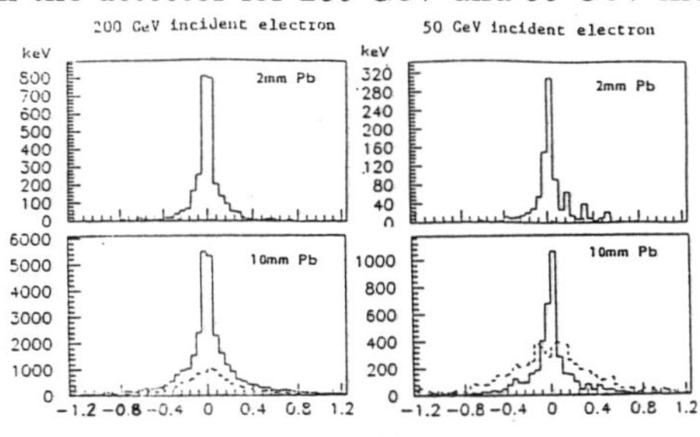

Figure (3)
Energy deposited in SF by synchrotron photons produced in a 5T-1M magnet at a distance x=1.5 mm from the electron shower center and projected onto the y-axis (line C of Fig 2).

The data from this prototype detector has just been collected at Fermilab. The electromagnetic showers are well defined and the results will be presented in a future report.

Acknowledgments

We would like to thank Cathy Ishida, Michael Esaili, Donald Humbert, and Debra Ruszkowski for their significant contributions to this project.

References

1.) The preforms were made by Alan Bross *et al.* at FNAL. These were then drawn, with an EMA coating, by Bicron.
2.) R.E. Ansorge, *et al.*, "Performance of a Scintillating Fibre Detector for the UA2 Upgrade", CERN preprint EP/87-63 (1987).

STATUS OF SCINTILLATING
FIBER CALORIMETRY

M. Sivertz

UC San Diego
Physics Department B-019
La Jolla, California, 92093

ABSTRACT

Preliminary results are presented on the performance of a lead/scintillating fiber calorimeter prototype. Energy resolution, linearity, uniformity, and electron/pion separation are reported on.

1. INTRODUCTION

The SPACAL collaboration has built and tested 37 calorimeter modules in a variety of sizes and configurations.[1] The design is primarily based on scintillating fibers of 1 mm diameter threaded through a honeycomb of lead sheets. The modules were tested at the UA2 test beam at the CERN SPS with electrons, pions and muons between the energies of 10 and 150 GeV. The most recent tests were conducted on a stack of 20 modules. Each module was hexagonal in cross section, 86 mm apex to apex and 2 m in length. The assembled stack had dimensions of 34 cm by 37 cm; insufficient to fully contain an hadronic shower. Figure 1 shows a schematic of the beam ends of the 20 modules.

The fibers were SCSN38, manufactured by Kyowa Gas Co. The lead/fiber ratio was 4 to 1 by volume. Most of the beam tests were performed with the module axis tilted at 3 degrees from the beam to avoid channelling effects seen in earlier tests.

2. ENERGY RESOLUTION AND LINEARITY

Figure 2 shows the results of energy resolution tests. Energy resolution for electrons was measured to be $\sigma/E = 15\%/\sqrt{E} + 1\%$ constant term. For pions $51\%/\sqrt{E} + 2\%$ was the observed energy resolution. It must be noted that this was achieved while containing only about 85% to 90% of the hadronic shower. Improvements are expected when the shower is fully contained.

Linearity for electromagnetic showers was maintained within $\pm 1.6\%$. Uniformity was ensured by the fabrication technique of the stack which had no cracks or seams. Questions concerning uniformity will need to be addressed again when projective modules are tested.

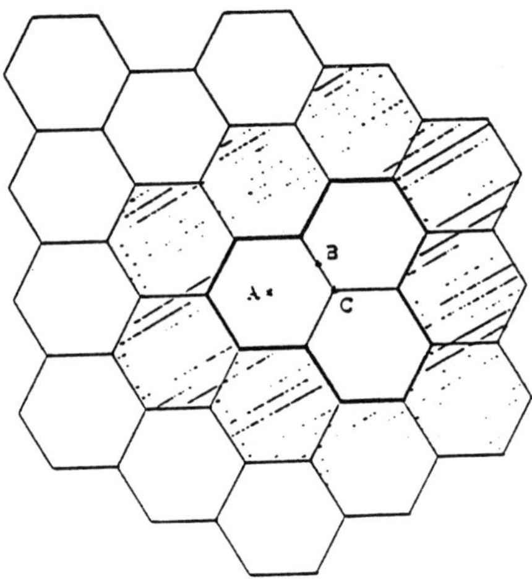

Fig. 1 shows a schematic of SPACAL calorimeter. Also shown is an outline of a typical central cluster of 3 towers surrounded by a ring of 9 outer towers. Energy containment requires the fraction deposited in the outer ring to be less than about 3 % of the total energy in the 12 towers.

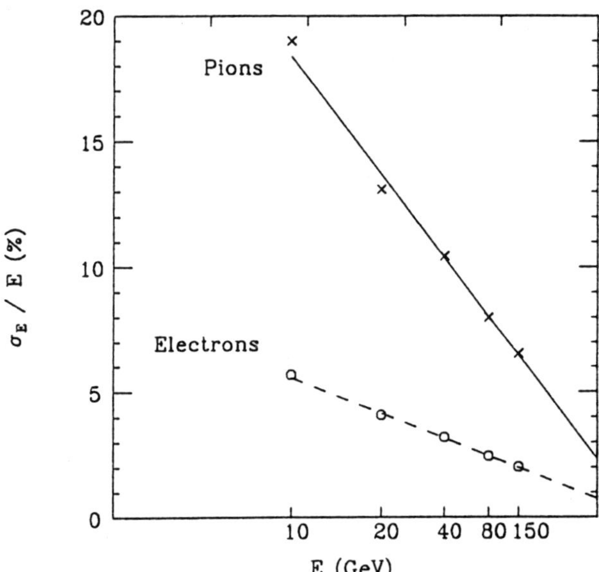

Fig. 2 Energy resolution as a function of beam energy, for electrons and pions. Also shown are the results of fits $\frac{\sigma}{E} = \frac{15\%}{\sqrt{E}} + 1\%$ and $\frac{51\%}{\sqrt{E}} + 2\%$.

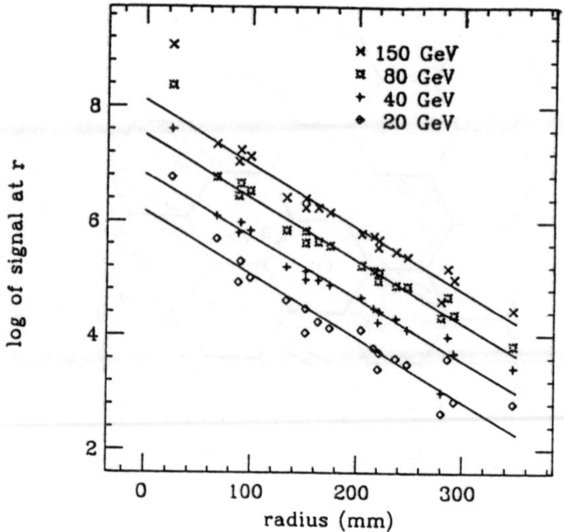

Fig. 3 Pion shower profile for 4 beam energies. The log of the energy deposited is plotted versus the distance from the shower center. Also shown are fits to the exponential shape, excluding the central point.

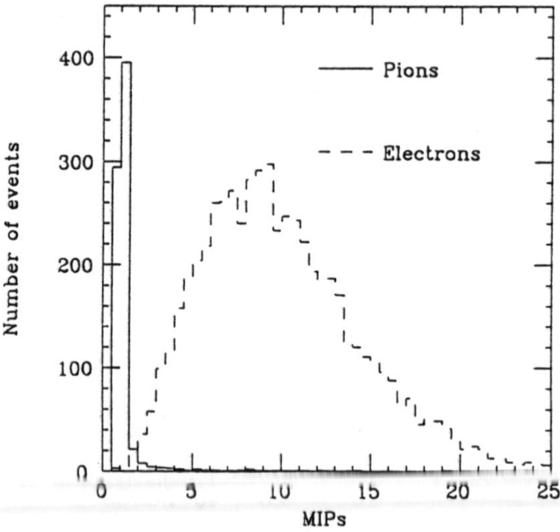

Fig. 4 Electron and pion response in the preradiator counter. 80 GeV beams of electrons and pions. Horizontal scale is in units of minimum ionizing particle response.

3. HADRONIC SHOWER SHAPE

The shape of hadronic showers was measured by directing the beam into the corner of the calorimeter stack allowing us to sample the shower up to an effective radius of 50 cm. Figure 3 shows the shape of pion induced showers as a function of beam energy. There is an electromagnetic core from neutral pions surrounded by a broad hadronic peak. The hadronic part falls off exponentially with distance. The $1/e$ radius of the hadronic part only is about 90 mm, and is independent of energy over the range tested. Using the measured shower shape, it is possible to correct for losses and estimate the degree of compensation for the calorimeter. Estimates of the e/h ratio show results close to 1.0, but await a conclusive test this summer with a larger prototype.

4. ELECTRON/PION SEPARATION

Electron identification is performed in this calorimeter without the benefit of depth segmentation. There are two elements to make use of. The first is a 1 cm thick scintillator plate which is mounted behind 2.6 radiation lengths of lead and tungsten. A cut at 3 times the typical minimum ionizing response passes 98% of the electrons while allowing less than 6% of the pions to pass. Figure 4 shows a comparison of the electron and pion response in this preradiator.

The second element of the electron ID is the fraction of the shower energy which is contained in 3 neighboring modules. A sum over the twelve closest modules was compared to the three central modules, as illustrated in figure 1. If greater than 97% of the observed signal was in the central three modules, an electron trigger would be generated. Alternatively, if there was more than 3% leakage beyond the three module boundary, the event would be considered a pion. Figure 5 shows the containment for electrons and pions. At cut at 0.97 loses only 3% of the electrons while rejecting all but 0.48% of the pions. Combining the two cuts makes it possible to achieve an electron efficiency of about 95% with a pion rejection of 1.7×10^{-4}.

5. SUMMARY

The SPACAL calorimeter has performed very well in measurements of energy resolution, and linearity. Initial indications are that upcoming tests in the summer of 1990 will confirm our beliefs that a lead/fiber ratio of 4 to 1 will achieve compensation and excellent energy resolution will be observed for hadronic showers. Tests next year of modules with projective geometry will prove very interesting. Specifically the uniformity of response across module boundaries and electron/pion separation results that were achieved with parallel geometry modules must be duplicated with the projective geometry modules.

6. REFERENCES

1. D. Acosta *et al.*, Results of Prototype Studies for a Spaghetti Calorimeter, Submitted to NIM 2/1/90.

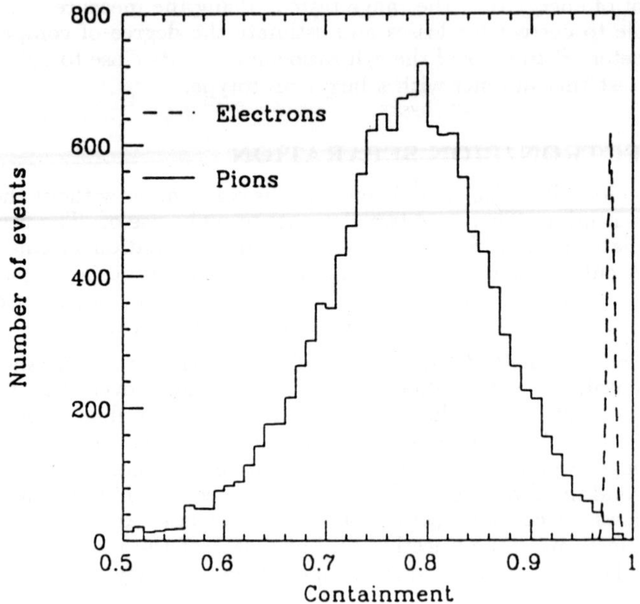

Fig. 5 Containment of electron and pion showers. The containment is defined as the ratio of the sum of three towers central to the shower to the sum of the 12 central towers as illustrated by example in Figure 1. The pions in this sample have already been required to have 1.5 MIPs in the preradiator counter.

Progress in Warm Liquid Calorimetry at PENN

Robert Hollebeek
University of Pennsylvania

1990 DPF Conference, Houston

Ionization Calorimeters

The direct collection of ionization produced in electromagnetic and hadronic calorimeters has been used for many years in high energy physics. Examples include calorimeters using primarily liquid Argon or Xenon for sampling, the liquified noble gases being chosen to eliminate absorption of the direct ionization by the sampling medium. This technology has now been used in many recent large detectors including Mark II, SLD, and D0.

In most of the examples to date, the motivation for using this technique has been a combination of two factors; calibration stability and ease of segmentation in the construction of the calorimeter. The good calibration stability of these devices comes about because the direct ionization yield in the detector depends only on fixed properties of the liquid (i.e. density, Z, de/dx etc.) The detector's calibration scale is therefore very stable. Of course, care must be taken to ensure that the liquid remains pure, and that the amplifier gains can be calibrated with direct charge injection; problems which have been shown to be solvable on a large scale in present detectors. Eliminated in this technique are the problems of variable quantum efficiency, spectral shifts, phototube gain variation, non-uniform light collection, sampling medium aging or radiation damage and other problems which affect techniques which rely on indirect measurements of calorimeter ionization.

Small segmentation in a calorimeter is important for electron identification, jet invariant mass measurement, detection of leptons close to a jet, and heavy quark isolation requirements. Often the desire to employ small segmentation in the design of a calorimeter must be mediated by the realities of mechanical design or by the relative cost of more electronics or readout channels. Relative to scintillator techniques however, liquid ionization has very little problem accommodating both the longitudinal and transverse segmentation requirements encountered in conceptual SSC detector designs.

Increased segmentation can become difficult to achieve when the extra room needed to get the signal out of a device is a significant fraction of the size of the calorimeter segment. An obvious point in favor of liquid ionization devices is that the mechanical space taken up in the design by the extra wire required for a new channel in a ionization device is small compared to the space taken by wave guides, light shifters, and even fiber readout in a scintillator design. In contrast to many current designs for calorimeters utilizing scintillators and/or fibers, longitudinal segmentation is not a problem for liquid ionization devices. This gives added confidence that the calorimeter electron identification will not be compromised. The cost of the additional electronics needed to support longitudinal segmentation is reasonable.

In summary, what started out 15 years ago as a difficult technology (liquid Argon detectors) involving as much art as science in the purification of noble gases and the construction of detectors, has progressed to a relatively standard and well understood technique with good prospects for extension to SSC scale calorimeters. Experience has shown however that two major draw-backs may continue to hinder the use of this technique. The first problem is that since the noble gas sampling medium lacks hydrogen, conversion of some of the hadronic energy flow carried by neutrons (a relatively large fraction of the total energy flow with significant fluctuations and hence significant contributions to the resolution) via slow neutron collisions which produce slow (but high ionizing) protons, is missing. The presence of this effect is now believed to be important in determining the ultimate resolution of a hadron calorimeter. Calorimeters which do have this mechanism are called "compensating" which refers to the balance between the yield expected for electromagnetic and hadronic shower components. Since the typical hadronic shower itself contains both electromagnetic (π^0) and hadronic components with large fluctuations between the two, this balance is important for the optimization of the calorimeter resolution. Tests have now shown that by adjusting the sampling medium thickness or composition so that this balance is achieved, the resolution can be improved by as much as factors of two to three over calorimeters built in the past with typically $100\%/\sqrt{E}$ hadronic resolution.

A further problem encountered in recent ionization calorimeters has been the need for a cryogenic envelope around the calorimeter. As physics focuses more and more on weak interaction physics involving W and Z production as well as future signatures for W$\prime$, Z$\prime$, supersymmetry, Higgs, top etc., the desire both to maintain adequate missing energy resolution for $W \to e\nu, \mu\nu, \tau\nu$ signatures as well as to reconstruct or at least constrain W and Z decays into hadronic channels becomes increasingly important. The need to provide cryogenic isolation for the calorimeter in the case of liquid noble gases conflicts directly with these desires because the vacuum vessel introduces significant dead space into the design which in turn affects both the overall calorimeter resolution and the missing energy resolution. This problem becomes particularly difficult for most detectors in the region between the central calorimeter and the plug or end cap region. The introduction of a magnet into the design makes the problem all the more complex. The most recent example, D0, paid particular attention to this problem, but nevertheless has significant problems in the region $0.8 \leq |\eta| \leq 1.2$ which is right in the middle of the range important for lepton, W, and Z detection from heavy sources ($|\eta| \leq 2$.) Attempts to improve the situation by using chambers in front of these problem regions (i.e. crack chambers) have been made both in conceptual designs (Liquid Argon SSC designs) and in present detectors (CDF). Experience with CDF has shown however that these regions are typically ignored in physics analysis because the loss in resolution, increase in systematics, and increased analysis problems tend to be so great as to not be worth the minor increase in acceptance. Nevertheless, the lost acceptance can be relatively more important for multi-body final states containing for example 4-leptons (an interesting Higgs signature) or multi-jets (for W signatures).

Room Temperature Ionization Calorimetry

By replacing the ionization medium used in the calorimeter with a material which is a liquid at room temperature, all of the advantages of liquid calorimetry can be retained while eliminating many of the difficulties. Several candidate liquids have been known now for a few years thanks primarily to the efforts of the UA1 group at CERN. The most attractive liquid is TMS or Tetra Methyl Silane. It is a very spherical molecule with a tetrahedral arrangement of four methyl (CH_3) groups surrounding a silicon atom. Silanes are chemical compounds which have silicon replacing carbon and are used as surficants, inerting compounds, biological implants, bonding agents and many other uses. TMS is the first by-product in the production of silanes, and is available in very large quantities. Its primary usage is in minute quantities used in NMR spectroscopy (its 12 fold degenerate C-H bond defines the zero for NMR).

The major draw back to TMS is shared by all of the other candidate compounds and that is its flammability. Because of the risk of a Uranium fire in the UA1 calorimeter which is situated in close proximity to a populated area (Meyrin), the UA1 group chose to use TMP (tetra Methyl Pentane) $4(CH_3)\,C_2$ instead of TMS. The disadvantage of TMP is that it has a lower drift velocity. The faster drift velocity in TMS is important for SSC detectors and results in a factor of 2-3 improvement in signal to noise for fixed measurement and shaping times relative to TMP. Minor additional advantages of TMS relative to TMP and other candidate liquids include the ability of the gas phase to withstand high electric fields without breakdown (due to the high vapor pressure) and ease of purification (due to low boiling point relative to common contaminants).

The recent problems encountered by the UA1 collaboration in their attempts to produce the first large scale TMP hadron calorimeter have been the subject of many discussions among physicists. The problems include production delays with the boxes used to isolate the liquid from the reused metal in the old gondolas, low yield·for HV feed thrus and high voltage breakdown within some of the constructed modules. None of these problems appear to be due to fundamental problems with liquids themselves, most are related to the detailed design of the boxes used to contain the liquid. The HV breakdown problem was not seen in prototype modules, and is now believed to be due to contaminants introduced during the filling procedure.

Because of the UA1 experience with the difficulty of producing the liquid containment boxes, and because of the difficulty of extrapolating the "box per gap" calorimeter design to SSC-scale hadron calorimeters, more recent efforts in warm liquid calorimetry have focused on what is now known as a "swimming pool" design in which the metal radiators are immersed directly in the liquid. This is in fact what is always done in liquid Argon detectors. In this approach, increased care needs to be taken in the selection of materials for the calorimeter construction since more of these materials come into direct contact with the liquid itself.

PENN Warm Liquid Program

At Penn, we have concentrated on the use of TMS because of the significant advantages which it offers in the construction and operation of the calorimeter if the safety problems can be solved. The greatest advantage is due directly to the fast drift velocity, which makes this detector one of the fastest radiation-hard, compensating detectors known. The primary goals of our research have been:

1. to develop diagnostic tools which can determine the degree of contamination of liquid samples and identify the contaminants

2. to investigate the available purification techniques, determine which are most effective, and simplify the process so that one can economically produce large volumes of liquid

3. to test the compatibility of the liquid with materials to be used in calorimeter construction.

The purification setup which we have constructed at Penn is presently used for TMS, but is general enough to handle other liquids including liquid Argon in small quantities. A diagram of the system is shown in figure 1. The vacuum pump is a turbomolecular pump which is used because it pumps heavy molecules with high speed, and it is a clean pump in the sense that there is very little backstreaming of heavy compounds. A large gate valve isolates the pump from the rest of the system when necessary. The system has several types of vacuum gauges including thermocouple gauges, a cold cathode gauge and an ionization guage which reads down to about 10^{-11}. The system regularly pumps to less than 10^{-7}. The primary tool for determining the quality of the vacuum and the cleanliness of the system is the residual gas analyzer. This device is a quadropole mass-spectrometer with a resolution of about 0.1 atomic mass units. It is used both to determine the cleanliness of the containers after bakeout or pumping and to analyze samples as they are purified or samples from the detectors. TMS spectra from this device are shown in figure 2. The spectrum has a small peak at mass 88 which is the molecular weight of TMS. The strongest peak is 73 which corresponds to the loss of one methyl group. Isotopic abundance ratios of the peaks can be used to recognize the contributions from TMS and other compounds particularly because of the characteristic values of these abundances for Silicon.

Inlet samples of gases, TMS, residual contaminants from the purification procedures and gas phase from the detectors can now be sent to a gas chromatograph which has just been installed in the last few months. A sample trace from a TMS sample is shown in figure 3. The trace (which comes from a flame ionization detector FID) shows that the TMS from the vendor is 99.96% pure but

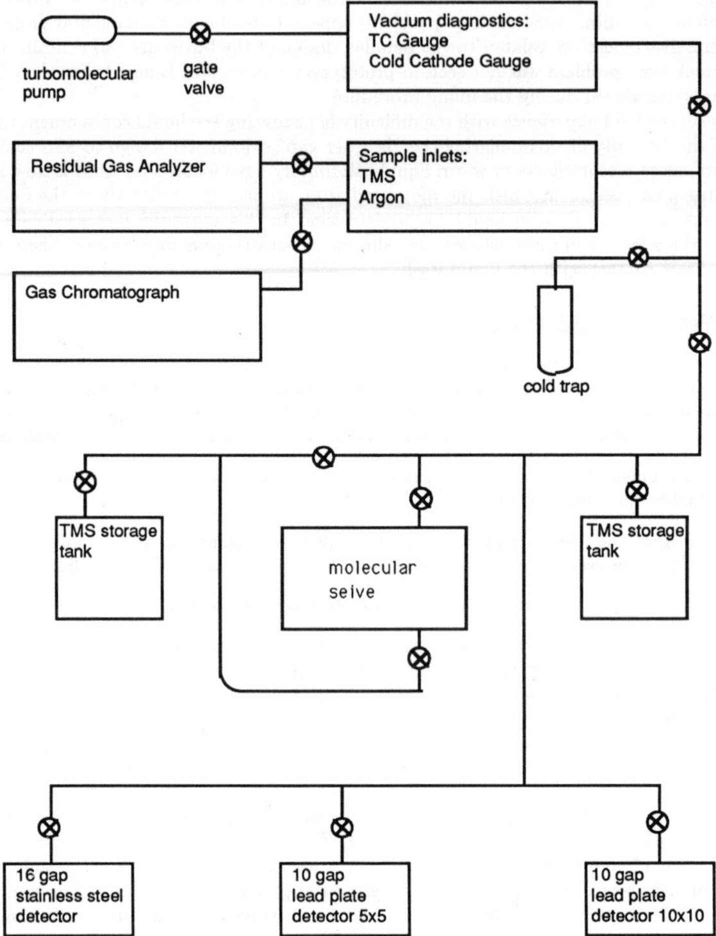

Figure 1: Vacuum layout for the PENN TMS system

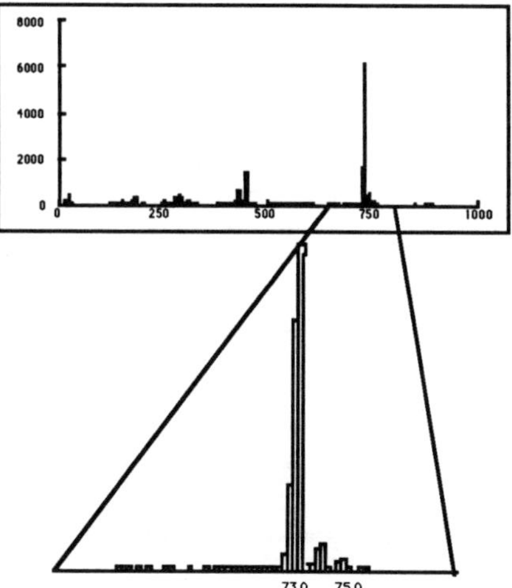

Figure 2: TMS sample Mass Spectrometer signal.

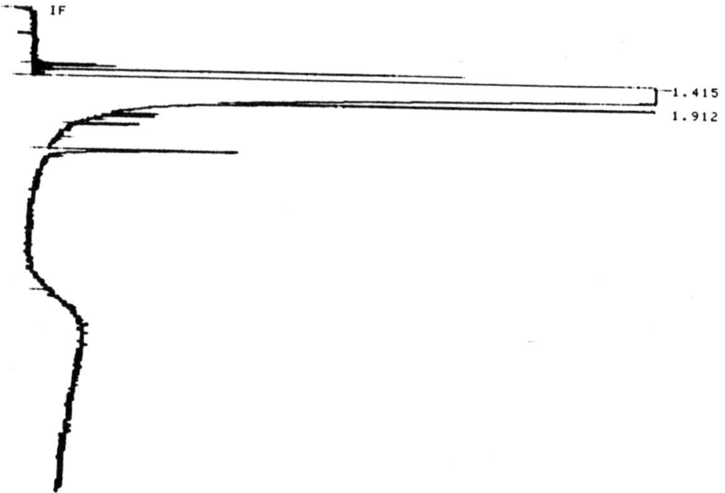

Figure 3: TMS sample Gas Chromatograph trace.

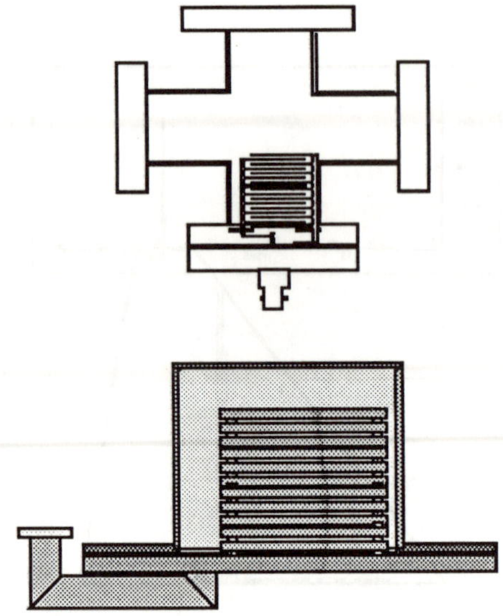

Figure 4: 16 gap stainless and the 10x10cm lead plate detectors.

contains as many as six other compounds. One of these compounds has been identified as Toluene. The gas chromatograph has two parallel detectors: the FID and an electron capture device. This latter detector consists of a large radioactive Nickel source and an ionization detector, and can therefore measure directly the absorption of electrons by the contaminants. In this way, we hope to be able to separate out those contaminants which affect the electron lifetime in the detector from those which do not so that we can concentrate on the former.

During the past year, we have been using three detectors (see figure 4). The first is a 16 gap stainless steel stack with 1.6mm gaps. This is the third in a series of such modules and contains some materials (copper, macor, silver solder connections) which have been determined not to significantly affect electron lifetime in previous tests. This module is used as a normalization module. The input capacitance of the stack is about 120pf.

The second detector is a 10 gap stack of 5x5 cm lead plates, each 1/4 in thick. The stack is constructed form lead-antimony alloy used for construction of liquid argon modules. A similar detector has 10x10cm plates which gives it roughly the size and input capacitance of an SSC hadronic calorimeter tower. These detectors were the first ones to demonstrate the feasibility of operating a detector with TMS in direct contact with lead plates. The cleaning procedure used for the plates has been a standard chemical wash sequence using high megohm water, acetone and alcohol. Auger surface analysis of lead plates treated in this way has since shown that the surface is as clean as is usually achieved with baked stainless steel.

Both detectors have had the liquid in contact with the plates for long enough to measure the time dependence of the charge yield from the TMS. The yield for the larger detector as a function of time is shown in figure 5. Since it has been shown with this series of tests that it is possible to achieve reasonable charge yield in this configuration, it is possible to contemplate calorimeter designs for TMS and TMP which are quite similar in their construction techniques to those used in liquid argon detectors.

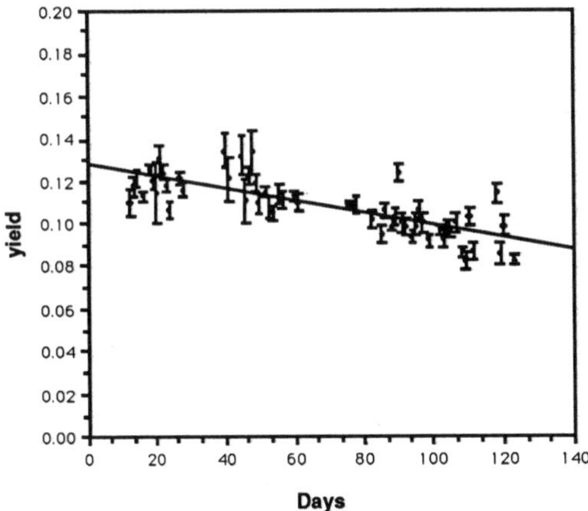

Figure 5: Time dependence of the yield from the big detector with TMS in direct contact with the lead plates.

Tests performed during the past year include:

1. Continued studies of the yield versus voltage curve for samples in the stainless detector. Comparisons are made between the shape of this curve and the Onsager theory to extract electron lifetime and charge recombination radius (see figure 6). Measurements at high voltage are used to compare the electron yield achieved from different batches of the liquid each with slight modifications to the purification procedure.

2. Measurements of the yield in the small lead detector and studies of the relative calibration of the amplifier systems on the lead and stainless detectors.

3. Yield measurements on the big lead detector as a function of time until a degradation could be measured. The results of these measurements were shown previously in figure 5. The measured degradation looks small enough that a modest amount of cleaning of the liquid should be sufficient to keep a detector functioning for years. Studies continue to try to find the origin of the effect. There are at least three possibilities. The detector is sealed with a 20cm diameter Viton ring which has a potential leak rate sufficient to explain the effect. Future modules will incorporate a standard copper gasket and conflat flange to eliminate this possibility. It is also possible that the contaminant is leaching slowly from the lead plates, or, the contamination may be the result of water and oxygen desorption from the stainless steel module walls. For the latter case, consultation with the surface experts in the condensed matter group at PENN has indicated that it may be useful to design the vacuum walls from Aluminum instead of stainless. While more difficult to weld, the Aluminum provides much better vacuum desorption properties than the stainless. This is fortunate since in the EG&G design for a full size SSC calorimeter in which we participated, it was found that replacing some of the structural materials with Aluminum was desirable from the standpoint of minimizing the amount of dead material seen by particles interacting in the calorimeter.

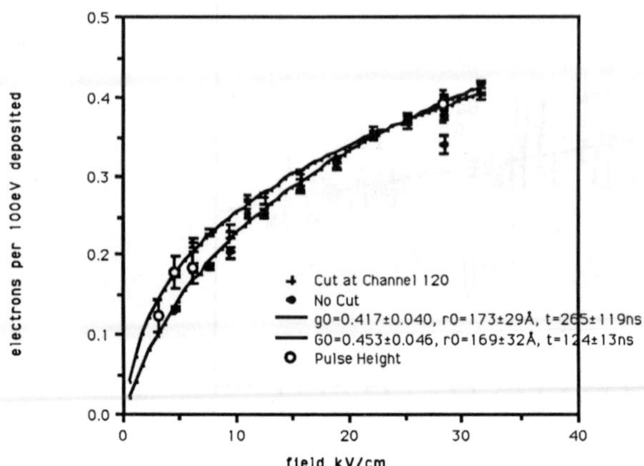

Figure 6: Comparison of the yield vs HV for Onsager theory.

4. Comparisons were made between several preamps and post amplifiers in order to minimize the noise. This is particularly important in the case of the full scale lead detector which has an input capacitance of almost 1 nf. This large input capacitance makes it difficult to design low noise high speed electronics for the calorimeter front end. The best performance to date has been achieved using a Rel-Labs preamp with standard spectroscopy amplifiers.

5. During the past year we installed the mass spectrometer system and have been using it to collect spectra of the TMS samples as they come from the vendor and after they have been purified. Work is underway on software to extract peaks from these spectra and to use the splitting patterns and isotopic abundance tables to identify contaminants.

6. A gas chromatograph was attached to the system within the past month with a standard detector and a detector which is specifically sensitive to compounds which will affect electron lifetime in the liquid.

7. The electronics has been modified to allow the digitization of the pulses using a Lecroy waveform digitizer. Individual pulses are then sent to the computer (which is a MacIIC using a Micron interface and the MacSys software from CERN) which then stores the data and performs pulse averaging. Pulses are fit to extract the lifetime of the electrons in the liquid. An example of the pulse shape and fit is shown in figure 7. The electronics response function must be known in order to extract the electron lifetime from this type of data. This is done by running the detector at high voltage. At 4500V for example, the drift time in our detector is 58ns which represents almost an impulse input for the amplifiers. Electronics rise and fall times are adjusted to fit the high voltage data (see figure 8).

8. Modifications of the basic purification procedure have been tried including multiple passes through the molecular sieve, high speed versus slow speed transfers, and low temperature versus high temperature distillation.

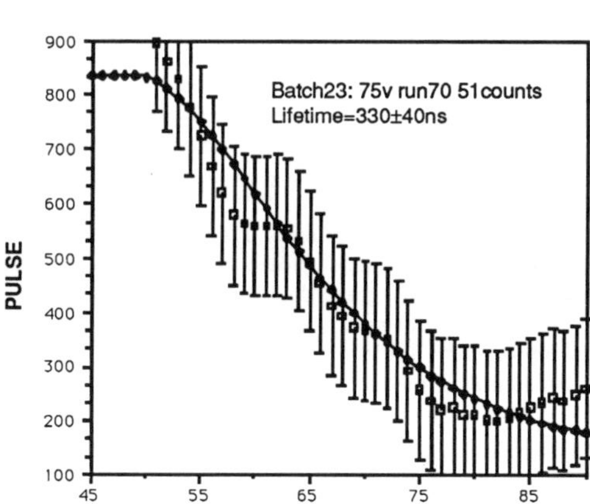

Figure 7: Average pulse waveform with a fit for the electron lifetime.

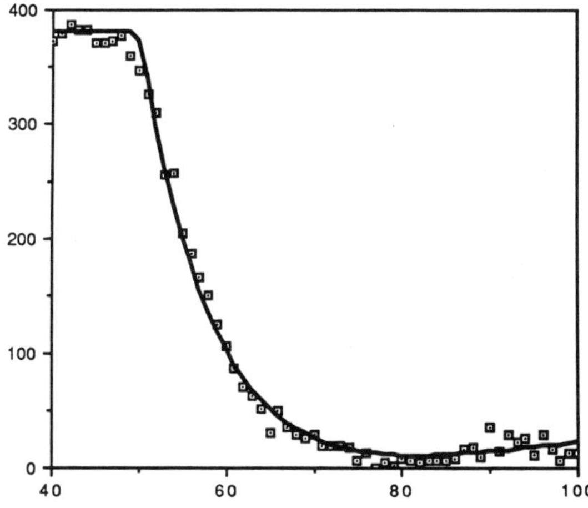

Figure 8: Average pulse waveform at high voltage with a fit for the electronics response.

Plans for Next Year

In addition to the continuing efforts to optimize the purification steps for TMS and to use the mass spectrometer and gas chromatograph as diagnostic tools, during the coming year, an Aluminum version of the containment vessel for the lead detector will be constructed, and the old vessels will be used to further test materials for compatibility with TMS including G10 and Ceramic circuit board materials, a prototype stack using Kapton foils, and other materials to be used in the construction of future hadron calorimeters. Under the Large Subsystem R&D program for Room Temperature liquids, we will also be testing materials to be used in the construction of the test beam module (TBM) to be constructed under that proposal. Also as part of that effort, we are designing and testing front-end electronics and readout systems for the TBM.

Simple methods are now available for the purification of TMS, and the operation of this liquid in the presence of lead calorimeter plates brings us much closer to being able to design TMS detectors using well-proven liquid Argon technology. A great deal of progress has thus been made in the last two years in the study of room temperature liquids for ionization calorimetry, and this new technology continues to be an interesting candidate for future calorimeter designs requiring fast, compensating and radiation hard sampling media.

Fast Warm Liquid Calorimetry *

D. DiBitonto, T. Pennington, M. Timko, and L. Wurtz

The University of Alabama
Tuscaloosa, Alabama 35487

S.-Y. Lee, K.-Y. Ling, H.-C. Liu,
and P.M. Van Peteghem

Texas A&M University
College Station, Texas 77843

Abstract

We report measurements on a fast, ultra-sensitive charge preamplifier for warm liquid calorimetry that is manufactured in an industrial BiFET process with 5 Megarad (1 Gigarad) radiation hardness. The measured noise level of this device is less than 3.5 nV/$\sqrt{\text{Hz}}$ with a 37 (10) nsec risetime.

I. Introduction

Warm liquid ionization media have recently attracted interest as a competitive technology for calorimetry in high energy physics experiments. Mainly in response to the inherent limitations of speed, hermiticity, and accessibility of cryogenic technology, warm liquid calorimetry offers the advantages of faster detector response time, flexible modularity in detector design, and the ability to measure electromagnetic and hadronic energy with equal detector response[1] (compensation). The choice of detector technology for hadron calorimetry at the SSC depends largely on the design considerations of detector medium speed, segmentation, and radiation hardness. Warm liquid hydrocarbon and silane media such as tetramethylpentane (TMP) and tetramethylsilane (TMS) appear to satisfy most detector requirements for radiation hardness[2,3] and segmentation (either as longitudinally segmented modular vessels, or as "swimming pool" type detectors).

---

* work supported by the Department of Energy, contract DE-AS05-81ER40039

With the development of fast, radiation-hard charge preamplifers for warm liquid calorimetry[4),5)], fast detector response is possible despite the relatively slow drift velocities in these liquids. The current signal generated within these liquid ionization chambers contains information on the total charge deposited at the <u>beginning</u> of this pulse[6)]. Fast front-end amplifiers can now be designed with matching bandwidth which can therefore respond to this initial current pulse with an output amplitude proportional to the initial (total) charge. To maintain the intrinsically fast response of the front-end electronics (≤ 10 nsec), the preamplifier must be mounted directly on the detector. Detector capacitance, in turn, must be kept as low as possible, either by the choice of small electrode size, or by series connection of much larger anodes.

II. <u>Circuit Design</u>

A monolithic charge preamplifier manufactured in a radiation-hardened industrial BiFET technology has recently been developed for warm liquid calorimetry[5)] with a 37 nsec risetime, 3.5 nV/$\sqrt{\text{Hz}}$ noise, and 5 Mrad radiation hardness. To compensate for the increase in flicker noise[7)] ($1/f_T$) from lower transistor β value after exposure to radiation, this original circuit has been modified and manufactured in another industrial BiFET process with higher radiation hardness (order 1 Gigarad). This circuit is shown in Figure 1.

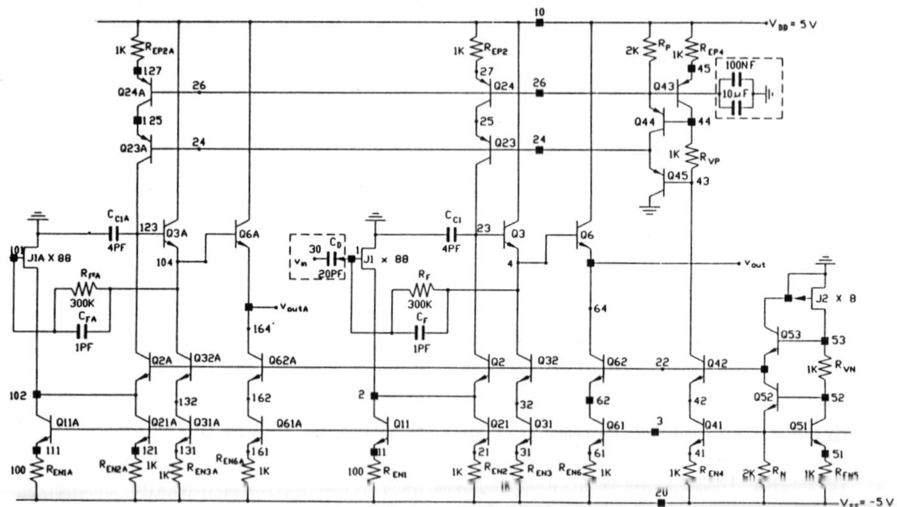

Figure 1. Circuit schematic of integrated low-noise BiFET charge preamplifier.

III. Test Results

The circuits of Figure 1 were evaluated with a 1 GHz digitizing oscilloscope (model HP54111D) and read out with an HP 9000 workstation. An HP8082A pulse generator provided a 1 nsec risetime reference. The output from this amplifier is shown in Figure 2. Tests are presently underway to evaluate the performance characteristics of this device after exposure to γ rays (order 1 Gigarad) and high neutron fluency (order $10^{14}\mathrm{cm}^{-2}\mathrm{yr}^{-1}$).

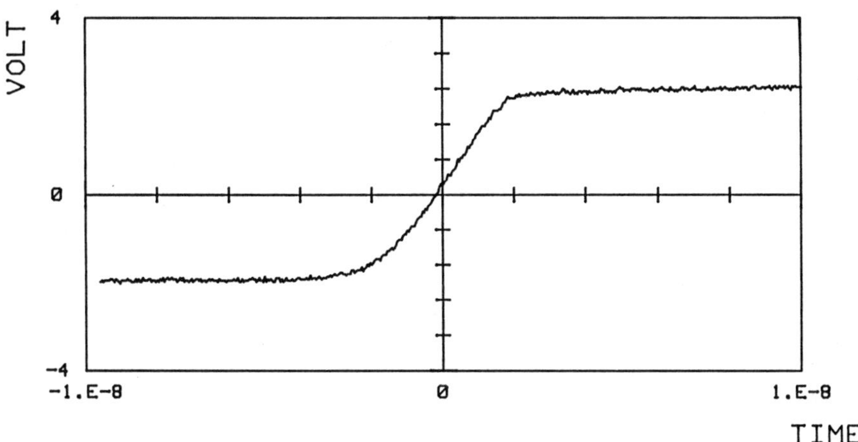

Figure 2. Output response of BiFET charge preamplifier to square wave input.

References

1. R. Wigmans, NIM A259, 389 (1987).
2. R. Holroyd, SSC-SR-1035, p. 335, June, 1988;
   R. Holroyd and D. Anderson, NIM A236, 294 (1985).
3. D. Groom, SSC-SR-1033, June 10,1988.
4. D. DiBitonto et al., NIM A279, 100 (1989).
5. D. DiBitonto et al., Proceedings of the Workshop on Calorimetry for the SSC, Tuscaloosa, Alabama, March 13-17, 1989.
6. W.J. Willis and V. Radeka, NIM 120, 221 (1974).
7. P.M. Van Peteghem et al., Proceedings of the ESSCIRC, San Francisco, September, 1989.

Performance of a Silicon Microstrip Detector in a High Radiation Environment

C. S. Mishra and C. N. Brown
Fermilab, Batavia, Illinois

J. Kapustinsky, M. J. Leitch,P. L. McGaughey,
J. C. Peng, W. Sailor, and K. Holzscheiter
Los Alamos National Laboratory, Los Alamos, New Mexico

M. Sadler
Abilene Christian University,Abilene,Texas

The performance of a silicon microstrip detector has been studied in a high rate environment using electron, pion, and proton beams. The pulse height, time response, and leakage current have been studied as a function of particle fluence up to a total integrated flux of about 4×10^{14} protons/cm^2.

Silicon microstrip vertex detectors are a major new component of Fermilab experiment 789[1], designed to study the rare decays of the B and D mesons in a very high radiation environment. Several other experiments are also using silicon detectors as precision tracking devices. There are also several proposals to use silicon detectors for central tracking at the SSC. The radiation levels in these experiments will be very high and it is important to understand the performance of these detectors and their associated electronics in such an environment. Because of these proposals, in recent years there has been a lot of activity studying radiation damage to particle detectors[2].

We have studied the performance of a silicon microstrip detector (SMD) at the Los Alamos Meson Physics Facility to determine the effects of high fluxes of ionizing radiation. The details of this measurement will be published in a forthcoming publication[3]. A high intensity 0.8 GeV proton beam was used to study the effects of radiation damage on pulse height, timing resolution and leakage current. The detector was exposed for several different periods, at rates varying from 10^8 to 10^9 proton/sec/cm^2 over several hours, up to a total integrated flux of 4×10^{14}. In the pion test channel (beam momentum ranging from 500 to 700 MeV/c^2), rate effects on the pulse height and timing resolution were studied.

The SMD, build by Micron Semiconductors, was 5 cm $\times$ 5cm, approximately 300 micron thick, and made of N-type silicon with ion implanted P type strips at 50 micron pitch. Eight strips in the central region of the detector were instrumented with Fermilab ASIC preamplifiers. These preamps are a bipolar-current amplifying type developed on an ASIC by the Research Division[4] at Fermilab. The outputs of these preamps were digitized using CAMAC based ADC's, TDC's, and scalers. The ADC gate and TDC start were provided by the accelerator beam gate. The time interval between beam pulses was 360 ns and each beam spill lasted for $625\mu s$ at 60 Hz. The duty factor of the accelerator was about 3% with an average silicon strip occupancy of about 10%. Five strips throughout the detector volume were also instrumented to measure any increase in leakage current due to radiation damage. All the other strips were grounded.

The proton beam was monitored throughout the run with a beam profile monitor which recorded the beam spot size and position, and an ion chamber which measures the beam intensity. Eight 1 cm $\times$ 1 cm aluminum activation foils were also placed over the active area of the detector to determine the total integrated flux seen at different locations on the detector. The fluence varied from 10 Mrad at the center of the detector to 100 Krad at the top and bottom.

The pulse height and time response of the detector, generated by the beam gate "Start", and the discriminated silicon "Stop" were monitored throughout this measurement. A sharp reduction in proton peak pulse height was observed between 5.5×10^{12} and 1.4×10^{13} protons/cm^2. The loss in the amplitude of the proton pulse height could be recovered by increasing the bias voltage. Further reduction in pulse height was observed at higher fluences. Some of this loss in pulse height is also due to a reduction in preamp gain caused by a high level of leakage current. The silicon output was DC coupled to the preamp input. Radiation damage to the preamp at its location, about $24''$ from the beam axis, was very small. The loss in amplification of the preamp due to radiation damage was measured to be about 10% for a fluence of 10^{13} protons/cm^2 on the preamplifier. At present, we are attempting to correct our silicon pulse height data for the gain loss due to the increase of leakage current. An increase in charge sharing between adjacent strips was observed at higher fluences. No significant change in the time response of the detector occured as a result of the irradiation. Hence there was no severe degradation of the drift time within the silicon, or of the shape of the leading edge of the pulse.

An increase of two orders of magnitude in the reverse bias current was noted after irradiation of the central strips. Fig. 1 shows the increase of the leakage current as a function of bias voltage at different particle fluences. The damage constant ($\alpha =$ Increase in leakage current/Particle fluence) calculated at 90 V bias voltage and 20° C from this data is 1.8×10^8 nA/cm, which is in good agreement with other measurements[5]. At a particle fluence of 3.75×10^{14} the SMD exhibits a resistive characterstic rather than a diode characterstic. During the irradiation a large increase in the leakage current was observed when the detector bias was on, compared to when it was off. This rapid increase in the bias-on leakage current decayed at a faster rate than the annealing rate for radiation induced leakage currents measured with the bias off. This effect is believed to be due to the buildup of charge at trapping centers in the SiO_2 layer.

After the proton beam exposure, the detector was transfered to a temperature controlled box to study the dependence of the leakage current on temperature and to study annealing at room temperature. The leakage current as a function of bias voltage at different temperatures is shown in Fig. 2. There is an order of magnitude drop in leakage current for every 20° C drop in temperature. The leakage current of the detector was also monitored to study room temperature annealing. A 50% decrease in the leakage current was noted after 120 days. The annealing rate seems to decrease as a function of time. The room temperature annealing data is shown in Fig. 3.

Summary

In conclusion, we observe that irradiation causes a Silicon Microstrip detector's performance to degrade due to an increase in leakage current, which is roughly proportional to the fluence. After irradiation the SMD required a higher bias voltage for full charge collection. The leading edge of the silicon pulse is uneffected by radiation. Such detectors can be operated successfully above 10^{13} minimum-ionizing particles/cm^2, perhaps even up to 10^{14}. We also find that cooling apparently is more effective method than annealing for continuing to use a radiation damaged detector.

We thank the Los Alamos Meson Physics Facility staff for providing beam for this study and the Fermilab Research Division for providing the preamps. This work was supported by the U.S. Department of Energy.

References

1. Fermilab Proposal 789, D. M. Kaplan and J. C. Peng, Spokesperson, Oct 1987.

2. M. Gilchriese, Editor, Radiation Effects at the SSC, SSC-sr-1035, 1988.

3. C. S. Mishra, J. Kapustinsky et al., To be submitted to Nucl. Instr. and Meth.

4. D. Christian et al. IEEE Trans. on Nucl. Sci NS-36(1988) 507 ; T. Zimmerman, presented at 1989 Nucl. Sci. Sympo.

5. T. Ohsugi et al., Nucl. Instr. and Meth. A265 (1988) 105 and reference therein.

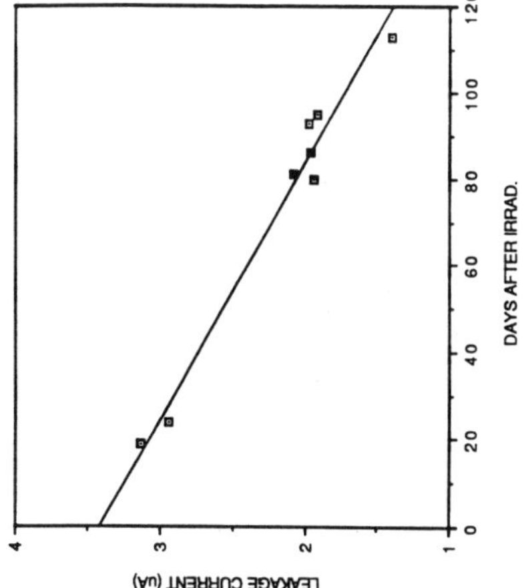

Fig. 3. Leakage current as a function of time elapsed after the exposure at 20° C temperature and 90 volts bias.

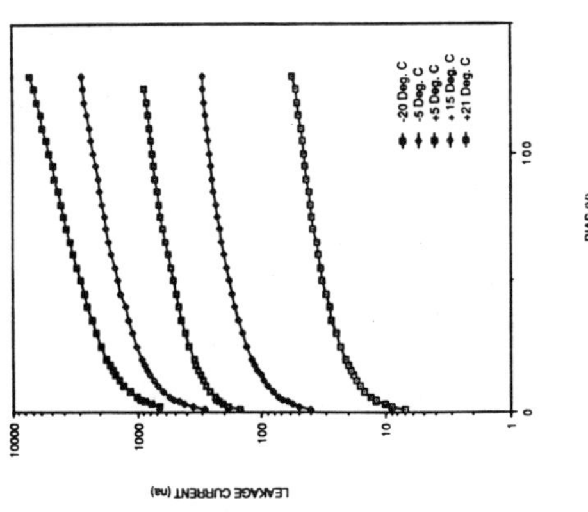

Fig. 2. Temperature dependence of the leakage current after irradiation.

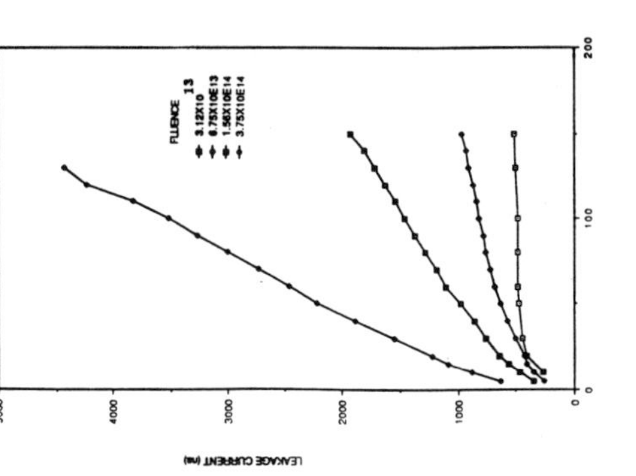

Fig. 1. Leakage current dependence on the bias voltage at several fluences.

Graded Potential Drift Chambers
With Applications to AMY Forward Tracking and the SSC

Alan F. Sill, AMY Collaboration, University of Rochester

We report on new construction methods for high-precision drift cells that provide a linear voltage-to-distance relationship over large regions within the cells, and hence lead to a linear time-to-distance calibration. These methods are being used to construct a pair of forward angle drift chambers to cover the angular range from 14° to 30° in the AMY experiment at the $e^+ e^-$ collider Tristan, and are discussed in relation to their possible use in SSC experiments.

As a result of an effort to develop rectangular-celled drift chambers with comparatively large cells (and hence lower instrumentation costs) for use in the forward angular direction at AMY, we have developed a set of drift cells that are based on the use of sheet films of resistive plastic in order to provide a linear voltage-to-distance relation within each cell of the chambers. They are formed from thin (80 μm) sheets of high resistance (10^{12} Ω per square) resistive-Kapton[1] plastic, preprinted with parallel lines of conductive epoxy paint, which are then laminated to 2 mm thick panels of nonconductive polymethacrylimide[2] foam.

The conductive traces, which are 2 mm in width, are printed at 1.5 cm intervals directly onto the high resistance Kapton. These correspond to the centers and the sides of the 3 cm wide cells. Low-resistance side walls, made of a high conductivity (200 Ω per square) resistive Kapton laminated to 1 mm thick Rohacell foam, are used to complete the sides of each cell. The conductive traces in contact with the cell walls are placed at negative high voltage, and the center strips are placed at ground potential. The high-resistance Kapton acts as a continuous, electrically noise-free sheet resistor between adjacent conductive traces to form a uniform potential gradient from the sides to the centers of the cells. Distortion of the equipotentials within the cell by the sense wire, which is placed at positive high voltage, is minimal, as shown in Figure 1. These cells consequently exhibit a uniform electric field throughout most of their width, leading in turn to an extremely linear time-to-distance relationship.

This technology is being used to construct the forward tracking chambers (FTCs) for the AMY experiment located at the e^+e^- storage ring TRISTAN. The chambers will consist of approximately 2500 drift cells to be housed within the magnetic field of the detector to provide tracking in the endcap regions. They will cover approximately the 14° to 35° angular range with respect to the beam line. The chambers are round, with outside diameters ranging up to 1.3 meters, and have a circular central cutout approximately 50 centimeters in diameter as clearance for the beam line and final focusing magnets of the accelerator. The cells are stacked together in layers, which in the case of the AMY FTCs have an internal-gas thickness of 1.5 cm. There are 21 layers of cells per endcap, with the cells arranged in sets of three staggered layers per orientation in each of three orientations. The wires in each set of three layers differ in orientation by 60 degrees from those in the preceding and following sets of layers.

We built several prototypes using other methods of construction, including the multiple parallel strip method which involves the use of an external resistor chain. The resistive kapton method appears to be superior in terms of construction time and effort, as well as overall complexity (since it gets rid of the external voltage divider and field shaping traces or feed throughs), at roughly comparable cost.

It's well known that tracking chambers for muon detection at the SSC will need to have significantly improved resolution compared to those used in previous detectors, on the order of 100 to 150 μm in typical SSC detector designs. Drift chambers that have active field shaping stand the best chance of achieving such resolution in cells large enough (on the order of a few cm) to allow designs that are reasonably cost-efficient, considering the large amount of muon tracking real estate that needs to be covered. In addition, high technology materials and methods, such as the use of composites and ultralight and reinforced plastics, may be needed in place of the traditional aluminum extrusions in order to achieve the needed strength-to-weight ratios, rigidity, and economy of production that will be needed to reach this level of precision with chambers of the size needed for use at the SSC.

To summarize, a method of achieving precise linear voltage-to-distance relations within cells on the order of one to many centimeters in size has been developed and applied to a detector for forward tracking inside of a magnetic field in AMY. This method may also have significant advantages for many different kinds of future chambers, including those of interest for possible use at the SSC.

REFERENCES

1. XC series resistive Kapton^{TM} polyimide film, DuPont Corporation, Wilmington DE 19880

2. Rohacell^{TM} polyimide/acrylic alloy foam, Röhm GmbH Chemische Fabrik, Darmstadt, W. Germany

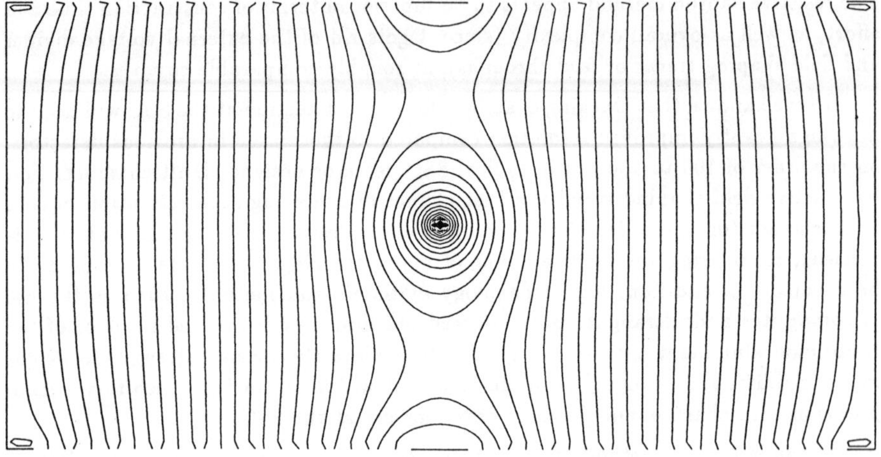

Figure 1. Equipotentials within an individual cell of the AMY Forward Tracking Chambers. The drift field is at or above 1.0 kV/cm throughout most of the cell in this example, which is plotted for the case that the cell sides are held at −1.3 kV and the sense wire is at +1.3 kV. The central 2 mm strips on the top and bottom walls are shown at ground potential.

THE KNIFE-EDGE CHAMBER
A Track Chamber for SSC Detectors

E.F. Barasch, T.J.V. Bowcock, M.M. Drew, S.M. Elliott, B. Lee,
P.M. McIntyre, Y. Pang, M. Popovic, D.D. Smith, J. Wahl
Department of Physics, Texas A&M University, College Station, TX 77843

ABSTRACT

The design for a new technology for particle track detectors is described. Using standard IC fabrication techniques,
a pattern of microscopic knife edges and field-shaping
electrodes can be fabricated on a silicon substrate. The
knife-edge chamber uniquely offers attractive performance for
the track chambers required for SSC detectors, for which no
present technology is yet satisfactory. Its features include:
excellent radiation hardness (10 Mrad), excellent spatial
resolution($\sim$20 μm), short drift time (20 ns), and large pulse
height (1 mV).

The Superconducting Super Collider presents formidable challenges
to the detector technology required to instrument its experiments. The
challenges are particularly distressing in the central magnetic
spectrometer[1] which is the heart of most collider detectors. Figure 1[1]
shows simulations of two "typical" interesting events as they would be
recorded in a conventional track chamber: a Higgs boson which decays
into 2 Z bosons; and a 2-jet quark scattering with a momentum transfer
of 1 TeV. Such events contain $\geq$200 charged particles, half of which
curl partly or fully within the chamber volume. On average there are
1.6 pp collisions in each bunch crossing, and the resolving time of most
track chambers would indiscriminately record tracks from several consecutive crossings. In total some 20,000 signals must be processed from
each event, and the (few) interesting tracks must be extracted from the
(many) irrelevant backgrounds. The demand upon resolution, speed, and
radiation hardness for tracking devices is one of the great challenges
for SSC detector design.

The Accelerator Research Laboratory at Texas A&M University is
developing a new technology for precision track chambers in which the
anode plane of a multi-wire proportional chamber is replicated in min-
iature on a silicon substrate. The heart of the design is a array of
microscopic knife-edge electrodes which can be fabricated in large
arrays on a silicon substrate using conventional IC process techniques.[2]
The knife-edge array can be configured with field-shaping electrodes and
suitably biased to produce an extended region of high field near the
knife-edge surface, as shown in Figure 2.

The knife-edge chamber has a number of advantageous properties for
track detection at the SSC. A knife-edge spacing of 50 μm can be
readily achieved. Operating in an appropriate gas (Xe/ethane/ethanol or
Xe/CF$_4$), the knife-edge chamber can produce well-controlled gas gain

~10⁴ and saturated drift velocity ~50 mm/μsec. With a 0.5 mm gas-filled sensitive region, the chamber should produce ~1 mV pulses, and should be capable of a 92% efficiency, spatial resolution of ~20 μm, and an resolving time of ~20 ns. The only elements of the device structure which are vulnerable to radiation damage are the gas medium and front-end readout electronics. The stability against gas aging effects should be 100 times better than in a straw tube chamber, because the charge collected per unit length of knife edge is 100 times less than in a straw tube.[3] The stability against leakage currents from radiation damage of readout electronics should be 10 times better than that of silicon microstrip detectors, because the charge collected for a minumum-ionizing particle is 10 times greater. This combination of properties would appear to make the knife-edge chamber an attractive technology for tracking at the SSC.

1. <u>FABRICATION OF THE KNIFE-EDGE CHAMBER</u>. The pattern of knife edges is fabricated on a silicon wafer using orientation-dependent etching (ODE). When a silicon surface is exposed to an appropriate alkaline wet-etch solution, the etch rate is ~50 times faster in the 100 and 110 crystal directions than in the 111 direction. Knife-edge fabrication begins with a wafer of silicon (resistivity ~.01 Ω cm), cleaved on the 100 axis, with a thickness ~0.2 mm. The surface is coated with photoresist, and a pattern of lines is lithographically defined on the resist. The silicon surface is then etched in an ODE solution until the desired knife-edge pattern is produced. The resist pattern is then removed by a polishing etch. The silicon surface is oxidized at high temperature to diffuse a SiO_2 insulating layer into a thickness of ~2 μm. A layer of gold is then deposited. A layer of photoresist is deposited, and a second pattern of lines is exposed lithographically. The gold is etched to produce the signal and field-shaping electrodes.

2. <u>ELECTRIC FIELD GEOMETRY</u>. The knife-edge geometry provides the possibility to replicate on a microscopic scale the field geometry of a multiwire proportional chamber. Gas atoms are ionized by high energy charged particles traversing the chamber. The planar electric field between cathode and anode drifts the electrons towards the anode plane. The field-shaping electrodes are biased so as to produce a convergence of electric field upon each knife-edge. Figure 3 shows the electric field distribution. The field-shaping electrodes make it possible to locally control the gas gain of the chamber. A negative electrode bias can be used to enhance the electric field near the knife-edge and extend the high field region further from the knife-edge than would be obtained by simply miniaturizing a conventional MWPC.

3. <u>PROPORTIONAL CHAMBER OPERATION</u>. The knife-edge chamber operates as a detector just as a microscopic multi-wire proportional chamber. Indeed an entire channel would fit within the diameter of the very <u>wire</u> of a typical MWPC. For use as a track chamber in a collider detector, the chamber would be oriented face-on to the track direction. The chamber could be operated essentially as a hodoscope - no timing or pulse height digitization is required.

If the knife-edge chamber were used for track detection in a magnetic spectrometer, the drift electrons would be deflected by the static magnetic field. For a field strength of 2 Tesla and a xenon-based gas, the ExB deflection angle is $\theta_B \simeq 30°$.[4] Dispersion of the charge packet at the anode plane can be eliminated by tilting the chambers at the same angle θ_B with respect to the direction of high-momentum tracks.

The response time of the knife-edge chamber is determined by three considerations: the maximum drift time for ionization electrons; the RLC response of the readout lines; and the response of the readout electronics. The chamber depth g = 0.5 mm produces a maximum drift time $T = g/V_d \cos \theta_B = 15$ nsec. The readout electrode for each knife edge has a capacitance C, inductance L, and resistance R:

$$C = \epsilon \cdot \ell \cdot w/d = 30 \text{ pF} \qquad R = \rho\ell/wt = 300 \text{ } \Omega \qquad L - \mu_o \ell d/w = 0.4 \text{ } \mu H$$

The transmission line properties are

$$Z = \sqrt{L/C} = 110 \text{ } \Omega \qquad RC = 9 \text{ ns} \qquad \sqrt{LC} = 3.5 \text{ ns.}$$

The response is thus compatible with a synchronous 64 MHz readout, with ~10% baseline remaining at the next clock cycle.

As the drifting electrons approach the vicinity of the knife-edge, the increasing electric field initiates an electron avalanche. The avalanche process is characterized by a transport coefficient $\alpha(E)$, usually called the first Townsend coefficient.[5] The gas multiplication is $M = \exp\left[\int_\infty \alpha dr\right]$. The gas multiplication may be estimated from the electric field dependence E(r) near the knife-edge. Using $\alpha(E)$ for pure Xe, a gas multiplication $\underline{M = 10^4}$ is obtained. This is probably an underestimate for the xenon/ethane/ethanol mixture.

<u>Pulse height and efficiency.</u> The number of primary interactions producing ionization electrons in Xe/ethane/ethanol is[6] $n_p = 44$/cm. The electrons ionized in primary interactions typically have sufficient energy to ionize several additional electrons before being thermalized; the total ionization is $n_t = 300$/cm. In the path length $g/\cos \theta_B$, there are $N_p = 2.5$ primary interactions and $N_t = 17$ total ionizations in a knife-edge chamber. The corresponding pulse height at the readout electronics is

$$V = MN_t \text{ } e/C = 1 \text{ mV.}$$

It is instructive to compare this signal with that in a silicon microstrip detector of comparable spatial resolution. In a 200 μm thick depletion layer of silicon, a total of 20,000 electron-hole pairs are produced by a minimum-ionizing particle. One knife-edge chamber thus detects a charge signal ten times larger than that in a comparable silicon microstrip detector.

The efficiency ϵ of a single knife-edge chamber is the probability of there being at least one primary ionization:

$$\epsilon = 1 - e^{-2.5} = 92\%.$$

The trade-off of efficiency against resolving time will determine the choice of chamber gas thickness for a given application.

The spatial resolution of the knife edge chamber is determined by transverse diffusion, track angle, and delta rays. The transverse diffusion of drifting electrons produces a charge distribution with a width $\delta = \sqrt{Dg}$. The coefficient D has been measured for various Xe/ethane mixtures[6], and has the approximate value

$$D = 2 \times 10^{-4} \text{ cm}.$$

Over a drift length of 0.5 mm, the ionization packet thus diffuses with a width $\delta = 34$ μm. By analogy with centroid determination in spark chambers and ionization chambers, the centroid of this distribution can provide a measure of the track coordinate to an accuracy which is considerably smaller than the spread width. A useful resolution $\sigma_x \leq 20$ μm thus appears to be achievable.

Delta rays produce a fluctuating, non-symmetric error to the position measurement. For example, a minimum-ionizing particle can eject a 3 keV electron, having a practical range $\delta = 100$ μm, with a probability of 2×10^{-4}. This is about the same as the statistical probability of a $4\sigma_x$ measurement error ($e^{-8} = 3 \times 10^{-4}$). The small gas thickness of the knife-edge chamber thus helps to reduce the occurrence of δ-rays so that they do not significantly degrade the measurement resolution.

<u>READOUT ELECTRONICS</u>. The readout electronics for each knife-edge chamber can be mounted on a single chip which is bonded to the readout electrode lines. Bonding can be accomplished using either 0.5 μm wire bond or bump bonding.

Figure 4 shows the functional schematic of the readout electronics. All readout electronics is envisioned to be synchronous with the 64 MHz SSC clock. A front-end bifet amplifier buffers each input signal to a comparator, the threshold level of which is generated by an on-board DAC, and is shared among all channels. The comparator outputs are applied to a silicon gate array which serves two functions. First, outputs from manageable widths of the chamber may be OR'd to produce a fast hodoscope signal for use in constructing a trigger decision. Second, the pattern of hits is pipelined through sequential CMOS memory for a number of beam crossings equal to the trigger decision time, eg. 64 cycles = 1 μsec. Events that are not triggered are cleared from memory as they emerge from the pipeline. When a trigger decision is made, the synchronous clock is stopped and all data from the desired event is encoded to describe each consecutive sequence of hits as a centroid position and width. This information emerges as a random-length string of words with a lead word giving event number and record length in the usual fashion. The compacted data is read out on an appropriate data bus. In this way all high-speed, synchronous electronics is mounted locally on each knife-edge chamber; only trigger-selected, compacted data is communicated for higher level processing.

ACKNOWLEDGEMENTS

This paper is dedicated to Michael J. Neumann, 1907-1989, a pioneer in the development of wire chambers for high energy physics and an inspiring teacher to generations of young physicists. It is a pleasure to acknowledge stimulating conversations with Muzaffer Atac (Fermilab),

George Blanar (LeCroy), Andrea Palounek (LBL), and Peter Winoker (Sandia).

Similar detector concepts are being developed by Renaldo Bellazzino (INFN-Pisa) and Aki Mahi (KEK).

This work is supported by the U.S. Department of Energy, contracts No. DE-AC01-85ER40236 and DE-AS05-81ER40039, and by the Texas Center for Energy and Mineral Resources.

REFERENCES

1.   G.G. Hanson, B.B. Niczyporuk, and A.P.T. Palounek, "Tracking Simulation and Wire Chamber Requirements for the SSC." Proc. DPF Summer Study: Snowmass '88, High Energy Physics in the 1990's, Snowmass, Colo., June 27-July 15, 1988.

2.   D.J. Campisi, R. Green, and H. Gray, "Vacuum Integrated Circuits," in Proc. 1986 Int. Electron Devices Meeting, p. 776.
     D.J. Campisi and H. Gray, "Microfabrication of field emission devices for vacuum integrated circuits using orientation dependent etching," in Proc. Mat. Res. Soc. Meeting, 1986.

3.   M. Atac, "Wire Chamber Aging and Wire Material," Vertex Detectors, F. Villa, ed. Plenum, New York (1988).

4.   A. Breskin, G. Charpak, F. Sauli, M. Atkinson, and G. Schultz, Nucl. Instr. Meth. 124, 189 (1975).

5.   V. Palladino and B. Sadoulet, "Application of the Classical Theory of Electrons in Gasses to Multiwire Proportional and Drift Chambers." CERN, April 1974.

6.   F. Sauli, "Principles of Operation of Multiwire Proportional and Drift Chambers." Experimental Techniques in High Energy Physics. T. Ferbel, ed., Addison-Wesley. Menlo Park. 1987.

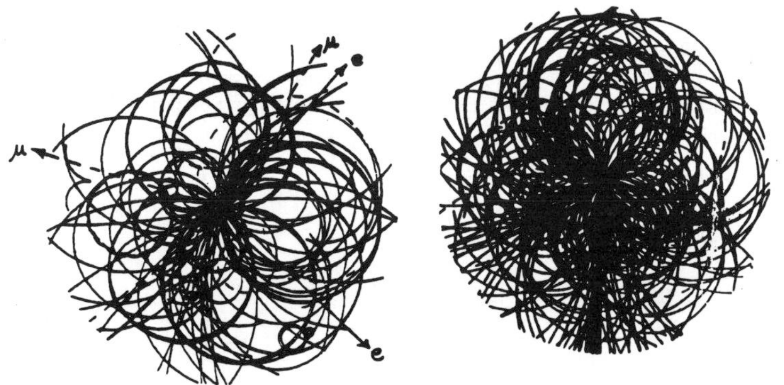

Figure 1.   Simulations of two "interesting" events as they would be recorded in a magnetic spectrometer (B = 2 T):
a) Higgs boson: pp → H → ZZ
        └ e⁺e⁻
        └ μ⁺μ⁻
b) a 2-jet event corresponding to quarks scattering with a momentum transfer of 1 TeV. (from Ref. 1).

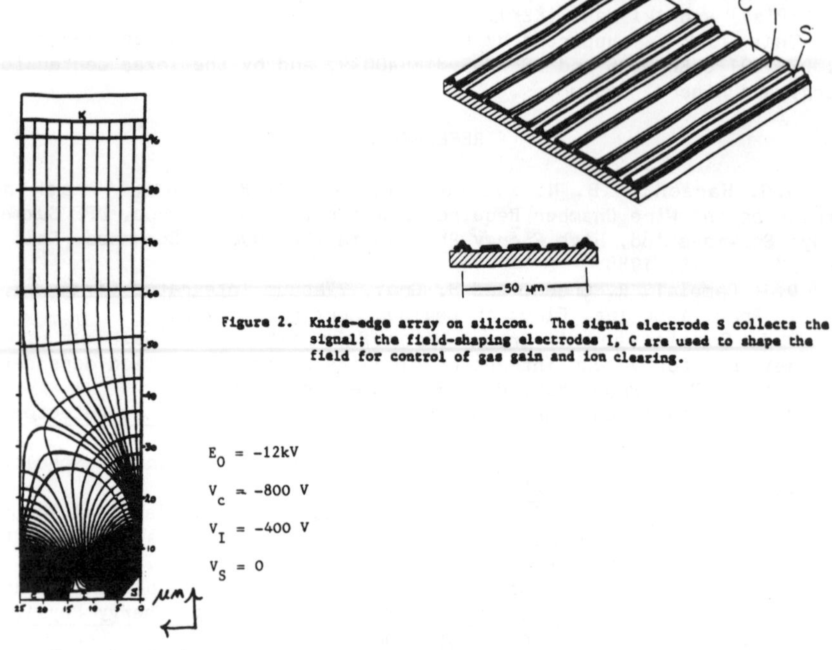

Figure 2. Knife-edge array on silicon. The signal electrode S collects the signal; the field-shaping electrodes I, C are used to shape the field for control of gas gain and ion clearing.

$$E_0 = -12\text{kV}$$

$$V_C = -800\ \text{V}$$

$$V_I = -400\ \text{V}$$

$$V_S = 0$$

Figure 3. Calculated electric field distribution near the knife-edge array, with electrode potentials set for optimum gas gain.

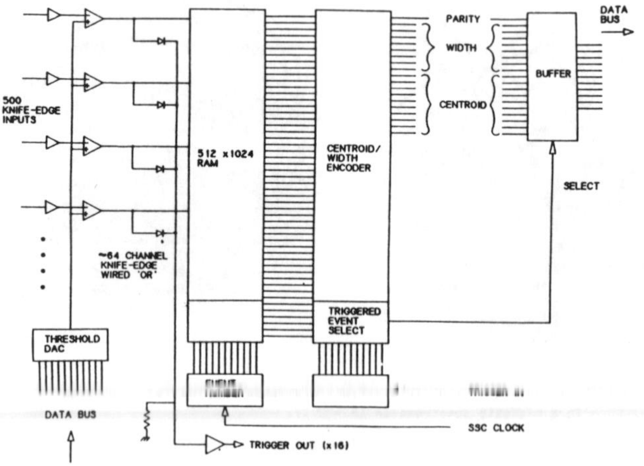

Figure 4. Functional schematic of readout electronics.

ACOUSTIC CHARGE TRANSPORT DELAY LINES

R. J. Davisson, W. M. Dougherty, H. J. Lubatti, R. J. Wilkes
Physics Department FM-15, University of Washington
Seattle, WA 98195

D. Fleisch, R. Kansy, G. Pieters
Electronic Decisions Inc., 1776 E. Washington St., Urbana, IL 61801

Presented by
H. J. LUBATTI

ABSTRACT

Properties of acoustic charge transport delay lines which make them good candidates for SSC/LHC "pipeline" applications are described. Results of tests on existing devices are reported.

1. INTRODUCTION

Data rates expected at the SSC and LHC will require analog storage of signals during which pretriggers can be formed[1]. In this paper we describe Acoustic Charge Transport (ACT) delay lines, which have delay times, sampling rates, band pass and dynamic range necessary for SSC and LHC "pipeline" applications. These devices which are based on acoustic charge transport in GaAs are capable of providing delay times of the order of microseconds for large bandwidth analog signals and permit non-destructive preview sampling of the signal. A brief description of the device and the results of measurements which demonstrate the delay line, pulse shaping capabilities, and resolution of the device are given below. A more detailed discussion of our work which includes more information on ACT technology is given in ref. [2].

The ACT delay line, shown schematically in Figure 1, repeatedly samples the analog signal. The discrete samples represented as discrete charge packets proceed sequentially down a linear transport channel formed in a high purity n-type GaAs epitaxial layer. Charge injection and extraction are accomplished using ohmic input and extraction contacts (IC and EX in Fig. 1). The superposition of the piezoelectric potential formed in the GaAs by a continuously applied surface acoustic wave (SAW) and the depletion potentials creates a series of three-dimensional potential wells which propagate through the transport channel at the SAW velocity, 2864 m/sec. Charge packets are injected into these potential wells from the input signal which is sampled at the SAW frequency. The devices which we have tested operate with SAW frequencies of 360 MHz and have an effective sampling aperture of approximately 10% of the SAW frequency. Transfer efficiencies $> .9999$ have been obtained. Devices with delay times of 1 to 3 μsec have been achieved.

Nondestructive (NDS electrodes shown in Fig. 1) sampling of the charge packets propagating in the transport channel is accomplished by electrodes above the surface of the transport channel which capacitively sense the charge packets propagating below. The NDS readout can be used to form pretriggers and preprocess the signal by providing weighted sums over many packets. After transport through the length of the channel, the electron packets are swept out of the wave by the large electric fields associated with the bias of the extraction electrode (EX in Fig. 1).

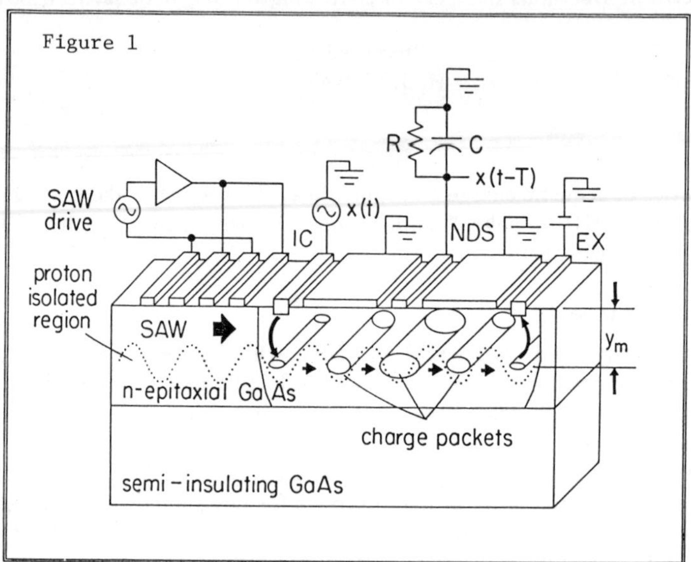

At the present early stage in the development of ACT technology, commercial effort has thus far not been directed to the needs of high energy physics. The specific devices required do not yet exist. However, such devices are well within the current capabilities of the technology. Our confidence that ACT devices will prove useful to the needs of particle physics is based on:

1. The demonstrable fact that the basic mechanism for confining and transporting charge indeed works;
2. The good fortune that one can integrate on the same die both the ACT device and its ancillary circuits; and
3. The coincidence that the acoustic velocity of approximately three microns per nanosecond provides an excellent match between the physical scale of microcircuits and the time scale dealt with in high energy physics

2. EVALUATION OF ACT DELAY LINES

A series of tests were performed to evaluate some of the device types described above. In particular we were interested in the band-pass and signal processing capabilities of ACT's when processing the fast pulses associated with particle physics detectors. In order to perform these tests we used off-the-shelf devices. These devices were designed for RF signal processing; they generally have gains less than unity, and in some devices attenuated response to frequencies below 20 MHz. In all the measurements discussed in this section a 225 MHz low pass filter was used at the output in order to attenuate the SAW drive signal and its harmonics.

The overall effect of the ACT on energy resolution is a primary concern for applications which involve calorimetry signals. We therefore studied the effect of an ACT delay line on the resolution of a NaI ^{60}Co spectrum. A 10 μCurie source illuminated a sodium iodide crystal which was coupled to a photomultiplier. The output of the photomultiplier was fed through the ACT and signals from the ACT were then amplified and analyzed by a LeCroy qVt. Spectra were obtained for both a non-destructive tap approximately half-way along the delay line (480 ns) and the destructive sensing tap at the end of the delay line (900 ns). For purposes of comparison, spectra were taken with the ACT replaced by an appropriate attenuator such that the input signal to the qVt had approximately the same dynamic range.

The pulse height spectra are shown in Figure 2. The typical NaI ^{60}Co energy spectrum is apparent: the two peaks at 1.17 and 1.33 MeV and the Compton shoulder. The spectra obtained from the destructive and non-destructive sensing taps are shown in Figures 2a and b, respectively. The corresponding spectra obtained with the ACT delay line bypassed are shown in Figure 2c,d.

Figure 2

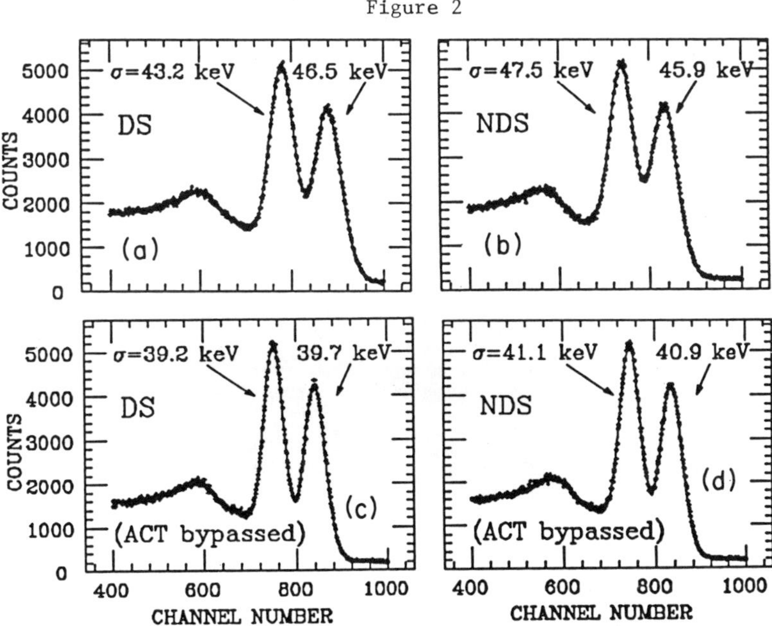

Different amounts of external gain were used in the DS and NDS cases in order to place the peaks in approximately the same qVt channels. We obtained the position and widths of each peak by fitting the spectrum above channel 700 with two Gaussians and an exponential. Using the known energy of the two peaks we thus obtained a calibration. Further, assuming that the resolution function of the ACT is Gaussian, we obtain its width by comparing the spectrum of signals passing through the ACT to those where the ACT is bypassed. We find a resolution $\sigma \simeq 22$ KeV for both the destructive sensing and non-destructive sensing taps.

To test the signal processing capabilities inherent in ACT devices we used a prototype Programmable Transversal Filter with 64 non-destructive sensing taps. The weight coefficient for each tap is determined by a digital word which is stored on-chip and loaded by means of a computer "interface." Figure 3 shows the output of tap 11 with weight +1 for a long square input pulse. One sees only the differentiated leading edge. The overshoot at the trailing edge is off-scale. This pulse propagated through the device and could be sensed at any of the 64 taps. Since this differentiated pulse has an exponentially decaying tail, it is ideal for performing pulse shaping tests. We were able to clip this pulse by subtracting a delayed signal with weight commensurate to the exponential decay of the amplitude. For example, Figures 4a,b,c show clipped pulses where we have added to tap 11 with weight 1 several nearby taps with weights between 0 and -1.0. The taps on this particular device are separated by 7.7 ns. The shortest pulse that can be achieved by this technique is obtained by subtracting adjacent taps. Figure 4d illustrates the results for taps 11 and 12. In the PTF, even and odd taps have weighting circuits on opposite sides of the device. In the early prototype device used, the two sides were not properly balanced. Thus for the results shown in Figure 4d, it was necessary to weight the output of tap 11 by +.65 and add to it the output of tap 12 with a weight of -1. This imbalance will not be present in production devices.

Figure 3

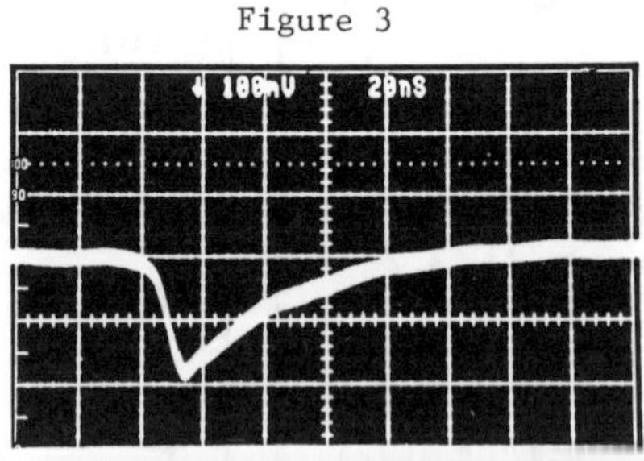

Tap II, weight + I.O
(used as input pulse)

Figure 4

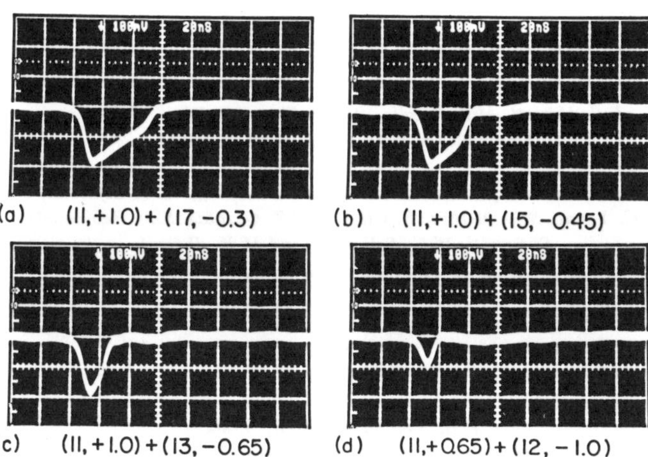

3. SUMMARY

Based on our initial study we conclude that ACT delay lines have a potential for application to SSC/LHC data acquisition systems. There are still some technical issues which must be resolved. These include power per channel, dynamic range, and radiation hardness. We plan to design a multi-channel ACT (8-16 channels) with power dissipation less than 200 mW per channel and dynamic range greater than 60 dB per channel. We also plan to perform radiation hardness tests using both gamma rays and neutrons.

ACKNOWLEDGEMENTS

We wish to thank Phil Berger and Mike Ingram of EDI for their very helpful assistance with the measurements and B. J. Hunsinger for his continued interest and support. This research was supported in part by the Department of Energy (grant DE-FG06-90ER40541), Electronic Decisions Inc., and the National Science Foundation (grant PHY-8918557).

REFERENCES

1. T.J. Devlin, A. Lankford and H.H. Williams, in *Proc. Summer Study on the Physics of the SSC*, eds. R. Donaldson and J. Marx, (AIP, New York, 1986) p. 499. See also R. Davisson, W. Dougherty and H. Lubatti, "Buffering SSC Data", U. of Wash. report VTL-PHY-116, February 1989 (unpublished).
2. R.J. Davisson, W.M. Dougherty, H.J. Lubatti, R.J. Wilkes, D. Fleisch, R. Kansy, G. Pieters, "Evaluation of Acoustic Charge Transport Delay Lines for SSC/LHC Applications," U. of Wash. preprint UWSEA PUB 90-6, to be published.

Hadronic Shower Punchthrough and TeV Muon dE/dx

W.K. Sakumoto, P. de Barbaro, A. Bodek, H.S. Budd, and B.J. Kim
University of Rochester, Rochester, NY 14627

F.S. Merritt, M.J. Oreglia, H. Schellman, and B.A. Schumm
University of Chicago, Chicago, IL 60637

K.T. Bachmann, R.E. Blair, C. Foudas, B. King, W.C. Lefmann,
W.C. Leung, S.R. Mishra, E. Oltman, P.Z. Quintas, S.A. Rabinowitz,
F. Sciulli, W.G. Seligman, and M.H. Shaevitz
Columbia University, New York, NY 10027

R.H. Bernstein, F.O. Borcherding, M.J. Lamm, W. Marsh,
K.W. Merritt, P.A. Rapidis, and D. Yovanovitch
Fermi National Accelerator Laboratory, Batavia, IL 60510

P.H. Sandler and W.H. Smith
University of Wisconsin, Madison, WI 53706

Abstract

We have measured the longitudinal particle punchthrough probability from shower cascades produced by hadrons incident on the iron-scintillator calorimeter of the CCFR neutrino detector. Measurements of the dE/dx energy loss in Fe of high energy cosmic ray muons (up to 1 TeV) incident on the same detector are presented and compared against calculations.

The design of muon detectors for the superconducting supercollider (SSC) requires knowledge on two topics: particle punchthrough from the calorimeter in front of the muon detector and the interactions of TeV energy muons with the components of the muon detector. Consider a muon detector that consists of absorber plates interspersed with particle tracking chambers. A large energy loss muon interaction within the absorber produces a shower of secondary particles that can penetrate into the tracking chamber region. How often these interactions occur for TeV muons is an important measurement for the proper design of an SSC muon detector.

Both the measurements on hadronic shower punchthrough and TeV muon dE/dx were taken using the CCFR neutrino detector [1] (the Lab E detector at Fermilab). The detector is a 690 ton iron-scintillator calorimeter followed by a 420 ton toroidal muon spectrometer. Shower punchthrough measurements are extracted from a series of hadron beam energy calibration runs of the calorimeter and measurements of hadron induced muon production rates taken in 1984 and 1987. Measurements of the longitudinal punchthrough probabilities of particles as a function of depth into the iron-scintillator calorimeter from showers produced by 15, 25, 40, 50, 70, 100, 200, and 300 GeV hadrons have been reported [2,3]. In Reference [3], particle punchthrough rates for 70 GeV tagged pions and kaons are also presented as well as muon production rates from 40, 70, and 100 GeV hadrons. Attempts at reproducing the longitudinal punchthrough rates with shower Monte Carlos [4] have been fairly successful.

The TeV muon dE/dx measurements are from cosmic ray exposures of the CCFR neutrino detector taken in 1987 and 1988. The trigger selected muons that were nearly horizontal. To ensure proper reconstruction of the muon momentum by the toroidal spectrometer, only muons that traverse the detector in the direction of the accelerator beam and through the full length of the muon spectrometer are used. The fractional momentum resolution is $\sigma/p \sim 0.11\sqrt{1 + (p/900)^2}$, where p is in GeV. In the final event sample, there are 9443 events, with 236, 95, and 43 events between 500 to 800 GeV, 800 to 1000 GeV, and 1000 to 2000 GeV respectively. Above 2000 GeV, there are 19 events. As resolution errors are expected to be large in this region, they are not used. The differential muon energy spectrum falls approximately as $1/E^2$ and steepens above 400 GeV. Muon energy losses are measured in the CCFR detector's calorimeter, which has 84 scintillation counters, one every 10.3 cm of Fe.

Figure 1 shows the muon dE/dx (average energy loss) as a function of the muon momentum. The measurement in each momentum bin is the total energy deposited by the muons in the bin divided by the total path length in iron traversed by those muons. The solid curve is a theoretical prediction [5], and it is consistent with the measurement. In iron, the energy loss from stochastic processes (bremsstrahlung, e^+e^- pair production, and photonuclear interactions) is predicted to become significant and eventually dominate the dE/dx at muon energies above a hundred GeV. Energy losses from these stochastic processes appear as occasional shower cascades in the calorimeter. These cascades are superimposed above the ionization energy deposited by the muon in the calorimetry counters. Energy losses under a few GeV are difficult to measure because their shower cascades are not readily distinguishable from the ionization background. The stochastic energy loss is obtained from the measured shower cascade energy by subtracting its average ionization energy loss from it.

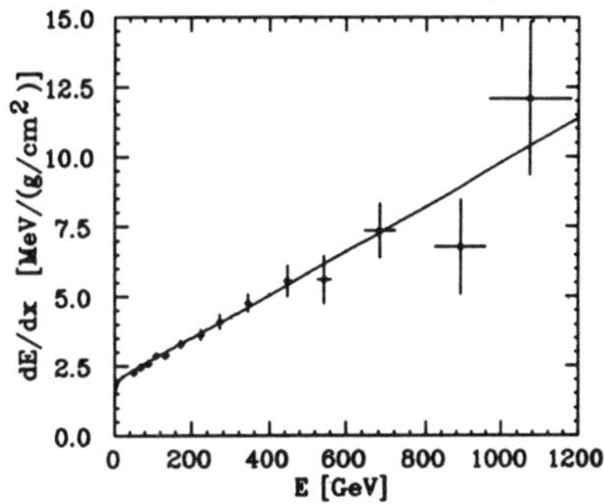

Figure 1: Muon dE/dx. The solid curve is the theoretical prediction.

Figure 2 shows the differential, stochastic energy loss spectrum for very energetic muons and for energy losses larger than 2 GeV. The solid curve is a prediction [5] for the shape of the spectrum. There is reasonable agreement between the measurement and shape predictions for all muon energies. From visual scans of a sample of cosmic ray event pictures, we observe that with increasing muon energies, events contain more visible shower cascades. Nearly all the shower cascades are electromagnetic

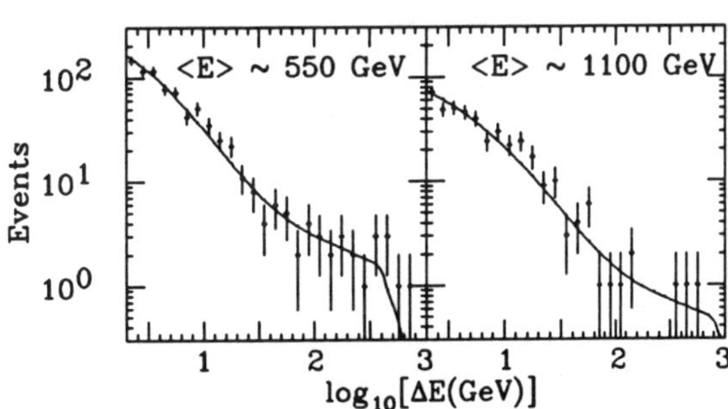

Figure 2: The differential muon energy loss spectrum for muon energies in the range 400 to 800 GeV ($< E > \sim 550$) and 800 to 2000 GeV ($< E > \sim 1100$). The solid curves are the predicted shapes that have been normalized to the data.

(the cascades do not penetrate through more than $\sim$30 cm of Fe). The calculated [5] interaction lengths for producing a muon energy loss in excess of 2 GeV via stochastic processes for 10, 100, 1000, and 10000 GeV muons are 47000, 3900, 280, and 61 cm of Fe respectively. For energy losses under 2 GeV, which are not as readily observable in our calorimeter, the corresponding interaction lengths are 330, 78, 34, and 21 cm. The implication is that a muon tracking chamber sandwiched between iron (hadron) absorbers will see more activity for increasing muon energies.

The CCFR calorimeter has 42 drift chambers, one every 20.6 cm of Fe. They are instrumented with multi-hit TDCs and flash ADCs. For two particles to be resolved by the flash ADCs, they must be separated by at least 2 mm. We use the drift chambers with the flash ADC electronics to measure the hit multiplicities as a function of the muon energy. The cosmic ray muons are put into momentum bins of 40-80, 80-180, 180-400, 400-800, and 800-2000 GeV. (The mean muon momenta of each bin are: 58, 119, 259, 546, and 1132 GeV). The fraction of events in each bin where the flash ADCs are able to detect multiple particles per drift cell are 14, 16, 20, 24, and 28% respectively. In the 40-80 GeV momentum bin, the multiplicity drops about two orders of magnitude after four hits, while for the 800-2000 GeV momentum bin, the multiplicity drops the same amount after nine hits. The resolution of the drift chamber (from the multi-hit TDCs) does degrade with increasing hit multiplicities observed in the flash ADCs.

In conclusion, we find that our measurements on average muon energy loss in iron and on the differential energy loss spectrum are consistent with the latest theoretical predictions [5]. Thus, the theoretical differential cross sections on bremsstrahlung, e^+e^- pair production, and photonuclear interactions can be used in conjunction with shower Monte Carlos to aid in the design of muon detectors that utilize large amounts of hadron absorber [6].

References

[1] W.K. Sakumoto et.al.,"Calibration of the CCFR Target Calorimeter," University of Rochester preprint: UR-1142, Mar 90. To be published in NIM-A.

[2] F.S. Merritt et.al., Nucl. Instr. and Meth. **A245** (1986) 27.

[3] P.H. Sandler et.al.,"Hadron Shower Punchthrough and Muon Production by Hadrons for 40, 70, and 100 GeV," Univ. of Wisc., Madison preprint: WISC-EX-89-306. Submitted to Phys. Rev. D.

[4] H. Fesefeldt et.al.,"Tests of Punch Through Simulations," Aachen, Tech. Hochshc., III Phys. Inst. preprint: PITHA 89/07-REV.

[5] All theoretical predictions in this paper are based on recent cross-section formulae summarized in: W. Lohmann, et.al., "Energy Loss of Muons in the Energy Range 1-10000 GeV." CERN preprint: CERN-85-03, Mar 1985.

[6] J.J. Eastman and S.C. Loken, "Muon Energy Loss at High Energy and Implications for Detector Design," Lawrence Berkeley Lab preprint: LBL-24039, 1987.

Bottom Collider Detector at the SSC

An intermediate and low Pt detector designed for B physics.

BCD Collaboration[1]

presented by

Paul Lebrun
Fermi National Accelerator Laboratory
P.O. Box 500, Batavia, Illinois 60501

Abstract : The opportunities to study in detail CP violation in the decay of B mesons at the SSC are discussed. Emphasis is placed on global detector design and on the technological developments necessary for this experiment. The R&D program of the Bottom Collider Detector group is reviewed.

I. The ultimate goal of BCD and the SSC

The physics goal of this experiment is the complete and thorough study of CP violating modes of the B meson[2,3]. Measurements of CP violation in the B - $\overline{\text{B}}$ system can determine C-K-M angles with little strong interaction uncertainty. By the study of several decay modes the C-K-M matrix elements can be overconstrained. Thus, this study is a clean and unique test of the standard model. Since very good mass resolution is required to distinguish between final states, only B mesons that decay into all charged final states are considered. Most interesting from a theoretical standpoint are the decay of a neutral B meson to a CP eigenstate, such as $B \to \psi K_S$ or $B \to \pi^+ \pi^-$. The study of CP violation in these systems can be accomplished by measurement of an asymmetry[4]

$$A_{CP} = (\Gamma (B \to f) - \Gamma (\overline{B} \to f)) / (\Gamma (B \to f) + \Gamma (\overline{B} \to f))$$

The particle-antiparticle character of the parent B must be "tagged" by the observation of the second B in the interaction. These modes may have asymmetries as large as 30% but small branching ratios, typically less than 10^{-5}. In addition, the reconstruction and tagging efficiencies will never be 100%. Therefore, large yields of B mesons are required. As an illustrative example, the observation of CP violation requires, conservatively, at least 1000 $B \to \psi K_S$ decays fully reconstructed and tagged via a detached semi-leptonic decay. If one assumes a single detached lepton as a trigger, we present in Table 1 reconstruction efficiencies and branching ratios. Combining these losses, the required number of produced B's decaying to $\psi\, K_s$ is 2.6 10^7. Assuming an optimistic value for the $B \rightarrow \psi\, K_s$ branching ratio of 5 10^{-4} , one needs 510^{10} B's produced.

At the Fermilab collider, the cross section for B pair production is estimated to be 40 to 50 μb[5]. At the SSC , this cross section is expected to be as high as 500 μb, that is, 10^6 times the $\Upsilon(4s)$ cross section at CESR and DORIS. At the SSC, one out of 200 events is likely to contain a B pair. This ratio is identical to the one observed in charm photoproduction at Fermilab fixed target energy, where very clean and rich samples of charm decay have been observed. This makes the SSC a very attractive place to study high sensitivity B physics. In addition to the cross section argument, the B lifetime is sufficiently long that the decay vertex is separated from the beam collision point. Finally, the rapidly advancing technology of solid state detectors, integrated circuits, fast RICH detectors, and high speed data acquisition systems will make a dedicated B experiment at the SSC feasible. The goal of this paper is to define the detector architecture and the areas where R&D is required to carry out this ambitious project.

Table 1 : Assumed efficiencies and branching ratios for
a CP violation study in the $\psi\, K_S$ system.

Detachment cuts at each secondary vertex (related to finite vertex resolution)	0.50
Tracking and reconstruction efficiency for the tagging lepton	0.30
Momentum cut in triggering on this lepton	0.30
Branching ratio for B to electron	0.12
Tracking and reconstruction efficiency for $\Psi\, K_S$	0.30
Branching ratio for Ψ to $e^+ e^-$	0.07
Branching ratio for K_S to $\pi^+ \pi^-$	0.69

II. The BCD detector at the SSC

The architecture for a B-physics detector is largely determined by the characteristics of the production of the B-mesons. The dominant process for B-meson production at the SSC is gluon-gluon fusion[5]. The B's are predicted to have an average P_t of approximately the mass of the b-quark ($\cong$ 5 GeV/c) and to be broadly distributed in pseudorapidity η : roughly over 5 to 6 units at the SSC energy. Finally, the B-$\bar{B}$ pairs are produced with strong positive correlations in pseudorapidity ($\Delta\eta \cong 1.5$) and therefore direction. Table 2 shows the acceptance as a function of pseudorapidity coverage and energy[6]. This table demonstrates the large gain in acceptance when the central region is included.

The BCD solution to the problem of covering both the forward and central regions is to use a large dipole magnet with its field transverse to the beam with both forward and barrel tracking elements as shown in fig 1. Other important factors are good momentum resolution for low P_t charged tracks, precision vertexing and good charged particle identification. An overview of the key detector components is given in the following list[2,7] :

1. The dipole magnet configuration optimizes the detection and momentum measurement for tracks of intermediate P_t over a wide rapidity gap. Assuming a single hit resolution of 50 to 100 microns in the tracking system, a one Tesla field makes possible a mass resolution of 0.3%, required to separate B_d from B_s and to set a narrow mass window around the B as a rejection against combinatoric background.

2. The vertex detector is designed to find and locate accurately, in 3 dimensions, secondary vertices of B particles with high efficiency. Such a vertex reconstruction capability is not only needed to measure the time dependence of the asymmetry A_{CP} but also to reject combinatoric background by many orders of magnitude. At least 3 hits per track must be recorded with an accuracy approaching 5 microns for each charged track suspected to come from a B-decay. As described in the R&D section in more detail, we are currently investigating a design based on double sided silicon strip detectors and a design based on pixels.

3. The straw tracking system is used as the primary means of pattern recognition and in the determination of the charged track momenta. We are currently pursuing a low-mass straw-tube design with the tubes operated as drift chambers. By using dimethylether gas, resolutions of 35 microns can be achieved at atmospheric pressure. These tubes will be glued together into 'superlayers' of 8 layers for mechanical support. There are typically 75 to 100 hits recorded along each track.

4 Particle identification is important for reducing the combinatoric background, especially in modes such as $B \to K^+ \pi^-$ or $B \to p\,\bar{p}$. The design presently incorporates RICH counters, and a time of flight system which is a very efficient and simple system for tracks of momenta less than 2. GeV/c. Finally, at pseudorapidity greater than 1.5, a muon detector will enrich the semileptonic decay sample we will be able to study.

5. Electron identification is essential for both triggering and tagging the identity of the B meson in the CP violation studies. The electron P_t spectrum from semileptonic B decay peaks around 1 GeV/c, making identification difficult because sources of non-prompt electrons such as conversions and Dalitz decay also have a very soft spectrum. By measuring the energy deposition in the silicon wafers used in the vertex detector, we hope to identify conversions occurring in the beam pipe or in the vertex detector itself. In order to reject Dalitz decay, one has to carefully fit the electron track and measure the closest distance of approach to the primary vertex. Significantly detached electron tracks presumably come from heavy quark decays and are worth triggering upon.

6. The small angle fiber system is designed to measure tracks at rapidities beyond those covered by the central detector. It provides a minimum bias trigger, a luminosity measurement, and a fast method of determining the longitudinal location of the primary vertex to roughly 1 cm[8] .

7. Triggering and Data acquisition system : It is planned to operate the detector at luminosities of $\cong 10^{32}$ cm^{-2} sec^{-1}. At the SSC, this corresponds to 10^7 interactions per second, each producing about 1 Mbyte of information (assuming data compression at the front end), or about 10^{13} bytes per second. The trigger consists of two levels, a prompt 'hardware' trigger and a non-prompt 'software' trigger. The BCD strategy emphasizes the latter one. The event data remains stored on the subdetector elements in buffers until a prompt trigger decision is received.   A fast, dedicated hardware trigger will make an early decision based on a very limited amount of information, such as whether or not an electromagnetic shower matches a charged track in a front end preconverter.   The non-prompt trigger and the corresponding data acquisition system described in more detail in the R&D section consist of 'event building' switches embedded in a large farm of parallel processors capable of detecting prompt electrons and/or detached vertices. We anticipate this farm must be able to process 1 Trillion instructions per second.

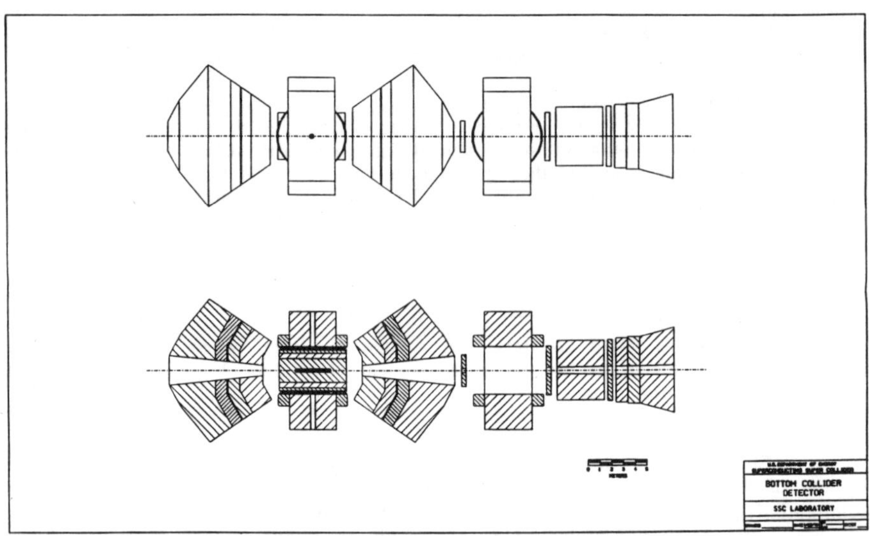

Fig. 1: The BCD detector layout at SSC.

III. Computer simulation and analysis.

A few decades ago, heavy quark experiments could have been planned based on previous experience and simple geometrical acceptance calculations. This is no longer the case. New technology - often immature - and overall detector costs need to be assessed in more detail. Computer simulations and analysis of this simulated data do not necessarily provide the ultimate correct answer, although they are often the only practical means by which one can plan for a reasonable detector strategy and define trade-offs in choosing between technologies. In this section, we describe a few items which have strongly influenced design decisions for Future Collider Initiatives. Although this list is by no means complete, we intend to cover the major topics.

A. Software engineering considerations

- Global detector strategy and B production models:

Transverse momentum and rapidity distributions, as well as correlations between the 2 B's are primordial ingredients to sound detector architecture. ISAJET and PYTHIA have been used extensively to define the rapidity coverage for BCD at the SSC[9] . Tranverse momentum distributions dictate the fate of "stiff" track triggers described below. These programs also allows us to estimate the loss of yield due to P_t trigger biases. Particle identification systems are sensitive to these distributions as well.

Particularly relevant is the correct simulation of the track multiplicity in minimum bias events, as well as the correct yield for 'mini-jet' events, in which a higher multiplicity and perhaps a higher heavy quark content are expected. These multiplicities dictate to a large extent the granularity of the subdetector elements. A specific study has been made for the silicon vertex detector (see below in the vertex section).

- Detector simulation and mechanical engineering:

Once the overall detector architecture is set, engineering CAD/CAM studies start in parallel with geometrical layouts in an early-stage detector simulation package. The BCD Monte-Carlo package is based on the GEANT3 system[10] . Often a detailed volume shape description is required to assess the physics reach of a given subdetector. For instance, a simple cylindrical Beryllium beam pipe design of a "reasonably optimistic" thickness of 300 microns implies that silicon vertex detectors which cover a rapidity range above roughly 3 units are useless because of multiple scattering in tracks which graze the pipe. Therefore, more complicated volume shapes must be tried. Flexible and powerful 3 dimensional volume modeling with HEP knowledge software is required. If performed on 32-bit machines, this ray tracing problem should be resolved using double precision arithmetic.

GEANT3 is unfortunately not suitable for serious CAD/CAM engineering studies. Detector mechanical designs are currently based on the Intergraph package at the SSC laboratory. Flexible interfaces between these 2 software environments - the physicist vs the engineer - are currently under study. Given the large amount of tracks to simulate and the complex volume structure layouts, such interfaces will required many man-years of effort.

• Detailed, fast,   detector response simulation:

GEANT3 provides adequate tools to generate simulated detector digitisations. Nevertheless, these are general tools, and quite a bit of code must be generated to simulated in detail the localization, ionization, amplification and discrimination of simulated electronic signals (about 1,000 lines of FORTRAN code for the BCD vertex detector). Since these detectors are quite specialized, there is no easy way to devise useful general tools.

• Calorimetry:

Electromagnetic calorimetry is now understood and can be simulated very accurately using either GEANT3 or EGGS4[11]. Fortunately, hadron calorimetry is not required for B physics, where accurate simulation is still somewhat of an issue.

B. Physics issues and   BCD simulation : selected topics.

• Search for a 'level 1' trigger:

BCD has studied a triggering scheme based on the fact that tracks coming from B decays tend to have a high transverse momentum, of the order of the B mass. In this scheme, it is required that at least one track is "stiff"; i.e., has a transverse momentum greater than a few GeV/c. The merit factor of such a trigger has not been well measured over a wide rapidity range at Tevatron energies. Therefore, one has to rely on Monte-Carlo calculations, which are very much model-dependent. Using ISAJET, 'minimum bias' events are easily rejected while B-B jets are characterized by higher P_t tracks. Asking for at least one track with a P_t greater than 3.0 GeV/c, one retains more than 95% of the B events and rejects up to 98% of minimum bias events. Unfortunately, this 'minimum bias' option in ISAJET does not reproduce the unbiased transverse energy distribution measured by CDF[12].   In order to better match the data, the minimum P_t parameter in ISAJET must be raised. When one does this, the merit factor drops from approximately 50 to an uninteresting value of 2 or 3. The rapidity gap explored by CDF (+- 1 units, roughly) is somewhat too narrow to reach a definite conclusion, however, and in any event, it would be unwise to base a data acquisition strategy on this trigger alone.

Therefore, we are also investigating other triggering schemes, based on particle identification and vertexing. BCD aims at detecting both B particles in the event. In order to measure CP violation, we select the events in which a B meson decays semileptonically; the lepton tag is then used to determine the particle/antiparticle characteristic of the other B meson. Moreover, very promising decay modes include the J/Ψ, a particle conveniently detected through it's decay mode to a lepton pair. Therefore, a lepton trigger for BCD does makes sense. Let us consider the electron channel which, unlike the muon, could be accessible at lower momenta.

This electron is identified by (i) a large energy deposition in the electromagnetic calorimeter, (ii) a matching track in a pre-radiator, (iii) a matching track in the TRD system and (iv) a matching track in the main tracking system. In addition, a link to the microstrip detector might be required in order to discriminate against photon conversion occurring in the tracking system. In order to perform the tracking quickly, the knowledge of the position of the primary vertex is also required, with an accuracy of a few mm. This is no longer a simple "hardwired" trigger, dealing with only one subdetector element. Special purpose processors have been used to trigger on transverse energy deposition in calorimeters (E705[13], UA1[14,15], D0[16]) but such a complete electron trigger as this is requires at least 3 complex analyses - such as tracking - to work well.

Finally, in these thorough Monte-Carlo studies, we are considering a vertex trigger based on the silicon vertex detector alone. This is clearly the most "physical" trigger: the very unique signature of a heavy quark event is the presence of at least 2 secondary decays. This type of trigger has been used very successfully at a fixed target charm experiment at CERN (WA82 [17]), where a merit factor of 50 to 100 was easily obtained. A similar trigger will be used by E789 at FNAL[18]. In these schemes, one only has to assess if track(s) are *not* pointing to the primary vertex position to establish the presence of secondary vertices. The presence of a magnetic field undoubtedly complicates the algorithm, although it allows us to measure the momentum of the tracks and therefore to correctly estimate the error due to multiple scattering.

None of these triggering schemes are easy to implement in real time and therefore are not likely to be suitable for a 'prompt'(few μsec) level 1 trigger. But they will certainly be used in the asynchronous processor farm.

• Performance of the Silicon Vertex detector:

In a simplified version of the simulation, the hit arrays are generated for each track and the resolution is computed only for tracks with angle of incidence into the silicon wafer of less than 45°. Using the correct assignment of hits to the track, a straight line is fit to these points (in this particular study, we turned the magnetic field off). The primary vertex is found first, and the pool of tracks failing to be associated with the primary is used to seed secondary vertices. The channel B $\to \pi^+ \pi^-$ was studied because it is most difficult in terms of combinatoric background. Because we found that most of the background comes from B decays, we generated a sample of 150,000 events of B-$\overline{\text{B}}$ pairs, in which the B mesons were *not* decaying to $\pi^+ \pi^-$. In figure 2, the K_S and the $D^0 \to \pi^+ \pi^-$ clearly emerge from the background with virtually no count in the mass bin of interest.

This simulation has been expanded to include various physics processes besides multiple scattering. The detector layout is based on the silicon strip design described in the R&D section. The simulation includes the exact path length calculation in the silicon active region, energy loss fluctuation in the silicon, charge correlation between one side of the wafer and the other one (we assumed double sided technology[19]), charge diffusion between neighboring strips, detection inefficiency, sparse readout based on charge division and electronic noise at the preamplification level. The Monte-Carlo data is saved on disk and is reconstructed using a prototype of the future BCD analysis program. The silicon hits are first used to characterize 'clusters' of strips, one cluster presumably generated by a single track. One assumes that the STRAW tracks are reconstructed prior to vertex analysis and the silicon clusters are matched to these tracks on a wafer by wafer basis. These cluster positions are then fitted to a helix, fixing the momentum and estimating the slope and the intercept at the vertex position. We are currently planning to upgrade this simulation by implementing the pixel detector and run a detailed analysis similar to the one presented in figure 2.

Already, a number of important conclusions have emerged from these analyses:

- 3-D vertex reconstruction is needed to untangle the B decay tracks from the underlying event, over a complete 4π geometric acceptance[20].

- The resolution of the intercept of a given track originating from the primary vertex at this primary vertex location has a most probable value of 25 μm. The average is about 40 to 80 μm, depending on the momentum of the track and the number of measurements included in the track fitting. This kind of resolution will allow us to reconstruct with good efficiency secondary vertices located 100 μm away from the primary vertex transverse to the beam.

- Only tracks with transverse momentum greater than about 500 MeV/c are de-facto used in the final vertex location fitting; low momenta tracks are automatically 'downweighted' because of multiple scattering.

- All resolution plots exhibit a flat tail which is attributed to pattern recognition failures in matching tracks to silicon clusters. These clusters are contaminated by nearby hits from unrelated tracks in the event. Table 3 shows the probability of having such a contamination as a function of various production models and experimental conditions. Substantial degradation of the vertex resolution is expected when this probability rises significantly above a few percent. This analysis justifies the need for shorter strips (of the order of 5 mm to 1 cm) or a pixel detector at SSC energy.

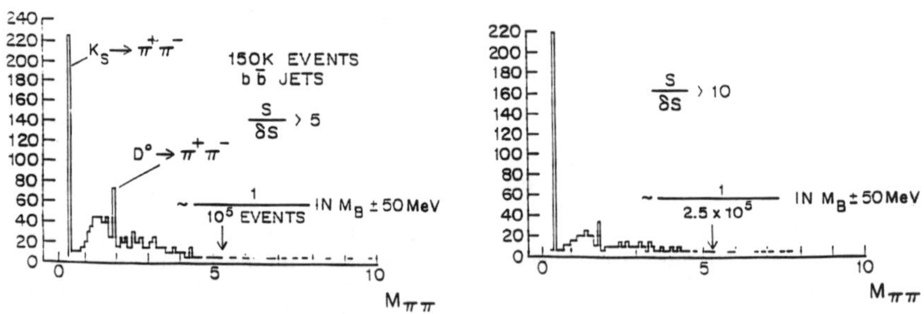

Fig. 2: Invariant mass of 2 tracks from secondary vertices, for 2 detachment significance. The sample is $\bar{b}b$ jets.

- STRAW tracking simulation:

Similar studies are underway on the STRAW tracking system. In a global version, all the 'superlayers ' (made of 8 layers of straws) are implemented with the entrance and exit points in the superlayer recorded. In an independent package, at a very detailed and local level the resolution of a single straw can be studied as a function of the electronic performance and gas composition.

Pattern recognition studies have begun, in which all 'mini-vectors' defined at the superlayer levels are combined to find tracks. A constant magnetic field is currently assumed. These found tracks can then be used to extrapolate into the silicon detector and form the base for pattern recognition studies in the silicon.

- Conclusion:

What has been described above can be summarized as complex software demanding many years of physicist effort and large amounts of CPU time, of the order of many hundreds of MIPS per year. These preliminary results are encouraging, but clearly more work is required to get to the final design of the detector.

Table 3   : Silicon cluster contamination probabilities.

Conditions :   Silicon   wafer type [*]:

	Barrel, 'ϕ' strips	Barrel, 'z' strips	Disk central	Disk F/B
✓ 2 TeV c.m., minimum bias, No magnetic field, Inf. STRAW accuracy	1.5 %	1.55%	2.1%	2.1%
✓ 2 TeV c.m., minimum bias, No magnetic field, 100 μm STRAW accuracy	1.5 %	1.55%	2.1%	2.1%
✓ 2 TeV c.m. B-B jet, P_t > 1.GeV/c, No magnetic field, 100 μm STRAW accuracy	2.1 %	2.25%	3.5%	3.5%
✓ 40 TeV c.m. Jets, P_t >4.5 Gev/c, Magnetic field = 1.0 T, Inf. STRAW accuracy	16.25%	12.6%	14.6%	16.5%
✓ 40 TeV c.m. Mini jets, P_t >4.5 GeV/c, No magnetic field. Outer layers only 100 μm STRAW accuracy	5.8%	7.25%	9.9%	11.%
✓ 40 TeV c.m. Mini jets, P_t >4.5 GeV/c, Magnetic field = 1.0 T, Silicon outer layers only 100 μm STRAW accuracy	11.8%	13.6%	14.9%	17.3%

* The 'ϕ' strips measure the track position in the R,ϕ plane, at a fixed R, counting the strip numbers in the ϕ direction (i.e., the strips run parallel to the beam). The 'z' strips measure the tracks along the beam. Because of the symmetry in ϕ for the Disks, there is no dependance on strip orientation. The central region sees a multiplicity slightly different than the high rapidity (forward/backward) ones, despite the fact that the spacing along the z axis has been adjusted to provide a constant number of measurements across the entire rapidity gap $-4.5 < \eta < 4.5$.

IV. The Required R&D Program: Vertexing, Tracking, Particle identification and Data Acquisition.

A three phase R&D program[21]   has begun at Fermilab and at the SSC laboratory that will study issues associated with building such a powerful vertex detector, the STRAW tracking system and the ambitious data acquisition system. Later on, this program will be extended to particle identification.   The Fermilab Physics Advisory Committee approved the first two years of this program in January, 1989. This work involves hardware and electronics development that aims towards a system test at the collider.

• Vertex Detector

The requirements of large geometric acceptance and 3D measurements lead to a novel detector design(fig 3.a). All parts of the detector are outside the vacuum vessel. The detector is logically spaced into two regions. The 'central' region covers most of the interaction region with a combined geometry of equally spaced silicon wafer planes ('disks') and segmented barrels. The forward/backward region covers the outer limits of the interaction region with disks spaced at equal intervals in pseudorapidity.   The inner (outer) barrel is about at 1.5 (13.5 cm) radius .   The silicon detectors are self supporting in structures called modules (fig 3.b) housed in a beryllium gutter.   A full scale model of such a structure has been built to study its thermal and mechanical properties[22]. In order to limit multiple scattering and therefore the amount of material, double sided technology has been chosen[19].

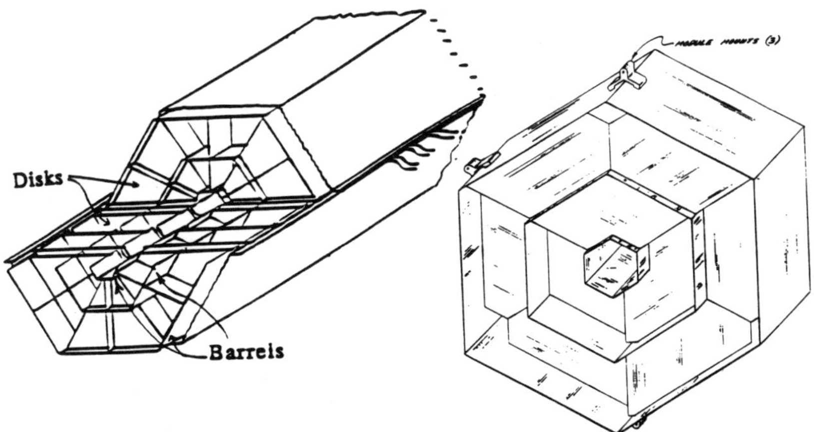

Fig. 3: The Silicon Vertex detector (a) The modules being housed in the Berylium gutter (b) Detailed layout of 1 central module.

The need for efficient 3-D track reconstruction over 4π solid angle creates a technical challenge which is illustrated in figure 4. A substantial number of tracks at grazing incidence contribute not only to pattern recognition failure, but also to a possible degradation of the position resolution. A 45° angle of incidence track has a path length of 71 μm in 200 μm thick silicon. Thus, the electronics must be able to detect signals of about 6,000 electrons (corresponding to 70 μm path length) with good signal to noise ratio, roughly 10. The design goal for the VLSI readout chip is therefore a 600 electron noise level with an input capacitance of 5 picofarads, dissipating roughly 1 milliwatt per channel. The amplification and storage must be achieved in roughly 50 ns in to run at 10^{**32} luminosity without appreciable deadtime. Finally, this electronics must be able to sustain about 1 Megarad[23].

As shown in Table 3, pixel detectors are probably required to decrease the 2 track confusion probability. Detectors based on CCD readouts are certainly too slow; we are currently studying a unique 'smart' pixel readout scheme[24].

• Tracking

The excessive cost of silicon detectors at large radius leads to a design in which the vertex detector acts only as a vernier on the track, while the primary pattern recognition and the momentum measurements are made in a gas tracking system. Straw tubes have several advantages over large 'central' drift chambers : (i) the small drift distance (about 4 mm) allows high rates, (ii) there are no massive end plates, (iii) the tracking can be extended to higher rapidity with minimal dead regions and (iv) isolated broken wires do not impair detection in a big section of the detector. At Princeton University, members of the collaboration are currently studying the characteristics of these devices (gas gain, temperature effects, and so forth). The readout system is being developed as a project associated with the SSC generic R&D program[25]. We plan to use electronics developed by H. Williams *et al.* as part of their program[26]. The VLSI TDC work of J. Watase and Y. Arai is also of great interest to the BCD tracking effort.

• RICH detectors

The stringent space constraints in the central as well as in the forward region impose restrictions on the radiator material. A preliminary, conceptual, design in the GEANT framework has begun and its response to ISAJET events has been investigated. Pad size optimization is in progress. A lot of work remains to be done in this area.

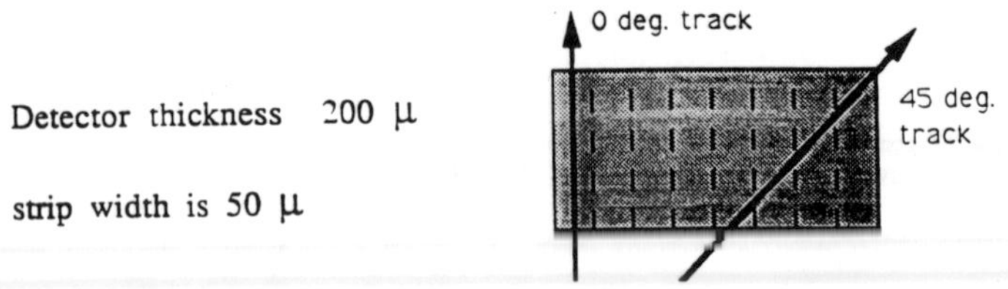

Fig 4: Tracks at small incident angles produce small signals spread over a large number of strips.

- Triggering and Data Acquisition

BCD emphasizes the concept of 'software triggers', using large amounts of computing power as soon as the data is available. Shown in fig. 5 is the generalized system architecture. The data flows from the subdetector elements into local storage buffers. Each segment of information is time stamped. Upon receipt of a prompt trigger - which could be as simple as detecting that an interaction took place -, corresponding data is transferred from the detector to an event builder. This device handles parallel streams of data to a processor farm. Only a small fraction of the events are written on sequential mass media for further analysis.

The Barrel Switch Event Builder, based on the principle of a telephone switch, can build events at the rate of several hundred thousand per second. This is being developed by the group of E. Barsotti *et al.* at Fermilab[27]. In the absence of an efficient prompt trigger, we are considering rejecting events prior to the complete event building phase. This can be achieved by inserting more than one sequence of switch/farms into the data path. The overall CPU capability of such a farm is anticipated to be in the order of 1 Tip[28]. Since this software trigger must be capable of identifying particles or reconstructing vertices with good accuracy, the analysis software running on this farm will be as complex as the HEP programs currently running offline.

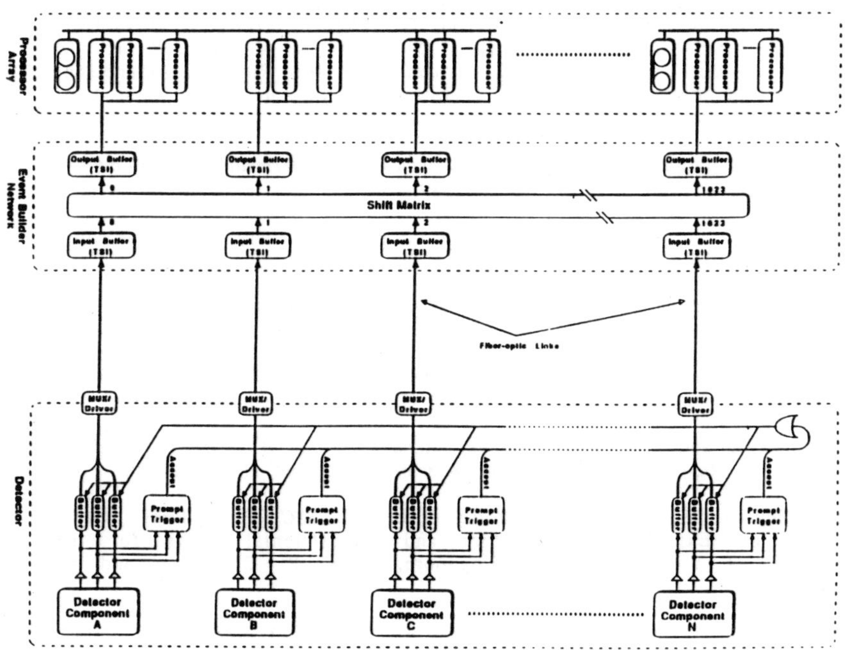

Fig 5: The Architecture of the BCD Data Acquisition system. Data is flowing from bottom to the top of the picture.

V.   Conclusion

The goal of the BCD group is to study CP violation in the B system. The SSC is the ultimate B factory: It will produce in excess of 10^{12} $B\bar{B}$ pairs per year and 1 in every 200 events is likely to contain a bottom meson antimeson pair. The challenge for the experimentalist is to build a detector capable of sustaining these high rates while retaining good granularity and very good tracking resolution. An R&D program is well underway to address these technical challenges.

References

[1]   The menbers of the BCD collaboration are:
H. Castro, B. Gomez, F. Rivera, J.C. Sanabria, *Universidad de los Andes;* E.Barsotti, M. Bowden, S. Childress, H. Jostlein, P. Lebrun, J. Morfin, L. A. Roberts, R. Stefanski, L. Stutte, C. Swoboda, *Fermilab;* P. Avery, J. Yelton, *University of Florida;* K. Lau,*University of Houston;* R. Burnstein, H. Rubin, *Illinois Institute of Technology;* E. McCliment, Y. Onel,*University of Iowa;* G. Alverson, W. Faisler, D. Garelick, M. Glaubman, I. Leedom, S. Reucroft, D. Kaplan, *Northeastern University;* S. E. Willis *Northern Illinois University;* G. R. Kalbfleisch, P Skubic, J. Snow,*University of Oklahoma;* N. S. Lockyer, R. Van Berg,*University of Pennsylvania;* D. Judd, D. Wagoner, *Prairie, View A&M University;* C. Lu, D. R. Marlow, K. T. McDonald, M. V. Purohit,*Princeton University;* A. Lopez, A. Mendez, *Universidad de Puerto Rico;* B. Hoeneisen, *Universidad San Francisco de Quito* ; S. Dhawan, P. E. Karchin, W. Ross, A. J. Slaughter, E. Wolin, *Yale University* .

[2] BCD collaboration, *Bottom Collider Detector, an Intermediate and Low* P_t *Detector for the SSC*, SSC-240 (October 1, 1989).

[3] N. Lockyer, *B factory with Hadron Colliders*, University of Pensylvania preprint UPR# 0175E.

[4] F. Gilman, *Predictions for CP Violation*, SLAC-PUB-5018, August 1989.

[5] K. Foley *et al.*, *Bottom and Top Physics*, Proceedings of the Workshop on Experiments, Detectors and Experimental Areas for the Supercollider (World Scientific, Singapore, 1988, R. Donaldson and G. Gilchriese, editors, p. 701.

[6] N. Lockyer and J. Morfin, *How the Acceptance of B - Decay Products Depends on the Rapidity Coverage of the Detector*, BCD note, Oct. 24 1989 .

[7] K. Foley *et al.,A Beauty Spectrometer for the SSC*, Proceedings of the Workshop on Experiments, Detectors and Experimental Areas for the Supercollider (World Scientific, Singapore, 1988, R. Donaldson and G. Gilchriese, editors, p. 701.

[8] H. Fenker, I. Leedom and S. Reucroft, *A PSF Vertex Detector for BCD*, Northeastern Univiersity, BCD memo.

[9] N. Lockyer and E. Wang, *Interfacing Monte Carlo Event Generators to Simulation Packages*, Proceedings of the Workshop on Detector Simulation for the SSC, ANL, AUGUST 24-28 1987, P. 325.

[10] R.Brun, *et al*, *GEANT3 User's Guide*, CERN DD/EE/84-1, May 1986.

[11] W. R. Nelson, H. Hirayama and D Rogers,. SLAC -Report-265 (1985).

[12] F. Abe *et al.*, Phys. Rev. Letter, Vol. 61, 1819 (1988); and N. Lockyer, private communication (1988).

[13] G. Zioulas *et al.*, IEEE Trans. on Nucl Science, Vol 36, 375 (1989).

[14] S. A. Baird *et al.*, IEEE Trans. on Nucl Science, Vol 36, 364 (1989).

[15] G. A. Baird *et al.*, IEEE Trans. on Nucl Science, Vol 36, 370 (1989).

[16] M. Abolins *et al.*, IEEE Trans. on Nucl Science, Vol 36, 384 (1989).

[17] M. I. Adamovich *e t al*, Charm Hadroproduction with an Impact Parameter Trigger, CERN - EP 89-123 Sept 1989, presented at the 1989 Int. Symp. on Heavy Quark Decay Physics, Ithaca, N.Y., June 13-17 1989.

[18] D. M. Kaplan *et al.*, E798 FNAL Proposal P-789 (1988).

[19] G. R. Kalbfleisch and P. Skubic, *Considerations Regarding Double Sided Solid State Detectors*, U. of Oklahoma preprint (Feb. 1989); G.R. Kalbfleish, P.L. Skubic and M. A. Lambrecht, *Results on Signal Correlation from a 'Micron' Doubled-sided Mini-Strip Detector*, U. of Oklahoma preprint (Aug. 1988).

[20] L. Roberts, *Monte Carlo Simulation of Silicon Vertex Detector for Bottom Collider Dectector*, Fermilab preprint FN-488 (1988).

[21] H. Castro *et al*, *Proposal for Research and Development : Vertexing, Tracking and Data Acquisition for a Bottom Collider Detector*, submitted to the Fermilab PAC (Jan 2, 1989) .

[22] H. Mulderick, N. Michels and H. Jöstlein , *Mechanical and Thermal Behavior of a Prototype Support Structure for a Large Silicon Vertex Detector (BCD)*, Fermilab TM-1612 (1989).

[23] T. Kondo, *Radiation Damage of Silicon Devices*, from Future Directions in Detector Research and Development for Experiment at PP Colliders (Snowmass, Colorado, July 5-7, 1988).

[24] E. Arens *et al* , *SSC Detector R&D Proposal : Development of Technology for Pixel Vertex Detector*, (1989).

[25] C. S. Lu *et al.*, *Proposal to the SSC Laboratory for Research and Development of a Straw-Tube Tracking System*, Princeton U. preprint DOE/ER/3072-56 (1989).

[26] L. Callewaert *et al*, IEEE Trans. on Nucl Science, Vol 36, 446 (1989).

[27] M. Bowden *et al*, *A High-throughput Data Acquisition Architecture Based on Serial Interconnects*, Fermilab preprint (1989).

[28] L. D. Gladney *et al.*, *Proposal to the SSC Laboratory for a Parallel Computing Farm*, (1989).

LIST OF PARTICIPANTS

Abashian, Alexander
Virginia Polytechnic Institute

Adams, David
Rice University

Adkins, Gregory S.
Franklin & Marshall College

Albright, Carl H.
Northern Illinois University

Aldinger, Randolph R.
Gettysburg College

Allen, Roland E.
Texas A&M University

Altarelli, Guido
CERN

Amidei, Dan
Fermilab

Andersson, Bo
SUNY at Stony Brook

Andrews, Thomas B.
Brooklyn, NY

Appel, Jeffrey A.
Fermilab

Appelquist, Thomas
Yale University

Arenton, Michael
University of Virginia

Arnowitt, Richard
Texas A&M University

Atwood, David
Brookhaven National Laboratory

Auge, E.
L.A.L. Orsay

Badalyan, Alla
University of Minnesota

Baer, Howard A.
Florida State University

Baggett, Neil
SSC Laboratory

Baker, Oliver K.
CEBAF

Bardeen, William A.
Fermilab

Barklow, Timothy
SLAC

Barletta, William A.
UCLA

Barnett, Bruce
Johns Hopkins University

Battiston, Roberto
University of Perugia, Italy

Baum, Joel L.
H.S. Truman

Baur, Ulrich
University of Wisconsin

Bean, Alice
University of California, Santa Barbara

Berger, Edmond L.
Argonne National Laboratory

Berley, David
National Science Foundation

Bern, Z.
Los Alamos National Laboratory

Bernard, Claude
University of California, Santa Barbara

Betke, Siegfried
University of Heidelberg

Bharucha, Cyrus
Rice University

Bird, Fred
CERN

Blaylock, Guy
CERN

Bloom, Elliot D.
SLAC

Bogart, Christopher W.
University of Colorado

Bonner, Billy
Rice University

Borchard, Regina
Martin Marietta

Bowles, Thomas J.
Los Alamos National Laboratory

Braaten, Eric
Northwestern University

Branson, James G.
University of California, San Diego

Brock, Raymond
Michigan State University

Brown, Chuck
Fermilab

Brown, Stanley
Physical Review Letters

Bugg, William
University of Tennessee

Bunce, Gerry
Brookhaven National Laboratory

Burgess, Cliff
McGill University

Burke, Jack W.
Southwestern Laboratories, Inc.

Butler, Joel
Fermilab

Cahill, Kevin E.
University of New Mexico

Cahn, Robert
Lawrence Berkeley Laboratory

Caldwell, David O.
University of California, Santa Barbara

Carroll, Alan S.
Brookhaven National Laboratory

Carson, Larry J.
University of Minnesota

Cassel, David
Cornell University

Casteel, Sharon
Rice University

Chan, Lai-Him
Louisiana State University

Chao, Alex
SSC Laboratory

Chen, Chia-Chu
Texas A&M University

Chen, Wei
Oklahoma School of Science and Mathematics

Cheng, Ta-Pei
University of Missouri

Chiou, Cheng-nan
Rice University

Chlouber, Dean
Alvin, TX

Cho, Hing Tong
University of Oklahoma

Christ, Norman
Columbia University

Clark, Robert B.
Texas A&M University

Corcoran, Marj
Rice University

Comyn, Martin
TRIUMF

Conway, John S.
University of Wisconsin

Cornell, Laird R.
SSC Laboratory

Coupal, David
SLAC

Cousins, Robert
UCLA

Cox, Brad
University of Virginia

Cox, Paul H.
Texas A&I University

Cranshaw, Jack
Rice University

Crawford, Glen
Cornell University

Cremaldi, Lucien M.
University of Mississippi

Crisler, Michael B.
Fermilab

Culy, Steven W.
University of Colorado

Cunningham, John D.
University of Notre Dame

Damiano, Gary
SSC Laboratory

Darling, Chris
Yale University

Davenport, Forest
University of North Carolina, Asheville

Day, George
SSC Laboratory

Delchamps, Stephen W.
Fermilab

Delfino, Manuel
Universidad Autonoma de Barcelona

Deshpande, N.G.
University of Oregon

DeTar, Carleton
University of Utah

DiBitonto, Daryl
University of Alabama

Diehl, H. Thomas
Rutgers University

1086

Dixon, Lance
SLAC

Dombeck, Thomas
SSC Laboratory

Domokos, Gabor
Johns Hopkins University

Doncheski, Michael A.
Penn State University

Dorfan, Jonathan
SLAC

Dougherty, William M.
University of Washington

Drinkard, John J.
University of California, Santa Cruz

Duboscq, Jean
University of California, Santa Barbara

Duchovni, Ehud
Weizmann Institute

Duck, Ian
Rice University

Dunwoodie, William M.
SLAC

Duryea, Jeff
University of Minnesota

Eichten, Estia
Fermilab

Eisner, Alan M.
University of California, IIRPA

Elias, John E.
Fermilab

Engel, Jean-Pierre
University of Strasbourg

Errede, Steven
University of Illinois

Erwin, Albert
University of Wisconsin

Everett, Allen E.
Tufts University

Feldman, Gary
SLAC

Ferris, Michael L.
SAES Getters

Findeisen, Charles
University of Wisconsin

Firestone, Alexander
Iowa State University

Fisher, John C.
Carpinteria, CA

Flaugher, Brenna
Rutgers University

Forbush, Michael
Texas A&M University

Fortney, L.R.
Duke University

Foudas, Costas
University of Wisconsin

Fox, Rosemary
Pasadena, CA

Franklin, Jerrold
Temple University

Frederiksen, Soren G.
Ohio State University

Gaisser, Thomas K.
Bartol Research Institute

Gandhi, Raj
University of Arizona

Garren, A.
SSC Laboratory

Garwin, Charles A.
Santa Monica, CA

Gary, J. William
University of Heidelberg

Gaussiran, Tom
Rice University

Gettner, Marvin
Northeastern University

Giacomelli, Giorgio
Universita di Bologna

Gibaut, Duncan
Carnegie-Mellon University

Gilchriese, Murdock G.
SSC Laboratory

Glover, E.W.N.
Fermilab

Golowich, Eugene
University of Massachusetts

Gonsalves, Richard J.
State University of New York at Buffalo

Gotow, Kazuo
Virginia Polytechnic Institute

Gottlieb, Steven
Indiana University

Green, Daniel
Fermilab

Groom, D.
Lawrence Berkeley Laboratory

Gyulassy, Miklos
Lawrence Berkeley Laboratory

Haber, Howard
University of California, Santa Cruz

Hagelin, John S.
Maharishi International University

Haim, Dan
University of Illinois

Haines, Todd
University of Maryland

Hall, Lawrence J.
University of California, Berkeley

Handler, Thomas
University of Tennessee

Hands, Simon
University of Illinois

Harigel, Gert
CERN

Harral, Brian
Johns Hopkins University

Harris, Robert M.
Lawrence Berkeley Laboratory

Hart, Edward L.
University of Tennessee

Hartill, D.
Cornell University

Harton, John L.
University of Wisconsin

Haymaker, Richard W.
Louisiana State University

Heilig, Steven J.
Grinnell College

Heltsley, B.
Cornell University

Herbst, Ingo
University of Wuppertal

Hewett, JoAnne
University of Wisconsin

Hill, Christopher T.
Fermilab

Ho, Yiu-Huang M.
University of Rochester

Hobbs, Frayne
SSC Laboratory

Hochberg, David
Vanderbilt University

Hofer, H.
ETH Zürich

Hoffman, Cyrus M.
Los Alamos National Laboratory

Hogan, Gary E.
Los Alamos National Laboratory

Hollebeek, Robert
University of Pennsylvania

Holmes, S.
Fermilab

Hong, Jiangtao
California Institute of Technology

Hou, George W.S.
Paul Scherrer Institute

Hu, Kangping
Virginia Polytechnic Institute

Huang, Yunxiang
University of Houston

Hurst, Roger B.
INFN

Huson, Fredrich R.
Texas A&M University

Ibañez-Meier, Rodrigo
Rice University

Incandela, Joseph
University of Milan

Isgur, Nathan
University of Toronto

Ito, Mark M.
Princeton University

Izen, Joseph M.
University of Illinois

Jackson, J. D.
Lawrence Berkeley Laboratory

Jaffe, Robert
Massachusetts Institute of Technology

Johnson, Denis P.
I.I.H.E. ULB-VUB, Belgium

Johnson, Robert P.
University of Wisconsin

Johnstad, Harold
SSC Laboratory

Jones, Lorella M.
University of Illinois

Judd, Dennis
Prairie View A&M University

Jung, Chang Kee
SLAC

Kahn, Stephen
Brookhaven National Laboratory

Kamal, A. N.
University of Alberta

Karagiannis, Evangelos
University of California, Los Angeles

Karlen, Dean
Carleton University

Kass, Richard
Ohio State University

Kazimi, RTeza
Texas A&M University

Keung, Wai-Yee
University of Illinois at Chicago

Kilcup, Greg
Brown University

Kirk, Thomas B.
Argonne National Laboratory

Kiskis, Joe
University of California, Davis

Klebanov, I.
Princeton University

Ko, Winston
University of California, Davis

Kofler, Richard R.
University of Massachusetts

Komamiya, Sachio
SLAC

Koshiba, Masatoshi
University of California, Riverside

Kostelecky, Alan
Indiana University

Kovesi-Domokos, S.
Johns Hopkins University

Kral, J. Frederic
Lawrence Berkeley Laboratory

Kroll, J.
University of Chicago

Kroll, Norman M.
University of California, San Diego

Kuhlen, Michael
California Institute of Technology

Kunszt, Zoltan
ETH-Zürich

Kuti, Julius
University of California, San Diego

Lander, Richard
University of California at Davis

Lankford, Andrew J.
SLAC

Larson, K. Dean
University of New Mexico

Lassila, Kenneth E.
Iowa State University

Lebrun, Paul
Fermilab

Lederman, Leon
University of Chicago

Lee, Bo
Texas A&M University

Lee, Chen-Han
Louisiana State University

Lee, S.Y.
Brookhaven National Laboratory

Lesikar, James D. II,
Computer Sciences Corporation

Lewis, David A.
Iowa State University

Lewis, Jonathan D.
Cornell University

Li, Ling-Fong
Carnegie-Mellon Unuversity

Li, Mingyang
University of Houston

Li, Yue Kuan
University of South Carolina

Lichtenberg, Don
Indiana University

Linde, Andrei
P.N. Lebedev Inst. of Physics

Lopez, Jorge L.
Texas A&M University

Love, George
Southwestern Laboratories

Lowe, James
University of New Mexico

Lowery, Bruce
University of Texas at Dallas

Lubatti, Henry J.
University of Washington

Ludlam, Thomas W.
Brookhaven National Laboratory

Lueking, Lee H.
Fermilab

Lundberg, Wayne R.
Wright State University

Luth, Vera
SLAC

Ma, Ernest
University of California, Riverside

Machida, Shinji
University of Houston

Mackenzie, Paul
Fermilab

Maeshima, Kaori
University of California, Davis

Maettig, Peter
CERN

Magill, Stephen R.
University of Illinois at Chicago

Mandula, Jeffrey E.
Department of Energy

Mani, Sudhindra
University of Pittsburgh

Marinari, E.
University of Rome

Marleau, Luc
Universite Laval

Marzia, Rosati
McGill University

Matthews, James
University of Michigan

Mattingly, Alan
Rice University

McBride, Patricia
Harvard University

McDonald, Kirk
Princeton University

McIntyre, Peter
Texas A&M University

McLerran, Larry
University of Minnesota

McMahon, Timothy
Purdue University

Meier, Karlheinz
CERN

Melese, Philip
Rockefeller University

Meshkov, Sydney
NIST

Metheny, Justine
Tufts University

Miettinen, Hannu E.
Rice University

Migdal, Alexander A.
Princeton University

Miller, James
Boston University

Miller, Dean F.
University of Oklahoma

Miller, Marshall
University of Chicago

Milton, Kimball A.
University of Oklahoma

Mishra, Chandra Shekhar
Los Alamos National Laboratory

Mitchell, Joseph W.
CERN

Mohapatra, Pramoda K.
University of Colorado

Month, Melvin
Brookhaven National Laboratory

Morfin, Jorge G.
Fermilab

Morris, M.
Cornell University

Morrison, Douglas R. O.
CERN

Mouthuy, Thierry
CERN

Mutchler, Gordon
Rice University

Nagashima, Yorikiyo
Osaka University

Nagel, Darragh
Los Alamos National Laboratory

Napier, Austin
Tufts University

Narayanan, Rajamani S.
University of California at Davis

Nelson, Donald John
Beaverton, OR

Neuenschwander, Dwight E.
Southern Nazarene University

Ng, John N.
TRIUMF

Nicholson, John L.
U.S. Department of Energy

Noble, Anthony J.
University of British Columbia

Nodulman, Larry
SSC Laboratory

Oakes, Robert
Nortwestern University

O'Fallon, John R.
Department of Energy

Ohnuma, Sho
University of Houston

Olness, Fredrick I.
University of Oregon

Olsen, Stephen L.
University of Rochester

Orgeron, Joseph A.
University of Texas at Dallas

Orie, Never
SSC Laboratory

Orr, J. Richie
Fermilab

O'Shaughnessy, Kathy
SLAC

Owens, Joseph F.
Florida State University

Padgett, Michael
Pennzoil

Pal, Palash B.
University of Oregon

Palmer, Stephen
Arizona Chemical

Pan, Yi-Bin
University of Wisconsin

Patrick, James
Fermilab

Peccei, Roberto D.
UCLA

Peggs, Steve
Fermilab

Peng, Yingcai
Louisiana State University

Perez, Miguel Angel
CINVESTAV, IPN, Mexico

Perl, Martin L.
SLAC

Petcher, Donald N.
Florida State University

Petridis, Athanasios N.
Iowa State University

Phillips, Gerald C.
Columbia Scientific Industries

Piilonen, Leo
Virginia Polytechnic Institute

Pinsky, Larry
University of Houston

Pinsky, Stephen S.
Ohio State University

Pitman, Dale
SLAC

Plasil, Frank
Oak Ridge National Laboratory

Pois, Heath
Vanderbilt University

Polchinski, Joe
University of Texas

Pordes, Stephen H.
Fermilab

Powers, Shannon
University of Utah

Price, Larry
Argonne National Laboratory

Price, LeRoy
University of California, Irvine

Purohit, Milind V.
Princeton University

Pyrlik, Jorg
University of Houston

Quevedo, F.
Los Alamos National Laboratory

Quigg, Chris
Lawrence Berkeley Laboratory

Raab, Johannes
CERN

Raby, Stuart
Ohio State University

Radeztsky, Scott A.
University of Wisconsin

Raffelt, Georg
University of California, Berkeley

Rankin, Patricia
University of Colorado

Raparia, Deepak
University of Houston

Reay, Neville W.
Ohio State University

Reed, James F.
McNeese State University

Rembiesa, Peter J.
The Citadel

Reucroft, S.
Northeastern University

Richter, Burton
SLAC

Rizzo, Thomas G.
University of Wisconsin

Roberts, Jabus
Rice University

Robinson, Jerry
McGill University

Rohaly, Timothy F.
University of Pennsylvania

Roland, Kaj
Niels Bohr Institute

Ronan, Michael T.
Lawrence Berkeley Laboratory

Rondio, Ewa
Warsaw University

Ross, William
Yale University

Roy, Probir
University of Texas

Ruchti, Randy
University of Notre Dame

Russell, Patrick
Princeton University

Ruth, Ronald D.
SLAC

Rutherford, John P.
University of Arizona

Sadoulet, Bernard
University of California, Berkeley

Sakumoto, Willis K.
University of Rochester

Sandweiss, Jack
Yale University

Sanger, P.
SSC Laboratory

Sarcevic, Ina
University of Arizona

Sato, Asao
Texas Accelerator Center

Schellman, Heidi
Fermilab

Schierholz, G.
DESY/HLRZ

Schlein, Peter
UCLA

1100

Schmidt, Michael P.
Yale University

Schmitt, Michael
Harvard University

Schneekloth, Uwe
Massachusetts Institute of Technology

Schultz, Jonas
University of California, Irvine

Schwarz, John H.
California Institute of Technology

Schwitters, Roy
SSC Laboratory

Sciulli, Frank
Columbia University

Seckel, David
Bartol Research Institute

Seidel, Sally
University of Toronto

Shafi, Qaisar
Bartol Research Institute

Sharma, V.
University of Wisconsin

Shen, Benjamin C.
University of California, Riverside

Sher, Marc
College of William and Mary

Shrauner, J. Ely
Washington University

Siegel, Steven
Boise State University

Sill, Alan F.
University of Rochester

Simhony, Menahem
Hebrew University, Israel

Simonov, Yuri
ITEP, Moscow

Simpson, J.
Argonne National Laboratory

Singh, Simon L.
University of Cambridge

Singh, Vandana
Louisiana State University

Sivertz, Michel
University of California, San Diego

Skrinsky, A.
USSR Acad. of Sciences, Novosibirsk

Slaughter, Jean
Yale University

Smith, A.J. Stewart
Princeton University

Smith, Frank D. Jr.
Cartersville, GA

Smith, Lesley L.
University of Kansas

Snow, Gregory
University of Michigan

Soni, Amarjit
Brookhaven National Laboratory

Soper, Davison E.
University of Oregon

Spalding, Jeff
Fermilab

Spang, Karsten
Niels Bohr Institute

Stancu, Ion
Rice University

Stanfield, Kenneth C.
Fermilab

Steele, Jeff
California Institute of Technology

Stefanski, Ray
SSC Laboratory

Stephenson, Gerard J. Jr.
Los Alamos National Laboratory

1102

Sterner, Kevin L.
Virginia Polytechnic Institute

Stevenson, Paul
Rice University

Stiening, R.
SSC Laboratory

Strausz, Steven C.
University of Washington

Stroynowski, Ryszard
California Institute of Technology

Stuart, David
University of California, Davis

Stubbins, Calvin
Franklin & Marshall College

Stuckey, William K.
National Geographic

Studenmund, Bill
Rice University

Sulak, Lawrence R.
Boston University

Sun, Chih Ree
SUNY at Albany

Swenson, Charles Allen
Texas Accelerator Center

Takach, Stephen F.
Yale University

Takeda, Seishi
KEK, Japan

Tanaka, Mitsuyoshi
Brookhaven National Laboratory

Tangherlini, Frank R.
College of the Holy Cross

Tata, Xerxes R.
University of Hawai Manoa

Tesarek, R.
Duke University

Thaler, Jon J.
University of Illinois

Thompson, Murray A.
University of Wisconsin

Thresher, John
CERN

Tipler, Frank J.
Tulane University

Toki, Walter
SLAC

Toll, John S.
University of Maryland

Tollestrup, Alvin
Fermilab

Tompkins, J.
SSC Laboratory

Tompkins, Perry
Texas A&M University

Tran, Huan Minh
Texas A&M University

Traynor, Mike
Rice University

Treille, Daniel
CERN

Trost, Hans-Jochen
Argonne National Laboratory

Turkot, Frank
Fermilab

Tzanakos, George
University of Athens

Ullman, Jack D.
Lehman College, CUNY

Uy, Zenaida E. S.
Millersville University

van Berg, R.
University of Pennsylvania

VanDalen, Gordon J.
University of California, Riverside

Van Kooten, Richard
SLAC

Vasek, James
Rice University

Vinson, Jon
Rutgers University

Visser, Matt
Washington University

Voyvodic, Louis
Fermilab

Wagner, Robert G.
Argonne National Laboratory

Wagoner, David E.
Prairie View A&M University

Walker, William D.
Duke University

Wallraff, W.
Physical. Inst. Aachen

Walsh, Jon
University of Pennsylvania

Wang, Angel T.M.
University of South Carolina

Wasserman, Eric
Harvard University

Watts, Terence L.
Rutgers University

Webb, Robert C.
Texas A&M University

Weber, Frederick
University of Wisconsin

Webster, Medford S.
Vanderbilt University

Weinberg, Steven
University of Texas

White, Christopher
University of Minnesota

White, D. Hywel
Los Alamos National Laboratory

White, James T.
Texas A&M University

White, Sharon L.
University of Oregon

Wicklund, Eric J.
California Institute of Technology

Wilcox, Judd
University of California, Davis

Wilcox, Walter
Baylor University

Willeke, Ferdinand
DESY

Williams, P. K.
U.S. Department of Energy

Wilson, Jeffrey R.
University of Illinois

Wilson, Scott R.
University of South Carolina

Witherell, Michael
University of California, Santa Barbara

Wojcicki, Stanley
Stanford University

Wolbers, Stephen
Fermilab

Woodside, Jeffrey
Oklahoma State University

Worris, Mark A.
Cornell University

Wu, Sau Lan
University of Wisconsin

Wu, Yu
Texas A&M University

Wyslouch, Boleslaw
CERN

Xu, Jianping
Rice University

Yamamoto, Hitoshi
University of Chicago

Yamanouchi, Taiji
Fermilab

Yamin, Peter
Brookhaven National Laboratory

Yao, Weiming
Purdue University

Yegneswaran, Amrit S.
CEBAF

Yoshida, R.
Northwestern University

Young, Glenn R.
Oak Ridge National Laboratory

Zachos, Cosmas
Argonne National Laboratory

Zeller, Michael E.
Yale University

Zhu, Qiuan
Rice University

Zhu, Xiquan
University of Alberta

Zielinski, Marek
University of Rochester

AUTHOR INDEX